CHEESE:
CHEMISTRY, PHYSICS AND MICROBIOLOGY

Volume 1
General Aspects

Second Edition

CHEESE:
CHEMISTRY, PHYSICS AND MICROBIOLOGY

Volume 1

General Aspects

Second Edition

Edited by

P.F. FOX

Department of Food Chemistry
University College
Cork, Republic of Ireland

CHAPMAN & HALL

London · Glasgow · New York · Tokyo · Melbourne · Madras

Published by Chapman & Hall, 2–6 Boundary Row, London SE1 8HN, UK

Chapman & Hall, 2–6 Boundary Row, London SE1 8HN, UK

Blackie Academic & Professional, Wester Cleddens Road, Bishopbriggs, Glasgow G64 2NZ, UK

Chapman & Hall Inc., 29 West 35th Street, New York NY10001, USA

Chapman & Hall Japan, Thomson Publishing Japan, Hirakawacho Nemoto Building, 6F, 1-7-11 Hirakawa-cho, Chiyoda-ku, Tokyo 102, Japan

Chapman & Hall Australia, Thomas Nelson Australia, 102 Dodds Street, South Melbourne, Victoria 3205, Australia

Chapman & Hall India, R. Seshadri, 32 Second Main Road, CIT East, Madras 600 035, India

First edition 1987
Second edition 1993

© 1993 Chapman & Hall

Typeset in Times Roman by Variorum Publishing Limited, Rugby
Printed in Great Britain by Galliard (Printers) Ltd, Great Yarmouth

ISBN 0 412 53500 9   0 412 58230 9 (Set)

A catalogue record for this book is available from the British Library

Library of Congress Cataloging-in-Publication data
Cheese: chemistry, physics, and microbiology / edited by P.F. Fox. — 2nd ed.
  p. cm.
  Includes bibliographical references and index.
  Contents: v. 1. General aspects — v. 2. Major cheese groups.
  ISBN 0 412 58230 9 (set). — ISBN 0 412 53500 9 (v. 1). — ISBN 0 412 53510 6 (v. 2)
  1. Cheese. 2. Cheese — Varieties. I. Fox, P.F.
SF271.C43 1993
637'.3 — dc20   92-42748
                  CIP

ISBN 0 412 53500 9 (v. 1)
ISBN 0 412 53510 6 (v. 2)
ISBN 0 412 58230 9 (set)

Preface to the Second Edition

The first edition of this book was very well received by the various groups (lecturers, students, researchers and industrialists) interested in the scientific and technological aspects of cheese. The initial printing was sold out faster than anticipated and created an opportunity to revise and extend the book.

The second edition retains all 21 subjects from the first edition, generally revised by the same authors and in some cases expanded considerably. In addition, 10 new chapters have been added: Cheese: Methods of chemical analysis; Biochemistry of cheese ripening; Water activity and the composition of cheese; Growth and survival of pathogenic and other undesirable microorganisms in cheese; Membrane processes in cheese technology, in Volume 1 and North-European varieties; Cheeses of the former USSR; Mozzarella and Pizza cheese; Acid-coagulated cheeses and Cheeses from sheep's and goats' milk in Volume 2. These new chapters were included mainly to fill perceived deficiencies in the first edition.

The book provides an in-depth coverage of the principal scientific and technological aspects of cheese. While it is intended primarily for lecturers, senior students and researchers, production management and quality control personnel should find it to be a very valuable reference book. Although cheese production has become increasingly scientific in recent years, the quality of the final product is still not totally predictable. It is not claimed that this book will provide all the answers for the cheese scientist/technologist but it does provide the most comprehensive compendium of scientific knowledge on cheese available.

Each of the 31 chapters is extensively referenced to facilitate further exploration of the extensive literature on cheese. It will be apparent that while cheese manufacture is now firmly based on sound scientific principles, many questions remain unanswered. It is hoped that this book will serve to stimulate further scientific study on the chemical, physical and biological aspects of cheese.

I wish to thank sincerely all the authors who contributed to the two volumes of this book and whose co-operation made my task as editor a pleasure.

P.F. Fox

Preface to the First Edition

Cheese manufacture is one of the classical examples of food preservation, dating from 6000–7000 BC. Preservation of the most important constituents of milk (i.e. fat and protein) as cheese exploits two of the classical principles of food preservation, i.e.: lactic acid fermentation, and reduction of water activity through removal of water and addition of NaCl. Establishment of a low redox potential and secretion of antibiotics by starter microorganisms contribute to the storage stability of cheese.

About 500 varieties of cheese are now produced throughout the world; present production is $\sim 10^7$ tonnes per annum and is increasing at a rate of $\sim 4\%$ per annum. Cheese manufacture essentially involves gelation of the casein via isoelectric (acid) or enzymatic (rennet) coagulation; a few cheeses are produced by a combination of heat and acid and still fewer by thermal evaporation. Developments in ultrafiltration facilitate the production of a new family of cheeses. Cheeses produced by acid or heat/acid coagulation are usually consumed fresh, and hence their production is relatively simple and they are not particularly interesting from the biochemical viewpoint although they may have interesting physico-chemical features. Rennet cheeses are almost always ripened (matured) before consumption through the action of a complex battery of enzymes. Consequently they are in a dynamic state and provide fascinating subjects for enzymologists and microbiologists, as well as physical chemists.

Researchers on cheese have created a very substantial literature, including several texts dealing mainly with the technological aspects of cheese production. Although certain chemical, physical and microbiological aspects of cheese have been reviewed extensively, this is probably the first attempt to review comprehensively the scientific aspects of cheese manufacture and ripening. The topics applicable to most cheese varieties, i.e. rennets, starters, primary and secondary phases of rennet coagulation, gel formation, gel syneresis, salting, proteolysis, rheology and nutrition, are reviewed in Volume 1. Volume 2 is devoted to the more specific aspects of the nine major cheese families: Cheddar, Dutch, Swiss, Iberian, Italian, Balkan, Middle Eastern, Mould-ripened and Smear-ripened. A chapter is devoted to non-European cheeses, many of which are ill-defined; it is hoped that the review will stimulate scientific interest in these minor, but locally important, varieties. The final chapter is devoted to processed cheeses.

It is hoped that the book will provide an up-to-date reference on the scientific aspects of this fascinating group of ancient, yet ultramodern, foods; each chapter is extensively referenced. It will be clear that a considerably body of scientific knowledge on the manufacture and ripening of cheese is currently available but it will be apparent also that many major gaps exist in our knowledge; it is hoped that this book will serve to stimulate scientists to fill these gaps.

I wish to thank sincerely the other 26 authors who contributed to the text and whose co-operation made my task as editor a pleasure.

P.F. Fox

Contents

List of Contributors

T.M. COGAN
National Dairy Products Research Centre, Teagasc, Moorepark, Fermoy, Co. Cork, Republic of Ireland.

D.G. DALGLEISH
Department of Food Science, University of Guelph, Guelph, Ontario, Canada N1G 2W1.

B. FOLTMANN
Institute of Biochemical Genetics, University of Copenhagen, 2A Oster Farimagsgade, DK-1353 Copenhagen K, Denmark.

P.F. FOX
Department of Food Chemistry, University College, Cork, Republic of Ireland.

M.L. GREEN
8 Harwich Close, Lower Earley, Reading RG6 3UD, UK.

T.P. GUINEE
National Dairy Products Research Centre, Teagasc, Moorepark, Fermoy, Co. Cork, Republic of Ireland.

C. HILL
National Dairy Products Research Centre, Teagasc, Moorepark, Fermoy, Co. Cork, Republic of Ireland.

K.R. LANGLEY
Institute of Food Research (Reading Laboratory), Earley Gate, Whiteknights, Reading RG6 2EF, UK.

J. LAW
Department of Microbiology, University College, Cork, Republic of Ireland.

A. MARCOS
Department of Food Science and Technology, University of Córdoba, E-14005 Córdoba, Spain.

R.J. MARSHALL
University of North London, Holloway Road, London N7 6DB, UK.

J.-L. Maubois
Laboratoire de Recherches Laitières, Institut National de la Recherche Agronomique, Rennes Cedex, France.

P.L.H. McSweeney
Department of Food Chemistry, University College, Cork, Republic of Ireland.

V.V. Mistry
Dairy Science Department, South Dakota State University, Brookings, South Dakota 57007, USA.

J.H. Prentice
Rivendell, 3 Millbrook Dale, Axminster, Devon EX13 7TF, UK.

E. Renner
Dairy Science Section, Justus Liebig University, Bismarckstrasse 16, D-6300 Giessen, Germany.

L.B. Smith
Department of Food Science and Nutrition, University of Minnesota, St. Paul, Minnesota 55108, USA.

J. Wallace
Department of Food Chemistry, University College, Cork, Republic of Ireland.

P. Walstra
Department of Food Science, Wageningen Agricultural University, Bomenweg 2, 6703 HD Wageningen, The Netherlands.

M.G. Wilkinson
National Dairy Products Research Centre, Teagasc, Moorepark, Fermoy, Co. Cork, Republic of Ireland.

E.A. Zottola
Department of Food Science and Nutrition, University of Minnesota, St. Paul, Minnesota 55108, USA.

1

Cheese: An Overview

P. F. Fox

Department of Food Chemistry, University College, Cork, Republic of Ireland

1 HISTORICAL

Cheese is the generic name for a group of fermented milk-based food products, produced in a great range of flavours and forms throughout the world. From humble beginnings, i.e. simply as a means of conserving milk constituents, cheese has evolved to become a food of *haute cuisine* with epicurean qualities, as well as being highly nutritious. Sandine & Elliker[1] suggest that there are more than 1000 cheese varieties. Walter & Hargrove[2] describe more than 400 varieties and list the names of a further 400, while Burkhalter[3] classified 510 varieties (although some are listed more than once).

It is commonly believed that cheese evolved in the 'Fertile Crescent' between the Tigris and Euphrates, in Iraq, some 8000 years ago. The so-called 'Agricultural Revolution' occurred here with the domestication of plants and animals. Presumably, Man soon realized the nutritive value of milk produced by his domesticated animals and contrived to share the mother's milk with her offspring. Unfortunately, milk is also a rich source of nutrients for bacteria which contaminate the milk, some species of which utilize milk sugar, lactose, as a source of energy, producing lactic acid as a by-product. Bacterial growth and acid production would have occurred during storage or during attempts to dry milk in the prevailing warm, dry climate to produce a more stable product—air-drying of meat, and probably fruits and vegetables, appears to have been practised as a primitive form of food preservation at this period in Man's evolution. When sufficient acid has been produced, the principal proteins of milk, the caseins, coagulate, i.e. at their isoelectric points, to form a gel entrapping the fat; thus the world's first fermented dairy foods were produced accidentally. Similar products are still popular in the Middle East and throughout the world.

The first fermented dairy foods were produced by a fortuitous combination of events—the ability of a group of bacteria, now known as the lactic acid bacteria, to grow in milk and to produce just enough acid to reduce the pH of milk to the isoelectric point of the caseins, at which these proteins coagulate. Neither the lactic acid bacteria nor the caseins were designed for this function. The caseins

1

were 'designed' to be enzymatically coagulated in the stomachs of neonatal mammals, the gastric pH of which is around 6, i.e. very much higher than the isoelectric point of the caseins. The ability of lactic acid bacteria for ferment lactose, a sugar specific to milk, is frequently encoded on plasmids, suggesting that this characteristic was acquired relatively recently in the evolution of these bacteria. Their natural habitats are vegetation and/or the intestines, from which they presumably colonized the teats of mammals, contaminated with lactose-containing milk; through evolutionary pressure, these bacteria acquired the ability to ferment lactose.

When an acid milk gel is broken, e.g. accidentally by movement of the storage vessels or intentionally by breaking or cutting, it separates into curds and whey. It would have been realized quickly that the acid whey is a pleasant, refreshing drink for immediate consumption while the curd could be consumed fresh or stored for future use. It was probably soon realized that the shelf-life of the curd could be extended greatly by dehydration and/or by adding salt; heavily-salted cheese varieties (e.g. Feta and Domiati) are still widespread throughout the Middle East.

This is the presumed origin of one of the principal families of cheeses, the acid cheeses, modern members of which include Cottage cheese, Cream cheese, Quarg and Quesco Blanco. While lactic acid, produced *in situ*, is believed to have been the original milk coagulant, an alternative mechanism was also recognized from an early date. Many proteolytic enzymes can modify the milk protein system, causing it to coagulate under certain circumstances. Enzymes capable of causing this transformation are widespread in nature, e.g. bacteria, moulds, plant and animal tissues, but the most obvious source would have been animal stomachs. It would have been observed that the stomachs of slaughtered young animals frequently contained curd, especially if the animals had suckled shortly before slaughter; curd would also have been observed in the vomit of human infants. Before the development of pottery (*c.*5000 BC), storage of milk in bags made from animals skins was probably common (as it still is in many countries). Stomachs from slaughtered animals provided ready-made, easily-sealed containers; under such circumstances, milk would extract coagulating enzymes from the stomach tissue, leading to its coagulation during storage.

The properties of rennet curds are very different from those produced by isoelectric (acid) precipitation, e.g. they have better syneresis properties which make it possible to produce low-moisture cheese curd without hardening. Rennet curd can, therefore, be converted to a more stable product than acid curds and rennet coagulation has become predominant in cheese manufacture and is the starting material for the vast majority of modern cheese varieties. Although animal rennets were probably the first enzyme coagulants used, rennets produced from a range of plant species, e.g. figs and thistles, appear to have been common in Roman times. However, plant rennets are not suitable for the manufacture of long-ripened cheese varieties and gastric proteinases from young animals became the standard rennets until shortage of supply made it necessary to introduce 'rennet substitutes', as will be discussed later.

While the coagulation of milk by the in-situ production of lactic acid was, presumably, accidental, the use of rennets to coagulate milk was intentional. It was, in fact, quite an ingenious invention—if the conversion of milk to cheese by the use of rennets were discovered today, it would be hailed as a major biotechnological discovery!

The advantages accruing from the ability to convert the principal constituents of milk to cheese would have been apparent from the viewpoints of storage stability, ease of transport and, probably eventually, as a means of diversifying the human diet. Cheese manufacture accompanied the spread of civilization throughout the Middle East, Egypt, Greece and Rome. There are several references in the Old Testament to cheese, e.g. Job (1520 BC) and Samuel (1170–1017 BC), in the tombs of Ancient Egypt and in classical Greek literature, e.g. Homer (1184 BC), Herodotus (484–408 BC), Aristotle (384–322 BC). Apparently, cheese was prescribed for the diet of Spartan wrestlers in training. Cheese manufacture was well established by the time of the Roman Empire and was a standard item in the rations issued to Roman soldiers. Cheese must have been popular with Roman civilians also and demand must have exceeded supply since the Emperor Diocletian (AD 284–305) was forced to fix a maximum price for cheese. Many Roman writers, e.g. Cato (about 150 BC), Varro (about 40 BC), Columella (AD 50), Pliny (AD 23–89), wrote at some length on cheese manufacture and quality and on the culinary uses of cheese; Columella, in particular, gave a detailed account of cheese manufacture in his treatise on agriculture, *De re Rustica*.

Movements of Roman armies and administrators would have spread the consumption of cheese throughout the then known world. Although archaeological evidence suggests that cheese may have been manufactured in pre-Roman Britain, the first unequivocal evidence credits the Romans with the establishment of cheesemaking in Britain. Palladius wrote a treatise on Roman–British farming in the 4th century AD, including a description of and advice on cheesemaking. Cheesemaking practice appears to have changed little from the time of Columella and Palladius until the 19th century.

The great migrations of peoples throughout Europe after the fall of the Roman Empire must have promoted the further spread of cheese manufacture, as must the Crusaders and other pilgrims of the Middle Ages. However, the most important agents contributing to the development of cheese 'technology' and to the evolution of cheese varieties were probably the monasteries and the feudal estates. In addition to their roles in the spread of Christianity and in the preservation and expansion of knowledge during the Dark Ages, the monasteries made considerable contributions to the advancement of agriculture in Europe and to the development and improvement of food commodities, notably wine and cheese. Many of our current well known cheese varieties were developed in monasteries, e.g. Wensleydale (Rievaulx Abbey, Yorkshire), Port du Salut or Saint Paulin (Monastery de Notre Dame du Port du Salut, Laval, France), Fromage de Tamie (Abbey of Tamie Lac d'Annecy, Geneva), Maroilles (Abbey Moroilles, Avesnes, France), Trappist (Maria Stern Monastery, Banja Luka, Bosnia). The inter-monastery movement of monks would have contributed to the spread of cheese varieties and probably to the development of new hybrid varieties.

The great feudal estates of the Middle Ages were self-contained communities. The conservation of surplus food produced in summer for use during winter was a major activity on such estates and undoubtedly cheese represented one of the more important of these conserved products, along with cereals, dried and salted meats, dried and fermented fruits and vegetables, beer and wine. Cheese probably represented an item of trade when amounts surplus to local requirements were available. Within these estates, individuals acquired specific skills which were passed on to succeeding generations. The feudal estates evolved into villages and some, later, into larger communities.

Because monasteries and feudal estates were essentially self-contained communities, it is readily apparent how several hundred distinct varieties of cheese could have evolved from essentially the same raw material, milk or rennet curd, especially under conditions of limited communication. Traditionally, many cheese varieties were produced in quite limited geographical regions, especially in mountainous areas. The localized production of certain varieties is still apparent and indeed is preserved for those varieties with a designation of Appellation d'Origine. Regionalization of certain cheese varieties is particularly marked in Spain and Italy, where the production of many varieties is restricted to very limited regions. Almost certainly, most cheese varieties evolved by accident because of a particular set of local circumstances, e.g. a peculiarity of the local milk supply, either with respect to chemical composition or microflora, an 'accident' during storage of the cheese, e.g. growth of mould or other microorganisms. Presumably, those accidents that led to desirable changes in the quality of the cheese would have been incorporated into the manufacturing protocol; each variety would thus have undergone a series of evolutionary changes and refinements.

Cheesemaking remained an art rather than a science until relatively recently. With the gradual acquisition of knowledge on the chemistry and microbiology of milk and cheese, it became possible to direct the changes involved in cheesemaking in a more controlled fashion. Although few new varieties have evolved as a result of this improved knowledge, the existing varieties have become better defined and their quality more consistent. Considering the long history of cheesemaking, we might be inclined to the idea that what we have come to regard as standard varieties have been so for a long time. However, although the names of many current varieties were introduced several hundred years ago (Table I), these cheeses were not standardized; for example, the first attempt to standardize the well known English varieties, Cheddar and Cheshire, was made by John Harding in the mid-19th century. Prior to that, 'Cheddar cheese' was that produced in a particular area in England around the village of Cheddar, Somerset, and probably varied considerably depending on the manufacturer and other factors. It must also be remembered that cheese manufacture was a farmstead enterprise until the mid-19th century—the first cheese factory in the US was established near Rome, NY, in 1851 and the first in Britain at Longford, Derbyshire, in 1870. Thus, there were thousands of cheese manufacturers and

TABLE I
First Recorded Date for some Major Cheese Varieties[a]

Gorgonzola	897	Cheddar	1500
Schabzieger	1000	Parmesan	1579
Roquefort	1070	Gouda	1697
Maroilles	1174	Gloucester	1783
Schwangenkäse	1178	Stilton	1785
Grana	1200	Camembert	1791
Taleggio	1282	St Paulin	1816

[a] From Scott.[4]

there must have been great variation within any one general type. When one considers the very considerable inter-factory, and indeed intra-factory, variations in quality and characteristics which still occur today in well-defined varieties, e.g. Cheddar, in spite of the very considerable scientific and technological advances, one can readily appreciate the variations that must have existed in earlier times.

Some major new varieties, notably Jarlsberg and Maasdamer, have been developed recently as a consequence of scientific research. Many other varieties have evolved very considerably, even to the extent of becoming new varieties, as a consequence of scientific research and the development of new technology—notable examples are (US) Quesco Blanco, Feta produced by ultrafiltration and various forms of Quarg. There has been a marked resurgence of farmhouse cheesemaking in recent years; many of the cheeses being produced on farms are not standard varieties and it will be interesting to see if some of these evolve to become new varieties.

A major source of variation in the characteristics of cheese resides in the species from which the milk was produced. Although milks from several species are used in cheese manufacture, the cow is by far the most important, while sheep, goat and buffalo are commercially important in certain areas. There are very significant interspecies differences in the composition of milk which are reflected in the characteristics of the cheeses produced from them. Major interspecies differences of importance in cheesemaking are the concentration and types of caseins, concentration of fat and especially the fatty acid profile, concentration of salts, especially of calcium. There are also significant differences in milk composition between breeds of cattle and these also influence cheese quality, as do variations due to seasonal, lactational and nutritional factors and of course the methods of milk production, storage and collection.

The final chapter in the spread of cheese throughout the world resulted from the colonization of North and South America, Oceania and Africa by European settlers who carried their cheesemaking skills with them. Cheese has become an item of major economic importance in some of these 'new' countries, notably the US, Canada, Australia and New Zealand, but the varieties produced are mainly of European origin, modified in some cases to meet local requirements. It is not certain whether or not cheeses were manufactured in these regions before colonization by Europeans but in most cases, probably not.

For further information on the history of cheese, the reader is referred to References 4–7. For references on Roman agriculture, see White.[8]

2 CHEESE PRODUCTION AND CONSUMPTION

World cheese production was ~12·5 × 10^6 tonnes in 1988 and has increased at an average annual rate of ~4% over the past 20 years.[9] Europe, with a production ~6 × 10^6 tonnes p.a., is by far the largest producing block; North American and USSR, with populations approximately similar to Europe, produce approximately 50% and 12% of European production, respectively (Table II).

Cheese consumption varies widely between countries, even within Europe; it is noteworthy that with the exception of Israel, no Asian, African or South American country is listed among the top 25 cheese-consuming countries (Table III).[10] Cheese consumption in most countries for which data are available has increased considerably since 1970.

TABLE II
Cheese Production, '000 tonnes, in the Leading Countries in 1988[9]

Belgium	91·8	Australia	182·5
Denmark	259·6	New Zealand	128·4
France	1291·0	Japan	83·0
Germany	1008·0		
Irish Republic	78·7	Total Pacific	393·9
Italy	660·0		
Luxembourg	3·4	Czechoslovakia	144·0
Netherlands	572·0	Hungary	54·3
United Kingdom	298·8	Poland	472·5
Greece[a]	206·0	USSR	890·0
Portugal	40·0	Other Eastern Europe	597·4
Spain	184·5		
		Total Eastern Europe	2158·1
Total EEC	4693·8		
		India	146·0
Austria	83·0	Zimbabwe	1·2
Finland	86·6	Uruguay	16·2
Norway	77·7	Israel	73·4
Sweden	123·0	South Africa	36·2
Switzerland	127·6	China	0·1
Iceland	3·4	Other countries	1479·7
Total other Western Europe	501·3	Total World	12 423·6
Canada	251·2		
USA	2527·2		
Total North America	2778·4		

[a] From Ref. 9a.

TABLE III
Consumption per Head of Cheese (kg per caput) 1988[10]

Country	Hard cheeses	Fromage frais	Total
France	15·1	6·9	22·0
Greece[a]	20·7	0·9	21·6
Italy	13·3	4·6	17·9
Federal Republic of Germany	9·7	7·7	17·4
Israel	3·8	13·1	16·9
Iceland	10·2	6·2	16·4
Sweden	15·0	0·9	15·9
Belgium	12·2	3·6	15·8
Switzerland	13·5	1·4	14·9
Netherlands	13·4	1·3	14·7
Norway	13·1	0·2	13·3
Bulgaria	12·8	0·4	13·2
Denmark	11·8	0·9	12·7
Czechoslovakia	7·0	5·6	12·6
USA	10·7	1·8	12·5
Finland	10·0	2·2	12·2
Canada	10·6	1·2	11·8
Poland	3·6	8·0	11·6
Luxembourg	—	—	10·6
Austria	6·7	3·6	10·3
Hungary	4·3	5·0	9·3
Australia	8·4	0·8	9·2
New Zealand	8·4	—	8·4
United Kingdom	7·4	0·8	8·2
USSR	3·1	3·3	6·4
Spain	—	—	5·2
Ireland	—	—	5·0
South Africa	1·5	0·1	1·6
Japan	1·1	0·05	1·1
India	—	0·2	0·2

[a] From International Dairy Federation Bulletin 160 (1983).

Thus, while cheese manufacture is practised world-wide, it is apparent from Tables II and III that cheese is primarily a product of European countries and those populated by European emigrants. With a few exceptions, notably Egypt, cheese is of relatively little importance in Asia, Africa and Latin America where diets are based much more strongly on plant than on animal products and where no tradition of dairying exists. However, cheese in some form is produced in most countries throughout the world and some interesting minor varieties are produced in 'non-dairying' countries (see Chapter 14, Volume 2).

3 CHEESE SCIENCE AND TECHNOLOGY

Cheese is the most diverse group of dairy products and is, arguably, the most academically interesting and challenging. While many dairy products, if properly manufactured and stored, are biologically, biochemically and chemically very stable, cheeses are, in contrast, biologically and biochemically dynamic, and, consequently, inherently unstable. Throughout manufacture and ripening, cheese production represents a finely orchestrated series of consecutive and concomitant biochemical events which, if synchronized and balanced, lead to products with highly desirable aromas and flavours but when unbalanced, result in off-flavours and odours. Considering that, in general terms, a basically similar raw material (milk from a very limited number of species) is subjected to a manufacturing protocol, the general principles of which are common to most cheese varieties, it is fascinating that such a diverse range of products can be produced. No two batches of the same variety, indeed no two cheeses, are identical.

A further important facet of cheese is the range of scientific disciplines involved: study of cheese manufacture and ripening involves the chemistry and biochemistry of milk constituents, fractionation and chemical characterization of cheese constituents, microbiology, enzymology, molecular genetics, flavour chemistry, rheology and chemical engineering.

It is not surprising, therefore, that many scientists have become involved in the study of cheese manufacture and ripening. A voluminous scientific and technological literature has accumulated, including several textbooks[4-7,11-21] and chapters in many others. However, these textbooks deal mainly with cheese technology; this book concentrates on the more scientific aspects of cheese.

The more general aspects of cheese manufacture, i.e. molecular properties of rennets, coagulation mechanism, curd syneresis, starters, salting, rheology, an overview of the biochemistry of ripening, pre-concentration by ultrafiltration, nutritional aspects and analytical techniques, which apply, more or less, to most cheese varieties, are considered in the first volume of this text. The second volume deals with specific aspects of the principal families of cheese.

The principal objective of this introductory chapter is to provide an integrated overview of cheese manufacture and to provide some general background for the more detailed later chapters that follow.

4 OUTLINE OF CHEESE MANUFACTURE

Although some soft cheese varieties are consumed fresh, i.e. without a ripening period, production of the vast majority of cheese varieties can be sub-divided into two well defined phases, manufacture and ripening:

$$\text{Milk} \xrightarrow{\text{Manufacture}} \begin{array}{c}\text{fresh}\\ \text{cheese}\\ \text{curd}\end{array} \xrightarrow{\text{Ripening}} \begin{array}{c}\text{Mature}\\ \text{cheese}\end{array}$$

The manufacturing phase might be defined as those operations performed during the first 24 h, although some of these operations, e.g. salting and dehydration, may continue over a longer period. Although the manufacturing protocols for individual varieties differ in detail, the basic steps are common to most varieties;

these are: acidification, coagulation, dehydration (cutting the coagulum, cooking, stirring, pressing, salting and other operations that promote gel syneresis), shaping (moulding and pressing) and salting.

Cheese manufacture is essentially a dehydration process in which the fat and casein in milk are concentrated between 6- and 12-fold, depending on the variety. The degree of hydration is regulated by the extent and combination of the above five operations, in addition to the chemical composition of the milk. In turn, the levels of moisture and salt, the pH and the cheese microflora regulate and control the biochemical changes that occur during ripening and hence determine the flavour, aroma and texture of the finished product. Thus, the nature and quality of the finished cheese are determined to a very large extent by the manufacturing steps. However, it is during the ripening phase that the characteristic flavour and texture of the individual cheese varieties develop.

5 SELECTION AND PRE-TREATMENT OF CHEESE MILK

Cheese manufacture commences with the selection of milk of high microbiological and chemical quality. The importance of microbiological quality will not be considered here; suffice it to say that milk of the highest quality should be used, with particular attention being paid to *Clostridium tyrobutyricium* in milk to be used for Dutch and Swiss cheese varieties. Obviously, cheesemilk must be free of antibiotics. The importance of the chemical quality of cheesemilk can best be treated in the chapters dealing with the rennet coagulation of milk and subsequent chapters on curd tension, gel syneresis and cheese texture.

In modern commercial practice, milk for cheese is normally cooled to 4°C immediately after milking and may be held at about this temperature for several days on the farm and at the factory. Apart from the development of an undesirable psychrotrophic microflora, cold-storage causes physico-chemical changes (e.g. shifts in calcium phosphate equilibrium and dissociation of some micellar caseins) which have undesirable effects on the cheesemaking properties of the milk; these changes will be discussed at some length in several subsequent chapters.

Although raw milk is still used in both commercial and farmhouse cheesemaking, most cheesemilk is now pasteurized, usually immediately before use. Pasteurization alters the indigenous microflora and facilitates the manufacture of cheese of more uniform quality, but unless due care is exercised, it may damage the rennet coagulability and curd-forming properties of the milk, as will be discussed in later chapters.

Even when properly pasteurized, Cheddar cheese (and probably other varieties) made from pasteurized milk develops a less intense flavour and ripens more slowly than raw milk cheese. Several heat-induced changes, e.g. inactivation of indigenous milk enzymes, killing of indigenous microorganisms, denaturation of whey proteins and their interaction with micellar κ-casein, perhaps even shifts in salt equilibria and destruction of vitamins, could be responsible for these changes. Until now it has not been possible to establish which of these factors was principally responsible for the differences in quality between raw and pasteurized milk cheeses. However, using microfiltration it is possible to remove more than 99·9% of the indigenous microorganisms from milk without inducing any of the other changes mentioned above.

In a recent study, McSweeney et al.[22] compared the quality of Cheddar made from raw, pasteurized or microfiltered milk. The cheeses from pasteurized and microfiltered milk were of good and equal quality; although the raw milk cheese was downgraded because the flavour was atypical, its flavour was much more intense and developed much faster than that of the raw milk cheeses. Polyacrylamide gel electrophoretograms of the three cheeses did not differ throughout ripening but the rate of formation of water soluble N was much faster in the raw milk cheese and PAGE and RP-HPLC showed that soluble peptides in the raw milk cheese differed markedly from those in the other cheeses. Other chemical differences between the cheeses are being sought. The number of lactobacilli was about 10-fold higher in the raw milk cheese than in the others and the species of lactobacilli also differed. These preliminary results indicate that the indigenous microorganisms probably play a significant role in flavour development in raw milk Cheddar cheese, although the flavour is atypical for current factory-made Cheddar.

Pasteurization of cheesemilk minimizes the risk of cheese serving as a vector for food-poisoning or pathogenic microorganisms, so that even high quality raw milk may be unacceptable for cheese manufacture. Inoculation of pasteurized milk with selected organisms from the indigenous microflora of raw milk probably warrants investigation.

Thermization (~65°C × 15 s) of cheesemilk on arrival at the factory is common or standard practice in some countries. The objective is to control psychrotrophs and the milk is normally pasteurized before cheesemaking.

Not more than 75% of the total protein in milk is recovered in rennet cheeses. Obviously, a considerable economic advantage would accrue if at least some of the whey proteins could be incorporated into the cheese. In recent years, ultrafiltration (UF) has been investigated extensively as a means of accomplishing this, with considerable success in the case of soft cheese, especially Feta and Quarg, but with less success for hard and semi-hard varieties. The application, and associated problems, of UF in cheese manufacture was reviewed by Lawrence[23] and is comprehensively reviewed in Chapter 13, Volume 1.

An alternative approach is to heat denature the whey proteins (e.g. 90°C × 1 min) to induce their interaction with the casein micelles. Normally, such severe heat treatments are detrimental to the renneting properties of milk but the effects can be offset by acidification or supplementation with calcium (see Chapters 3 and 4, Volume 1). In the author's experience, yield increases of up to 8% can be achieved by this approach, while retaining acceptable quality. However, to the author's knowledge, the technique is not used commercially except for Quarg, e.g. the thermo-Quarg process (see Chapter 13, Volume 2).

6 ACIDIFICATION

One of the basic operations in the manufacture of most, if not all, cheese varieties is a progressive acidification throughout the manufacturing stage, i.e. up to 24 h, and for some varieties during the early stages of ripening also, i.e. acidification commences before and transcends the other manufacturing operations. Acidification is normally via in-situ production of lactic acid, although pre-formed acid or acidogen (usually gluconic acid-δ-lactone) are now used to directly acidify curd for some varieties, e.g. Mozzarella, UF Feta and Cottage. Until relatively recently, the indigenous microflora of milk was relied upon for acid production. Since this was probably a mixed microflora, the rate of acid production was unpredictable and the growth of undesirable bacteria led to the production of gas and off-flavours. It is now almost universal practice to add a culture (starter) of selected lactic acid-producing bacteria to pasteurized cheesemilk to achieve a uniform and predictable rate of acid production. For cheese varieties that are cooked to not more than 40°C, a starter consisting of *Lactococcus lactis* subsp *lactis* and/or *Lc. lactis* subsp *cremoris* is normally used while a mixed culture of *Str. salavarius* var *thermophilus* and *Lactobacillus* spp (*Lb. bulgaricus, Lb. helviticus, Lb. casei*) or a *Lactobacillus* culture alone is used for varieties that are 'cooked' to higher temperatures, e.g. Swiss and hard Italian varieties.

Originally, and in many countries still, mixed-strain mesophilic starters were used. Because the bacterial strains in these starters may be phage-related (i.e. subject to infection by a single strain of bacteriophage) and also because the strains in the mixture may be incompatible, thereby leading to the dominance of one or a few strains, the rate of acid production by mixed-strain starters is variable and unpredictable, even when the utmost care in their selection and handling is exercised. To overcome these problems, single-strain mesophilic starters were introduced in New Zealand about 1935. Unfortunately, many of the fast acid-producing, single-strain starters produced bitter cheese, the cause(s) of which will be discussed in later chapters. This problem was resolved by using selected pairs of fast and slow acid producers. Defined-strain mesophilic starters are widely used in many countries, frequently consisting of a mixture (cocktail) of two to six selected, phage-unrelated strains which give very reproducible rates of acid production if properly selected and maintained. The use of defined-strain thermophilic starters is, as yet, rather limited.

The science and technology of starters have become highly developed and specialized; Chapter 6, Volume 1, is devoted to these developments. Other reviews on starters include References 24–32.

Acid production at the appropriate rate and time is the key step in the manufacture of good quality cheese (excluding the enzymatic coagulation of the milk, which is a *sine qua non* for rennet cheese varieties). Acid production affects several aspects of cheese manufacture, many of which will be discussed in more detail in later chapters, i.e.:

1. Coagulant activity during coagulation.
2. Denaturation and retention of the coagulant in the curd during manufacture and hence the level of residual coagulant in the curd; this influences the rate of proteolysis during ripening, and may affect cheese quality.
3. Curd strength, which influences cheese yield.
4. Gel syneresis, which controls cheese moisture and hence regulates the growth of bacteria and the activity of enzymes in the cheese; consequently, it strongly influences the rate and pattern of ripening and the quality of the finished cheese.
5. The rate of pH decrease determines the extent of dissolution of colloidal calcium phosphate which modifies the susceptibility of the caseins to proteolysis during manufacture and influences the rheological properties of the cheese, e.g. compare the texture of Emmental, Gouda, Cheddar and Cheshire cheese.
6. Acidification controls the growth of many species of non-starter bacteria in cheese, especially pathogenic, food poisoning and gas-producing microorganisms—in fact properly-made cheese is a very safe product from the public health viewpoint. In addition to producing acid, many starter bacteria produce probiotics that also restrict or inhibit the growth of non-starter microorganisms.

Mesophilic *Lactococcus* spp are capable of reducing the pH of cheese to 4·9–5·0 and *Lactobacillus* spp to somewhat lower values, perhaps 4·6. Thus, the natural ultimate pH of cheese curd falls within the range 4·6–5·1. However, the period required to attain the ultimate pH varies from ~5 h for Cheddar to 6–12 h for Blue, Dutch and Swiss varieties. The differences arise from the amount of starter added to the cheesemilk (0·2–5%), the cooking schedule which may retard the growth of the starter microorganisms and the rate of subsequent cooling of the curd. The pH of Blue cheese curd is 6·1 at draining and decreases smoothly to 5·1 within ~6 h. The pH of Gouda curd also decreases smoothly whereas that of Cheddar is irregular due to the higher cooking temperature used in the latter which retards acid development during cooking.

Acidification is more complex in Emmental and Gruyère cheeses which are cooked to high temperatures (53–56°C) and in which the starter (*Lb. helveticus* or *Lb. casei* and *Str. thermophilus*) grow mainly after pressing. The curd is placed in the press at ~50°C and cools during pressing at a markedly faster rate at the periphery than at the centre; hence, the rate of starter growth and of acid development vary throughout a cheese.[33] This results in a lactic acid gradient from the periphery to the centre of the cheese which approaches equilibrium, due to diffusion, as the cheese ages. The concentration of lactic acid and pH influence the development of *Propionibacteria shermanii* and hence the amount of CO_2 produced. Variations in pH probably also influence the functional properties of the cheese proteins and hence cheese texture. Both the rate of production and the volume of CO_2 and the rheological properties of the protein matrix control eye development, number and distribution, which is an essential feature of these cheese varieties.

The level and method of salting have a major influence on pH changes in cheese. The concentration of NaCl in cheese (commonly 0·7–4%, i.e. 2–10% salt in the moisture phase) is sufficient to halt the growth of starter bacteria. Some varieties, mostly of British origin, are salted by mixing dry salt with the curd towards the end of manufacture and hence the pH of curd for these varieties must be close to the ultimate value (~pH 5·1) at salting. However, most varieties are salted by immersion in brine or by surface application of dry salt; as will be discussed in Chapter 7, Volume 1, salt diffusion in cheese moisture is a slow process and thus there is ample time for the pH to decrease to ~5·0 before the salt concentration becomes inhibitory. The pH of the curd for most cheese varieties, e.g. Swiss, Dutch, Tilsit, Blue, etc., is 6·2–6·5 at moulding and pressing but decreases to ~5 during or shortly after pressing and before salting. The significance of various aspects of the concentration and distribution of NaCl in cheese will be discussed in Chapter 7, Volume 1.

In a few special cases, e.g. Domiati, a high level of NaCl (10–12%) is added to the cheesemilk, traditionally to control the growth of the indigenous microflora. This concentration of NaCl has a major influence, not only on acid development, but also on rennet coagulation, gel strength and syneresis (cf. Chapter 9, Volume 2).

7 COAGULATION

The essential characteristic step in the manufacture of all cheese varieties involves coagulation of the casein component of the milk protein system to form a gel which entraps the fat, if present. Coagulation may be achieved by:

1. limited proteolysis by selected proteinases;
2. acidification to pH ~4·6;
3. acidification to pH values >4·6 (perhaps 5·2) in combination with heating.

The majority of cheeses are produced by enzymatic (rennet) coagulation. With a few exceptions, such as Serra de Estrela (Portugal) in which a plant proteinase, from the cardoon flowers of *Cynara cardunculus*, is used, acid (aspartate) proteinases of animal or fungal origin are used. Chymosin from the stomachs of young animals (calves, kids, lambs, buffaloes) was used traditionally as rennets but limited supplies of such rennets (due to the increasing trend in many countries to slaughter calves at an older age than previously), together with a worldwide increase in cheese production, has led to a shortage of calf rennet, and consequently rennet substitutes (usually bovine or porcine pepsins and less frequently, chicken pepsin, and the acid proteinases from *Mucor miehei* and less frequently *M. pusillus* or *Endothia parasitica*) are now used widely for cheese manufacture in many countries with more or less satisfactory results. The calf chymosin gene has been cloned in *K. lactis*, *E. coli* and *A. niger* and chymosin from these organisms is now commercially available. Reviews on rennet substitutes include References 34–41.

The molecular and enzymatic properties of calf chymosin and other acid proteinases used as rennets are reviewed in detail in Chapter 2, Volume 1.

Although it appears to have been recognized since 1917 (see Ref. 42) that milk is not coagulated by rennet at low temperatures, Berridge[42] is usually credited with clearly demonstrating that the rennet-catalysed coagulation of milk occurs in two phases: a primary enzymatic phase and a secondary non-enzymatic phase. The primary phase has a temperature coefficient (Q_{10}) of ~2 and occurs down to 0°C, while the secondary phase has a Q_{10} of ~16 and occurs very slowly or not at all at temperatures < ~15°C. The two phases can thus be readily separated by performing the primary phase below 15°C; if cold-renneted milk is warmed, coagulation occurs very quickly. Cold renneting, followed by rapid warming, forms the basis of attempts to develop methods for the continuous coagulation of milk. Normally, the two phases of rennet coagulation overlap to some extent, the magnitude of overlap being quite extensive at low pH, high temperatures and in milks concentrated by ultrafiltration (see Ref. 43).

The primary phase of rennet action appears to have been recognized, in general terms, by Hammersten (1880–90) who observed the formation of small peptides during renneting. Views on rennet coagulation were extended by Linderstrom-Lang in the 1920s but a full explanation of the process had to await the isolation of the casein micelle-protective protein, κ-casein, by Waugh & von Hippel.[44] These workers showed that the protective capacity of κ-casein was destroyed on renneting and Wake[45] demonstrated that κ-casein is the only milk protein hydrolysed during the primary phase of rennet action. Only one peptide bond, Phe105–Met106, is hydrolysed,[46] resulting in the release of the hydrophilic C-terminal segment of κ-casein [the (caseino) macropeptides, some of which are glycosylated]. The unique sensitivity of the Phe–Met bond of κ-casein, hydrolysis of which occurs optimally at pH 5·1–5·5[47] has been the subject of extensive study since 1965 and this work is reviewed in Chapter 3, Volume 1.

Since κ-casein is the principal factor stabilizing the casein micelles (see Refs 48, 49), its hydrolysis destabilizes the 'residual' (para-casein) micelles, which coagulate in the presence of a critical concentration of Ca^{2+} at temperatures > ~20°C, i.e. the secondary, non-enzymatic phase of rennet coagulation.

Although the precise mechanism of coagulation has not yet been described, the kinetics of the process can be described by the theory of Smoluchowski for the slow aggregation of hydrophobic colloids. Hydrolysis of κ-casein during the primary phase of rennet action reduces the zeta potential of the residual micelles to about 50% of that of the native micelles and destroys the steric stabilization caused by the protruding C-terminal segments of κ-casein ('hairs'). Presumably, loss of these stabilizing factors is the major cause of coagulation, although a critical concentration of Ca^{2+} and a minimum temperature (~20°C) are required for coagulation. Reduction of the colloidal calcium phosphate (CCP) content of the casein micelles prevents coagulation unless the $[Ca^{2+}]$ is increased.[50] This is perhaps unexpected since CCP-free milk is unstable to Ca^{2+}; the Ca sensitivity of renneted CCP-free milk has not been investigated. Perhaps disruption of the micellar structure by removal of CCP alters the 'conformation' of the micelles such that a protein gel network cannot be formed.

Recent work on the secondary phase of rennet coagulation is reviewed in Chapters 3 and 4, Volume 1.

The visual coagulation of milk is really only the start of the gelation process which continues for a considerable period thereafter. Although these post-coagulation changes determine many of the critical cheesemaking properties of the gel, e.g. curd tension (which influences cheese yield) and syneresis properties (which determine the moisture content and hence the ripening profile of the cheese), it is perhaps the least well understood phase of the cheesemaking process. The recent literature on aspects of the post-visual coagulation phase is reviewed in Chapters 4 and 5, Volume 1.

8 POST-COAGULATION OPERATIONS

A rennet gel is quite stable if maintained under quiescent conditions but if it is cut or broken, syneresis occurs rapidly, expelling whey. The rate and extent of syneresis are influenced, *inter alia*, by milk composition, especially $[Ca^{2+}]$ and [casein], pH of the whey, cooking temperature, rate of stirring of the curd–whey mixture and of course, time (see Chapter 5, Volume 1). The composition of the finished cheese is to a very large degree determined by the extent of syneresis and since this is readily under the control of the cheesemaker, it is here that the differentiation of the individual cheese varieties really begins, although the composition of cheesemilk, the amount and type of starter and the amount and type of rennet are also significant in this regard. The unique manufacturing schedules for the specific varieties are not considered in this text and the interested reader should consult appropriate texts.[2,4,11-21] Some chemical and physico-chemical aspects of the manufacture of the major cheese families are discussed in Volume 2.

The last manufacturing operation is salting. While salting contributes to syneresis, it should not be used as a means of controlling the moisture content of cheese. Salt has several functions in cheese which are described in Chapter 7, Volume 1. Although salting should be a very simple operation, quite frequently it is not performed properly, with consequent adverse effects on cheese quality.

As indicated previously, cheese manufacture is essentially a dehydration process. With the development of ultrafiltration as a concentration process it was obvious that this process would have applications in cheese manufacture, not only for standardization of cheesemilk with respect to fat and casein, but more importantly for the preparation of a concentrate with the composition of the finished cheese, commonly referred to as 'pre-cheese'. Standardization of cheesemilk by adding UF concentrate (retentate) is now common in some countries but the manufacture of pre-cheese has to date been successful commercially for only certain cheese varieties, most notably Feta and Quarg. Undoubtedly, the use of ultrafiltration will become much more widespread in cheese manufacture (see Chapter 13, Volume 1). The potential of microfiltration to control the indigenous microflora of cheesemilk is likely to attract industrial attention in the immediate future.

9 RIPENING

Some cheeses are consumed fresh and as indicated in Table III, fresh cheeses constitute a major proportion of the cheese consumed in some countries; the principal fresh cheeses are described in Chapter 13, Volume 2. However, most cheese varieties are not ready for consumption at the end of manufacture but undergo a period of ripening (curing, maturation) which varies from about three weeks to more than two years, the duration of ripening being generally inversely related to the moisture content of the cheese. Many varieties may be consumed at any of several stages of maturity, depending on the flavour preferences of consumers and economic factors.

Although curds for different cheese varieties are recognizably different at the end of manufacture (mainly as a result of compositional and textural differences arising from differences in milk compositional and processing factors), the unique characteristics of the individual cheeses develop during ripening, although in most cases the biochemical changes that occur during ripening, and hence the flavour, aroma and texture of the mature cheese, are largely predetermined by the manufacturing process, that is by composition, especially moisture, NaCl and pH, by the type of starter and in many cases by secondary inocula added to, or gaining access to, the cheesemilk or curd.

During ripening an extremely complex set of biochemical changes occurs through the catalytic action of the following agencies:

1. coagulant;
2. indigenous milk enzymes, especially proteinase and lipase, which are particularly important in cheese made from raw milk;
3. starter bacteria and their enzymes;
4. secondary microflora and their enzymes.

The secondary microflora may arise from the indigenous microflora of milk that survive pasteurization or gain entry to the milk after pasteurization, e.g. *Lactobacillus, Pediococcus, Micrococcus,* or they may arise through the use of a secondary starter, e.g. *Propionibacteria* in Swiss cheese, *Penicillium roqueforti* in Blue varieties, *P. caseicolum* in Camembert or Brie or the cheese may acquire a surface microflora from the environment during ripening, e.g. *Brevibacterium linens* in Tilsit, Limburger, etc. In many cases, the characteristics of the finished cheese result from the metabolic activity of these microorganisms.

The primary biochemical changes involve glycolysis, lipolysis and proteolysis but these primary changes are followed and overlapped by a host of secondary catabolic changes, including deamination, decarboxylation and desulphurylation of amino acids, β-oxidation of fatty acids and even some synthetic changes, e.g. esterification.

It is not possible to review the biochemistry involved in the ripening of all individual cheese varieties. Instead, an overview of the principal ripening reactions (proteolysis, glycolysis and lipolysis) is presented in Chapter 10, Volume 1, and the general rheological properties of cheese are reviewed in Chapter 8, Volume 1. More detailed discussions of specific aspects of the ripening of the principal families of cheese, Cheddar, Dutch, Swiss, Iberian, Italian, Balkan, North European, mould-ripened, bacterial surface-ripened, high-salt varieties, fresh (acid) cheeses and processed cheeses are given in Volume 2. An attempt is made in Chapter 14, Volume 2, to collate information on a number of minor but interesting non-European cheese varieties. In most cases, these varieties have not been well studied, microbiologically, biochemically or physically, and consequently the treatments are very superficial in comparison with those for the major European varieties. Nevertheless, it was considered that some of these minor varieties warranted attention; perhaps some interest will be generated in the study of these cheeses as a result.

While most people consume cheese principally for its organoleptic qualities, it must be remembered that cheese is a very valuable source of nutrients, especially protein, calcium and phosphorus. The nutritional aspects of cheese are considered in Chapter 15, Volume 1.

As in most areas of science and technology, fast, accurate and easy analytical methods are a key to success in the study of cheese ripening. Numerous methods have been developed to monitor and assess cheese ripening and many of these are reviewed in Chapter 9, Volume 1.

10 CLASSIFICATION OF CHEESE

A considerable international trade exists in the principal cheese varieties, many of which are produced in several countries but which may not be identical. To assist international trade, to provide nutritional information and perhaps for other reasons. e.g. research, a number of attempts have been made to develop a classification scheme for cheese varieties.

The difficulties in classifying cheese varieties were discussed by Schulz[51] who reviewed earlier attempts to do so. Schulz[51] was critical of these earlier schemes because they relied excessively on knowledge of the manufacturing process. He proposed a modified scheme consisting primarily of five groups based essentially on moisture content (moisture in fat-free cheese): dried (< 40%), grated (40–49·9%), hard (50–59·9%), soft (60–69·9%) and fresh (70–82%). Four of these groups (fresh, soft, hard, grated) were each sub-divided into two sub-groups (i.e. eight sub-groups) based on whether or not the cheeses were pressed and/or cooked. An interesting development was the subdivision of each of the eight sub-groups into six sub-sets (a–f) on the basis of the concentration of calcium in the fat-free, NaCl-free solids: >2·5%, 2·1–2·5%, 1·6–2·0%, 1·1–1·5%, 0·6–1·0%, <0·6%, which reflects the rate and extent of acidification.

Davis[7] discussed the problems encountered in attempting to classify cheese and suggested a number of possible schemes. One scheme (Table IV) was based on the rheological properties, or, more likely, on the moisture content. In fact, most schemes include similar criteria. In a second scheme, cheeses were classified[7] primarily into hard, semi-hard and soft which were subdivided on the basis of the principal characteristic microflora, e.g. normal lactic starter, propionibacteria, surface slime, surface mould or interior mould. Burkhalter[3] classified 510 varieties by the criteria summarized in Table V, again using moisture content as the primary criterion.

Scott[4] also classified cheeses primarily on the basis of moisture content, i.e. hard, semi-hard and soft, and sub-divided these groups on the basis of cooking (scalding) temperature and/or secondary microflora (Table VI).

A classification scheme suggested by P. Walstra (pers. comm.) is shown in Table VII. Interesting innovations are the use of the water : protein ratio rather than moisture content as the primary criterion for classification and replacement of cooking temperature by starter-type, i.e. mesophile, thermophile, which is essentially equivalent to the use of cooking temperature used by other authors.

Walter & Hargrove[2] suggest that there are probably only about 18 distinct types of natural cheese, no two of which are made by the same method, i.e. they differ with respect to: setting the milk, cutting, stirring, heating, draining, pressing and salting of the curd and curing of the cheese. He lists the following varieties as typical examples of the 18 types: Brick, Camembert, Cheddar, Cottage, Cream, Edam, Gouda, Hand, Limburger, Neufchatel, Parmesan, Provolone, Romano, Roquefort, Sapsago, Swiss, Trappist and whey cheeses. The authors acknowledged the imperfection and incompleteness of such a classification scheme and indeed a cursory glance at the list of examples highlights this. Edam and Gouda, for example, are clearly quite similar varieties while high-salt varieties, like Feta and Domiati, are omitted.

TABLE IV
Suggested Classification of Cheeses Based on Rheological Properties[a]

Type	Moisture, %[b]	pV	pM	pS
Very hard	<25	>9	>6·3	>2·3
Hard	25–36	8–9	5·8–6·3	2–2·3
Semi-hard	36–40	7·4–8	<5·8	1·8–2
Soft	>40	<7·4	<5·8	>1·8

pV = viscosity factor, logarithmic scale.
pM = elasticity factor, logarithmic scale.
pS = springiness factor, logarithmic scale.

[a] From Ref. 7.
[b] Suggested moisture levels appear to be very low.

TABLE V

Classification of Cheese According to Source of Milk, Moisture Content, Texture and Ripening Agents[a]

1. COWS' MILK

1.1 *Hard (<42% H_2O)*	1.2 *Semi-hard/semi-soft (43–55% H_2O)*	1.3 *Soft (>55% H_2O)*	1.4 *Fresh, rennet*	1.5 *Fresh, acid*	1.6 *Fresh*
1.1.1 Grating cheese (extra hard)	1.2.1 Small round openings	1.3.1 Blue veined			
1.1.2 Large round openings	1.2.2 Irregular openings	1.3.2 White surface mould			
1.1.3 Medium round openings	1.2.3 No openings	1.3.3 Bacterial surface smear			
1.1.4 Small round openings	1.2.4 Blue veined	1.3.4 No rind			
1.1.5 Irregular openings					
1.1.6 No openings					

2. SHEEP'S MILK
   Hard; semi-hard; soft; blue-veined; fresh.

3. GOATS' MILK

4. BUFFALO'S MILK

[a] Modified from Ref. 3; unless otherwise stated, the cheeses are internal bacterially ripened.

TABLE VI

Classification of Cheese according to Moisture Content, Cooking Temperature and Secondary Microflora[a]

Hard cheese (moisture content 20–42%)

Low scald Ns	Medium scald Ns	High scald Ns or Pr	Plastic curds Ns or Pr
Edam (N)	Cheddar (UK)	Grana (Parmesan) (I)	Scamorza (I)
Gouda (N)	Gloucester (UK)	Emmental (Sw)	Provolone (I)
Cantal (F)	Derby (UK)	Gruyère (Sw)	Caciocavallo (I)
Fontina (I)	Leicester (UK)	Beaufort (F)	Mozzarella (I)
Cheshire (UK)	Svecia (S)	Herrgardsost (S)	Cecil (USSR)
	Dunlop (UK)	Asiago (I)	Kaaseri (Gr)
	Turunmaa (FL)	Sbrinz (Sw)	Kashkaval (B)
			Perenica (C)

Semi-hard cheese (moisture content 44–55%; low scald)

Ns	Sm	Bv
St Paulin (F)	Herve (Bel)	Stilton (UK)
Caerphilly (UK)	Limburg (Bel)	Roquefort (F)
Lancashire (UK)	Romadur (G)	Gorgonzola (I)
Trappist (H)	Munster (F)	Danablu (D)
Providence (F)	Tilsit (G)	Mycella (D)
	Vacherin- Mont d'Or (Sw)	Wensleydale (UK)
	Remoudou (Bel)	Blue Vinny (UK)
	Srainbuskerkase (G)	Gammelost (Nor)
	Brick (USA)	Adelost (S)
		Tiroler-Graukäse (G)
		Edelpitzkäse (A)
		Aura (Ice)
		Cabrales (E)

Soft cheese (moisture content >55%; very low or no scald)

Sm or Hm	Hm	Ns	Un, Ac
Brie (F)	Camembert (F)	Colwich (UK)	Coulommier (F)
Bel Paese (I)	Carre d'est (F)	Lactic (UK)	York (UK)
Maroilles (F)	Neufchatel (F)	Bondon (F)	Cambridge (UK)
	Chaource (F)		Cottage (UK)
			Quarg
			Petit Suisse (F)
			Cream (UK)

Pr = propionic acid bacteria.
Ns = normal lactic starter of milk flora.
Sm = smear coat (*Brevibacterium linens*).
Hm = surface mould (*Penicillium candidum* or *P. camemberti*).
Bv = blue veined internal mould (*P. glaucum* or *P. roqueforti*).
Ac = acid coagulated.
Un = normally unripened, fresh cheese.

[a] Modified from Scott.[4]

TABLE VII

Classification of Cheese Types on the Basis of Type of Ripening and Moisture Content (Ratio Water/Protein W/P) (from Walstra, pers. comm.)

W/P	log W/P	PRIMARY STARTER					
		Mesophile	—	Thermophile	SECONDARY STARTER	Mesophile	Mesophile
		Fresh		Propioni bacteria	Corynebacteria	White mould	Blue mould
-5		Quarg					
-4	0.6	Cottage					
-3		Feta[a,b]					
			Meshanger				
	0.4		Quesco blanco[b,c]			Camembert	
-2			Butterkäse		Munster		Lymeswold
			Caerphilly[d]			Chévre[b]	
			St. Paulin				
			Mozzarella[e]				
-1.5	0.2		Amsterdammer		Port Salut		Roquefort[b]
					Tilsiter		
			Gouda				
			Jarlsberg				
-1.25			Cheddar[d]		--- x ---		--------- Gorgonzola
				Emmentaler			
			Provolone[e]				
-1			Leidse				
-0.8	0		Gruyère --------- Parmigiano --------- x ---------				Stilton[d]

[a] Stored in brine.
[b] Usually not made from cows' milk.
[c] Normally curd prepared with acid.
[d] Curd salted before pressing.
[e] Pasta filata.
x Means that the factor concerned is appropriate.

Walter & Hargrove[2] suggested an alternative classification into eight families, generally similar to those of Burkhalter:[3]

1. Very hard (grating):
    1.1 Ripened by bacteria: Asiago (old), Parmesan, Romano, Sapsago, Spalen.
2. Hard:
    2.1 Ripened by bacteria, without eyes: Cheddar, Granular, Caciocavallo.
    2.2 Ripened by bacteria, with eyes: Emmental, Gruyère.
3. Semi-soft:
    3.1 Ripened principally by bacteria: Brick, Munster.
    3.2 Ripened by bacteria and surface microorganisms: Limburger, Port du Salut, Trappist.
    3.3 Ripened principally by blue mould in the interior: Roquefort, Gorgonzola, Danablu, Stilton, Blue Wensleydale.
4. Soft:
    4.1 Ripened: Bel Paese, Brie, Camembert, Hand, Neufchatel.
    4.2 Unripened: Cottage, Pot, Bakers, Cream, Ricotta, Mysost, Primost.

While the published classification schemes appear to be inadequate, I am unable to offer a significantly improved classification scheme. However, I would like to suggest classification of cheeses into super-families, based on the coagulation agent:

1. Rennet cheeses: most of the major international varieties;
2. Acid cheeses: e.g. Cottage, Quarg, Quesco Blanco, Cream;
3. Heat/acid: e.g. Ricotta, some forms of Quesco Blanco, Sapsago, Ziger, Schottenziger;
4. Concentration/crystallization: e.g. Mysost.

Obviously, the classification schemes of Davis,[7] Walter & Hargrove[2] and Burkhalter[3] and others can be applied to the rennet cheeses but are not really applicable to the other three super-families since most are high-moisture, soft cheese and most are, normally, not ripened. It would appear reasonable that the very heavily salted varieties (which are normally stored in brine or brined whey), including Domiati, Feta, Brinza, Lightvan, Bulgarian pickled white cheese, should be classified as a separate sub-family of rennet cheeses or within any of the above classification schemes.

The FAO/WHO has published standards for about 30 major cheese varieties in various editions of Code of Quality Standards for Cheese, forming part of the Joint FAO/WHO Codex Alimentarius. Some of these standards are included by Mair-Waldburg.[16]

A further set of standards arises from the Stresa Convention of 1951 on the use of 'Appellation d'Origine' for cheese varieties. Under the Convention, certain cheese, e.g. Roquefort, Stilton, Manchego, Grana, Parmigiano, Reggiano, Gruyère de Comte, can be produced only within certain regions, sometimes by a defined protocol and frequently from milk of a defined species, e.g. sheep or goat. Several other varieties may be produced outside the country or region of origin, e.g. Emmental, Gouda, Gruyère, Camembert, but the name of the producing country should be included. Cheese defined by the Stresa Convention are listed by Mair-Waldburg.[16]

Davis[7] suggested the possibility of classifying cheese according to the extent of chemical breakdown during ripening and expressed the view that it might be possible within a few years (from 1965) to classify cheese on the basis of chemical fingerprints; 25 years later it is still not possible to do so. Interestingly, Schulz[51] used residual casein as a criterion for classification.

An obvious problem encountered when attempting to chemically fingerprint cheeses arises from the fact that ripening cheese is a dynamic system and therefore the age at which cheeses are fingerprinted creates a major problem of definition. Within any particular variety there is considerable variability with respect to any particular characteristic. At present, insufficient information is available, even on the major varieties, to permit such a chemical fingerprinting. However, it seems worth while to speculate on some possible methods and criteria that might be useful for chemical fingerprinting.

10.1 Glycolysis

As discussed in Chapter 10, Volume 1, all the lactose in cheese is normally converted to lactate, generally the L-isomer. In most internal bacterially ripened cheeses, L-lactate is isomerized to a racemic (DL) mixture at a rate which is influenced by the population of non-starter lactic acid bacteria. Notable exceptions to the above are Swiss-type cheeses, in which most of the lactate is converted to propionate, acetate and CO_2 by the propionic acid bacteria, and surface mould varieties, in which the lactate is converted to CO_2 and H_2O. Racemization or metabolism of lactate may give an estimate of the age of young cheese but is probably not useful as a basis for cheese classification.

The pH of most cheese varieties is initially within the range 4·6–5·2. During ripening, the pH of most varieties increases somewhat but the change is most marked in surface or internal mould-ripened cheeses in which the pH increases to 6·5–7·5 due to the utilization of lactate and/or production of NH_3.

Thus, it would appear that the products of lactose metabolism do not provide a useful basis for a classification scheme that is more meaningful than schemes already available based on other criteria, i.e. cheese with eyes or internal or surface mould ripened.

10.2 Lipolysis

Most cheese varieties undergo relatively little lipolysis, notable exceptions being some Italian and Blue varieties. While lipolysis in blue cheeses is of no additional value in cheese classification, it may represent an additional criterion by which to classify hard and/or extra-hard varieties. The ratios of selected free fatty acids may also be useful, e.g. Marcos & Esteban (Chapter 6, Volume 2) discuss the ratio of C_{10}/C_{12} in some Spanish cheese varieties.

Products of fatty acid catabolism, principally by methyl ketones and lactones, would appear to be of little value as criteria for classification. Methyl ketones are produced at significant concentrations only in Blue cheeses and are therefore of no special value for classification. At present, there is very little information on the concentration of lactones in cheeses.

10.3 Proteolysis

Proteolysis is probably the most complex of the three primary events in cheese ripening and hence it might be expected to hold the greatest potential as a basis for chemical classification. Although proteolysis has been extensively studied in several varieties, there are, unfortunately, few intervarietal comparisons. Proteolysis in cheese in general is discussed in Chapter 10, Volume 1, and in most of the chapters in Volume 2 on the individual varieties.

Several indices of proteolysis might be useful for cheese classification.

10.3.1 Polyacrylamide Gel Electrophoresis (PAGE)

As discussed in Chapter 10, Volume 1, α_{s1}-casein is the principal substrate for rennets during ripening; it is converted to α_{s1}-I casein initially and later to other peptides. Ledford et al.[52] compared the gel electrophoretograms of a limited number of cheeses and a larger number were compared by Marcos et al.[53] Electrophoretograms prepared in our laboratory of a number of varieties are shown in Fig. 1. The results of these three studies show that there are substantial differences in the gel electrophoretograms of cheeses of different varieties but with some exceptions they are not unique to the variety. In the above studies only one sample of each variety was analysed and it is not known how representative of the variety that sample was. Another obvious problem is that the electrophoretic patterns change as the cheese matures so that PAGE may not be very useful in the classification of cheese unless its age is known, and even then, the rate of ripening varies with temperature and cheese composition, especially moisture and salt.

β-Casein is not extensively hydrolysed in bacterially ripened cheeses although in some varieties, e.g. Emmental and Gouda, a considerable amount of γ-casein is formed via the action of plasmin, the activity of which appears to be dependent on the cooking temperature.[54] Perhaps the ratio of $\beta:\gamma$-caseins might be useful as the basis for classification. The data of Marcos et al.[53] and in Fig. 1 indicate that there are considerable intervarietal variations in the ratio of $\beta:\gamma$, although precisely which electrophoretic bands represent the γ-caseins requires careful definition. A possible source of error in defining γ-casein on PAGE gels may arise from the fact that the cell-wall proteinase of some strains of mesophilic starters preferentially hydrolyses β-casein to peptides with mobilities similar to the γ-caseins. A second major problem is that $\beta:\gamma$ ratio changes as

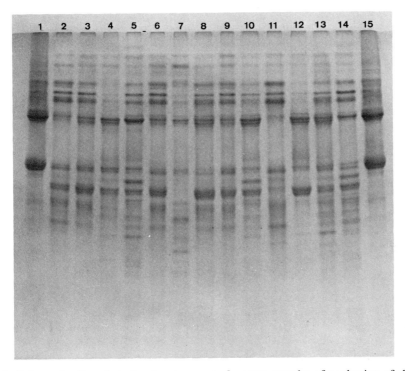

Fig. 1. Polyacrylamide gel electrophoretograms of mature samples of a selection of cheese varieties. 1. Na-caseinate; 2. Appenzell; 3. Beaufort; 4. Brie; 5. Cheddar; 6. Comté; 7. Danish Blue; 8. Emmental; 9. Fontina; 10. Gouda; 11. Parmesan; 12. Port Salut; 13. Regato; 14. Svecia; 15. Na-caseinate.

the cheese matures; in fact Haasnoot et al.[55] found that the β:γ ratio, which decreased with cheese age, was the best indicator of the age of mature Gouda cheese, the ratio a_{s1}:a_{s1}-I being a better index of maturity for young cheeses.

In internal and surface mould-ripened cheeses, β-casein is completely degraded after development of the mould, but this is of no special value for classification purposes, since the presence of mould is normally used to classify such varieties.

It has been our experience in studies on Cheddar cheese during ripening that PAGE of water soluble N (WSN) is more discriminating than electrophoretograms of whole cheese and this appears also to be the case between varieties as shown in Fig. 2 for the WSN (or pH 4·6 sol. N) from the cheeses in Fig. 1. Again, it must be emphasized that these electrophoretograms are for just one sample at one time point and therefore may not be typical but it appears that this approach is promising and warrants further study.

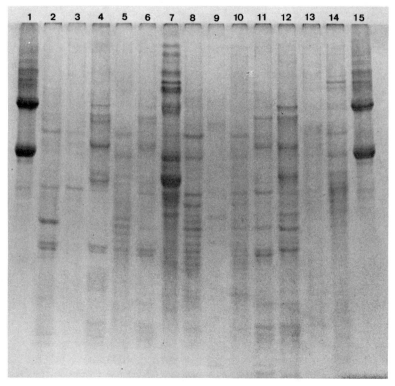

Fig. 2. Polyacrylamide gel electrophoretograms of the water-soluble nitrogen from the cheeses in Fig. 1. Sample codes as for Fig. 1.

Other forms of electrophoresis, e.g. SDS-PAGE, isoelectric focusing and two-dimensional electrophoresis have been applied to cheese to only a limited extent and probably warrant greater study, especially for intervarietal comparisons.

10.3.2 FPLC and HPLC

These techniques, especially RP-HPLC, are now being widely applied to the study of cheese ripening (see Chapter 9, Volume 1). However, there are very few intervarietal comparisons. RP-HPLC profiles of the WSN of Emmental, Gruyère and Appenzaller, shown in Chapter 3, Volume 2, show clear differences between these cheeses. Intervarietal differences between Edam, Gouda, Cheddar, Swiss and Parmesan are also apparent from the data of Smith & Nakai.[56] However, there are significant differences between cheeses of the same variety, e.g. HPLC showed marked differences between the WSN of Cheddar cheeses made with individual single strain starters (Fig. 3) and McSweeney et al.[22] found substantial differences between the RP-HPLC profiles of WSN from Cheddar cheeses made from raw or pasteurized milks (Fig. 4). RP-HPLC profiles of the cheese varieties used in Figs 1 and 2 show marked intervarietal differences (Fig. 5). Again, it

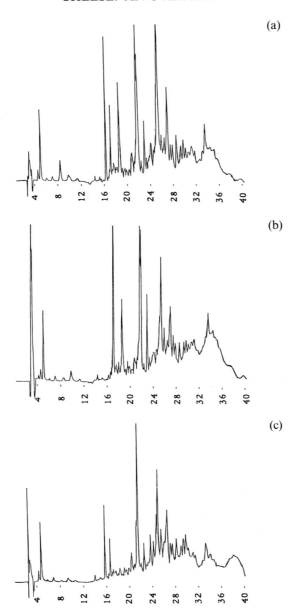

Fig. 3. HPLC chromatograms of WSN fractions of 6-months-old Cheddar cheeses made with *L. lactis* subsp. *lactis* strains UC317 (a), JL3601 (b) or JL521 (c) (from Ref. 57).
Column: ODS Ultrasphere C18 R.P. (5μm) 250 mm × 4·6 mm
                  (Alltech Associates, Carntath Lanes, United Kingdom)
Eluent: – Solvent A: 0·1% Trifluoroacetic acid (TFA) in double distilled water.
              – Solvent B: 0·1% TFA in acetonitrile: H$_2$O (50:50). Samples (lyophilized WSN)
                were dissolved in Solvent A (4 mg/ml). The samples were eluted at 100% A for
                5 min; a gradient of 0–100% B was then commenced (3·3% B/min) and 100% B
                was maintained for 5 min at the end of the run (solvent flow rate: 1 ml/min).

P. F. FOX

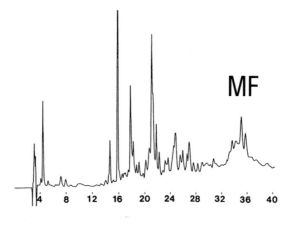

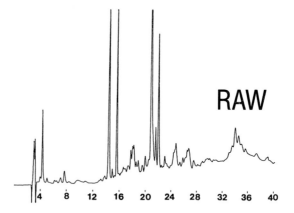

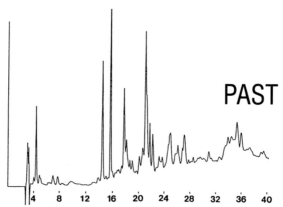

Fig. 4. Reverse phase HPLC chromatogram of the water-soluble nitrogen fraction of 6-months-old Cheddar cheeses manufactured from raw, pasteurized (Past.) or microfiltered (MF) milk. Chromatographic conditions: see Fig. 3.

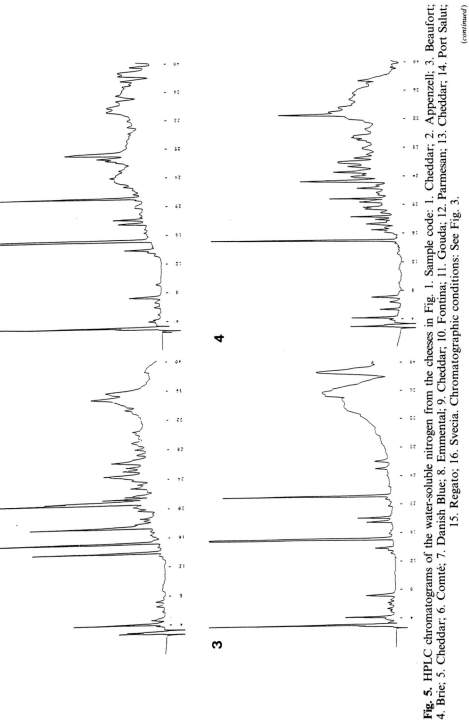

Fig. 5. HPLC chromatograms of the water-soluble nitrogen from the cheeses in Fig. 1. Sample code: 1. Cheddar; 2. Appenzell; 3. Beaufort; 4. Brie; 5. Cheddar; 6. Comté; 7. Danish Blue; 8. Emmental; 9. Cheddar; 10. Fontina; 11. Gouda; 12. Parmesan; 13. Cheddar; 14. Port Salut; 15. Regato; 16. Svecia. Chromatographic conditions: See Fig. 3.

(continued)

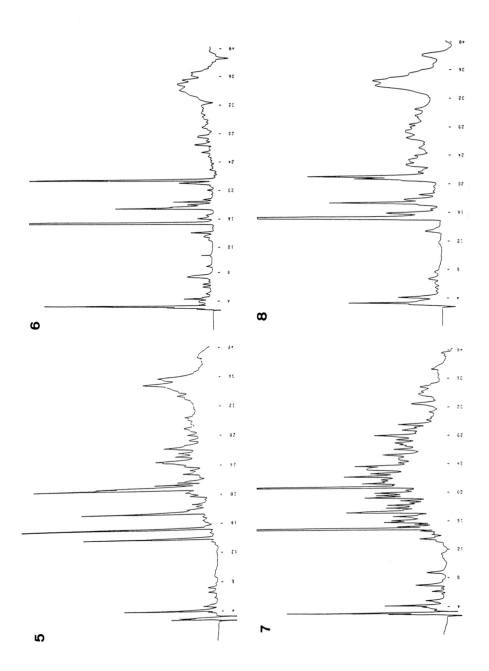

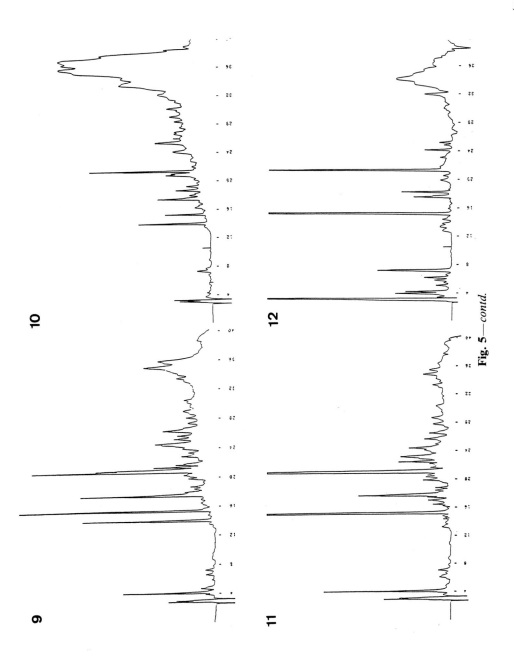

Fig. 5—contd.

(continued)

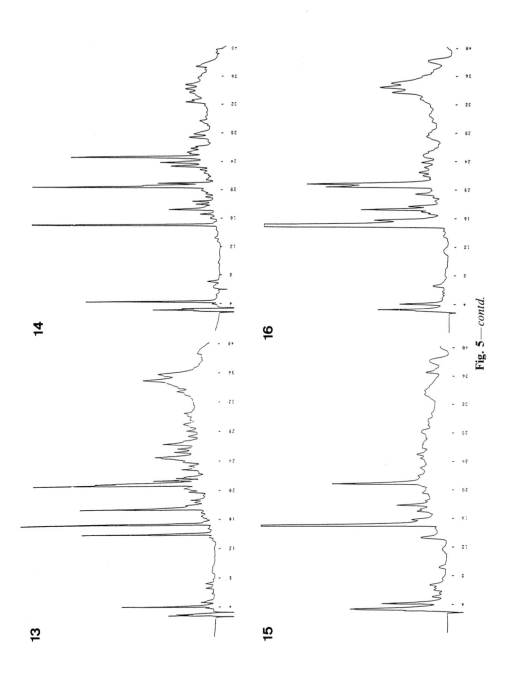

Fig. 5—*contd.*

must be emphasized that the reproducibility of these profiles is unknown but the approach appears to merit further study.

FPLC on ion-exchange columns has been applied to the fractionation of cheese in only a few studies. FPLC on Mono-Q columns using pH 8·5 buffers containing 4·5 M urea resolves the water insoluble N of cheese, or whole cheese, very effectively (Breen & Fox, unpublished data). As far as we are aware, FPLC has not been used in a comparative study; it is likely that it is less effective than PAGE.

Soluble tyrosine and tryptophan appear to be indices of the hydrolysis of β-casein rather than of α_{s1}-casein, since the concentrations of these amino acids are higher in the former than in the latter (see Chapter 6, Volume 2). Thus, the concentration of tyrosine, e.g. A_{280}, in soluble N may be a useful index for cheese classification, as the data of Marcos et al.[53] suggest, but analysis of more cheeses from any particular variety is necessary before the value of this index can be assessed.

10.3.3 Soluble N

The proportion of the total N soluble in water, at pH 4·6, in 2 or 12% TCA, 5% PTA, 30 or 70% ethanol increases during ripening (see Chapter 10, Volume 1). Perusal of the literature indicates intervarietal differences with respect to the proportion of N soluble in any of these precipitants, e.g. the proportion of soluble N in blue cheese is very high. It should be possible to classify cheese based on the proportion of total N soluble in any of these precipitants, but insufficient data are available at present for a wide range of varieties to permit such a classification. However, such an approach is probably less discriminating than PAGE or HPLC.

A high percentage of total N is converted to NH_3 in some varieties, e.g. surface mould or surface smear-ripened cheeses; this also may be a useful criterion for the classification of some varieties.

The free amino acid profile of cheese may also be a useful basis for classification although Leu, Phe, Glu appear to be the principal free amino acids in most varieties (see Chapter 10, Volume 1). Notable exceptions are Swiss-type cheeses which contain high concentrations of free proline. There are insufficient data on a range of cheeses at present by which to assess the value of the free amino acid profile as the basis for classification.

The volatiles in many cheese varieties have been analysed, mainly with the objective of identifying the compounds responsible for cheese flavour. However, head-space analysis of cheese may also be a useful method for the objective chemical classification of cheese, e.g. Aishima & Nakai[58] found 118 large to medium peaks (plus about 80 minor peaks) in the gas chromatograms of Cheddar, Gouda, Edam, Swiss and Parmesan and claimed that it should be possible to distinguish between these varieties by stepwise discriminant analysis of these chromatograms. However, as with previous analytical methods, information on the intravarietal variation in gas chromatograms of cheese head-space and the influence thereon of cheese composition, age and ripening conditions is not known.

The water activity (a_w) of cheese ranges from about 0·88 to >0·99. A major role is played by a_w in determining the safety of cheeses from the public health viewpoint and in controlling the type and depth of ripening. It can be determined relatively easily and data are already available for several cheese varieties. The classification of cheeses based on a_w would appear to be a viable possibility (see Chapter 11, Volume 1, for a review on a_w of cheese).

In summary, while it is not yet possible to classify cheeses based on chemical fingerprinting of their ripening patterns, considerable progress has been made in recent years and modern techniques, especially PAGE, RP-HPLC and GC, hold considerable potential but insufficient data are available at present. In preparing for the second edition of this book, I decided to circulate a questionnaire (Appendix) to contacts in the principal dairying countries, requesting a large body of information on which a classification scheme might be based. Unfortunately, only about half of the questionnaires were returned. Data for many of the principal cheesemaking countries are missing which makes it impossible to attempt the planned classification. Perhaps in the third edition of this book, the task can be completed and by then sufficient detailed information on proteolysis in a sufficient number of samples of a wide range of cheese varieties will be available to permit a more detailed chemical classification of cheeses.

REFERENCES

1. Sandine, W.E. & Elliker, P.R., 1981. *J. Agr. Food Chem.*, **18**, 557.
2. Walter, H.E. & Hargrove, R.C., 1972. *Cheeses of the World*. Dover Publications Inc., New York.
3. Burkhalter, G., 1981. *Catalogue of Cheese*. International Dairy Federation, Doc. 141.
4. Scott, R., 1986. *Cheesemaking Practice*. Elsevier Applied Science Publishers, London.
5. Cheke, V., 1959. *The Story of Cheesemaking in Britain*. Routledge & Kegan Paul, London.
6. Squire, E.H. (ed.), 1937. *Cheddar Gorge: A Book of English Cheeses*. Collins, London.
7. Davis, J.G., 1965. *Cheese, Vol. 1, Basic Technology; Vol. 2, Bibliography*. Churchill Livingstone, London.
8. White, K.D., 1970. *Roman Farming*. Thames and Hudson, London.
9. International Dairy Federation, 1988. *The World Market for Cheese*, IDF Bulletin, 243.
9a. Anifantakis, E.M., 1991. *Greek Cheeses:* a tradition of centuries. National Dairy Committee, Athens.
10. International Dairy Federation, 1990. *Consumption Statistics for Milk and Milk Products*, IDF Bulletin, 246.
11. Sammis, J.L., 1948. *Cheesemaking*. Cheesemaker Book Co., Madison, Wisconsin, USA.
12. Van Slyke, L.L. & Price, W.V., 1949. *Cheese*. Orange Judd, New York.
13. Kosikowski, F.V. & Mocquot, G., 1958. *Advances in Cheese Technology*. FAO Studies 38, Food Agricultural Organisation, Rome.
14. Simon, A.L., 1956. *Cheeses of the World*. Faber & Faber, London.
15. Layton, J.A., 1973. *The Cheese Handbook*. Dover Publications Inc., New York.
16. Mair-Waldburg, H., 1974. *Handbook of Cheese; Cheeses of the World A to Z*. Volkwirtschaftlicher Verlag GmbH, Kempton Allgan, Germany.

17. Davis, J.G., 1967. *Cheese. Vol. III. Manufacturing Methods; Vol. IV, Bibliography.* Churchill Livingstone, London.
18. Eekhof-Stork, N., 1976. *World Atlas of Cheese.* Paddington Press Ltd, London.
19. Kosikowski, F.V., 1977. *Cheese and Fermented Milk Foods.* Edwards Bros Inc., Ann Arbor, Michigan.
20. Cantin, C., 1976. *Guide Pratique des Fromages.* Solar Editeur, Paris.
21. Eck, A. (ed.), 1984. *Le Fromage.* Lavoisier, Paris.
22. McSweeney, P.L.H., Lucey, J.A., Jordon, K., Cogan, T.M. & Fox, P.F., 1991. *J. Dairy Sci.*, **74** (Suppl. 1), 91.
23. Lawrence, R.C., 1989. *The Use of Ultrafiltration Technology in Cheesemaking,* Bulletin 240, International Dairy Federation.
24. Lloyd, G.T., 1971. *Dairy Sci. Abstr.*, **33**, 411.
25. Reiter, B., 1972. *J. Soc. Dairy Technol.*, **26**, 3.
26. Stadhouders, J., 1974. *Milchwissenschaft*, **29**, 329.
27. Lawrence, R.C., Thomas, T.D. & Tarzaghi, B.E., 1976. *J. Dairy Res.*, **43**, 141.
28. Sandine, W.E., 1977. *J. Dairy Sci.*, **60**, 322.
29. Daly, C., 1983. *Ir. J. Food Sci. Technol.*, **7**, 39.
30. Daly, C., 1983. *Antoine von Leeuwenhoek, J.*, **49**, 297.
31. Accolas, J.P. & Auclair, J., 1983. *Ir. J. Food Sci. Technol.*, **7**, 27.
32. Auclair, J. & Accolas, J.P., 1983. *Antoine van Leeuwenhoek J.*, **49**, 313.
33. Accolas, J.P., Veaux, M., Vassal, L. & Mocquot, G., 1978. *Le Lait*, **58**, 118.
34. Sardinas, J.L., 1972. *Adv. Appl. Microbiol.*, **15**, 39.
35. Ernstrom, C.A., 1974. In *Fundamentals of Dairy Chemistry*, 2nd edn, ed. B.H. Webb, A.H. Johnson & J.A. Alford. AVI Publishing Co. Inc., Westport, CT, p. 662.
36. Nelson, J.H., 1975. *J. Dairy Sci.*, **58**, 1739.
37. Sternberg, M., 1976. *Adv. Appl. Microbiol.*, **20**, 135.
38. Green, M.L., 1977. *J. Dairy Res.*, **44**, 159.
39. Martens, R. & Naudts, M., 1978. International Dairy Federation Annual Bulletin, Doc. 180, p. 51.
40. De Koning, P.J., 1979. International Dairy Federation, Doc. 126, p. 11.
41. Phelan, J.A., 1985. PhD Thesis, National University of Ireland.
42. Berridge, N.J., 1942. *Nature*, **149**, 194.
43. Fox, P.F., 1983. In *Developments in Food Proteins—3. Proteins*, ed. B.J.F. Hudson. Elsevier Applied Science Publishers, London, p. 69.
44. Waugh, D.F. & Von Hippel, P.H., 1956. *J. Am. Chem. Soc.*, **78**, 4576.
45. Wake, R.G., 1959. *Aust, J. Biol. Sci.*, **12**, 479.
46. Delfour, A., Jolles, J., Alais, C. & Jolles, P., 1965. *Biochim. Biophys. Res. Commun.*, **19**, 452.
47. Humme, H.E., 1972. *Neth. Milk Dairy J.*, **26**, 180.
48. Schmidt, D.F., 1982. In *Developments in Dairy Chemistry—1. Proteins*, ed. P.F. Fox. Elsevier Applied Science Publishers, London, p. 61.
49. McMahon, D.J. & Brown, R.J., 1984. *J. Dairy Sci.*, **67**, 499.
50. Pyne, G.T. & McGann, T.C.A., 1962. *Proc. 16th Intern. Dairy Congr. (Copenhagen)*, A, 611.
51. Schulz, M.E., 1952. *Milchwissenschaft*, **7**, 292.
52. Ledford, R.A., O'Sullivan, A.C. & Nath, K.R., 1966. *J. Dairy Sci.*, **49**, 1098.
53. Marcos, A., Esteban, M.A., Leon, F. & Fernandez-Salguero, J., 1979. *J. Dairy Sci.*, **62**, 892.
54. Farkye, N.Y. & Fox, P.F., 1990. *J. Dairy Res.*, **57**, 413.
55. Haasnoot, W., Stouten, P. & Venema, D.P., 1989. *J. Chromat.*, **483**, 319.
56. Smith, A.M. & Nakai, S., 1990. *Can. Inst. Food Sci. Technol. J.*, **23**, 53.
57. Law, J.M., 1991. PhD Thesis, National University of Ireland.
58. Aishima, T. & Nakai, S., 1987. *J. Food Sci.*, **52**, 939.

Appendix: Questionnaire on Characteristics of Cheese Varieties

VARIETY

1. Type of milk
   cow, goat, sheep, buffalo
   raw or pasteurized

2. Type of starter
   Lactococcus lactis ssp. *lactis*
   L. lactis ssp. *cremoris*
   L. lactis ssp. *diacetylactis*

   Str. thermophilus
   Lactobacillus (species)

   Others

   Defined, Undefined, 'Natural'

3. Coagulant
   rennet (type)
   acid
   acid/heat

4. Setting temperature

5. Cooking temperature

6. pH at whey draining

7. Curd pressed

8. Salting:
   mixed with curd
   surface brining
              dry

9. Curd stretched   Yes
   (pasta filata)   No

10. Composition, %   fat
                      moisture
                      fat in dry matter
                      protein
                      salt

11. pH   1 day
                     mature

12. Consistency   Hard
                     Semi hard
                     Semi soft
                     Soft

13. Ripening: Duration, months
                     Temperature
                     Humidity

14. Flavour: mild
                     sharp
                     piquant

15. Interior: large eyes
                     small eyes
                     irregular openings
                     close texture

16. Secondary microflora
                     Propionibacteria
                     Blue mould
                     White mould
                     Surface smear
                     Dry rind
                     Others

17. Indices of maturity
    (typical values for mature cheese)

    pH
    % of total N soluble in water or
       at pH 4·6
    % of total N soluble in 12% TCA
    % of total N soluble in PTA
    % of total N as free amino acids
    α-casein completely hydrolysed
    β-casein completely hydrolysed
    Free fatty acids, meq/kg
    Lactic acid, %
    Lactic acid isomer (D,L)
    Does lactic acid undergo change?

18. Physical appearance: shape
                           weight (kg)
                           height
                           diameter

19. Volume of Production
    (tonnes/annum)

20. Does the cheese have Appellation
    d'origine status?

21. Synonyms: related varieties
    (use separate page if necessary)

22. Any other relevant information
    (use separate page if necessary)

2

General and Molecular Aspects of Rennets

BENT FOLTMANN

Institute of Biochemical Genetics, University of Copenhagen, Denmark

1 INTRODUCTION

Cheesemaking and fermentation represent the first examples of applied biochemistry and biology. Whereas living microorganisms are used in fermentation processes, the clotting of milk for cheesemaking has always required soluble enzymes. The milk-clotting enzyme from the fourth stomach of the calf was one of the first enzymes of which purification was attempted, and Deschamps[1] suggested the name chymosin, derived from the Greek word for gastric liquid 'chyme'. This designation was later used in continental European languages, whereas in English the name rennin, derived from rennet, was used.[2] Misunderstandings often occurred between rennin and renin from the kidneys, and therefore the designation chymosin was recently adopted in English[3] and it is now used in the recommended international enzyme nomenclature.[4]

Calf chymosin is still the prevailing milk coagulant used in cheesemaking but due to the shortage of calf stomachs, animal rennets are sometimes fortified by addition of pepsins. Furthermore, proteases of microbial origin are widely used for cheesemaking, and cloned calf chymosin, produced in microorganisms, is now also in industrial production.

In some countries, vegetable rennets are traditionally used and bacterial proteases may also have milk-clotting properties (reviews, Refs 5–8), but compared to animal and fungal rennets little is known about vegetable and bacterial rennets and hence they will not be considered further in this review.

The proteases in animal and fungal rennets all belong to the group of aspartic proteases (previously called acid proteases). In the international enzyme classification,[4] this group carries the number EC 3.4.23. Table I presents a list of proteases of major importance for rennets.

The aim of this review is to give a survey with representative references, mainly from 1970 to 1990. Further references and references to the older literature may be found in reviews on the following topics: microbial rennets,[7,8] aspartic proteases,[13–16] gastric proteases,[17,18] chymosin,[3,12] pepsin,[10,17–21] rennets and cheese.[22–25]

37

TABLE I
Nomenclature and Sources of Major Proteases in Rennets

	IUB-name and number[4]	Other names	Sources
Pepsin	Pepsin A EC 3.4.23.1	Pepsin II[9]	Ruminants Pigs Chicken
Gastricsin	Gastricsin EC 3.4.23.3	Pepsin I[9] Parapepsin II[10] Pepsin B[11] Pepsin C	Ruminants Pigs
Chymosin	Chymosin EC 3.4.23.4	Rennin[12]	Ruminants
		Trade names	
M. miehei protease	EC 3.4.23.6	Rennilase (Novo) Hanilase (Chr. Hansen) Fromase (Wallerstein) Marzyme (Miles)	*Mucor miehei*
M. pusillus protease		Emporase (Dairyland) Meito (Meito Sangyo) Noury (Vitex)	*Mucor pusillus* var. Lindt
E. parasitica protease		Sure curd Suparen (Pfizer)	*Endothia parasitica*

Notes:
Pepsin: Chicken pepsin is tentatively classified together with the mammalian pepsins though the evolutionary and physiological relationships are not yet firmly established. Suffixes of Roman numerals have been used to characterize the individual gastric proteases in their order of elution by ion-exchange chromatography on DEAE-cellulose.
Gastricsin: The mammalian gastricsins form a separate group of gastric proteases that are different from both the predominant pepsin and chymosin. The designation pepsin C was previously used in the IUB-nomenclature.
Chymosin: The mammalian chymosins are neonatal proteases, see also Section 6.
Fungal proteases: Though these proteases show large differences in structures and specificities, all microbial aspartic proteases have the same EC number.

2 PREPARATION, PURIFICATION AND CHARACTERIZATION

2.1 Preparation

2.1.1 Animal Rennets
Frozen calf stomachs are now sold in bulk, but large-scale extraction procedures based on frozen starting material have not been published. The traditional method of preparation from dried or salted vells is described by Placek *et al.*[26]

The stomachs are cut into small pieces and mixed with Excelsior. Counter-current extraction is carried out with a 10% NaCl solution. The crude extracts consist of a mixture of zymogens and active enzymes. After activation by addition of acid to a pH between 2 and 4·6, and pH is adjusted to 5·5–5·7 (see also Sections 4 and 5), the salt concentration is increased to about 20% NaCl and preservatives, e.g. sodium benzoate or sodium propionate, are added. Finally, the rennet is filtered and the activity is adjusted ready for use.

For research purposes, several small-scale extraction procedures have been published. If fresh or frozen material is available, it is convenient to dissect the mucosa from the muscular sheet before extraction. Extraction may take place after passage through a meat grinder or directly in a blender. Ultrasound extraction has also been described.[27]

For preparation of zymogens, it may be advantageous to perform the extraction under neutral or weakly alkaline conditions (bicarbonate[12] or phosphate buffers of pH 7·3–7·5).[28] Under such conditions, active enzymes are denatured, whereas the zymogens are stable (Section 4). If no buffers are added, the buffering capacity of a fresh mucosal extract will generally result in a pH about 6. For small-scale extraction, salt solutions containing up to 0·5 M NaCl have been used[29] but in other preparations of gastric zymogens, salt has not been added during the extraction and for purification, the zymogens and enzymes have been adsorbed on DEAE-cellulose directly from the extraction liquid,[30] or after activation and dialysis.[31]

2.1.2 Microbial Rennets
Different strains of *Mucor* are often used for the production of microbial rennets. Today, taxonomists prefer the name *Rhizomucor* for this genus, but the traditional designation is used in this review.

Best yields of the milk-clotting protease from *Mucor pusillus* are obtained from semi-solid cultures containing 50% of wheat bran, whereas *Mucor miehei* and *Endothia parasitica* are well suited for submerged cultivation. From the former, good yields of milk-clotting protease may be obtained in a medium containing 4% potato starch, 3% soybean meal and 10% barley. During growth, lipase is secreted together with the protease; the lipase activity is destroyed by treatment at low pH values before the preparation can be used as cheese rennet.[8,32]

2.1.3 Recombinant rennets
Due to the shortage of calf stomachs and the economic value of cheese rennet, calf chymosin was one of the first mammalian enzymes which was cloned and expressed in microorganisms. Many different laboratories have cloned the gene for calf prochymosin in *E. coli* and analysed the structure of the gene as well as the properties of the recombinant chymosin (representative references, 33–40). By expression in *E. coli* the synthesized proenzyme is found mainly as insoluble inclusion bodies, consisting of reduced prochymosin as well as molecules which are interlinked by disulphide bridges.[41] After disintegration of the cells, inclusion bodies are harvested by centrifugation. The individual laboratories have reported

some differences in the procedure for renaturation of prochymosin from the inclusion bodies, but all follow the same general scheme: the insoluble precipitate is dissolved under denaturing conditions, e.g. in 8–9 M urea at pH 8–10, and the solution is made 0·5 M with respect to NaCl. Salt and urea are removed by dialysis, and pH adjusted to 7·5–8·0. After centrifugation, the supernatant is purified by chromatography.[37-40] The enzymic properties of the recombinant *E. coli* chymosin are indistinguishable from those of native calf chymosin.[40,42,43] The enzymes were found to be identical by immunodiffusion in gels,[43] but a slight difference was observed by enzyme-linked immunosorbent assay.[40]

The gene for prochymosin has also been cloned in yeast (*Saccharomyces cerevisiae*); the levels of expression correspond to 0·5–2% of total yeast protein.[44,45] In yeast, about 20% of the prochymosin can be released in soluble form which can be activated directly; the remaining 80% is still associated with cell debris. The experiments of Mellor *et al.*[44] also comprised cloning of chymosin without the propart. Though the level of chymosin m-RNA was similar to that of prochymosin m-RNA, no milk-clotting activity was observed in clones containing the chymosin gene only. The results suggest that the propart is essential for correct folding of the peptide chain.

The zymogen for the aspartic protease from *M. pusillus*, also called Mucor rennin, has likewise been cloned and expressed in yeast.[46] Its conversion to active form was investigated[47] (see also Section 5), and it was observed that the secretion of *M. pusillus* protease from recombinant yeast was dependent on glycosylation of the enzyme.[48]

Compared with yeast, the filamentous fungi will in general secrete larger quantities of proteins into the culture medium. Furthermore, filamentous fungi will secrete the major part of heterologous proteins with correct folding of the peptide chain and the correct pairing of disulphide bridges. The gene for *M. miehei* protease has been expressed in *Aspergillus oryzae*.[49] Prochymosin has been expressed in *Kluyveromyces lactis*,[50] *Aspergillus nidulans*,[51] *Aspergillus niger*[52] and *Tricoderma reesei*.[53] Most of the reported yields of the published model systems are about 10–40 mg of enzyme per litre of culture medium but Christensen *et al.*[49] obtained up to 3·3 g/l. In an industrial production the yields will probably be of the latter magnitude.

Several cheesemaking experiments have been carried out with recombinant chymosin[54-57] and the general aspects of recombinant chymosin have been dealt with in a report by Teuber.[58] The conclusion of this is that in cheesemaking experiments no major difference could be detected between cheeses made with cloned chymosin or the natural enzyme. Table II gives a summary of recombinant chymosins which are approaching or have marketing and legal acceptance; the references are to GRAS petitions (abbreviation: *G*enerally *R*egarded *A*s *S*afe).

2.2 Purification

Calf chymosin may be crystallized after repeated precipitations from solutions saturated with NaCl.[12] The crystals are rectangular plates but unless the

TABLE II

Recombinant Chymosin Preparations at or Approaching Legal and Commercial Acceptance (Modified from Ref. 58)

Source of DNA	Producing microorganism	Producing company, brand name	Refs
1. Calf abomasum	*Kluyveromyces lactis*	Gist-Brocades MAXIREN	59
2. Calf abomasum	*Aspergillus niger*	Genencor/Chr.Hansen CHYMOGEN	60
3. Synthetic	*Escherichia coli*	Pfizer CHY-MAX	61

solutions are seeded with crystals from a previous preparation, it is often difficult to obtain crystallization. Today crystallization of a protein is not regarded as a criterion of homogeneity, and chromatographic or electrophoretic analyses have shown that crystalline chymosin may contain at least three different components;[12,62,63] the two major components are designated chymosin A and B. It should be noted that crystalline chymosin is by far the most stable form of the enzyme. At this Institute, we have a batch of crystalline chymosin suspended in 4 M NaCl, and stored at $-10°C$ to $-20°C$; when samples of these crystals are dissolved, we have for 30 years observed no significant decrease in the ratio of the milk-clotting activity relative to absorbency at 278 nm.

Ion-exchange chromatography on columns of DEAE-cellulose or DEAE-Sephadex has been used successfully for both preparative purification and analytical characterization of extracts of gastric mucosa, of solutions of fungal proteases and of recombinant chymosins as well as commercial rennets. The milk-clotting proteases are stable at pH 5–6 and buffers within this pH range are used in most of the chromatographic fractionations. Typical examples of conditions used for ion-exchange chromatography are summarized in Table III. By fast protein liquid chromatography similar results have been obtained.[68,68a]

Affinity chromatography using inhibitors like pepstatin,[69] ε-amino capronyl-D-phenylalanine methyl ester, gramicidin or bacitracin[70,71] has been used successfully for the preparation of chymosin and other aspartic proteases. Affinity chromatography with inhibitors is especially suitable for the isolation of enzymes from crude extracts. Though different binding constants occur, these methods will generally not discriminate among the individual proteases. However, by a dye-ligand affinity chromatography with combinations of hydrophobic and electrostatic interactions, good separation of chymosin and pepsin in calf rennet was obtained.[72]

2.3 Quantitation and Identification

The essential property of a rennet is its milk-clotting activity. The milk-clotting process is described in detail in Chapter 3 of this volume. At this Institute, we

TABLE III
Examples of Chromatographic Systems used for Purification and Characterization of Rennets on Columns of DEAE-ion Exchangers

Equilibration buffer	Elution	Comments	Refs
0.1 M sodium phosphate, pH 5·8	Linear gradient of the equilibration buffer and 0·4 M sodium phosphate, pH 5·5	Optimum conditions for separation of prochymosin and the individual components of chymosin	12
0·05 M sodium phosphate, pH 5·8	Linear gradient of the equilibration buffer and 0·5 M NaCl in the same buffer	Fractionation of chymosin and pepsin in bovine rennet	11
0·02 M piperazine, pH 5·3	Linear gradient of 0·15 M NaCl and 0·6 M NaCl in the equilibration buffer	Fractionation of chymosin and pepsin in bovine rennets, also used for analytical separation of fungal proteases and pepsin	64 65
0·05 M citrate buffer, pH 4·0	Isocratic	Initial purification of *M. miehei* protease	66
0·05 M sodium acetate, pH 5·0	Linear gradient of the equilibration buffer and 0·5 M KCl in the same buffer	Final purification of *M. pusillus* protease	67

use radial diffusion in agarose-skim milk gels[73] for a semiquantitative and rapid screening of the milk-clotting activity in preparations or chromatographic fractionations; clotting tests with 10 ml of reconstituted skim milk in bifurcated glass tubes are used for accurate determinations.[3] Our coagulation or chymosin unit (CU) is defined from chromatographically purified chymosin B: a solution of freshly prepared chymosin B with an absorbency at 278 nm of 1·00 contains 100 CU/ml; this corresponds to 143 CU/mg of chymosin B.[3] As the chromatographic fractionation of chymosin is very reproducible, this unit could be used by others also.

The milk-clotting activity of a solution is a measure of the total amount of milk-clotting proteases. Whether the question concerns a sample of rennet or an evaluation of a purification procedure, the activity test must be followed up by analyses that characterize the enzymic composition of the samples.

Indirect information about the contents of the individual enzymes may be obtained by utilizing the differences in milk-clotting activities at different pH values,[11] or by selective denaturation of the individual components,[74] preferably combined with chromatographic fractionation.[65] In an IDF report, de Koning[75] has suggested a systematic procedure for identification of milk-clotting enzymes. This scheme involves tests for enzymic activity and inactivation, as well as several types of electrophoresis. Separations by ion-exchange chromatography or electrophoresis provide direct information about the enzyme profile of rennets. Traditional ion-exchange chromatography is rather time-consuming and a single experiment gives only a limited separation of the individual components. Thus, under the conditions of Garnot et al.,[64] chymosin and gastricsin co-elute. The method of Garnot et al. has been modified for fast protein liquid chromatography.[68] Another method for fast protein liquid chromatography of rennet has also been described by Rauch et al.,[76] but this fractionation takes places at pH 8·3 which will inactivate most of the milk-clotting enzymes and thus it is not possible to distinguish between active enzymes and inert material.

Electrophoretic separations are rapid and several samples may be tested simultaneously. Samples of rennets have been evaluated by polyacrylamide gel electrophoresis or isoelectric focusing[75] followed by staining of the proteins. Such methods give a detailed picture of the protein content, but with crude samples they do not discriminate between enzymic components and inert proteins; however, if the gels are developed as zymograms, only the enzymatically active components, show up. Detection of proteases after incubation with haemoglobin followed by precipitation and staining of undigested proteins[77] has been widely used in clinical investigations on pepsins and pepsinogens (review, Ref. 78). The method is also suitable for detection of pepsins in rennets, but due to its weak general proteolytic activity it is difficult to observe chymosin in haemoglobin zymograms. The milk-clotting proteases may be detected by pouring skim-milk on the top of the electrophoresis gel;[79] a more accurate method that allows permanent documentation has recently been described.[80] As the precipitates consist mainly of casein, this method is designated a 'caseogram'. Figure 1 illustrates caseograms of different milk-clotting enzymes after electrophoresis in agar gel; it should be pointed out that the caseogram detection works well after all kinds of non-denaturing gel electrophoresis, including immunoelectrophoresis.

Immunochemical methods represent efficient tools for characterization and identification of proteins (review, Ref. 82). Early applications of immunochemical evaluations of rennets were reviewed by de Koning.[83] Using monospecific antisera, quantitative determinations may be obtained by radial diffusion or by rocket immunoelectrophoresis; good agreement between immunochemical determinations and other analytical methods has been reported.[11,84] Selective precipitation by monospecific antisera against chymosin or pepsin, followed by determination of the residual activity, has also been used.[85] There are, however, certain reservations which must be considered in an evaluation of the immunochemical determinations. As pointed out in the following section, the gastric proteases are highly homologous proteins, and pepsins from different species may have common antigenic determinants. Such common antigenicity may not be expressed in all antisera but partial immunochemical identity has been observed among pepsins from different species.[86] Hence, if one wants to determine bovine pepsin in the presence of porcine pepsin, the common antigen binding sites in anti-(bovine

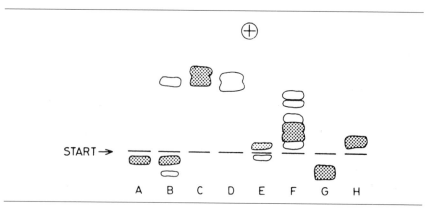

Fig. 1. Tracings of caseograms of different rennets. Electrophoresis was performed in a 1% agar gel, 0·05 M sodium acetate, pH 5·0, at 15 V/cm for 1·5 h. All samples were diluted to a milk-clotting activity of approx. 0·1 CU/ml; 5 μl was used in each slot. A: calf chymosin (the apparent cationic mobility is due to electro-osmotic flow in the system); B: calf rennet (a faint zone is observed below the chymosin band, which may correspond to degradation products, probably the chromatographic fraction designated chymosin C cfr.,[12] a small amount of pepsin is also observed); C: bovine rennet (pepsin only); D: porcine pepsin; E: lamb rennet; F: chicken pepsin; G: *M. miehei* rennet; H: *M. pusillus* rennet.[80,81]

pepsin) should be removed by adsorption with porcine pepsin prior to use. Conversely, anti-(porcine pepsin) should be purified by adsorption with bovine pepsin before determination of porcine pepsin by immunochemical methods. No cross-reactivity has been observed among bovine chymosin, pepsin or gastricsin by gel immunoelectrophoresis. In this connection, it may be noted that most analyses of rennets disregard gastricsin, though the content in bovine rennets may be up to 10% of the content of pepsin.[11] Furthermore, it should be observed that the milk-clotting enzymes may undergo denaturation in such a way that activity is lost but at least some of the antigenic determinants are retained. Thus, some types of rennet do not show proportionality between milk-clotting activity and immunological tests.[87]

As regards the microbial rennets, partial immunochemical identity has been observed between *M. miehei* and *M. pusillus* proteases.[88]

3   STRUCTURE

3.1 Amino Acid Compositions and Molecular Weights

The amino acid compositions of aspartic proteases are generally characterized by a high content of dicarboxylic and hydroxy amino acids and a low content of basic amino acids.[12–21]

Molecular weights from 30 400 to 40 000 have been reported for chymosin and pepsin.[3,10] After determination of the primary structures, the molecular weights of calf chymosin and pig pepsin turned out to be 35 600 and 34 600, respectively.[89,90] Among the fungal proteases, a molecular weight of about 38 000 has been reported for the protein moiety of the glycosylated *M. miehei* protease.[91] This was confirmed by the amino acid sequence from which a molecular weight of 38 700 was determined.[92] For *E. parasitica* protease,[93] a molecular weight of 37 500 was reported, whereas 33 800 has been calculated from the amino acid sequence.[94] If one considers the structural homology among these enzymes, the general feature appears that all the proteases in question consist of about 325–360 amino acid residues, and, depending on their amino acid compositions, the molecular weights add up to between 33 000 and 38 000.

3.2 Primary Structure

When the manuscript for the first edition of this book was finished in 1986, 12 complete amino acid sequences of aspartic proteases were known. Today, more than 30 complete sequences are published or under publication. Figure 2 shows some selected sequences that illustrate the homology among proteases used in cheese-making and related proteases. All the gastric proteases are synthesized as zymogens which are converted into active enzymes after cleavage of an N-terminal propart consisting of 42 to 48 residues (see also section 5). Nucleotide sequencing of c-DNA indicates that this is also the case for the fungal enzymes.[92,108,110] Thus, the numbering of residues is a problem in alignment of amino acid sequences from this superfamily of proteins. As the active site aspartic acid residues are often characterized by their positions in the peptide chain of porcine pepsin, this system of numbering is used in the alignment. In order to obtain some consistency, the residues of the proparts are numbered relative to porcine pepsinogen with the prefix 'p'. This numbering systems creates some shortcomings in the ultimate N-termini of proparts and in the region between the C-terminal ends of the proparts and the N-termini of the active enzymes, but no ideal solution seems possible at the moment.

The overall homology among the aspartic proteinases is anchored in clusters of invariant or highly conserved residues. Most of these residues are located in, or close to, the active site cleft. Outside these regions it is precarious to attempt a general alignment based on the amino acid sequences only. In Fig. 2, gaps in the alignment are preferentially introduced where the consensus folding[96] suggests bends or random coil in the tertiary structure. This fits well with the observed identities, but it does not prove that all the aligned residues have topologically equivalent positions in the tertiary structure (see next section).

In porcine pepsin, Tang *et al.*[90] reported an Ile-residue at position 230. This was not found by Moravek & Kostka[99] nor by X-ray crystallography[111] or c-DNA sequencing;[112] consequently, this residue is deleted and further numbering is changed accordingly. Nucleotide sequencing also indicates the presence of isoenzymes, as Tyr was found instead of Asp at position 242.

The sequence of human pepsinogen illustrates the relationship among the mammalian pepsinogens; about 80% of identity is observed with porcine

```
Fold.      bbbbbb     aaaaaaaaa aaaaaaaa      aaaaa            b
           p1         p10       p20       p30       p40        1
            . *         .         .         .         .
P PgA      LVKVPLVRKKSLRQNLIKNGKLKDFLKTHKHNPASKYFPEAAAL-----IG
H PgA      IMYKVPLIRKKSLRRTLSERGLLKDFLKKHNLNPARKYFPQWEAPTL---VD
G PgA      SIHRVPLKKGKSLRKQLKDHGLLEDFLKKHPYNPASKYHPVL-----TATES

R PgC      SLLRVPLRKMKSIRETMKEQGVLKDFLKTHKYDPGQKYHFGNFGDY---SVL
H PgC      AVVKVPLKKFKSIRETMKEKGLLGEFLRTHKYDPAWKYRFGDL------SVT
S PgC      AVVKVPLKKFKSIRETMKEKGLLGEFLRTHKYDPAWKYHFGDL------SVS

B Pch      AEITRIPLYKGKSLRKALKEHGLLEDFLQKQQYGISSKYSGF-------GEVA
O Pch      AEITRIPLYKGKPLRKALKERGLLEDFLQKQQYGVSSXYSGF-------GEVA
G EPg      /DGITRLPLERGKKLREILREKGLLHHFLQHHRYDIGTKFPHAFPD---VLTVV
            .         .         .         .         .
P JP                                                          AASG
E PP                                                          STG

MMP        /RPVSKQSESKDKLLALPLTSVSRKFSQTKFGQQQLAEKLAGLKFPSEAAADGS
MPP        /RPVSKQSDADDKLLALPLTSVNRKYSQTKHGQQA-AEKLGGIKAF--AEGDGS
            .         .         .         .         .
Com Gas        PL    K   R       G L  FL          K
Pep/Fun      K       K L   L                K         K
```

Fig. 2. Alignment of amino acid sequences of proteases from rennets and some related proteases.

Fold: Folding of the peptide chains as determined by X-ray crystallography. Propart according to James and Sieleckie.[95] Active enzymes, consensus folding (secondary structure) according to Overington et al.[96] 'b' indicates β-strands and 'a' helices. * marks location of introns (see Section 6). Prefix 'p' indicates propart numbering starting from the N-terminus of porcine pepsinogen A. Numbering starts again from the N-terminus of porcine pepsin A. Known N-termini of active enzymes are underlined. / indicates that the sequence is derived from nucleotide sequencing. The sequences are expressed in the single letter code A:Ala; C:Cys; D:Asp; E:Glu; F:Phe; G:Gly; H:His; I:Ile; K:Lys; L:Leu; M:Met; N:Asn; P:Pro; Q:Gln; R:Arg; S:Ser; T:Thr; V:Val; W:Trp; X:Unknown; Y:Tyr; –:gap. The vertebrate proteases have S–S bridges from C45 to C50, C206 to C210, and C249 to C282, RMP and RPP have bridges from C45 to C50 and C249 to C282. EPP and PJP have an S–S bridge from C249 to C282 only.

P PgA, PnA: porcine pepsinogen and pepsin.[90, 97–99] H PgA, H PnA: human pepsinogen and pepsin.[100] G PgA, G PnA: chicken (galline) pepsinogen and pepsin.[101] R PgC, R PnC: rat progastricsin and gastricsin[102] H PgC, PnC: human progastricsin and gastricsin.[103] S PgC, S PnC: monkey (simian) progastricsin and gastricsin;[104] B Pch, B Ch: bovine prochymosin and chymosin.[89] O Pch, O Ch: sheep (ovine) prochymosin and chymosin.[105] G EPg, G EPn: chicken embryonic pepsinogen and pepsin.[106] PJP: *Penicillium janthinellum* protease.[107] EPP: *Endothia parasitica* protease.[94] MMP: *Mucor miehei* protease.[92,110] MPP: *Mucor pusillus* protease.[108,109] Com Gas: Residues that are common in all gastric enzymes. Pep/Fun: Propart residues that are identical in porcine pepsinogen and the zymogens for the *Mucor* proteases. Predom: predominant residues found in at least 50% of all known structures. At such positions substitutions generally occur with residues which have the same chemical character as the predominant residues. Com all: residues that are common in all known sequences of aspartic proteases. For further comments see the text.

```
Fold.    bbb bb        bbbbbbb    bbbbbbb    bbb
                  10         20         30         40          50
                  .*                                            *  53
P PnA   DEPLENY--LDTEYFGTIGIGTPAQDFTVIFDTGSSNLWVPSVYC-SSLACSDH
H PnA   EQPLENY--LDMEYFGTIGIGTPAQDFTVVFDTGSSNLWVPSVYC-SSNACTNH
G PnA   YEPMTNY--MDASYYGTISISIGTPQQDFSVIFDTGSSNLWVPSIYC-KSSACSNH

R PnC   YEPM-AY--MDASYFGEISISIGTPPQNFLVLFDTGSSNLWVSSVYC-QSEACTTH
H PnC   YEPM-AY--MDAAYFGEISISIGTPPQNFLVLFDTGSSNLWVPSVYC-QSQACTSH
S PnC   YEPM-AY--MDAAYFGEISISIGTPPQNFLVLFDTGSSNLWVPSVYC-QSQACTSH

B Ch    SVPLTNY--LDSQYFGKIYLGTPPQEFTVLFDTGSSDFWVPSIYC-KSNACKNH
O Ch    SVPLTNY--LDSQYFGKIYLGTPPQEFTVLFDTGSSDFWVPSIYC-KSNACKNH
G EPn   TEPLLNT--LDMEYYGTISIGTPPQDFTVVFDTGSSNLWVPSVSC-TSPACQSH

P JP    VATNTPTA-NDEEYITPVTIG--GTTLNLNFDTGSADLWVFSTEL-PASQQSGH
EPP     SATTTPIDSLDDAYITPVQIGTPAQTLNLDFDTGSSDLWVFSSET-TASEVDGQ

MMP     VDTPGYYDFDLEEYAIPVSIGTPGQDFLLLFDTGSSDTWVPHKGCTKSEGCVGS
MPP     VDTPGLYDFDLEEYAIPVSIGTPGQDFYLLFDTGSSDTWVPHKGCDNSEGCVGK

Com Gas  P        D  Y G I  GTP Q F  V FDTGSS  WVPS YC  S AC   H
Predom   P        D  Y     IGTP Q F    FDTGSS  WVPS  C      C  H
Com all            Y       G          DTGS   W
```

```
Fold.         aaa   bbb  bbbbbb    bbb bbbbbbbb      bbbb bb
              60          70         80         90          100
                                               *
P PnA   NQFNPDDSSTFEAT-SQELSITYGT-GSMTGILGYDTVQV-----GGISDTNQI
H PnA   NRFNPEDSSTYQST-SETVSITYGT-GSMTGILGYDTVQV-----GGISDTNQI
G PnA   KRFDPSKSSTYVST-NETVYIAYGT-GSMSGILGYDTVAV-----SSIDVQNQI

R PnC   ARFNPSKSSTYYTE-GQTFSLQYGT-GSLTGFFGYDTLTV-----QSIQVPNQE
H PnC   SRFNPSESSTYSTN-GQTFSLQYGS-GSLTGFFGYDTLTV-----QSIQVPNQE
S PnC   SRFNPSESSTYSTN-GQTFSLQYGS-GSLTGFFGYDTLTV-----QSIQVPNQE

B Ch    QRFDPRKSSTFQNL-GKPLSIHYGT-GSMQGILGYDTVTV-----SNIVDIQQT
O Ch    QRFDPRKSSTFQNL-GKPLSIRYGT-GSMQGILGYDTVTV-----SNIVDIQQT
G EPn   QMFNPSQSSTYKST-GQNLSIHYGT-GDMEGTVGCDTVTV-----ASLMDTNQL

P JP    SVYNPSATGKE-L-SGYTWSISYGDGSSASGNVFTDSVTV-----GGVTAHGQA
EPP     TIYTPSKSTTAKLLSGATWSISYGDGSSSSGDVYTDTVSV-----GGLTVTGQA

MMP     RFFDPSASSAFKAT-NYNLNITYGT-GGANGLYFEDSIAIGDITVTKQILAYVD
MPP     RFFDPSSSSTFKET-DYNLNITYGT-GGANGIYFRTSITVGGATVKQQTLAYVD

Com Gas  F P  SST           YG  G    G  G DT  V        I    Q
Predom     P  SSTY     G    SI YG  GS  G    DT          GG   Q
Com all             G          Y       G
```

Fig. 2—contd.

(continued)

```
Fold.    bbbbbb    aaa     bbbb  aaa          aaaaaaa
          110       120        130           140
           .         .          .             .          147
P PnA    FGLSETEPGSFLYYAPFDGILGLAYPSISAS---GATPVFDNLWD---QGLVS
H PnA    FGLSETEPGSFLYYAPFDGILGLAYPSISSS---GATPVFDNIWN---QGLVS
G PnA    FGLSETEPGSFFYYCNFDGILGLAFPSISSS---GATPVFDNMMS---QHLVA

R PnC    FGLSENEPGTNFVYAQFDGIMGLAYPGLSSG---GATTALQGMLG---EGALS
H PnC    FGLSENEPGTNFVYAQFDGIMGLAYPALSVD---EATTAMQGMVQ---EGALT
S PnC    FGLSENEPGTNFVYAQFDGIMGLAYPTLSVD---GATTAMQGMVQ---EGALT

B Ch     VGLSTQEPGDVFTYAEFDGILGMAYPSLASE---YSIPVFDNMMN---RHLVA
O Ch     VGLSTQEPGDVFTYAEFDGILGMAYPSLASE---YSVPVFDNMMD---RRLVA
G EPn    FGLSTSEPGQFFVYVKFDGILGLGYPSLAAD---GITPVFDNMVN---ESLLE

P JP     VQAAQQISAQFQQDTNNDGLLGLAFSSINTVQPQSQTTFFDTVKSS-----LA
EPP      VESAKKVSSSFTEDSTIDGLLGLAFSTLNTVSPTQQKTFFDNAKAS-----LD

MMP      NVRGPTAEQSPNADIFLDGLFGAAYPDNTAMEAEYGSTYNTVHVNLYKQGLIS
MPP      NVSGPTAEQSPDSELFLDGIFGAAYPDNTAMEAEYGDTYNTVHVNLYKQGLIS

Com Gas  GLS   EPG     Y   FDGI G   P
Predom   F                 DGILGLAYP           FDN      Q L
Com all                    G
```

```
Fold.      bbbbb       bbbb          bbbb          bbbb
            150       160     170      180          190
             .         .       .        .            .      195
P PnA    QDLFSVYLSSNDD-SGSVVLLGGIDSSYYTGSLNWVPVSV----EGYWQITLD
H PnA    QDLFSVYLSADDQ-SGSVVIFGGIDSSYYTGSLNWVPVTV----EGYWQITVD
G PnA    QDLFSVYLSKGE--TGSFVLFGGIDPNYTTKGIYWVPLSA----ETYWQITMD

R PnC    QPLFGVYLGSQQGSNGGQIVFGGVDKNLYTGEITWVPVTQ----ELYWQITID
H PnC    SPVFSVYLSNQQGSSGGAVVFGGVDSSLYTGQIYWAPVTQ----ELYWQIGIE
S PnC    SPIFSVYLSDQQGSSGGAVVFGGVDSSLYTGQIYWAPVTQ----ELYWQIGIE

B Ch     QDLFSVYMDRDG--QESMLTLGAIDPSYYTGSLHWVPVTV----QQYWQFTVD
O Ch     QDLFSVYMDRSG--QGSMLTLGAIDPSYYTGSLHWVPVTV----QKYWQFTVD
G EPn    QNLFSVYLSREP--MGSMVVFGGIDESYFTGSINWIPVSY----QGYWQISMD

P JP     QPLFAVALKH---QQPGVYDFGFIDSSKYTGSLTYTGVDNS---QGFWSFNVD
EPP      SPVFTADLGY---HAPGTYNFGFIDTTAYTGSITYTAVSTK---QGFWEWTST

MMP      SPLFSVYMNT--NSGTGEVVFGGVNNTLLSGDIAYTDVMSRYGGYYFWDAPVT
MPP      SPVFSVYMNT--NDGGGQVVFGGVNNTLLGGDIQYTDVLKSRGGYFFWDAPVT

Com Gas   F  Y             G  D    TG    W P          YWQ
Predom    FSVYL           FGG D S Y G      PV         W
Com all
```

Fig. 2—*contd.*

```
Fold.   bbbb      bb     bbbbb      bbb aaaaaaa      bb
          200            210       220       230        240
        *         .            .         .         .      . 243
P PnA   SITMDG-ETIACS-GGCQAIVDTGTSLLTGPTSAIANIQSDI-GASENS-DG
H PnA   SITMNG-EAIACA-EGCQAIVDTGTSLLTGPTSPIANIQSDI-GASENS-DG
G PnA   RVTVGN-KYVACF-FTCQAIVDTGTSLLVMPQGAYNRIIKDL-GVSS---DG

R PnC   DFLIGDQASGWCSSQGCQGIVDTGTSLLVMPAQYLSELLQTI-GAQEGE-YG
H PnC   EFLIGGQASGWCS-EGCQGIVDTGTSLLTVPQQYMSALLQAT-GAQEDE-YG
S PnC   EFLIGGQASGWCS-EGCQAIVDTGTSLLTVPQQYMSALLQAATGAQEDE-YG

B Ch    SVTISG-VVVACE-GGCQAILDTGTSKLVGPSSDILNIQQAI-GATQNQ-YG
O Ch    SVTISG-AVVACE-GGCQAILDTGTSKLVGPSSDILNIQQAI-GATQNQ-YG
G EPn   SIIVNK-QEIACS-SGCQAIIDTGTSLVAGPASDINDIQSAV-GANQNT-YG

PJP     SYTAGS-QSG----DGF SGIADTGTTLLLLBDSVVSQYYSQVSGAQQDSNAG
EPP     GYAVGSGTFKS---TSIDGIADTGTTLLYLPATVVSAYWAQVSGAKSSSSVG

MMP     GITVDGSAAVRFS-RPQAFTIDTGTNFFIMPSSAASKIVKAALPDATETQQG
MPP     GVKIDGSDAVSFD-GAQAFTIDTGTNFFIAPSSFAEKVVKAALPDATESQQG

                .            .              .        .
Com Gas           C      CQ I DTGTS L    P              G       G
Predom        G              I DTGTSLL   P S            GA      G
Com all                       D G
```

```
Fold.   bbb        bbbbb  bbbbb aaaabb b           bbbb
          250      260       270                   280
        *        .          .             *      . 284
P PnA   EMVISCSSIDSLPDIVFTINGVQYPLSPSAYILQDD-----------DSCTS
H PnA   DMVVSCSAISSLPDIVFTINGVQYPVPPSAYILQSE-----------GSCIS
G PnA   E--ISCDDISKLPDVTFHINGHAFTLPASAYVLNED-----------GSCML

R PnC   EYFVSCDSVSSLPTLSFVLNGVQFPLSPSSYIIQED-----------NFCMV
H PnC   QFLVNCNSIQNLPSLTFIINGVEFPLPPSSYILSNN-----------GYCTV
S PnC   QFLVNCNSIQNLPTLTFIINGVEFPLPPSSTILNNN-----------GVCTV

B Ch    EFDIDCDNLSYMPTVVFEINGKMYPLTPSAYTSQDQ-----------GFCTS
O Ch    EFDIDCDSLSSMPTVVFEINGKMYPLTPYAYTSQEE-----------GFCTS
G EPn   EYSVNCSHILAMPDVVFVIGGIQYPVPALAYTEQNG----------QGTCMS

PJP     GYVFDCSS--SLPDFNVSISGYTATV-PGSLINYGPS-------GNGSTCLG
EPP     GYVFPCSA--TLPSFTFGVGSARI-VIPGDYIDFGPI------STGSSSCFG

MMP     -WVVPCASYQNSKSTISIVMQKSGSSSDTIEISVPVSKM-LPVDQSNETCMF
MPP     -YTVPCSKYQDSKTTFSLDLQKSGSSSDTIDVSVPISKMLLPVDKSGETCMF

                .            .            .              .
Com Gas         C        P      F     G                        C
Predom  Y       C        LP     F     G                        C
Com all
```

Fig. 2—*contd.*

(continued)

```
Fold.     bbb                 bbb aaaa  bbbbbb      bbbbbb
          290                 300       310         320
            .                   .         .           .        326
P  PnA    GF EGMDVPTSSGE-L--WILGDVF IRQYYTVF DRAN-NKVGLAPVA
H  PnA    GF QGMNLPTESGE-L--WILGDVF IRQYF TVF DRAN-NQVGLAPVA
G  PnA    GF ENMGTPTELGE-Q--WILGDVF IREYYVIF DRAN-NKVGLSPLS

R  PnC    GLESISLTSESGQPL--WILGDVF LRSYYAIF DMGN-NKVGLATSV
H  PnC    GVEPTYLSSQNGQPL--WILGDVF LRSYYSVYDLGN-NRVGF ATAA
S  PnC    GVEPTYLSAQNSQPLVYWILGDVF LRSYYSVYDLSN-NRVGF ATAA

B  Ch     GF QS-----ENHSQK--WILGDVF IREYYSVF DRAN-NLVGLAKAI
O  Ch     GF QG-----ENHSHQ--WILGDVF IREYYSVF DRAN-NLVGLAKAI
G  EPn    SF QN------SSADL--WILGDVF IRVYYSIF DRAN-NRVGLAKAI

P JP      GIQSNSGI---GF----SIF GDIF LKSQYVVF DSDG-PQLGF APQA
EPP       GIQSSAGI---GI----NIF GDVALKAAF VVF NGATTPTLGF ASK

MMP       IILP-----NGGN-Q--YIVGNLF LRF F VNVYDF GN-NRIGF APLASAYENE
MPP       IVLP-----DGGN-Q--F IVGNLF LRF F VNVYDF GK-NRIGF APLASGYENN

Com Gas                       WILGDVF R Y    D    N N VG
Predom    G                   ILGDVF R  Y VFD   N NRVG  A
Com all                       G
```

Fig. 2—*contd.*

pepsinogen and a similar relationship is found between the known sections of bovine pepsinogen and the corresponding sections of porcine pepsinogen.[113-115]

The primary structure of chicken pepsin[101] is apparently more closely related to that of mammalian pepsins than to those of the gastricsins or chymosins. In Table I and Fig. 2, chicken pepsin is tentatively placed together with the mammalian pepsins. However, more knowledge about the structures of other avian gastric proteases is required before a final classification of the avian gastric proteases is established.

The primary structures of the gastricsins are exemplified with three structures from man, monkey and rat. The primate gastricsins show 95% of mutual identity, corresponding to the close evolutionary relationship. A partial structure of bovine gastricsin indicates about 80% of identity with the primate gastricsins (Harboe and Foltmann, unpublished). The identity between the primate gastricsins and that from rat is about 70%, whereas the identity between the gastricsins and pepsins is about 50%.

At least two isoenzymes have been found for bovine chymosin; at position 243, chymosin A has Asp while B has Gly.[89] The difference in charge distribution apparently results in a different binding of SDS to the two izoenzymes. By SDS-PAGE the mobility of chymosin A is retarded more relative to that of chymosin B than may be accounted for by the increase of molecular weight due to the substitution of Gly to Asp.[68a]

Chymosins from different species show pronounced immunochemical cross-reactions.[86] Corresponding to this, the amino acid sequences of lamb and pig chymosin show respectively about 90% and 85% of identity to the amino acid sequence of calf chymosin.[68a]

Chicken embryonic forestomach contains a protease for which the corresponding c-DNA has been sequenced; the derived amino acid sequence shows about 50% of identity with both pepsin and chymosin, but a matrix calculation with specific attention to gaps indicates a closer relation to chymosin.[102]

As one might expect, the vertebrate gastric proteases are more closely related to each other than to the proteases of fungal origin. Though the overall identity of distantly related proteins is limited, the amino acid sequences of the fungal proteases clearly demonstrate homology with the gastric proteases.

P. janthinellum protease is not used in the dairy industry, but its structure is included for comparison and because this enzyme has been of great importance for elucidation of the tertiary structure of the aspartic proteases. The *E. parasitica*, *P. janthinellum* and *M. miehei* proteases show no immunochemical cross-reactions. The partial immunochemical identity between *M. pusillus* and *M. miehei* proteases is consistent with similarities in their primary structures.

In addition to the sequences shown in Fig. 2, complete structures for the following aspartic proteases have been published: monkey pepsin;[116] mouse submandibular renin,[117,118] mouse kidney renin,[119] human kidney renin;[120–122] cathepsin D from pig,[123] mouse,[124] and man;[125] human cathepsin E;[126] *Rhizopus chinensis* protease,[127,128] *Rhizopus niveus* protease,[129] *Irpex lacteus* protease,[130] *Candida albicans* protease[131] and from *Saccheromyces cerevisiae*: protease A,[132] Bar 1 protease,[133] and protease 3.[134] The two latter proteases form a group which has an extended C-terminal sequence, and these two proteases are only distantly related to the other fungal proteases.

3.3 Tertiary Structure

Much progress has recently been made in determination of the tertiary structures of the aspartic proteases. The latest published results are summarized in Table IV. The R-factor is a measure for the difference between the molecular model and the observed data. For proteins, R-factors around 0·15 are quite satisfactory, but for interpretation of the finer details of the structures there are minor differences

TABLE IV
X-ray Crystallographic Structures of Pepsin, Chymosin and Fungal Aspartic Proteases

Enzyme	Resolution (Å)	R-factor (Å)	Refs
Porcine pepsin	1·8	0·174	111
Porcine pepsin	2·3	0·190	135
Porcine pepsin	2·3	0·171	136
Bovine chymosin	2·3	0·165	137
P. janthinellum protease	1·8	0·136	107
R. chinensis protease	1·8	0·143	138
E. parasitica protease	2·1	0·178	139

among the papers. In spite of this all agree that the homology of the folding of the peptide chains is even more pronounced than the homology that is observed by comparison of the primary structures. Overington *et al.*[96] have compared the structures of bovine chymosin, porcine pepsin and the fungal proteases from *R. chinensis, P. janthinellum* and *E. parasitica*. The consensus structure of α-helices and β-structures is indicated above the alignment in Fig. 2. As also seen from Fig. 3, the molecule consists mainly of β-sheets with short stretches of α-helices. The overall shape of an aspartic protease molecule is bilobal with an N-terminal

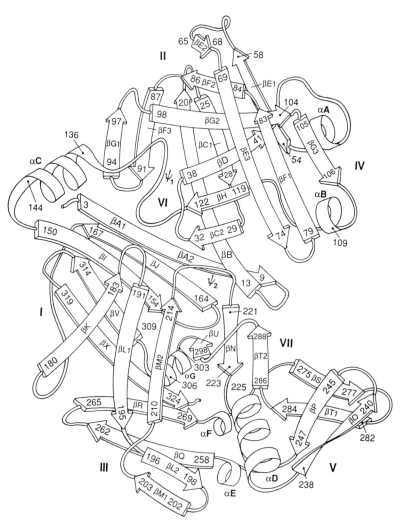

Fig. 3. Schematic diagram of the folding of the peptide chain in porcine pepsin. Numbers of residues that define the approximate boundaries for elements of secondary structures are indicated. The most prominent β-sheets are labelled with Roman numerals as used in Refs 107, 111, 138. Drawing by Collins and Abad-Zapatero, modified from Ref. 136. Permission to reproduce the figure is gratefully acknowledged.

(residues 1–176) and a C-terminal (residues 177–326) domain separated by a cleft which runs perpendicular to the largest diameter. There is an internal symmetry in the molecule so that the two lobes are related by an approximate two-fold axis.[140] A central β-sheet involves three antiparallel strands from each of the lobes; this six-stranded sheet forms, in effect, a wall upon which the rest of the molecule resides.

It is noteworthy that the cleft and its immediate surroundings comprise most of the amino acid residues which are non-conditionally conserved in the primary structures. This applies especially to two aspartic acid residues (numbers 32 and 215) and their neighbouring residues. These are located in the so-called psi-shaped structures, i.e. two bends from the N-terminal and the C-terminal domains with side-chains of Asp-32 and Asp-215 pointing out towards the cleft. Behind each bend run the peptide chains with the conservative residues around Gly-122 and Gly-302. The internal symmetry is especially pronounced in this part of the molecule.

While this paper was in print, a second report on the tertiary structure of bovine chymosin has been published;[201] with a resolution of 2·2 Å and R-factor of 0·17, the results are consistent with the other structures presented in Table VI.

3.4 Post-translational Modifications

Several mammalian pepsins occur in both unphosphorylated and phosphorylated forms. In porcine and bovine pepsins the phosphorylated forms are predominant. After determination of the amino acid sequence of porcine pepsin, Ser-68 was identified at the site of phosphorylation.[90] Bovine pepsin may contain up to three phosphate groups[141,142] but their locations in the structure have not been determined. The presence of phosphate and sialic acid has been reported in bovine gastricsin[143] but structural analyses were not made. Chicken pepsin is glycosylated at Asn-68 and the carbohydrate moiety may be sulphated.[101] Calf chymosin has two potential N-glycosylation sites (Asn-251 and Asn-295); unpublished observations by Hayenga and collaborators indicate that a small percentage of native calf chymosin may be glycosylated, and when recombinant chymosin is expressed in filamentous fungi a larger amount of glycosylation may occur. Glycosylation of the fungal rennet enzymes apparently depends on the culture conditions. The commercial enzyme from *E. parasitica* is not glycosylated. The protease from *M. pusillus* has three potential N-glycosylation sites at positions 72, 101 and 171. Baudys *et al.*[109] observed glycosylation at Asn-171, but Aikawa *et al.*[48] did not observe glycosylation in the native protease. The latter authors found that Asn-72 and Asn-171 are glycosylated when the protease is expressed in recombinant yeast and that glycosylation is important for secretion of the enzyme by yeast. Furthermore, it was reported that the glycosylation reduces the milk-clotting activity of the enzyme.[48] The protease from *M. miehei* has the same potential glycosylation sites of those of *M. pusillus*, and the commercial product is glycosylated at Asn-72 and Asn-171.[92]

4 PROPERTIES

4.1 General Proteolytic Activity and Stability

All the milk-clotting enzymes have pH optima for general proteolysis at acidic conditions, but it must be recalled that for these proteases the pH optimum is not an absolute value. The observed values for the pH of optimum activity depend on experimental conditions such as denaturation of substrates, ionic strength of the solutions, duration and temperature of the experiments, and the methods used to follow the progress of the proteolysis like, for instance, the conditions for precipitation of undigested substrate. Hence, though the general trends are the same in different experiments, a detailed comparison among these proteases is possible only with one common set of conditions. With these reservations in mind, the following values for pH optima for general proteolytic activities may be stated: pepsin, pH 2;[9,10,19–21] gastricsin, broad optimum about pH 3;[9,21] chymosin, pH 3–4;[9,12] fungal proteases, pH 3–4.[67,93,144] When tested with haemoglobin as substrate and at the pH of optimum activity, solutions of equal milk-clotting activity against bovine milk show the following approximate ratios of general proteolytic activities; calf chymosin/bovine pepsin/porcine pepsin; 1/3/6.[9,12]

As regards stability, the observed limits for pH stability may also depend on the experimental conditions like ionic strength of solutions, temperature, duration of the experiment and concentration of enzymes. Pepsin and chymosin undergo autolysis at the pH of optimum proteolytic activity[12] whereas the fungal proteases are stable down to pH 2·5.[66,67,93] In the neutral or alkaline ranges, pig pepsin and calf chymosin lose activity at pH values above 6·5, but the rate of inactivation is different for the two enzymes and pig pepsin has a sharp drop in stability at pH 7.[9,12] Bovine and chicken pepsins are stable at pH 7 but are inactivated at pH 8.[9,145] Inactivation of bovine gastricsin occurs at pH values about 7·5.[9] The mammalian gastric proteases show different susceptibilities to inactivation by urea, the pepsins being rather stable. This has been used for differentiation among the bovine enzymes.[74] For the *M. miehei* protease, a considerable loss of activity was observed at pH 9 after 1 h at 38°C,[66] but at 20°C both the *M. miehei* and the *M. pusillus* proteases withstood pH 9 for 1 h without observable loss of activity.[65]

The native proteases from *M. miehei* and *M. pusillus* are rather thermostable at pH 5·9–6·6, as used in cheesemaking, and this may have undesirable effects for ripening of the cheese and for further use of the whey. In order to circumvent this, the enzymes may be treated with peroxide reagents[146] and such rennets are marketed with the trade names 'Rennilase TL' and 'Rennilase XL'. The technological aspects of the heat stability of the milk-clotting enzymes have been reviewed by Garnot.[147]

The presence of other proteins often has a stabilizing effect on proteases. Therefore, it is difficult to draw definitive conclusions from stability experiments in aqueous solutions as to the stability of proteases in cheese or whey, but in general the same trends are found. Ernstrom and collaborators[148] investigated

TABLE V
Residual Activity of Milk-Clotting Proteases after Pasteurization, Expressed as Percent of Initial Activity

pH	Calf rennet		M. pusillus protease	M. miehei protease		Modified M. miehei protease
	H	L	H	H	L	H
6·6	0^a	$<2^b$	0^a	18^a	5–10^b	0^c
6·4	—	7–14^b	—	—	17–22^b	—
6·3	10^b	—	—	—	—	—
6·2	0^a	20–27^b	0^a	—	29–39^b	—
6·0	0^a	—	0^a	40^a	—	0^c
5·9	—	32–39^b	—	—	38–47^b	—

Refs [a](148), [b](149), [c](150).
Pasteurization: H: 74°C and 15 s; L: 62°C and 30 min. (modified from Ref. 147).

the stability of milk-clotting enzymes in whey over the pH range 5·2–7·0 and heat treatment at 68·3 and 73·9°C. Of six rennets, the protease from *M. miehei* was the most heat stable followed in order by *M. pusillus* protease, calf rennet, bovine pepsin, *E. parasitica* protease and porcine pepsin. With decreasing pH, the heat stability of all enzymes increased with the exception of *E. parasitica* protease on which pH had little effect. Using HTST pasteurization, all measurable activity of *M. miehei* protease was destroyed above 79·5°C and pH 5·4, 76·6°C and pH 5·8 or 73·9°C and pH 6·0. A summary of inactivation of rennets under conditions used for pasteurization of whey is presented in Table V. The results clearly reflect the pH-dependence of the stability. It is seen that with the exception of native *M. miehei* rennet, all the milk-clotting enzymes can be inactivated by pasteurization of the whey provided that pH is kept high enough. A comparative investigation which mimics the conditions for Swiss cheesemaking has shown that the modified protease from *M. miehei*, 'Rennilase XL', has a thermostability which is similar to that of the protease from *E. parasitica*. Furthermore, it was found that the stability of calf chymosin is greatly reduced by raising the temperature from 51 to 55°C and increasing pH from 6·5 to 6·75.[151]

4.2 Catalytic Mechanisms and Specificity

During investigations on the enzymic properties of porcine pepsin, Knowles and collaborators found that two groups with pK_a values about 1·5 and 4·5 participated in the catalytic mechanism.[152] It has also been observed that pepsin and related proteases are inhibited after reaction with diazoacetyl-DL-norleucine methyl ester (abbreviated DAN)[153] or 1,2-epoxy-3-(p-nitrophenoxy)-propane (abbreviated EPNP).[154] Determination of the primary structures showed that the inhibition with DAN is due to esterification of Asp-215 whereas the inhibition

with EPNP preferentially occurs by esterification of Asp-32. The importance of
these two residues is consistent with their location in the molecule. In the primary structure, both are surrounded by highly conservative residues. All X-ray
crystallographic investigations show that Asp-32 and Asp-215 are located in the
middle of the cleft, and are interwoven in a network of hydrogen bonds. Furthermore, it has been found that different types of peptide inhibitors all bind in the
cleft (e.g. Refs 14, 155 156). This means that the cleft is an extended substrate
binding site and that the two aspartic acid residues constitute the active centre as
such; however, the catalytic mechanism itself is not well elucidated. Asp-32 is
buried in the cleft in such a way that its side-chain cannot come in direct contact
with the scissile peptide bond, and in spite of many attempts, it has not been
possible to observe any covalent intermediate during the enzymic catalysis. Hence,
it has been suggested that a nucleophilic attack of a negative charge from the
two aspartic acid residues on the carbonyl group of the scissile peptide bond is
mediated via a water molecule that is hydrogen bonded between the two aspartic
acid residues.[17,157-59] Different mechanistic models are discussed in Refs 14–16; the
question may not be finally settled, but all agree that a covalent intermediate does
not occur.

A characteristic feature of peptide hydrolysis by aspartic proteases is the
extended subsite specificity on either side of the scissile bond. In a long series of
experiments, Fruton *et al.* showed that the length of the substrate peptide chain
was of great importance for the catalytic efficiency of pig pepsin[19,20] and subsequent investigations have shown that the same holds true for chymosin[25] and
for other aspartic proteases also. The individual subsites are numbered relative to the scissile peptide bond according to Schechter & Berger[160] (see also
Tables VI and VII). Dunn and collaborators have carried out systematic investigations on synthetic chromophoric substrates for aspartic proteases,[161] and by
variations of residues at positions P_2 and P_3 they have mapped differences in the
specificities of porcine pepsin, calf chymosin and *E. parasitica* protease.[162]

TABLE VI

Model of Porcine Pepsin Residues in Contact with Residues of an Ideal Pepsin Heptapeptide Substrate[111]

Substrate	Subsite	Residue numbers
P_4 Asn	S_4	Thr-12,Ser-219,Leu-220,(Glu-244,Met-245), (Leu-276,Asn-278)[a]
P_3 Ile	S_3	Glu-13,Ile-30,Phe-111,Phe-117,Ser-219
P_2 Glu	S_2	Thr-218,Thr-222,Glu-287,Met-289,Gly-76,Thr-77
P_1 Phe	S_1	Ile-30,Asp-32,Tyr-75,Thr-77,Phe-111,Leu-112,Phe-117, Ile-120,Gly-217
P'_1 Tyr	S'_1	Tyr-189,Ile-213,Val-291,Thr-293,Leu-298,Ile-300
P'_2 Val	S'_2	Gly-34,Ser-35,Asn-37,Ile-73,Tyr-75,Ala-130
P'_3 Leu	S'_3	Ile-128,Gly-188,Tyr-189

[a] The detailed interactions for S_4 will depend on the conformations of loops at residues
240 and 280 when the substrate is bound.

TABLE VII
Subsites of *E. parasitica* Protease as Defined by Interactions with a Reduced Inhibitor[155]

Inhibitor	Subsite	Residues
P_4 Pro	S_4	Asp-12,Thr-219
P_3 Thr	S_3	Asp-12,Ala-13,Asp-77,Thr-219
P_2 Glu	S_2	Tyr-75,Gly-76,Asp-77,Thr-218,Tyr-222
P_1 Phe$_{(red.)}$	S_1	Asp-30,Asp-32,Tyr-75,Asp-77,Ser-79,Phe-111,Leu-120, Asp-215,Gly-217
P'_1 Phe	S'_1	Asp-32,Gly-34,Gly-76,Ile-213,Asp-215,Thr-218,Ile-297, Ile-299,Ile-301
P'_2 Arg	S'_2	Gly-24, Ile-73,Ser-74,Leu-128,Thr-130,Phe-189
P'_3 Glu	S'_3	Ser-74,Gly-76

Furthermore, it was also shown that ionic interaction may occur between porcine pepsin and substrates with basic amino acids.[163] In order to explain the enhancement of the catalytic efficiency by increased substrate peptide length, Pearl[164] suggested that the binding energy of the substrate is transferred to the scissile peptide bond and thereby facilitates cleavage.

The interactions between the aspartic proteases and their substrates have to a wide extent been investigated with derivatives of pepstatin which is a putative transition stage analogue inhibitor (review, Ref. 165). This inhibitor includes the abnormal amino acid, 3-hydroxy-4-amino-6-methylheptanoic acid, also called 'statine'. In complex with an aspartic protease, statine acts as a dipeptide unit which occupies both S_1 and S'_1 subsites. Other inhibitors which have been used to investigate subsite specificities of aspartic proteases included peptides with a reduced peptide bond, hydroxy isosteres, keto isosteres and ketone hydrates (see Refs 15, 155, 165 for further references). The positions of the inhibitors have been determined by X-ray crystallography (e.g. Refs 14, 15, 156), and a substrate peptide chain may also be modelled into the cleft.[111] Tables VI and VII show the subsite residues as determined for an inhibitor and a substrate; several of the residues have the same sequence numbers, but due to the flexibility of the enzymes and substrates, the detailed interactions will depend on the amino acid sequences of both.

Though all aspects are not clarified, we are able to conclude that the binding cleft may accommodate a substrate peptide chain with at least four residues before and three residues after the scissile peptide bond. The extended binding site, with possible cooperative interaction of the individual subsites, opens the possibility of a high degree of specificity, but the molecular details of the binding of natural substrates are not yet worked out for any of the milk-clotting proteases.

Digestion of the B-chain of oxidized insulin has been used in several investigations on the specificities of these enzymes; although minor differences are observed, the overall pattern of degradation is quite similar for all.[12,144,166-168] The influence of ionic strength and composition of buffers have been investigated by degradation of ribonuclease with bovine and porcine pepsins.[168] In this case, the

activity of bovine pepsin is lower and more dependent on ionic strength than
that of porcine pepsin. Calf chymosin does not cleave ribonuclease.[12] Cleavage
of α- and β-caseins by chymosin and pepsin has also been investigated.[169-173] The
experiments show that the milk-clotting proteases preferentially cleave between
amino acid residues with apolar side-chains. Cleavages after Asp or Glu have
also been observed, but under acidic conditions the carboxylic acids in the side-
chains become protonated and thus are able to fit into an apolar binding site.

The clotting of bovine milk with chymosin depends primarily on a limited
proteolysis involving cleavage of the bond Phe-105–Met-106 in κ-casein. Porcine
pepsin, gastricsin and *M. miehei* protease all cleave the same bond, whereas *E.
parasitica* protease cleaves the preceding Ser–Phe bond.[174] Several studies have
been carried out with synthetic peptides that mimic this peptide bond and its neighbouring amino acid sequence in κ-casein. These will be dealt with in Chapter 3
of this book. Here it should be mentioned that the tryptic fragment of κ-casein
His-98–Lys-111 is cleaved with a K_{cat}/K_m value which is about 75 times larger
than the value for the peptide His-102–Ile-108.[25,175] This means that interactions
of residues outside the cleft also influence the catalytic efficiency.

Chymosin A has an aspartic acid residue at position 243, whereas chymosin B
has glycine at this position. It is seen from Table VI that residue 243 is located
close to subsite S_4 in porcine pepsin. Interaction between Asp-243 in chymosin A
and His-102 of κ-casein may explain why the milk-clotting activity of chymosin
A is about 20% higher than that of chymosin B.[12] A third chymosin fraction
from chromatography on DEAE-cellulose is designated chymosin C. This is a
mixture of several components, all of which have not been analysed in detail.
Among other components, chymosin C includes a degradation product of chymosin A, designated C-2. The general proteolytic activity of this degradation
product is 50% of that of chymosin A but its milk-clotting activity is only 25%.[12]
Experiments with cloned chymosin A[176] have shown that the degradation consists
of an autolytic excision of the tripeptide Asp–Glu–Phe (residues 243–245); the
resulting change in activity and specificity again illustrates the importance of
residues which are located in or near subsite S_4. Site-directed mutagenesis of
chymosin has also contributed to elucidation of the significance of some of the
individual residues in chymosin. Beppu and collaborators[177] changed Lys-220 to
Leu as found in pepsin; this displaced the optimum pH for proteolytic activity
to the acidic side along with increased activity for degradation of acid-denatured
haemoglobin. Exchange of Val-111 to Phe or Ser, and Phe-112 to Tyr also
changed kinetic parameters. If Tyr-75 was changed to Phe, the mutant enzyme
was still active, though with K_{cat}/K_m values which were 2–20% of that of the
native enzyme. This is most interesting since Tyr-75 has been found as an invariant residue in all aspartic proteases sequenced so far. In another investigation,[178]
the mutation of Val-111 to Phe was examined by X-ray crystallography of the
mutant chymosin. The results confirm the importance of residue 111 which par-
ticipates both in subsite S_1 and S_3 (see also Table VI).

Other modifications included Asp-304 and Thr-218.[179] Asp 304 is involved in a
network of hydrogen bond interactions with Asp-215. Substitution of Asp 304

with Ala results in an increase in the pH for optimum activity, as does replacement of Thr-218 with Ala.

It should be observed that relative to the primary structure of porcine pepsin A, the *E. parasitica* protease has several gaps between residues 200 and 300. In this region the alignment numbering is not the same in Refs 94 and 139. In a following report on inhibitor binding in *E. parasitica* protease[202] residues 297, 299, and 301 of Table VII were not observed as a part of the S'₁ binding site.

5   ACTIVATION OF ZYMOGENS

All the vertebrate gastric proteases are secreted as inactive precursors. Manufacturers of cheese rennets ensure that the maximum milk-clotting activity is developed in the rennet before it is sold. Hence, the activation process is not of importance in cheesemaking, but from a biochemical point of view, the activation of the gastric zymogens has interesting aspects.

The zymogens for the vertebrate extracellular proteases are activated by limited proteolysis, which, generally, is brought about by other proteases, but the activation of zymogens of the gastric proteases may be initiated by the action of hydrogen ions only. From experiments on the activation of prochymosin[12] it was suggested that at neutral pH the zymogens are stabilized in an inactive conformation though electrostatic interactions between basic amino acids in the propart and dicarboxylic amino acids in the enzyme moiety of the molecule. By lowering the pH, the dicarboxylic acids are protonated and the electrostatic interactions are weakened. Thus, the first step in the activation consists of a conformational change to the active form without cleavage of a peptide bond. The proteolytic cleavage occurs as a second step and may occur as bimolecular or monomolecular reactions.

The primary structure of the zymogens (Fig. 2) shows: (1) that the majority of the basic amino acid residues are located in the N-terminal region of the peptide chains, and (2) that the distribution of these amino acids is homologous. Lys or Arg may substitute each other at the positions p3 and p8, but the positive charge of a basic amino acid is retained, and Arg-p13 is common in all vertebrate zymogens. Of the gastric zymogens, the structure of porcine pepsinogen only has been solved by X-ray crystallography.[135,180] However, there is a high degree of identity in the amino acid sequences of the proparts, and since enzymes with about 30% of identity in the primary structures have almost the same tertiary structure, it is most likely that all the proparts also have the same folding in principle.

The tertiary structure of porcine pepsinogen shows salt bridges from Lys-p3 to Asp-171 and from Arg-p13 to Asp-11; the two Asp residues are conserved in all known gastric enzymes and the salt bridges illustrate the importance of electrostatic interactions for stabilization of the zymogen molecule. The tertiary structure further shows that the first six residues of pepsinogen form a part of the same large, centrally located six-stranded antiparallel β-sheet as do the N-termini of the active aspartic proteases. The rest of the propart forms a flattened domain which covers the active site cleft; the partly helical structure is indicated in Fig. 2. From the third of the helices, the side-chain of Lys-p36 points down in the cleft and the ε-amino group forms salt bridges to Asp-32 and Asp-215. From Pro-p39 to Pro-5, the folding is irregular, and corresponding to this, large variations are seen in the alignment of the primary structures. In the folding of pepsinogen, residues Leu-6 to Thr-12 form a fourth helix, but from residue Glu-13 the folding of pepsinogen is almost the same as that of pepsin. It is most noteworthy that in the zymogen, the Ile residue which becomes the N-terminus in pepsin is located 40 Å away from its final position. It is still an unsolved question at which stage of the activation process the six N-terminal residues of pepsinogen are exchanged with those of pepsin in the central six-stranded β-sheet.

Alignment of the nucleotide-derived amino acid sequences of fungal aspartic proteases suggested that these are also synthesized as zymogens.[92] Such are not secreted under normal conditions, but a zymogen has been isolated after the M. pusillus protease was expressed in yeast, and it was activated in a process analogous to the activation of pepsinogen.[47]

The main features of the activation process are the same for all the gastric zymogens, but depending on pH, ionic strength, concentration of the zymogen, the amino acid sequence of the propart and the specificity of the active enzyme, the activation process may follow different pathways, and the individual pathways may partly overlap. Furthermore, under physiological conditions and during the production of cheese rennet, different zymogens and active enzymes are present. These will probably also interact with each other. Thus, the system is very complex and all details have not been elucidated.

When pure prochymosin is activated at pH 2, the first cleavage occurs between Phe-p25 and Leu-p26, the process is predominantly intermolecular but may also occur as an intramolecular reaction.[12,181] The resulting intermediate has been designated pseudochymosin. Nothing is known about the folding of the remaining fragment of the propart in pseudochymosin, but for 20 h at pH 2 and 0°C, no further proteolysis was observed. However, at pH 4–5.5, the limited proteolysis proceeds. When a solution of pseudochymosin was transferred to pH 5.5 only chymosin was found after 20 h at 25°C.

Activation of prochymosin has also been studied with genetically modified zymogens. The sequence Leu–Gln–Lys (residues p26 to p28) was substituted with Pro–Arg; this prevented the formation of pseudochymosin at pH 2, and almost 50% of the milk-clotting activity was obtained without any proteolytic processing; at pH 4.5, the activation was normal with cleavage of the propart.[182] In another series of experiments,[183] eight residues from Tyr-p37 to Ala-2 were replaced with a single serine residue; at pH 2, the formation of pseudochymosin was observed though at a slower rate than for authentic prochymosin; at pH 4.5, activation was extremely slow and cleavage occurred between Ser-p35 and Lys-p36.

During the production of cheese rennet, activation may take place at about pH 4·6.[26] At this pH only a minor part of the zymogen molecules are in an active conformation but still sufficient to start the activation process. Thereafter, the activation is mainly autocatalytic in the sense that the rate is determined by molecules which have undergone limited proteolysis.[12] On activation at pH between 2 and 4·7, a combination of the different pathways may occur simultaneously; the resulting mixtures have not been analysed.

The activation of porcine pepsinogen at pH 2 has been described in several papers. The initial rapid conformational change has been investigated by stopped-flow measurements of fluorescence[184] and the formation of the substrate binding site has been determined.[185] In solutions of 0·1 mg/ml at pH 2, porcine or bovine pepsinogens are first cleaved between Leu-p18 and Ile-p19; under these conditions the reaction is predominantly monomolecular.[186] In contrast to the activation of prochymosin, the degradation of pseudopepsin to pepsin proceeds at pH 2. At higher initial concentrations of porcine pepsinogen, cleavage after Phe-p25 has been observed and the activation may also occur in a single step that removes the entire propart.[187] The latter reactions seem to be bimolecular. The propart peptides from the pepsinogens have pronounced inhibitory effects[188] and complexes between the peptides and intermediates or active enzyme behave as single components on ion exchange chromatography.[187] At pH 2, chicken pepsinogen is cleaved at the same site as prochymosin; the propart peptide has inhibitory properties.[189] During preparation of chicken pepsinogen a non covalent complex between the entire propart and cleaved chicken pepsin was observed.[203]

6   BIOLOGY OF MAMMALIAN GASTRIC PROTEASES

Cloning experiments and nucleotide sequencing have provided new insight in the structure and organization of genes for the gastric proteases. Moir et al.[36] reported that the genes for the individual prochymosins represent different alleles at a single locus. Two alleles (chymosin A and B) are predominant, but other forms have been isolated by chromatography of extracts from stomachs of individual calves.[190] Lu et al.[114] have determined partial sequences of two different but closely related genes for bovine pepsin. Furthermore, they observed a gene which is closely related to that of chymosin, but a corresponding functional protease has not been isolated. As regards introns, it is most interesting that the genes for calf prochymosin,[33] human pepsinogen[100] and mouse prorenin, EC 3.4.23.15,[119] have a total of eight introns located at identical positions in their nucleotide sequences. This means that the location of introns appears more conservative than the identity observed in the primary structures of the genes. The significance of this has not yet been elucidated, but it is interesting that several of the suggested insertions in the alignment of the amino acid sequences (Fig. 2) are located in the immediate vicinities of the observed introns.

Pepsinogen, progastricsin and prochymosin are secreted by the chief cells in the fundic part of the gastric mucosa.[78,191] These are located mainly in the lower half of the oxyntic glands which are tubular structures with a narrow lumen oriented almost perpendicular to the muscular sheet surrounding the stomach. The acid-producing parietal cells are located mainly above the chief cells. This means that the zymogens of the gastric proteases are secreted directly into a highly acidic liquid, and under physiological conditions the initial conformational change and limited proteolysis presumably occur already in the lumen of the glands. This is important for the clotting of milk in the stomachs of young mammals: milk has a rather high buffering capacity, so after suckling the contents of the lumen may have a pH up to 5 or 6[192,193] and at such pH, it would take hours to activate the zymogens.

Until a few years ago it was often assumed that the production of chymosins was characteristic for young ruminants (e.g. Ref. 12) but electrophoretic and immunochemical investigations have shown that the production of gastric proteases changes from young to adult mammals, and that several species have gastric proteases which show partial immunochemical identity with calf chymosin.[86]

Quantitative analyses of the ontogeny of gastric proteases were carried out with pigs.[194] The results show that chymosin is produced in large amounts from birth with a rapid decrease after the first week of life. Pepsin is virtually absent during this period but a rapid increase in the production of pepsin starts after about two weeks. In fetal calves, chymosin is present from the sixth month of gestation.[195] Investigations on the production of chymosin and pepsin in calves showed in principle the same pattern of secretion as found for pigs, though with the difference that in cattle the production of chymosin never comes to a complete stop.[196]

To illustrate the physiological significance of chymosin it should be recalled, firstly, that piglets and calves, as well as other mammals in which chymosin-like enzymes have been observed, also have a postnatal uptake of immunoglobulins from colostrum,[86,197] and secondly, that pepsin may cleave the immunoglobulins in the hinge region between the Fab and the Fc fragments. The degradation of immunoglobulins by pepsin or chymosin has not been investigated under physiological conditions but its appears likely that secretion of pepsin would damage the uptake of immunoglobulins in newborn mammals whereas the weaker general proteolytic activity of chymosin will cause less damage. Man has predominantly placental transfer of IgG, and the reports on the existence of human chymosin are conflicting. Örd et al.[198] found a pseudogene for human prochymosin in which two internal stop codons prevent the synthesis of a functional protease. Furthermore, their results indicated that no other chymosin gene was present in their human genomic libraries. Investigations at this Institute are in agreement with the genomic analyses: we have analysed 30 samples of human gastric mucosa obtained at autopsy from the 28th week of gestation to two weeks after birth, and we have not been able to detect chymosin-like proteases, neither by immunoelectrophoresis against anti-calf chymosin, nor by zymograms after agar gelelectrophoresis. This is in contrast to results from Henschel et al.;[199] by single radial diffusion, they tested gastric juice from 17 infants against anti-calf chymosin, some unspecific precipitation took place, but five of the samples apparently gave distinct immunoprecipitates.

The physiological significance of clot formation has not been investigated. The significance may be that the newborn, toothless mammal has the advantage of taking liquid food that solidifies as a rubber-like clot in the stomach. The clot may then act as a mechanical stimulus for the subsequent reactions in the digestive tract. In this context, it is interesting that the clotting activity of pig chymosin against porcine milk is about six times that against bovine milk and, conversely, the clotting activity of calf chymosin against porcine milk is about half of the activity against bovine milk.[194] This means that in spite of a close structural relationship between the two chymosins, an adaptation between the specificity of the chymosins and the structures of the caseins has occurred.

Finally, it should be mentioned that the aspartic proteases represent an interesting example on the evolution of proteins in general. The structures of the molecules show a pronounced internal two-fold symmetry. At an early stage it was suggested[140] that these proteases have evolved by gene duplication and fusion of a gene which originally coded for a protein corresponding to one domain of the present day aspartic proteases. This monomer was supposed to form a noncovalent proteolytically active dimer. Eventually, such a protease has been found in retroviruses. The possibility of designing inhibitors for the protease as a treatment for AIDS has given rise to a large number of papers in this field; for an introduction see Ref. 200.

REFERENCES

1. Deschamps, J.B., 1840. *J. Pharm.*, **26**, 412.
2. Lea, A.S. & Dickinson, W.L., 1890. *J. Physiol. Lond.*, **11**, 307.
3. Foltmann, B., 1970. *Methods in Enzymology*, Vol. 19, ed. G.E. Perlmann & L. Lorand. Academic Press, New York, p. 421.
4. IUB Enzyme Nomenclature Recommendations 1984, Suppl. 1, 1986. *Eur. J. Biochem.*, **157**, 1.
5. Veringa, H.A., 1961. *Dairy Sci. Abstr.*, **23**, 197.
6. Sardinas, J.L., 1972. *Adv. Appl. Microbiol.*, **15**, 39.
7. Sardinas, J.L., 1976. *Process Biochem.*, **11**(4), 10.
8. Sternberg, M., 1976. *Adv. Appl. Microbiol.*, **20**, 135.
9. Antonini, J. & Ribadeau Dumas, B., 1971. *Biochemie*, **53**, 321.
10. Ryle, A.P., 1970. *Methods in Enzymology*, Vol. 19, ed. G.E. Perlmann & L. Lorand. Academic Press, New York, p. 316.
11. Rothe, G.A.L., Axelsen, H.H. Jøhnk, P. & Foltmann, B., 1976. *J. Dairy Res.*, **43**, 85.
12. Foltmann, B., 1966. *C.R. Trav. Lab. Carlsberg*, **35**, 143.
13. Hofmann, T., 1974. *Adv. Chem. Ser.*, **136**, 146.
14. James, N.M.G. & Sieleckie, A.R., 1987. In *Biological Macromolecules and Assemblies* Vol. 3, ed. F.A. Jurnak & A., McPherson. John Wiley, New York, p. 414.
15. Davies, D.R., 1990. *Annu. Rev. Biophys. Chem.*, **19**, 189.
16. Fruton, J.S., 1987. In *Hydrolytic Enzymes.*, ed. A. Neuberger & K. Brocklehurst. Elsevier Science Publishers, Amsterdam, p.1.
17. Foltmann, B., 1981. *Essays in Biochemistry*, Vol. 17, ed. P.N. Campbell & R.D. Marshall. Academic Press, New York, p. 52.
18. Foltmann, B., 1986. In *Molecular & Cellular Basis of Digestion*, ed. P. Desnuelle, H. Sjöström & O. Noren. Elsevier Biomedical Press, Amsterdam, p. 491.
19. Fruton, J.S., 1971. In *The Enzymes*, 2nd edn, ed. P.D. Boyer. Academic Press, New York, p. 119.

20. Fruton, J.S., 1976. *Adv. Enzymol.*, **44**, 1.
21. Tang, J., 1970. *Methods in Enzymology*, Vol. 19, ed. G.E. Perlmann & L. Lorand. Academic Press, New York, p. 406.
22. Scott, R., 1979. In *Topics in Enzyme and Fermentation Biotechnology*, Vol. 3, ed. A. Wiseman. John Wiley, New York, p. 103.
23. Brown R.J. & Ernstrom, C.A., 1988. In *Fundamentals of Dairy Chemistry*, 3rd edn, ed. N.P. Wong, R., Jenness, M. Keeney & E.H. Marth. Van Nostrand Reinhold Comp. New York, p. 609.
24. Green, M.L., 1977. *J. Dairy Res.*, **44**, 159.
25. Visser, S., 1981. *Neth. Milk Dairy J.*, **35**, 65.
26. Placek, C., Bavisotto, V.S. & Jadd, E.C., 1960. *Ind. Engng Chem.*, **52**(1), 2.
27. Kim, S.M. & Zayas, J.F., 1989. *J. Food Sci.*, **54**, 700.
28. Meitner, P.A. & Kassell, B., 1971. *Biochem, J.*, **121**, 249.
29. Harboe, M., Andersen, P.M., Foltmann, B., Kay, J. & Kassell, B., 1974. *J. Biol. Chem.*, **249**, 4487.
30. Foltmann, B. & Jensen, A.L., 1982. *Eur. J. Biochem.*, **128**, 63.
31. O'Leary, P.A. & Fox, P.F., 1975. *J. Dairy Res.*, **42**, 445.
32. Aunstrup, K., 1980. In *Microbial Enzymes and Bioconversions*, ed. A.H. Rose. Academic Press, London, p. 50.
33. Hidaka, M., Sasaki, K., Uozumi, T. & Beppu, T., 1986. *Gene*, **43**, 197.
34. Nishimori, K., Kawaguchi, Y., Hidaka, M., Uozumi, T. & Beppu, T., 1982. *Gene*, **19**, 337.
35. Harris, T.J.R., Lowe, P.A., Lyons, A., Thomas, P.G., Eaton, M.A.W., Millican, T.A., Patel, T.P., Bose, C.C., Carey, N.H. & Doel, M.T., 1982. *Nucleic Acid Res.*, **10**, 2177.
36. Moir, D., Mao, J., Schumm, J.W., Vovis, G.F., Alford, B.L. & Taunton-Rigby, A., 1982. *Gene*, **19**, 127.
37. Emtage, J.S., Angal, S., Doel, M.T., Harris, T.J.R., Jenkins, B., Lilley, C. & Lowe, P.A., 1983. *Proc. Nat. Acad. Sci. USA*, **80**, 3671.
38. Marston, F.A.O., Lowe, P.A., Doel, M.T., Schoemaker, J.M., White, S. & Angal, S., 1984. *Biotechnology*, **2**, 800.
39. McCaman, M.T., Andrews, W.H. & Files, J.G., 1985. *J. Biotechnology*, **2**, 177.
40. Kawaguchi, Y., Kosugi, S., Sasaki, K., Uozumi, T. & Beppu, T., 1987. *Agric. Biol. Chem.*, **51**, 1871.
41. Schoemaker, J.M., Brasnett, A.H. & Marston, F.A.O., 1985. *EMBO J.*, **4**, 775.
42. Meisel, H., 1987. *Milchwissenschaft*, **42**, 787.
43. Meisel, H., 1988. *Milchwissenschaft*, **43**, 71.
44. Mellor, J., Dobson, M.J., Roberts, N.A., Tuite, M.F., Emtage, J.S., White, S., Lowe, P.A., Patel, T., Kingsman, A.J. & Kingsman, S.M., 1983. *Gene*, **24**, 1.
45. Goff, C.G., Moir, D.T., Kohno, T., Gravius, T.C., Smith, R.A., Yamasaki, E. & Taunton-Rigby, A., 1984. *Gene*, **27**, 35.
46. Yamashita, T., Tonouchi, N., Uozumi, T. & Beppu, T., 1987. *Mol. Gen Genet.*, **210**, 462.
47. Hiramatsu, R., Aikawa, J., Horinouchi, S. & Beppu, T., 1989. *J. Biol. Chem.*, **264**, 16862.
48. Aikawa, J., Yamashita, T., Nishiyama, M., Horinouchi, S. & Beppu, T., 1990. *J. Biol. Chem.*, **265**, 13955.
49. Christensen, T., Woeldike, H., Boel, E., Mortensen, S.B., Hjortshoej, K., Thim, L. & Hansen, M.T., 1988. *Biotechnology*, **6**, 1419.
50. Hollenberg, C.P., DeLeeuw, A., Das, S. & van den Berg, J.A., 1983. European Patent Application 0096 430.
51. Cullen, D., Gray, G.L., Wilson, L.J., Hayenga, K.J., Lamsa, M.H., Rey, M.W., Norton, S. & Berka, R.M., 1987. *Biotechnology*, **5**, 369.
52. Ward, M., Wilson, L.J., Kodama, K.H., Rey, M.W. & Berka, R.M., 1990. *Biotechnology*, **8**, 435.

53. Harkki, A., Uusitalo, J., Bailey, M., Penttilä, M. & Knowles, J.K.C., 1989. *Biotechnology*, **7**, 596.
54. Hicks, C.L., O'Leary, J. & Bucy, J., 1988. *J. Dairy Sci.* **71**, 1127.
55. Green, M.L., Angal, S., Lowe, P.A. & Marston, F.A.O. 1985. *J. Dairy Res.*, **52**, 281.
56. Koch, M., Prokopek, D. & Krusch, U., 1986. *Kieler Milchw. Forschungsber.*, **38**, 193.
57. Prokopek, D., Meisel, H., Frister, H., Krusch, U., Reuter, H., Schlimme, E. & Teuber, M., 1988. *Kieler Milchw. Forschungsber.*, **40**, 43.
58. Teuber, M., 1991. *IDF Bulletin*, No. 251.
59. Gist-Brocades Ltd. GRAS Affimation Petition 1989, 9 G 0349.
60. Genencor Inc. GRAS Affimation Petition 1989, 9 G 0352.
61. Pfizer Inc. GRAS Affimation Petition 1988, 8 G 0337.
62. De Koning, P.J. & Draaisma, J. Th. M., 1973. *Neth. Milk Dairy J.*, **27**, 368.
63. Ernstrom, C.A., 1958. *J. Dairy Sci.*, **41**, 1663.
64. Garnot, P., Thapon, J.L., Mathieu, C.M., Maubois, J.L. & Ribadeau Dumas, B., 1972. *J. Dairy Sci.*, **55**, 1641.
65. O'Leary, P.A. & Fox, P.F., 1974. *J. Dairy Res.*, **41**, 381.
66. Ottesen, M. & Rickert, W., 1970. *C.R. Trav. Lab. Carlsberg*, **37**, 301.
67. Arima, K., Yu, J. & Iwasaki, S., 1970. *Methods in Enzymology*, Vol. 19, ed. G.E. Perlmann & L. Lorand. Academic Press, New York, p. 446.
68. Berankova, E., Sajdok, J., Rauch, P. & Kas, J., 1988. *Neth. Milk Dairy J.*, **42**, 337.
68a. Foltmann, B., 1992. *Scand. J. Clin. Lab. Invest.*, **52**(210), 65.
69. Kobayashi, H. & Murakami, K., 1978. *Agric. Biol. Chem.*, **42**, 2227.
70. Stepanov, V.M., Lavrenova, G.L., Adly, K., Gonchar, M.V., Balandina, G.M., Slavinskaya, M.M. & Strongin, A.Ya., 1976. *Biochemia*, **41**, 294.
71. Stepanov, V.M., Rudenskaya, G.N., Gaida, A.V. & Osterman, A.L., 1981. *J. Biochem. Biophys. Meth.*, **5**, 177.
72. Subramanian, S., 1987. *Prep. Biochem.*, **17**, 297.
73. Lawrence, R.C. & Sanderson, W.B., 1969. *J. Dairy Res.*, **36**, 21.
74. Douillard, R., 1971. *Biochemie*, **53**, 447.
75. De Koning, P.J., 1974. *IDF Ann. Bulletin*, Document 80.
76. Rauch, P., Berankova, E., Valentova, O., Sajdok, J. & Kas, J., 1988. *J. Chromatography*, **438**, 451.
77. Uriel, J., 1960. *Nature, Lond.*, **188**, 853.
78. Kreuning, J., Samloff, I.M., Rotter, J.I. & Eriksson, A.W. (eds), 1985. *Pepsinogens in Man*. Alan R. Liss, Inc., New York.
79. Shovers, J., Fossum, G. & Neal, A., 1972. *J. Dairy Sci.*, **55**, 1532.
80. Foltmann, B., Szecsi, P.B. & Tarasova, N.I., 1985. *Anal. Biochem.*, **146**, 353.
81. Foltmann, B., Tarasova, N.I. & Szecsi, P.B. 1985. In *Aspartic Proteinases and their Inhibitors*, ed. V. Kostka. W. de Gruyter, Berlin, p. 491.
82. Axelsen, N.H. (ed.), 1983. *Handbook of Immunoprecipitation in Gel Techniques*, *Scand. J. Immunol.*, **17**, suppl. 10.
83. De Koning, P.J., 1972. *IDF Ann. Bulletin*, part IV, 1.
84. Rothe, G.A.L., Harboe, M.K. & Martiny, S.C., 1977. *J. Dairy Res.*, **44**, 73.
85. Berankova, E., Rauch, P. & Kas, J., 1989. *J. Dairy Res.*, **56**, 631.
86. Foltmann, B. & Axelsen, N.H., 1980. In 'Trends in Enzymology', *FEBS Proceedings*, Vol. 60, p. 271.
87. Harboe, M.K., 1985. In *Aspartic Proteinases and their Inhibitors*, ed. V. Kostka. W. de Gruyter, Berlin, p. 537.
88. Etoh, Y., Shoun, H., Beppu, T. & Arima, K., 1979. *Agric. Biol. Chem.*, **43**, 209.
89. Foltmann, B., Pedersen, V.B., Jacobsen, H., Kauffman, D. & Wybrandt, G., 1977. *Proc. Nat. Acad. Sci. USA*, **74**, 2321.
90. Tang, J., Sepulveda, P., Marciniszyn, J., Chen, K.C.S., Huang, W.Y., Tao, N., Liu, D. & Lanier, J.P., 1973. *Proc. Nat. Acad. Sci.*, *USA*, **70**, 3437.
91. Rickert, W.S. & Elliott, J.R., 1973. *Can. J. Biochem.*, **51**, 1638.
92. Boel, E., Bech, A.-M., Randrup, K., Draeger, B., Fiil, N.P. & Foltmann, B., 1986. *Proteins: Structure, Function, and Genetics*, **1**, 363.

93. Whitaker, J.R., 1970. *Methods in Enzymology*, Vol. 19, ed. G.E. Perlmann & L. Lorand. Academic Press, New York, p. 436.
94. Barkholt, V., 1987. *Eur. J. Biochem.*, **167**, 327.
95. James, M.N.G. & Sielecki, A.R., 1986. *Nature*, **319**, 33.
96. Overington, J., Johnson, M.S., Sali, A. & Blundell, T.L., 1990. *Proc. Roy. Soc. (London)*, Ser. B, **241**, 132.
97. Ong, E.B. & Perlmann, G.E., 1968. *J. Biol. Chem.*, **243**, 6104.
98. Pedersen, V.B. & Foltmann, B., 1973. *FEBS Lett.*, **35**, 255.
99. Moravek, L. & Kostka, V., 1974. *FEBS Lett.*, **43**, 207.
100. Sogawa, K., Yoshiaki, F.K., Mizukami, Y., Ichihara, Y. & Takahashi, K., 1983. *J. Biol. Chem.*, **258**, 5306.
101. Baudys, M. & Kostka, V., 1983. *Eur. J. Biochem.*, **136**, 89.
102. Ichihara, Y., Sogawa, K., Morohashi, K., Fujii-Kuriyama, Y. & Takahashi, K., 1986. *Eur. J. Biochem.*, **161**, 7.
103. Hayano, T., Sogawa, K., Ichihara, Y., Fujii-Kuriyama, Y. & Takahashi, K., 1988. *J. Biol. Chem.*, **263**, 1382.
104. Kageyama, T. & Takahashi, K., 1986. *J. Biol. Chem.*, **261**, 4406.
105. Pungercar, J., Strukelj, B., Gubensek, F., Turk, V. & Kregar, T., 1990. *Nucleic Acids Res.*, **18**, 4602.
106. Hayashi, K., Agata, K., Mochii, M., Yasugi, S., Eguchi, G. & Mizuno, T., 1988. *J. Biochem. (Tokyo)*, **103**, 290.
107. James, M.N.G. & Sielecki, A.R., 1983. *J. Molec. Biol.*, **163**, 299.
108. Tonouchi, N., Shoun, H., Uozumi, T. & Beppu, T., 1986. *Nucleic. Acids. Res.*, **14**, 7557.
109. Baudys, M., Fouling, S., Pavlik, M., Blundell, T. & Kostka, V., 1988. *FEBS Lett.*, **235**, 271.
110. Gray, G.L., Hayenga, K., Cullen, D., Wilson, L.J. & Norton, S., 1986. *Gene*, **48**, 41.
111. Sieleckie, A.R., Federov, A.A., Boodhoo, A., Andreeva, N.S. & James, M.N.G., 1990. *J. Molec. Biol.*, **214**, 143.
112. Lin, X., Wong, R.N.S. & Tang, J., 1989. *J. Biol. Chem.*, **264**, 4482.
113. Harboe, M.K. & Foltmann, B., 1975. *FEBS Lett.*, **60**, 133.
114. Lu, Q., Wolfe, K.H. & McConell, D.J., 1988. *Gene*, **71**, 135.
115. Rasmussen, K.T. & Foltmann, B., 1971. *Acta Chem. Scand.*, **25**, 3873.
116. Kageyama, T. & Takahashi, K., 1986. *J. Biol. Chem.*, **261**, 4395.
117. Panthier, J-J., Foote, S., Chambraud B., Strosberg, A.D., Corvol, P. & Rougeon, F., 1982. *Nature, Lond.*, **298**, 90.
118. Misono, K.S., Chang, J.-J. & Inagami, T., 1982. *Proc. Nat. Acad. Sci. USA*, **79**, 4858.
119. Holm, I., Ollo, R., Panthier, J.-J. & Rougeon, F., 1984. *EMBO J.*, **3**, 557.
120. Imai, T., Miyazaki, H., Hirose, S., Hori, H., Hayashi, T., Kageyama, R., Ohkubo, H., Nakanishi, S. & Murakami, K., 1983. *Proc. Nat. Acad. Sci. USA*, **80**, 7405.
121. Miyazaki, H., Fukamizu, A., Hirose, S., Hayashi, T., Hori, H., Ohkubo, H., Nakanishi, S. & Murakami, K., 1984. *Proc. Nat. Acad. Sci. USA*, **81**, 5999.
122. Hobart, P.M., Fogliano, M., O'Connor, B.A., Schaefer, I.M. & Chirgwin, J.M., 1984. *Proc. Nat. Acad. Sci. USA*, **81**, 5026.
123. Shewale, J.G. & Tang, J., 1984. *Proc. Nat. Acad. Sci., USA*, **81**, 3703.
124. Grusby, M.J., Mitchell, S.C. & Glimcher, L.H., 1990. *Nucleic Acids Res.*, **18**, 4008.
125. Faust, P.L., Kornfeld, S. & Chirgwin, J.M., 1985. *Proc. Nat. Acad. Sci. USA*, **82**, 4910.
126. Azuma, T., Pals, G., Mohandas, T.K., Couvreur, J.M. & Taggart, R.T., 1989. *J. Biol. Chem.*, **264**, 16748.
127. Delaney, R., Wong, R.N.S., Meng, G.-Z., Wu, N.-H. & Tang, J., 1987. *J. Biol. Chem.*, **262**, 1461.
128. Takahashi, K., 1987. *J. Biol. Chem.*, **262**, 1468.

129. Horiuchi, H., Yanal, K., Okazaki, T., Takagi, M. & Yano, K., 1988. *J. Bact.*, **170**, 272.
130. Kobayashi, H., Sekibata, S., Shibuya, H., Yoshida, S., Kusakabe, I. & Murakami, K., 1989. *Agric. Biol. Chem.*, **53**, 1927.
131. Lott, T.J., Page, L.S., Boiron, P., Benson, J. & Reiss, E., 1989. *Nucleic Acids Res.*, **17**, 1779.
132. Dreyer, T., Halkier, B., Svendsen, I. & Ottesen, M., 1986. *Carlsberg Res. Commun.*, **51**, 27.
133. MacKay, V.L., Welch, S.K., Insley, M.Y., Manney, T.R., Holly, J., Saari, G.C. & Parker, M.L., 1988. *Proc. Nat. Acad. Sci., USA*, **85**, 55.
134. Egel-Mitani, M., Flygenring, H.P. & Hansen, M.T., 1990. *Yeast*, **6**, 127.
135. Cooper, J.B., Khan, G., Taylor, G., Tickle, I.J. & Blundell, T.L., 1990. *J. Molec. Biol.*, **214**, 199.
136. Abad-Zapatero, C., Rydel, T.J. & Erickson, J., 1990. *Proteins: Structure, Function and Genetics*, **8**, 62.
137. Gilliland, G.L., Winborne, E.L., Nachman, J. & Wlodawer, A., 1990. *Proteins: Structure, Function and Genetics*, **8**, 82.
138. Suguna, K., Bott., R.R., Padlan, E.A., Subramanian, E., Sheriff, S., Cohen, G.H. & Davies, D.R., 1987. *J. Molec. Biol.*, **196**, 877.
139. Blundell, T.L., Jenkins, J.A., Sewell, B.T., Pearl, L.H., Cooper, J.B., Tickle, I.J. Veerapandian, B. & Wood, S.P., 1990. *J. Molec. Biol.*, **211**, 919.
140. Tang, J., James, M.N.G., Hsu, I.N., Jenkins, J.A. & Blundell, T.L., 1978. *Nature, Lond.*, **271**, 618.
141. Martin, P., 1984. *Biochemie*, **66**, 371.
142. Martin, P. & Corre, C., 1984. *Anal. Biochem.*, **143**, 256.
143. Martin, P., Tricu-Cuot, P., Collin, J.C. & Ribadeau Dumas, B., 1982. *Eur. J. Biochem.*, **122**, 31.
144. Rickert, W., 1970. *C.R. Trav. Lab. Carlsberg*, **38**, 1.
145. Bohak, Z., 1970. *Meth. Enzymology*, **19**, 347.
146. Branner-Jörgensen, S., Schneider, P. & Eigtved, P., 1986. US Patent 4591565.
147. Garnot, P., 1985. *IDF Bulletin*, No. 194, 2.
148. Ernstrom, C.A., Thunell, R.K. & Duersch, J.W., 1979. *J. Dairy Sci.*, **62**, 373.
149. Harper, W.J. & Lee, J., 1975. *J. Food Sci.*, **40**, 282.
150. Olesen, T., 1984. *Milchwirtschaftliche Berichte*, **80**, 183.
151. Garnot, P. & Molle, D., 1987. *J. Food Sci.*, **52**, 75.
152. Knowles, J.R., 1970. *Phil. Trans. Roy. Soc. Lond. B*, **257**, 135.
153. Rajagopalan, T.G., Stein, W.H. & Moore, S., 1966. *J. Biol. Chem.*, **241**, 4295.
154. Tang, J., 1971. *J. Biol. Chem.*, **246**, 4510.
155. Cooper, J., Foundling, S., Hemmings, A., Blundell, T.L., Jones, D.M., Hallett, A. & Szelke, M., 1987. *Eur. J. Biochem.*, **169**, 215.
156. Cooper, J., Foundling, S.I., Blundell, T.L., Boger, J., Jupp, R.A. & Kay, J., 1989. *Biochemistry*, **28**, 8596.
157. Antonov, V.K., Ginodman, L.M., Kapitannikov, Yu.V., Barshevskaya, T.N. Gurova, A.G. & Rumsh, L.D., 1978. *FEBS Lett*, **88**, 87.
158. Pearl, L. & Blundell, T.L., 1984. *FEBS Lett.*, **174**, 96.
159. James, M.N.G., Sieleckie, A.R., Haykawa, K. & Gelb, M.H., 1992. *Biochemistry*, **31**, 3872.
160. Schechter, I. & Berger, A., 1967. *Biochem. Biophys. Res. Comm.*, **27**, 157.
161. Dunn, B.M., Jimenez, M., Parten, B.F., Valler, M.J., Rolph, C.E. & Kay, J., 1986. *Biochem, J.*, **237**, 899.
162. Dunn, B.M., Valler, M.J., Rolph, C.E., Foundling, S.I., Jimenez, M. & Kay, J., 1987. *Biochim. Biophys. Acta*, **913**, 122.
163. Pohl, J. & Dunn, B.M., 1988. *Biochemistry*, **27**, 4827.
164. Pearl, L., 1985. In *Aspartic Proteinases and their Inhibitors*, ed. V. Kostka. W. de Gruyter, Berlin, p. 189.

165. Rich, D.H., 1985. *J. Med. Chem.*, **28**, 263.
166. Oka, T., Ishino, K., Tsuzuki, H., Morihara, K. & Arima, K., 1973. *Arg. Biol. Chem.*, **37**, 1177.
167. Sternberg, M., 1972. *Biochim. Biophys. Acta*, **285**, 383.
168. Pedersen, U.D., 1977. *Acta Chem. Scand. B*, **31**, 149.
169. Pelissier, J.P., Mercier, J.C. & Ribadaeu Dumas, B., 1974. *Ann. Biol. Anim. Biochim. Biophys.*, **14**, 343.
170. Visser, S. & Slangen, K.J., 1977. *Neth. Milk Dairy J.* **31**, 16.
171. Carles, C. & Ribadeau Dumas, B., 1984. *Biochemistry*, **23**, 6839.
172. Mulvihill, D.M. & Fox, P.F., 1979. *Milchwissenschaft*, **34**, 680.
173. Mulvihill, D.M., Collier, T.M. & Fox, P.F., 1979. *J. Dairy, Sci.*, **62** 1567.
174. Drøhse, H.B. & Foltmann, B., 1989. *Biochim. Biophys. Acta*, **995**, 221.
175. Visser, S., van Rooijen, P.J. & Slangen, Ch. J., 1980. *Eur. J. Biochem.*, **108**, 415.
176. Danley, D.E. & Geoghegan, K.F., 1988. *J. Biol. Chem.*, **263**, 9785.
177. Beppu, T., Suzuki, J., Sasaki, K., Sasao, Y., Hamu, A., Kawasaki, H., Nishiyama, M. & Horinouchi, S., 1989. *Protein Engineering*, **2**, 563.
178. Strop, P., Sedlacek, J., Stys, J., Kaderabkova, Z., Blaha, I., Pavlickova, L., Pohl, J., Fabry, M., Kostka, V., Newman, M., Frazao, C., Shearer, A., Tickle, I.J. & Blundell, T.L., 1990. *Biochemistry*, **29**, 9863.
179. Mantafounis, D. & Pitts, J., 1990. *Protein Engineering*, **3**, 605.
180. James, M.N.G. & Sieleckie, A.R., 1986. *Nature, Lond.*, **319**, 33.
181. Pedersen, V.B., Christensen, K.A. & Foltmann, B., 1979. *Eur. J. Biochem.*, **94**, 573.
182. McCaman, M.T. & Cummings, D.B., 1986. *J. Biol. Chem.*, **261**, 15345.
183. McCaman, M.T. & Cummings, D.B., 1988. *Proteins: Structure, Function and Genetics*, **3**, 256.
184. Auer, H.E. & Glick, D.M., 1984. *Biochemistry*, **23**, 2735.
185. Glick, D.M., Auer, H.E., Rich, D.H., Kawai, M. & Kamath, A., 1986. *Biochemistry*, **25**, 1858.
186. Christensen, K.A., Pedersen, V.B. & Foltmann, B., 1977. *FEBS Lett.*, **76**, 214.
187. Kageyama, T. & Takashashi, K., 1983. *J. Biochem. (Tokyo)*, **93**, 743.
188. Dunn, B.M., Deyrup, C., Moeshing, W.G., Gilbert, W.A., Nolan, R.J. & Trach, M.L., 1978. *J. Biol. Chem.*, **253**, 7269.
189. Keilova, H., Kostka, V. & Kay, J., 1977. *Biochem, J.*, **167**, 855.
190. Donnelly, W.J., Carroll, D.P., O'Callaghan, D.M. & Walls, D., 1986. *J. Dairy Res.*, **53**, 657.
191. Andrén, A., Björck, L. & Claesson, O., 1982. *J. Physiol.*, **327**, 247.
192. Decuypere, J.A., Bossuyt, R. & Henderickx, H.K., 1978. *Br. J. Nutr.*, **40**, 91.
193. Cranwell, P.D., Noakes, D.E. & Hill, K.J., 1976. *Br. J. Nutr.*, **36**, 71.
194. Foltmann, B., Jensen, A.L., Lønblad, P., Smidt, E. & Axelsen, N.H., 1981. *Comp. Biochem. Physiol.*, **68B**, 9.
195. Pang, S.H. & Ernstrom, C.A., 1986. *J. Dairy Sci.*, **69**, 3005.
196. Andrén, A., Björck, L. & Claesson, O., 1980. *Swedish, J. Agric. Res.*, **10**, 123.
197. Brambell, F.W.R., 1970. *The Transmission of Passive Immunity from Mother to Young*. North-Holland, Amsterdam.
198. Örd, T., Kolmer, M., Villems, R. & Saarma, M., 1990. *Gene*, **91**, 241.
199. Henschel, M.J., Newport, M.J. & Parmar, V., 1987. *Biol. Neonate*, **52**, 268.
200. Blundell, T.L., Lapatto, R., Wilderspin, A.F., Hemmings, A.M., Hobart, P.M., Danley, P.E. & Whittle, P.J., 1990. *Trends in Biochem. Sci.*, **15**, 425.
201. Newman, M., Safro, M., Frazao, C., Khan, G., Zdanov, A., Tickle, I.J., Blundell, T.L. & Andreeva, N., 1991. *J. Mol. Biol.*, **221**, 1295.
202. Veerapandian, B., Cooper, J.B., Sali, A. & Blundell, T.L., 1990. *J. Mol. Biol.*, **216**, 1017.
203. Baudys, M., Pichova, I., Pohl, J. & Kostka, V., 1985. In *Aspartic Proteases and their Inhibitors*, ed. V. Kostka. W. de Gruyter, Berlin, p. 309.

3

The Enzymatic Coagulation of Milk

D.G. DALGLEISH

*Department of Food Science, University of Guelph,
Guelph, Ontario, Canada N1G 2W1*

1 INTRODUCTION

After milk has been treated with chymosin or other milk coagulating enzymes, there is little apparent reaction for some time, and then the milk coagulates rapidly. This phenomenon, which is the first step of cheesemaking, results from two processes, the first being the attack on κ-casein, which stabilizes the casein micelles, by the proteolytic enzymes (chymosin, pepsin or microbial proteinases) contained in the rennet, and the second being the subsequent clotting of the micelles which have been destabilized by this enzymatic attack. These processes have been described as the primary and secondary stages of the renneting reaction. This sequential description of the reaction is, however, somewhat over-simplified, since it is not necessary for the enzymatic reaction to be complete before aggregation of partly renneted micelles can occur. Indeed, the milk may begin to clot well before the enzymatic cleavage of the κ-casein is complete.[1,2,3]

Since the two processes, i.e. the enzymatic hydrolysis of κ-casein and the subsequent aggregation, occur simultaneously during the later stages of the enzymic reaction, this chapter will deal with both. Until the primary enzymic phase has at least partly occurred, no clotting can occur, and therefore it is necessary to define the behaviour of the enzymes responsible for the hydrolysis of κ-casein. However, the enzymatic reaction cannot be considered in isolation from the effects which it has on the casein micelles, both in chemical and in physical and mechanistic terms, and these also will be described. There is in fact a third stage of the reaction which involves the changes in the properties and structure of the rennet curd once it has been formed. However, this chapter will describe only the mechanisms leading to coagulation (i.e. the primary and secondary stages of clotting).

To understand the coagulation reaction, it is important to define a suitable model for the particles which are being clotted, namely the casein micelles. From the point of view of the renneting reaction, these particles, which contain about 80% of the protein in cows' milk, may be considered as being approximately spherical, composed of several thousand individual molecules of all four types of caseins (α_{s1}, α_{s2}, β, κ) and containing calcium phosphate in one of its insoluble states. The different caseins are not evenly distributed throughout the particle; in particular, the κ-casein appears to be located mainly at the surface of the micelle,[4-6] probably along with α_s-caseins,[7] so that it can exercise a stabilizing effect upon the native micelles and prevent them from coagulating. The stabilizing effect arises because κ-casein, as shown by its peptide sequence,[8] is divided into two distinct regions, namely the hydrophobic para-κ-casein (residues 1–105) and the hydrophilic macropeptide or glycomacropeptide (CMP or GMP, residues 106–169). In its natural position on the surface of the micelles, the κ-casein is probably linked to the remainder of the micelle via the hydrophobic para-κ-casein moiety of the molecule, allowing the macropeptide to protrude from the surface into the surrounding solution. This hydrophilic moiety interacts with the solvent to stabilize the micelles.[9-15]

Successful milk-clotting enzymes split κ-casein at the junction between the para-κ-casein and macropeptide moieties,[16] i.e. in bovine κ-casein, at the Phe_{105}–Met_{106} bond. When this occurs, the macropeptide diffuses off into the serum, its stabilizing influence is lost, and the micelles can begin to coagulate once sufficient of their κ-casein has been hydrolysed.[15] These reactions are shown schematically in Fig. 1.

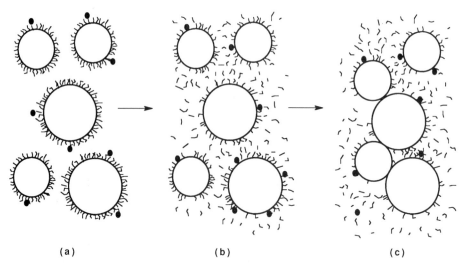

(a)   (b)   (c)

Fig. 1. Schematic diagram of the attack by chymosin (shown as small filled circles) on casein micelles. Three different points in the reaction are illustrated. In (a) the κ-casein coat of the micelles is intact, and chymosin has just been added; (b), some time later, much of the κ-casein has been hydrolysed and a proportion of the GMP is in solution, but sufficient remains to prevent aggregation; (c), at a later time still, nearly all of the κ-casein has been hydrolysed and the micelles have started to aggregate.

The enzymes which specifically cleave the Phe–Met bond of bovine κ-casein all belong to the group of acid proteinases (see previous chapter). The enzymes of this class traditionally used in the manufacture of cheeses are chymosins and pepsin, the former extracted from the stomachs of calves, kids or lambs, and the latter from adult cattle. However, in recent years, the use of enzymes of the same class from microorganisms or from other higher animals has been developed, for economic, religious or cultural reasons. More recently still, chymosin has been produced by microorganisms using recombinant DNA techniques, and this form of pure material is becoming of increasing importance. In the main, this chapter will deal with bovine chymosin and pepsin, since they are the best studied and the most specific for bovine milk, but some consideration will be given to the non-bovine proteinases.

Much of the literature on rennet clotting considers the activity of the proteinases. It should be pointed out that it is deceptively simple to relate the activity of the proteinases to the rate at which they clot milk. Since the clotting is a two-stage process, any variation in conditions can affect the two stages of the reaction differently. For example,[17] milk does not clot at temperatures below 15°C: this is caused, not by a loss of enzymatic activity, but by the slow coagulation rate of renneted micelles at or below this temperature. Also,[18] milks are found to clot differently at different pH values and concentrations of Ca^{2+}. In the discussion which follows, measurements of clotting time have not been used on their own as indicators of enzymatic activity, but details of the separate enzymatic and coagulation stages are considered. Also, since the effects of additives or processing conditions on the overall clotting time cannot on their own be construed as evidence for specific changes in either the primary or secondary stages of the enzymatic reaction, attempts have been where possible made to describe their effects upon individual steps in the clotting process.

2 MEASUREMENT OF THE REACTION OF ACID PROTEINASES WITH κ-CASEIN

Since the primary stage of the reaction involves the hydrolysis of κ-casein, it is important to be able to estimate quantitatively the extent and rate of attack of the acid proteinases on κ-casein, when the protein is either on its own in solution or in its natural state as a component of casein micelles. Unfortunately, there is no method which is simple and rapid enough to be used as a 'real-time' analysis which can be performed during the course of a renneting process.

The most obvious method for following the reaction is of course to measure the disappearance of κ-casein. This, however, can be complicated since κ-casein contains a number of components. There are two genetic variants of the protein (a problem which can be avoided in the laboratory by using milk from individual animals), and in addition, each of the genetic variants contains fractions which are glycosylated to different extents.[19-21] Thus, on anion exchange column chromatography,[22,23] or on electrophoresis on starch or polyacrylamide gels, the κ-casein appears as a number of fractions, which complicates its estimation. Until modern methods of chromatography became available, it was generally found to be more convenient to study the kinetics by analysing the formation of the products of the reaction, namely the macropeptide or the para-κ-casein.

Para-κ-casein, which is insoluble in most buffers, may be solubilized in solutions containing urea, and its amount has been estimated by electrophoresis on polyacrylamide gels[3] or cellulose acetate strips,[2] by staining the bands and photometrically scanning the para-κ-casein band. Since all of the glycosylated residues of κ-casein are found in the macropeptide moiety, para-κ-casein appears as a single band on electrophoresis. A further advantage is that the positively charged para-κ-casein migrates in the opposite direction to the other (negatively-charged) caseins, allowing simple identification and quantification. An alternative method which has been employed to quantify para-κ-casein depends on the insolubility of the peptide in the absence of Ca^{2+}; proteolysis of milk followed by the chelation of the Ca^{2+} present has been used to assess the formation of para-κ-casein.[24] It is also possible to precipitate para-κ-casein[25] specifically because of its higher isoelectric point (>pH 5·0) compared with that of native κ-casein (pH 4·5).

Quantification of the macropeptide liberated during renneting is somewhat more complex. All of the proteins in milk, with the exception of the macropeptide, and of β-lactoglobulin,[26] are precipitated by 2% trichloroacetic acid (TCA).[27] Therefore, all of the proteinaceous material in the supernatant of renneting milk after treatment with TCA is macropeptide. The solubility of the macropeptide in solutions of TCA depends on the degree of glycosylation; as the concentration of TCA is increased, the less glycosylated peptides are precipitated while the more highly glycosylated forms remain soluble.[21] The amount of macropeptide measured therefore depends on the concentration of TCA used to precipitate the caseins, and on the relative amounts of the differently glycosylated fractions present in the macropeptide mixture. This variation has, in fact, been used to differentiate between non-glycosylated and highly glycosylated fractions, by comparing the different amounts of peptides soluble in 2% and 12% TCA.[28] Even when it has been separated, estimation of macropeptide is somewhat hampered, since it lacks aromatic amino acids, and does not absorb in the near UV. Measurement of the absorbence of the peptide at about 215 nm or determination of the nitrogen content of the TCA filtrate requires extensive dialysis to remove traces of interfering materials.

A much more rapid estimation of the macropeptide can be achieved using high-performance liquid chromatography (HPLC),[29] which obviates the need for dialysis. An alternative rapid method is to use fluorescence techniques. Flurescamine (4-phenylspiro[furan-2(3H),1-phthalan]-3,3'dione) binds to amino groups in proteins,[30] especially to the N-termini, and in the process becomes fluorescent. During renneting, new N-termini (of GMP) are produced, and so the fluorescence of samples taken during the course of the reaction and treated with fluorescamine increases.[31] Treatment of the milk with TCA is still required to separate the macropeptides, since the background fluorescence arising from the other proteins in milk is otherwise too high to allow accurate estimation of the small increase in fluorescence caused by the increasing proteolysis by rennet.

HPLC can be used to estimate amounts of κ-casein and possibly the formation of para-κ-casein.[32] Another rapid method which requires no extensive prior treatment of the samples is fast protein liquid chromatography (FPLC). Using this method, it is possible to estimate, relatively rapidly, the amounts of the different forms of κ-casein in renneting samples.[33] These two methods have the advantage that they measure directly the loss of the protein. However, there is still no method for quantifying the formation of products in real time, since all estimates depend on the analyses of samples taken during the reaction and treated with TCA or dispersed in dissociating solvents.

It is possible to make approximate estimates of enzyme activity by observing the clotting time of a standardized sample of reconstituted skim milk powder under defined conditions.[34] An approximately linear relationship exists between the enzyme activity and the time required for the sample to clot under defined conditions.[35-37] Such methods do not allow any detailed analysis of the kinetics of the enzymatic reaction itself, since the measured clotting time is compounded of the time required for a considerable proportion of κ-casein hydrolysis and the time taken by the renneted micelles to form detectable aggregates. However, under strictly defined conditions, the activities of different enzyme preparations can be compared using such measurements of clotting time.

The potential activity of the acid proteinases, rather than their specific action on κ-casein, can be studied using synthetic peptides, especially those which possess a chromophoric group whose absorbence changes when the peptide is cleaved enzymatically. The peptide, Leu–Ser–Phe(NO$_2$)–Nle–Ala–Leu–OMe, is a suitable peptide, the hydrolysis of which can be followed spectrophotometrically,[38] and the use of other similar peptides has been proposed.[39] Alternatively, peptides containing phenylalanine rather than nitro-phenylalanine may be used, with the reaction being followed by estimation of the products using ninhydrin.[40] These peptide substrates can also be used to measure general proteolytic activity: this can then be compared with milk-clotting activity to define the usefulness of the enzyme as a milk-clotting agent, where high clotting activity must be linked to generally low proteolytic activity.

Since commercial rennets often contain varying proportions of chymosin and pepsin; it is often desirable to determine the separate amounts of the two enzymes. This has been attempted in several ways, and can be achieved by a combination of chromatographic separation of the enzymes and the subsequent analysis of the enzyme activity using the synthetic peptide described above.[41,42] These measurements can be related, if care is taken, to a standard milk-clotting assay, so that the assay using the reference peptide need not always be used.[42] It is also possible in principle to estimate the amounts of active enzymes in rennets using immuno-electrophoresis or immunodiffusion techniques.[43-45] Although qualitative identification of the enzymes is possible using the appropriate antisera,[44] the occurrence of an immunological reaction does not guarantee that the enzyme is active.[43,45] The techniques must therefore be used with care.

3 OVERALL KINETICS OF THE PROTEOLYTIC REACTION

Since the breakdown of the κ-casein substrate is essentially a single-step enzyme-catalysed reaction, it seems reasonable to suppose that the kinetics of the proteolysis should obey the standard Michaelis–Menten formulation for the kinetics of such reactions. In this, the instantaneous rate of the reaction (i.e. the rate at which substrate (S) is converted into product), v, is given by the relation:

$$v = -\frac{d[S]}{dt} = \frac{V_{max} \cdot [S]}{(K_m + [S])} \tag{1}$$

In this equation, V_{max} is the maximum rate which will occur at infinite concentration of substrate (i.e. it depends on the concentration of the enzyme) and K_m is the dissociation constant for the enzyme-substrate complex. The renneting reaction has been analysed in this way in a number of studies.[2,3,46-50] The reaction of synthetic peptides with chymosin has also been shown to obey the Michaelis–Menten behaviour.[51] However, it should be noted that overt behaviour of this type is not always observed, because of the relation between the concentration of substrate and the constant K_m. It can be seen from eqn (1) that if [S] is much larger than K_m, a reaction which is zero-order with respect to the concentration of substrate will be observed, at least in the early part of the reaction. Conversely, if $K_m \gg [S]$ the reaction becomes apparently first order, i.e.

$$-\frac{d[S]}{dt} = \frac{V_{max} \cdot [S]}{K_m} \tag{2}$$

Various analyses of the reaction have been made using both isolated and micellar κ-casein as substrates, to give values of K_m in the range 3×10^{-6} to about 5×10^{-4} mole 1^{-1}. In milk, the reaction appears to be approximately first order,[48] and fitting the time-course of the κ-casein proteolysis gave values of K_m which were greater than the substrate (i.e. κ-casein) concentration, which is of the order of $1-1.5 \times 10^{-4}$ M in unconcentrated milk. Lower values of K_m would not fulfil this condition, and would tend to predict a reaction with kinetics closer to zero-order. Thus, to describe the reaction which occurs in milk, it seems to be necessary to use either a first-order formulation or a Michaelis mechanism with a relatively high K_m. The decision as to whether or not the reaction is to be considered as a Michaelis reaction must depend on the method in which the results are analysed. Properly, the reaction should be studied during its early stages, using a variety of enzyme and substrate concentrations. However, a second method of analysing the Michaelis kinetics is to use the integrated form of eqn (1) over the time course of the reaction:

$$K_m \ln\left(\frac{[S]_0}{[S]}\right) + ([S]_0 - [S]) = V_{max} \cdot t \tag{3}$$

where $[S]_0$ is the initial concentration of κ-casein. This has been used in some studies,[2,3] and it is these which give the highest values of K_m (in milk).

It is possible that the Michaelis–Menten mechanism may not be the correct formulation to use to describe the reaction. By using eqns (1) and (3), an implicit assumption is made that enzyme and substrate are able to equilibrate at all times. This in turn implies that both enzyme and substrate are mobile throughout the solution; such an assumption may also have implications for the mechanism of clotting after proteolysis (see below, Section 8). The proteinases, being relatively small in size compared with the casein micelles, are free to move through the solution, but this is hardly true of the κ-casein which forms part of the casein micelles, and which must be constrained to move through the solution as the micelles move, i.e. very slowly in comparison to the enzyme.[52] Moreover, the individual molecules of κ-casein do not move through solution independently, since they are clumped together in the casein micelles. It is in principle possible for the proteinase molecules to bind to the micelles and to create patches of para-κ-casein by attacking adjacent κ-casein molecules one after the other by a 'catch-and-razor' mechanism,[53] rather than producing randomly distributed individual molecules of para-κ-casein, although studies of the aggregation behaviour appear to preclude this.[54]

The fact that κ-casein is naturally incorporated into casein micelles suggests that the kinetics of proteolysis of micellar and soluble κ-casein might be different. It may be significant in this context that the lower values of K_m which have been measured have mainly derived from studies of isolated κ-casein or of peptides,[23,46,49,51] and the higher values have been measured mainly using casein micelles in dispersions or in milk.[2,3,48] It has indeed been shown[23] that aggregation of κ-casein increases the K_m. The possibility of non-random proteolysis should, at least, be borne in mind as a possible reaction pathway.

In eqn (1), the value of V_{max} can be replaced by the expression $k_{cat}.[E]$, where $[E]$ is the concentration of the enzyme and k_{cat} is the catalytic constant. Measured values for this vary, as do the values for K_m. The highest value, 216 s^{-1}, was found for the action of chymosin on reconstituted skim milk powder;[48] lower values of about 60 s^{-1} have been found for peptides,[51] and 30 s^{-1} and 25–40 s^{-1} for isolated κ-casein.[23,46] The highest value was obtained from the same experiments as those which gave high values of K_m. Since this result was obtained using milk and the others used κ-casein or peptides, it appears once again that there may be significant differences between the kinetics of chymosin action in milk and in model systems.

The inverse dependence of the clotting time of milk on the concentration of rennet is well known, and this variation is mainly attributable to the effect of enzyme concentration on the rate of proteolysis. The rate of the enzymatic reaction will increase linearly with the concentration of enzyme, which accords with either a Michaelis–Menten or a first-order mechanism.[48] Temperature also affects the clotting time, and although much of this variation can be attributed to the change in the rate of aggregation of renneted micelles, at least some can be attributed to the enzymatic reaction. For isolated κ-casein[46,55] there appears to be little change in K_m with temperature, and k_{cat} approximately doubles between 25 and 40°C. In milk, the same appears to be true for the value of k_{cat}, at least.[48] Certainly, the κ-casein of milk can be hydrolysed by rennet even at 4°C,[2] at which temperature the clotting reaction does not occur.[17]

4   SENSITIVITY OF THE PHE–MET BOND IN PEPTIDES AND IN κ-CASEIN AND FACTORS WHICH AFFECT THE RATE OF HYDROLYSIS

To clot casein micelles effectively, bovine κ-casein must be split in the region of the Phe_{105}-Met_{106} bond.[16] Cleavage at other points may, and in fact does, occur subsequently, but any general proteolysis of the caseins during the clotting reaction is a disadvantage, leading to increased losses of soluble peptides. The acid proteinases used in cheesemaking are therefore all highly specific for this bond. This specificity may arise from factors such as the sequence of peptides about the sensitive bond (i.e. the primary structure of the substrate), the conformation of κ-casein (i.e. its secondary and tertiary structure), and even its state of aggregation (quaternary structure). The first of these will define specific interactions between individual amino acids in the substrate and in the enzyme, and certain residues close to the sensitive bond appear to be important in this respect. However, the conformation of κ-casein must be such that the sensitive bond is located in a region of the protein readily accessible to the enzyme.

The important part of the sequence of bovine κ-casein appears to be the residues between 97 and 129 of the protein, and changes in this region can cause changes in the reactivity of the sensitive bond.[56,57] The Phe–Met dipeptide is not split by chymosin,[58] nor is the bond hydrolysed when it is in tri- or tetra-peptides.[59,60] However, incorporation of the bond into the pentapeptide, Ser–Leu–Phe–Met–Ala–OMethyl, allows it to be hydrolysed, and when the serine and leucine are exchanged to give the correct sequence appearing in κ-casein, the rate is enhanced.[59,61] Thus, some residues in the immediate vicinity of the sensitive bond function to hold the substrate in its correct orientation in the active site of the enzyme. Further lengthening of this pentapeptide shows that amino acids added at either end, towards the N-and C-terminals of the protein, also influence the reactivity of the sensitive bond.[60,62,63] The effect of elongating the chain towards the C-terminal by three amino acids to give the peptide Ser_{104}-Lys_{111} increases the catalytic ratio, k_{cat}/K_m, and incorporation of the N-terminal Leu_{103}, gives a further increase of about 600-fold in the catalytic ratio. Incorporation of His_{102} and Pro_{101} leads to further increases in the activity of chymosin towards the peptide substrates.[64]

The model peptides just described have sequences identical to that of κ-casein. Other substrates have been used to determine the importance of either the phenylalanine or the methionine on the efficiency of the peptides as substrates. Norleucine is isosteric with methionine and can be used as a replacement; its incorporation[63] increases the ratio k_{cat}/K_m by a factor of about three. Conversely, the replacement of phenylalanine by p–nitrophenylalanine[62] reduces the catalytic efficiency of the enzyme on the peptide substrate by about three. Synthetic peptides suggest that a good substrate should possess hydrophobic residues at positions 103 and 108 (Leu and Ile in the natural protein). The hydroxyl group of Ser_{104} is strongly involved in both the binding and catalysis, and the two proline residues at positions 109 and 110 are also important.[65] The shorter synthetic substrates do not show the same k_{cat} and K_m as are found for κ-casein, but a peptide derived from tryptic hydrolysis and containing residues 98–112 of κ-casein gives a rate of hydrolysis similar to intact κ-casein,[51] so that the presence of the residues at positions 98–102 (His–Pro–His–Pro–His) is also important in the enzyme/substrate binding: K_m for the longer peptide is considerably smaller than that of the peptide 103–112.

The sequence 98–102 contains the only histidine residues in κ-casein, and their importance has been demonstrated by studies of the photo-oxidation of histidyl and other residues of κ-casein, which results in a much reduced sensitivity of the protein to renneting.[66,67] It may be that only one of the histidyl residues performs an essential function and others may play a role in the aggregation of para-κ-casein, although it has been shown that the whole sequence 98–102 is important in determining the susceptibility of the Phe–Met bond to the enzymatic attack.[68] Iodomethylation of the methionine in intact κ-casein did not apparently reduce the sensitivity of the Phe–Met bond,[66] so that, as found for the synthetic substrates, it appears that in κ-casein the residue in position 106 needs not be methionine, so long as the sequence around it is correct. This is also confirmed by the rennetability of both human and porcine κ-caseins, which do not possess a Phe–Met bond.[69] Human κ-casein has a Phe–Ile bond in the sensitive position,[70,71] and is attacked by calf chymosin, although somewhat more slowly than bovine κ-casein. On the other hand, κ-casein from rat milk, which has a Phe–Leu bond in the sensitive region[71,72] is very resistant to attack by calf rennet,[73] although the peptides 98–112 in cow and rat κ-caseins have 53% homology. The insensitivity may arise because the sequence of rat κ-casein shows a large deletion in the position occupied by residues 80–90 in the bovine protein, or because there may be a phosphorylated threonine residue close to the sensitive bond, or because lysines at positions 111 and 112 in the bovine casein are replaced by asparagine and glutamic acid in the rat casein;[72] lysine 111 is itself important in defining the rate of the enzymatic reaction.[68]

The binding of the various substrates to the proteolytic enzymes is significantly governed by the interaction of specific residues with particular sites on the enzyme. However, increases in the efficiency of the binding will be possible if the three-dimensional structure of the κ-casein presents the relevant peptides to the enzyme in an 'optimum' conformation. Many of the small peptide model substrates will be structureless, because they are too short to fold significantly: this is not the case in the intact protein. No detailed definite structural information is available for κ-casein, but the use of calculations which aim to predict the conformations of proteins has suggested that the Phe–Met bond is situated in a region of the protein which can form a projecting β-structure,[74] which may also interact with the enzyme by participating in β-sheet formation.[75] An alternative prediction of the secondary structure suggests that the Phe–Met bond is situated on a β-turn which would also render it accessible to enzymatic attack.[76] A very recent calculation of the 3-dimensional structure of κ-casein[77] suggests that the conformation of the protein in the region of the Phe–Met bond is compact, but is found in a relatively accessible portion of the molecule. These calculations also suggest that the para-κ-casein has a fairly ordered conformation, while the GMP has a much less defined conformational state.

There are also naturally-occurring variations in the structure of κ-casein which may be used to estimate the effect of some structural alterations remote from the Phe–Met bond (at least in terms of primary structure) on the susceptibility towards hydrolysis. A number of studies have been made on the effect of glycosylation and it has been suggested that glycosylated forms of κ-casein are hydrolysed somewhat more slowly than the non-glycosylated forms.[23,48,78,79] The effect is not necessarily large, and other studies have failed to find evidence that glycosylation has a significant effect.[3,33] A second natural modification is the difference between the A and B genetic variants of κ-casein where the modifications consist of replacement of Ile for Thr at position 136 and Ala for Asp at position 148, both being some distance from the site of enzymatic attack. This modification seems to have a small effect on the rate of hydrolysis.[33] It is agreed that milk containing κ-casein B clots more rapidly than milk containing κ-casein A.[80–82] However, there are differences in protein composition, micelle size distribution and in gel structure and syneresis[82,83] associated with the two variants, so the specific influence of the genetic variant upon only the hydrolysis is not yet clear.

A further possible source of variation in the rate of hydrolysis of the sensitive bond may be related to the size of the micelles in which the κ-casein is to be found. From measurements of clotting times on suspensions of micelles of different sizes, the clotting times of the largest and smallest micelles appeared longer than those of medium size.[84] Since it is known that the rate of aggregation of fully-renneted micelles does not depend on the micellar size,[85] it is possible that the κ-casein in micelles of different sizes is hydrolysed at different rates. However, evidence for this is based on measurements of the clotting time only, and may reflect factors such as the concentration of micelles in different preparations,[86,87] and the extent of aggregation required to give visible clotting, as well as possibly different rates of attack of the enzyme on the κ-casein.

The proteolytic activities of the acid proteinases depend on both pH and ionic strength, which makes comparisons of rate constants difficult, since in many cases measurements have been made under different conditions. Measurement of the maximal activity of the proteinases is hampered by the precipitation of the substrate caseins at pH values below 5. It has been suggested that the pH optimum for the attack of chymosin on κ-casein is in the range 5·0–5·5,[88] but in milk it appears that the optimum is in the region of pH 6·0.[89] Sufficient activity of chymosin remains at the natural pH of milk (about 6·7) to allow clotting of the milk, but some other acid proteinases lose their activity at greater or lesser rates around this pH value. The ionic strength of the renneting medium is also important in defining the activity of the proteinase. This may be because the enzyme and substrate are both negatively charged and tend to repel each other: this can be overcome by increasing the ionic strength. If the ionic strength is increased too far, however, it will interfere with specific charge interactions which are essential for enzyme activity, and consequently the activity will decrease. The activity will therefore go through a maximum as the ionic strength is increased.[90,91]

The concentration of specific ions, notably Ca^{2+}, has an effect on the rate of the enzymatic reaction. It is an established fact that the addition of Ca^{2+} to milk accelerates the overall clotting process, principally because of the effect on the aggregation stage of the reaction.[92] It has been reported that these ions also have an effect on the enzymatic reaction, increasing its rate,[93] although on the other hand, it has been claimed that concentrations of Ca^{2+} above 8 mM decrease the enzymatic activity.[94] A third study has reported[95] that addition of $CaCl_2$ (up to 1·8 mM) to milk causes no change in the enzymatic rate. It is possible that these different observations relate to the types of experiments which were attempted, since in some [94] the milk was diluted into a buffer solution containing Ca^{2+} while in others[95] the solution of Ca^{2+} was added to milk. It has also been noted[96] that anions affect the action of chymosin, mainly it is thought, because of their size. Thus, inhibition increases in the range $Cl^- < Br^- < NO_3^- < SCN^-$; these ions also affect the rate of the coagulation reaction.

As described in an earlier section, the values for K_m and k_{cat} for the attack of chymosin on the rennet-sensitive bond are subject to some uncertainty, although this may be partly because of different experimental conditions. Garnier[46] quotes 33 μM for K_m, which is in good agreement with the 27 μM suggested by Visser et al.,[51] but is higher than some of the values given by Castle & Wheelock[47] which were in some cases as low as 6 μM, and Vreeman et al.,[23] who found values in the range 3–9 μM, depending on the level of glycosylation. All of these values are considerably lower than the values of 500 μM given by Dalgleish[2] and 283 μM given by Chaplin & Green,[3] and the high value defined by van Hooydonk et al.[48] Values of 66 μM and 48 μM have been reported by Azuma et al.[49] and Carles & Martin,[50] respectively. According to Garnier,[46] the K_m was not dependent on temperature between 25 and 40°C, so that the discrepancies cannot be reconciled by postulating that K_m is temperature-dependent. All of the values, except those of Visser et al.,[51] were obtained at pH values between 6·6 and 7·0, so that the differences cannot all be explained by variations in pH. The values obtained by Garnier[46] and Visser et al.[51] for k_{cat} are in very good agreement, being 36 s^{-1} at 35°C and pH 6·9 and approximately 65 s^{-1} at 30°C and pH 4·7, respectively. This is reinforced by the value of 68·5 s^{-1} obtained[50] at 30°C and pH 6·2. These results are of interest, because they do not show the expected variations with temperature and pH which are shown by the overall enzyme activity.[88,89]

5 CLONED CHYMOSINS AND ACID PROTEASES OTHER THAN CHYMOSIN

Since the world supply of calf rennet has been found to be inadequate to supply the needs of cheesemakers, it has been necessary to develop the use of other acid proteinases as rennet substitutes. Only acid proteinases possess suitable activities, since, although proteinases such as trypsin cause milk to clot, they also degrade the milk clot almost as soon as it has formed. Until recently, two types of chymosin substitutes were available: the pepsins (obtained from the stomachs of ruminants and other species), and the acid proteinases of bacterial or fungal origin. However, recently the gene for chymosin has been cloned and expressed in microorganisms (e.g. yeasts or *E. coli*), to allow the production of rennets containing pure chymosin only.[97] These products of cloning are available and being used commercially. A number of tests have shown that, as expected, they optimize cheese yield and behave otherwise in the same way as chymosin extracted from the stomachs of calves.[98–101] It is too early to determine whether the availability of cloned chymosins will make any appreciable difference to the pattern of coagulant use, as described below, or whether the cloned chymosin will simply be used to replace veal rennet.

Among the more traditional chymosin substitutes, which have been used or studied, are pepsins from cow, sheep, goat, pig and chicken, and the acid proteinases of *Mucor miehei*, *M. pusillus* and *Endothia parasitica*. The most important criterion to be applied to these enzymes in relation to cheesemaking is that their general proteolytic activity should be low in comparison with their activity towards the rennet-sensitive bond of κ-casein. In addition to individual proteinases, mixtures of different proteinases can be used as rennet substitutes. The distribution of use of rennet substitutes appears to be variable worldwide; for example, it was estimated that in 1974 about 60% of the cheese manufactured in the United States was produced using microbial proteinases,[102] whereas in Europe the microbial rennets are less widely used. A number of reviews are available on the different rennets and rennet substitutes,[102–106] and on their structures.[97]

All of the acid proteinases used have the ability to hydrolyse κ-casein at approximately the required position, but they are not identical in their properties. For example, they have different pH optima, and indeed different stabilities and pH-dependence of activity.[107] Pig pepsin is unstable above pH 6·0,[108,109] compared to chymosin (pH 6·7),[110] *Mucor miehei* proteinase (pH 6·7)[111] and chicken pepsin (pH 8·0).[112] Since pH affects the enzymes differently, it may in principle be necessary to modify the cheesemaking process to accommodate this, if mixtures of proteinases are used,[113] but differences may be sufficiently small to render the adjustment unnecessary. A further difference between the coagulants is their heat stability, which is important if the cheese whey is to be processed. Chymosin shows maximum activity at about 40°C, and at higher temperatures shows evidence of heat-denaturation, as does pepsin, whereas for some *M. miehei* protease inactivation occurs at temperatures about 65°C.[105] The order of inactivation of various acid proteinases is pig pepsin, bovine pepsin, chymosin, *M. pusillus* proteinase and *M. miehei* proteinase, and all are rendered more heat stable by decreasing pH.[114-116] All of the enzymes can be inactivated by prolonged heating, but care must be taken that the whey proteins are not also denatured by too high a temperature during this process. There are different proteinases from *M. miehei*,[117] some of which are heat-labile,[118] and these tend to be more widely used.

Comparisons of the proteolytic and clotting activities of the enzymes show that almost all the rennet substitutes compare unfavourably with chymosin. The clotting to proteolysis ratios of chymosin (40·9) and bovine pepsin (4·3), demonstrate the greatly enhanced general proteolysis caused by pepsin.[104] In a study using synthetic substrates, bovine pepsin was found to be considerably more proteolytic than chymosin; *Mucor* proteinases were somewhat less proteolytic than pepsin but considerably more so than chymosin.[119] However, the heat-labile *Mucor* proteinase has been reported to compare very favourably with chymosin in respect of its proteolytic activity.[120] Reconstituted *M. pusillus* proteinase powder, rather than enzyme extract, has been reported to be less proteolytic than chymosin[121] while all other preparations tested were more proteolytic, both on κ-casein and on α_s- and β-caseins.

It is evident that, since the proteolytic activities of non-chymosin acid proteinases are almost invariably greater than that of chymosin, there may be problems associated with long-maturing cheeses. The ideal solution is to use as coagulants mixtures of proteinases which contain enzymes less proteolytic and more proteolytic than chymosin. Of the available enzymes, pig pepsin is the only one which is potentially less proteolytic than chymosin, since it is readily denatured in the cheesemaking process.[122] Mixed proteinases have been studied[123,124] and were used as the so-called Fifty-Fifty coagulants which found commercial use.[113] In general, however, the enzymes used as coagulants are chosen so as to minimize their proteolytic activities. Most commercial calf rennets do in fact contain mixtures of chymosin and bovine pepsin in varying amounts.

6 DO IMMOBILIZED PROTEASES COAGULATE MILK?

Economically, there would be obvious benefits to be derived from the possession of a reusable clotting agent. Therefore, a good deal of research has been performed to immobilize chymosin or pepsin on solid supports, with the aim of using these immobilized enzymes in column reactors to perform the proteolysis of κ-casein. An ideal process would consist of cooling milk to prevent aggregation[17] and passing it through a column of the immobilized proteinase, to convert its κ-casein to para-κ-casein. Warming the milk would then allow a curd to form. There have been a number of claims that such reactors are practicable.

However, there are good reasons why immobilized proteinases will not be efficient in causing milk clotting. It has been demonstrated experimentally that the efficiency with which κ-casein is hydrolysed by preparations of pepsin, coupled to dextrans of different sizes, decreases with the size of the dextran-pepsin conjugate.[25] Extrapolation to the limit where the pepsin is completely immobilized, indicates that the activity of the protease will be very low. Immobilized enzymes can be successfully used with small, rapidly diffusing substrates, which move quickly around the immobilized enzyme; in contrast, the casein micelles are large particles with slow rotational and translational diffusion. In normal milk clotting, the chymosin is small in relation to the casein micelles, and it is this which diffuses rapidly and causes rapid reaction. So, when the enzyme is immobilized, and the substrate has only slow translational and rotational diffusion, it is to be expected that proteolysis will be slow and inefficient, as has been found experimentally. This has been confirmed by an experiment which showed that only limited hydrolysis of κ-casein occurred even when the proteinase was linked to the immobile support via a flexible link,[126] and it is possible that only the soluble fraction (i.e. that fraction not forming part of the casein micelles) of κ-casein was hydrolysed by this form of immobilized chymosin. Similarly, it has been shown that very prolonged treatment of milk with immobilized chymosin was required to destroy sufficient κ-casein to allow aggregation of the casein micelles.[2] Since the reactors must be run at a low temperature to prevent coagulation of the milk, and it is known that some β- and κ-caseins dissociate from the micelles into the soluble phase, this mobilization will render these caseins more susceptible to attack by the immobilized proteinase.

Since a number of reports have been published of apparently successful coagulation of milk by immobilized proteinases, it must be explained, in view of the arguments given above, how sufficient hydrolysis of κ-casein can occur in such reactors to cause treated milk to coagulate. The simplest explanation is that proteolysis of cold-solubilized κ-casein occurs, and that in addition some of the 'immobilized' chymosin is released from the column and acts on the milk in its soluble form. Only small amounts of proteinase which have been dissociated in this way will be necessary to cause clotting when the milk is warmed. That such desorption of the proteinases can occur from hydrophobic carriers has been demonstrated[127] and indeed desorption of only partly immobilized proteinases has been estimated to underlie many of the apparently successful results.[128,129] It is common for enzyme activity to be irreversibly lost during extensive use of the reactors,[130] and it is probable that some or all of this loss of activity is attributable to desorption of the originally immobilized proteinases. The loss of enzyme activity in this way will depend partly on the method by which the enzyme is coupled to its support, but also on the materials present in the solutions being used, which may enhance the tendency to desorption.

Thus, dissociation of the proteinases from their supports is probably the main reason why clotting has been observed in many studies. At normal temperatures and pH values, extensive proteolysis of the κ-casein is required before the micelles will aggregate,[2,3] and so it is likely that any experiments where coagulation occurs at low extents of proteolysis, and at normal temperature and pH, must be doubted.[131,132] This is confirmed by the observation that apparently fully immobilized chymosin hydrolysed only limited amounts of κ-casein and no clotting was observed.[126] It has been shown that milk can clot when it has been mixed with ultracentrifugate which has been passed through a column of apparently immobilized pepsin.[131] This observation can most readily be explained by assuming that a quantity of the enzyme had been released into the solution, since no other mechanism for such clotting can be suggested.

Therefore, although it is in principle possible to achieve clotting via very careful use of reactors involving immobilized enzymes, the general opinion must be that this cannot be practicable in anything larger than a laboratory experiment. There are no commercial reactors available, and this must reflect the uncertainties in the mechanism, in the efficiency, and the overall practicability of the methods.

7 CLOTTING REACTIONS AND THE PRETREATMENT OF MILK

Heating milk prior to renneting renders it difficult to clot using normal renneting procedures.[133] The effects are essentially reversible until the milk has been heated to temperatures which cause the denaturation of β-lactoglobulin (approximately 75°C for a few minutes or 90°C for 1 minute).[134] Several possible reasons for this loss of clotting may be advanced: the denaturation of β-lactoglobulin and its interaction with κ-casein may cause the κ-casein to be less accessible and/or susceptible to rennet action;[135] conversely, the enzyme may hydrolyse the κ-casein but the renneted micelles may be unable to clot because of the denatured β-lactoglobulin which is bound to their surfaces (the interaction of β-lactoglobulin and κ-casein occurs via formation of disulphide bonds, so that the β-lactoglobulin can bind only to the para-κ-casein moiety, which contains both cysteine residues of κ-casein). Alternatively, the heating may cause more far-reaching changes in the micellar structure itself such that neither of these need be considered as primary causes of the loss of rennetability of the milk. It is, for example, possible that heating causes changes in the distribution of calcium phosphate in the micelles and the serum[136] which render even renneted micelles less capable of aggregating.[137] Such a structural factor may be involved because the effect of heating (providing that it has not been too severe) can be almost totally reversed by decreasing the pH of the milk to below 6 and then readjusting it to pH 6·3 or even higher before adding rennet.[138,139] The effect of changing the pH in this way is to partly dissociate the calcium phosphate of the micelles as the pH is lowered,[140,141] followed by its reformation (and possible partial reconstitution of the micelles) as the pH is raised again, although not all of the dissociated Ca^{2+} is restored to the micelles as the pH is increased.[139] Caseins may be lost from the micelles as the result of decreasing pH,[142] depending on the temperature.[143] The major casein dissociated is β-casein, but if the milk is maintained at 30°C, very little pH-induced dissociation of the caseins occurs. The effect of heating can also be reversed, at least partially, by adding $CaCl_2$, and this suggests that heating interferes with the coagulation mechanism of the renneted micelles, which is highly sensitive to Ca^{2+}, as well as with the enzymatic reaction.

The enzymatic reaction appears to be inhibited in milk which has been heated[144] but inhibition does not appear to be complete, as gauged by the release of peptides by the enzyme.[134,137,145,146] The inhibition appears to prevent a portion (between one-third and one-half) of the κ-casein from being hydrolysed after skim milk has been heated for 1 h at 90°C or casein/β-lactoglobulin mixtures have been heated for short times at 85°C.[144] The effect has been studied over a range of temperatures,[137,146] and it has been demonstrated that both the rate and the extent of the hydrolysis of κ-casein are affected by heating, although different authors provide different estimates of the magnitude of the effects. It may be concluded, therefore, that the enzyme is prevented from hydrolysing sufficient κ-casein on the surfaces of the micelles to make them aggregable, since it is known[2,3] that rennet must hydrolyse a considerable proportion of the κ-casein before aggregation at a measurable rate can occur. The adjustment of pH described above may, by altering the structure of the micelles, render them more susceptible to aggregation at limited extents of hydrolysis of the κ-casein, or may, by reforming the micelles, re-establish the susceptibility of the κ-casein to rennet. Although the primary cause of inhibition is expected to be the reaction of β-lactoglobulin with κ-casein, the effect of adding reducing agents is small, and the enzyme still cannot hydrolyse all of the κ-casein.[147] It seems that α-lactalbumin also plays a part in the effect of heat, since it also inhibits the release of peptides by rennet.[137,148] There remains, however, much to be explained on the mechanism of the renneting reaction in heated milk.

Cooling milk before renneting tends to increase its clotting time[149] and this has been shown to arise at least in part from the decrease in the rate of the renneting reaction.[48] Changes in the casein micelles occur during cooling, e.g. dissociation of some β- and κ-caseins, and solution of some micellar calcium phosphate; which of these may be responsible for the change in clotting time has not been established. This decrease can be almost completely reversed by pasteurization or by holding at high temperatures but recovery is not complete.

A further factor which influences the renneting of milk is homogenization. After the homogenization process, casein micelles are attached to the surface of the fat globules, and some of the micelles may be partly dissociated before adsorption.[150] The change in the surroundings of κ-casein and the structures of the adsorbed casein micelles may, therefore, induce changes in behaviour during renneting. Qualitatively, this does not seem to be the case: a study in which the kinetics of rennet-induced aggregation of skim and homogenized milks were compared[151] showed that maximal rates of aggregation were attained at the same time, suggesting that rennet acted similarly on both systems, although the particles aggregated rather differently.[152] This presumably reflects the fact that the micelles are relatively intact when bound to the surfaces of the fat globules in the homogenized milk.[153] A different situation has been found to obtain if κ-casein, rather than casein micelles, is bound directly to the fat globule surfaces: in this case, the action of the rennet can be considerably altered, presumably because it binds to the surface in such a conformation that the susceptible bond it not available to the enzyme.[154]

Concentration by ultrafiltration is an important factor which affects the renneting of milk. Its effect is to decrease the rate of the renneting reaction.[48] It is claimed that this results from the retardation of the diffusion of the enzyme because of the increased volume fraction of the casein micelles. However, the clotting of the concentrated milks is enhanced so that the overall clotting process appears faster.[86,87,155,156]

8 COAGULATION OF THE RENNETED CASEIN MICELLES

Before they are renneted, the casein micelles in milk show no tendency to aggregate. This stability arises from either or both of two possible mechanisms. First, the micelles carry negative charges on their surfaces, partly, but not completely, as a result of the micellar surface being coated with macropeptide regions of κ-casein.[5,7] Because of the repulsion of these like charges, DLVO theory will potentially explain the micellar stability.[157-159] This principle cannot, however, completely account for all aspects of micellar stability,[160] since the densities of surface charge appear to be rather too low to prevent aggregation. An alternative mechanism for stabilizing the micelles is that of steric stabilization, in which the macropeptide moieties of the κ-casein project into the solution from the surfaces of the micelles (Fig. 1). Because the macropeptides are flexible and hydrophilic, the aggregation of the casein micelles is prevented because the 'hairy' outer layers of the micelles cannot interpenetrate. This latter cause of stabilization has been favoured by a number of authors.[9-12,14,15,161]

During renneting, some of the charge on the surface of the casein micelles is lost. Measurements of the electrophoretic mobility of the particles[157,162-164] show that approximately half of the charge is lost, and this in itself will make the particles coagulate more readily. Moreover, the charge decreases approximately proportionately with the extent of proteolysis of κ-casein;[12] the relationship is not linear, since the rate of change of the surface charge with the extent of renneting is faster towards the end of the reaction. The hydrodynamic radii of the micelles decrease by approximately 5 nm during renneting[11] and this decrease also appears to be linear with the change in surface charge.[12] Because of the extensive hydration of the surface layer,[10] the true decrease in the radii may be as much as 12 nm.[12] It has also been shown by electron microscopy that the number average diameter of the particles decreases, although by a smaller amount than found by the other investigations which used photon correlation spectroscopy, during the early stages of renneting.[165] The viscosity of the milk also decreases during this period before aggregation starts,[165-167] as does its turbidity.[168] The decrease in micellar radius, which seems to have the same value irrespective of the initial radii of the micelles, is consistent with the loss of the 'hairy' macropeptide surface of the particles. However, κ-casein stabilizes even when the surface layer is not extended: the surface layer may be collapsed by treatment with ethanol, but the micelles are still stable until renneted.[13] Therefore, the changes which are brought about in micelles by renneting involve possible losses of both charge stabilization and steric stabilization. The experimental measurements, especially the change in radius caused by renneting, strongly suggest that the κ-casein must be largely on or near to the surface of the micelle, a theory supported by other evidence.[4,5,7] Para-κ-casein is insoluble, and so renneting of isolated κ-casein causes precipitation almost immediately. The kinetics of the aggregation have been described[168,169] and have been shown to describe the reaction in detail.[170] However, the aggregation behaviour of isolated κ-casein during renneting does not serve as an adequate model for the clotting of casein micelles. As soon as an individual molecule of κ-casein has been hydrolysed by chymosin, it is free to aggregate; the effect of chymosin is therefore to provide a steady stream of monomeric material for a subsequent aggregation reaction.[169] This can indeed be seen when κ-casein is hydrolysed by pepsin.[170]

In casein micelles, proteolysis of a small number of κ-casein molecules will have much less effect on the aggregation properties, since the micelles contain many hundreds or even thousands of such molecules. The para-κ-casein produced in the micelles by renneting can only aggregate when the whole micelle is capable of aggregating, and it is this which causes the lag stage before aggregation is observed.[161,171] A short lag phase is seen when isolated κ-casein is renneted, but the origin of this is different: the concentration of particles capable of aggregation is low at the start of the reaction, and then increases rather rapidly.[169] Renneted micelles appear to be incapable of aggregating until about 60–80% of their κ-casein has been destroyed, after which the concentration of micelles capable of aggregating increases rapidly.[2] This behaviour can be explained either by the loss of surface charge during renneting or by loss of steric stabilization; both of the predicted curves have much the same overall shape, and it is not easy to conclude definitively that only one mechanism is operative, although the model based on steric stabilization appears to provide a reasonable fit to observations.[13,15,161] The action of rennet can be seen as providing 'hot spots' via which the micelles can aggregate, these reactive areas being produced by removal of κ-casein from a sufficiently large area.[160,172,173] This area has been suggested to be the space occupied by about 20 molecules of κ-casein.[15]

During the last 20% of the proteolytic reaction both the concentration of micelles which are capable of aggregation and the rate at which they can aggregate increase rapidly, as the last of the stabilizing surface is removed (Fig. 2). Finally, when the micelles have been completely denuded of their κ-casein macropeptide, a limiting rate for the aggregation is reached, and the micelles aggregate by a Smoluchowski[174] mechanism, i.e. the growth of particle weight with time is linear.[169] This certainly applies in dilute solution, where most of the measurements have been made;[85,92] it is possible that in the more concentrated dispersion represented by milk, the aggregation mechanism may be rather different because the micelles are fairly close to each other (within one or two diameters). In such cases, Smoluchowski theory is not strictly applicable, although it may be so in practice. Most theoretical treatments of the renneting reaction[161,169,171] have assumed that Smoluchowski kinetics apply. In milk during normal cheesemaking, the aggregation is very rapid, and it is demonstrable that

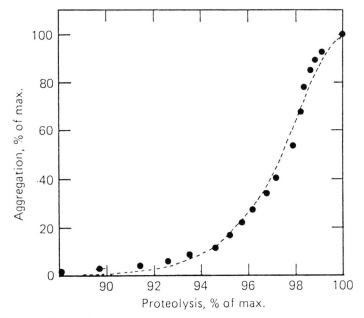

Fig. 2. The experimentally-observed increase in aggregation rate of renneted casein micelles as a function of the extent of breakdown of κ-casein (adapted from Ref. 2).

not all of the κ-casein has been hydrolysed before clotting takes place at 30°C,[93] so that the micelles are not aggregating at their maximum rate during clotting at the temperatures normally used for cheesemaking, nor is it necessarily true that all of the casein micelles in the milk actually participate in the formation of the initial curd.

The aggregation rate of renneted micelles is unaffected by the concentration of rennet,[175] or by the size of the micelles.[85] It is, however, very sensitive both to the concentration of Ca^{2+} present in the solution and to temperature.[92,175] Milk itself will not clot at less than about 15°C,[17] and this is a direct consequence of the slowness of the aggregation reaction at that temperature. The Arrhenius plot for the coagulation reaction is non-linear, being curved very sharply at the low-temperature end (Fig. 3). However, at temperatures of above 45°C the aggregation is very efficient indeed, and approaches the theoretical maximum rate at which particles can collide (Fig. 3). To some extent, the effect of temperature below 45°C can be overcome by increasing the concentration of Ca^{2+} present in the solution; above 45°C, the concentration of Ca^{2+} has little effect on the reaction because it is already near its maximum rate.[92] These results for temperature apply equally to measurements made when micelles are suspended in synthetic buffers[92] or in ultrafiltrate from the same milk.[175]

It is probable that more than one mechanism is involved in the coagulation, because a number of factors affect the rate of aggregation. In the absence of repulsive forces, van der Waals' attraction may be sufficient to hold the micelles

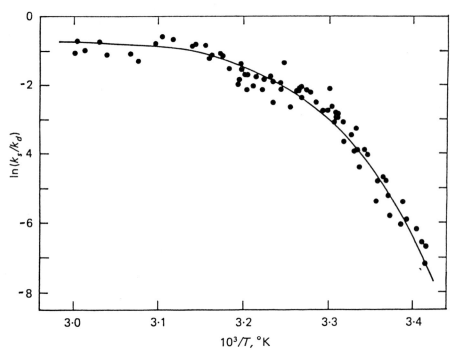

Fig. 3. The increase of the rate constant for the aggregation of completely renneted micelles, as a function of temperature, shown as an Arrhenius plot. Casein micelles were pre-renneted at 0°C before the aggregation rates were measured. k_s is the observed rate constant and k_d is the maximum possible rate constant calculated from the number of collisions (adapted from Ref. 92).

together, but a number of other forces may act as well. From the temperature-dependence of the rate of aggregation or coagulation,[176] it is plausible to assume that hydrophobic interactions (which become stronger with increasing tempera-ture) may be important. However, because the rate of aggregation changes with the concentration of Ca^{2+}, these ions must play a part in the aggregation[93] which is incompatible with the aggregates being held together only by hydrophobic interactions.[92] Furthermore, the rate of aggregation is decreased when the ionic strength is increased, suggesting that specific ion-pair formation may be impor-tant.[93,168,177,178]

An additional factor in the aggregation appears to be the state of the micellar calcium phosphate. It is well established that a decrease in pH leads to a de-crease of the rennet coagulation time[89] but most of this is probably caused by the increase in enzyme activity as the pH is lowered.[88,89] However, it has also been shown that pH does exert some effect on the rate of coagulation of the ren-neted micelles.[176] The rate of aggregation increases as the pH decreases,[179] and the extent of proteolysis required for aggregation decreases markedly.[89,180,181]

These results, however, may depend on the method of acidification, how long the acidified milk is stored, and the effect of the buffer into which milk is diluted, since the composition of the micelles changes markedly with pH.[141,142] The increase in rate has been suggested[162] to arise from the increase in the activity of Ca^{2+} as the pH is lowered, but an alternative explanation may be found by considering the effect of pH on the micellar (or colloidal) calcium phosphate.[182] It has been suggested[183] that dissolution of micellar calcium phosphate may in fact lead to a decrease in the efficiency with which micelles coagulate. As the pH is lowered, more calcium phosphate dissolves, but in milk this only serves to increase the concentration of Ca^{2+}. The effects may therefore tend to cancel out, to give only a small pH-dependence of the aggregation.

The fact that micellar charge is important in the clotting of the renneted micelles is confirmed by the effect of cations such as Ca^{2+} on the clotting time[93] and the effect of polyions. Lysozyme appears to bind to casein micelles, and to catalyse clotting,[184] probably by acting as a polycation.[158] Other polycations which have similar effects are cetyltrimethylammonium bromide[162] and salmine.[93] Polycations can cause milk to clot in the absence of rennet,[185] and so it is not surprising that subcritical amounts can enhance the rate of coagulation.[186] The effect is produced by materials which bind to the casein micelles,[187] and the important factor appears to be the relative charge concentration of polycations.[188] Since the mean sizes and the surface potentials of the casein micelles appear to be largely unaltered by the binding of the ionic materials, it is possible that they bind in the interiors of the micelles. Binding may also occur via hydrophobic interactions, and it appears that the casein, rather than the calcium phosphate, is the primary binding site.[189] Additives such as proline show the importance of hydrophobic interactions,[187] although they may also disrupt the micelles, and additives which decrease the clotting time tend to be of this type or positively charged and strongly bound to the micelles.[187] Additives which decrease the clotting time tend to remain bound to the micelles after coagulation, and most (with the notable exception of sodium dodecyl sulphate) do not affect the rate of proteolysis by rennet.

All of these results tend towards a view that the enhancement of the rate of coagulation arises from the neutralization of negative charge within the micelles, diminishing the charge repulsions, and allowing hydrophobic interactions to occur, once some of the stabilizing κ-casein has been hydrolysed by rennet.

9   MECHANISM AND CALCULATION TO DESCRIBE THE OVERALL COAGULATION PROCESS

When rennet is added to milk, there is no apparent action for some time, and then the milk is seen to clot rapidly. No matter what means are used to monitor the reaction, whether measurement of the visual clotting time, or turbidity[168] or viscosity,[166] or electron microscopy,[165,190] coagulation of the milk only occurs

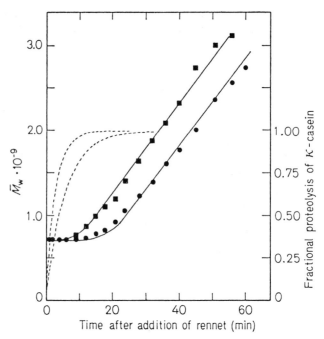

Fig. 4. Lag stage and growth of molecular weight during renneting of casein micelles. Two sets of results are shown, using different concentrations of rennet, the higher concentration giving the shorter lag-time. The two broken lines show the extent of breakdown of κ-casein in the two reactions (adapted from Ref. 85).

after a period during which no aggregation occurs (Fig. 4). This is simply the result of the coagulation reaction being partly sequential on the enzymatic reaction.

In the interests of defining the rate at which milk is converted from a suspension to a gel, and thus allowing the rennet clotting of milk to be carefully controlled, several attempts to construct a usable mathematical model for the overall process, involving both enzymatic and coagulation stages, have been made. Most of the models require a certain simplification of the known mechanisms in order to provide analytical solutions, but otherwise it is possible to use computer-generated algorithms to attempt to describe the reactions in numerical terms.

The object of these models is to allow values for such factors as rennet concentration, temperature and the concentration of milk to be entered into an equation which will predict the rate of curd formation. The earliest attempt to describe the kinetics of the clotting process in this way was made in the 1870s by Storch and Segelcke (see Foltmann, Ref. 36). This simply stated that the clotting time was inversely related to the concentration of rennet used to clot the milk, i.e.

$$CT = k/[E] \tag{4}$$

where k is a constant.

To a first approximation, this equation is valid,[35] but a further refinement was postulated by Holter[191] and rearranged by Foltmann[36] to give the familiar equation:

$$CT = \left(\frac{k}{[E]} \right) + A \qquad (5)$$

where k and A are constants. This preserves the observed linear dependence of the clotting time on the inverse of the enzyme concentration. In such a form, the equation is acceptable for use with a wide range of enzyme concentrations.[37] It may be taken as a test of any more advanced model of the reaction that it reduces to the Holter formulation, which is amply supported by empirical experimental evidence. However, the relationship given in eqn (5) is not descriptive, in the sense that the particular mechanistic significance of the two constants, A and k, is not established, and indeed they depend on the composition of the milk. More recent models have sought to describe the overall reaction by considering the details of the individual reactions involved. Although several approaches to the problem have been made, which come to somewhat different conclusions, they all possess some aspects in common.

The first of these model calculations was established by Payens et al. in a series of papers,[87,159,168,169,175,192,193] where micelles were considered to aggregate in exact proportion to the amount of κ-casein destroyed. Thus, in the early stages of the reaction, proteolysis can only have created small concentrations of micelles which can aggregate. Since the growth of the particles is a bimolecular process, aggregation can occur only slowly at this stage of the reaction. As more proteolysis occurs, the number of coagulating micelles, and hence the rate of aggregate formation, will increase. Thus, there will be a lag period during which no great effect of the enzyme on the average particle size will be seen. On more sensitive criteria of particle size, the size (or, more precisely, the particle weight) will decrease by a small amount because of the loss of the macropeptides from the micelles. The increase in aggregation during renneting simply arises from the increase in the number of aggregating units, as κ-casein is hydrolysed by rennet. This model described well the kinetics of coagulation of isolated κ-casein treated with pepsin,[170] but also predicted that in milk a characteristic time (which may be identified as the clotting time) could be defined by the simple relation:

$$t_c = \sqrt{\frac{2}{k_s \cdot V}} \qquad (6)$$

where k_s is the rate constant for flocculation, and V is the enzymatic velocity (i.e. a constant rate at which the κ-casein was assumed to be destroyed and the casein micelles became capable of aggregation). This equation states that the clotting time will be inversely proportional to the square root of the enzyme concentration, rather than to the enzyme concentration itself (and is therefore not in agreement with experimental observation). The major problem with the model was that, to describe the coagulation of milk, the rate constant for the flocculation of renneted micelles had to be assumed to be very low compared with the known rates which have been measured directly.[85,92,175] This is essentially because the value of k_s which must be used in the calculations represents an average over all values and concentrations of aggregating micelles, from the low value at the start of the reaction to the high value at the end.

Considerations such as this led to a re-estimation of the model, where renneted casein micelles were considered as multi-functional particles, whose rate constants for interaction increased according to the number of functional 'hot spots' via which aggregation could occur, and in which it was possible to use a 'time-averaged' value for the rate at which proteolysis of κ-casein occurred, i.e. the rate at which functionalities were produced on micelles.[193] Once again, the model predicted a square root dependence of the clotting time on the enzyme concentration, which arose as a consequence of the estimates of the rates at which flocculating micelles were produced.

This model of the reaction highlighted a number of important factors, among which are: (1), how much κ-casein must be hydrolysed before a micelle can aggregate; (2), whether the aggregation rate constant of micelles (i.e. k_s) is constant or variable, and if the latter, how the variation arises; (3), how the reactions of κ-casein hydrolysis and aggregation can be linked in time in such a way as to allow calculation of the overall reaction. While the first two of these are matters of the chemistry of the reactions, the third may simply be a matter of defining mathematical functions which have meaningful analytical solutions.

Having observed that micelles did not aggregate until most of their κ-casein had been hydrolysed (Fig. 2), Dalgleish[171] developed a model which described the dependence of the aggregation tendency of micelles upon the extent of the proteolytic reaction by giving k_s a value of zero until a defined critical extent of proteolysis had occurred on any particular micelle (i.e. micelles became capable of aggregating in a catastrophic way). In micelles where more than the critical degree of proteolysis had been achieved, k_s was given a maximal value. The proportions of micelles which had achieved the critical state at any defined extent of κ-casein breakdown could be simply computed using probability theory. The complete mathematical description of the system was relatively complex because of the necessity of incorporating the mathematics of the Michaelis mechanism into the calculation of the extent of κ-casein hydrolysis, but it was possible to employ computer simulation to show that the Holter relationship could be reproduced by the calculation.

Moreover, the calculation could be simplified by making the assumptions that (1), the transition between all micelles being stable and all being capable of aggregation could be approximated by a step function, at a defined degree of proteolysis of κ-casein, and (2), the observed coagulation time occurred at a defined degree of polymerization of the casein micelles. This gave an equation for clotting time:[86]

$$t_c \sim \frac{K_m}{V_{max}} \ln\left(\frac{1}{(1-\alpha_c)}\right) + \frac{\alpha_c \cdot s_0}{V_{max}} + \frac{1}{2k_s c_0}\left(\frac{M_{crit}}{M_0} - 1\right) \tag{7}$$

where α_c is the critical extent of conversion of κ-casein, s_0 is the original concentration of κ-casein, c_0 is the number concentration of casein micelles, and M_0 and M_{crit} are average micellar particle weights of casein micelles initially and at the point where clotting is observed. A similar formulation has been used by Carlson et al.[181] Since for any set of conditions, most of the terms in eqn (7) are constant, the equation can be seen to reduce to the Holter equation (5) (since of course V_{max} is proportional to the concentration of rennet).

In this mechanism, no reason was given for the sudden change in aggregation of the micelles; the model was simply based on experimental observation of the rate at which the casein micelles aggregated when renneted to different extents. However, it is possible to explain the change in the value of k_s by postulating an energy barrier which prevents aggregation of the original micelles, but which is lowered by the action of rennet.[161] In this model, the native casein micelles were considered to be sterically stabilized, this stabilization being lost as the renneting proceeded. In this way, it was possible to define an energy function (stability ratio) for interactions between average micelles at any stage in the reaction, this stability ratio (W) being related to the extent of proteolysis at time, t, by:

$$W_t = W_0 \exp(-C_m \cdot V \cdot t) \tag{8}$$

where C_m is a constant and V is the rate of the enzymatic reaction. Like Payens's formulation, the kinetics of the enzymatic reaction were simplified to give linear proteolysis of κ-casein. At the start of the reaction, rennet has negligible effect upon the aggregation since W_t is large at that time, and calculation shows that effective aggregation cannot take place until about 70% of the κ-casein has been destroyed. At this point, the ability of the micelles to aggregate increases rapidly with the extent of proteolysis, and Smoluchowski aggregation can take place, at an increasing rate, because the effective rate constant for the aggregation increases. This is in accordance with measurements of this rate.[1,2,3,181] The description of the reaction in this way provided the first attempt to link the interaction energies of renneted micelles[157,159] with their changing rates of aggregation during renneting. However, the difficulty with the model lay in its simplification of Michaelis kinetics to allow calculation to be made.

This model was based on the destruction of steric stabilization by rennet, although the equation used for the stability factor (8) in fact is valid for any form of stabilizing interaction energy, whether by charge or steric means. It is, however, possible to envisage the onset of aggregation in a purely geometrical way.[15] If the micelles are stabilized by a thick layer of κ-casein, and can only react when large enough gaps in this layer have been made, then from the known dimensions of the casein micelles, and of their surface layers,[12] it is possible to calculate from geometric principles how large the gaps must be. A scheme of reaction showing the encounters of partially renneted micelles is shown in Fig. 5, from which it can be seen that the micelles must not only possess gaps in their surface of the appropriate size, but must also collide in the proper orientation. On these purely geometrical grounds it is possible to describe the increase in aggregation rate with the extent of κ-casein hydrolysis, since as the extent of

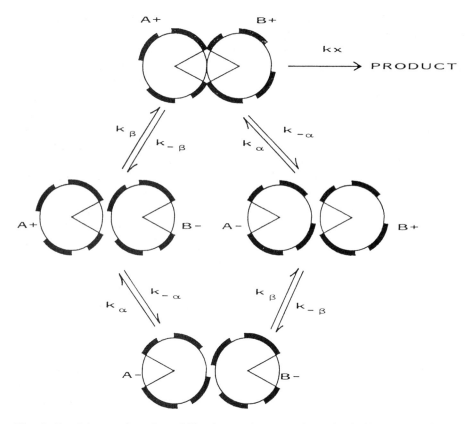

Fig. 5. Breakdown of steric stabilization and aggregation of micelles. Chymosin has partly destroyed the stabilizing layers, to leave some micelles with gaps large enough to cause aggregation (these gaps are identified as segments on the micelles). However, the micelles must approach each other in such an orientation that a collision may lead to aggregation. As the number of gaps of sufficient size increases, so will the chances of a successful collision. The diagram is simplified in that there would be many more gaps, of smaller size, because of random attack by the chymosin (reproduced from Ref. 15, with permission).

proteolysis increases, so do both the sizes of the gaps in the micellar surface and the chances of collision in a reactive orientation. This model of the reaction explains the change in aggregation of micelles with the extent of proteolysis but does not explain the changes in the critical values of α with changing temperature or pH.

In the reaction mechanisms described above, explicit analytical solutions for the coagulation time could not be calculated, because of the complexity of the reaction; only by simplifying the Michaelis kinetics could analytical solutions be found. However, a calculation involving less drastic modification of the enzyme kinetics can be made. According to van Hooydonk et al.,[48] the hydrolysis of κ-casein can be regarded as a first-order reaction in milk, and the high values of

K_m found by Dalgleish[2] and Chaplin & Green[3] suggest also that a pseudo-first order reaction occurs. Van Hooydonk & Walstra[52] accepted and used this in calculating the overall reaction; however, they found that in their formulation, the exponential form of the stability factor, W, as in equation (8), did not permit a full analytical solution. They therefore used an approximation where W was linearly related to the amount of κ-casein remaining in the micelles at any time, which is only true over a limited period of the reaction, and so the calculation did not cover the whole of the experimental time-scale from rennet addition to clotting. However, it has been found possible to calculate analytical solutions where the first-order kinetics of κ-casein hydrolysis are combined with the stability criteria of the casein micelles.[54] This applies both in the case in which an energy barrier between casein micelles is postulated, of the exponential form used by Darling & van Hooydonk,[161] and also in the model in which the purely geometrical approach to micellar aggregation is used. In either of these cases, the analytical solution is very complex and unfortunately cannot be reduced to a simple form. The models did, however, when entered into a computer, show the appropriate behaviour, e.g. in relationship between coagulation time and enzyme concentration. It may be possible to use these calculations as the basis for future models.

All of the models describe the overall reaction as consisting of a lag-stage followed by rapid aggregation. In two of the models, the lag time is explained as arising from the time required for aggregable micelles to be formed, giving the impression of a primary (enzymatic) stage followed by a secondary (coagulation or flocculation) stage. However, all of the models show that the onset of coagulation will occur before the exhaustion of the κ-casein substrate of the enzyme. None of the models is complete, inasmuch as they cannot explain all of the changes which can be caused in the coagulation process by such factors as temperature, pH, ionic strength, and concentrations of different ions such as Ca^{2+}. What they have achieved is a rationale for the reaction in fairly general terms, within which framework more detailed investigations of the behaviour of the casein micelles and the enzyme-substrate reactions can be made.

10   CONCLUSION

We have considered in this chapter the factors which influence rennet clotting, and the reactions which are believed to occur during the process. Clotting times can be measured in a variety of ways, ranging from simple observation of the formation of visible particles to more sophisticated methods.[194] The model calculations which have been used to understand the coagulation have in fact given some insights and produced some discussion.[195,196,197] However, the models have their limitations: once the aggregation of the renneted micelles is well advanced, it is almost certain that the description of the system in terms of Smoluchowski kinetics cannot be accurate, and more complex theories of gel formation and properties must be used. Some attempts have been made to consider the reaction in this way[198,199] to describe such factors as gel strength and syneresis. These have led to partial success, and although the theories can be criticized in terms of whether they are applicable to the milk system,[200] they have made some attempt to explain gel formation and properties.

To change from considering the properties of renneting milk as those of an aqueous suspension to considering renneting milk as a gel requires changes in the types of experiments which should be done, and also the development of different theoretical backgrounds within which to discuss the results. The point of division between the two regimes is one which requires further study, since there are as yet no certain theoretical developments, and very few experimental techniques[201] which will allow the details of the suspension-gel transition to be investigated successfully.

REFERENCES

1. Green, M.L., Hobbs, D.G., Morant, S.V. & Hill, V.A., 1987. *J. Dairy Res.*, **45**, 413.
2. Dalgleish, D.G., 1979. *J. Dairy Res.*, **46**, 643.
3. Chaplin, B. & Green, M.L., 1980. *J. Dairy Res.*, **47**, 351.
4. McGann, T.C.A., Donnelly, W.J., Kearney, R.D. & Buchheim, W., 1980. *Biochim. Biophys. Acta*, **630**, 261.
5. Donnelly, W.J., McNeill, G.P., Buchheim, W. & McGann, T.C.A., 1984. *Biochim. Biophys. Acta*, **789**, 136.
6. Rollema, H.S., Brinkhuis, J.A. & Vreeman, H.J., 1988. *Neth. Milk Dairy J.*, **42**, 233.
7. Dalgleish, D.G., Horne, D.S. & Law, A.J.R., 1989. *Biochim. Biophys. Acta*, **991**, 383.
8. Mercier, J.-C., Brignon, G. & Ribadeau Dumas, B., 1973. *Eur. J. Biochem.*, **35**, 222.
9. Holt, C., 1975. In *Proceedings of the Conference on Colloid & Surface Science, Budapest*, ed. E. Wolfram. Akademiai Kiado, Budapest, p. 641.
10. Walstra, P., 1979. *J. Dairy Res.*, **46**, 317.
11. Walstra, P., Bloomfield, V.A., Wei, G.J. & Jenness, R., 1981. *Biochim. Biophys. Acta*, **669**, 258.
12. Holt, C. & Dalgleish, D.G., 1986. *J. Colloid Interf. Sci*, **114**, 513.
13. Horne, D.S., 1984. *Biopolymers*, **23**, 989.
14. Horne, D.S., 1986. *J. Colloid Interf. Sci.*, **111**, 250.
15. Dalgleish, D.G. & Holt, C., 1988. *J. Colloid Interf. Sci.*, **123**, 80.
16. Jollès, J., Alais, C. & Jollès, P., 1968. *Biochim. Biophys. Acta*, **168**, 591.
17. Berridge, N.J., 1942. *Nature*, **149**, 194.
18. Jakob, E. & Puhan, Z., 1986. *Schweiz. Milchw. Forschung*, **15**, 27.
19. Pujolle, J., Ribadeau Dumas, B., Garnier, J. & Pion, B., 1966. *Biochem. Biophys. Res. Commun.*, **25**, 285.
20. Schmidt, D.G., Both, P. & de Koning, P.J., 1966. *J. Dairy Sci.*, **49**, 776.
21. Armstrong, C.E., MacKinlay, A.G., Hill, R.J. & Wake, R.G., 1967. *Biochim. Biophys. Acta*, **140**, 123.
22. Vreeman, H.J., Both, P., Brinkhuis, J.A. & van der Spek, C., 1977. *Biochim. Biophys. Acta*, **491**, 93.
23. Vreeman, H.J., Visser, S., Slangen, C.J. & van Riel, J.A.M., 1986. *Biochem. J.*, **240**, 87.
24. Lawrence, R.C. & Creamer, L.K., 1969. *J. Dairy Res.*, **36**, 11.
25. Bingham, E.W., 1975. *J. Dairy Sci.*, **58**, 13.
26. Fox, K.K., Holsinger, V.H., Posati, L.P. & Pallansch, M.J., 1967. *J. Dairy Sci.*, **50**, 1363.

27. Wake, R.G., 1957. *Aust, J. Sci.*, **20**, 147.
28. Hindle, E.J. & Wheelock, J.V., 1970. *J. Dairy Res.* **37**, 389.
29. van Hooydonk, A.C.M. & Olieman, C., 1982. *Neth. Milk Dairy J.*, **36**, 153.
30. Udenfriend, S., Stein, S., Bohlen, P., Dairman, W., Leimgruber, W. & Weigele, M., 1972. *Science*, **178**, 871.
31. Beeby, R., 1980. *N.Z.J. Dairy Sci. Technol.*, **15**, 99.
32. Humphrey, R.S. & Newsome, L.J., 1984. *N.Z.J. Dairy Sci. Technol.*, **19**, 197.
33. Dalgleish, D.G., 1986. *J. Dairy Res.*, **53**, 43.
34. Berridge, N.J., 1952. *Analyst*, **77**, 57.
35. Brown, R.J. & Collinge, S.K., 1986. *J. Dairy Sci.*, **69**, 956.
36. Foltmann, B., 1959. *Proc. XV Int. Dairy Congr. (London)*, Vol. 2, p. 655.
37. McMahon, D.J. & Brown, R.J., 1983. *J. Dairy Sci.*, **66**, 341.
38. Raymond, M.N., Bricas, E., Salesse, R., Garnier, J., Garnot, P. & Ribadeau Dumas, B., 1973. *J. Dairy Sci.*, **56**, 419.
39. Raymond, M.N. & Bricas, E., 1979. *J. Dairy Sci.*, **62**, 1719.
40. de Koning, P.J., van Rooijen, P.J. & Visser, S., 1978. *Neth. Milk Dairy J.*, **32**, 232.
41. Martin, P., Collin, J.-C., Garnot, P., Ribadeau Dumas, B. & Mocquot, G., 1981. *J. Dairy Res.*, **48**, 447.
42. Collin, J.-C., Martin, P., Garnot, P., Ribadeau Dumas, B. & Mocquot, G., 1981. *Milchwissenschaft.*, **36**, 32.
43. Andren, A. & de Koning, P.J., 1982. In *Use of Enzymes in Food Technology*, ed. P. Dupuy. Technique et Documentation Lavoisier, Paris, p. 275.
44. Collin, J.-C., Muset de Retta, G. & Martin, P., 1982. *J. Dairy Res.*, **49**, 221.
45. Garnot, P. & Molle, D., 1982. *Le Lait*, **62**, 671.
46. Garnier, J., 1963. *Biochim. Biophys. Acta*, **66**, 366.
47. Castle, A.V. & Wheelock, J.V., 1972. *J. Dairy Res.*, **39**, 15.
48. van Hooydonk, A.C.M., Olieman, C. & Hagedoorn, H.G., 1984. *Neth. Milk Dairy J.*, **37**, 207.
49. Azuma, N., Kaminogawa, S. & Yamauchi, K., 1984. *Agric. Biol. Chem.*, **48**, 2025.
50. Carles, C. & Martin, P., 1985. *Arch. Biochem. Biophys.*, **242**, 411.
51. Visser, S., van Rooijen, P.J. & Slangen, C., 1980. *Eur. J. Biochem.*, **108**, 415.
52. van Hooydonk, A.C.M. & Walstra, P., 1987. *Neth. Milk Dairy J.*, **41**, 19.
53. Brinkhuis, J. & Payens, T.A., 1985. *Biochim. Biophys. Acta*, **832**, 331.
54. Dalgleish, D.G., 1988. *J. Dairy Res.*, **55**, 521.
55. Carlson, A., Hill, C.G. & Olson, N.F., 1987. *Biotech. Bioeng.*, **29**, 582.
56. Beeby, R., 1976. *J. Dairy Res.*, **43**, 37.
57. Hill, R.D. & Hocking, V.M., 1978. *N.Z.J. Dairy Sci. Technol.*, **13**, 195.
58. Voynick, I.M. & Fruton, J.S., 1971. *Proc. Nat. Acad. Sci. (US)*, **68**, 257.
59. Hill, R.D., 1968. *Biochem. Biophys. Res. Commun.*, **33**, 659.
60. Schattenkerk, C. & Kerling, K.E.T., 1973. *Neth. Milk Dairy J.*, **27**, 286.
61. Hill, R.D., 1969. *J. Dairy Res.*, **36**, 409.
62. Raymond, M.N., Bricas, E. & Mercier, J.-C., 1973. *Neth. Milk Dairy J.*, **27**, 298.
63. Visser, S., van Rooijen, P.J., Schattenkerk, C. & Kerling, K.E.T., 1977. *Biochim. Biophys. Acta*, **481**, 171.
64. Visser, S., van Rooijen, P.J., Schattenkerk, C. & Kerling, K.E.T., 1976. *Biochim. Biophys. Acta*, **438**, 265.
65. Visser, S., 1981. *Neth. Milk Dairy J.*, **35**, 65.
66. Hill, R.D. & Laing, R.R., 1965. *J. Dairy Res.*, **32**, 193.
67. Kaye, N.M.C. & Jollès, P., 1978. *Biochim. Biophys. Acta*, **536**, 329.
68. Visser, S., Slangen, C.J. & van Rooijen, P.J., 1987. *Biochem, J.*, **244**, 553.
69. Chobert, J.-M., Mercier, J.-C., Bahy, C. & Hazé, G., 1976. *FEBS Lett.*, **72**, 173.
70. Fiat, A.-M. & Jollès, P., 1977. *C.R. Acad. Sci., Paris*, **D284**, 393.
71. Brignon, G., Chtourou, A. & Ribadeau Dumas, B., 1985. *FEBS Lett.*, **188**, 48.
72. Thompson, M.D., Dave, J.R. & Nakhasi, H.L., 1985. *DNA*, **4**, 263.

73. Kotts, C. & Jenness, R., 1976. *J. Dairy Sci.*, **59**, 816.
74. Raap, J., Kerling, K.E.T., Vreeman, H.J. & Visser, S., 1983. *Arch. Biochem. Biophys.*, **221**, 117.
75. Jenkins, J.A., Tickle, I., Sewell, T., Ungaretti, L., Wollmer, A. & Blundell, T.L., 1977. *Adv. Exp. Med. Biol.*, **95**, 43.
76. Loucheux-Lefebvre, M.-H., Aubert, J.-P. & Jollès, P., 1978. *Biophys. J.*, **23**, 323.
77. Farrell, H.M., Jr., Brown, E.M. & Kumosinski, T.F., 1990. *Proceedings*, 23 *Int. Dairy Congr. Montreal*, vol. 2, 1526.
78. Sinkinson, G. & Wheelock, J.V., 1970. *Biochim. Biophys. Acta*, **215**, 517.
79. Addeo, F., Martin, P. & Ribadeau Dumas, B., 1984. *Milchwissenschaft*, **39**, 202.
80. Schaar, J., 1984. *J. Dairy Res.*, **51**, 397.
81. Aaltonen, M.-L. & Antila, V., 1987. *Milchwissenschaft*, **42**, 490.
82. Pagnacco, G. & Caroli, A., 1987. *J. Dairy Res.*, **54**, 479.
83. McLean, D.M. & Schaar, J., 1989. *J. Dairy Res.*, **56**, 297.
84. Ekstrand, B., Larsson-Raznikiewicz, M. & Perlmann, C., 1980. *Biochim. Biophys. Acta*, **630**, 361.
85. Dalgleish, D.G., Brinkhuis, J. & Payens, T.A.J., 1981. *Eur. J. Biochem.*, **119**, 257.
86. Dalgleish, D.G., 1980. *J. Dairy Res.*, **47**, 231.
87. Payens, T.A., 1984. *J. Appl. Biochem.*, **6**, 232.
88. Humme, H.E., 1972. *Neth. Milk Dairy J.*, **26**, 180.
89. van Hooydonk, A.C.M., Boerrigter, I.J. & Hagedoorn, H.G., 1986. *Neth. Milk Dairy J.*, **40**, 297.
90. Payens, T.A.J. & Both, P., 1980. In *Biochemistry: Ions, Surfaces, Membranes*, ed. M. Blank, *ACS Adv. Chem. Ser.*, **188**, 129.
91. Payens, T.A.J. & Visser, S., 1981. *Neth. Milk Dairy J.*, **35**, 387.
92. Dalgleish, D.G., 1983. *J. Dairy Res.*, **50**, 331.
93. Green, M.L. & Marshall, R.J., 1977. *J. Dairy Res.*, **44**, 521.
94. Bringe, N.A. & Kinsella, J.E., 1986. *J. Dairy Res.*, **53**, 371.
95. van Hooydonk, A.C.M., Hagedoorn, H.G. & Boerrigter, I.J., 1986. *Neth. Milk Dairy J.*, **40**, 369.
96. Bringe, N.A. & Kinsella, J.E., 1986. *J. Dairy Sci.*, **69**, 965.
97. Foltmann, B., 1987. In *Cheese: Chemistry, Physics & Microbiology* Vol. 1, ed. P.F. Fox. Elsevier Applied Science, p. 33.
98. Hicks, C.L., O'Leary, J. & Bucy, J., 1988. *J. Dairy Sci.*, **71**, 1127.
99. Meisel, H., 1988. *Milchwissenschaft*, **43**, 71.
100. Bines, V.E., Young, P. & Law, B.A., 1989. *J. Dairy Res.*, **56**, 657.
101. Pszczola, D.E., 1989. *Food Technol.*, **43**, 84.
102. Sternberg, M., 1976. *Adv. Appl. Microbiol.*, **20**, 135.
103. Green, M.L., 1977. *J. Dairy Res.*, **44**, 159.
104. de Koning, P.J., 1978. *Dairy Industries International*, **43**, 7.
105. Scott, R., 1979. In *Topics in Enzyme and Fermentation Biotechnology*, **3**, 103.
106. Kay, J. & Valler, M.J., 1981. *Neth. Milk Dairy J.*, **35**, 281.
107. Hofmann, T., 1974. In *Food-Related Enzymes*, ed. J.R. Whitaker, *ACS Adv. Chem. Ser.*, **136**, 146.
108. Fox, P.F., 1969. *J. Dairy Res.*, **36**, 427.
109. Fruton, J.S., 1971. In *The Enzymes*, vol 3., ed. P. Boyer. Academic Press, London, p. 119.
110. Foltmann, B., 1966. *C.R. Trav. Lab. Carlsberg*, **35**, 143.
111. Alais, C. & Lagrange, A., 1972. *Le Lait*, **52**, 407.
112. Bohak, Z., 1969. *J. Biol. Chem.*, **244**, 4638.
113. Jespersen, N.J.T. & Dinesen, V., 1979. *J. Soc. Dairy Technol.*, **32**, 194.
114. Duersch, J.W. & Ernstrom, C.A., 1974. *J. Dairy Sci.*, **57**, 590.
115. Hyslop, D.B., Swanson, A.M. & Lund, D.B., 1979. *J. Dairy Sci.*, **62**, 1227.
116. Thunell, R.K., Duersch, J.W. & Ernstrom, C.A., 1979. *J. Dairy Sci.*, **62**, 373.

117. Lagrange, A., Paquet, D. & Alais, C., 1980. *Int. J. Biochem.*, **11**, 347.
118. Wallace, D.L., 1981. In *Proceedings from the Second Biennial Marschall International Cheese Conference.* Marschall Products, p. 289.
119. Martin, P., Raymond, M.-N., Bricas, E. & Ribadeau Dumas, B., 1980. *Biochim. Biophys. Acta*, **612**, 410.
120. Ramet, J.P. & Weber, F., 1981. *Le Lait*, **61**, 458.
121. Philippos, Sh.G & Christ, W., 1976. *Milchwissenschaft*, **31**, 349.
122. Green, M.L. & Foster, P.M.D., 1974. *J. Dairy Res.*, **41**, 269.
123. Phelan, J.A., 1973. *Dairy Industries*, **38**, 418.
124. Green, M.L. & Stackpoole, A., 1975. *J. Dairy Res.*, **42**, 297.
125. Chaplin, B. & Green, M.L., 1982. *J. Dairy Res.*, **49**, 631.
126. Beeby, R., 1979. *N.Z.J. Dairy Sci. Technol.*, **14**, 1.
127. Voutsinas, L.P. & Nakai, S., 1983. *J. Dairy Sci.*, **66**, 694.
128. Green, M.L. & Crutchfield, G., 1969. *Biochem, J.*, **115**, 183.
129. Carlson, A., 1983. *Diss. Abs. Int. B*, **43**, 3671.
130. Taylor, M.J., Olson, N.F. & Richardson, T., 1979. *Process Biochem.*, **14**, 10.
131. Hicks, C.L., Ferrier, L.K., Olson, N.F. & Richardson, T., 1975. *J. Dairy Sci.*, **58**, 19.
132. Lee, H.J., Olson, N.F. & Richardson, T., 1977. *J. Dairy Sci.*, **60**, 1683.
133. Morrissey, P.A., 1969. *J. Dairy Res.*, **36**, 333.
134. Wheelock, J.V. & Kirk, A., 1974. *J. Dairy Res.*, **41**, 367.
135. Dalgleish, D.G., 1990. *Milchwissenschaft*, **45**, 491.
136. Fox, P.F., 1981. *J. Dairy Sci.*, **64**, 2127.
137. van Hooydonk, A.C.M., De Koster, P.G. & Boerrigter, I.J., 1987. *Neth. Milk Dairy J.*, **41**, 3.
138. Banks, J.M., Stewart, G., Muir, D.D. & West, I.G., 1987. *Milchwissenschaft*, **42**, 212.
139. Singh, H., Shalabi, S.I., Fox, P.F., Flynn, A. & Barry, A., 1988. *J. Dairy Res.*, **55**, 205.
140. Davies, D.T. & White, J.C.D., 1960. *J. Dairy Res.*, **27**, 171.
141. Dalgleish, D.G. & Law, A.J.R., 1989. *J. Dairy Res.*, **56**, 727.
142. Snoeren, T.H.M., Klok, H.J., van Hooydonk, A.C.M. & Damman, A.J., 1984. *Milchwissenschaft*, **39**, 461.
143. Dalgleish, D.G. & Law, A.J.R., 1988. *J. Dairy Res.*, **55**, 529.
144. Damicz, W. & Dziuba, J., 1975. *Milchwissenschaft*, **30**, 399.
145. Wilson, G.A. & Wheelock, J.V., 1972. *J. Dairy Res.*, **39**, 413.
146. Reddy, I.M. & Kinsella, J.E., 1990. *J. Agric. Food Chem.*, **38**, 50.
147. Shalabi, S.I. & Wheelock, J.V., 1977. *J. Dairy Res.*, **44**, 351.
148. Shalabi, S.I. & Wheelock, J.V., 1976. *J. Dairy Res.*, **43**, 331.
149. Qvist, K.B., 1979. *Milchwissenschaft*, **34**, 467.
150. Oortwijn, H., Walstra, P. & Mulder, H., 1977. *Neth. Milk Dairy J.*, **31**, 134.
151. Robson, E.W. & Dalgleish, D.G., 1984. *J. Dairy Res.*, **51**, 417.
152. Robson, E.W. & Dalgleish, D.G., 1987. In *Food Emulsion and Foams*, ed. E. Dickinson. Royal Society of Chemistry Special Publication No. 58, p. 64.
153. Green, M.L., Marshall, R.J. & Glover, F.A., 1983. *J. Dairy Res.*, **50**, 341.
154. Dickinson, E., Whyman, R. & Dalgleish, D.G., 1987. In *Food Emulsions and Foams*, ed. E. Dickinson. Royal Society of Chemistry Special Publication No. 58, p. 40.
155. Garnot, P. & Corre, C., 1980. *J. Dairy Res.*, **47**, 103.
156. Mehaia, M.A. & Cheryan, M., 1983. *Milchwissenschaft*, **38**, 708.
157. Green, M.L. & Crutchfield, G., 1971. *J. Dairy Res.*, **38**, 151.
158. Green, M.L., 1973. *Neth. Milk Dairy J.*, **27**, 278.
159. Payens, T.A.J., 1978. *Farad. Disc. Chem. Soc.*, **65**, 164.
160. Payens, T.A.J., 1979. *J. Dairy Res.*, **46**, 291.
161. Darling, D.F. & van Hooydonk, A.C.M., 1981. *J. Dairy Res.*, **48**, 189.

162. Pearce, K.N., 1976. *J. Dairy Res.*, **43**, 27.
163. Darling, D.F. & Dickson, J., 1979. *J. Dairy Res.*, **46**, 441.
164. Dalgleish, D.G., 1984. *J. Dairy Res.*, **51**, 425.
165. Guthy, K., Auerswald, D. & Buchheim, W., 1989. *Milchwissenshaft*, **44**, 560.
166. Scott Blair, G.W. & Oosthuizen, J.C., 1960. *J. Dairy Res.*, **28**, 165.
167. Guthy, K. & Novak, G., 1977. *J. Dairy Res.*, **44**, 363.
168. Payens, T.A.J., 1977. *Biophys. Chem.*, **6**, 263.
169. Payens, T.A.J., Wiersma, A.K. & Brinkhuis, J., 1977. *Biophys. Chem.* **6**, 253.
170. Hyslop, D.B., Richardson, T. & Ryan, D.S., 1979. *Biochim. Biophys. Acta*, **566**, 390.
171. Dalgleish, D.G., 1980. *Biophys. Chem.*, **11**, 147.
172. Payens, T.A.J., 1982. *J. Dairy Sci.*, **65**, 1863.
173. Green, M.L. & Morant, S.V., 1981. *J. Dairy Res.*, **48**, 57.
174. von Smoluchowski, M., 1917. *Z. Physik. Chem.*, **92**, 129.
175. Brinkhuis, J. & Payens, T.A.J., 1984. *Biophys. Chem.*, **19**, 75.
176. Kowalchyk, A.W. & Olson, N.F., 1977. *J. Dairy Sci.*, **60**, 1256.
177. Knoop, A.-M. & Peters, K.-H., 1976. *Milchwissenschaft*, **31**, 338.
178. Slattery, C.W., 1976. *J. Dairy Sci.*, **59**, 1547.
179. Kim, B.Y. & Kinsella, J.E., 1989. *Milchwissenschaft*, **44**, 622.
180. Pierre, A., 1983. *Le Lait*, **63**, 217.
181. Carlson, A., Hill, C.G. & Olson, N.F., 1987. *Biotech. Bioeng.*, **29**, 601.
182. Shalabi, S.I. & Fox, P.F., 1982. *J. Dairy Res.*, **49**, 153.
183. Roefs, S.P.F.M., Walstra, P., Dalgleish, D.G. & Horne, D.S., 1985. *Neth. Milk Dairy J.*, **39**, 119.
184. Bakri, M. & Wolfe, F.H., 1971. *Canad. J. Biochem.*, **49**, 882.
185. Di Gregorio, F. & Sisto, R., 1981. *J. Dairy Res.*, **48**, 267.
186. Green, M.L. & Marshall, R.J., 1979. *J. Dairy Res.*, **46**, 365.
187. Marshall, R.J. & Green, M.L., 1980. *J. Dairy Res.*, **47**, 359.
188. Green, M.L., 1982. *J. Dairy Res.*, **49**, 87.
189. Green, M.L., 1982. *J. Dairy Res.*, **49**, 99.
190. Green, M.L., Hobbs, D.G. & Morant, S.V., 1978. *J. Dairy Res.*, **45**, 405.
191. Holter, H., 1932. *Biochem. Z.*, **255**, 160.
192. Payens, T.A.J. & Wiersma, A.K., 1980. *Biophys. Chem.*, **11**, 137.
193. Payens, T.A.J. & Brinkhuis, J., 1986. *Colloids Surf.*, **20**, 37.
194. van Hooydonk, A.C.M., 1987. Thesis, Agricultural University of Wageningen, p. 109.
195. van Hooydonk, A.C.M. & Walstra, P., 1987. *Neth. Milk Dairy J.*, **41**, 293.
196. Payens, T.A.J., 1987. *Neth. Milk Dairy J.*, **41**, 289.
197. Dalgleish, D.G., 1988. *Neth. Milk Dairy J.*, **42**, 341.
198. Johnston, D.E., 1984. *J. Dairy Res.*, **51**, 91.
199. Merin, U., Talpaz, H. & Fishman, S., 1989. *J. Dairy Res.*, **56**, 79.
200. Kumosinski, T.F., Brown, E.M. & Farrell, H.M. Jnr., 1991. *J. Dairy Sci.*, **74**, 2879.
201. Horne, D.S. & Davidson, C., 1990. *Milchwissenschaft*, **45**, 712.

4

Secondary (Non-enzymatic) Phase of Rennet Coagulation and Post-coagulation Phenomena

Margaret L. Green

Formerly of AFRC Institute of Food Research, Reading Laboratory, Earley Gate, Whiteknights Road, Reading, RG6 2EF, UK*

&

Alistair S. Grandison

Department of Food Science and Technology, University of Reading, Whiteknights, Reading, RG6 2AP, UK

1 INTRODUCTION

Cheesemakers often state that a good curd is required to make a quality cheese, one which is highly acceptable in both flavour and texture. This implies that the initial stages of casein aggregation to form a network influence the characteristics of the cheese. The mechanisms involved in network formation have been probed over the last 15 years, microscopic methods and rheological techniques being especially useful. It has been shown that the structure of the curd determines its properties and, thus, the retention of fat and moisture on which cheese yield and composition depend. The curd structure is also a direct precursor of the cheese structure, which must be the basis of its texture.

The formation of the curd is itself influenced by the composition and treatment of the milk. Ideally, the milk should coagulate quickly after addition of rennet to produce a firm curd which drains well but retains a high proportion of the fat. Much progress has been made during the last few years in elucidating the factors which influence these processes, mainly because of improvements in methods for measuring the different stages of cheesemaking in isolation.

In this chapter, the formation of curd and its conversion to cheese and the effects of compositional and processing factors will be reviewed. This will involve discussion of the methods now available for studying the formation, properties and structure of curd and cheese and the way in which their application

*Present address: 8 Harwich Close, Lower Earley, Reading, RG6 3UD

has led to some understanding of the mechanisms involved. From this basis, explanation of the diverse influences of factors in the milk and the manufacturing process will be attempted. Emphasis will be given to work published during the last 10 years and the review will extend and update earlier ones.[1-3]

2 EXPERIMENTAL METHODS

2.1 Curd Formation

Gels formed by the action of rennet in milk are viscoelastic materials constantly changing with time. A period roughly equal to the rennet clotting time (RCT) is usually allowed from the onset of visual coagulation for the gel to become sufficiently firm prior to cutting.[3] The development of the gel during this period and the strength at cutting are related to the yield and quality of the finished cheese. Soft curds lead to casein and fat losses in the whey, while if the curd is too hard it is difficult to cut and likely to shatter. Many techniques have been devised to measure curd tension both as research tools and, more particularly, to standardize commercial cheese manufacture. Curd firmness at cutting has been shown to vary widely during commercial Cheddar manufacture[4] and several workers have attempted to develop a rheometer for quality control in the factory, preferably in the vat.

To obtain information on milk gel structure and to allow comparison with gels of other materials, absolute rheological measurements are required. These are now beginning to be made. For a time-dependent gel system, Bohlin et al.[5] argue that a dynamic test is most suitable. They described an oscillating coaxial-cylinder rheometer which permits continuous measurement of the loss and storage moduli, G' and G", and the phase shift at constant shear strain. Other authors[6,7] have described the application of similar apparatus to rennet coagula, and extensive studies have been carried out on the effects of chemical and physical changes on these measurements.[8-11] Tokita et al.[12] used a rather simpler torsion pendulum apparatus to measure G' and G" continuously at a fixed frequency. A variable frequency rheometer was used for acid milk gels.[13] G' can also be determined from the displacement of a column of curd in a U-tube by an applied pressure.[14] A stress relaxation modulus has been determined with an Instron Food Tester (Instron Corp., Canton, Mass., USA) from the relaxation of a gel with time after a small displacement.[15]

However, most measurements of curd firmness are empirical. This is satisfactory only if the instrumental output can be related to some useful property of the milk gel, or for strictly comparative measurements. If any alteration is made in the conditions, such as varying the casein concentration or using homogenized milk, the output may not necessarily relate to the extent of micellar aggregation.[16] Further, different instruments do not usually give comparable outputs (Fig. 1), because they measure different aspects of the gel's properties. In fact, the outputs of different instruments may move in opposite directions

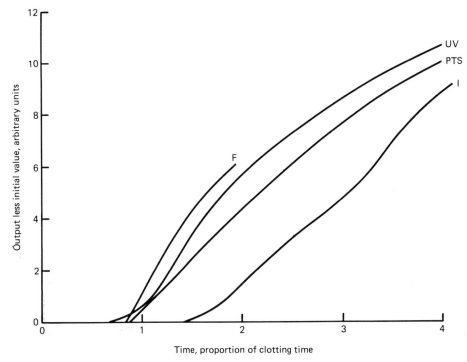

Fig. 1. Typical curves of increase in curd firmness with time. The Ultra Viscoson (UV) and pressure transmission system (PTS) were used on pasteurized whole and skim milk, respectively, as described by Marshall *et al.*[17] The Formograph (F) was used as recommended by the manufacturers and the Instron (I), as described by Storry and Ford,[28] on pasteurized whole milk.

when the conditions are changed. For instance, the curd firming rate increased with temperature when measured with a vibrating reed viscometer or an oscillating diaphragm apparatus[17] but was either unaffected or declined with temperature when measured as yield force with an Instron.[18] This makes it difficult to compare the effect of using different milks if curd firmness measurements were made with different instruments. Thus, it would seem hazardous to attempt to use such measurements to derive fundamental information.

The empirical measurement of curd formation has become much easier in recent years with the development of a number of instruments enabling reproducible measurements and giving continuous outputs. Most of those which have become available since Prokopek's[19] review of the subject and which seem to be generally useful are described below. Of the instruments incorporating a slow oscillating cylinder or probe, an upgraded version of the Plint torsiometer, incorporating transducers to measure the angle and the movement, has been described.[20] However, the Formograph (Foss & Co., Hellerup, Denmark) is based on the same principle as the torsiometer and lactodynamograph but is a simpler and cheaper instrument. It involves the oscillation of stainless steel pendula and measures 10 samples simultaneously. The manufacturers claim that it parallels curd formation in the vat. This apparatus has become increasingly popular for research purposes (e.g. Politis and Ng-Kwai-Hang[21]), and a recent report[22] demonstrated a significant relationship between curd firmness and cheese yield using the Formograph. Variations of this apparatus are a slowly oscillating sphere attached to an Instron probe and suspended in a sample of milk[23] and an oscillating vertical steel disc.[24] The latter instrument can be used in a cheese vat.

The hydraulically-operated oscillating diaphragm apparatus, developed by Vanderheiden,[25] measures the rigidity of the curd by its ability to transmit a pressure wave. It can be used directly in the vat and preliminary trials suggest that it would be useful in indicating the optimum cutting time for Cheddar curd.[26] An electronically-operated version has been developed.[27] It was used successfully in preliminary trials in the vat, and also compared with a vibrating-reed viscometer (Ultra Viscoson 1800, Bendix Corp. Inc., Lewisburg, W. Virginia, USA).[17] This and another instrument of the same type (Unipan 505, Unipan Scientific Instruments, Warsaw, Poland) vibrates much more rapidly than those described previously. They are particularly useful for detecting the early stages of aggregation, before visible coagulation (Fig. 1). A more recent instrument, the Gelograph R (Gel Instrumenti AG Rüschlikon, Germany), involving damping the vibrations of a needle by the curd, appears to be available, but the authors have not seen the instrument or an independent assessment of its effectiveness.

Most other methods for the continuous assessment of gel firmness in current use involve measuring the force required to break the curd. Storry and Ford[28] used the traverse of a wire probe through undisturbed curd by means of an Instron. Matěj[29] used a spherical sensor, electrically polarized to avoid coagulum adhesion, to travel spirally through the curd and Kulkarni & Vishweshwariah[30] derived curd strength from the force and time required to move a knife through curd.

A promising novel development is the use of temperature measurements of electrically heated wire entrapped in the coagula, which relate to the thermal conductivity.[31] This could permit non-destructive, continuous measurement during cheesemaking, although extreme caution must be applied in relating thermal conductivity data to absolute rheological data. However, it has been reported that results from the hot-wire apparatus correlate reasonably well with data from a lactodynamograph.[32]

2.2 Syneresis of Curd

Syneresis, which is curd shrinkage and loss of whey, occurs very readily and so is extremely sensitive to the experimental conditions. The purpose of its measurement in model experiments may be either to assess the effect of a variable on whey loss during cheesemaking or to obtain information about the structure of curd or the mechanisms involved in its shrinkage. For the former purpose, it is desirable to simulate the conditions in the cheese vat as closely as possible, but comparative measurements are likely to be acceptable. For the latter purpose, absolute measurements are to be preferred.

Most of the methods available have been reviewed by Pearse & Mackinlay.[33] One category comprises those involving physical separation of the curd from the whey. This is perhaps most readily achieved by forming and cutting identical gels in a number of vessels, the contents of which are separated by a sieve at intervals.[34] However, the necessity to conserve time and materials often demands that all points on a syneresis-time curve are derived from a single sample of curd and whey.[35] Such measurements are simple to perform under a range of conditions and appear to give satisfactory comparative results. However, the whey loss is likely to be overestimated as the curd is not supported by the whey. Tracer dilution methods overcome this disadvantage but there are problems in mixing and diffusion into the curd if added materials are used. These have been overcome by Pearse *et al.*[36] using fat globules as an endogenous tracer. These workers describe a suitable apparatus for stirring the curd and whey, consisting of six vials held on a wheel which is rotated at 20 rpm in a water bath. The method is equally suitable for following syneresis in whole milk, skim milk, or reconstituted milk gels,[33] and can give an absolute measure of whey production. Alternatively, syneresis can be measured directly by following the one-dimensional shrinkage of undisturbed curd.[37]

2.3 Moisture Content of Curd and Cheese

Modern, semi-continuous methods of cheese production require a rapid method for determining the moisture content of curd. A method for various cheeses using a specific microwave oven was reported to be very reproducible.[38] However, a more thorough collaborative study indicated that drying of samples in microwave ovens can be beset by a number of problems,[39] the most serious of which are the influence of the salt content of the sample and the variability between ovens, even of the same model, with respect to optimal power setting and sample positioning. Extensive calibration against the British Standards Institution standard method of drying in an electrically-heated oven is recommended.

2.4 Curd Fusion

A direct method for determining the stickiness of cheese grains has been described.[40] It involves pressing two cups containing cheese curd together and measuring the force required to separate them. In a more involved method, the frequency of holes, originally containing whey, in electron micrographs of curd at the required stage has been measured (R.J. Marshall, pers. comm.). A photographic method for detecting curd granule junctions in cheese has been described.[41] It should be adaptable for use with cheddared curd.

The flow of curd during pressing has been measured from the extent of penetration of cheese mass into the openings of perforated moulds,[42] but the authors have no details of the methodology.

No method has been described for measuring salt uptake by curd, and it is not known whether curd samples differ in this respect. However, there has been speculation that differences may exist and the view has been expressed that such measurements are desirable.

2.5 Curd and Cheese Structure

During the last few years, a number of experimental and theoretical advances have been made which appear to be useful in studying curd and cheese structure. It is too early yet to assess the value of some of them, because they have not yet been applied extensively enough. In fact, some have only been used with other protein gel systems, although all appear to the authors to hold promise for use with curd or cheese. Different methods can often be used to give complementary information on a system.

Now that methods are beginning to be available for determining the rheological properties of curd (Section 2.1) and cheese (Chapter 8) in absolute terms, it is timely that theoretical advances are being made to enable the results to be interpreted in structural terms. Bohlin et al.[5] suggest that the complex modulus of rigidity of curd is a measure of the network density and that the phase shift between the applied deformation and resultant force reflects the elastic and viscous contributions to the linkage. Johnston[43] has applied polymer cross-linking theory to rheological measurements of renneted curd. This also suggests that the shear modulus is a measure of the number of cross-links in the gel. Haddad[44] has developed a theoretical analysis for the steady-state deformation of fibrous systems, which takes the microstructure into account and gives information on the distribution of internal stresses.

Information on curd and cheese structure has been derived most usually from microscopic methods. The transition of curd to cheese was first studied by conventional transmission and scanning electron microscopy (TEM, SEM)[45,46] involving dehydration of the specimens. However, the conclusions reached appeared to be reliable in that Green et al.[46] found that the structures observed on TEM and SEM examination accorded with those observed by light microscopy (LM) of the same samples, but frozen rather than dehydrated. A wider range of preparation methods for EM samples has been developed more recently and applied within the same study. Taneya et al.[47] compared the structures of processed cheese using two different TEM methods. The first involved sectioning of fixed, dehydrated and stained sections. In the second, the sample was frozen and fractured before preparing a replica of the surface which was examined (Fig. 2). Kalab & Modler[48] examined an acid curd product using two SEM and two TEM methods. All involved dehydration of the specimens, but the fat was retained in one method of each type but not in the other. SEM was most useful in showing the interrelations of the protein, fat and aqueous phases, whereas TEM showed the detail of the protein phase. Rousseau[49] followed the conversion of curd into Saint Paulin cheese both by conventional SEM involving dehydration and critical point drying and by cold-stage SEM after rapid freezing and holding at −160°C. The conventional method was more useful for observing the protein phase, with the cold-stage being more suitable for observing the fat globules (Fig. 3).

The cold-stage technique has the advantage that the sample undergoes no treatment with solvents or chemicals so that its chemical composition is unaltered. Brooker[50] took advantage of this in his study of mineral movements during the ripening of Camembert cheese. Frozen samples were examined by X-ray spectroscopy in the SEM to reveal the Ca and P locations in relation to the background cheese structure at different ripening stages. This technique is semi-quantitative but can be made quantitative.[51]

Quantitative information can be derived from micrographs obtained by conventional TEM and LM. Tombs[52] deduced the mean pore size and strand thickness in protein gels from measurements on transmission electron micrographs. Green et al.[46] described methods for assessing the fat/protein interfacial area from transmission electron micrographs and the coarseness of the protein matrix from

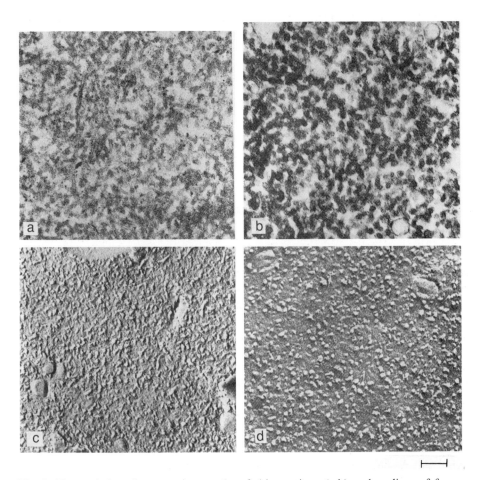

Fig. 2. Transmission electron micrographs of thin sections (a,b) and replicas of freeze-fracture surfaces (c,d) of hard-type (a,c) and soft-type (b,d) processed cheeses (from Ref. 47). (Scale bar = 100 nm).

light micrographs, both applicable to curd and cheese. However, the freezing used to fix the curd for LM probably distorts the true structure[53] so the results should be interpreted with caution. The use of water-soluble fluorochromes to stain cheese components for examination by light fluorescence microscopy[54] may be more suitable for quantitative studies.

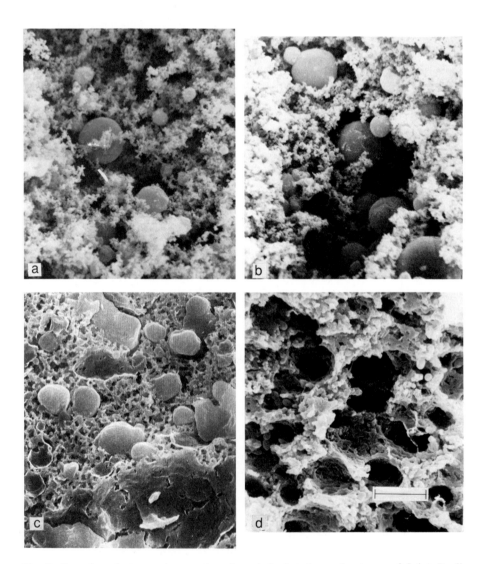

Fig. 3. Scanning electron micrographs of curd during the early stages of Saint Paulin cheesemaking, (a) before cutting; (b,c) after hooping; (d) after pressing. (a,b,d) fixed in glutaraldehyde and osmium tetroxide, dehydrated and critical-point dried; (c) examined on a cold stage after fast freezing and fracturing (from Ref. 49). (Scale bar = 5 μm).

Ultrasonics, which should give information on the size and distribution of the fat particles, has been applied to the study of cheese structure.[55] Small angle X-ray scattering combined with computer modelling has been used on globular protein gels.[56] It has shed light on the assembly of the gel, and the strand thickness and degree of uniformity of the network. In principle, such methods should be applicable to cheese curds.

Proton nuclear magnetic resonance (NMR) can be used to give information on the extent and strength of water-binding in materials, and has been applied to both curd[57] and cheese.[58] An NMR method has also been described for determining the mean size and variance of distribution of liquid droplets in a solid phase, and has been applied to the fat in cheese.[58]

3 MECHANISMS INVOLVED IN THE FORMATION OF CURD AND ITS CONVERSION TO CHEESE

3.1 Curd Assembly

The curd starts to form at about the visually-observed clotting time (RCT) (Fig. 1) and the process can be followed quantitatively thereafter by measuring the increase in firmness. It is rather slow, and therefore can also be followed by electron microscopy of samples taken at intervals. It is characterized by a steady aggregation of the rennet-treated casein micelles. Chains of micelles are formed at first.[59] By the RCT, these have begun to link into a loose network (Fig. 4a). The network then extends and becomes more differentiated, with the chains of micelles aligning together (Fig. 4b). During this time, the linkages between the micelles also appear to strengthen. Initially, many micelles are joined by bridges (Fig. 4a), but later these appear to contract, bringing the micelles into contact and eventually causing partial fusion. These observations suggest that the increase in curd firmness is due to increases in both the number and strength of linkages between micelles.

Further light has been shed on the process by means of rheological measurements, reviewed by Walstra & van Vliet.[2] Curd formation can be described by a single first-order[12] reaction, as expected if it is a continuation of the initial aggregation of the casein micelles (Chapter 3). The gel formed is made up of linked particles and behaves differently from one consisting of cross-linked, randomly-coiled molecules.[2] It shows very low yield stress (i.e. little capacity to flow) and low breaking stress (i.e. brittleness). Both the storage modulus (G'), related to the elastic deformation, and the loss modulus (G"), related to the viscous deformation, increase with time, reflecting an increase in firmness. The moduli depend on the number, strength and relaxation time of the bonds between the particles in the network. Probably, there is first an increase in the number of junctions between particles and then an increase in the number of bonds per junction. This accords with the effect of pressure on curd formation.[60] High pressures retard the start of curd firming but accelerate curd firming if applied during the process,

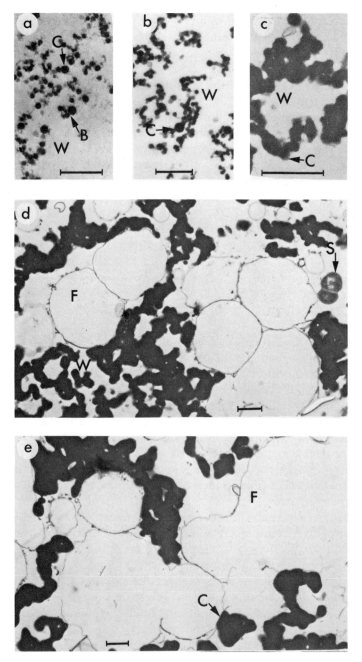

Fig. 4. Thin section electron micrographs of curd during the early stages of normal Cheddar cheesemaking. (a) At about 1·3 × clotting time, 15 min after renneting; (b) just before cutting, 40 min after renneting; (c) at maximum scald, 1·7 h after renneting; (d) 2·7 h after renneting; (e) at pitching, 3·5 h after renneting. C, casein; F, fat; W, whey; S, starter bacteria; B, bridge between micelles. Scale bar = 1 μm. (Modified from Ref. 45)

suggesting that a volume-increasing step is followed by a volume-decreasing one. There may be a change in the chemical nature of the bonds formed with time, probably in the direction of increasing strength at longer times. It has been noted that the firming rate of curd formed from small micelles is faster than that formed from large ones at the same casein concentration.[61] This may indicate, as expected intuitively, that the same number of cross-links would be formed more readily by a large number of small particles than by fewer large ones.

The presence of fat in the milk decreases the initial shear modulus and exerts a plasticizing action enabling greater deformation of the gel,[62] as expected if fat limits, but also directs and supports, the casein strands without inhibiting their movement in relation to one another.

The chemical nature of the cross-links is not yet entirely clear, but the phosphoryl side chains of casein, especially β-casein, are probably involved.[63] These may be linked by Ca^{2+} bridges, since the increase in the elastic modulus of the curd is directly related to Ca^{2+} concentration.[64] In cheese, α_{S1}-casein appears to play the major structural role,[65] but this is much affected by the early stages of hydrolysis to α_{S1}-I-casein. Thus, it seems likely that groups near the N-terminal end, probably in the region of residues 14–24 may be involved in linking α_{S1}-casein into the network.

The structure of the gel is related to its process of formation. There is a tendency for faster firming rates to lead to gels with coarser network structures.[66] For instance, Ca depletion in the milk reduced the firming rate and the gel network was finer, while higher temperatures increased the firming rate and gave coarser gel networks. However, the relationship between firming rate and gel structure is not maintained if the casein micelle composition or structure is altered, for instance by acidification or the addition of chemicals adsorbed by casein.

If a rennet gel is left undisturbed, the shear modulus continues to increase for a period of hours.[43] This must reflect a continuing increase in the number and/or strength of the links between casein micelles. This has been observed directly by microscopy over a 24-h period as an increase in the amount of contact between micelles.[67] The casein strands became shorter and thicker and eventually fused into large masses. Similar changes in the casein were observed in Cheddar cheesemaking (Fig. 4); the casein micelles lost their individuality and steadily fused into masses of increasing size.[45] There is also a steady increase in the size of the casein particles during Cottage cheesemaking.[68] Thus, casein aggregation is not terminated by cutting, but continues throughout cheesemaking. It is now clear that this process is fundamental to the conversion of milk to cheese and that its manipulation is essential for the control of cheese manufacture.

3.2 Syneresis

There has been a surge of interest in the mechanism of syneresis in the last few years. This is reflected in the publication of several reviews.[69,70,33] It is the subject of Chapter 5, so will be covered only in sufficient detail here to enable its role in cheesemaking to be appreciated.

TABLE I
Factors Influencing the Formation, Syneresis and Fusion of Curd

Factors	Curd formation	Syneresis	Curd fusion
Curd treatment			
Lower pH	I[17]	I[33]	—
Increase temperature	I[17]	I[33]	—
Milk treatment			
Store cold	D[71]	D[71]	—
Grow psychrotrophs	D[72]	D[72]	—
Homogenize	D[16]	D[16]	D[16]
Increase pasteurization temperature	D[73]	D[74]	D[73]
Increase fat content	D[62]	D[33]	—
Add $CaCl_2$	I[17]	I[33]	I[75]
Chemical modification of casein			
Block—NH_2	D[75]	D[76]	—
Limited dephosphorylation of β-casein	D[33]	D[33]	—

I = increase; D = decrease.

The term 'syneresis' strictly covers the loss of whey from the curd particles. However, as this is not accompanied by a change in casein hydration, it is not a simple physical process. In fact, it involves rearrangement of the casein network in the curd, with strands being broken and reformed to form a highly cross-linked, more compact structure (Fig. 4c).

If it is driven by casein aggregation, whey expulsion from the curd should show a positive pressure. This has been verified, but the pressure is very low, only about 1 Pa.[2] If whey expulsion was prevented by blocking shrinkage, the permeability of the curd increased with time.[37] This reflects the formation of a more compact structure in the curd. The time course of syneresis can be explained by a theoretical model including a low endogenous pressure combined with gradually increasing external pressure and permeability. The loss of whey and contraction of the casein network appear to start at the outside of the curd particle, so, if the process is too rapid initially, a relatively impermeable 'skin' can be formed.

The interactions between casein micelles leading to syneresis are incompletely understood. Although some are probably non-specific hydrophobic interactions, others are very specific. They may not be identical to the interactions occurring during curd formation, though considerable overlap would be expected. Limited dephosphorylation of β-casein severely inhibits loss of whey from the curd and also causes some inhibition of curd formation.[33] This suggests that the β-casein phosphate cluster has an important role, either by participating directly in ionic interactions or by holding the β-casein molecule in a defined orientation at the micelle surface. However, the same workers have shown that blocking sulphydryl interactions does not affect curd formation or syneresis.

If curd formation and syneresis are two aspects of the same basic process of casein aggregation, they should respond similarly to changes in conditions. A number of variables do indeed change the two parameters in the same direction (Table I), and they are also affected to a similar extent. Increased acidity, temperature and pasteurization temperature and modification of casein amino groups tend to have a large effect on both curd formation and syneresis, while fat content, added $CaCl_2$, cold storage, psychrotroph growth and homogenization tend to have less effect on both parameters.

Increasing the fat content of the milk probably also decreases the rate of whey loss by physical means. As well as limiting casein aggregation, fat globules may act as 'plugs', blocking the flow of whey through channels in the curd.

3.3 Subsequent cheesemaking stages

Once started, casein aggregation and whey loss continue throughout cheesemaking and the early stages of ripening. These processes lead to extensive compaction and fusion of the curd and reduction in moisture levels. Variations in the cheesemaking procedure appear to have similar influences on curd fusion to those they have on curd formation and syneresis (Table I). Further, the junctions between curd granules are more deficient in fat globules than the bulk of the granule, as expected if fusion is dependent on protein interactions. However, fusion requires flow of the curd, which requires pressure in the later stages of cheesemaking.[2] Pressure at this stage does not necessarily cause a reduction in moisture content. Aggregation of casein and whey exudation have been shown to occur over at least the first two months of ripening of Ras cheese, a Kashkaval-type of hard cheese.[77]

The manufacture of most types of cheese involves, as a last step, salt flux through the curd. This appears to occur by diffusion, in that it accords with Fick's law, but the diffusion coefficient is about 0.2 cm^2/day, 20% of the value in pure water.[78,79] This discrepancy has been explained by the observation that salt uptake is accompanied by water loss, and the supposition that the salt and water flux would occur through tortuous pores in the cheese.[78] Pulsed field NMR measurement of the diffusion coefficient of water in cheese, 0.35 cm^2/day,[58] is consistent with these observations. In accord with the proposed mechanism, the rate of salt flux was increased when the moisture content of the cheese was raised.[78] Further, when the fat content of the cheese was increased, the fluxes of both salt and water were decreased and the difference between them was reduced. This could be explained if the role of the fat is to block some of the pores through which the fluxes occur and to prevent shrinkage of the cheese matrix, analogous to the proposed mechanism for the effect of fat on syneresis. Aspects of salt diffusion in cheese are considered in Chapter 7.

3.4 Influence of Curd Structure and Composition on Curd Properties and Cheese Structure

It is now well established that there is a direct influence of curd structure on the structure and texture of the derived cheese. Both the curd and Cheddar cheese formed with bovine pepsin have more open, looser structures than those formed with calf rennet.[80,81] Fat losses during cheesemaking were higher with pepsin, indicating that the curd with the more open structure retained less fat. When Cheddar cheese was made from milk concentrated by ultrafiltration (UF), both the curd and the cheese had an abnormally coarse protein network.[46] The network became coarser as the concentration factor (CF) of the milk was increased, the relative differences being the same throughout cheesemaking into the mature cheese. Less fat was recovered and the moisture in non-fat solids (MNFS) was lower in the cheeses from the more concentrated milks,[82] presumably because the curds with the coarser protein network were less effective in retaining fat and moisture. This accords with the observations of Sone et al.[83] on whey protein gels. Stronger protein interactions in forming the network were associated with a coarser structure and reduced water holding capacity.

The question arises as to why curds vary in the coarseness of the protein network. Knoop & Peters[84] suggested that casein micelles would tend to link together to form chains rather than clumps under ideal conditions. This is because, for negatively-charged renneted micelles, the electrostatic repulsive forces against lateral interaction between a single particle and a chain or between two chains are greater than against end-to-end addition.[85] However, the mean free path of micelles decreases in concentrated milk in proportion to the CF, from about one micelle diameter in native milk. Thus, as the concentration of casein micelles in the milk increases, it may be that the conditions for aggregation become progressively less ideal with a greater tendency to lateral aggregation and, consequently, a coarser protein network is formed.

Even with no change in the concentration factor of the milk there is a tendency for fast-forming curds to have coarser casein networks.[66] This may be because lower energy barriers to aggregation increase the tendency to lateral interactions between particles, thus increasing the probability of clump rather than chain formation.

The foregoing discussion has emphasized the importance of casein aggregation and whey loss as determinants of cheese structure and composition. During cheesemaking, the casein is compacted in the curd, water is lost and fat globules are entrapped and compressed (Figs 3, 4). Inevitably, changes in the relative concentrations of the components affect these processes. Higher fat concentrations in milk result in higher MNFS in cheese, reflecting the inhibitory effect of fat on syneresis. Lower casein levels also give higher MNFS in cheese, suggesting that casein aggregates more slowly in low-casein milks. Coarser curds tend to lose more fat, presumably due to less effective entrapment.

The physical and chemical conditions in the curd throughout cheesemaking are chosen to ensure that the young cheese has the correct composition and structure to develop the required flavour and texture. The manipulation of cheesemaking variables to give different characteristics in the cheese has been reviewed by Lawrence et al.[86] These authors point out that cheese types can be broadly distinguished by their pH values and concentrations of minerals. Both these factors are largely controlled by the acidity at whey draining which determines the extent of removal of Ca, inorganic phosphate (Pi) and water, with dissolved components such as lactose, from the casein. Cheeses which are highly acid and have a low mineral content, such as Cheshire, tend to have a structure consisting of smaller protein aggregates, which may affect the perceived texture. Further, the acidities at both coagulation and whey draining influence the extent to which the enzymes of the coagulant and the milk are retained in the cheese and, thus, participate in ripening. A possible synthesis of the known manufacturing influences on the composition and structure of cheese is summarized in Fig. 5.

In Cottage cheesemaking, a period of hours is generally allowed before cutting, so that coagulation occurs under more acid conditions than in the manufacture

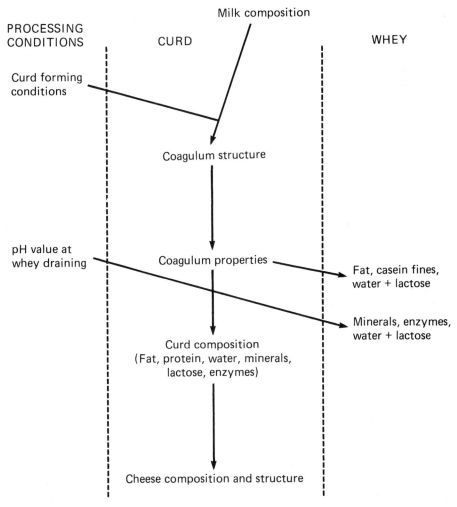

Fig. 5. Factors influencing the composition and structure of curd and cheese.

of most other types of cheese. Acid coagulation of milk results in less aggregation of the casein than is obtained with rennet.[87] Probably, the acid coagulation which occurs in Cottage cheesemaking also results in slower and less complete aggregation of casein than is obtained in the manufacture of other cheese types. If so, this would account for the high stability of the curd particles against fusion and loss of moisture.

4   FACTORS AFFECTING MILK COMPOSITION OR DISTRIBUTION OF COMPONENTS

4.1 Composition

The suitability of milk for cheesemaking is affected by its composition. This, in turn, depends on a number of factors including the husbandry of the animals and technological treatments of the milk prior to cheese manufacture.

Different breeds of cow produce milks of widely differing composition. The concentrations of fat and protein decrease in the order: Jersey, Guernsey, Ayrshire, British Friesian, British Holstein.[88] Further marked changes in gross composition can arise from selective breeding within a specific breed.[89]

The contents of fat, protein and lactose in milk fall slightly as the age of the cow increases,[90] but this is not relevant to a bulk milk supply where the age ranges of herds cannot be manipulated readily. Stage of lactation can exert a large effect on milk composition and is especially important for a milk supply where the contributing herds are predominantly synchronized to either spring or autumn calving. The most dramatic changes occur in early lactation when levels of fat, casein, Ca and P fall markedly.[91] However, in late lactation the concentrations of casein[92] and fat[93] tend to rise and to be accompanied by considerable changes in the mineral composition.[94] In late lactation, the amount of plasmin-like proteinase secreted into the milk also tends to rise[95] and this causes degradation of part of the β-casein to γ-caseins and proteose peptones.

Seasonal changes in bulk milk composition arise from a combination of lactational and dietary effects. In the UK, the most dramatic variations result from sudden dietary changes, especially during the spring when the cows are put out to pasture and to a lesser extent when they are brought inside for the winter.[88,96] The composition of the diet is a major determinant of milk composition and is the subject of much literature. The effects of nutrition on milk protein were reviewed by Rogers & Stewart[90] and Thomas[97] and the dietary effects on milk fat by Storry[98] and Sutton.[99]

Milking interval or frequency may affect the composition at a single milking, but is unlikely to affect a bulk supply.[100] Milk composition is also affected by climate[101] and excitement or stress.[90]

The presence of diseased animals in a herd can result in compositional changes in the milk which are detrimental to its use for cheesemaking. Although other diseases, such as ketosis, may affect the milk, mastitis is by far the most important disease to the dairy industry. Changes in milk composition due to mastitis result from both changes in the composition of the secreted milk and through the post-secretory action of enzymes derived from somatic cells, blood plasma and the mammary gland epithelium. These changes have been extensively reviewed by Kitchen[102] and Munro et al.[103] Increased somatic cell counts are associated with losses in milk total solids content[104] and depressed concentrations of lactose.[105] Total milk fat is not greatly altered,[106] but changes have been reported in the composition of the fat fraction, for example increased free fatty acid levels,[107] during mastitis. While total protein levels remain constant during mastitic infec-tions, the proportions of proteins which are synthesized in the mammary gland (α_{S1}^-, α_{S2}^-, β- and κ-caseins, β-lactoglobulin, α-lactalbumin) are reduced, while the increased permeability of the mammary gland results in increased levels of immunoglobulins and serum albumin. Mastitis also causes an increase in the ratio, soluble/micellar casein,[108,109] which may result in increased casein proteolysis. The changes in the permeability of the udder also affect the mineral composition of the milk. Most significantly for cheesemaking, the total[110] and micellar[111] con-centrations of Ca and P are considerably reduced during mastitis.

Milk composition may also be altered deliberately by ultrafiltration (UF) of the milk. Fat and protein are completely retained in the retentate by the mem-brane, but vitamins and salts are only partly retained, the extent depending on the proportion bound to the protein.[112] Thus, concentration by UF leads to changes in the relative amounts of the various components in the milk and in the bound/free ratio for macromolecular-bound materials. Further, at high total solids levels, there is a perceptible change in the ratio of the aqueous and par-ticulate phases, so that the concentration of lactose is reduced in the whole milk though unaltered in the milk serum. Treatment of milk in a UF-plant may also cause some disruption of the fat globules, rendering the fat susceptible to lipase action unless this enzyme is destroyed by heating. This problem can be worse during reverse osmosis (RO), when higher pressures are required as the mem-branes are permeable only to water.

The final concentration of water-soluble components in a UF-retentate can be influenced by additional treatments. Acidification of the initial milk, which causes dissociation of both salts[113] and vitamins,[114] can be used to reduce the concen-tration of these components in the retentate. Diafiltration, i.e. the addition and removal of extra water during the UF process, can be used to reduce the final concentration of any or all of the water-soluble components in the retentate.[115]

4.2 Distribution of Components

Storage of milk at refrigeration temperatures allows proteolysis and lipolysis to occur by the action of psychrotrophic bacteria and indigenous milk enzymes, as well as by somatic cells. Law[116] has reviewed the action of psychrotrophic en-zymes in milk. Cold storage, at 4 or 7°C for up to 48 h, also causes the caseins, particularly β-casein, Ca and Pi to dissociate from the micelles.[117] Solubilization increases the susceptibility of the casein to proteolysis in the milk. The process is partially reversed by prolonged storage, and completely reversed by heat treatment, 60°C/30 min or 72°C/1 min.[117] There is also a change in the micelle size distribution after holding milk at 4°C for 48 h, with a decrease in the proportion of small micelles.[118] It has been shown that micelle size affects the coagulation of milk, with a greater proportion of smaller micelles being associated with greater coagulum strength at the same casein concentration.[61,119]

Heat treatments in excess of pasteurization, in the range 90°C/15 s to 140°C/4 s, can be used as preludes to cheesemaking. These conditions cause partial denaturation of most of the whey proteins,[120] which then participate in SS/SH interchange reactions to precipitate as aggregates or bind to the κ-casein on the outside of the casein micelles.[121] Such heating also reduces the concentrations of soluble Ca and Pi, although these recover, at least partially, on subsequent storage in the cold.

Milk may also be homogenized prior to cheesemaking. The fat globules become smaller and the total surface area larger, to an extent depending on the pressure.[122] The new fat surface becomes covered with milk proteins, mostly casein micelles and submicelles.

5 MILK COMPOSITIONAL AND TREATMENT FACTORS AFFECTING CHEESEMAKING PROPERTIES

The most important considerations to a cheese manufacturer are the yield and characteristics of the product. The yield depends on the milk composition, especially the casein and fat contents and the efficiency of conversion of milk to cheese. The characteristics of importance are the composition and structure of the product. These are determined by the composition and treatment of the milk and the variables during cheese manufacture. It is by the deliberate control and manipulation of these factors that different cheese varieties are made. There may be influences on the structure of the curd, which then determines the cheese structure. The properties of the curd may be affected, thus influencing the retention of fat or water in the finished product and hence its composition. Alternatively, the retention of minor components such as rennet, minerals or starter in the curd can have large influences on the rate and type of proteolysis during ripening and the uptake of salt. In this section and the next, we shall attempt to explain the mechanisms underlying the influence of most of the variables normally encountered on the cheesemaking process and the structure of the product.

5.1 Breed and Genetics

Different species of ruminant produce milks of widely differing renneting properties,[123,124] reflecting their different compositions. While particular properties of milks from non-bovine species may be exploited in the manufacture of specific cheeses, this discussion will concentrate on cows' milk.

The major effect of breed on the cheesemaking properties of milk probably results from variations in casein content.[123,125,126] Curd firmness and structure are directly affected by the casein concentration.[123] The casein and fat in the milk, being the major solid components, also largely determine cheese yield.[127,128] In fact, various equations have been published for predicting cheese yield from the fat and casein (or protein) levels in the milk.[128–130] As expected from the casein and fat contents, the yields of Cheddar cheese decrease in the order: Jersey, Guernsey, Ayrshire, Holstein.[131] Genetic improvement in cows in the UK has tended to be concerned with increasing milk yield by replacing traditional English breeds by the Friesian and, more recently, by the North American Holstein.[132] With the recent increases in the proportion of milk being used for processing, where the composition is important, questions must be raised as to whether the higher yielding Holstein–Friesian is the optimal breed or whether reintroduction of more Channel Island characteristics may be economically beneficial to the dairy industry.[133]

One possibility for improving the properties of milk with respect to cheese-making lies in the fact that individuals secrete genetically determined variants of the caseins and β-lactoglobulin. Even though the variants differ by only a few amino acid substitutions, these differences cause significant changes in renneting and cheesemaking properties. Some of these changes may result directly from physical changes in the milk caused by differences in the physical properties of the protein molecules, but in most cases the variant seems to be genetically linked to some other composition factor and therefore just acts as a marker trait.

Several authors have noted the effect of the different κ-casein variants in milks during cheesemaking. The B and AB variants are associated with shorter clotting times and higher curd firmness than the A variants.[134–137] Schaar[138] found that the rennet-to-cut time was reduced by more than 30% and fat recovery from the milk was slightly greater when cheese was made from milk containing κ-casein B rather than the A variant, but there was little difference in the yield or composition of the finished cheeses. However, Mariani et al.[139] reported that the B variant resulted in improved texture of Parmesan cheese. The B variant of κ-casein is therefore desirable, at least for the renneting stage of cheesemaking. This is probably not a direct effect but due to associated compositional factors. In particular, milk containing κ-casein B has higher contents of casein, Ca and P and lower contents of citrate than milk containing κ-casein A.[139] Each of these characteristics will tend to favour the association of casein into the micellar form and promote micelle stability. In agreement with this, El-Negoumy[140] observed that casein micelles were more stable in milks containing the B variant rather than the A variant of κ-casein. The B variant of β-casein has also been reported to be associated with higher curd firmness than the A variant.[134] This may be due to differences in the distribution between serum and micelles or to linkage between the genes for the B variants of β- and κ-casein. In complete contrast to these latter studies, Marziali & Ng-Kwai-Hang[141] reported that coagulation properties were unrelated to the variants of κ- and β-casein, while the overall concentration of κ-casein was the major determinant of curd firmness.

Milk containing α_{S1}-casein A gave slower acid development than that containing the B or BC variants, while curd firmness decreased in the order of variants, BC, B, AB, AA.[142] Grandison et al.[143] and Remeuf & Lenoir[144] found that curd firmness is closely related to the content of α_{S1}-casein in milk and Creamer et al.[65] showed that the hydrolysis of α_{S1}-casein correlated with changes in the rheological properties of young cheeses. Therefore, it seems likely that α_{S1}-casein is basic to the formation of the network in the curd, analogous to its known structural role in the micelle, and that variants of differing physical properties may cause direct changes in curd rheology.

Different genotypes of β-lactoglobin also seem to be important in cheese-making although this whey protein has no direct role in curd formation. Milk containing the B variant produced firmer curds than that containing the AA or AB variants[141] and greater cheese yield and dry matter content were obtained with β-lactoglobulin B milk compared with that containing the A variant, with the protein loss in the whey decreasing in the order: AA, AB, BB.[138] These results can be explained by the observation that the proportion of casein in the protein fraction of milks decreases in the order: β-lactoglobulin BB, AB, AA.[138,145]

Thus, there seems to be potential to the cheese industry in including considerations of milk protein genotypes in future breeding programmes. Further research is needed, however, to check that the improvements in cheesemaking associated with the different variants are not linked with undesirable traits.

5.2 Mastitis

Inclusion of high levels of mastitic milk in the milk supply is clearly detrimental to renneting properties and cheese yield and quality.[129] Very severe mastitis results in doubling of the clotting time and halving of the curd firmness of the milk[103] but as little as 10% mastitic milk in the milk supply has been shown to prolong curd formation and give unsatisfactory curd.[146] Curd properties appear to deteriorate in proportion to the somatic cell counts,[147,148] and therefore to the compositional defects in the milk. Probably, the reduced casein content of the milk and alterations in the mineral balance leading to higher than normal casein solubility are both influential, since curd formation from mastitic milk could be improved, while still remaining sub-normal, by dialysis against normal milk.[149]

As expected, the formation of weak gels from mastitic milk is associated with high moisture retention and excessive fat losses in the whey.[129] These high fat losses and the low casein concentration in the milk were also found to reduce cheese yields by about 5%.[150] Grandison & Ford[151] reported that even very modest increases in somatic cell counts resulted in textural changes in Cheddar cheese, even though no decrease in yield could be detected. With treatment of the mastitis, the cheese yield returned to normal levels. There may also be some potential for increasing cheese yields by improving cow health further, since milks of very low cell counts (5×10^4/ml) gave better yields of curd than milks with counts typical of bulk milk supplies (5×10^5/ml).[109] An alternative approach suggested by Kosikowski & Mistry[152] is to supplement high somatic cell count milks with ultrafiltration retentate prepared from low count milks. Supplemented milks, despite their high counts, clotted normally to produce high yields of good quality cheese.

5.3 Environmental, Seasonal, Lactational and Dietary Effects

Most cheesemakers are aware of seasonal variations in the cheesemaking properties of milk. These result from variations in milk composition, which in turn result mainly from lactational and dietary effects.

Variations in the clotting times of milks during lactation appear to result mainly from variations in the pH value, with the more acid milk clotting more rapidly.[153,154] Curd firmness tends to be highest very early in lactation and declines to a fairly constant level within a few weeks, in parallel with the fall in casein concentration in the milk.[91,124] However, the rate of syneresis increases during this period, perhaps related to the fall in fat concentration in the milk.[91] Syneresis tends to be slow in late lactation milks[155] and to be accompanied by high moisture levels in the derived cheeses.[156] This has been traced to proteolysis in the milk.[157]

Dietary factors can have a marked effect on the renneting behaviour of milk. Several authors[96,158,159] have noted that curd firmness is greater during pasture feeding than when animals are kept indoors. Curd firmness also increases with the proportion of concentrate in the diet.[160] These changes resulted from variations in the casein concentrations in the milks. The effects of diet on clotting times and syneresis were less well defined. It has also been suggested that climatic changes affect renneting properties[161] and cheese quality,[101] but it seems probable that these effects are mediated through dietary changes, particularly resulting from compositional changes in the pasture.

The resultant effect of these variables on seasonal changes in a bulk milk supply is difficult to assess. It is dependent on the proportion of the contributing herds which are autumn or spring calving combined with the timing of putting the animals out to pasture and returning them indoors as well as the basic composition of pasture and winter rations on the individual farms. Generally, the effect of bulking the milk from a number of herds is to smooth out the extremes of composition and properties so that problems such as slow-renneting milks may be removed by bulking. However, seasonal variation in cheesemaking properties does still exist in bulked supplies. Olson[162] reported that cheese yield varied with seasonal changes in (fat + protein) concentration and that slight deviations from this were due to changes in the proportions of milk proteins. Milks that produce good curds give better yields and quality[129,163] whereas weak coagula result in cheese of lower quality and increased moisture.[156,164] Probably, maintenance of a high casein concentration throughout the year is the most important consideration for good yield and quality of cheese.

The composition of the fat fraction of the diet probably influences cheesemaking only in extreme cases. Cheese made from milk containing 19% linoleic acid in the milk fat, resulting from feeding protected sunflower seed oil, had poor texture and flavour.[165] Churning of fat during milk handling can also cause reduced yields.[162]

It is still not clear exactly how changes in composition affect the cheesemaking potential of milk, because much of the data are conflicting. However, the main positive factors appear to be high casein concentration, low pH and high micellar calcium phosphate.[166,167] Of these, the last is the most complex and difficult to determine. There is now reasonable agreement on its composition[168,169] but, as there appear to be direct links between the Ca of the calcium phosphate and casein phosphate groups,[170] there is no clear distinction from casein-bound Ca. Further, as the same factors appear to favour association of Ca with Pi to form micellar calcium phosphate and with casein phosphate groups, these two effects cannot be separated at present.

A clear understanding of the relation between milk composition and cheesemaking potential really depends on precise knowledge of the components present. A start has been made in this direction by the development of computer models describing the distribution of minerals in a milk ultrafiltrate among the various possible complexes and the influence thereon of pH and temperature.[171,172] Now that the micellar calcium phosphate composition is established,[168,169] extension of this to the casein micelle requires information on the various ionic binding sites in casein complexes.

However, in qualitative terms, it is clear that compositional factors favourable for cheesemaking are also those that promote both association of casein into micelles and coagulum formation. Both reactions are also facilitated by the binding of a number of polycations to casein.[173] Thus, the association of caseins to micelles and of para-casein to a coagulum may operate by similar mechanisms.

5.4 Bovine Somatotropin (BST)

BST is a natural hormone which stimulates milk production and is found in minute quantities in all cows' milk. Developments in biotechnology have permitted the manufacture of commercial quantities of substances with the same properties as natural BST, which can be used to increase milk yields of dairy cows by 10–20%. Although these may not yet be marketed in the EC, licences have been granted in some countries and several studies have been undertaken to determine the effects of BST administration on both milk composition and cheesemaking properties.[174-176] In general, it is concluded that milk from BST-treated cows is not significantly different from milk from untreated animals with respect to cheesemaking. Piva et al.[177] concluded that the use of BST for milk production for cheesemaking is economically viable.

5.5 Cold Storage

Storage of the milk may affect cheesemaking by both the physical effect of casein solubilization from the micelles and hydrolysis of casein and fat by enzymes in the milk, mainly from psychrotrophic bacteria but also from somatic cells or blood. In some cases it is difficult to distinguish between the two effects.

Milk stored at 4°C or 7°C, having increased soluble casein, shows slower clotting, higher losses of fat and curd fines into the whey, weaker curds and lower curd yield than that stored at 10–20°C.[117] This suggests that soluble casein may not be incorporated into the curd matrix, but the results are confounded by bacterial growth and proteolysis in the milks stored at the higher temperatures. The effects of cold storage are at least partially reversed by holding the milk at 60–65°C, which reduces the clotting time and improves curd properties.[178]

5.6 Psychrotroph Action and Thermization

The action of enzymes from psychrotrophic bacteria during cold storage can cause lipolysis and proteolysis in the milk. The main effects noted on the cheesemaking properties are rancid off-flavours due to lipolysis and reduced yields resulting from proteolysis. In Cottage cheesemaking, yield losses only become significant after the total bacterial count has reached 10^6/ml and are related to increased non-casein N formed by proteolysis.[179] When the bacterial count had reached 10^8/ml, clotting occurred more slowly and Cottage cheese curds were softer.[180] This level of bacterial growth also caused serious losses in yield. Psychrotroph growth in milk also causes significant losses in Cheddar cheese yield.[181] However, the numbers of psychrotrophic bacteria occurring in stored raw milks under commercial conditions are unlikely to cause significant losses in yield.[116,138] High psychrotrophic counts also cause the micelle size distribution to change[182] which may influence the coagulation properties of the milk.

Reduced cheese yields due to extended cold storage may be avoided by on-farm heat treatment of milk prior to cooling and storage. Heating to 74°C for 10 s was reported to result in 5% more Cottage cheese after storage of the milk for 7 days at 3°C.[183]

Reports of yield increases with Cheddar cheese were not substantiated by a detailed study in which it was shown that heating milk at 65°C for 15 s prior to storage kept psychrotroph levels acceptable over the subsequent 3 days.[184] Longer storage times or lower heat treatments led to losses of yield or quality. Yields of Edam and Limburg cheeses were reportedly increased 1·5–2% by heating to 73°C for 30 s, cooling to 8–10°C and reheating to 67°C for 30 s before cooling overnight and pasteurization.[185] Thus, it seems that thermization of milk before cold storage can help to keep psychrotrophs below dangerous levels, especially if the storage temperature is at 6–7°C rather than 2°C.[184] Probably, the higher heat treatments used result in improved microbiological quality rather than cause whey protein–casein interactions, as suggested by Dzurec and Zall.[183]

5.7 Carbon Dioxide Addition

Another possible means of improving the keeping quality of raw milk by reducing the growth of psychrotrophs is to add about 30 mM CO_2 to milk, decreasing the pH value to about 6·0.[186] The CO_2 can be removed by sparging with nitrogen, but this would seem to be unnecessary where the milk is to be converted to cheese as acid production is required and all the CO_2 is driven off by the time of pitching, so as to have no effect on the later stages of cheesemaking (J.S. King, pers. comm.).

In model experiments with milk containing CO_2 but without starter, using rennet levels to give the same RCT, both curd formation and whey loss were faster than with the controls, but the differences were entirely accounted for by the difference in pH value.[66] Reducing the pH value from 6·6 to 6·2 by either CO_2 or lactic acid had no perceptible effect on curd structure. As expected from these findings, milk preserved with CO_2 created no problems in pilot-scale Cheddar cheesemaking trials, provided allowance was made for the faster curd formation by cutting the curd at the optimal point (L.A. Mabbitt & J.S. King, pers. comm.). This might cause difficulties in factory-scale cheesemaking, because the concentration of CO_2 in the milk in the vat at renneting is likely to be variable, so that the optimal time for cutting the curd may vary between vats. Cutting too early or too late may cause fat losses or excessive moisture retention, respectively.[187] However, one advantage would be that the amount of rennet required could be reduced with a greater proportion retained than normal.

5.8 High Heat Treatment

Heating milk above pasteurization temperatures causes loss of coagulability without necessarily affecting the enzymic action of rennet,[73] presumably due to steric hindrance by the β-lactoglobulin bound to the surface of the casein micelles. The curd from such milk also shows reduced syneresis, which has been shown to be due to β-lactoglobulin binding.[74] These processes are not a simple function of (time $\times$ temperature), since whey proteins undergo other interactions in the milk at high temperatures. The cheesemaking parameters appear to be affected more at higher heating temperatures,[188] even if the level of whey protein denaturation is kept constant (R.J. Marshall, pers. comm.).

Acidification and Ca^{2+} addition can be used to facilitate the formation of curd from high-heat-treated milks.[189] This suggests that the best types of cheese to make from such milks may be those in which the curd is formed under acid conditions. In agreement with this, satisfactory Mozzarella cheese was made from milk heated for 2 s at up to 130°C by allowing coagulation to occur at pH 5·6;[190] there was a 3–4% increase in cheese yield. Cheshire cheese, which is a crumbly, semi-hard variety, was also made satisfactorily from high-heat-treated milk.[73] In this instance, heat treatment was at 97°C for 15 s and coagulation occurred at pH 6·4. The curd tended to retain moisture and fuse poorly, which correlated with slower aggregation of the casein masses observed by electron microscopy. To overcome this, it was scalded at a higher temperature than normal and the manufacturing time extended to encourage curd fusion. By this means, a product of normal composition and properties could be prepared but with an increase in yield in terms of dry matter of 4–5%.

In contrast, most attempts to prepare satisfactory Cheddar cheese, a variety containing lower moisture and requiring coagulation at a higher pH value, from

high-heat-treated milk have not been successful. However, Banks[191] overcame
the problem by acidification to pH 6·2 of milk heated at 90°C for 60 s. Such
milk could be coagulated using 0·9 times the normal rennet level and converted
into Cheddar cheese with 4–7% increased dry matter yield over pasteurized milk.
The MNFS was within the optimal range and the product was acceptable although somewhat inferior to the control. This may relate to the observation by
Creamer et al.[192] that Cheddar cheese made from acidified milk contains a lower
than normal level of Ca and has a brittle, crumbly body. Poznanski et al.[193] also
patented a method for making several varieties of cheese in increased yield from
milk heated at 92°C for 15 s.

An alternative to heating the milk to be used for cheesemaking is to recover
the proteins from whey for addition to the next batch of cheesemilk. A number
of methods have been tried, including recovery of the proteins by heat treatment,[194] by heating after concentration[195] or as a complex with carboxymethyl
cellulose.[196] However, although yields were higher in all instances, the moisture
content of the cheeses containing the extra whey proteins was significantly above
normal, often causing flavour and texture defects.

5.9 Homogenization

When milk is homogenized, the rate of aggregation of the casein particles is
reduced, the bimolecular rate constant (k) being about 10^{-2} of that for skim
milk.[197] This is unlikely to be due to the change in size of the casein particles,
since k is little affected by this factor. Curd is also formed more slowly from
homogenized milk and the curd synereses more slowly,[198] so that the final cheese
retains more moisture than normal. In curds formed from concentrated milk,
homogenization also reduces the ability of the curd particles to fuse.[16]

Homogenization of the milk also influences the structure of the curd[16] and young
cheese[199] formed from it. Both have a less coarse protein network than normal;
this may contribute to the improvement in soft cheese texture observed with
homogenized milk.[200]

It seems likely that the effects of homogenization of the milk on curd formation, structure and properties all stem from the same basic mechanism. There is
no evidence for a change in the surface of the casein particles, since the electrophoretic mobility is unaffected by homogenization.[201] However, many micelles
are linked to fat particles in homogenized milk[122] without altering the micelle
size distribution. This may reduce the total surface area of the casein, reducing
the amount available for interaction and so slowing it.

5.10 Membrane Concentration

The increase in efficiency of cheesemaking and yield of cheese theoretically possible
with concentrated milk has stimulated interest in UF-concentration. Three different approaches have been made: concentration up to two-fold has been used with
traditional cheesemaking methods to overcome milk variability, concentration

three to six-fold has been used with success for some cheese varieties but usually involves non-traditional methods and equipment, and concentration 8 to 10-fold is required to form a 'pre-cheese' which can be converted to cheese without syneresis. In the present context, the middle range is of most interest.

The concentration of milk by UF has little effect on the clotting time[202] and the aggregation rate of micelles does not appear to be much affected by concentration.[16] However, the curd firmness, measured by empirical methods, increases markedly with concentration. This apparent discrepancy probably results from the higher total solids content of the curds from concentrated milks compared with unconcentrated ones, reducing the relative ease of movement of the strands in the protein network and thus increasing the viscosity and rigidity.

As the concentration of fat and protein in the milk increases, the viscosity of the curd increases faster than the elastic modulus.[203] The curd structure also tends to change with the protein network becoming coarser.[46] This is probably responsible for the poor retention of fat and moisture by curds from concentrated milks.[82] Further, the structure of the curd leads directly to that of the cheese, so that cheeses from more concentrated milks tend to have coarser protein networks. The atypical structures are perceived as textural defects, cheeses from more concentrated milks tending to be more crumbly, grainy and dry.[204]

The structure of the curd from concentrated milks can be improved by reducing the aggregation rate via lower temperature[204] or Ca concentration.[64] These conditions also reduce the rate of syneresis. However, the mechanisms by which the curd structure is controlled are not yet clear. Aggregation of the casein micelles in concentrated milks starts when a smaller proportion of κ-casein has been hydrolysed than for normal milk[205] and can occur at lower temperatures[206] which may lead to a change in the balance of small and large aggregates.[207] Alternatively, the decreased mean free path of casein micelles as the milk becomes more concentrated may reduce the proportion of the energetically-favoured end-to-end linking of chains of micelles and increase lateral interactions, thus creating coarser protein strands.[1]

Technological approaches to the problems of cheesemaking with UF-concentrated milk have tended to be rather successful. Special curd forming and handling equipment has been devised enabling hard and semi-hard cheeses such as Cheddar, Havarti and Mozzarella to be manufactured successfully. These approaches are discussed in detail in Chapter 13.

UF-concentration and high heat treatment (Section 5.8) tend to have opposite effects on cheesemaking parameters. This suggests that combining the two processes may be desirable. Certainly, the yields of cheese should be greater than from either process alone, as denaturation of the whey proteins should increase the proportion retained in the curd from UF-concentrated milk. Some success has been reported. Prokopek *et al.*[208] prepared Camembert cheese of satisfactory consistency from four-fold concentrated milk heated at 80°C for 1 min and coagulated at 16–18°C. Rao & Renner[209,210] used four-fold concentrated milk, heated at 76°C for 5 min for Cheddar cheesemaking. They reported improvements in yield and flavour, though not in texture, even though the moisture content was higher than normal. In a more detailed study,[211] it was found that the proportion of whey protein denatured increased with concentration factor, the rate of curd formation from concentrated milks was approximately halved by heating at 90°C for 15 s, and syneresis from the gel was considerably retarded. The recovery of milk solids in Cheddar cheese made from 2·3-fold concentrates heated at 90°C was comparable with that from five-fold pasteurized concentrates.[212] The fusion of the curd and, consequently, the texture of the cheese was poorer when the heated concentrate was used, but the difference was small. Thus, the results suggested that high heat-treated concentrates may be suitable for Cheddar cheesemaking, and that less concentration may be required to achieve the same yields as from pasteurized concentrates.

A study has been made of the use for Cheddar cheesemaking of milk in which all the solids have been concentrated by reverse osmosis (RO).[213,214] It was found that a 15% reduction in milk volume gave satisfactory results provided a few changes were made in the cheesemaking technology and the RO plant did not damage the fat. Increased retention of fat, casein and whey solids was achieved.

5.11 Reconstitution and Recombination

Whole milk powder or skim milk powder and cream or anhydrous milk fat reconstituted in clean water to the same TS and fat/casein ratio as whole milk with added $CaCl_2$ can be used reasonably successfully for cheesemaking provided the powders are prepared under low heat conditions so that interaction of whey proteins with the casein is minimized. Both Camembert[215] and soft brine cheeses[216] tended to have a more open structure than normal, with the casein less evenly distributed, giving more rapid loss of whey and penetration of brine. This was probably due to relatively poor aggregation of the casein resulting from heat damage during the drying process, since the problem was accentuated with milks dried under high heat conditions.[215]

In a thorough study of cheesemaking from recombined milk prepared by homogenizing anhydrous milk fat into reconstituted skim milk powder with $CaCl_2$ addition Gilles & Lawrence[217] found that coagulation, firming and syneresis were all slower than normal and the curds tended to retain moisture, probably due to the effects of both heat treatment and homogenization. The typical fibrous structures of Cheddar and Mozzarella cheeses were not developed, perhaps because the reduced aggregation potential of the casein decreased the tendency of the curd to flow during the texturization steps. However, Dutch cheeses, made by a process not involving texturization, could be produced satisfactorily from recombined milks.

One possible way to simplify cheesemaking from reconstituted milk would be to carry out some of the normal cheesemaking processes before the milk is dried. In a study of the effects of acidification and renneting of the milk prior to drying on its subsequent cheesemaking properties, Ehsani et al.[218] found that curd giving satisfactory rates of firming and whey loss could be prepared.

As most of the water used in reconstitution is removed by syneresis, the addition of as little as possible seems a worthwhile aim. However, use of recombined milk containing more than 25% total solids (2 × normal concentration) was not successful in Ras-type cheesemaking,[219] apparently because of a reduced ripening rate.

A quite different approach to the use of dried milk fractions is to combine them to form a mix of the same composition as required in the final cheese, similar to the use of 'pre-cheese' for making soft cheeses from milk concentrated by ultrafiltration. Davis[220] first tried this approach, mixing skim milk powder, sodium caseinate and ripened cream to the composition of hard, semi-hard and soft cheeses, though no really satisfactory product was prepared. It was developed further by Ali[221] who prepared a soft cheese of the Domiati type which developed acceptable flavour and a typical conchoidal texture on maturation in brine. The changes in structure throughout cheesemaking and ripening were followed by electron and light microscopy. In the first stage, combination of skim milk powder with anhydrous milk fat, casein micelles were adsorbed onto the globule surface as in homogenized milk. This cream was ripened and sodium caseinate was added to give a mixture having a denser matrix, with a fuzzy background of caseinate. When rennet was added to this, aggregation occurred at only about 10% of the expected rate. After 1 h, the caseinate was less evenly dispersed in the matrix, appearing to have moved towards the micelles, and a firm curd was formed in about 6 h. Some whey expulsion started after 15–20 h and this was increased by the addition of NaCl. In the 1-day old cheese, the caseinate and casein micelles were coalesced into aggregates containing entrapped fat, with whey-filled spaces between them. During ripening, the protein masses broke down into smaller units still separated by whey-filled areas. The structure of the ripened cheese prepared by the new method was comparable with that of the authentic product.

5.12 Alteration of Fat Levels

Milk for Cheddar cheesemaking is frequently standardized to within a narrow range of casein/fat ratios (typically $0.68–0.72:1$) to optimize quality and yield.[222] This tends to involve a reduction in the proportion of fat and a slightly higher proportion of the fat tends to be retained in the cheese.[223] Thus, a smaller proportion of the fat escapes from the curd into the whey, perhaps because it is more effectively entrapped. Johnston & Murphy[62] have shown that the fat present in curd acts as a plasticizer and inhibits the formation of cross-links between the casein chains. The weaker and more porous protein network would tend to lose fat more readily.[187]

When the fat in milk is reduced so as to halve that in the cheese, the low-fat product tends to be harder, more crumbly and less smooth than normal. Probably there is increased cross-linking within the curd, carried through into the cheese, together with reduced plasticizer. Raising the moisture level in an attempt to overcome the defects leads to weak body and encourages undesirable flora and poor flavour. Replacement of the fat by whey proteins has been more successful. Semi-hard Dutch cheeses with 20% fat were made successfully from a 'pre-cheese' prepared by UF-concentration followed by evaporation.[224] This enabled the whey proteins to be incorporated quantitatively in an undenatured form. As they were not hydrolysed during ripening, they behaved as inert fillers. This caused the microstructure to be relatively open and the cheeses to be smoother, creamier and less rubbery than usual for low-fat cheeses. A similar approach, but using five-fold UF-concentrated milk, which still requires some whey drainage, has been used for low-fat Cheddar cheese.[225] However, the cheeses produced, even with added proteases and lipases,[226] received low texture and flavour scores.

An alternative approach has been to replace fat by water bound to hydrocolloids. This has shown some success with Ras cheese, using milk containing 2% fat and adding 0·02% carrageenan or 0·1% carboxymethylcellulose.[227] The additives induced a softer, smoother texture.

5.13 Addition of Non-milk Materials that Influence the Cheesemaking Properties of Milk and Cheese Characteristics

CaCl$_2$ is frequently used as an additive in cheesemaking because it stimulates coagulation, curd firming and whey loss (Table I). This is probably because it binds to the casein micelles[228] in such a way as to reduce the repulsive forces between them, perhaps by promoting hydrophobic interactions.[173] In this circumstance, all the stages of cheesemaking that depend on casein aggregation would probably be stimulated.

Many other positively-charged materials have similar effects to Ca^{2+}, their efficacy being dependent on the extent of binding and on the charge.[229] It appears that their actions also mimic that of Ca^{2+}.[173] In fact, some cationic polyelectrolytes can almost completely coagulate native casein,[230] probably because of the extent of their binding and charge neutralizing abilities. Addition of soy protein to milk reduced the strength of the coagulum produced by rennet.[231] Various modifications of the additive, including chemical changes to side chains[232] and hydrolysis,[233] gave no real improvement. However, adding Ca^{2+} and lowering the pH of coagulation slightly stimulated both coagulum formation and syneresis, but not to the levels obtained with milk alone.[234] As expected, incorporation of 20% soy protein into Domiati cheese made from a mixture of cows' and buffaloes' milks gave a high moisture product.[235]

As mentioned previously (Section 5.12), incorporation of polysaccharides might be used to increase moisture levels in cheese. Some materials, such as modified starches and carrageenans, did not affect coagulation, but others, such as xanthan and low methoxyl pectin, slightly accelerated both coagulation and curd firming.[236] This suggests that study of their action is needed before such materials are added to cheese milks, so that a proper selection can be made.

Incorporation of phospholipids into cheese curd does not appear to involve entrapment within the casein network. Using electron microscopy, Kirby et al.[237]

found lecithin-derived liposomes to be situated close to the fat-protein interface, with none within the casein matrix.

6   Influence of Variables During Cheesemaking on Curd and Cheese Structure

6.1 Curd Formation

The structure of the initial coagulum is very influential in determining that of the cheese (Section 3.4). This means that curd formation is a vitally important step in the manufacture of good cheese.

For Cottage cheesemaking, the coagulum can be formed using either acid alone or by adding low levels of rennet as well as acid. The enzyme-set curd tended to be more integrated and the particles were larger after cutting.[238] However, the acid-set curd drained more readily, giving denser, firmer particles.

For some cheese types, such as Domiati, NaCl is added to the milk before it is coagulated. This caused the pH value to decrease[239] and the soluble Ca, Mg and P levels to rise.[240] The curd formed more slowly[239] and syneresis was considerably inhibited.[240] Therefore, the NaCl addition may help to retain whey in the curd.

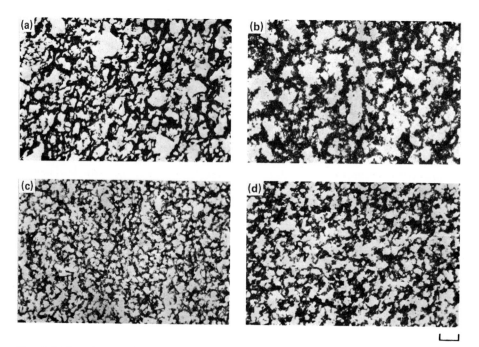

Fig. 6. Light micrographs of curds from milk concentrated 3·25-fold by UF. Conditions of formation: (a) Renneted at 30°C; (b) renneted at 38°C; (c) pre-renneted at 5°C then raised to 30°C; (d) pre-renneted at 5°C then raised to 38°C. Scale bar = 20 μm.

TABLE II
Effect of Temperature and Pre-renneting at 5°C on the Rate of Formation, Structure and Properties of Curd from Milk Concentrated 3·25-fold by Ultrafiltration

	Renneted at curd forming temperature		Pre-renneted at 5°C, then warmed	
Temperature, °C	30	38	30	38
Temperature at which viscosity started to rise, °C	—	—	14	14
Maximum curd firming rate, arbitrary units	55	86	62	123
Coarseness of protein network, arbitrary units	3·8	6·0	3·2	3·4
Fat loss in whey, % of milk fat	2·9	7·1	2·6	4·9

The nature of the coagulant also affects the structure and properties of the curd. The main action is probably on the extent of proteolysis during manufacture and ripening. The more proteolytic enzymes appear to cause the curd to be formed more slowly.[241] They also give a coagulum with a looser, more open structure,[80,242] which can lead to reduced fat[80] and water[242] retention. The yields of cheese may also be reduced, presumably due to hydrolysis of the curd matrix. Coagulants showed similar effects in milk concentrated by ultrafiltration, with casein micelles being linked into a more extensive network in gels formed by rennet than in those formed by fungal or bacterial coagulants.[243]

As the temperature at which the curd is formed is raised, the casein micelles aggregate more readily and extensively.[244] As expected, the protein network became coarser as the temperature of curd formation was raised for both normal and UF-concentrated milks (Fig. 6).[66] More fat tended to be lost in the whey from the coarser curds (Table II).

When the milk was pre-renneted at 5°C, so as to effect the enzymic reaction without coagulation, and then warmed by plunging into a bath at the required temperature, the curd formed more rapidly than normal (Table II).[66] However, with milk concentrated 3·25-fold by UF, the protein network was less coarse than when formed by normal renneting at the same temperature (Fig. 6) and the fat losses in the whey were reduced (Table II). This unexpected result can probably be explained by the slowness of the heating process. Curd formation actually started well before the final temperature was reached, probably under the minimum conditions required for aggregation. Thus, the views that the initial stages of formation strongly influence the curd structure, and that curds formed under less favourable conditions have a less coarse protein network, are consistent with these observations.

6.2 Curd Handling

Studies of the effect of firmness at cutting on curd properties and cheese yields have generally concluded that the cutting time has little influence unless it falls

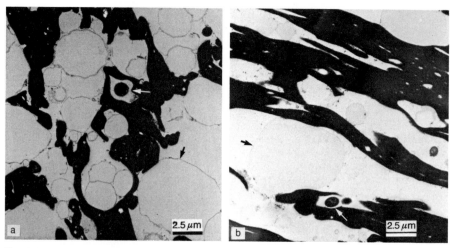

Fig. 7. Transmission electron micrographs of cheddared curd showing dark protein matrix, fat globules (black arrow) and bacteria (white arrow). (a) Section cut across the direction of applied pressure; (b) section cut along the direction of applied pressure (kindly provided by Dr M. Kalab).

outside the range of about 85–200% of normal. The amount of fat in the whey decreased and the moisture content of the cheese increased linearly with increase in the cutting time.[187] The loss of fat from soft curd was explained by the inability of a poorly-developed network to retain it, and the retention of moisture by the difficulty of a very rigid highly cross-linked curd subsequently forming sufficient cross-links to expel the required amount of whey. The healing time, the interval between cutting and stirring, has been shown to be important for both cheese yield and composition.[245,246] Fat was retained better and cheese moisture was lower when longer healing times, 10–15 min, were used. This probably occurred because the conditions permitted cross-linking and syneresis throughout the curd grain rather than accelerating moisture loss from the outside of the particle. If this occurs, the casein network can compact at the particle surface, creating an impermeable film and trapping moisture within the curd grain. Too rapid a rise in temperature at cooking can also lead to this defect. It probably explains why intermediate cooking rates and temperatures appear to be optimal for preparing firm, low-moisture Cottage cheese.

In Cheddar cheesemaking, the high cooking temperature liquefies the fat, so that it can flow in the curd.[247] During cheddaring, the protein chains become oriented, causing the fat globules, still containing liquid fat, to elongate in the same direction and aggregate (Fig. 7).[248] The protein and fat are also oriented in Mozzarella and Provolone cheeses, which are stretched at temperatures above the melting point of the fat, but not in Edam and Gouda cheeses, where flow during manufacture is not induced.

Pre-pressing, the piling of curd in the vat, is important in giving the characteristic open structure of certain semi-hard cheeses.[249] During this stage, the curd flows[42] presumably aiding fusion of the grains. This is also assisted by vacuum pressing, which removes air and whey from between the grains.

6.3 Cheese Ripening

Ripening of both soft and hard cheeses is accompanied by disintegration of the casein network to form a smoother, more homogeneous structure.[250-252] This may be affected by the Ca present, perhaps because of cross-linkage of the protein. The extent of proteolysis is also important. In one soft cheese, the softening has been correlated with α_{s1}-casein hydrolysis by residual rennet.[253] This results in reduced hydrophobic interactions between molecules, which can cause the protein network to weaken.[65]

In practice, an important factor controlling proteolysis in cheese is the amount and type of salt present. This subject has been of interest in recent years because of the nutritional significance of salt in the diet. In general, it has been found that the NaCl content can be reduced significantly. Cheddar cheese was reported to ripen satisfactorily with 1·6% NaCl or equivalent ionic strength 1:1 mixtures of NaCl and KCl.[254] Alternatively, good quality Cheddar cheeses containing 1% NaCl were produced from milk concentrated 1·5–1·9-fold by UF.[255] The salt content also affects the textural properties of Mozzarella cheese, the meltability and stringiness increasing at low salt concentrations.[255]

In mould-ripened soft cheeses, the fungal enzymes are responsible for much of the protein breakdown. For instance, in Camembert proteolysis has been shown to occur more rapidly near to the fungal hyphae.[256] Protein breakdown is extensive, leading to the formation of NH_3, which raises the pH of the surface of the cheese to ~7.[257] This rise in pH precipitates the calcium phosphate at the surface causing the migration of Ca and phosphate from the bulk of the cheese to the surface.[50] The combination of increased pH and low Ca level causes partial solubilization of the protein matrix and, thus, softening of the cheese.

6.4 Cheese Processing

In processed cheesemaking, cheese is heated with a mixture of salts, mostly phosphates and citrates. The salts sequester the Ca, allowing hydration, swelling and solubilization of the protein, control of pH and emulsification of fat.[258] Thus, a uniform, gel-like structure is obtained, suitable for slicing and melting. Differences in properties may be caused by structural differences. Various electron microscopic techniques were used to show that harder, less meltable products had longer protein strands than softer ones.[47]

7 CONCLUSION

Although it is now possible to describe the conversion of milk to cheese at the microscopic level in some detail, the forces involved still remain to be understood sufficiently to enable full control of the process. Much data have been collected on the effect of composition and processing on the cheesemaking properties of milk, but the complex interaction of multiple components has sometimes been difficult to disentangle. A coherent theory linking the composition, cheesemaking potential and optimum processing conditions of a milk would be of great benefit. However, we are beginning to be able to specify the factors in a milk that determine the efficiency of its conversion to cheese. This offers the possibility of either producing milk of the optimal composition or adjusting the composition and properties by processing prior to cheesemaking. The essential aim is to optimize the recovery of solids, the utilization of equipment and the quality of the product.

REFERENCES

1. Green, M.L., 1984. In *Advances in the Microbiology and Biochemistry of Cheese and Fermented Milk*, ed. F.L. Davis & B.A. Law. Elsevier Applied Science Publishers, London, p. 1.
2. Walstra, P. & van Vliet, T., 1986. *Neth. Milk Dairy J.*, **40**, 241.
3. Fox, P.F., 1987. *Dairy Ind. Int.*, **52**(7), 11.
4. Richardson, G.H., Okigbo, L.M. & Thorpe, J.D., 1985. *J. Dairy Sci.*, **68**, 32.
5. Bohlin, L., Hegg, P.O. & Ljusberg-Wahren, H., 1984. *J. Dairy Sci.*, **67**, 729.
6. Zoon, P., van Vliet, T. & Walstra, P., 1988. *Neth. Milk Dairy J.*, **42**, 249.
7. Dejmek, P., 1987. *J. Dairy Sci.*, **70**, 1325.
8. Zoon, P., van Vliet, T. & Walstra, P., 1988. *Neth. Milk Dairy J.*, **42**, 271.
9. Zoon, P., van Vliet, T. & Walstra, P., 1988. *Neth. Milk Dairy J.*, **42**, 295.
10. Zoon, P., van Vliet, T. & Walstra, P., 1989. *Neth. Milk Dairy J.*, **43**, 17.
11. Zoon, P., van Vliet, T. & Walstra, P., 1989. *Neth. Milk Dairy J.*, **43**, 35.
12. Tokita, M., Hikichi, K., Niki, R. & Arima, S., 1982. *Biorheology*, **19**, 209.
13. van Vliet, T. & Dentener-Kikkert, A., 1982. *Neth. Milk Dairy J.*, **36**, 261.
14. Scott Blair, G.W. & Burnett, J., 1958. *J. Dairy Res.*, **25**, 297.
15. Johnston, D.E., 1984. *Milchwissenschaft*, **39**, 405.
16. Green, M.L., Marshall, R.J. & Glover, F.A., 1983. *J. Dairy Res.*, **50**, 341.
17. Marshall, R.J., Hatfield, D.S. & Green, M.L., 1982. *J. Dairy Res.*, **49**, 127.
18. Storry, J.E. & Ford, G.D., 1982. *J. Dairy Res.*, **49**, 469.
19. Prokopek, D., 1978. *Deutsche Milchwirtschaft*, **29**, 534.
20. Gervais, A. & Vermeire, D., 1983. *J. Texture Studies*, **14**, 31.
21. Politis, I. & Ng-Kwai-Hang, K.F., 1988. *J. Dairy Sci.*, **71**, 1740.
22. Aleandri, R., Schneider, J.C. & Buttazzoni, L.G., 1989. *J. Dairy Sci.*, **72**, 1967.
23. Ramet, J.P., El Mayda, E. & Weber, F., 1982. *Lait*, **62**, 511.
24. Richardson, G.H., Okigbo, L.M. & Thorpe, J.D., 1985. *J. Dairy Sci.*, **68**, 32.
25. Vanderheiden, G.H., 1976. *CSIRO Fd. Res. Q.*, **36**, 45.
26. Kowalchyk, A.W. & Olson, N.F., 1978. *J. Dairy Sci.*, **61**, 1375.
27. Hatfield, D.S., 1981. *J. Soc. Dairy Technol.*, **34**, 139.
28. Storry, J.E. & Ford, G.D., 1982. *J. Dairy Res.*, **49**, 343.
29. Matej, V., 1984. Cited from *Dairy Sci. Abstr.*, **46**, 200.
30. Kulkarni, S. & Vishweshwariah, L., 1983. *J. Food Sci. Technol. (Mysore)*, **20**, 214.
31. Hori, T., 1985. *J. Food Sci.*, **50**, 911.
32. Cindio, B., Grasso, G., Spagna Musso, S., Matteo, M. & Mincione, B., 1987. *Scienza e Tecnia Lattiero-Casearia*, **38**, 201.
33. Pearse, M.J. & Mackinlay, A.G., 1989. *J. Dairy Sci.*, **72**, 1401.
34. Lawrence, A.J., 1959. *Aust. J. Dairy Technol.*, **14**, 166.
35. Marshall, R.J., 1982. *J. Dairy Res.*, **49**, 329.

36. Pearse, M.J., Mackinlay, A.G., Hall, R.J. & Linklater, P.M., 1984. *J. Dairy Res.*, **51**, 131.
37. van Dijk, H.J.M. & Walstra, P., 1986. *Neth. Milk Dairy J.*, **40**, 3.
38. Pieper, H., Stuart, J.A. & Renwick, W.R., 1977. *J.A.O.A.C.*, **60**, 1392.
39. Barbano, D.M. & Della Valle, M.E., 1984. *J. Food Protect.*, **47**, 272.
40. Ilyushkin, V.S., Lepilkina, O.V. & Tabachnikov, V.P., 1983. Cited from *Food Sci. Technol. Abstr.*, **15**, P1887.
41. Kalab, M., Lowrie, R.J. & Nichols, D., 1982. *J. Dairy Sci.*, **65**, 1117.
42. Ogorodnikov, E.N. & Tabachnikov, V.P., 1985. Cited from *Food Sci. Technol. Abtrs.*, **17**, 8P83.
43. Johnston, D.E., 1984. *J. Dairy Res.*, **51**, 91.
44. Haddad, Y.M., 1984. *J. Colloid Interface Sci.*, **100**, 143.
45. Kimber, A.M., Brooker, B.E., Hobbs, D.G. & Prentice, J.H. 1974. *J. Dairy Res.*, **41**, 389.
46. Green, M.L., Turvey, A. & Hobbs, D.G., 1981. *J. Dairy Res.*, **48**, 343.
47. Taneya, S., Kimura, T., Izutsu, T. & Buchheim, W., 1980. *Milchwissenschaft*, **35**, 479.
48. Kalab, M. & Modler, H.W., 1985. *Food Microstructure*, **4**, 89.
49. Rousseau, M., 1988. *Food Microstructure*, **7**, 105.
50. Brooker, B.E., 1987. *Food Microstructure*, **6**, 25.
51. Brooker, B.E., 1990. *Food Structure*, **9**, 9.
52. Tombs, M.P., 1974. *Faraday Disc.*, **57**, 158.
53. Hermansson, A.-M. & Buchheim, W., 1981. *J. Colloid Interface Sci.*, **81**, 519.
54. Shimmin, P.D., 1982. *Aust. J. Dairy Technol.*, **37**, 33.
55. Orlandini, I. & Annibaldi, S., 1983. *Scienza e Tecnica Lattiero-Casearia*, **34**, 20.
56. Clark, A.H. & Tuffnell, C.D., 1980. *Int. J. Peptide Protein Res.*, **16**, 339.
57. Lelievre, J. & Creamer, L.K., 1978. *Milchwissenschaft*, **33**, 73.
58. Callaghan, P.T., Jolley, K.W. & Humphrey, R.S., 1983. *J. Colloid Interface Sci.*, **93**, 521.
59. Green, M.L., Hobbs, D.G., Morant, S.V. & Hill, V.A., 1978. *J. Dairy Res.*, **45**, 413.
60. Ohmiya, K., Fukami, K., Shimizu, S. & Gekko, K., 1987. *J. Food Sci.*, **52**, 84.
61. Niki, R. & Arima, S., 1984. *Jpn. J. Zootech. Sci.*, **55**, 409.
62. Johnston, D.E. & Murphy, R.J., 1984. *Milchwissenschaft*, **39**, 585.
63. Yun, S.E., Ohmiya, K. & Shimizu, S., 1982. *Agric. Biol. Chem.*, **46**, 443.
64. Casiraghi, E.M., Peri, C. & Piazza, L., 1987. *Milchwissenschaft*, **42**, 232.
65. Creamer, L.K., Zoerb, H.F., Olson, N.F. & Richardson, T., 1982. *J. Dairy Sci.*, **65**, 902.
66. Green, M.L., 1987. *J. Dairy Res.*, **54**, 303.
67. Knoop, A.M., 1977. *Deutsche Milchwirtschaft*, **28**, 1154.
68. Glaser, J., Carroad, P.A. & Dunkley, W.L., 1980. *J. Dairy Sci.*, **63**, 37.
69. Blanc, B., 1983. In *Progress in Food Engineering,* ed. C. Cantarelli & C. Peri. Forster-Verlag AG, Kusnacht, Switzerland, pp. 667–679.
70. Walstra, P., van Dijk, H.J.M. & Guerts, T.J., 1985. *Neth. Milk Dairy J.*, **39**, 209.
71. Knoop, A.M. & Peters, K.H., 1978. *Proc. 20th Intern. Dairy Congr., Paris,* p. 808.
72. Lelievre, J., Kelso, E.A. & Stewart, D.B., 1978. *Proc. 20th Intern. Dairy Congr., Paris,* p. 760.
73. Marshall, R.J., 1986. *J. Dairy Res.*, **53**, 313.
74. Pearse, M.J., Linklater, P.M., Hall, R.J. & Mackinlay, A.G., 1985. *J. Dairy Res.*, **52**, 159.
75. Beeby, R., Hill, R.D. & Snow, N.J., 1971. In *Milk Proteins*, Vol. 2, ed. H.A. McKenzie. Academic Press, New York, p. 421.
76. Wallace, G.M. & Aiyar, K.R., 1970. *Proc. 18th Intern. Dairy Congr., Sydney*, Vol. 1E, p. 48.
77. Omar, M.M., 1984. *Food Chem.*, **15**, 19.
78. Geurts, T.J., Walstra, P. & Mulder, H., 1974. *Neth. Milk Dairy J.*, **28**, 102.
79. Guinee, T.P. & Fox, P.F., 1983. *J. Dairy Res.*, **50**, 511.

80. Eino, M.F., Biggs, D.A., Irvine, D.M. & Stanley, D.W., 1976. *J. Dairy Res.*, **43**, 113.
81. Eino, M.F., Biggs, D.A., Irvine, D.M. & Stanley, D.W., 1979. *Canad. Inst. Food Sci. Technol. J.*, **12**, 149.
82. Green, M.L., Glover, F.A., Scurlock, E.M.W., Marshall, R.J. & Hatfield, D.S., 1981. *J. Dairy Res.*, **48**, 333.
83. Sone, T., Dosako, S. & Kimura, T., 1984. *Snow Brand Milk Prod. Co. Rep. Res. Lab.*, **79**, 371.
84. Knoop, A.M. & Peters, K.H., 1975. *Kiel. Milch. Forsch.*, **27**, 227.
85. Thomas, I.L. & McCorkle, K.H., 1971. *J. Colloid Interface Sci.*, **36**, 110.
86. Lawrence, R.C., Gilles, J. & Creamer, L.K., 1983. *N.Z.J. Dairy Sci. Technol.*, **18**, 175.
87. Knoop. A.M. & Buchheim, W., 1980. *Milchwissenschaft*, **35**, 482.
88. Crabtree, R.M., 1984. British Society of Animal Production, Occasional Publication No. 9, p. 35.
89. Mulholland, J.R., 1984. British Society of Animal Production, Occasional Publication No. 9, p. 27.
90. Rogers, G.L. & Stewart, J.A., 1982. *Aust. J. Dairy Technol.*, **37**, 26.
91. Grandison, A.S., Ford, G.D., Millard, D. & Owen, A.J., 1984. *J. Dairy Res.*, **51**, 407.
92. Ng-Kwai-Hang, K.F., Hayes, J.F., Moxley, J.E. & Monardes, H.G., 1982. *J. Dairy Sci.*, **65**, 1993.
93. Phelan, J.A., O'Keeffe, A.M. & Keogh, M.K., 1982. *Irish J. Food Sci. Technol.*, **6**, 1.
94. Keogh, M.K., Kelly, P.M., O'Keeffe, A.M. & Phelan, J.A., 1982. *Irish J. Food Sci. Technol.*, **6**, 13.
95. Donnelly, W.J. & Barry, J.G., 1983. *J. Dairy Res.*, **50**, 433.
96. Grandison, A.S., Ford, G.D., Owen, A.J. & Millard, D.J., 1984. *J. Dairy Res.*, **51**, 69.
97. Thomas, P.C., 1984. British Society of Animal Production, Occasional Publication No. 9, p. 53.
98. Storry, J.E., 1981. In *Recent Advances in Animal Nutrition*, ed. W. Haresign. Butterworth, London, p. 3.
99. Sutton, J.D., 1984. British Society of Animal Production, Occasional Publication No. 9, p. 43.
100. Dodd, F.H., 1984. British Society of Animal Production, Occasional Publication No. 9, p. 77.
101. Zannoni, M., 1984. *Industria del Latte*, **20**, 99.
102. Kitchen, B.J., 1981. *J. Dairy Res.*, **48**, 167.
103. Munro, G.L., Grieve, P.A. & Kitchen, B.J., 1984. *Aust. J. Dairy Technol.*, **39**, 7.
104. Ashworth, U.S., Forster, T.L. & Luedecke, L.O., 1967. *J. Dairy Sci.*, **50**, 1078.
105. Renner, E., 1975. *IDF Bull.*, **85**, 53.
106. Needs, E.C. & Anderson, M., 1984. *J. Dairy Res.*, **51**, 239.
107. Randolph, H.E. & Erwin, R.E., 1974. *J Dairy Sci.*, **57**, 865.
108. Sharma, K.K. & Randolph, H.E., 1974. *J. Dairy Sci.*, **57**, 19.
109. Ali, A.E., Andrews, A.T. & Cheeseman, G.C., 1980. *J. Dairy Res.*, **47**, 393.
110. Bogin, E. & Ziv, G., 1973. *Cornell Veterinarian*, **63**, 666.
111. Singh, L.N. & Ganguli, N.C., 1975. *Milchwissenschaft*, **30**, 17.
112. Green, M.L., Scott, K.J., Anderson, M., Griffin, M.C.A. & Glover, F.A., 1984. *J. Dairy Res.*, **51**, 267.
113. Brule, G., Maubois, J.L. & Fauquant, J., 1974. *Lait*, **54**, 600.
114. Salter, D.N., Scott, K.J., Slade, H. & Andrews, P., 1981. *Biochem. J.*, **193**, 469.
115. Sutherland, B.J. & Jameson, G.W., 1981. *Aust. J. Dairy Technol.*, **36**, 136.
116. Law, B.A., 1979. *J. Dairy Res.*, **46**, 573.

117. Ali, A.E., Andrews, A.T. & Cheeseman, G.C., 1980. *J. Dairy Res.*, **47**, 371.
118. Ali, A.E., 1979. PhD thesis. University of Reading, England.
119. Ford, G.D. & Grandison, A.S., 1986. *J. Dairy Res.*, **52**, 129.
120. Farah, Z., 1979. *Milchwissenschaft*, **34**, 484.
121. Parry, R.M., 1974. In *Fundamentals of Dairy Chemistry*, ed. B.H. Webb, A.H. Johnson & J.A. Alford. AVI Publishing Co. Inc., Westport, p. 603.
122. Walstra, P. & Jenness, R., 1984. *Dairy Chemistry and Physics*, John Wiley, New York, p. 266.
123. Storry, J.R., Grandison, A.S., Millard, D., Owen, A.J. & Ford, G.D., 1983. *J. Dairy Res.*, **50**, 215.
124. Rao, R.V., Chopra, V.C., Stephen, J. & Bhalerao, V.R., 1964. *J. Food Sci. Technol. (Mysore)*, **1**, 19.
125. Jauranm, B.T., Vajayayalakshmi, B.T. & Nair, P.G., 1980. *Indian J. Dairy Sci.*, **33**, 17.
126. Tervala, H.L., Antila, V., Syvajarvi, J. & Lindstrom, U.B., 1983. *Meijeritieteellinen Aikakauskirja*, **41**, 24.
127. Howells, J.C., 1982. *Dairy Ind. Int.*, **47**(3), 13, 15, 17, 19, 22.
128. Banks, J.M., Muir, D.D. & Tamime, A.Y., 1984. *Dairy Ind. Int.*, **49**(4), 14.
129. Davis, J.G., 1965. *Cheese*, Vol 1. J. & A. Churchill Ltd, London.
130. Callanan, T. & Lewis, K., 1983. In *Proceedings of IDF Symposium: Physico-Chemical aspects of dehydrated protein-rich milk products.* Helsingor, Denmark, p. 160.
131. Custer, E.W., 1979. *J. Dairy Sci.*, **62**(Suppl. 1), 48.
132. Wilson, P.N. & Lawrence, A.B., 1984. British Society of Animal Production, Occasional Publication No. 9, p. 95.
133. Kindstedt, P.S., Duthie, A.H. & Nilson, K.M., 1984. *Cultured Dairy Prod. J.*, **19**(1), 20, 23.
134. Feagan, J.T., Bailey, L.F., Hehir, A.F., McLean, D.M. & Ellis, N.J.S., 1972. *Aust. J. Dairy Technol.*, **27**, 129.
135. Morini, D., Losi, G., Castagnetti, G.B. & Mariani, P., 1979. *Scienza e Tecnica Lattiero-Casearia*, **30**, 243.
136. Schaar, J., 1984. *J. Dairy Res.*, **51**, 397.
137. Jakob, E. & Puhan, Z., 1986. *Milchwissenschaft*, **107**, 833.
138. Schaar, J., 1985. *J. Dairy Res.*, **52**, 429.
139. Mariani, P., Losi, G., Russo, V., Castagnetti, G.B., Grazia, L., Morini, D. & Fossa, E., 1976. *Scienza e Tecnica Lattiero-Casearia*, **28**, 208.
140. El-Negoumy, A.M., 1972. *J. Dairy Res.*, **39**, 373.
141. Marziali, A.S. & Ng-Kwai-Hang, K.F., 1986. *J. Dairy Sci.*, **69**, 1793.
142. Sadler, A.M., Kiddy, C.A., McGann, E. & Mattingly, W.A., 1968. *J. Dairy Sci.*, **51**, 28.
143. Grandison, A.S., Ford, G.D., Millard, D. & Anderson, M.J., 1985. *J. Dairy Res.*, **52**, 41.
144. Remeuf, F. & Lenoir, J., 1985. *Revue Laitière Française*, **446**, 32.
145. McLean, D.M., Graham, E.R.B. & Ponzoni, R.W., 1984. *J. Dairy Res.*, **51**, 531.
146. Abdel-Galil, H. & Nassib, T.A., 1980. *Assiut Vet. Med. J.*, **7**, 149.
147. Butkus, K.D., Butkene, V.P. & Potsyute, R.Y., 1973. *Prikladnaya Biokhimiya i Mikrobiologiya*, **9**, 473.
148. Politis, I. & Ng-Kwai-Hang, K.F., 1988. *J. Dairy Sci.*, **71**, 1740.
149. Erwin, R.E., Hampton, O. & Randolph, H.E., 1972. *J. Dairy Sci.*, **55**, 298.
150. O'Leary, J. & Leavitt, B., 1982. *Kentucky Agric. Exp. Stn. Prog. Rep.*, No. 264, p. 25.
151. Grandison, A.S. & Ford, G.D., 1986. *J. Dairy Res.*, **53**, 645.
152. Kosikowski, F.V. & Mistry, V.V., 1988. *Milchwissenschaft*, **43**, 27.
153. White, J.C.D. & Davies, D.T., 1958. *J. Dairy Res.*, **25**, 267.
154. McDowell, A.K.R., Pearce, K.N. & Creamer, L.K., 1969. *N.Z.J. Dairy Sci. Technol.*, **4**, 166.

155. O'Keeffe, A.M., Phelan, J.A., Keogh, K. & Kelly, P., 1982. *Irish J. Food Sci. Technol.*, **6**, 39.
156. O'Keeffe, A.M., 1984. *Irish J. Food Sci. Technol.*, **8**, 27.
157. Donnelly, W.J., Barry, J.G. & Buchheim, W., 1984. *Irish J. Food Sci. Technol.*, **8**, 121.
158. Zienkiewicz-Skulmowska, T., Michalakowa, W., Michalak, W., Sinda, H., Jasinska, L. & Goszczynski, J., 1978. *Prace i Materially Zootechniczne*, **16**, 85.
159. Chapman, H.R. & Burnett, J., 1972. *Dairy Ind.*, **37**, 207.
160. Bartsch, B.D., Graham, E.R.B. & McLean, D.M., 1979. *Aust. J. Agric.*, **30**, 191.
161. Fossa, E., Pecorari, M. & Mariani, P., 1984. *Industria del Latte*, **20**, 87.
162. Olson, N.F., 1977. *Dairy Ind. Int.*, **44**(4), 14.
163. Bynum, D.G. & Olson, N.F., 1982. *J. Dairy Sci.*, **65**, 2281.
164. Banks, J.M. & Muir, D.D., 1984. *Dairy Ind. Int.*, **49**(9), 17.
165. Ahmad, N., 1978. *Aust. J. Dairy Technol.*, **33**, 50.
166. El-Shibiny, S. & Abd-El-Salam, M.H., 1980. *Egyptian J. Dairy Sci.*, **8**, 35.
167. Shalabi, S.I. & Fox, P.F., 1982. *J. Dairy Res.*, **49**, 153.
168. Holt, C., 1982. *J. Dairy Res.*, **49**, 29.
169. Chaplin, L.C., 1984. *J. Dairy Res.*, **51**, 251.
170. Holt, C., Hasnain, S.S. & Hukins, D.W.L., 1982. *Biochim. Biophys. Acta*, **719**, 299.
171. Lyster, R.L.J., 1981. *J. Dairy Res.*, **48**, 85.
172. Holt, C., Dalgleish, D.G. & Jenness, R., 1981. *Anal. Biochem.*, **113**, 154.
173. Green, M.L., 1982. *J. Dairy Res.*, **49**, 87.
174. Auberger, B., Lenoir, J. & Remeuf, F., 1988. *Technique Laitière & Marketing*, **1030**, 14.
175. Desnouveaux, R., Montigny, H., Trent, J.H., Schockmel, L. & Biju-Duval, R., 1988. *Technique Laitière & Marketing*, **1030**, 17.
176. Kindstedt, P.S., Rippe, J.K. & Pell, A.N., 1988. *J. Dairy Sci.,* **71**(Suppl. 1), 96.
177. Piva, G., Masoero, F. & Lazzari, A., 1989. In *Use of Somatotropin in Livestock Production*, ed. K. Sejrsen, M. Vestegard & A. Neimann-Sorensen. Elsevier Applied Science Publishers, London, p. 307.
178. Reimerdes, E.H., Perez, S.J. & Ringqvist, B.M., 1977. *Milchwissenschaft*, **32**, 154.
179. Aylward, E.B., O'Leary, J. & Langlois, B.E., 1980. *J. Dairy Sci.*, **63**, 1819.
180. Yan, L., Langlois, B.E., O'Leary, J. & Hicks, C., 1983. *Milchwissenschaft*, **38**, 715.
181. O'Leary, J., Hicks, C.L., Aylward, E.B. & Langlois, B.E., 1983. *Proc. 6th Int. Congr. of Food Sci. and Technol.*, Vol. 1, ed. J.V. McLoughlin & B.M. McKenna. Boole Press Ltd, Dublin, p. 150.
182. Burlingame-Frey, J.P. & Marth, E.H., 1984. *J. Food Protect.*, **47**, 16.
183. Dzurec, D.J. & Zall, R.R., 1982. *J. Dairy Sci.*, **65**, 2296.
184. Banks, J., 1990. *J. Soc. Dairy Technol.,* **43**, 35.
185. Bochtler, K., 1982. *North Eur. Dairy J.*, **48**, 127.
186. King, J.S. & Mabbitt, L.A., 1982. *J. Dairy Res.*, **49**, 439.
187. Mayes, J.J. & Sutherland, B.J., 1984. *Aust. J. Dairy Technol.*, **39**, 69.
188. Humbert, G. & Alais, C., 1975. *Rev. Lait Fr.*, **337**, 793.
189. Humbert, G. & Alais, C., 1976. *Rev. Lait Fr.*, **344**, 407.
190. Schafer, H.W. & Olson, N.F., 1975. *J. Dairy Sci.*, **58**, 494.
191. Banks, J.M., 1988. *J. Soc. Dairy Technol.*, **41**, 37.
192. Creamer, L.K., Lawrence, R.C. & Gilles, J., 1985. *N.Z.J. Dairy Sci. Technol.*, **20**, 185.
193. Poznanski, S., Smietana, Z., Jakubowski, J. & Rymaszewski, J., 1979. *Acta Alimentaria Polonica*, **5**, 125.
194. Banks, J.M. & Muir, D.D., 1985. *J. Soc. Dairy Technol.*, **38**, 27.
195. Brown, R.J. & Ernstrom, C.A., 1982. *J. Dairy Sci.*, **65**, 2391.
196. Baky, A.A.A., Elfak, A.M., El-Ela, W.M.A. & Farag, A.A., 1981. *Dairy Ind. Int.*, **46**(9), 29.
197. Robson, E.W. & Dalgleish, D.G., 1984. *J. Dairy Res.*, **51**, 417.

198. Emmons, D.B., Lister, E.E., Beckett, D.C. & Jenkins, K.J., 1980. *J. Dairy Sci.*, **63**, 417.
199. Knoop, A.M. & Peters, K.H., 1972. *Milchwissenschaft*, **27**, 153.
200. Jameson, G.W., 1983. *CSIRO Fd. Res. Q.*, **43**, 57.
201. Dalgleish, D.G., 1984. *J. Dairy Res.*, **51**, 425.
202. Dalgleish, D.G., 1980. *J. Dairy Res.*, **47**, 231.
203. Culioli, J. & Sherman, P., 1978. *J. Texture Studies*, **9**, 257.
204. Green, M.L., 1985. *J. Dairy Res.*, **52**, 555.
205. Garnot, P. & Corre, C., 1980. *J. Dairy Res.*, **47**, 103.
206. van Leeuwen, H.J., 1982. *Proc. 21st Intern. Dairy Congr. Moscow*, Vol. 1(1), p. 457.
207. Dalgleish, D.G., 1981. *J. Dairy Res.*, **48**, 65.
208. Prokopek, D., Knoop, A.M. & Buchheim, W., 1976. *Kieler Milch. Forschung.*, **28**, 245.
209. Rao, D.V. & Renner, E., 1988. *Milchwissenschaft*, **43**, 708.
210. Rao, D.V. & Renner, E., 1989. *Milchwissenschaft*, **44**, 351.
211. Green, M.L., 1990. *J. Dairy Res.*, **57**, 549.
212. Green, M.L., 1990. *J. Dairy Res.*, **57**, 559.
213. Barbano, D.M. & Bynum, D.G., 1984. *J. Dairy Sci.*, **67**, 2839.
214. Bynum, D.G. & Barbano, D.M., 1985. *J. Dairy Sci.*, **68**, 1.
215. Peters, K.-H. & Knoop. A.M., 1975. *Milchwissenschaft*, **30**, 205.
216. Omar, M.M. & Buchheim, W., 1983. *Food Microstructure*, **2**, 43.
217. Gilles, J. & Lawrence, R.C., 1981. *N.Z.J. Dairy Sci. Technol.*, **16**, 1.
218. Ehsani, R., Bennasar, M. & Tarodo de la Fuente, B., 1982. *Lait*, **62**, 276.
219. Abdel Baky, A.A., Abo El Ella, W.M., Aly, M.E. & Fox, P.F., 1987. *Food Chem.*, **26**, 175.
220. Davis, J .G., 1980. *Dairy Ind. Int.*, **45**(10), 7.
221. Ali, M., 1984. PhD thesis, University of Reading, UK.
222. McIlveen, H. & Strugnell, C.K., 1990. *Dairy Ind. Int.*, **55**(11), 17.
223. Banks, J.M., Muir, D.D. & Tamime, A.Y., 1984. *J. Soc. Dairy Technol.*, **37**, 83.
224. de Boer, R. & Nooy, P.F.C., 1980. *North Eur. Dairy J.*, **46**, 52.
225. McGregor, J.U. & White, C.H., 1990. *J. Dairy Sci.*, **73**, 314.
226. McGregor, J.U. & White, C.H., 1990. *J. Dairy Sci.*, **73**, 571.
227. El-Neshawy, A.A., Abdel Baky, A.A., Rabie, A.M. & Ashour, M.M., 1986. *Food Chem.*, **22**, 123.
228. Green, M.L. & Marshall, R.J., 1977. *J. Dairy Res.*, **44**, 521.
229. Marshall, R.J. & Green, M.L., 1980. *J. Dairy Res.*, **47**, 359.
230. Di Gregorio, F. & Sisto, R., 1981. *J. Dairy Res.*, **48**, 267.
231. Lee, Y.H. & Marshall, R.T., 1979. *J. Dairy Sci.*, **62**, 1051.
232. Lee, Y.H. & Marshall, R.T., 1984. *J. Dairy Sci.*, **67**, 263.
233. Mohamed, M.O. & Morris, H.A., 1988. *J. Food Sci.*, **53**, 788.
234. Mohamed, M.O., May, A.T. & Morris, H.A., 1988. *J. Food Sci.*, **53**, 798.
235. Metwalli, N.H., Shalabi, S.I., Zahran, A.S. & El-Demerdash, O., 1982. *J. Food Technol.*, **17**, 297.
236. Olsen, R.L., 1989. *J. Dairy Sci.*, **72**, 1695.
237. Kirby, C.J., Brooker, B.E. & Law, B.A., 1987. *Int. J. Food. Sci. Technol.*, **22**, 355.
238. Bishop, J.R., Bodine, A.B. & Janzen, J.J., 1983. *Cult. Dairy Prod. J.*, **18**(3), 14.
239. Ramet, J.P., El-Majda, E. & Weber, F., 1983. *J. Texture Studies*, **14**, 11.
240. Gouda, A., El-Zayat, A. & El-Shabrawy, S.A., 1985. *Deutsche Lebensm. – Rundschau*, **81**, 216.
241. McMahon, D.J. & Brown, R.J., 1985. *J. Dairy Sci.*, **68**, 628.
242. Gouda, A. & El-Shabrawy, A.A., 1987. *Chem. Mikrobiol. Technol. Lebensm.*, **10**, 129.
243. Gavaric, D.Dj., Caric, M. & Kalab, M., 1989. *Food Microstructure*, **8**, 53.
244. Knoop, A.M., 1976. *Deutsche Molkerei-Z.*, **97**, 1092, 1167.
245. Riddell-Lawrence, S. & Hicks, C.L., 1989. *J. Dairy Sci.*, **72**, 313.

246. Mayes, J.J. & Sutherland, B.J., 1989. *Aust. J. Dairy Technol.*, **44**, 47.
247. Hall, D.M. & Creamer, L.K., 1972. *N.Z.J. Dairy Sci. Technol.*, **7**, 95.
248. Kalab, M., 1977. *Milchwissenschaft*, **32**, 449.
249. Hansen, R., 1984. *North Eur. Dairy J.*, **50**, 196.
250. Stanley, D.W. & Emmons, D.B., 1977. *Canad. Inst. Food Sci. Technol. J.*, **10**, 78.
251. De Jong, L., 1978. *Neth. Milk Dairy J.*, **32**, 1, 15.
252. Omar, M.M., 1984. *Food Chem.*, **15**, 19.
253. De Jong, L., 1977. *Neth. Milk Dairy J.*, **31**, 314.
254. Fitzgerald, E. & Buckley, J., 1985. *J. Dairy Sci.*, **68**, 3127.
255. Jameson, G.W., 1987. *Food Technol. Aust.*, **39**, 99.
256. Knoop, A.M. & Peters, K.H., 1971. *Milchwissenschaft*, **26**, 193.
257. Le Graet, Y., Lepienne, A., Brule, G. & Ducruet, P., 1983. *Lait*, **63**, 317.
258. Caric, M., Gantar, M. & Kalab, M., 1985. *Food Microstructure*, **4**, 297.

5

The Syneresis of Curd

P. WALSTRA

Department of Food Science, Wageningen Agricultural University,
Wageningen, The Netherlands

1 INTRODUCTION

Gels formed from milk by renneting or acidification under quiescent conditions may subsequently show syneresis, i.e. expel liquid (whey), because the gel (curd) contracts. Under quiescent conditions, a rennet-induced gel may lose two thirds of its volume, and up to 90% or even more, if external pressure is applied. Often, syneresis is undesired, e.g. during storage of products like yoghurt, sour cream, cream cheese or quark; hence, it is useful to know under what conditions syneresis can be (largely) prevented. In making cheese from renneted or acidified milk, syneresis is an essential step. Consequently, it is useful to understand and quantitatively describe syneresis as a function of milk properties and process conditions, particularly when new methods or process steps are introduced in cheesemaking. This concerns several aspects:

— regulation of water content of the cheese implies controlling syneresis;
— the rate of syneresis affects the method of processing, and thereby the equipment and time needed, and the losses of fat and protein in the whey;
— rate of syneresis in relation to other changes (e.g. acidification, proteolysis, inactivation of rennet enzymes) affects cheese composition and properties;
— the way in which syneresis of curd grains proceeds may affect the propensity of the grains to fuse into a continuous mass during shaping and/or pressing;
— differences in syneresis throughout a mass of curd cause differences in composition of the cheese between loaves of one batch and between sites in one loaf;
— after a cheese loaf has been formed, it may still show syneresis and hence moisture loss.

(Note: throughout this chapter we will use the word 'moisture' for any liquid that may move through curd or cheese; it is thus generally an aqueous solution and not just water.)

Thus, the importance of syneresis is obvious. Accordingly, numerous research reports have been published, providing many important data on the influence of various factors on the rate, and sometimes on the end-point, of syneresis. However, the results vary considerably according to the conditions during the test method employed and are difficult to interpret. During the last 10 years, syneresis has been studied as a rate process, taking the basic phenomena occurring into account.[1-5] This has considerably enhanced our understanding and the results obtained in these studies will be treated before discussing investigations or a more practical nature.

2   GEL FORMATION AND PROPERTIES

The properties of casein micelles under various conditions and the gels formed by their aggregation will be discussed briefly in as far as is needed for understanding of the syneresis of these gels.

2.1 The Casein Micelle

We will largely proceed on the model of the casein micelle proposed before (see e.g. Refs 6–8). This implies that the casein molecules are present in small aggregates (submicelles, 15–20 nm in diameter), each containing different casein species and having a predominantly hydrophobic core and a predominantly hydrophilic outer layer. The submicelles are clustered into approximately spherical aggregates (micelles), for the most part 50–300 nm in diameter, and with interstitial moisture; they are probably kept together by the undissolved or colloidal calcium phosphate. There is, however, a dynamic equilibrium between casein and calcium phosphate in the micelles and in solution. Also, submicelles can go in and out of the micelles. Most of the κ-casein of the micelles is at the outside and the strongly hydrophilic C-terminal part of these molecules apparently sticks out from the micelle surface as a flexible chain that perpetually changes its conformation by Brownian motion,[7] thereby causing steric repulsion. The micelles are thus said to be 'hairy'. They also have a negative charge, causing electrostatic repulsion between them. Steric and electrostatic repulsion provide complete stability of the micelles against aggregation under physiological conditions.

The micelles may change considerably due to changes in their environment. At low temperatures, a considerable part of the casein, especially β-casein, goes into solution and additional 'hairs' of partly protruding β-casein molecules are presumably formed. A small part of the micellar calcium phosphate also goes into solution. The micelles attain a higher voluminosity (they swell). These changes are reversible, although it is not quite certain that the micelles exactly regain their native structure after cooling and rewarming. At high temperatures, the amount of micellar phosphate increases somewhat. At temperatures high enough for serum proteins to denature, association of denatured serum proteins with the micelles occurs, to an extent greatly dependent on pH: the lower the pH, the stronger the association.

Lowering the pH causes considerable change. Some trends are illustrated in Fig. 1; in as far as it concerns properties of a rennet gel, these are discussed later. The main change is that micellar calcium and phosphate go into solution, thereby loosening the bonds keeping the micelles together. This leads to dissolution of casein, especially at low temperature. At still lower pH, electrostatic bonds between positive and negative groups on the caseins keep the micelles together, and at the isoelectric pH these bonds are quite strong, again. In fact, the casein particles at this pH are very different from the micelles at physiological conditions, although their size distribution has not changed greatly.[9] It should further be noted that a lower pH in milk leads to a higher calcium ion activity, which also lowers the negative charge on the micelles. At pH 4·6 and not too low a

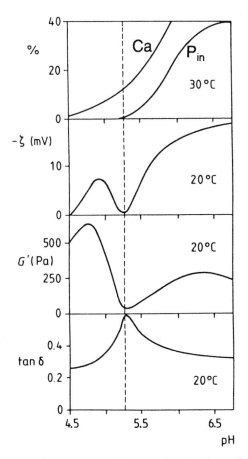

Fig. 1. The proportion of calcium (Ca) and inorganic phosphate (P_{in}) in and the electrokinetic potential (ζ) of casein micelles, as well as the dynamic shear modulus (G, frequency $1\,s^{-1}$) and the loss tangent (tan δ, frequency $0·01\,s^{-1}$) of rennet-induced skim milk gels, as a function of pH. From Ref. 8.

temperature, the casein particles aggregate: electrostatic repulsion is now absent and the κ-casein hairs, which provide steric repulsion, are lost also (they are presumably 'curled up').

Addition of calcium at a constant pH to milk causes the negative charge on the micelles to diminish and the amount of micellar phosphate to increase. This decreases the stability of the micelles and high levels of added calcium cause their aggregation.

2.2 Renneting

During the renneting of milk, the proteolytic enzymes in the rennet (mainly chymosin) hydrolyse the κ-casein molecules into para-κ-casein and soluble caseinomacropeptides (the C-terminal region), thereby largely removing the hairs and greatly reducing steric and electrostatic repulsions. The micelles can now approach one another closely and it is observed that they flocculate, i.e. remain close together. It is not yet certain what forces keep the paracasein micelles together since van der Waals attraction most probably is insufficient to achieve this, at least for the smaller micelles. Both electrostatic interactions (salt bridges) and hydrophobic bonding have been held responsible;[10] see further Ref. 11.

The kinetics of renneting are intricate since two reactions are involved. The enzymic reaction is essentially first order and the flocculation can be described, in principle, by Smoluchowski kinetics.[12] The caseinomacropeptide segments are removed from the micelles one by one (each micelle contains in the order of 10^3 κ-casein molecules and the number of micelles is roughly 10^2 times the number of chymosin molecules normally added to cheese milk). Consequently, the reactivity of the micelles, i.e. the probability that micelles that meet each other become flocculated, at first remains low and strongly increases as a greater proportion of the κ-casein has been hydrolysed (see also Fig. 2). The reactivity is roughly an inverse exponential function of the concentration of unhydrolysed κ-casein molecules on the micelles. As long as less than about 70% is hydro-

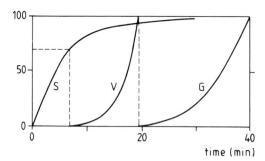

Fig. 2. Changes that occur in milk after adding rennet. Degree of hydrolysis of κ-casein (S), aggregation of paracasein micelles as measured by viscosity (V) and shear modulus (G) of the gel formed as a percentage of the values after 40 min, as a function of time.

lysed, the flocculation rate is virtually zero, at least at physiological pH and 30°C. If the pH is lowered, the enzymic reaction becomes much faster and, moreover, flocculation starts at a lower proportion of hydrolysed κ-casein molecules.[13] It appears that at low pH, the chymosin becomes adsorbed onto the micelles and this causes the hydrolysis of the κ-casein to be not quite random any more. Presumably, a chymosin molecule now often makes a 'bare' patch on the micelle before becoming desorbed and diffusing away, to find another (or possibly the same) micelle on which to act. At such a bare spot, the micelle is reactive. This implies that at a lower pH, flocculation starts at a stage where less κ-casein has been hydrolysed.

The reactivity of fully renneted micelles, i.e. those that are fully converted into paracasein micelles, depends little on pH, increases with Ca^{2+} concentration, decreases with increasing ionic strength (NaCl), and increases markedly with temperature, especially from 15 to 30°C.[14] Above 50°C, the flocculation rate becomes almost independent of temperature, being roughly equal to that predicted by Smoluchowski's equation for diffusion-controlled coagulation.[14] The temperature dependence is often taken as indicative of hydrophobic bonds being responsible for the reaction between the paracasein micelles,[10,14] but this is difficult to reconcile with the observation that the shear modulus of a rennet-induced skimmilk gel increases markedly with decreasing temperature.[15] Consequently, it is far more likely that with decreasing temperature only the activation free energy for flocculation increases, presumably because of protrusion of β-casein chains.

2.3 Gel Formation

After a while, flocculation leads to the formation of a gel (see Fig. 2). Microscopically, one can observe that aggregates are formed, at first irregular, but often somewhat threadlike; these grow to form large tenuous flocs, until they start to touch and form a continuous network.[1,16,17] Electron microscopy reveals (e.g. Refs 18–20) that the network can be described as consisting of strands of micelles, one to four micelles in thickness and some 10 micelles long, alternated by thicker nodes of micelles and leaving openings up to 10 μm in diameter.

The observation that the flocculation of particles leads to formation of a gel, as well as some properties of the gel formed, can readily be explained by the theory of 'fractal' aggregation.[21,22] This theory has been applied successfully to the flocculation of casein particles by Bremer and coworkers.[23–26] Assuming random aggregation of particles and of aggregates already formed (called cluster–cluster aggregation), computer simulations show the aggregates formed to be stochastic fractals, i.e. structures that are on average scale invariant at scales larger than that of the primary particles (radius a). The number of particles in an aggregate or floc is given by

$$N_p = (R/a)^D \tag{1}$$

where R is the radius of the floc and D the fractal dimensionality, which is always smaller than three. This implies that the floc becomes ever more tenuous as it becomes larger, and the computer simulations show indeed rarefied structures, consisting mainly of long irregular strands of particles, which are in most places only one particle thick. Equation (1) has been shown to hold remarkably well over a wide range of R and under many conditions, both in simulations and in experiments; colloidal interaction forces and geometrical constraints determine the value of D. The number of particles that could be present in a floc if the particles were closely packed, obviously is

$$N_a = (R/a)^3 \tag{2}$$

This implies that the average volume fraction of particles in a floc is given by

$$\phi_{floc} \equiv N_p/N_a = (R/a)^{D-3} \tag{3}$$

The average volume fraction of the flocs thus decreases during flocculation, and when it has reached the volume fraction of particles in the system ϕ (for paracasein micelles at 30°C about 0·09), the flocs fill the total space available and a gel has formed. It also follows that the average radius of the flocs at the moment of gelation is given by

$$R_{gel} = a\phi^{1/(D-3)} \tag{4}$$

In the above derivation, it has been implicitly assumed that flocculation proceeds undisturbed. But if the liquid is stirred during flocculation, gel formation may be hindered. Another disturbance may be appreciable sedimentation of the flocs occurring before a gel can be formed. The casein micelles in milk are small enough, and differ little enough in density from the milk serum, for sedimentation to be negligible. It may thus be assumed that under normal renneting conditions, gel formation occurs unhindered.

If equal-sized spherical particles flocculate in Brownian motion and if each encounter leads to lasting contact (so called diffusion–limited cluster–cluster aggregation), the fractal dimensionality turns out to be about 1·8. Several deviations from this simplest model, for instance a situation (as during renneting) in which only a certain small proportion of the encounters of particles leads to their lasting contact (so called chemically limited aggregation), or rearrangements occurring in the floc structure, lead to higher D values. Moreover, during gel formation, the flocs interpenetrate to some extent and this also causes a higher dimensionality. One type of change that certainly does occur in the flocs is a rearrangement of just-flocculated particles in such a way that each particle will touch more than two other particles; this leads, in principle, to strands of a thickness of about three particles rather than one.[22] This is in agreement with microscopical observations on casein gels.[26] Such a rearrangement does not detract from the fractal nature of the flocs or the gel formed from the flocs; it only means that the value of a in the above equations becomes somewhat higher (by a factor of about 3⅓) and that D becomes somewhat larger.

From different types of experiments on various casein gels it was found that $D = 2·3 \pm 0·1$.[23,25] Assuming the radius of paracasein micelles to be 55 nm and their volume fraction in milk to be 0·09, it is calculated that the average radius of the flocs at the onset of gelation is about 2·5 μm and that these flocs contain several thousand paracasein micelles. There is, however, considerable spread in these values within one gel, and the gel is thus fairly inhomogeneous: see Fig. 3a. The average pore size in the gel is of the order of R_{gel}, but some pores are larger. Average pore size is related to the permeability B in the equation of Darcy

$$v = (B/\eta)\nabla p \tag{5}$$

which relates the superficial velocity, v, of a liquid of viscosity η flowing through the gel due to a pressure gradient ∇p. The permeability of a 'fractal' gel is given by

$$B = \text{const.} \ a^2 \ \phi^{2/(D-3)} \tag{6}$$

The constant is not easily calculated; it is much smaller than unity. For $D = 2\cdot3$, the power of ϕ is about $-2\cdot9$ (in agreement with experiments), which implies that the permeability of the gel depends strongly on the initial ϕ and thus on casein concentration. A similar strong dependence on ϕ holds for some other properties and for the size of the flocs at the onset of flocculation.

Above, it has been tacitly assumed that skim milk is renneted. In the presence of fat globules, flocculation and gel formation proceed somewhat differently, but not greatly. The pores in the gel of paracasein micelles are roughly large enough (about 4 μm) and sufficient in number (about $2\cdot10^{16}$ m^{-3}) to accommodate the fat globules (average diameter: about 3·4 μm, number of globules larger than 1 μm: $3\cdot10^{15}$ m^{-3}).[7] Nevertheless, the pore size distribution in the gel is, of course, somewhat influenced by the presence of the fat globules, and most fat globules are entrapped in the gel.

2.4 Rheological Characteristics

The discussion will be based mainly on the extensive results of Zoon et al.[12,27–30] A rheological characteristic of a gel is its modulus, i.e. the ratio of the stress applied to it over the resulting strain (relative deformation). Mostly, the dynamic shear modulus, G, is determined (which implies that the deformation type is simple shear) as a function of the frequency of deformation, ω. Most gels are visco-elastic materials and these are characterized by two parameters. The storage modulus, G', is a measure of the true elastic property of the gel, the loss modulus, G'', of the viscous property; G''/ω can be seen as a viscosity. We further have $G^2 = G'^2 + G''^2$. In these dynamic measurements, the material is brought under an oscillating small strain, and G' and G'' can be determined separately, each as a function of ω; the time scale of the deformation is about ω^{-1}. Values of G' are shown in Fig. 1. The moduli were observed to depend generally on ω and to increase steeply with ϕ, in agreement with the theory of fractal gels.[25,31]

An important parameter is the loss tangent (tan $\delta \equiv G''/G'$), as it is a measure of the preponderance of viscous (or liquid-like) over elastic (or solid-like) properties of the gel. It is related to the relaxation of bonds in the gel during its deformation, and it therefore mostly increases with increasing time scale

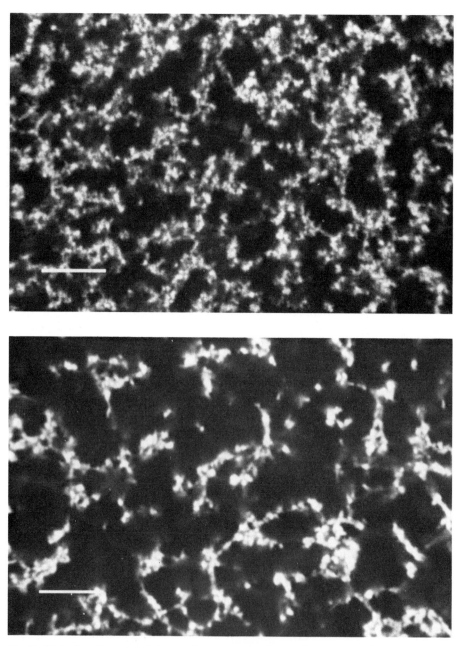

Fig. 3. Optical sections, made by confocal scanning laser microscopy in fluorescent mode, of rennet-induced skim milk gels, aged for 1 h (top) or 18 h at 30°C. The bars indicate 10 μm. From Bremer.[26]

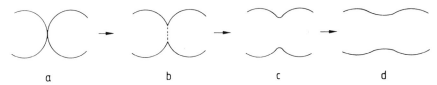

Fig. 4. Schematic picture of the change in conformation of flocculated paracasein micelles during ageing of the gel. From Ref. 32.

(decreasing oscillation frequency); this is because in general a greater proportion of the bonds that are under stress can relax when the time scale is longer. For rennet milk gels at physiological pH and 30°C, tan $\delta \approx 0.6$ at $\omega = 10^{-3}\,\mathrm{s}^{-1}$, i.e. at conditions as relevant for syneresis. This implies that a rennet gel has a significant viscous component in its rheological behaviour. In accordance with this, it is observed that its relaxation time, i.e. the time needed for the stress to decrease to 1/e of its initial value if a certain small deformation is applied to the material, is only of the order of 1 min. The loss tangent does not depend on casein concentration and is virtually independent of the age of the gel once formed.

The modulus of the gel strongly increases after it is formed: see Fig. 2. Presumably, the increase is due to two phenomena. One is that additional junctions are formed between casein particles, partly because there are strands of particles that are attached to the gel at only one end, partly because additional casein particles and small clusters thereof become incorporated into the gel. The latter situation will always occur to some extent during formation of a particle gel, but more strongly during normal renneting, since at the moment gel formation not all casein micelles have been fully transformed into paracasein micelles. Nevertheless, one would expect this phenomenon to be complete after, say, twice the time needed for gel formation, but the increase in modulus goes on for about 6 h.[27] The other phenomenon is illustrated in Fig. 4, which is derived from electron microscopical studies.[33] Any 'junction', by which is meant a contact region between two original micelles, must contain several bonds, and the number of bonds per junction increases on ageing. One may say that the micelles more or less fuse and after some hours the original particles making up the gel can no longer be distinguished. If no starter is added and the proteolytic enzymes of milk have been inactivated, the increase in modulus goes on for about 24 h.[27] The lower the temperature, the slower and the longer lasting is the increase in modulus. As mentioned, the increase in the number of bonds does not lead to a significant change in the loss tangent.

For deformations (in shear) larger than about 3%, the rheological behaviour of rennet milk gels becomes non-linear, and one can no longer speak of a modulus. In practice, the stresses applied are often too large for linear behaviour. Figure 5 shows what happens when a relatively large stress is applied.[30] After the instantaneous (elastic) response, the deformation soon becomes virtually viscous, i.e. $d\gamma/dt$ is constant. After some, often fairly long, time, the deformation rate increases and eventually becomes infinite: the gel fractures. Fracture does not mean falling

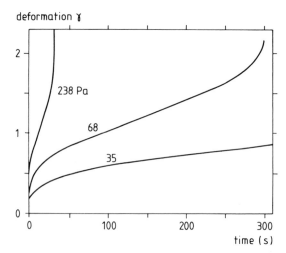

Fig. 5. Deformation in shear (γ) of a gel of renneted skim milk as a function of time, applying a constant stress. Temperature: 30°C, pH: 6·65, gel aged for 3·5 h. The stress applied is indicated near the curves. At 35 Pa, fracture occurred after 1350 s. From Zoon *et al.*[30]

into pieces, but rupture of the gel matrix only: the cleft formed immediately fills with whey. Presumably, local fracture occurs already in an early stage; the small cracks formed slowly increase in size and number, and coalesce until a fracture plane throughout the whole test piece has formed. This implies that long before macroscopic fracture, the gel structure has been altered markedly, which has been confirmed in loading–unloading experiments.[30] Note that the shear at fracture is very large; values between 1 and 5 have been obtained,[2,30] according to conditions. It is seen (Fig. 5) that a higher stress leads to smaller γ at fracture and to a much shorter time needed for fracture to occur. In other words, at a shorter time scale, the fracture stress is higher. Like the modulus, the fracture stress increases with ageing of the gel. The results of experiments at large deformations depend on the type of test applied (e.g. creep or dynamic), but the same trends are observed.

Temperature has a big effect on gel properties.[15,30] One should, however, distinguish between temperature of renneting and of measuring rheological properties. If renneting is at a lower temperature, gel formation is much slower and the modulus of the gel may consequently be smaller when measured at the same time, but this is not a true representation of the effect of temperature on gel properties. Lowering the temperature of a formed rennet gel generally causes a very brief decrease in modulus, but the latter subsequently starts to increase to reach a constant higher level after, say, 1 h. At $\omega = 10^{-3}$ s^{-1}, the storage modulus at 20°C is about 2·4 times that at 30°C. Figure 6 shows the effect of temperature on the loss tangent and it is seen that a rennet gel is much more solid-like at lower temperatures. The permeability tends to be higher at higher temperature; this will be considered later. At large deformations, a higher temperature causes a larger deformation and a lower stress at fracture.

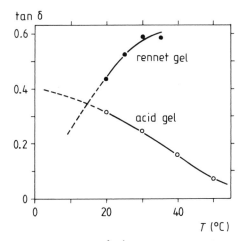

Fig. 6. Loss tangent (tan δ, frequency 10^{-3} s^{-1}) as a function of temperature for skim milk gels made by acidification[34] or renneting.[15]

Some effects of acidity[29,34] are shown in Fig. 1. Again, one should distinguish the pH of renneting from that at measurement, since renneting at a lower pH causes quicker gelation. Figure 1 gives results obtained several hours after renneting at the pH values indicated, and it is seen that the storage modulus at first increases with decreasing pH, to decrease again at still lower pH values; the loss tangent continues to increase. At large deformations, the effects of acidity are not great;[30] the fracture stress is somewhat higher for a lower pH, if determined at the same time scale.

2.5 Acid Gels

The casein particles at pH 4·6 are rather different from those at physiological pH, as is illustrated in Fig. 1. They are very prone to aggregation (except at low temperature), they contain no undissolved inorganic phosphate, and they have a (presumed) voluminosity at 30°C of about 3·4 ml/g^{-1} (leading to $\phi \approx 0.08$ in skim milk). They form a gel at temperatures above about 10°C and the gel properties have been extensively studied by Roefs *et al.*,[35,36] including those of gels made by combined action of acid and rennet.[34] The gels are in many respects quite comparable to rennet gels. They are also of a fractal nature and have roughly the same fractal dimensionality and thereby about the same dependence of modulus and of permeability on casein concentration.[23,24] The absolute value of the permeability is also roughly the same, as is the pore size distribution. The rheological properties are, however, rather different. A summary of properties is given in Table I; those relating to syneresis will be discussed later.

Two kinds of acid gel can be distinguished. Type 1 is made by adding acid to pH 4·6 in the cold (e.g. 2°C) and then quiescently warming the material, thereby preventing velocity gradients that would disturb gel formation. The voluminosity

TABLE I

Properties of Skim Milk Gels Obtained by Renneting (Aged for about 1 h) and by Acidification (Aged for 6–16 h). Acid gels are of Type 1 (Obtained by Cold Acidification and Subsequent Warming) or of Type 2 (Obtained by Slow Acidification with Glucono Delta Lactone at 30°C). Approximate Results at 30°C. From Various Sources.

Property	Rennet gel	Acid gel Type 1	Acid gel Type 2
pH	6·65	4·6	4·6
G' at $\omega = 0{\cdot}01$ rad.s^{-1} (Pa)	32	180	20
$\tan \delta$ at $\omega = 0{\cdot}01$ rad.s^{-1}	0·55	0·27	0·27
Fracture stress[a] (Pa)	10	100	100
Fracture strain[a] (—)	3·0	0·5	1·1
Permeability B (μm^2)	0·25	0·15	0·15
Fractal dimensionality[b]	2·23	2·39	2·36
dB/dt (nm^2.s^{-1})	20	<1	—
Initial syneresis rate[c]	15	<1	<1

[a] Loading time 10^3 s.
[b] From the relation between concentration and B.
[c] Arbitrary units.

of the casein particles at the onset of gel formation is now markedly higher than at the temperature where the measurements on the gels are made. This causes the particles to shrink after the gel has formed, which then leads to a straightening of the rather tortuous strands of particles originally formed. Type 2 gels are made as is common in practice, i.e. by slow and quiescent acidification at higher temperature, whether by bacterial action or by adding a substance that slowly hydrolyses to yield an acid (e.g. glucono delta lactone); in this way disturbance of gelation is also prevented. Here, the strands remain tortuous. It is seen in Table I that the difference leads to a marked difference in modulus, but that the fracture stress is about the same, whereas the fracture strain for Type 2 gels is much higher: the strands have to be straightened first by the deformation applied before the gel obtains roughly the same properties as a Type 1 gel. These differences have been explained, using fractal gel theory.[25] Note that the loss tangent does not differ, implying that the bonds involved are the same: only the geometrical structure differs.

Of greater importance are the differences between rennet and acid gels. Table I shows that acid gels are clearly more solid-like (smaller loss tangent), have a much higher fracture stress and a lower fracture strain than rennet-induced gels. In other words, they are stronger and more brittle. The fusion of the aggregated particles, as depicted in Fig. 4, is much slower in acid gels than in rennet-induced gels, as follows from electron microscopy.[35] In accordance with this, the modulus keeps increasing; it is roughly linear with log t, where t is the time after acidification.

The effect of temperature on the loss tangent is shown in Fig. 6; contrary to the trend for a rennet gel, tan δ increases with decreasing temperature, i.e. the gel becomes more liquid-like (At higher frequencies, i.e. shorter time scales, the dependence of tan δ on temperature is more like that of rennet gels.) The storage modulus is somewhat higher if the gel has been formed at a higher temperature, but clearly lower when the measurement is made at a higher temperature, as in rennet gels. The loss tangent is not dependent on temperature history.

Figure 1 suggests that pH $\approx 5\cdot2$ marks the border-line between acid and rennet gels; to be sure, this concerns, again, gels that have completely formed. In acid gels, most bonds presumably are protein–protein salt bridges, and these are most abundant at pH $4\cdot6$, implying that the number of bonds keeping the gel together decreases when increasing the pH. At physiological pH, the bonds may mostly be of the colloidal phosphate type, and lowering the pH leads to dissolution of colloidal phosphate, hence to a reduced number (or strength) of bonds. At pH $\approx 5\cdot2$ the resultant number of bonds presumably is at the minimum; see also Refs 8, 9 and 11. The relevance of these aspects for syneresis will be discussed below.

It may finally be mentioned that an interesting and somewhat different view on the role of calcium phosphate in micelles structure and stability, and on its dependence on pH, has been proposed by van Dijk.[37-39] This view may well provide a better understanding of the phenomena involved, but it does not materially alter the conclusions drawn in Sections 2 and 3.

3 MECHANISM OF SYNERESIS

Various mechanisms have been held responsible for syneresis; see an earlier review.[1] Summarizing, the following types of mechanisms were distinguished:

— A decrease in solvation or water binding of the material making up the gel. For a particle gel, this way of explaining syneresis does not appear suitable and there is no indication that an ongoing change in solvation is involved.
— Shrinkage of the building blocks of the gel, i.e. the (para)casein micelles in our case. This may happen when the pH is lowered or the temperature increased, but syneresis also occurs under constant conditions.
— Rearrangement of the network of (para)casein micelles. This is the main cause of syneresis.

The (para)casein particles in the gel form junctions with a limited number (mostly two to four) of others. (Strictly speaking, this is not true. As mentioned in Section 2.3, there is a rapid rearrangement into thicker strands, leading to a higher coordination number. However, one may use the same arguments by considering the 'particles' to be aggregates of, on average, three micelles.) However, the particles are expected to be reactive over their entire surface (or to contain numerous reactive sites smeared out over their surface) and in the network as described in Section 2, by far the greater part of the surface of each particle does not touch (form bonds with) another one. Rearrangement of the particles into a more compact network would thus increase the number of bonds and hence decrease

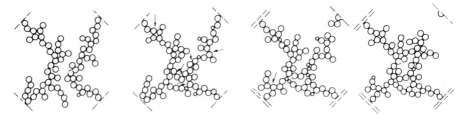

Fig. 7. Schematic representation of strands of paracasein micelles forming new cross links, leading to breaking of one of the strands (from Ref. 2).

the total free energy (the counteracting loss in conformational entropy is very small). But the particles cannot easily attain a more compact configuration because they are almost immobilized in the network. In other words, the network has to be deformed locally to form new junctions.

The mechanisms that may lead to rearrangement have also been reviewed[1] and it may be recalled here that long-range attractive forces cannot be responsible. Thermal motion of the strands may occasionally bring two particles in different strands close to each other so that a new junction is formed, especially shortly after renneting. This would lead to the build-up of a tensile stress in at least some of the strands. The fusion process illustrated in Fig. 4 may also cause such a stress to develop. As a result, strands may occasionally break, providing a possibility for the formation of more new junctions, thereby tending to make the network contract. These events are illustrated in Fig. 7. Even if no syneresis would follow, the changes mentioned would cause the strands of particles to become straightened, which implies that a normal rennet milk gel would be comparable to a Type 1 acid gel (see Section 2.5). This is indeed in agreement with the relation found between the modulus and the volume fraction of particles making up the gel.[25]

The propensity of the strands to break has been carefully studied by van Vliet et al.[40] They concluded that spontaneous breakage is possible if (1) the bonds in a junction can relax, and (2) the number of bonds in a junction is not too high. If the first condition is met, this is reflected in the loss tangent being fairly high at the time scale considered; the second is met if the strands are (locally) only one particle thick and the junction zones fairly small (small particles, little fusion). For normal (para)casein micelle gels, the critical loss tangent appears to be about 0·4, syneresis being stronger at a higher tan δ. To say it in other words: the activation free energy for breaking of bonds should be fairly low for syneresis to be possible. But also the activation free energy for bond formation should be fairly low, since otherwise no new junctions will be formed. It is at present not possible to calculate these energies, but it is fairly clear that both depend on temperature, pH and other variables. Finally, it is rather evident that breaking of strands occurs more readily when the strands are under a greater tensile stress.

An interesting, slightly different explanation of the origin of the syneresis pressure has been given by Bremer.[26] As mentioned, the build-up of the stress in the network is at least partly due to the fusion process illustrated in Fig. 4. This comes down to a rearrangement of sub-micelles, leading to a smaller contact area between paracaseinate and serum. It can be reasoned that this is to some extent analogous to a situation where two phases exist, with an interfacial free energy or interfacial tension between them. In such a case, this tension tries to lower the contact area, thereby leading to a shortening of strands in the network, if the rheological properties of the network material allow this. This may in some instances lead to local thinning and breaking of strands. Bremer calculated that the apparent interfacial tension would be of the order of $10^{-6}\,\text{N}\cdot\text{m}^{-1}$. It may be mentioned that this explanation is not really different from that given above, rather it is a different way of describing the same phenomena.

If the gel is formed undisturbed and completely sticks to the wall of the vessel in which it is formed (e.g. clean glass), it usually shows no syneresis, at least if the vessel is not too large and has vertical walls, and if the temperature is not too high (e.g. 30°C).[2] Apparently, the gel is now constrained and cannot skrink. Spontaneous syneresis is observed if the milk is renneted in a conical flask: presumably, the gel tears loose from the glass wall by gravity before it is fully set. Similarly, spontaneous syneresis may occur in a cylindrical glass if it is tilted slightly for a moment during setting. After the gel has become firmer, it can withstand a greater disturbance without exhibiting spontaneous syneresis. Usually, it does not show syneresis at the milk surface, either; presumably, the surface is covered by a thin lipid-rich layer which causes the contact angle between milk serum, air and the paracaseinate matrix (as measured in the serum) to be obtuse, so that capillary forces prevent the serum from leaving the matrix. As soon as the gel is cut or the surface (locally) wetted, syneresis occurs. This effect permits experimentally starting syneresis at any desired moment after a gel is formed.

These observations imply that in a constrained milk gel no syneresis occurs. However, the processes depicted in Fig. 7 will nevertheless occur; there is no reason to suppose they would not. This implies that at a local scale the gel network becomes more dense: this has been called microsyneresis.[1,2] At the same time, the network will become less dense at other sites; these changes are shown in Fig. 3. The surface-weighted average pore size will thus increase and it is indeed observed that the permeability of a constrained gel keeps increasing;[1,2] see Fig. 8. It may be argued that the rate of change of the permeability, dB/dt, is a measure of the tendency of the gel to exhibit syneresis.

Acid milk gels, if at a pH near 4·6 and at 30°C, show little or no syneresis. Table I shows that the permeability of an acid gel is somewhat lower than that of a rennet gel, but the difference is far too small to cause the large difference in syneresis behaviour. Neither is there a correlation with the modulus, but acid gels exhibit a much lower loss tangent (more solid-like behaviour) and a much higher fracture stress; because the configuration of the network is very similar, this implies that the strands in a rennet gel can break more readily than in an acid gel. It is also observed that an acid gel shows negligible dB/dt. It can be seen in Fig. 6 that the dependence of the loss tangent on temperature is very different in acid and rennet gels.

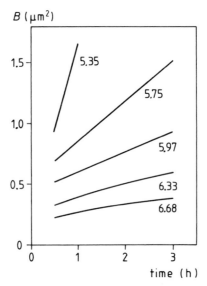

Fig. 8. Permeability of rennet-induced skim milk gels of various pH (indicated near the curves) as a function of time after renneting. Temperature 30°C. From van den Bijgaart.[5]

Until now only the inherent or endogenous tendency of a gel to show syneresis has been considered. Exerting a stress on the gel may be expected to speed up greatly the expulsion of whey, because of the increased pressure (see Eqn (7), below). Moreover, it may enhance syneresis by bringing strands of the network closer to each other and, perhaps more important, it will enhance breaking of strands, thereby providing a greater possibility for the number of junctions to increase. As was discussed above in relation to Fig. 5, deformation of the gel also causes local rupture of the network, thereby increasing its permeability.

4   ONE-DIMENSIONAL SYNERESIS AT CONSTANT CONDITIONS

In this section the detailed investigations by van Dijk and van den Bijgaart and coworkers[2-5] will be discussed. They studied horizontal slabs of renneted milk the top of which was moistened at a predetermined time after renneting, after which syneresis was followed by measuring the change in height h of the slab; whey could flow out only at the top. The milk has been brought to the desired pH and the apparatus was kept at a constant temperature. Examples of results are shown in Fig. 9. The diameter of the cylindrical slabs was much larger than their thickness (mostly 5 mm). In this way, one-dimensional syneresis under constant conditions could be determined. This is, of course, an oversimplification of the situation during actual curd making, but it allowed precise and unequivocal determination of syneresis under various conditions, providing insight into the processes occurring, and permitting the development of a workable mathematical model.[3]

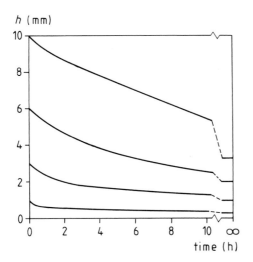

Fig. 9. The height of slabs of renneted skim milk of various initial height, h, as a function of time after initiation of syneresis. Temperature 30°C, pH ≈ 6·7. The values at infinite time are from extrapolation of log–log plots. From van Dijk.[2]

In parallel experiments, the permeability, B, and its change with time, dB/dt, were determined. Unless stated otherwise, the results pertain to renneted skim milk.

4.1 Modelling the Process

Syneresis comes down to the flow of liquid (whey) through the paracasein network and can thus be described by the equation of Darcy (eqn 5), which is conveniently written as:

$$v = (B/\eta) \cdot (p/l) \tag{7}$$

where v is the superficial velocity of the liquid in the direction of l, the distance over which the liquid has to flow. The pressure causing the syneresis can, in general, be written as:

$$p = p_s + p_g + p_e \tag{7a}$$

which terms are, respectively, the endogenous syneresis pressure, the pressure exerted by the network itself due to gravity, and any external pressure applied to the network. Results for $p_e = 0$ will be discussed first; note that p_g varies from zero at the top of the slab to a maximum of $gh\Delta\rho$ at the bottom, e.g. 1 Pa for a 1 cm slab (ρ = density).

Attempts to directly measure p_s failed: its value was too small. Only the order of magnitude could be estimated, and it was 1 Pa.[2] This is a very small pressure: it corresponds to the pressure exerted by a water 'column' of 0·1 mm and, as seen in Fig. 9, unaided syneresis is indeed very slow: it takes 7 h at 30°C for a 6 mm slab to be reduced to 3 mm. Since eqn (7) must hold, and since syneresis,

B, l and η can be measured, it is possible to indirectly determine p and thereby p_s, because p_g can also be calculated. The calculation is, however, very intricate, because:

— permeability increases with time (Fig. 8);
— permeability becomes smaller because of syneresis (see below);
— most likely, endogenous syneresis pressure also varies with ongoing syneresis;
— pressure due to gravity changes as well;
— the coordinates change with syneresis.

Consequently, most variables vary with time and location. Syneresis will start in the uppermost layer, thereby altering its permeability, etc., and progressively reaches deeper layers. A finite difference model was developed,[3] in which the slab was divided into parallel thin slices, to each of which eqn (7) and the equation of continuity were applied, to calculate the outflow of liquid in small time intervals. By inserting various values for p_s and comparing the computed results with the observed h as a function of time, the endogenous syneresis pressure could be derived.

By assuming p and B to be constant, which may be assumed to be the case at the very beginning of syneresis, an analytical solution can be found, which is mathematically equivalent to the solution of the diffusion equation. This would imply that h changes proportionally to $\sqrt{t}$ and directly yields the initial endogenous syneresis pressure, $p_{s,0}$. This proved not to be the case. The explanation probably is as follows.[5] An implicit assumption in the application of eqn (7) is that the network can without significant resistance comply with the outflow of whey. But the initial shrinkage rate of the outermost layer would then be very high (the proportionality with $\sqrt{t}$ even implies an infinite rate at $t = 0$) and that is clearly not possible. The extensional viscosity of the network will now determine the rate, implying that h changes linearly with t. In careful experiments, an initial proportionality with about $t^{3/4}$ was obtained. After a few minutes, the viscous resistance apparently becomes negligible (which is also in accordance with the observed stress relaxation time of a few minutes), but then B and probably p have already changed. Consequently, it was not easy to determine unequivocally the initial endogenous syneresis pressure, although, again, it turned out to be close to 1 Pa at 30°C and physiological pH.

There are further complications. Firstly, p_s is not constant. It changes with time after renneting, as is shown in Fig. 10. It should be realized that this figure gives the initial pressure, i.e., the abscissa gives the time when syneresis was started. It is seen that $p_{s,0}$ at first increases, which is logical because syneresis can occur only in a network and it takes some time for the network to develop. Later on, the pressure decreases slowly, which must be due to ongoing microsyneresis and to ongoing fusion of paracasein micelles (see Section 2.3 and Fig. 4). In a syneresing slab, the situation must be different, because after a certain shrinkage has occurred the gel cannot shrink any further. Several relations for the decrease in p_s with shrinkage were tried and they did not cause a great difference in the calculated results, provided that for a shrinkage to 30% of the initial volume ($i = 0.3$)

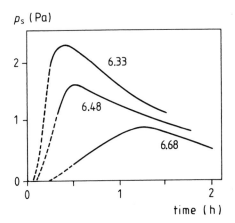

Fig. 10. The endogenous syneresis pressure (p_e) of rennet-induced skim milk gels as a function of the time elapsed after renneting when syneresis was initiated, at various pH (indicated near the curves). Temperature 30°C. The broken lines are assumed. From van den Bijgaart.[5]

$p_s = 0$ was predicted. A good agreement between observed and calculated syneresis could then be obtained, at least for a couple of hours after starting syneresis.

Secondly, the permeability should decrease with increasing shrinkage (besides increasing with time: Fig. 8). To estimate this dependence, B was determined in gels made from skim milk concentrated by ultrafiltration. However, a few experiments in which the permeability was determined in slabs of syneresed skim milk gels showed a different relation with concentration, especially at low i: see Fig. 11.

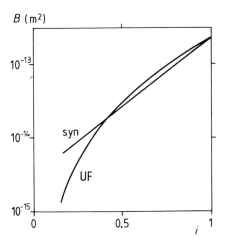

Fig. 11. The permeability (B) of skim milk gels made by renneting ultrafiltration retentates (UF) or by renneting skim milk and letting the gel synerese for a considerable time (syn). i is the volume of the gel relative to the volume of the skim milk. Approximate results. From van den Bijgaart.[5]

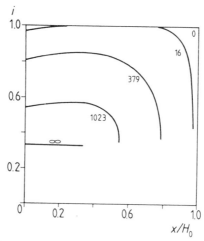

Fig. 12. Shrinkage profile of a slab of rennet curd at various times (indicated in minutes) after initiating syneresis. i = concentration factor, x/H_0 = relative distance from the bottom of the slab. Assumptions: thickness of slab: 10 mm; endogenous syneresis pressure: 1 Pa; syneresis stops at $i = \frac{1}{3}$. (Kindly computed by H.J.C.M. van den Bijgaart.)

It may be concluded that the syneresis pressure can be calculated from the Darcy equation, although not with great accuracy. Nevertheless, the fairly good agreement between observed and calculated overall syneresis lends confidence to the model. This implies that shrinkage profiles in a syneresing slab can be calculated, a result than cannot be obtained experimentally. Some examples are given in Fig. 12; the early decrease in i near the bottom of the slab is due to the effect of gravity, a situation that will mostly not occur in the early stage of curd making, because the curd particles float in the whey. The importance of the results is, however, that they show that very soon an outer layer of much higher concentration is formed, which has a much lower permeability and thus considerably impedes further syneresis.

4.2 Results

Some important results are given in Fig. 13. The variables, *temperature and pH*, affect tan δ (see Figs 1 and 6), dB/dt and $p_{s,0}$ in a similar way, in accordance with the ideas outlined in Section 3. All these trends would cause a faster syneresis at a higher temperature and a lower pH; this is indeed observed, but the change in syneresis is decidedly less than would be expected from a simple application of eqn (7). The reason for this discrepancy is that rapid syneresis also leads to the rapid formation of a highly shrunk outer layer, which implies that the permeability of that layer becomes very low, thereby slowing down further syneresis. In this way, the effect of any factor that strongly increases syneresis pressure is reduced, the more so because the endogenous syneresis pressure will eventually

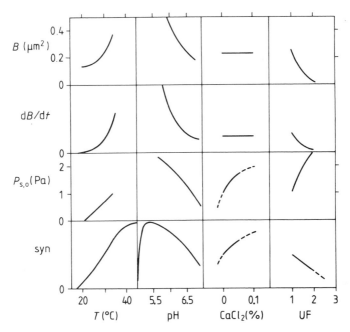

Fig. 13. Properties of rennet-induced skim milk gels. Permeability (B), rate of change of permeability (dB/dt), initial endogenous syneresis pressure ($p_{s,0}$), and approximate initial rate (syn, arbitrary scale) as a function of temperature (T), pH, added quantity of $CaCl_2$ and preconcentration by ultrafiltration (UF, degree of concentration). Mostly from van Dijk[2] and van den Bijgaart.[5]

relax (see Fig. 10). The close correlation between B and dB/dt is presumably due to the latter factor already causing an increase in B before the gel is firm enough to allow the estimation of B. Temperature has a very large effect: below 20°C, endogenous syneresis is virtually zero. This may be different at a low pH; Fig. 6 shows that for an acid gel tan δ increases with decreasing temperature, which suggests that syneresis is possible. Actual measurements at low pH and low temperature have not been made, but practical experience indicates that syneresis may indeed occur. Syneresis of acid gels (e.g. yoghurt) becomes much stronger if the pH drops still further, i.e. below about 4.

The relations for the effect of concentrating the milk by ultrafiltration are different. Naturally, B decreases with increasing concentration: the network becomes denser. The bonds remain of the same type, however, which is reflected in tan δ remaining constant.[27] Nevertheless, dB/dt decreases with increasing concentration, but this is due to B becoming smaller: the relative change in B, i.e. $d\ln B/dt$, increases somewhat with concentration. The initial endogenous syneresis pressure, $p_{s,0}$, increases with increasing concentration, presumably due to a more concentrated network having a higher concentration of strands that can be involved in syneresis. The overall result is that the rate of syneresis decreases with increasing concentration. The relative shrinkage rate—i.e. relative to one minus the volume fraction of paracasein particles in the gel—increases somewhat with increasing concentration.[5] It also appears that the rate of syneresis of a gel from preconcentrated milk (by UF) is higher than that of a gel of the same concentration but caused by syneresis; at least part of the explanation is presumably that in the latter case considerable relaxation of the internal stresses in the network has occurred, implying a lower syneresis pressure.

The effect of adding *calcium chloride* is somewhat puzzling: B and dB/dt are not affected (and neither is tan δ),[28] whereas $p_{s,0}$ and syneresis rate increase. Addition of $CaCl_2$ at a constant pH causes a slight increase in the Ca ion activity and a more important increase in the amount of micellar calcium phosphate.[7] Lowering the pH causes a marked increase in Ca ion activity and a decrease in micellar phosphate; it goes along with an increase in tan δ, dB/dt and $p_{s,0}$. For syneresis to occur, bonds (junctions) between paracasein micelles have to be broken as well as new bonds formed. Both reactions will have an activation free energy.[2] It may be so that adding $CaCl_2$ does not greatly affect the activation free energy for breakage, implying that microsyneresis will not be greatly affected, whereas it reduces the activation free energy for bond formation, going along with a higher endogenous syneresis pressure. This would be in accordance with the enhancement of the rate of flocculation of paracasein micelles due to the addition of $CaCl_2$.[41] It should be noticed, however, that the effect of adding $CaCl_2$ is rather variable and that the time elapsed between addition and doing the experiments also affects the results.[5] It would require painstaking investigations to settle these fine points.

Some *other variables* also (slightly) affect endogenous syneresis. The quantity of rennet added has very little effect, provided that the time elapsed after rennet addition has been sufficient to ensure almost complete hydrolysis of the κ-casein.

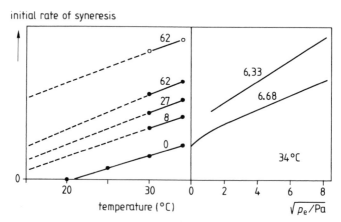

Fig. 14. Effects of temperature and external pressure (p_e, in Pa, indicated near the curves in the left-hand graph) on syneresis of renneted skim milk. pH was 6·68 (●) or 6·33 (○). From van den Bijgaart.[5]

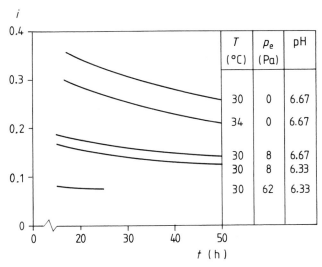

Fig. 15. Effects of temperature (T), external pressure (p_e) and pH on the shrinkage of renneted skim milk. Relative remaining volume (i) as a function of time (t) after renneting. Initial slab height 5 mm. From van den Bijgaart.[5]

Adding NaCl has very little effect, unless large quantities are added. Comparison of renneted milk with skim milk shows that the presence of fat globules causes a somewhat lower permeability; dB/dt is not affected, and syneresis is a little slower. As an example, after 5 h a slab had shrunk to 53% of its height, compared to 48% in the case of skim milk.[5]

Experiments on the influence of an *external pressure*, p_e, were performed by placing a porous disk on top of the synerising slab. Some results are shown in Fig. 14. It is seen that the effect is considerable and is about proportional to $\sqrt{p_e}$. The effect of external pressure cannot be seen as an amplification of syneresis: it is about additive to the endogenous syneresis. These results are in accordance with those of model calculations. Figure 14 also suggests that in the presence of an external pressure, the lowest temperature at which syneresis can occur is lower, the more so for a higher pressure. Although this has not been verified by experiments, the effect must exist at least to some extent.

Presumably, syneresis has an *endpoint*: eventually the system will be close-packed. Such an endpoint has, however, not been observed. Figure 15 gives some results up to 50 h syneresis. It is seen also that after a long time, the shrinkage is greater for a higher temperature, a lower pH and a higher external pressure. A higher fat content also causes somewhat less shrinkage after long times. To what extent these differences are due to variation in the parameters of the Darcy equation (mainly B and p) or to the resistance of the network to further compression (i.e. mainly its elongational viscosity at the stress applied), cannot be established easily.

5   SYNERESIS DURING CURD MAKING

After renneting has led to a gel of sufficient firmness, it is usually cut into pieces
to promote whey release. For most types of cheese, the mixture of curd and
whey is then stirred, often, part of the whey is removed, and it is fairly common
to increase the temperature of the mixture after some time (scalding or cooking),
which are all measures aimed at enhancing syneresis. Moreover, during this process of curd making, the pH decreases, again enhancing syneresis. An example
of the water content of the curd during the course of the process is given in
Fig. 16. Note how time, temperature, acidity and pressure affect the water content; the effect of pressure is seen when the curd is taken out of the whey
for moulding, by which action the pressure due to gravity increases by a factor
of about 30. All these effects are in qualitative agreement with the results of
Section 4.

In this section, the effect of several variables under conditions during actual
curd making, or conditions more or less mimicking these, will be considered.
This is because most published experiments were done in such a way, and methods
for estimating syneresis will be reviewed briefly. Some effects of milk composition and pretreatment will also be discussed.

It goes without saying that curd making is aimed at other things besides regulating the rate and extent of syneresis. The main aspect is that a higher moisture
content goes along with a higher sugar content, which, in turn, leads to a lower
pH; this can be modified by 'washing' the curd. A lower pH at the moment of

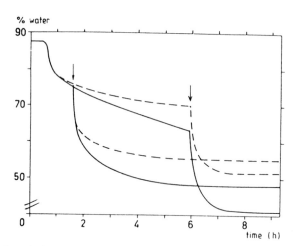

Fig. 16. Examples of changes in the water content of curd (determined by oven drying)
as a function of time after renneting. The gel was cut after 0·5 h and the curd and whey
mixture was continually stirred. At two moments (indicated by arrows), curd was removed from the whey and put into a cheese mould. Experiments with (———) and without (– – – –) added starter. Temperature of the whey was 32°C throughout, temperature in
the mould gradually fell to 20°C. (Recalculated from Refs 42 and 43).

separating curd and whey causes the cheese to contain less calcium phosphate. Other aspects are the limitation of the loss of curd fines, the inclusion and the activity of rennet in the cheese and, in some types, the killing of undesired micro-organisms (caused by scalding).

5.1 Methods for Estimating Syneresis

The ultimate result of syneresis is reflected in the water content of the cheese after pressing. Determining only this quantity yields, however, little understanding. It is much more interesting to follow syneresis during the curd making process, but it is not easy to do this unequivocally. The various methods and their pros and cons have been reviewed extensively[1] and only the salient points will be described here. The methods may be classified as follows:

1. Measuring the shrinkage of the curd, either the height of a slab (as discussed in Section 4) or the volume or mass of a slab or pieces of curd (in air or in whey). These methods are typically applied in laboratory experiments.
2. Determining the amount of whey expelled. This can be done in two ways:
   a. determination of the volume of whey drained off. The results strongly depend on conditions, especially the often imprecisely known external pressure. It also may be fairly uncertain how much interstitial whey is left between the curd grains. See also Section 6.
   b. determination of the degree of dilution of an added tracer. This method has an inherent uncertainty, inasmuch as the tracer may adhere onto or diffuse into the curd.
3. Determination of the dry matter content of curd pieces taken out of the whey. The main uncertainty is the unknown quantity of whey adhering to the curd particles; trying to remove the adhering whey may introduce the opposite error.
4. Determining the density of the curd grains by putting them in solutions of various density. This method is fairly crude, but is hardly biased if carefully executed.

All these methods can, of course, be executed under various conditions that affect syneresis—e.g. temperature, pH and effective pressure—and at various times after renneting or cutting.

Most authors have used method 2a, and methods 2b and 3 have also been fairly popular, especially in experiments involving stirring of the curd–whey mixture. Hardly ever have different methods been compared on the same curd. Figure 17 gives an example, and it is seen that the difference is considerable. It may be concluded that absolute values are hard to obtain and that most methods only provide trends. Even then, one has to be careful, since the method may not be linear. In relation to this it should be realized that at the beginning of syneresis a large amount of whey has to be removed for the moisture content of the curd to become appreciably lower, whereas at the end of the process the opposite is true; see Fig. 18.

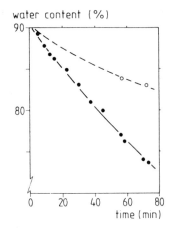

Fig. 17. Water content of curd from skim milk renneted and kept at 31°C as a function of time after cutting, determined from the concentration in the whey of added polyvinyl alcohol (●) and by oven drying of pieces of curd strained off (○). (From Ref. 43).

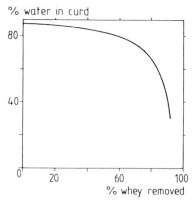

Fig. 18. Calculated relation between the water content of curd (from whole milk of 12·3% dry matter) and the quantity of whey (6·8% dry matter) expelled as % (w/w) of the original milk.

5.2 Rate Equations

As was discussed in Section 4.1, even in the simple case of one-dimensional syneresis under constant conditions, solution of eqn (7) in conjunction with the equation of continuity leads to complicated relations. This will be even more so for the situations considered here, where the geometric boundary conditions are more complicated and changing, and where the physico-chemical conditions affecting syneresis are not constant either. Nevertheless, some authors have tried to give simple analytical expressions for the process.

Kirchmeier[44] reported that the change in volume, V, of a piece of curd is, under 'constant conditions', given by

$$V = V_0 \exp(-Kt) \qquad (8)$$

where V_0 = original volume, t = time after starting syneresis and K would be a first order rate constant, linearly dependent on temperature. A similar relation, albeit with some 'extra' syneresis immediately after cutting, was observed by Marshall.[45] Apart from the lack of theoretical justification for eqn (8), it predicts that V approaches zero for very high t, which is clearly impossible.

Weber,[46] therefore, modified eqn (8) into

$$V = V_0 [0.15 + 0.85 \exp(-Kt)] \qquad (9)$$

where it was assumed that the curd eventually shrinks to 0.15 times its original volume (actually, Weber used mass rather than volume).

A further modification was made by Peri et al.[47] They introduced the final (relative) volume, V_∞, as a variable and obtained

$$V - V_\infty = [(V_0 - V_\infty) \exp(-Kt)] \qquad (10)$$

This would be a correct equation for a simple relaxation process where $1/K$ stands for the relaxation time. As we have seen, syneresis can certainly not be considered such a process; nevertheless, Peri et al. found a good agreement with their results, obtained under a fairly wide range of conditions. The good fit may have been due to eqn (10) containing two adjustable parameters.

Lawrence & Hill[48] found that the amount of whey expelled $(V_0 - V)$ from pieces of curd was proportional to $t^{1/2}$, and concluded that 'rate syneresis is substantially diffusion-controlled'. Such a conclusion had also been reached for syneresis in cross-linked polymer gels.[49] We have seen, however, that syneresis is pressure-controlled. Nevertheless, some other workers[50,51] also found an approximate proportionality with $t^{1/2}$ for rennet curd.

The results quoted above are incompatible: for $Kt \ll 1$, eqns (8–10) predict that $(V_0 - V)$ is proportional to t, not $t^{1/2}$; more generally, different authors find different relations. Neither can the results discussed in Section 4.2 be fitted to any of the relations given above. The only conclusion can be that, under constant conditions, the rate of syneresis $(-dV/dt)$ decreases as syneresis proceeds. This need not always be true for the relative rate of syneresis $(-d\ln V/dt)$, although this quantity also approaches zero eventually.

5.3 Effects of Cutting and Stirring

Cutting the renneted milk into pieces creates a free surface through which syneresis can occur: before cutting, the gel mostly sticks to the wall and its top surface does not show syneresis, unless it is wetted.[1] Modifying eqn (7) into

$$Q = Av = ABp/\eta l \qquad (11)$$

where Q = the volume flow rate of whey out of the curd and A is the surface area of the curd, it directly follows that cutting the gel into smaller pieces further

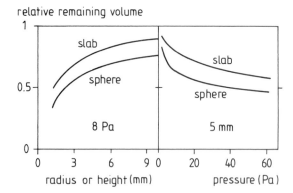

Fig. 19. Syneresis during 30 min compared in semi-infinite slabs and spheres of various height and radius, respectively, for various pressures. Calculated by van den Bijgaart.[5]

promotes (initial) syneresis rate. If the pieces have a size x, A is proportional to $1/x$ and l to x, hence Q to x^{-2}. Altogether, cutting greatly enhances syneresis and small pieces of curd shrink more than large ones. The latter implies that uneven cutting may cause local variation in moisture content and acidity in the fresh cheese.

In Section 4, a finite-difference method for calculation of one-dimensional syneresis from a slab was discussed. Van den Bijgaart[5] has used the same method to calculate syneresis from a sphere, i.e. in all radial directions. It turned out that for the same conditions, the initial syneresis rate was much higher—as would be expected—but that its time exponent (i.e. the slope when plotting log remaining volume against log time) was smaller, the more so for a higher pressure and a smaller radius. Some results are given in Fig. 19. It was observed that initial syneresis especially is much faster, and that for further shrinkage the differences become relatively less. These calculations involve several simplifications, of which the most important may be neglecting the rheo-logical properties: especially in the case of a sphere, the more dense outer layer (or 'skin') developing will increasingly resist its compression and thereby slow down further syneresis.

Stirring the mixture of already shrunk curd particles and whey markedly enhances syneresis; see for example, Fig. 20. The main factor may be prevention of sedimentation of the curd particles: although in a sedimented layer the pres-sure on the curd may be higher, the possibility of the whey flowing out of the curd layer soon becomes small, thereby strongly impeding syneresis; see further Section 6. Another factor is that stirring causes some pressure to be exerted on the curd grains and external pressure has a large effect (see Section 4.2 and below). Van den Bijgaart[5] has made some rough calculations. Stirring causes velocity gradients and consequently—according to Bernoulli's law—pressure differences. In laminar flow, these remain fairly small: they may amount to several Pa during curd making. Mostly, flow will be turbulent and pressures up to 160 Pa were

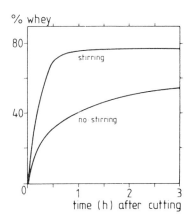

Fig. 20. The volume of whey expelled (as % of the original milk volume) from curd kept in the whey at 38°C as a function of time after cutting, with or without stirring (from Lawrence).[52]

calculated, although these exist only for short times. Collision of curd particles with each other or with the stirrer gives rise to brief pressure bursts of the order of 100 Pa, although average external pressure will probably be about 10 Pa. It has indeed been observed that more vigorous stirring[53] or removing more whey[52 54] (which causes more frequent collisions between curd grains, hence a higher average pressure) lead to somewhat more whey expulsion.

The intermittent deformation of the curd grains occurring during stirring may have another effect. As discussed in relation to Fig. 5, large deformation of renneted milk causes cracks to be formed in it. Experiments in which an amount of renneted milk between two concentric cylinders was brought temporarily under shear and the permeability determined before and afterwards,[4] yielded the following results. Up to shear of 0·35, B had altered little, a shear of about 0·7 caused an increase by, on average, 20%, and larger shear could cause a much higher permeability. It has also been observed[55] that an external pressure of the order of 100 Pa can in certain, not very well known, conditions cause several small cracks to appear at the outside of shrunken curd grains. (Perhaps the cracks are always formed if the local pressure is high enough, but are often sealed again.) To what extent these phenomena mitigate the strong inhibition of further syneresis due to the formation of a dense outer layer on the curd grains is unknown.

Practical conditions of modern curd making usually allow few opportunities to markedly affect syneresis by varying cutting, stirring, etc. For instance, the size to which the renneted milk is cut certainly has an effect on the water content of the cheese, but the effect is small,[56-60] at most some 1% water in the cheese.[60] The main reason presumably is that curd size cannot be varied greatly. If the initial particles are very large, they will inevitably be broken into smaller ones during stirring as long as they are still soft. If one tries to make very small particles, a considerable loss of curd fines occurs. Stirring for a longer time causes a lower

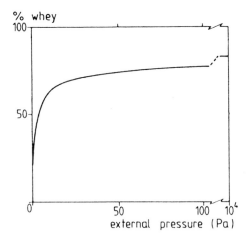

Fig. 21. The amount of whey expelled from curd (as % of the original milk volume) after 2 h at 30°C as a function of the external pressure applied to the curd. Approximate results, recalculated from Ref. 61 (100 Pa = 10^{-3} bar).

moisture content (see e.g. Fig. 16), but a certain minimum duration of stirring is needed to give the particles sufficient firmness. After that any longer stirring leads to a slope of, for instance, -0.04% water in the cheese per minute stirring for semi-hard cheese.[60] Consequently, other measures should be taken to influence the water content, especially varying the temperature; see Section 5.4.

After stirring, the curd particles are usually allowed to sediment. If they are sufficiently rigid (which implies mainly after they have lost sufficient whey), they will deform and fuse only to a limited extent in the sedimented layer, implying that any additional external pressure leads to a considerable loss of whey. This is illustrated in Fig. 21; the lower pressures in this graph were due to stirring curd and whey, the higher ones due to pressure exerted on the sedimented curd layer, either by the curd itself or by perforated plates lying on top. The factors affecting the expression of curd are further discussed in Section 6.

5.4 Effects of Other Process Variables

Numerous authors have studied the effects of product and process variables on syneresis rate, beginning with Sammis *et al.*[56] in 1910. Other extensive studies were by, successively, Wurster,[57] Koestler & Petermann,[50] Van der Waarden,[62] Thomé *et al.*[58] and Stoll.[51] Several others have studied one or a few variables. Harwalkar & Kalab[63] studied syneresis of set yoghurt.

As will be seen below, the results often vary somewhat. Results that are obviously in error in view of our present understanding have generally been omitted. But even then, differences in the individual milk samples, in the methods used and in the conditions employed, cause variation. Particularly, the stage at which syneresis is measured affects the results. Moreover, the effect of one variable may be influenced greatly by the level of another, and altering one factor often causes other conditions to change also.

5.4.1 Heat Treatment of the Milk

Heat treatment of milk to such an extent that serum proteins are denatured, increasingly diminishes the syneresis rate of renneted milk, according to many authors (Refs 51, 57, 59, 62, 64–66). See also Section 7, Fig. 30. Some found even a slight decrease caused by mild heat treatments,[51,64,67] but the others did not. Pearse et al.[66] found the decrease in syneresis to be almost linearly correlated with denaturation of β-lactoglobulin. Heat treatment of synthetic milk free of serum proteins hardly affected syneresis. Addition of κ-casein to milk diminished the detrimental effect of heating, presumably because β-lactoglobulin now reacted primarily with κ-casein in the serum during heating, thus affecting the casein micelles less.[66]

Syneresis of set yoghurt is reduced but not prevented by intense heat treatment.[63] Presumably, the increased voluminosity of the whey proteins in combination with their association with casein, causes a higher effective volume fraction of the protein network, thus diminishing the possibility of syneresis.

5.4.2 Homogenization of the Milk

Homogenization or recombination of milk significantly decreases syneresis rate (Refs 59, 68–72). This is related to the incorporation of micellar casein in the surface coat of the fat globules, which causes the fat globules to become part of the paracasein network, which, in turn, may hinder shrinking of the network. A comparable effect on syneresis was observed if milk had been concentrated by evaporation and diluted again before clotting,[73] which has a similar consequence for the fat globules.[74] If fat is homogenized into whey, so that the fat globules do not contain much casein in their surface layers, the detrimental effect of homogenization on syneresis is clearly less.[69]

5.4.3 Various Additions to the Milk

Additions meant to modify specific residues of the milk proteins, in order to study the clotting reaction, will not be considered here. Adding sugars, which are fairly unreactive, has been reported to cause no effect,[51,75] a slight decrease[62] or a slight increase[76] in syneresis rate. About the same holds for addition of up to 10% urea.[49,62,76]

In cheesemaking, some (~1 mM) $CaCl_2$ is frequently added to enhance coagulation. Most authors report that small additions (e.g. up to 10 mM) of $CaCl_2$ enhanced syneresis somewhat,[51,57,59,62,77] while others found little or no effect;[53,69,76] larger additions were generally found to reduce syneresis.[57,76,78,79] Van der Waarden[62] clearly showed that the main enhancing effect of $CaCl_2$ is due to its lowering the pH; if the pH was kept constant, addition of $CaCl_2$ caused syneresis to decrease, while $MgCl_2$ caused a marked increase.[51,62,80] Presumably, one has to consider two points: the calcium ion activity (enhancing syneresis) and the colloidal calcium phosphate (diminishing syneresis). Presumably, Mg^{2+} ions act much the same as Ca^{2+} ions, whereas Mg-phosphates are much more soluble than Ca-phosphates (addition of $MgCl_2$ may thus cause some dissolution of colloidal phosphate). Lowering the pH causes, of course, a dissolution of colloidal phosphate and an increase in Ca^{2+} activity. Addition of phosphate,[51,62] citrate,[51,62] oxalate[62] or EDTA[51] at constant pH, all caused decreased syneresis; these additions considerably reduce Ca^{2+} activity and adding phosphate also increases colloidal phosphate content. The salt equilibria in milk are intricate, depend on several conditions and often exhibit slow changes, as discussed by, for instance, Walstra and Jenness.[7] See further Section 4.2.

Increasing the ionic strength of milk with univalent ions (e.g. NaCl) has been reported to cause at first no change[76] or a slight increase in syneresis;[51] it tends to reduce the amount of colloidal phosphate and possibly the Ca^{2+} activity. A large increase in ionic strength causes a decrease in syneresis,[51,62,76] but then, milk with added salt coagulates very poorly on renneting. Addition of $AlCl_3$ causes syneresis to be less.[51]

In the production of set yoghurt, some gelatin is often added to enhance firmness and to reduce syneresis. Adding 0·5% gelatin to skim milk before acidification by yoghurt bacteria was found to be very effective in reducing syneresis after cutting the gel.[81] Presumably, the gelatin gel formed on cooling immobilizes the network of casein particles, thereby preventing their rearrangement and thus syneresis.

5.4.4 Coagulation

Most authors agree that rennet concentration has no effect on syneresis.[51,56,57,82] Others found that more rennet gave a slight increase[59,77,78] or a decrease[80] in syneresis, or observed an optimum concentration.[46] These effects should be considered in relation to the time of cutting.[46,51,82] As made clear by, for instance, Weber,[46] it is the stage of the coagulation process or the firmness of the curd at the moment of cutting that is the variable: if cutting is very late, syneresis may be somewhat less. It has also been observed that a higher coagulation temperature leads to slightly less syneresis;[60] this may be due to the cutting starting at an effectively later stage.

Whether milk is renneted by chymosin or pepsin makes no significant difference.[83] Renneting with proteolytic enzymes from *Mucor miehei* or *Endothia parasitica* caused somewhat slower syneresis, but curd firming was slower also, and when cutting 45 rather than 30 min after adding the rennet the syneresis rate was observed to be normal.[84]

Disturbance of the gel during setting may considerably enhance syneresis rate because it causes enlargement of the free surface[57] (see Section 5.3). But also when the free surface was kept constant, such disturbance caused syneresis rate to increase by 20–30% in some experiments.[85,86] This may have been due to the disturbance inducing a higher permeability.

Curd formed solely by acidification shows very little syneresis if left undisturbed (e.g. Refs 2, 87). Syneresis became progressively slower if the gel was left for a longer time after setting before it was cut or wetted.[2] In skim milk gelled by growth of yoghurt bacteria to a pH in the range of 3·8–4·5, however, cutting of the gel led to considerable syneresis, up to about 40% of the original volume being expelled as whey.[63,81] In milk clotted below about pH = 5, the presence of rennet was found to enhance syneresis considerably, the more so when the amount of added rennet was increased.[88] Presumably, this signifies a gradual change from an acid to a rennet-induced gel and is of importance in the production of fresh cheese types[46] and for the understanding of syneresis in acid milk products.

5.4.5 Temperature

Temperature greatly affects syneresis rate of rennet curd; some results are summarized in Fig. 22. All authors agree as to the trend and all results show that the rate of change of syneresis with temperature (Q_{10} or $-$ d ln V/dT) decreases with increasing temperature, but otherwise the results are fairly different. At 25°C, reported values of Q_{10} vary from about 2·5 to 15, at 45°C from about 1·1 to 1·5.

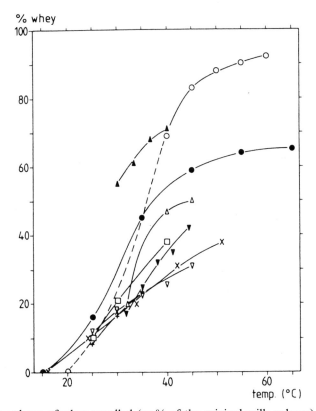

Fig. 22. The volume of whey expelled (as % of the original milk volume) from curd set, kept and treated at different (constant) temperatures. Most results were obtained 1 h after cutting. Recalculated from Sammis et al.[56] ▲; Wurster[57] ▽; Koestler and Petermann[50] ●; Gyr[78] ○; Lawrence[52] ▼; Stoll[51] △; Kirchmeier[44] ×; Kammerlehner[59] □; Marshall[45] +.

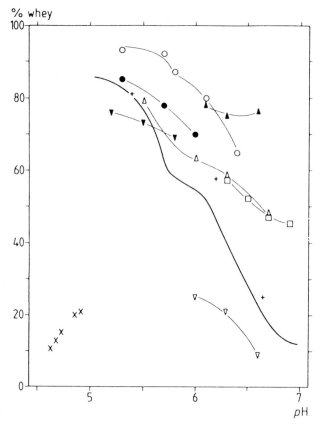

Fig. 23. The volume of whey expelled (as % of the original milk volume) from curd set, kept and treated at different pH; in some cases, pH decreased slightly during the experiment. Most results were obtained 1 h after cutting. Recalculated from Wurster,[57] heavy line = average of several experiments; Gyr[78] □; Cheeseman[76] ●; Stoll[51] Δ; Berridge & Scurlock[86] ▲; Patel et al.[53] ▼; Marshall[45] ▽; Pearse et al.[89] □; Weber[46] +. See text for the results of Emmons et al.[88] ×.

It appears that the rate at which temperature is changed (dT/dt) does not, as such, affect syneresis.[53,57] Keeping the milk for some time at a low temperature before renneting has been reported to have no effect,[90] a small detrimental effect on syneresis[64] or a considerable effect: holding for 20 h at 5°C reduced syneresis by about 30%.[59] Any detrimental effect of precooling is probably reversed by pre-warming the milk to a fairly high temperature before renneting, as is commonly done in cheesemaking to ensure normal setting.

For acid gels, the effect of temperature on syneresis appears not to have been determined, but it has been observed that in gels of skim milk soured by bacteria, considerable syneresis occurred after cutting the gel at 6°C[63] or 3°C.[81]

5.4.6 Acidity

If milk has been acidified to a lower pH before renneting, syneresis rate is faster. Some observations are summarized in Fig. 23; other authors have reported similar results.[50,56,58,62,77,79] Although the observed trends were mostly the same, there were, again, considerable quantitative differences. The deviating relation found by Berridge & Scurlock[86] is not due to inaccuracy but may be related to the different experimental set up (syneresis of a cylinder of curd attached to a grid, in air). The inflection points in the curves near pH 6 are also realistic. Stoll[51] observed that the effect of pH was relatively greater at lower temperature and in the absence of stirring, i.e. if syneresis was slower. There were no appreciable differences according to the acid used.[62]

If the pH falls during syneresis this may enhance syneresis rate to a greater extent than is found when the pH is previously brought to the same value,[42,51] because the building blocks of the protein network tend to shrink due to the change in pH. This is also exemplified in some results by Emmons et al.,[88] shown in Fig. 23. Here, the milk contained variable numbers of starter bacteria and the pH values indicated in the figure are those at the moment of cutting. A higher pH at that stage implies a greater drop in pH after cutting, hence more syneresis. Milk that has already been soured to a very low pH (e.g. 4·5) exhibits only weak syneresis, even after renneting.[56,88] However, as mentioned before, set yoghurt did exhibit syneresis after cutting and the amount of whey expelled did not vary significantly in the pH range of 3·8–4·5. As mentioned in Section 4.3, the temperature dependence of the syneresis in acid gels (pH < 5·2) may well be different from that at higher pH.

5.4.7 Washing of the Curd

Washing, i.e. adding water after part of the whey has been removed, has been reported to enhance syneresis,[59] and to give a slightly, possibly insignificantly, lower water content.[91] However, washing may coincide with a change in temperature and with a difference in the effectiveness of stirring, both of which affect syneresis. In studies in the author's laboratory,[42] either water or an equal quantity of whey at the same temperature was added at a certain stage during cheesemaking and the water content of the curd determined after various times. The water content of the curd to which water was added was up to two percentage units higher, but the difference could be fully explained by taking into account the difference in dry matter content of the moisture (liquid) in the curd. Hence, the osmotic effects of washing are negligible.

5.4.8 Ultrafiltration

Ultrafiltration of cheese milk and renneting the retentate allows the manufacture of curd in such a way that less, or even no, syneresis occurs. The latter, i.e. concentrating the milk to a composition roughly equal to that of the (unsalted) cheese to be made, is only feasible for soft-type cheese; it usually involves diafiltration also. For harder cheeses, partial ultrafiltration can be applied and an important point then is to what extent syneresis is affected. Some results were already given in Section 4.2, e.g. Fig. 13.

Extensive studies were done by Peri et al.,[47] applying eqn (10). They concentrated the milk up to 5·2-fold. The rate constant of the first order equation, which is thus a measure of the rate relative to the amount of whey yet to be removed, varied little with degree of concentration; clear correlations with pH or extent of diafiltration were not observed either. The extrapolated proportion of whey eventually expelled ($V_0 - V_\infty$) varied roughly linearly with the reciprocal of the degree of concentration. A lower pH resulted in a lower (extrapolated) final moisture content. The effects of diafiltration, pre-acidification and sequestering of Ca were also studied.[91]

Other workers obtained slightly different results;[71,72] see also Section 4.2. This may have been due to variation in the time elapsed after renneting before cutting. When renneting normal milk, about 2% of the κ-casein is still unhydrolysed at the moment of cutting, whereas this may be about 12% for a milk concentrated two-fold by ultrafiltration;[92] this proportion will be higher for a more concentrated milk. Consequently, the early stages of gel formation and syneresis probably proceed somewhat differently, depending on the moment of cutting.

5.5 Effect of Milk Composition

Milk composition may clearly affect syneresis, but the effect is usually not large. A higher fat content in the milk on average is accompanied by somewhat slower syneresis.[46,51,58,59,69,72,75,93,94] In practice, milk is usually standardized as to fat content. A higher casein content goes along with a slower absolute rate of syneresis, but a hardly different relative rate; see further above, under Ultrafiltration.

Minor components may have a larger influence and it must be presumed that the calcium ion activity is an especially important variable. For instance, separate milkings of individual cows may vary by a factor of three in syneresis rate,[50,58,59,95] but addition of some $CaCl_2$ greatly reduces the variation. Minor variation has been observed with the stage of lactation,[59,95] and this may possibly be related to the calcium ion activity also.

Milk from cows suffering severe mastitis exhibits poor clotting by rennet and somewhat diminished syneresis.[96,97,98] Extensive growth of pseudomonades in milk was shown to reduce syneresis markedly.[99] On the other hand, considerable proteolysis caused by plasmin activity hardly affected whey expulsion.[100]

The effect of protein composition has been studied as well. Pearse et al.[101] made milk with synthetic micelles of variable casein composition. The proportions of β- and κ-caseins clearly affected clotting time, but syneresis far less. Dephosphorylation of β-casein caused the clotting time to increase and syneresis rate to decrease. Interpretation of these results is very difficult without knowing such variables as micelle size and voluminosity, or loss tangent and permeability of the renneted milk. There also appears to be some correlation between syneresis and genetic variants of milk proteins, especially with the β-lactoglobulin variant.[102] This may, again, be due to differences in the calcium ion activity, which correlates with the genetic variants.

Other conditions being equal, renneted goats' milk exhibited greater syneresis than cows' milk, and ewes' milk syneresed less.[72] It may be noted that the latter usually has a clearly higher casein content.[7] The effect of several variables on syneresis of renneted cows' milk, as discussed above, was often different for either ewes' or buffaloes' milk.[65]

5.6 Concluding Remarks

The results presented in Section 5 (see also a review by Pearse & Mackinlay[103]) generally agree with those of Section 4, although only in a qualitative sense. In a practical situation, quantitative predictions on syneresis rate can hardly be made. Nevertheless, it may be concluded that the main variables affecting syneresis rate are:

— the geometrical constraints (surface area of curd, distance over which the whey has to flow);
— pressure applied to the curd (grains), where the relative effect is greatest in the low pressure range;
— pH in the case of rennet-induced gels;
— temperature for rennet-induced gels, where the relative effect is greatest in the low temperature range.

The effect of the other variables is generally small (with the exception of intense heat treatment) and tends to be relatively smaller when overall syneresis rate is higher. Stoll[51] observed, for instance, that stirring the curd–whey mixture almost eliminated differences caused by some variables observed when studying syneresis under quiescent conditions. This was explained by van den Bijgaart[5] from the overriding effect of the permeability of the outer layer of the curd grains. Any condition leading to very rapid syneresis also causes rapid development of a poorly permeable layer, which then markedly slows down any further syneresis. In a qualitative sense, this has been observed before, e.g. Ref. 50. Cheese makers speak of a 'skin' around the curd grains, and it is even assumed that very rapid initial syneresis may lead to an ultimately higher water content in the curd, as compared to a situation where syneresis proceeds more slowly. The author is, however, unaware of unequivocal proof for such a result.

6 DRAINAGE OF A CURD COLUMN

When the curd grains are sufficiently dry, they are usually allowed to sediment, into a 'bed' in the cheese vat or in a drainage pipe. The layer of curd grains compacts, more whey is expelled from the grains and the grains partly fuse to form a coherent mass. The compaction may be due to pressure exerted by the layer itself or by perforated plates laid on top. Effective pressure ranges from about 100 to 500 Pa. The compaction is either allowed to proceed for a considerable time, after which the curd mass is cut into small pieces ('milling', as for

TABLE II
Processes Occurring during Compaction and Drainage of a Curd–Whey Column

Process	Results in	Depends on
1. expulsion of whey from grains	a. lower whey content of grains b. grains less deformable	degree of concentration, temperature, pH, effective pressure, free surface area of grains
2. drainage of whey from column	a. closer packing of grains b. narrower pores	external pressure, pore size distribution (hence, grain size distribution and shape), geometrical constraints
3. deformation and fusion of grains	a. narrower pores b. smaller free surface area	deformability of grains (hence, degree of concentration, temperature, pH), external pressure, duration

Cheddar types), or after a short while blocks of curd are cut from it, which then are subject to moulding and pressing.

The phenomena occurring during compaction and drainage are complicated. At the very first, compaction occurs due to sedimentation and reorientation of the curd grains. Further events are summarized in Table II. The different processes are mutually dependent and especially the deformation of the grains and the expulsion of whey from them are coupled. Moreover, fusion of curd grains is greatly enhanced when they are deformed. By and large, processes 1 and 2 lead to a lower moisture content, whereas 3 impedes moisture loss. Fusion may proceed until the pores between the grains are no longer interconnected, when further drainage virtually stops. On the other hand, some initial fusion may promote drainage, as it may prevent further reorientation of the grains into a denser packing.

These conclusions may be true enough, but they are only qualitative and thus not very helpful. To arrive at quantitative relations, the processes were studied in some detail by Akkerman[55] for the case of Dutch-type cheese manufacture. He used whole milk, mostly without starter added, and made curd grains that had been left to synerese to roughly a quarter of their initial volume ($i \approx 0.25$). Temperature was mostly 35°C. Expression of single grains, fusion of a collection of grains, compaction of a column and change in pore size (distribution) in a column of grains were studied separately. Because of the many variables of importance, the intricacy of the processes and the experimental difficulties, the results are to some extent uncertain, but they clearly show quantitative trends.

Some results on *uniaxial expression of single grains* are shown in Fig. 24. The situation here is quite different from that leading to the results in Figs 14 and 15. A curd grain is now involved, which implies that the outer layer is much denser than the centre; see for instance Fig. 12. Moreover, the pressures are generally higher, the curd grain can deform sideways and most of the outflow of whey is

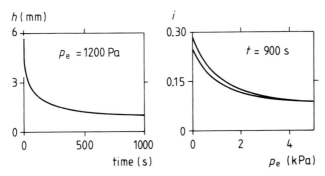

Fig. 24. Uniaxial compression of curd grains. h is the height of the curd grain, t duration of pressing, i the relative remaining volume and p_e the pressure applied. From Akkerman.[55]

in directions perpendicular to the applied force. It is seen that the grain shows an immediate, i.e. elastic, deformation, followed by a viscous one that becomes ever slower. If the pressure is released within a few seconds, the grain more or less regains its original shape, but after some minutes the deformation is permanent; this agrees with the average relaxation time of renneted milk of about 1 min found earlier.[30,104] Analysis of the results leads to the conclusion that, except in the very beginning, the permeability of the curd matrix is no longer the factor limiting the outflow of whey. The deformability of the grain, more precisely its effective biaxial elongational viscosity, soon becomes rate determining; the elongational viscosity will markedly depend on (decrease with) the stress applied. Akkerman defined a pseudo Poisson number, μ, as

$$\mu \equiv 0\cdot5(1 - \mathrm{d} \ln V/\mathrm{d}\, \epsilon_h) \tag{11}$$

where V is grain volume and ϵ_h is the relative deformation expressed as the Hencky strain; he found as an average over the first 15 minutes, $\mu \approx 0\cdot27$, almost independent of conditions. This implies that a (nearly) constant part of the decrease in height of the grain is due to shrinkage. The constancy of μ agrees with the deformability (i.e. the rheological properties) of the curd now being rate determinant. As shown in Fig. 24, the deformation, and thereby the expression of whey, depends on i, pressure and time. It is little dependent on grain size. The expression reasonably followed the relation

$$(i_t - i_\infty)/(i_0 - i_\infty) = \exp(-Kp_e\sqrt{t}) \tag{12}$$

with $i_\infty = 0\cdot1$ and where the rate constant $K \approx 4\cdot10^{-5}$ $\mathrm{Pa}^{-1}\cdot\mathrm{s}^{-0\cdot5}$. The effect of pH was not studied, but in view of the effect of pH on the rheological properties of renneted milk (see e.g. Fig. 1) and of cheese, it must be significant.

The *fusion of curd grains* in a curd–whey column was evaluated by determining the fracture stress when pulling two parts of the column, separated by a perforated plate, away from each other. The force divided by the contact area of the grains was taken as the fracture stress. Some results are given in Fig. 25, and it is seen that the pressure and time during which it is applied have a strong

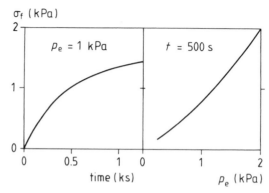

Fig. 25. Strength of fusion of curd grains in a column expressed as the fracture stress σ_f after pressing for various times (t) at various pressure (p_e); see text. From Akkerman.[55]

effect. At 34°C, the fracture stress obtained was about 60% higher than at 32°C, whereas at 36°C it was, maybe, somewhat lower again. The less the grains had shrunk prior to the experiment, the higher the fracture stress. The effect of pH was not studied, but it is known in practice that at a low pH, say near 5, fusion of curd grains is very difficult to achieve.

The *compaction of a curd column* was studied for radial drainage. The results strongly depended on the geometrical constraints, especially the radius of the column. The curd particles tend to stick to the wall and leakage along the wall is also of importance. The total pressure exerted was, following Schwartzberg *et al.*,[105] split up as

$$p_e = p_c + p_l + p_w \tag{13}$$

where part of the pressure is lost by friction to the wall (subscript w), and the remainder may not only be on the curd particles (subscript c), but also on the liquid (subscript l). As soon as the outflow of whey is hindered by a lack of interconnected pores, the pressure on the liquid increases rapidly. If the pores become completely disconnected, all the pressure is exerted on the liquid (except for p_w) or, in other words, the pressure is isotropic, and expression of the curd grains stops.

Some results are shown in Fig. 26. The total pressure exerted appears to be the dominant variable. For a low p_e, the expression increases strongly with pressure, roughly following eqn (12). But above a certain p_e, called the threshold pressure, any higher pressure leads to a progressively increasing p_l. Also the pressure loss at the wall p_w markedly increases with total pressure, being, for instance, proportional to $p_e^{2.5}$ Altogether, at a high p_e the effective pressure on the grains, p_c, soon becomes very small and expression (almost) stops. The other results in Fig. 26 speak for themselves; note the very strong decrease of the permeability of the curd column with time, despite p_e being fairly small.

Other variables affecting the drainage are the degree of concentration of the curd grains at the beginning, given as i_0, and the temperature. For a higher i_0

and a higher temperature, the initial rate of compaction is higher, but the threshold pressure mentioned above is lower; in other words, the highest pressure that can be applied for the drainage to proceed satisfactorily is smaller. Threshold pressures are mostly somewhere between 800 and 2000 Pa; they considerably depend on the geometry of the system. The presence of curd fines may strongly lower the drainage rate, as the fine particles tend to block the pores between the grains; the threshold pressure now is much lower. It must be assumed that the curd particle size (average and spread) has some influence, but not a lot within the range studied. The effect of pH was not estimated.

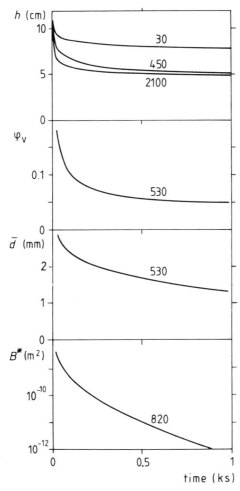

Fig. 26. Compression of a curd–whey column (radius 6 cm) under uniaxial compression and with radial drainage. Column height (h), volume fraction of pores (φ_v), volume-average apparent pore diameter ($\bar{d}$) and average radial permeability (B^*) calculated as a function of time after applying pressure (indicated near the curves, Pa). From Akkerman.[55]

Akkerman[55] has further compared the results obtained for single grains with the compaction of a curd column by developing a computer model of the process. Up to pressures of a few hundred Pa, the calculated results agreed reasonably well with the experimental ones. He also concluded that at a high p_e the curd grains can, in principle, initially be expressed quite fast, without the pressure on the liquid, p_l, becoming substantial. This opens up possibilities for improving the drainage process.

In the author's laboratory, some preliminary studies on axial drainage in a curd column had been performed under a wider range of conditions;[106] actually, there may have been considerable radial drainage as well. Curd was made from skim milk, using no starter. After cutting and stirring (and removing some whey), a column of curd and whey, 30 cm high, was taken; the curd sedimented almost immediately to a height of about 20 cm, after which pressure was applied via a perforated disc, and the curd column gradually compressed to a height of, for instance, 5 cm; the compression was allowed to proceed for 90 min. The final

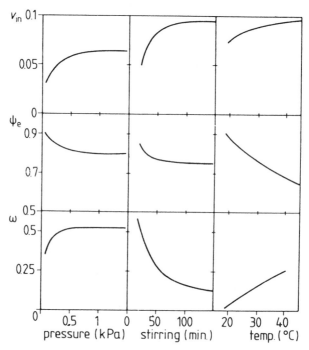

Fig. 27. Effect of some variables on the change in moisture in a column of curd grains in whey. v_{in} is the initial rate of moisture loss after applying pressure ($-d \ln V/dt$, where V = volume and t = time in min); ψ_e is the final fraction of moisture in the column; ω is the fraction of removed moisture that originated in the curd grains. Variables are external pressure applied (stirring time 20 min); time of (cutting and) stirring before applying pressure; temperature of stirring and compressing (stirring time 80 min). Unless indicated otherwise, temperature was 31°C and the external pressure was 0·4 kPa (approximate results from Ref. 106).

moisture content of the curd column was determined and the earlier values were calculated from the change in height and expressed as the mass fraction of moisture (moisture means liquid containing dissolved substances, i.e. whey in this case). The proportion of moisture between the grains was originally about 40%. Some results are shown in Fig. 27, which partly speaks for itself. In a qualitative sense, these results agree with those of Akkerman. Also here, very high pressures give little improvement. Note the large effect of temperature: at 20°C, very little whey is expelled from the curd grains, in accordance with the strong dependence of apparent viscosity[15] and of syneresis on temperature.

Some of the conclusions drawn in this section may in a sense be derived from earlier observations.[75,77,79,82,95,107] Especially interesting is the work of Scott Blair & Coppen,[108] who reasoned that firmer curd grains would permit faster drainage of whey from a mass of grains, and made use of this in devising a test method to determine the 'pitching point' of the curd, i.e. the moment at which the grains have lost sufficient moisture to stop stirring. A volume of curd and whey is put into a perforated cylinder and allowed to drain for a fixed time; now the 'superficial density' is determined, i.e. the weight of curd divided by the height of the curd column. They found a fair positive correlation between the water content and the superficial density, indicating that high moisture and thus soft grains deformed rapidly to close the channels between them, thereby greatly hindering further drainage. Firm (i.e. 'dry') grains permitted ongoing drainage, leading to a low superficial density, because the voids between particles now become filled with air. As is to be expected, other factors affect the draining rate: Scott Blair & Coppen[108] found for the same value of superficial density a range in water content of about 14 percentage units. There was a tendency for rapid initial syneresis, hence presumably the presence of a more or less rigid 'skin' around the curd grains, to lead to a lower superficial density. Likewise, curd at a lower pH (Cheshire as compared to Cheddar) tended to have a lower superficial density. It would be useful to study these and other variables in greater detail.

7 THE WATER CONTENT OF CHEESE

There obviously is a lowest possible water content of (freshly made) cheese: the paracasein particles have a given voluminosity. This aspect was reviewed earlier[1] and it was concluded that few hard conclusions can be drawn. The equilibrium voluminosity of paracasein micelles at room temperature and physiological pH was roughly estimated to correspond to 1·4 g water per g protein; this would come down to a water content of an unsalted full-cream cheese of about 40%. The voluminosity would be lower for a lower pH and for a higher temperature. The latter effect is considerable. It is also known that a cheese or curd may take up moisture when the temperature is lowered, at least at low pH (e.g. Ref. 109 on Herve cheese; results obtained in the author's laboratory on renneted milk ultrafiltration retentate at pH ≈ 5·2; observations in practice on Feta cheese kept in brine). In a cheese, the lowest possible water to protein ratio may be slightly lower for a lower fat content; in practice, a lower fat in dry matter content always goes along with a distinctly lower water to protein ratio, but this presumably is due to faster syneresis. Altogether, the final moisture content of most cheeses is primarily determined by the rate and duration of the processes causing whey expulsion, rather than by the equilibrium swelling state of the paracasein.

After curd making and drainage, one of the following procedures is usually applied.

1. Moulding the curd, followed by further drainage under its own weight; this is only applied for fairly soft cheese.
2. Moulding and pressing the curd; this is the common method for semi-hard and several hard cheeses.
3. Letting the curd rest for a considerable time to develop sufficient acidity (often while allowing the curd to flow: cheddaring) after which the coherent curd mass is cut into fairly small pieces (milling), salted, moulded and pressed.
4. Intensively working the already acidified curd (pH ≈ 5·3) at a quite high temperature, as is done in making pasta filata cheeses.

During several of these process steps the curd may lose considerable moisture. Merely taking the curd out of the whey allowing further whey to leak out, has already a marked effect; see e.g. Figs 16 and 28. Sometimes, the curd grains are worked after removal from the whey, which leads to a much drier cheese with an open texture (numerous small, irregularly shaped holes). The pasta filata treatment also causes appreciable moisture loss (high temperature and pressure), although the fairly large size of the lumps of curd formed has a mitigating effect (long distance and relatively small surface area).

Pressing of the curd mass is aimed at obtaining a coherent mass with a closed rind. The formation of a rind, i.e. an outer layer in which all the curd grains are fully fused with their neighbours, is greatly favoured by the possibility of rapid removal of moisture from the outer layer, for instance by application of a cloth around the curd mass.[16] The closed rind greatly reduces further expression of moisture. Figure 26 shows that the effective permeability of a drained mass of curd is about $10^{-12}\,m^2$, which still allows considerable flow of moisture under pressures of 10 to 100 kPa, which are common in practice. The permeability in the outer layer may be as low as $10^{-16}\,m^2$ or less, and even a layer of a few mm then makes a substantial barrier.

The effects of moulding, pressing and resting on the moisture content have been carefully studied by Geurts[110] and some results are in Fig. 28. It is seen that the lower the moisture content before pressing, the less the further moisture loss. The initial water content has other important consequences. First, consider the situation where it is high, say about 55% at the beginning of pressing. Now, pressing at an earlier stage or at a higher pressure leads to a higher water content (more precisely, a less reduced water content); the difference is of the order of 1% water. The explanation is that a closed rind is formed at an earlier stage or of a greater thickness. Pressing at a higher initial temperature or having a

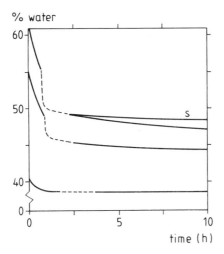

Fig. 28. Water content of a loaf of curd as a function of time after moulding. The dotted lines give the assumed course during pressing. Spherical loaves of about 22 cm diameter, except for one of 12 cm (designated S). From Geurts.[110]

larger loaf of curd lead to a lower water content. Although a higher temperature implies a softer curd, and thereby presumably easier rind formation, the over-riding effect seems to be the effect of temperature itself on syneresis; see e.g. Fig. 14. A smaller loaf will cool quicker and thereby lose less moisture; see Fig. 28.

These relations are rather different if the curd mass has a low water content at the beginning of pressing, say 40%. (Such a low water content can only be obtained by prolonged stirring at a high temperature and letting the pH decrease appreciably.) A higher pressure, a smaller loaf and a lower temperature all lead to a lower water content. Presumably, it takes a longer time to obtain a closed rind, the more so for a lower temperature, and before the rind is formed, moisture can be pressed out of the loaf.

Geurts[110] also studied the distribution of moisture in unsalted cheese. Some results are shown in Fig. 29. Apart from a thin outer layer, i.e. the rind, which has a slightly reduced water content, the lowest water content is at the centre. This is the region where the temperature has remained highest, especially in a large loaf. It was even observed that the temperature increased in the greater part of a large loaf, undoubtedly due to the heat generated by the starter bacteria. If an unsalted cheese is left to rest, the water content tends to become somewhat lower at the bottom side. It may thus be concluded that the moisture moves away from regions where the temperature is higher and/or the pressure higher than elsewhere. Soon, however, the process of fusion of curd grains becomes complete, say after two days,[111] and the permeability of the cheese mass becomes too low to permit appreciable transport of moisture.

In several types of cheese, the drained mass of curd is allowed to spread laterally for a considerable time ('cheddaring'). Olson & Price[112] found that this led

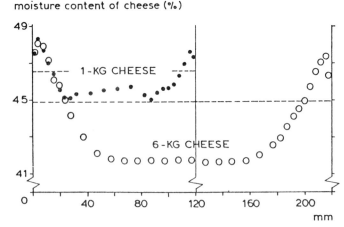

Fig. 29. Moisture (water) distributions in unsalted spherical cheeses (1 and 6 kg), moulded from the same curd, lightly pressed and kept for a few days. The broken lines indicate the average water content. From Geurts.[110]

to a higher moisture content (1–2% more water), as compared to curd kept for the same time but which was prevented from spreading. Although the cheddaring may have caused a slightly lower average temperature, the main cause for the differences was presumably that the flow of curd promoted deformation of the curd grains; hence, closing of pores between grains and hindering drainage of any moisture still leaving the grains due to syneresis.

The water content of the cheese must to a considerable extent depend on the amount of syneresis during curd preparation and the results on syneresis as given in Sections 4 and 5 indeed qualitatively agree with results on the water content of cheese (e.g. Refs 54, 56, 60, 113–115). Whether there is exact agreement is uncertain. The water content of cheese always shows considerable random variation (e.g. Ref. 116) and this makes exact comparisons difficult. During cheesemaking, conditions usually change, for instance pH, temperature and effective pressure acting on the curd, so that one has to take some kind of average. Moreover, the factors are interrelated; for instance, temperature affects the rate of acidification. The latter is also affected by oxygen content[117] and this may explain why some authors found a negative correlation between stirring rate of the curd–whey mixture and the final water content of the cheese;[118] presumably, faster stirring caused a higher oxygen content, hence slower acidification, and consequently less syneresis.

The main causes of discrepancy may be, however, the considerable effects of curd drainage (Section 6) and further treatment, such as pressing (this section). If these processes are kept constant, as is nearly always more or less the case during modern cheesemaking, the correlation between syneresis and final water content may be fairly good; one should then take into account the remark made earlier about the inhibiting effect of the formation of a dense outer layer around

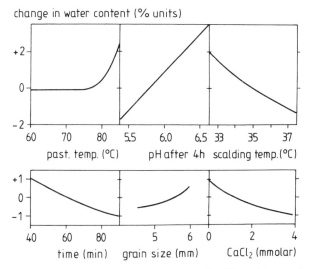

Fig. 30. The effect of some variables in treatment of milk and in curd making on the water content of unsalted Gouda type cheese, 5·5 h after renneting, other conditions being equal; time means time after cutting. The water content under standard conditions was about 46%. From Straatsma & Heijnekamp.[60]

the curd grains. Some interesting practical results for the case of semi-hard brine-salted cheese were obtained by Straatsma & Heijnekamp,[60] and some of these are summarized in Fig. 30. It is seen that most variables have, within the variation that can reasonably be applied in practice, a fairly small effect. Only the scalding (cooking) temperature and the acidity of the curd have a significant influence. The acidity depends primarily on type and quantity of starter added, any preacidification applied, temperature and duration of acid development. The manner in which the acidity was reached appeared to make little difference.

It may finally be mentioned that the water content of cheese also depends, of course, on salting (by pseudo osmosis[119]), on drying up and on proteolysis (which causes water to be converted into dry matter).

ACKNOWLEDGEMENTS

This review could not have been written without the work done by several people cooperating with the author in studying renneting, gel formation, syneresis and early stages of cheese making, notably J.C. Akkerman, H.J.C.M. van den Bij-gaart, L.G.B. Bremer, H.J.M. van Dijk, T.J. Geurts, A.C.M. van Hooydonk, T. van Vliet and P. Zoon. The extensive discussions with them are gratefully acknowledged. Some of them (J.A., H.v.d.B., H.v.D., T.G., T.v.V.) also gave useful comments on the manuscript.

188   P. WALSTRA

REFERENCES

1. Walstra, P., van Dijk, H.J.M. & Geurts, T.J., 1985. *Neth. Milk Dairy J.*, **39**, 209.
2. van Dijk, H.J.M., 1982. Doctoral thesis, Wageningen Agricultural University, Wageningen.
3. van Dijk, H.J.M., Walstra, P. & Schenk, J. 1984. *Chem. Eng. J.*, **28**, B43.
4. van Dijk, H.J.M. & Walstra, P., 1986. *Neth. Milk Dairy J.*, **40**, 3.
5. van den Bijgaart, H.J.C.M., 1988. Doctoral thesis, Wageningen Agricultural University, Wageningen.
6. Schmidt, D.G., 1982. In *Developments in Dairy Chemistry—I, Proteins*, ed. P.F. Fox. Elsevier Applied Science Publishers, London, pp. 61–86.
7. Walstra, P. & Jenness, R., 1984. *Dairy Chemistry and Physics*, Wiley, New York.
8. Walstra, P., 1990. *J. Dairy Sci.*, **73**, 1965.
9. Roefs, S.P.F.M., Walstra, P., Dalgleish, D.G. & Horne, D.S., 1985. *Neth. Milk Dairy J.*, **39**, 119.
10. Payens, T.A.J., 1979. *J. Dairy Res.*, **46**, 291.
11. van Vliet, T., Roefs, S.P.F.M., Zoon, P. & Walstra, P., 1989. *J. Dairy Res.*, **56**, 529.
12. van Hooydonk, A.C.M. & Walstra, P., 1987. *Neth. Milk Dairy J.*, **41**, 19.
13. van Hooydonk, A.C.M., Boerrigter, I.J. & Hagedoorn, H.G., 1986. *Neth. Milk Dairy J.*, **40**, 297.
14. Dalgleish, D.G., 1983. *J. Dairy Res.*, **50**, 331.
15. Zoon, P., van Vliet, T. & Walstra, P., 1988. *Neth. Milk Dairy J.*, **42**, 271.
16. Mulder, H., de Graaf, J.J. & Walstra, P., 1966. *Proc 17th Intern. Dairy Congr., Munich*, **D**, 413.
17. Henstra, S. & Schmidt, D.G., 1970. *Proc. 7th Intern. Congr. Electron Microscopy, Grenoble*, **1**, 389.
18. Knoop, A.M. & Peters, K.H., 1975. *Kieler Milchw. Forsch. Ber.*, **27**, 227.
19. Kalab, M. & Harwalkar, V.R., 1973. *J. Dairy Sci.*, **56**, 835.
20. Green, M.L., Hobbs, D.G., Morant, S.V. & Hill, V.A., 1978. *J. Dairy Res.*, **45**, 413.
21. Family, F. & Landau, D.P. (eds), 1984. *Kinetics of Aggregation and Gelation*, North-Holland, Amsterdam.
22. Meakin, P., 1988. *Adv. Colloid Interface Sci.*, **28**, 249.
23. Bremer, L.G.B., van Vliet, T. & Walstra, P., 1989. *J. Chem. Soc. Faraday Trans. 1*, **85**, 3359.
24. Walstra, P., van Vliet, T. & Bremer, L.G.B., 1990. In *Food Polymers, Gels and Colloids*, ed. E. Dickson. Royal Soc. of Chem., London, p. 369.
25. Bremer, L.G.B., Bijsterbosch, B.H., Schrijvers, R., van Vliet, T. & Walstra, P., 1990. *Colloids Surfaces*, **51**, 159.
26. Bremer, L.G.B., 1992. Doctoral thesis, Wageningen Agricultural University, Wageningen.
27. Zoon, P., van Vliet, T. & Walstra, P., 1988. *Neth. Milk Dairy J.*, **42**, 249.
28. Zoon, P., van Vliet, T. & Walstra, P., 1988. *Neth. Milk Dairy J.*, **42**, 295.
29. Zoon, P., van Vliet, T. & Walstra, P., 1989. *Neth. Milk Dairy J.*, **43**, 17.
30. Zoon, P., van Vliet, T. & Walstra, P., 1989. *Neth. Milk Dairy J.*, **43**, 35.
31. Bremer, L.G.B. & van Vliet, T., 1991. *Rheologica Acta*, **30**, 98.
32. Walstra, P. & van Vliet, T., 1986. *Neth. Milk Dairy J.*, **40**, 241.
33. Knoop, A.M. & Peters, K.H., 1975. *Kieler Milchw. Forsch. Ber.*, **27**, 315.
34. Roefs, S.P.F.M., van Vliet, T., van den Bijgaart, H.J.C.M., de Groot-Mostert, A.E.A. & Walstra, P., 1990. *Neth. Milk Dairy J.*, **44**, 159.
35. Roefs, S.P.F.M., de Groot-Mostert, A.E.A. & van Vliet, T., 1990. *Colloids Surfaces*, **50**, 141.
36. Roefs, S.P.F.M. & van Vliet, T., 1990. *Colloids Surfaces*, **50**, 161.
37. van Dijk, H.J.M., 1990. *Neth. Milk Dairy J.*, **44**, 65.
38. van Dijk, H.J.M., 1990. *Neth. Milk Dairy J.*, **44**, 111.

39. van Dijk, H.J.M., 1992. *Neth. Milk Dairy J.*, **46**, 102.
40. van Vliet, T., van Dijk, H.J.M., Zoon, P. & Walstra, P., 1991. *Colloid Polym. Sci.*, **269**, 620.
41. van Hooydonk, A.C.M., Hagedoorn, H.G. & Boerrigter, I.J., 1986. *Neth. Milk Dairy J.*, **40**, 369.
42. van de Grootevheen, J.G. & Geurts, T.J., 1977. *Het Instellen van het Vochtgehalte van Verse Ongezouten Wrongel*, Wageningen, (unpublished).
43. Kwant, P.R., Geurts, T.J. & van Dijk, H.J.M., 1980. *Een methode ter bepaling van het watergehalte van pas gesneden wrongel*, Wageningen (unpublished).
44. Kirchmeier, O., 1972. *Milchwissenschaft*, **27**, 99.
45. Marshall, R.J., 1982. *J. Dairy Res.*, **49**, 329.
46. Weber, F., 1984. In *Le Fromage*, ed. A. Eck. Lavoisier, Paris, p. 22.
47. Peri, C., Lucisano, M. & Donati, E., 1985. *Milchwissenschaft*, **40**, 650.
48. Lawrence, A.J. & Hill, R.D., 1974. *Proc. 19th Intern. Dairy Congr., New Delhi*, **1E**, 204.
49. Beltman, H., 1975. Doctoral thesis, Wageningen Agricultural University, Wageningen.
50. Koestler, G. & Petermann, R., 1936. *Landw. Jahrb. Schweiz*, **50**, 103.
51. Stoll, W.F., 1966. Doctoral thesis, University of Minnesota, St. Paul (Diss. Abstr. 27B(3), 851).
52. Lawrence, A.J., 1959. *Austr. J. Dairy Technol.*, **14**, 169.
53. Patel, M.C., Lund, D.B. & Olson, N.F., 1972. *J. Dairy Sci.*, **55**, 913.
54. Birkkjaer, H.E., Sørensen, E.J., Jorgensen, J. & Sigersted, E., 1961. *Beretn. Statens Forsøgsmejeri*, **128**.
55. Akkerman, J.C., 1992. Doctoral thesis, Wageningen Agricultural University, Wageningen.
56. Sammis, J.L., Suzuki, S.K. & Laabs, F.W., 1910. *Bur. Animal Industry, USDA*, Bull., **122**.
57. Wurster, K., 1934. *Milchw. Forsch.*, **16**, 200.
58. Thomé, K.E., Axelsson, I. & Liljegren, G., 1958. *Milk Dairy Res. Rep. Alnarp*, **53**.
59. Kammerlehner, J., 1974. *Deutsche Molkerei-Ztg.*, **95**, 306, 342, 377.
60. Straatsma, J. & Heijnekamp, A., 1988. *Kwantitatieve Invloed van Diverse Factoren op het Kaasbereidingsproces*. NIZO-Mededeling M21.
61. van Dijk, H.J.M., Walstra, P. & Geurts, T.J., 1979. *Neth. Milk Dairy J.*, **33**, 60.
62. van der Waarden, M., 1947. *Onderzoek naar de factoren die invloed hebben op de wei-uittreding uit gestremde melk*, Alg. Ned. Zuivelbond FNZ, The Hague (unpublished).
63. Harwalkar, V.R. & Kalab, M., 1983. *Milchwissenschaft*, **38**, 517.
64. Nilsen, K.O., 1982. *Meieriposten*, **71**, 123, 162.
65. Dimov, N.D. & Mineva, P., 1962. *Proc. 16th Intern. Dairy Congr., Copenhagen*, **BIV**, 817.
66. Pearse, M.J., Linklater, P.M., Hall, R.J. & Mackinlay, A.G., 1985. *J. Dairy Res.*, **52**, 159.
67. Siegenthaler, E. & Flückiger, E., 1964. *Schweiz. Milchztg.*, **90**, 766 (Wissensch. Beilage 96).
68. Vaikus, V., Lubinskas, V. & Mitskevichus, E., 1970. *Proc. 18th Intern. Dairy Congr., Sydney*, **1E**, 320.
69. Emmons, D.B., Lister, E.E., Beckett, D.C. & Jenkins, K.J., 1980. *J. Dairy Sci.*, **63**, 417.
70. Humbert, G., Driou, A., Guerin, J. & Alais, C., 1980. *Lait*, **60**, 574.
71. Green, M.L., Marshall, R.J. & Glover, F.A., 1983. *J. Dairy Res.*, **50**, 341.
72. Storry, J.E., Grandison, A.S., Millard, D., Owen, A.J. & Ford, G.D., 1983. *J. Dairy Res.*, **50**, 215.
73. Cheeseman, G.C. & Mabbitt, L.A., 1968. *J. Dairy Res.*, **35**, 135.
74. Mulder, H. & Walstra, P., 1974. *The Milk Fat Globule. Emulsion Science as Applied to Milk Products and Comparable Foods*. Pudoc, Wageningen.

75. Grandison, A.S., Ford, G.D., Owen, A.J. & Millard, D., 1984. *J. Dairy Res.*, **51**, 69.
76. Cheeseman, G.C., 1962. *Proc. 16th Intern. Dairy Congr., Copenhagen*, **BIV**, 465.
77. Lelievre, J. & Creamer, L.K., 1978. *Milchwissenschaft*, **33**, 73.
78. Gyr, A., 1944. *Ann. Paediatrici*, **163**, 314.
79. Tarodo de la Fuente, B. & Alais, C., 1975. *Chimia*, **29**, 379.
80. Kovalenko, M.S. & Bocharova, S.G., 1973. *Dairy Sci. Abstr.*, **35**, 372.
81. Modler, H.W., Larmond, M.E., Lin, C.S., Froehlich, D. & Emmons, D.B., 1983. *J. Dairy Sci.*, **66**, 422.
82. Lelievre, J., 1977. *J. Dairy Res.*, **44**, 611.
83. Andersson, H. & Andren, A., 1990. *J. Dairy Res.*, **57**, 119.
84. Gouda, A. & El-Shabrawy, S.A., 1987. *Chem. Mikrob. Technol. Lebensm.* **10**, 129.
85. Cheeseman, G.C. & Chapman, H.R., 1966. *Dairy Industries*, **31**, 99.
86. Berridge, M.J. & Scurlock, P.G., 1970. *J. Dairy Res.*, **37**, 417.
87. Aiyar, K.R. & Wallace, G.M., 1970. *Proc. 18th Intern. Dairy Congr.*, Sydney, **1E**, 47.
88. Emmons, D.B., Price, W.V. & Swanson, A.M., 1959. *J. Dairy Sci.*, **42**, 866.
89. Pearse, M.J., Mackinlay, A.G., Hall, R.J. & Linklater, P.M., 1984. *J. Dairy Res.*, **51**, 131.
90. Johnston, D.E., Murphy, R.J. & Whittaker, M.R., 1983. *J. Dairy Res.*, **50**, 231.
91. Casiraghi, E.M., Peri, C. & Piazza, L., 1987. *Milchwissenschaft*, **42**, 232.
92. van Hooydonk, A.C.M. & van den Berg. G., 1988. *Intern. Dairy Fed., Bull.* 225.
93. Beeby, R., 1959. *Austr. J. Dairy Technol.*, **14**, 77.
94. Johnston, D.E. & Murphy, R.J., 1984. *Milchwissenschaft*, **39**, 585.
95. Grandison, A.S., Ford, G.D., Millard, D. & Owen, A.J., 1984. *J. Dairy Res.*, **51**, 407.
96. Thomé, K.E. & Liljegren, G., 1959. *Proc. 15th Intern. Dairy Congr., London*, 3, 1922.
97. Kiermeier, F. & Keis, K., 1964. *Milchwissenschaft*, **19**, 79.
98. Kiermeier, F., Renner, E. & Djafarian, M., 1966/7. *Z. Lebensm. Unters. Forsch.*, **132**, 352.
99. Lelievre, J., Kelso, E.A. & Stewart, D.B., 1978. *Proc. 20th Intern. Dairy Congr., Paris*, **E**, 760.
100. Pearse, M.J., Linklater, P.M., Hall, R.J. & Mackinlay, A.G., 1986. *J. Dairy Res.*, **53**, 477.
101. Pearse, M.J., Linklater, P.M., Hall, R.J. & Mackinlay, A.G., 1986. *J. Dairy Res.*, **53**, 381.
102. McLean, D.M. & Schaar, J., 1989. *J. Dairy Res.*, **56**, 297.
103. Pearse, M.J. & Mackinlay, A.G., 1989. *J. Dairy Sci.*, **72**, 1401.
104. Zoon, P., Roefs, S.P.F.M., de Cindio, B. & van Vliet, T., 1990. *Rheol. Acta*, **29**, 223.
105. Schwartzberg, H.G., Huang, B., Abularach, V. & Zaman, S., 1985. *Lat. Am. J. Chem. Eng. Appl. Chem.*, **15**, 141.
106. Heerink, G.J. & Geurts, T.J., 1981. *Comprimeerbaarheid van een wrongelbed onder de wei.* Wageningen (unpublished).
107. Vas, K., 1931. *Milchw. Forsch.*, **11**, 519.
108. Scott Blair, G.W. & Coppen, F.M.V., 1940. *J. Dairy Res.*, **11**, 187.
109. Delbeke, R. & Naudts, M., 1970. *De Bereiding van Herve-kaas met Rauwe Melk*, **19** Meded. Rijkszuivelstation Melle.
110. Geurts, T.J., 1978. *Neth. Milk Dairy J.*, **32**, 112.
111. Luyten, H., 1988. Doctoral thesis, Wageningen Agricultural University, Wageningen.
112. Olson, N.F. & Price, W.V., 1970. *J. Dairy Sci*, **53**, 1676.
113. Whitehead, H.R., 1948. *J. Dairy Res.*, **15**, 387.
114. Whitehead, H.R. & Harkness, W.L., 1954. *Austr. J. Dairy Technol.*, **9**, 103.

115. Feagan, J.T., Erwin, L.J. & Dixon, B.D., 1965. *Austr. J. Dairy Technol.*, **20**, 214.
116. Straatsma, J., de Vries, E., Heijnekamp, A. & Kloosterman, L., 1984. *Zuivelzicht*, **76**, 956.
117. Gillies, A.J., 1959. *Proc. 15th Intern. Dairy Congr., London*, **2**, 523.
118. Kiermeier, F. & von Wüllerstorf, B., 1963. *Milchwissenschaft*, **18**, 75.
119. Geurts, T.J., Walstra, P. & Mulder, H., 1974. *Neth. Milk Dairy J.*, **28**, 46.

6

Cheese Starter Cultures

TIMOTHY M. COGAN & COLIN HILL

*National Dairy Products Research Centre, Moorepark, Fermoy,
Co. Cork, Republic of Ireland*

1 INTRODUCTION

The use of starter cultures (often simply called starters) containing lactic acid bacteria (LAB) is an essential requirement in the manufacture of most cheeses. These cultures are called starters because they initiate or 'start' the production of lactic acid, their primary purpose in cheese manufacture. Acid production, combined with heating and stirring of the curd/whey mixture, causes the casein curd to synerese and expel moisture (whey) from the coagulum to produce a product—cheese—with a much lower water content (87% down to 35–60%), a lower pH (6·6 down to 4·6–5·2), and consequently a much longer shelf-life than milk. Acid production during cheese manufacture has other effects besides gel syneresis, e.g. it affects the activity, denaturation and retention of the coagulant in the cheese, curd strength, the extent of dissolution of colloidal calcium phosphate and inhibits the growth of many species of pathogenic and defect-producing bacteria. These aspects are considered in other chapters in these volumes and will not be considered further here. In this chapter, we will attempt to provide a broad understanding of the physiology, metabolism, genetics and propagation of starter cultures. It is not intended to be an exhaustive review but we would hope to indicate the likely direction of future research.

2 HISTORICAL

The use of starter cultures to produce acid during cheese manufacture was practised long before it was appreciated that bacteria were involved. Milk was held at room temperature for several hours during which the indigenous LAB multiplied and produced lactic acid. The soured (coagulated) milk was used to inoculate the cheese milk. If the cheese was of good quality, the inoculum was transferred into fresh milk for further use. A variation of this process is still in use in Switzerland, Italy and France where small traditional production units exist. The whey from one day's cheese production is incubated at the appropriate temperature for use the following day. All starter cultures in use today probably originated in this way and were passed between cheesemakers and from one generation to the next. Such empirical methods of starter production were used until the end of the 19th century when Storch in Denmark and Conn in the US showed that a good flavoured ripened butter could be produced from cream which was soured with pure cultures of *Lactococcus lactis* subsp. *lactis* or *Lactococcus lactis* subsp. *cremoris*. However, the butter still lacked the truly fine flavour of the traditional product. The reason for this became apparent in 1919 when three groups of workers—Hammer & Bailey in the US, Storch in Denmark and Boekhout & de Vries in Holland—independently established that starter cultures capable of producing the finest flavoured butter were mixtures of different types of bacteria, one of which (lactococci) was responsible for lactic acid production and the other (leuconostocs) for flavour production. Initially, it was thought that volatile acids were responsible for the flavour but subsequently it was shown that diacetyl is the important aroma compound.[1,2]

3 TYPES OF CULTURE

Essentially two types of starter cultures are used: mesophilic with an optimum temperature of ~30°C and thermophilic with an optimum temperature of ~45°C. The choice of culture depends on the cheese being made, e.g. mesophilic cultures are used in the production of Cheddar, Gouda, Edam, Blue and Camembert, while thermophilic cultures are used for Swiss and Italian varieties. This choice is related to the method of manufacture since Swiss and Italian cheese are cooked to much higher temperatures (50–55°C) which the starter bacteria must be capable of withstanding. More information is available on mesophilic than on thermophilic cultures, which probably reflects the greater amounts of cheese made with mesophilic cultures.

The LAB involved in starter cultures and some of their more important distinguishing properties are shown in Table I. Recent studies in molecular taxonomy, especially nucleic acid hybridization data, comparison of 16S ribosomal RNA sequences and analysis of cell wall components has led to significant changes in the taxonomy of starter LAB.[3–6] The older names are also listed to facilitate reading the earlier literature. Growth at 10 and 45°C can be used to distinguish mesophilic from thermophilic cultures while microscopic observations, measurement of the amount and isomer of lactic acid produced and the ability to metabolize citrate (in the case of mesophilic cultures) can readily distinguish most of the species within these broad categories. The release of galactose from lactose by *Lactobacillus delbrueckii* subsp. *bulgaricus* and *Lb. delbrueckii* subsp. *lactis* but not *Lb. helveticus* can be used to distinguish the lactobacilli (see Section 4.1); *Streptococcus salivarius* subsp. *thermophilus* also releases galactose from lactose.[7–9] Leuconostocs can be distinguished from the other starter bacteria by their ability to metabolize sugars by the phosphoketolase pathway, and their inability

TABLE I
Taxonomy and Some Distinguishing Characteristics of the Lactic Acid Bacteria Found in Starter Cultures

Organism	Old name	Shape	Reduction of litmus in milk before coagulation	% Lactic acid produced in milk[a]	Isomer of lactate	Metabolism of citrate	NH₃ from arginine	Growth at 10 °C	40 °C	45 °C	Fermentation of[b] Glu	Gal	Lac
Streptococcus salivarius subsp. thermophilus	Str. thermophilus	Cocci	−	0·6	L	−	−	−	+	+	+	−	+
Lactobacillus helveticus	Unchanged	Rods	−	2·0	DL	−	−	−	+	+	+	+	+
Lactobacillus delbrueckii subsp. bulgaricus	Lb. bulgaricus	Rods	−	1·8	D	−	−	−	+	+	+	−	+
Lactobacillus delbrueckii subsp. lactis	Lb. lactis	Rods	−	1·8	D	−	−/+	−	+	+	+	+/−	+
Lactobacillus lactis subsp. cremoris	Str. cremoris	Cocci	+	0·8	L	−	−	+	−	−	+	+	+
Lactococcus lactis subsp. lactis	Str. lactis	Cocci	+	0·8	L	+/−	+	+	+	−	+	+	+
Leuconostoc lactis	Unchanged	Cocci	−	<0·5	D	+	−	+	−	−	+	+	+
Leuconostoc mesenteroides subsp. cremoris	Leuc. cremoris	Cocci	−	0·2	D	+	−	+	−	−	+	+	+

[a] These are approximate values; individual strains vary.
[b] Glu, glucose; Gal, galactose; Lac, lactose.

to grow in milk unless a stimulant, e.g. 0·3% (w/v) yeast extract, is added, and, in some cases, 1·0% (w/v) glucose.[10] Under these conditions *Leuconostoc* spp. will coagulate milk in less than 24 h at 30°C and produce ~0·6% lactic acid. However, *Leuc. lactis* grows relatively well in milk (Table I). The requirement for yeast extract probably reflects the inability to produce sufficient proteinase to hydrolyse milk proteins to the amino acids and small peptides required for growth (see Section 5.3) while the requirement for glucose reflects the inability to use lactose as an energy source.

Both mesophilic and thermophilic cultures can be further sub-divided into mixed (undefined) cultures, in which the number of strains is unknown, and defined cultures in which the number of strains is known.

3.1 Mixed Strain Mesophilic Cultures

Undefined or mixed-strain mesophilic cultures are descended from the good starters of the late 19th and early 20th centuries and are commonly used in the production of fermented dairy products in Northern Europe. They are composed mainly of *Lc. lactis* subsp. *cremoris*, although occasionally they also contain the closely related *Lc. lactis* subsp. *lactis.* These species are called the lactococci (old names: group N streptococci or lactic streptococci). Some mesophilic mixed cultures also contain a lactococcus which metabolizes citrate (Cit⁺) to CO_2 and flavour compounds (see Section 4.2). The old name for this organism was *Streptococcus diacetylactis*, but as the major difference between it and *Lc. lactis* subsp. *lactis* is possession of a plasmid which encodes the uptake of citrate, it is no longer recognized as a species or sub-species.[3] Thus, many mesophilic cultures are comprised of Cit⁻ and Cit⁺ lactococci; the former is involved in acid production and the latter in flavour formation. *Leuconostoc* spp. are also involved in producing flavour compounds from citrate in many mixed cultures but the species is not known. Galesloot & Hassing[11] and Garvie[4] believe that *Leuc. mesenteroides* subsp. *cremoris* is the principal species while Stadhouders[12] feels that *Leuc. lactis* is also involved. Depending on the nature of the flavour producers, mesophilic mixed cultures are classified as: L type, containing *Leuconostoc* spp.; D type, containing Cit⁺ lactococci (diacetylactis); DL type containing both and O type containing no flavour producers.

The flavour producers are often called aromabacteria, aroma producers, citrate utilizers or citrate fermenters. The term 'fermenter' is an unfortunate choice since it implies, incorrectly as it happens, that citrate is used as an energy source. The acid and flavour producers in mesophilic cultures comprise about 90 and 10% of the microflora, respectively. They are called mixed cultures not only because they contain different bacterial species but also because they contain different strains of the same species. The evidence for the latter observation is not definitive but single cell isolates from mixed cultures have been shown to contain different plasmids, to have different growth rates, and different phage sensitivities.[13-16] In addition, some strains lack proteinase (Prt⁻) and as a result do not grow well in milk.[17,18]

Starter cultures are particularly susceptible to attack by phage (see Section 6.3) which slows down, or, in extreme cases, prevents the production of lactic acid. In the Netherlands, mixed strain mesophilic cultures which have been sub-cultured commercially without protection against airborne phage are called P cultures (P for practice) to distinguish them from similar cultures which have been sub-cultured aseptically in the laboratory (L for laboratory).[16] This type of L culture should not be confused with L cultures which contain leuconostocs as flavour producers. P cultures are more phage resistant than L cultures and can contain high numbers of phage (up to 10^8/ml) without affecting the ability of the culture to produce acid. These are called 'own' phage to distinguish them from 'disturbing' phage which inhibit acid production. L cultures are dominated by one or more strains, and are, therefore, more susceptible to phage attack. The best of these P cultures have been collected and stored at $-80°C$ at the Netherlands Institute for Dairy Research (NIZO) where they are propagated under carefully controlled conditions and distributed to the cheese factories.

3.2 Defined Strain Mesophilic Cultures

Defined strain mesophilic cultures are mainly pure cultures of *Lc. lactis* subsp. *cremoris* and were first introduced commercially in New Zealand in 1934.[19] The first bacterium ever purified was *Lc. lactis* subsp. *lactis* by Lister in 1878 and it could be argued that this was, in fact, the first defined culture. The literature on defined cultures can be confusing as they have also been called mono cultures, single cultures, pairs, triplets and multiple cultures. Many of these names reflect the way in which they are used, i.e. as single cultures or as mixtures of two to six strains (most commonly two or three). They are propagated together only when bulk culture is being produced. The important differences between these cultures and traditional mixed cultures are that the number of strains is known and they do not contain flavour producers.

In the 1930s, open texture was a big problem in Cheddar cheese being made in New Zealand due to CO_2 production from citrate by the flavour producers in the mixed strain starters.[20] The acid producers in these starters were isolated and used individually (the single strain starter system) to make cheese. The cheese was close-textured but when these strains were used commercially, slow acid production occurred which was subsequently shown to be due to lysis of the starter cultures by phage (see Section 6.3). This problem was overcome by using pairs of phage-unrelated strains in four day rotations with a different pair being used on each of the four days.[19,21] The idea behind this was that phage numbers were sufficiently reduced on the fifth day, when the first pair was used again, to have little effect on growth and acid production. In this way, phage build-up in the plant was avoided. In addition to being phage unrelated, one of the strains in each pair grew and produced acid at the cooking temperature (a fast or temperature-resistant strain) while the other did not grow but continued to produce acid although at a slower rate (a slow or temperature-sensitive strain). The use of the slow strain reduced the development of bitterness (see Section 4.3.5) in the cheese caused by too much growth of the fast strain.

This starter system (rotation of pairs) was used very successfully for several decades but the amalgamation of smaller cheese plants into larger units put it under increased pressure. Phage became a much greater problem as greater volumes of starter were required, vats were filled several times daily and cheese was made 'by the clock' which left little time to nurse slow vats to the desired final acidities. This problem was solved by the development of multiple strain starters whose use has now spread from New Zealand[22] to Australia,[23,24] the US[25,26] and Ireland.[27] The key to their development was a simple test to predict the ability of a strain to withstand phage.[28] In this test, potential strains are screened for resistance to as large a mixture of phage as possible over several growth cycles (usually seven) in milk. Whey from the previous growth cycle is also included so that any phage present in low numbers have a chance to multiply and exert an inhibitory effect. These phage-resistant strains are then further screened for rates of acid production, inability to act as indicators of lysogenic phage, inability to produce bacteriocins against the other strains being used and lack of off-flavour production. Cultures which survive these screening tests have proved very useful commercially. Multiple strain cultures are used without rotation. Originally, combinations of six strains were used but, in the past few years, the number of strains has been reduced to two or three without an adverse affect on acid production or cheese quality. Despite the fact that multiple strain cultures are selected on the basis of phage resistance, they can still be attacked by phage. When this occurs, the offending strain is removed from the system and (bacterio)phage-insensitive mutants (BIMs) are isolated to replace it. This approach to strain replacement is practised in New Zealand, Australia, the US and Ireland.[19,23-29] Unfortunately, not all strains are amenable to isolation of BIMs for reasons which are not clear.

3.3 Thermophilic Cultures

Undefined or mixed thermophilic cultures are especially common in the small cheesemaking factories of France, Switzerland and Italy in which Gruyère, Emmental and Grana cheese are produced and are often called artisan cultures.[30] Very often, these cultures are produced by incubating whey from the previous day's production overnight at 40–45°C. The flora is composed mainly of *Str. salivarius* subsp. *thermophilus* (enterococci and lactococci may also be present) and several species of *Lactobacillus* e.g. *Lb. fermentum, Lb. helveticus, Lb. delbrueckii* subsp. *lactis* and *Lb. acidophilus.*[31,32] Low-acid natural whey cultures, in which *Str. salivarius* subsp. *thermophilus* predominated, have been used in the past but, more recently, high-acid whey cultures, consisting mainly of lactobacilli, are used.[33] Efforts have been made to refine these cultures. Emmental cheese has been prone to a defect, called the secondary fermentation, which is characterized by increased CO_2 production and blowing of the cheese after the propionic acid fermentation. This was thought to be due to further metabolism by propionic acid bacteria (PAB) (see Chapter 3, Volume 2) but it was subsequently shown to be due to the use of an excessively proteolytic *Lb. helveticus* which led to increased growth of the PAB and further production of CO_2.[34] The problem has been alleviated to some extent by careful selection of the starter cultures. Mixed cultures were collected from factories making good quality Emmental cheese and were intensively screened for acid production and proteolytic activity, storage stability and cheesemaking performance. Only 8 out of 186 cultures survived the selection process. They are propagated under carefully controlled conditions for distribution to the cheese factories.

In modern factories, defined strain thermophilic starters (single or multiple strain cultures) are generally used. In contrast to mesophilic defined cultures, these are used in rotation. Thermophilic cultures almost always contain *Str. salivarius* subsp. *thermophilus* and a thermophilic *Lactobacillus*: *Lb. delbrueckii* subsp. *lactis; Lb. delbrueckii* subsp. *bulgaricus* or *Lb. helveticus*. An exception is Beaufort cheese in which no *Streptococcus* is used. The species of *Lactobacillus* used is determined by the product being made: *Lb. helveticus* and *Lb. delbrueckii* subsp. *lactis* are used for Swiss cheese manufacture and *Lb. delbrueckii* subsp. *bulgaricus* and *Lb. lactis* subsp. *lactis* for yoghurt manufacture. Care should be taken in selecting the Lactobacillus for Swiss type cheese since many strains of *Lb. helveticus* are very proteolytic and can give rise to the secondary fermentation defect.[34] *Lb. helveticus* and a few strains of *Lb. delbrueckii* subsp. *lactis* can ferment the galactose produced from lactose (Gal^+) while *Lb. delbrueckii* subsp. *bulgaricus*, most strains of *Lb. delbrueckii* subsp. *lactis*, and *Str. salivarius* subsp. *thermophilus* do not (Gal^-).[7-9] Instead, galactose is excreted in proportion to the lactose taken up. Only Gal^+ strains of thermophilic lactobacilli should be used as starter cultures in cheese since they will ferment the galactose produced by *Str. salivarius* subsp. *thermophilus*. If Gal^- strains are used, galactose accumulates in the cheese and may act as an energy source for growth and off-flavour production by non-starter bacteria. Turner & Martley[35] suggest that many strains of *Lb. delbrueckii* subsp. *bulgaricus* used commercially in the US are Gal^+ and should be reclassified as *Lb. helveticus* and that this mistaken identification may explain why *Lb. delbrueckii* subsp. *bulgaricus* is said to be preferred for Swiss cheese manufacture in the US while *Lb. helveticus* and Gal^+ *Lb. delbrueckii* subsp. *lactis* are preferred in Europe. The streptococci and lactobacilli are propagated together for yoghurt manufacture and separately for cheese manufacture. The volume of inoculum used in the manufacture of Gruyère and Emmental cheese is only 3–10% of that used for Cheddar or Dutch type cheese. Consequently, the cultures are often added directly to the cheese milk in the vat. In this case these cultures are grown individually, concentrated by centrifugation or membrane filtration and frozen at $-30°C$ in small vials until required for direct inoculation.[36,37]

3.4 Media for Growth and Differentiation

Starter LAB are very fastidious and require several amino acids, vitamins and other nutritional factors for growth. Because of this, the media used for isolation and growth generally include yeast and/or beef extract, and one or more enzymatic digests of protein; tryptone, a tryptic digest of casein, is commonly used. These bacteria also require a fermentable carbohydrate for energy production and since their metabolism is fermentative, large amounts of lactate are produced. Therefore, media must be adequately buffered. Inorganic ions (Fe^{2+} Mg^{2+} Mn^{2+}) are also frequently added. In the past, the most common general purpose medium used to grow the lactococci was the unbuffered lactic agar of Elliker et al.[38] The current medium of choice is M17 which is buffered with β-glycerophosphate.[39] This medium is also very useful for growing Str. salivarius subsp. thermophilus[9,40] and for preparing and assaying phage for both this organism and the lactococci.[39] Media containing β-glycerophosphate are not suitable for growth of many thermophilic lactobacilli, especially Lb. delbrueckii subsp. bulgaricus and Lb. delbrueckii subsp. lactis and, in fact, M17 is used to enumerate Str. thermophilus selectively in yoghurt.[41] MRS[42] is the most useful general purpose medium for lactobacilli; streptococci and leuconostocs also grow well on this medium but it can be made selective for lactobacilli by reducing the pH to 5·4. Differential/selective media to distinguish Str. thermophilus from the thermophilic lactobacilli have been reviewed[43] but none of them works equally well for all strains. A provisional standard has been proposed by the IDF:[44] Str. thermophilus is enumerated on M17 agar (after aerobic incubation at 37°C for 48 h) and Lb. bulgaricus on MRS agar at pH 5·4 (after anaerobic incubation at 37°C for 72 h).

Numerous differential media have been proposed to distinguish the flavour and acid producers found in mesophilic starter cultures. Many contain calcium citrate: the flavour producers are detected as colonies surrounded by a clear zone due to metabolism of the insoluble calcium citrate. These media are opaque so that it is difficult to see the acid producers. Addition of 2,3,5-triphenyltetrazolium chloride (0·1 mg/ml) allows the acid producers to be easily visualized (Cogan, unpublished data). Of several media examined by Waes,[45] the KCA medium of Nickels and Leesment[46] was found to be the most useful. In mesophilic cultures, leuconostocs contain β-galactosidase (βgal) (see Section 4.2) while lactococci contain phospho-β-galactosidase (pβgal). Vogensen et al.[47] further modified KCA by adding the chromogenic substrate, 5-bromo-4-chloro-3-indolyl-β-D-galactopyranoside (X-gal), to detect βgal activity. In this medium, the Cit^+ lactococci are white in colour surrounded by a clear halo due to citrate utilization while leuconostocs are blue, due to the action of βgal on X-gal, and are also surrounded by a halo. Tetracycline (0·15 μg/ml) has been reported to be selective for Leuconostoc spp.[48] Two other promising media have been suggested.[49,50] In the first one, the Cit^+ lactococci are detected by their ability to metabolize citrate while Lc. lactis subsp. cremoris is distinguished from Lc. lactis subsp. lactis by its inability to produce NH_3 from arginine. The second medium[50] is very useful for the differential enumeration of Cit^+ and Cit^- lactococci but many Leuconostoc spp. grow poorly on it (Cogan, unpublished data). Potassium ferrocyanide and a mixture of sodium and ferric citrates are added to a milk/tryptone/glucose agar. The differential property of this medium depends on the precipitation of Fe^{3+} as 'Prussian blue'. Cit^- lactococci are white while Cit^+ lactococci are blue because the citrate, which prevents the precipitation of 'Prussian blue', has been utilized around the colonies.

4 METABOLISM

4.1 Lactose Transport

Starter LAB require a fermentable carbohydrate for energy production and growth. In milk, this is lactose, a disaccharide composed of galactose and glucose. Transport of lactose requires energy but the exact mechanisms used in all starter LAB have not been elucidated. *Str. salivarius* subsp. *thermophilus* uses a proton motive force (PMF)[51,52] while the lactococci use a group translocation system involving the phosphoenol pyruvate (PEP) phosphotransferase system (PTS). In the PMF system, lactose is transported against a concentration gradient whereas in the PTS system no gradient is established because lactose is transformed to lactose-P during transport. No information is available on the transport systems used by thermophilic lactobacilli or leuconostocs. The PTS system is found only in bacteria which ferment sugar by glycolysis[53] and so is unlikely to be found in leuconostocs which ferment sugar by the phosphoketolase pathway. A pH gradient (ΔpH) exists between the inside and outside of the cell with the inside being more alkaline (for a review see Ref. 54). This causes an electrical potential ($\Delta\psi$) to be set up at the membrane. $\Delta\psi$ and ΔpH collectively comprise the PMF. PMF systems frequently involve movement of a second molecule into (symport) or out of (antiport) the cell by specific membrane-located carrier proteins, called permeases. Antiport is the simultaneous transport of two compounds in opposite directions with the resultant concentration gradient in one transporting the other against its concentration gradient. The function of the efflux of galactose in the transport of lactose by *Str. salivarius* subsp. *thermophilus* is to drive the transport of lactose by an antiporter mechanism.[51]

In the PTS system, PEP is the energy source and lactose is transformed into lactose-P as it is transported into the cell.[55,56] This is a complex system requiring Mg^{2+} and four proteins, two of which, enzyme II and enzyme III, are membrane-associated and sugar-specific while the other two, enzyme I and the low molecular weight heat stable protein (HPr), are soluble proteins and common to all PTS systems (Fig. 1). Enzyme III is peripheral to and enzyme II an integral component of the cell membrane. Lawrence & Thomas[57] have suggested that the PTS transport system and pβgal are prerequisites for rapid lactose fermentation. This was supported by the findings of Farrow[58] and Crow & Thomas.[59] The former showed that starter strains which fermented lactose rapidly had high pβgal and no βgal activities while strains which fermented lactose slowly had high levels of βgal and low levels of pβgal. The latter workers showed that *Lc. lactis* subsp. *lactis* ATCC 7962 transports lactose by the PTS system but has no pβgal activity

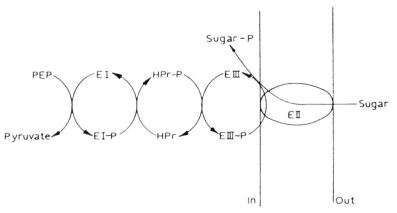

Fig. 1. The phosphoenol-pyruvate phosphotransferase system of sugar uptake in lactococci. Abbreviations: EI, enzyme I; EII, enzyme II; EIII, Enzyme III; HPr, heat stable phosphocarrier protein; ~ phosphorylated compound. EI and HPr are soluble proteins and are common to all sugar transport systems while EII and EIII are membrane-associated proteins, which are sugar specific (from Dills et al.).[56]

and grows slowly in milk. In contrast, *Str. salivarius* subsp. *thermophilus* does not have a PTS transport system nor pβgal and ferments lactose rapidly.[60]

In the PMF and PTS transport systems, lactose is transported as lactose and lactose-P, respectively. Lactose is initially hydrolysed by βgal to glucose and galactose while lactose-P is hydrolysed by pβgal to glucose and galactose-6-P. βgal and pβgal are easily measured using chromogenic substrates (ONPG and ONPGP, respectively) and their presence is a good indicator of the transport system being used. There are contradictory reports as to the presence of βgal and pβgal in *Str. salivarius* subsp. *thermophilus* and the thermophilic lactobacilli; some workers report βgal[9] only while others report the presence of both enzymes.[8,61] Where both activities are present, much greater amounts of βgal are found. In the case of the thermophilic lactobacilli, pβgal activity was an artefact caused by hydrolysis of ONPGP, the substrate for pβgal, to ONPG, the substrate for βgal, by a phosphatase.[8] It should also be remembered that both enzymes are assayed with artificial substrates and low activities do not necessarily mean that activities on the natural substrate are also low.

The transport of sugars other than lactose has also been studied. Galactose transport in *Str. salivarius* subsp. *thermophilus* occurs via an ATP-dependent permease but only in the presence of another energy source like sucrose.[8,62] Some evidence for a glucose-specific PTS system has been reported in thermophilic lactobacilli.[8] In the lactococci, glucose is transported by a PTS system and galactose by two sugar-specific systems, a low affinity PTS system and a high affinity permease system.[63,64] This has led to speculation[63,65] that when growing on low levels of galactose, uptake of the sugar is via the high affinity system and further metabolism by the Leloir pathway and that when growing on high levels of galactose, uptake is via the low affinity PTS system and further metabolism by the tagatose pathway.

4.2 Lactose Metabolism

4.2.1 Lactococci

In the lactococci, lactose-P transported by the PTS system is hydrolysed to glucose and galactose-6-P by pβgal. The former is fermented via the glyolytic pathway while the latter is metabolized through several tagatose derivatives of glyceraldehyde-3-P and dihydroxy acetone-P which then enter the terminal reactions of glycolysis (Fig. 2).[66] Lactose transport and the enzymes of the tagatose pathway are plasmid encoded in the lactococci.[67] Tagatose is a stereoisomer of fructose but separate enzymes are involved in the formation of fructose-6-phosphate and tagatose-6-phosphate, fructose-1,6-diphosphate (FDP) and tagatose-1,6,-diphosphate (TDP).[68] TDP aldolase has some activity on FDP but FDP aldolase has no activity on TDP. The terminal reactions of both glucose and tagatose metabolism are similar, and result in the reduction of pyruvate to lactate in an NADH-requiring reaction catalysed by LDH. Only L-lactate is formed. The reason for lactate formation is the need to regenerate stoichiometric amounts of NAD$^+$ to continue the fermentation. For this reason, lactococci contain large amounts of LDH, an allosteric enzyme, which is activated by FDP.[69]

L-lactate is generally the sole product of fermentation but when these bacteria are grown on galactose, maltose or low levels of glucose, other products of pyruvate metabolism besides L-lactate, e.g. formate, ethanol and acetate, are formed.[65,70] The pathways for formation of these products are shown in Fig. 3. Acetyl Co-A can be produced from the pyruvate-formate lyase (PFL) reaction or from the pyruvate dehydrogenase reaction. Both enzymes have been detected in lactococci, but under normal (anaerobic) growth conditions, PFL activity is responsible for acetyl CoA production (see Section 4.2.5). PFL has a lower affinity for pyruvate than LDH and is inhibited by triose phosphates while LDH is activated by FDP. Under conditions where the other products of pyruvate metabolism are formed, the concentrations of triose phosphates and FDP decrease, which results in higher PFL and lower LDH activities, respectively. The net result is that more pyruvate is metabolized to formate, acetate and ethanol at the expense of lactate. Pyruvate is also produced from citrate by Cit$^+$ lactococci but much of this is used for acetoin and diacetyl production (see Section 4.2). Theoretically, pyruvate produced from citrate could be used for lactate formation but the stoichiometry between lactate and NAD$^+$ production in glycolysis suggests otherwise.

PEP plays a pivotal role in determining the balance between sugar transport and metabolism. The regulation of sugar metabolism is complicated and essentially relies on the allosteric enzyme, pyruvate kinase (PK), and the intracellular concentrations of various effectors, especially FDP and iP.[68,71,72] FDP is a positive effector while iP is a negative effector of PK. In growing cells, the concentration of FDP is high while that of iP is low so that PK activity is maintained and pyruvate is produced. In contrast, during starvation, the FDP concentration is low and that of iP high. This results in no PK activity and a consequent increase in the concentration of PEP.

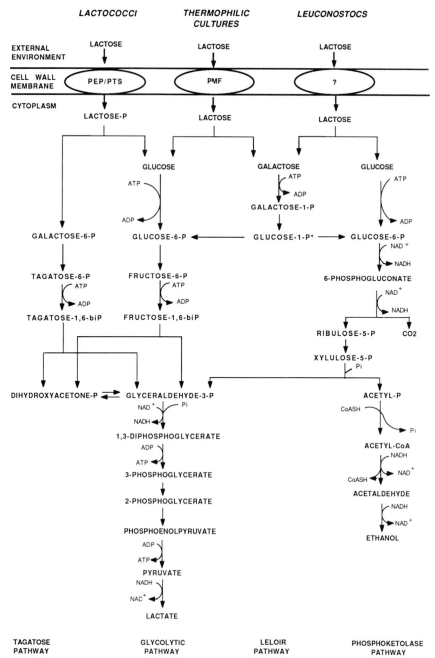

Fig. 2. Sugar metabolism by lactococci, leuconostocs and thermophilic cultures. Only some thermophilic cultures (*Lb. helveticus* and some strains of *Lb. delbrueckii* subsp. *lactis*) metabolize galactose (see text). These Gal⁺ strains, and probably leuconostocs also, metabolize galactose by the Leloir pathway. The subsequent metabolism of glucose-6-P is by the glycolytic pathway in the lactobacilli and by the phosphoketolase pathway in the leuconostocs.

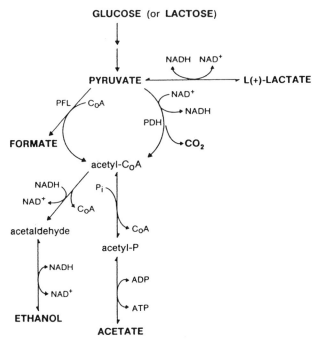

Fig. 3. Pyruvate metabolism in lactococci. PFL, pyruvate-formate lyase; PDH, pyruvate dehydrogenase (from Fordyce *et al.*).[70a]

4.2.2 Leuconostocs

In Leuconostocs, lactose is hydrolysed to glucose and galactose by βgal. Glucose is metabolized by the phosphoketolase pathway to equimolar concentrations of lactate, ethanol and CO_2 while galactose is probably metabolized by the Leloir pathway to glucose-6-P (Fig. 2), using the enzymes galactokinase (GK), galactose-1-P uridyl transferase and UDP-4-epimerase (Fig. 2). Lactate and ethanol are produced because of the need to regenerate NAD^+ to continue the fermentation. Unlike the lactococci, the D isomer of lactate is formed. Co-metabolism of glucose and citrate results in a switch from ethanol to acetate production and the production of more lactate than can be accounted for in terms of the glucose used.[73,74] This is due to conversion of all the pyruvate produced from both citrate and glucose to lactate to regenerate NAD^+ (See Section 4.3).

4.2.3 Streptococcus salivarius *subsp.* thermophilus

A review of carbohydrate metabolism in this organism has been published recently.[60] Transport of lactose in this organism is via a PMF system and hydrolysis of lactose is catalysed by βgal which is constitutively present in glucose-grown cells, but is increased two- and four-fold by lactose and galactose, respectively.[7,62] However, only the glucose moiety is subsequently metabolized; galactose is excreted in amounts equimolar with the lactose used.[9] Most strains of *Str. salivarius* subsp. *thermophilus* are considered to be Gal⁻, although 12 of 31 strains studied by Somkuti & Steinberg were Gal⁺.[75] Gal⁺ strains can be selected from Gal⁻ strains on lactose-limited chemostats, but these revert to Gal⁻ within 100 generations (10 subcultures).[7] Wild-type Gal⁺ strains also excrete galactose when grown on lactose but they will metabolize it in the presence of low (3·5 mM) levels of lactose.[9,62] No evidence for a tagatose pathway has been found but high levels of hexokinase, aldolase, glyceraldehyde-3-P dehydrogenase, PK and LDH activities are found, implying that glucose is fermented by the glycolytic pathway.[7] Low levels of the Leloir pathway enzymes, galactokinase (GK), galactose-1-P-uridyl transferase and UDP-glucose-4-epimerase, are found in cells metabolizing lactose and phosphorylation of galactose by GK appears to be the rate-limiting step in the utilization of galactose.[7,62] Three pieces of information support this contention:

1. GK activity in Gal⁻ strains growing on lactose, with one exception, is lower than in Gal⁺ strains.[62]
2. Gal⁺ strains growing on galactose contain 10 times more GK activity than when grown on lactose.[62]
3. Gal⁺ strains, selected from Gal⁻ strains growing in lactose-limited chemostats contain three to five times more GK activity than Gal⁻ strains growing on lactose.[7]

These data suggest that lactose and/or glucose repress GK activity and that while glucose can be readily metabolized, the capacity of the organism to metabolize the equivalent amount of galactose produced from lactose is exceeded and galactose is excreted.

There is very little information on the regulation of sugar metabolism in *Str. salivarius* subsp. *thermophilus*. βGal is induced by lactose and repressed by glucose;[75] repression can be lifted by cAMP which is found in high concentrations in these bacteria.[76] In contrast to the lactococci, the LDH of *Str. salivarius* subsp. *thermophilus* is not affected by FDP but PK is activated by FDP, fructose-6-P and glucose-6-P and inhibited by iP.[77] In addition *Str. salivarius* subsp. *thermophilus* does not use PEP in lactose transport. Consequently, the regulatory role of PK may not be as important in this bacterium as it is in the lactococci. Unlike lactococci, *Str. salivarius* subsp. *thermophilus* ferments sugars homofermentatively[78] even when growing on low concentrations of glucose or lactose. However, small amounts of formate (<1% of the total sugar metabolized) are produced from sugar.

4.2.4 Thermophilic Lactobacilli

These bacteria contain both βgal and pβgal but the latter was shown to be an artefact due to formation of ONPG, the substrate for βgal, from ONPGP, the substrate for Pβgal, by a phosphatase.[8] Thus, these bacteria probably contain only βgal. Only *Lb. helveticus* and a few strains of *Lb. delbrueckii* subsp. *lactis* metabolize galactose; all strains of *Lb. delbrueckii* subsp. *bulgaricus* and most strains of *Lb. delbrueckii* subsp. *lactis* do not.[8,35] The latter organisms, like

TABLE II
Salient Features of Lactose Metabolism in Starter Culture Organisms

Organism	Transport[a]	Cleavage enzyme	Pathway[b]	Products (mol/mol lactose)
Lactococcus	PTS	pβgal	GLY	4 L Lactate
Leuconostoc	?	βgal	PK	2 D Lactate + 2 Ethanol + 2 CO_2
Str. salivarius subsp. thermophilus	PMF	βgal	GLY	2 L Lactate[c]
Lb. delbrueckii subsp. lactis	PMF?	βgal	GLY	2 D Lactate[c]
Lb. delbrueckii subsp. bulgaricus	PMF?	βgal	GLY	2 D Lactate[c]
Lb. helveticus	PMF?	βgal	GLY	4 L (mainly) + D Lactate

[a] PTS, phosphotransferase system; PMF, proton motive force.
[b] GLY, glycolysis; PK, phosphoketolase.
[c] These species metabolize only the glucose moiety of lactose.

Str. salivarius subsp. *thermophilus*, excrete galactose in amounts equimolar to the lactose used when growing on lactose; some strains will metabolize galactose but only when low (4·0 mM) concentrations of lactose are present.[8] Whether this is due to an antiporter PMF transport system involving galactose remains to be established. Sugar metabolism by these organisms is by glycolysis and different isomers of lactate are formed which are useful in identification (Table I). The production of both isomers could be due to the presence of two stereo-specific LDHs or to one LDH and a racemase which transforms one isomer into the other. Racemases are found in very few strains of LAB but in none of those present in starter cultures. Consequently, production of L and D lactate is generally considered to be due to two specific LDHs.

A summary of the different mechanisms of transport and metabolism of lactose in starter LAB is given in Table II.

4.2.5 Aerobic Metabolism
Many starter LAB, when exposed to O_2, produce H_2O_2 which can inhibit their subsequent growth. Production of H_2O_2 is due to increased activity of several enzymes, including pyridine nucleotide oxidases and peroxidases, pyruvate oxidase and α-glycerophosphate oxidase. The inhibition can normally be prevented by addition of catalase (for a review see Condon[79]). The stoichiometry of product formation from sugars can be quite different when these organisms grow aerobically since some of these enzymes are alternative mechanisms for regenerating NAD^+. Leuconostocs growing aerobically produce acetate instead of ethanol due to increased synthesis of NADH oxidase and acetate kinase and lower activities of phosphotransferase and alcohol dehydrogenase.[80] It is interesting that metabolism of citrate has similar effects on product formation.[73,74]

Lactococci growing on complex media with glucose as the energy source are essentially homofermentative under both aerobic and anaerobic conditions. However, in defined media growth does not occur aerobically unless lipoic acid or acetate is present.[81] Lipoic acid is an integral part of the pyruvate dehydrogenase (PDH) enzyme complex which is involved in acetyl CoA synthesis. Aerobically, acetyl CoA is produced from the PDH complex (or acetate) while anaerobically, it is produced from PFL activity; this enzyme is sensitive to oxygen and is inactivated under aerobic conditions of growth. Acetate and lactate are the major products of sugar metabolism in the presence of excess lipoate while acetoin replaced acetate with limiting lipoate. Increased diacetyl production may be obtained by addition of hemin to Cit[+] lactococci.[82] All of these data show that it is possible for essentially homofermentative organisms to divert a major portion of their carbohydrate metabolism from C3 to C4 compounds by manipulating the growth conditions.

H_2O_2 coupled with SCN[-] and lactoperoxidase (LP) is a potent inhibitor of some starter lactococci. Both SCN[-] and LP are naturally present in milk and H_2O_2 can be produced by aeration of milk during growth.[83] LP is inactivated at ~80°C so that inhibition does not normally occur in milk used for starter production. The exact mechanism of inhibition is not understood but oxidation products of SCN[-], especially OSCN[-], are believed to be involved.

4.2.6 Lactate Levels in Cheese

The levels of lactate in Camembert, Swiss, Cheddar and Romano cheese after manufacture are 1·0, 1·4, 1·5 and 1·7%, respectively, with the L form the dominant isomer.[84-87] Information on Dutch type cheese is lacking. However, the level of lactose in Dutch cheese at pressing is ~1·4%[88] which theoretically would form ~1·4% lactic acid. Allowing for loss of lactate in the whey and during brining, a value of ~1·2% could be presumed. The lactate in cheese can be metabolized further. In Camembert cheese, it is transformed to CO_2 and H_2O by the surface moulds; in Swiss cheese, both L and D lactate are metabolized to propionate, acetate and CO_2 while in Cheddar cheese the L lactate can be converted to D lactate or to acetate by the non-starter LAB (NSLAB). These transformations are discussed in more detail in other chapters and have been reviewed recently by Fox et al.[89]

4.3 Citrate Metabolism

Even though the concentration of citrate in milk is low (8 mM), the metabolism of this acid is important in determining the texture and flavour of cheese: CO_2 production is responsible for eye formation in Dutch-type cheese while diacetyl and acetate production are responsible for the flavour and aroma of fresh (unripened) cheese. Diacetyl is also thought to be an important flavour compound in Cheddar cheese.[90] CO_2 production is sometimes undesirable as it is thought to be responsible for floating curd in cottage cheese and for openness in Cheddar cheese.[91,92] It is also produced from fermentation of lactose by *Leuconostoc* species.

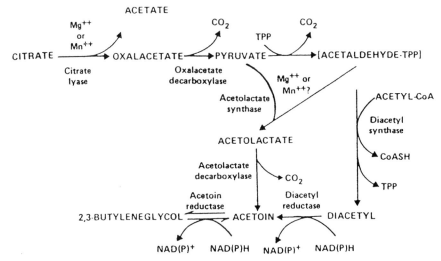

Fig. 4. Citrate metabolism in lactococci and leuconostocs.

Among starter LAB, only the flavour producers (Cit$^+$ *Lc. lactis* subsp. *lactis* and *Leuconostoc* spp.) in mesophilic cultures metabolize citrate. It is not used as an energy source but is metabolized rapidly in the presence of a fermentable carbohydrate by the pathway outlined in Fig. 4.[93–95] Citrate is transported by a plasmid-encoded permease,[50,96] and hydrolysed to oxalacetate by citrate lyase (CL).[97,98] The permease from Cit$^+$ lactococci has been cloned, sequenced and expressed recently in *E. coli* (see Section 5.4.2).[99,100] CL is constitutively present in Cit$^+$ lactococci[101] and inducible in *Leuconostoc* spp.[102] It requires acetylation by another enzyme, citrate lyase ligase, for activity:[103]

Diacetyl-Citrate Lyase + Acetate + ATP → Citrate Lyase + AMP + PPi
(Inactive enzyme)   (Active enzyme)

CL has an optimum pH of ~7·2 and the substrate is an equimolar complex of citrate and Mg^{2+} or Mn^{2+}. It has Km and V_{max} values of 990 mM and 670 mM/min with Mg-citrate as substrate and 94 mM and 470 mM/min with Mn-citrate.[97,98]

Oxalacetate is decarboxylated to pyruvate which in turn is decarboxylated to acetaldehyde-TPP. The latter can condense with acetylCoA to form diacetyl or with another molecule of pyruvate to form α-acetolactate (ALA) which is then decarboxylated to acetoin. It is not clear if pyruvate decarboxylase is involved in decarboxylating pyruvate to the acetaldehyde-TPP complex or whether ALA and diacetyl synthases catalyse both the decarboxylation and condensation reactions. It is also not clear whether the ALA synthase requires divalent cations (Mn^{2+} or Mg^{2+}), like that from other microorganisms.[104] The ALA synthase of *Leuc. lactis* does not require Mn^{2+} but the large amounts of this cation present in cell free extracts may mask any effect of additional Mn^{2+}.[105] The ALA decarboxylase gene from Cit$^+$ *Lc. lactis* subsp. *lactis* has been cloned recently and expressed in *E. coli*.[106]

ALA is an unstable molecule and easily decomposes chemically; non-oxidatively to acetoin and oxidatively to diacetyl.[107–109] This raises the question of whether or not diacetyl is formed chemically from ALA or enzymatically from acetylCoA, a contentious issue which has not been resolved. Support for the enzymatic formation of diacetyl from acetylCoA has come from studies which showed that ^{14}C-labelled diacetyl is formed from ^{14}C-labelled acetate.[94] The rationale behind these experiments is that in the absence of lipoic acid, Cit[+] lactococci can form acetyl-CoA only from acetate via acetate kinase and phosphotransacetylase. Thus, if diacetyl is labelled, acetylCoA must be involved in its synthesis; if ALA were involved, the diacetyl would have been unlabelled. Another argument in support of the enzymatic formation of diacetyl is that the redox potential is too low to allow oxidative decarboxylation of ALA to diacetyl.[110] This argument does not preclude the non-enzymatic formation of acetoin from ALA but ALA decarboxylase activity has been detected in Cit[+] lactococci.[95] Only one of 17 strains of Cit[+] lactococci was capable of excreting ALA and this was a special strain isolated from a commercial culture whose function is to do exactly that.[111] Leuconostocs do excrete ALA,[111] suggesting that acetoin may be produced chemically in these bacteria.

Because of its instability, ALA can interfere with the estimation of both diacetyl and acetoin.[109,111] Many of the processing steps (e.g. steam distillation) applied to cultures prior to measurement of these compounds result in decarboxylation of ALA. Thus, some values for acetoin and diacetyl in the literature may be overestimated.

Acetoin can also be produced from diacetyl by acetoin dehydrogenase (previously called diacetyl reductase) and 2,3-butylene glycol (BG) from acetoin by BG dehydrogenase (previously called acetoin reductase). Both enzymes require NADH or NADPH for activity and both substrates differ by only a single H atom which raises the question whether one or two enzymes are involved. These activities are catalysed by one enzyme in *Saccharomyces cerevisiae* and *Enterobacter aerogenes* and by two enzymes in *E. coli*.[112–114] In Cit[+] *Lc. lactis* subsp. *lactis*, two BG dehydrogenases have been purified and both have activity on diacetyl and acetoin.[115] Although activity on diacetyl is much higher than that on acetoin in a mixture of both substrates, acetoin is reduced preferentially. Both enzymes have similar MWs (170 kD). The optimum pHs of enzymes I and II were 10·0 and 8·5 when assayed with BG, 6·1 and 7·0 when assayed with acetoin and 5·8 and 6·1 when assayed with diacetyl, respectively. The products formed from diacetyl by enzyme I were acetoin and meso-BG while those of enzyme II were acetoin and optically active BG.[115] Two stereo-isomers of acetoin (L and D) and three isomers of BG (L, D and meso) are possible. The isomer of acetoin produced was not reported and perhaps enzymes I and II are responsible for the production of different isomers of acetoin.

The presence of citrate represses the synthesis of acetoin and BG dehydrogenase in Cit[+] lactococci which helps to explain why diacetyl and acetoin accumulate in these cultures.[101,115,116] Once citrate has been exhausted, increased synthesis of the enzymes occurs with consequent reduction in the levels of both diacetyl and acetoin. Reduction is more rapid when cultures are held at higher pH values.[115] This result suggests that these enzymes are on the outside of the cell wall or that their substrates are freely transportable across the cell wall.

The production of diacetyl and acetoin in milk by L, D and DL cultures is shown in Fig. 5. All three cultures produce lactic acid at the same rate but D

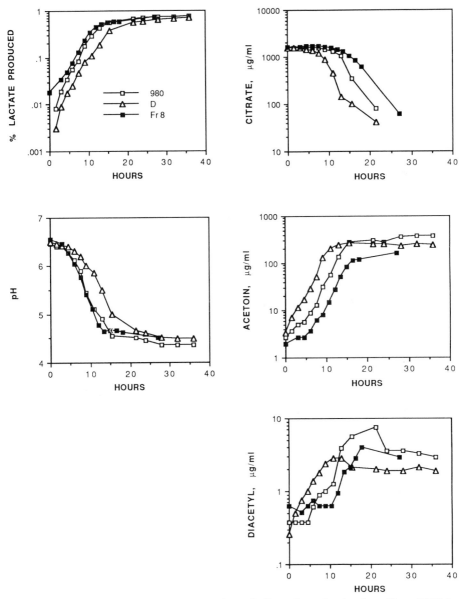

Fig. 5. A comparison of lactic acid, acetoin and diacetyl production in milk at 21°C by D, DL (980) and L (Fr8) mesophilic cultures.

and DL cultures utilize citrate and produce diacetyl and acetoin much more rapidly than L cultures, probably due to faster growth of Cit$^+$ lactococci than leuconostocs. Invariably, greater amounts of acetoin than diacetyl are produced by all cultures, which may be due to a limited rate of acetylCoA synthesis.

Pure cultures of Cit$^+$ lactococci and leuconostocs behave completely differently in their metabolism of citrate. The Cit$^+$ lactococci produce no diacetyl and little acetoin from sugars in the absence of citrate, even though the necessary enzymes and substrate (pyruvate) are present.[101,115,116] This is attributed to the need of the organisms to convert all the pyruvate to lactate in order to produce the NAD$^+$ needed to continue glycolysis. When the citrate is also metabolized, Cit$^+$ lactococci produce major amounts of acetoin and acetate and minor amounts of diacetyl and BG.[101,115] The presence of Cu^{2+}, Fe^{2+} and hemin also increase the production of diacetyl by Cit$^+$ lactococci, while simultaneously decreasing the uptake of citrate.[82] Subsequent experiments showed that diacetyl decreased the uptake of citrate.[117,118] In contrast, leuconostocs produce equimolar concentrations of lactate, ethanol and CO_2 from glucose and no ethanol, diacetyl, acetoin or 2,3BG from co-metabolism of glucose and citrate. Instead, greater amounts of acetate and lactate are produced than can be accounted for in terms of the citrate and sugar used.[73,74] The pyruvate produced from citrate and that from glucose are both reduced to lactate with increased production of lactate and NAD$^+$. This relieves pressure on alcohol dehydrogenase from also regenerating NAD$^+$ and instead, acetyl-P (Fig. 2) is converted to acetate:

$$\text{Acetyl-P} + \text{ADP} \rightarrow \text{Acetate} + \text{ATP}$$

The ATP produced also results in more rapid growth.[73] Thus, acetate arises from both citrate and sugar metabolism in this case. Leuconostocs will produce diacetyl and acetoin from citrate at low pH values for reasons that are not clear.[119] It may be due to regulation of ALA synthase which, at pH 5·4, is inhibited by many of the intermediates (6PG, 2PG, 3PG, PEP and ATP) of sugar metabolism.[105] Except for ATP which becomes more inhibitory, the inhibition is relieved at low pH, allowing ALA and, consequently acetoin, to be produced.

4.3.1 Citrate Metabolism in Cheese

Little information is available about the metabolism of citrate in cheese. Diacetyl is thought to contribute to Cheddar cheese flavour but few definitive data have been published.[90] Keen & Walker[120] followed the production of diacetyl, acetoin and BG in cheese; unfortunately, the disappearance of citrate was not reported. About 1 μg of diacetyl/g occurs in cheese made with Cit$^-$ lactococci. In cheese made with Cit$^-$ Lc lactis subsp. cremoris AM2 large amounts of acetoin (up to 80 μg/g) were produced initially, but the level decreased to zero as ripening continued and this was accompanied by a corresponding increase in the concentration of BG level. In cheese made with Cit$^-$ Lc. lactis subsp. cremoris HP, a low level (~30 μg/g) of acetoin was formed, which decreased very slowly during ripening; no BG was produced. When the cheese was made with Cit$^+$ Lc. lactis subsp. *lactis* DRC1, the level of diacetyl reached 11·3 μg/g 5 days of ripening after which it decreased to ~1·0 ppm after 100 days; the level of acetoin was constant at about 15 μg/g throughout ripening while the level of BG varied from 100 to 180 μg/g. It is unusual for Cit⁻ strains of lactococci to produce large amounts of acetoin. However, NSLAB (mesophilic lactobacilli and pediococci) can reach high levels (~10^8/g) in cheese and are able to metabolize citrate.[116,121–124] The numbers of these bacteria were not reported[120] but may be a complicating factor in interpreting the above data.

There is little information on the accumulation of acetate in cheese, even though it is also considered to contribute to flavour. Presumably, cheese made with Cit⁺ cultures would contain large amounts of acetate while cheese made with Cit⁻ cultures would not. Thomas *et al.*[122,123] found 3·5 to 24 mmol acetate/kg in 6 months-old Cheddar cheese while Kristofferson *et al.*[124] found 3·5 to 29 mmol/kg. Many NSLAB can metabolize lactate to acetate and the amounts produced depend on the number and type of NSLAB and on the oxygen transfer rates of the wrapping film.[123] Oxygen appears to be necessary for the formation of acetate by pediococci but not by lactobacilli.

4.4 Protein Metabolism

Starter LAB have limited abilities to synthesize amino acids, e.g. glutamate, methionine, valine, leucine, isoleucine and histidine are required by lactococci and many strains have an additional requirement for phenylalanine, tyrosine, lysine and alanine.[125] The amino acid requirements of *Str. salivarius* subsp. *thermophilus* are very similar to those of the lactococci: glutamate, valine, leucine and histidine are required in addition to cysteine and tryptophan.[126] The requirements of the leuconostocs are strain-dependent.[127] All require valine and glutamate and methionine stimulates most of them. The other 14 amino acids tested either stimulated or were required by one strain or another. No information appears to be available on the amino acid requirements of the thermophilic lactobacilli.

Milk contains insufficient amounts of amino acids and low molecular weight peptides to sustain the growth of LAB. To obtain the high cell numbers required in starters and cheese manufacture, starter LAB must have a proteinase system capable of hydrolysing the milk proteins to amino acids and peptides.[128] Casein is the major protein in milk and its open, largely random structure makes it more susceptible to proteolysis than the whey proteins, which have a high level of secondary and tertiary structure. There are four prerequisites for good growth of LAB in milk: (a) a proteinase system which hydrolyses the casein to oligopeptides, (b) a peptidase system which further hydrolyses the oligopeptides to amino acids and smaller peptides (c) transport systems for uptake of the amino acids and peptides, and (d) intracellular peptidases to hydrolyse the peptides to the constituent amino acids. Peptide uptake is essential for the growth of *Lc. lactis* subsp. *lactis* ML3 in milk.[129]

4.4.1 Proteinases

The proteinases of the lactococci have been more intensively studied than those of other starter LAB. They are encoded by plasmid DNA and numerous proteolytic systems have been described based on differences in activity, specificity and immunology.[130,131] Some aspects of the proteinases have been contentious and difficult to interpret; e.g. in many reports their cellular location is not clear since cell leakage was not studied simultaneously. In more recent work,[132-139] the association of proteinases with the cell surface or cell membrane has been indicated by data which show no release of proteinase into the supernatant and no release of intracellular marker enzymes (e.g. LDH) when proteinases have been assayed in intact cells or in protoplasts produced from lysozyme or phage lysin treated cells. Several proteinases have been purified.[137,140-142] All have high MWs (80–140 kD), pH optima of ~6·0, isoelectric points of ~4·5, are activated or stabilized by Ca^{2+} and inhibited by PMSF and DFP, implying that they are serine proteinases. They can be distinguished on the basis of their pH and temperature optima into two distinct types: PI having an optimum pH of 5·8, an optimum temperature of 40°C and an ability at act mainly on β-casein and PIII with an optimum pH of 5·4, an optimum temperature of 30°C and activity on both α- and β-caseins.[143] PII activity, with an optimum pH of 6·5 and an optimum temperature of 30°C, was subsequently shown to be an artefact. Strains contain either PI or PIII. Hugenholtz *et al.*[142] have shown that only one of four immunologically distinct proteinases was common to all 10 strains of *Lc. lactis* subsp. *cremoris* tested.

Genetic studies have considerably clarified the difficulties surrounding interpretation of the activity and specificities of the proteinases.[131] Nucleotide sequences of the proteinase genes from three lactococci, viz. Wg$_2$, SK11 and NCDO 763, have been determined and are highly conserved.[144-148] Slight differences in sequences are responsible for the differences in specificity of PI and PIII proteinases. The sequences surrounding the amino acids involved in the active site (aspartate, histidine and serine) are similar to those of the active site of subtilisin, the extracellular proteinase of *Bacillus subtilis*.[131] Like subtilisin, the proteinases of the lactococci are synthesized as pre-pro-proteins but, unlike this truly extracellular enzyme, the lactococcal proteinases are attached to the cell through an anchor at the C-terminal end of the molecule.[147] The anchor has a high concentration of proline residues which are thought to be involved in the attachment. A second gene (*prt*M), whose product is a lipoprotein, is involved in activation in a way which is not clear but may involve removal of the pre-region of the pre-pro-protein.[147,148] The production of proteinase is a plasmid-encoded property; in some strains the proteinase and lactose utilization genes are associated with the same plasmid. The biochemistry and genetics of the proteolytic systems of starter LAB have been comprehensively reviewed by Thomas & Pritchard[130] and Kok[131] (see Section 5.4.1).

There is limited information on the proteinases of other starter LAB with the exception of *Lb. helveticus* and *Lb. delbruecki* subsp. *bulgaricus*.[138,149-152] The helveticus proteinase could be liberated from cells by repeated washing in a

Ca^{2+}-free buffer and is inhibited by DFP and PMSF, but not EDTA, implying that it also is a serine proteinase. It has an optimum pH of 7·5 to 8·0. A symbiosis exists between *Lb. delbrueckii* subsp. *bulgaricus* and *Str. salivarius* subsp. *thermophilus* in thermophilic cultures.[43] The lactobacilli produce amino acids, especially glycine, histidine and valine, which stimulate the streptococci and presumably proteinase activity is involved in this process. The streptococci produce CO_2 and formate from sugar which stimulate the lactobacilli.[153,156]

4.4.2 Prt⁻ Strains

Many mixed starters contain strains which are able to hydrolyse milk proteins (Prt⁺) and other strains which are not (Prt⁻). The latter are missing the plasmid which encodes the proteinase and rely on the proteolytic ability of the Prt⁺ strains for growth. In the past, these strains were called 'slow coagulating variants' since they failed to coagulate milk quickly. Such variants were recognized over 50 years ago but the explanation for their slowness had to await the development of plasmid isolation techniques. It was determined that Prt⁻ strains grow as rapidly in milk as Prt⁺ strains but stop growing when the free amino acids and peptides are utilized.[157,158] Serial transfer increases the number of Prt⁻ strains;[18] after 100 transfers, Prt⁻ variants comprised 90–98% of the total number. The number of Prt⁻ variants can also be increased by maintaining the pH at relatively high values, e.g. pH 5·0 or 6·5. This finding has implications for growth in phage inhibitory media (see Section 6.3.7).

Cheesemaking with Prt⁻ variants has been advocated[159,160] Advantages claimed for their use include fewer problems with phage and antibiotic residues, decreased bitterness and greater cheese yields. The disadvantages are that greater amounts of inoculum are required and manufacturing times are increased. Cheese made with Prt⁻ variants had higher concentrations of pH 4·6 soluble N than cheese made with Prt⁺ strains,[160] which was attributed to stimulation of growth of the Prt⁻ variant by carry-over of the yeast extract used to produce the bulk culture. Stadhouders *et al.*[161] concluded that soluble N was the same in cheese made with Prt⁺ and Prt⁻ mutants but amino acid N was significantly lower in cheese made with Prt⁻ starters. They recommended that exclusively Prt⁻ variants should not be used for Gouda cheese. PTA soluble N was higher in Prt⁺ cheese than in Prt⁻ cheese. In the experiments of Faryke *et al.*,[162] levels of water soluble N were similar in Prt⁺ and Prt⁻ cheeses but the Prt⁺ cheese contained more PTA soluble N and received higher flavour scores. The consensus, therefore, appears to be that no particular advantage accrues to the exclusive use of Prt⁻ starter cultures.

4.4.3 Peptidases

The second step in casein utilization involves the degradation of the relatively large oligopeptides to smaller products by peptidases.[130,131] Carboxypeptidase activity has not been found in starter LAB but amino peptidases, di- and tri peptidases, an aryl peptidyl amidase and several which act on proline-containing peptides have been purified. The multiplicity of enzymes found may reflect the different substrates used for their detection. The consensus appears to be that LAB contain only a small number of exopeptidases. Both intra- and extra-cellular peptidases have been reported, although convincing proof of their location *in vivo* has not always been determined by monitoring suitable marker enzymes.

Casein has a high proline content and several investigators have looked for peptidases which act on proline-containing substrates. Five types have been found:[130,131] an aminopeptidase P, a proline iminopeptidase, an iminodipeptidase (prolinase), an imidodipeptidase (prolidase) and X-prolyl dipeptidyl amino-peptidase (XPDAP). Some of these may be located in the cell envelope but XPDAP is considered to be intracellular. The purified peptidases vary in MW from 36–95 kD and many are multimeric enzymes. All are either serine or metallo enzymes. XPDAPs have been purified from several starter LAB, including lacto-cocci, *Lb. delbrueckii* subsp. *lactis*, *Lb. delbrueckii* subsp. *bulgaricus*, *Lb. helveticus* and *Str. salivarius* subsp. *thermophilus*.[163–169]

4.4.4 *Amino Acid and Peptide Transport*

The third step in the utilization of casein involves the transport of the peptides and amino acids. Again, lactococci are the most studied group of starter LAB. Lactococci have separate transport systems for the uptake of peptides and amino acids, and these have been reviewed.[51,54] Peptides containing up to six amino acid residues can be transported and are then hydrolysed to the constituent amino acids by intracellular peptidases.

In the lactococci, methionine, the branched-chain acids (leucine, valine and isoleucine, all of which are required for growth), the neutral amino acids (serine, threonine, glycine and alanine) and the basic amino acid, lysine, are all trans-ported by a PMF system (for a review see Poolman *et al.*[54]). Proline must be taken up as a peptide. Glutamate, glutamine and asparagine are taken up by an ATPase catalysed reaction while arginine is taken up by a PMF (see Section 4.1) antiport mechanism involving ornithine. Arginine is subsequently metabolized by the deiminase pathway:

$$\text{Arginine} + H_2O \qquad \rightarrow \text{Citrulline} + NH_3$$
$$\text{Citrulline} + Pi \qquad \rightarrow \text{Ornithine} + \text{Carbamyl Phosphate}$$
$$\text{Carbamyl Phosphate} + ADP \rightarrow ATP + CO_2 + NH_3$$

The enzymes involved are arginine deiminase, ornithine transcarbamylase and carbamate kinase, respectively, and ATP is produced. One of the major proper-ties which distinguishes *Lc. lactis* subsp. *lactis* from *Lc. lactis* subsp. *cremoris* is the ability to metabolize arginine. All strains of *Lc. lactis* subsp. *cremoris* lack deiminase activity and some also lack transcarbamylase; all strains possess carbamate kinase activity.[170]

4.4.5 *Bitterness*

Bitterness is a problem in some varieties of cheese made with mesophilic cultures and is due to the production of bitter peptides, which contain predominantly hydrophobic amino acids residues, by rennet and starter bacteria. An early hypothesis[171-174] to explain the occurrence of bitterness suggested that all starter strains can cause bitterness but do not if their rate of multiplication during manufacture is controlled (e.g. by high cooking temperatures or the presence of phage) such that relatively low cell numbers are present in the curd. Bitterness develops due to increased hydrolysis of non-bitter to bitter peptides by high cell numbers.

A different hypothesis for the development of bitterness has emanated from more recent work.[175-177] Only some starter bacteria produce bitter peptides but all can degrade them. Salt decreases the permeability of the starter cells and increases the hydrophobic association between the bitter peptides. Both factors result in reduced availability of the peptides to the membrane-associated proteinases and their accumulation results in the production of a bitter cheese. The principal sources of bitter peptides are residues 84–89 and 193–209 of the C-terminal region of β-casein.[177]

4.5 Acetaldehyde

Acetaldehyde is produced by both thermophilic and mesophilic cultures but greater amounts are produced by the former (up to 20 μg/ml). Acetaldehyde is an important flavour component in yoghurt but causes an off-flavour in fermented milks made with mesophilic cultures, especially if the ratio of diacetyl to acetaldehyde decreases to 3:1; ideally this ratio should be 4:1.[178] In mesophilic cultures, acetaldehyde is produced mainly from threonine by threonine aldolase in which threonine is hydrolysed to glycine and acetaldehyde.[179] The only strain of lactococcus which has a requirement for glycine is a strain which is missing threonine aldolase. Thus, the main function of threonine aldolase is to provide glycine for growth. One of the roles of *Leuconostoc* spp. in mesophilic mixed cultures is the reduction of the acetaldehyde, produced by excessive numbers of lactic streptococci, to ethanol.

In thermophilic cultures, more acetaldehyde is produced by *Lb. delbrueckii* subsp. *bulgaricus* than by *Str. salivarius* subsp. *thermophilus* but because of the symbiotic relationship between these organisms more of it is produced when the cultures are grown together.[180] The source of acetaldehyde in thermophilic cultures has not been determined, but is most likely sugar. *Lb. delbrueckii* subsp. *bulgaricus* does have threonine aldolase activity[181] but produces little or no acetaldehyde from threonine.[182] The reason(s) for these conflicting results is not clear.

4.6 Measurement of Growth

Bacterial growth curves are normally constructed by plotting the log of absorbence, determined spectrophotometrically at 400–650 nm, or the log of dry weight, against time. Because of the opacity of milk, it is not possible to measure absorbence values; however, treatment with KOH and EDTA renders milk translucent and amenable to measurement of absorbence.[183,184] Lactic acid is easily measured by titration with NaOH and, as it is directly related to cell mass, it

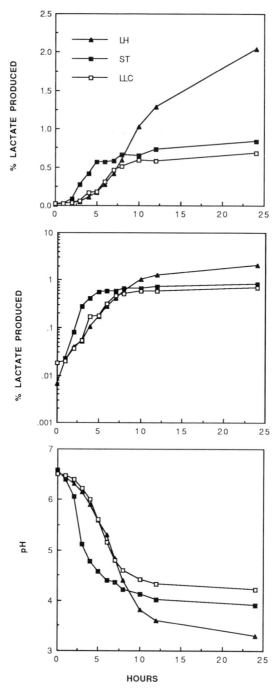

Fig. 6. Acid production by *Lb. helveticus* (LH) and *Str. salivarius* subsp. *thermophilus* (ST) is at 42°C and *Lc. lactic* subsp. *cremoris* (LLC) at 30°C in milk. The top graph is an arithmetic plot while the middle graph is a semi logarithmic plot of the data.

can also be used to construct growth curves. However, care should be taken in interpretation since, at high temperatures and salt concentrations, production of cell mass can become uncoupled from acid production.[184] In plotting growth curves it is important that the measurements are corrected for the inherent acidity of the uninoculated milk and that they are plotted semi-logarithmically rather than arithmetically. Typical results for mesophilic and thermophilic cultures are shown in Fig. 6. The data have been corrected for the inherent acidity and both arithmetic and logarithmic plots are shown. As long as the curves are plotted semi-logarithmically, the slope (*m*) of the linear portion of the plot is related to *k*, the specific growth rate, by the formula:

$$m = k/2 \cdot 303$$

Generation times (GT) or doubling times are related to *k* by the formula:

$$GT = \frac{0 \cdot 693}{k} = \frac{0 \cdot 301}{m}$$

The optimum growth temperatures of mesophilic and thermophilic cultures are ~30 and ~45°C, respectively. GTs for mesophilic cultures growing in milk at 30 and 21°C (the normal temperature for bulk starter production) are 1 and 2·3 h, respectively,[185] while for *Str. salivarius* subsp. *thermophilus* and *Lb. helveticus* they are ~30 min and 1 h, respectively, at 42°C (unpublished data).

pH values are easily measured and the decrease in pH as a result of lactic acid production is frequently used to measure growth. However, pH is not directly related to cell numbers or lactic acid production and cannot be used to calculate GTs. Large amounts of lactic acid are produced by starter LAB and, therefore, growth media must be adequately buffered. Milk, especially when the non-fat milk solids are increased to 10–12%, is well buffered and pH values of fully grown milk cultures typically fall from an initial value of 6·5 to a final value 4·5, 4·5 and 3·5 for mesophilic cultures, *Str. salivarius* subsp. *thermophilus* and thermophilic lactobacilli, respectively. Such cultures contain ~10^9 cells/ml.

5   GENETIC STUDIES ON LACTIC ACID BACTERIA

5.1 Introduction

The instability of a number of key industrial traits in lactococci has long been recognized but the scientific basis for this observation was only established in 1974 when it was demonstrated that lactococci contain plasmids,[186] relatively small DNA molecules which exist independently of the cell chromosome. Plasmids are not essential for cell growth but some carry genes which have a significant impact on the host cell. For example, while most lactococcal plasmids have no known function (cryptic plasmids), plasmid genes encode the enzymes necessary to hydrolyse casein, transport and metabolize lactose, transport citrate, produce bacteriocins, and resist phage attack. None of these traits is essential for the growth and survival of the cell *per se*, but they may be desirable commercially. The unstable nature of plasmids is directly responsible for the appearance of 'slow growing' variants of lactococci.

Since the initial discovery of plasmids in lactococci, and subsequently in other LAB, attention has been focused on the possibility of exploiting the techniques of molecular biology to understand and ultimately control the metabolism of these important industrial bacteria. Recently, genetic analysis has been extended to include the much larger bacterial chromosome. Most attention has been directed towards the lactococci, but other LAB are also being studied. One of the most exciting advances in this area has been the development of phage resistant starter cultures, and their acceptance and use by the cheese industry.

5.2 Systems for Genetic Analysis in Lactic Acid Bacteria

One of the prerequisites for the meaningful genetic analysis of a bacterium is the development of a host-vector system. It is unsatisfactory to conduct the analysis of lactococcal genes entirely within a well-defined genetic system such as *E. coli*, since the results obtained may not reflect the situation pertaining in the original host. Plasmid-free lactococcal strains are not commonly encountered, and most strains carry between four and seven individual plasmids (and, in most cases, many copies of each plasmid). The 'plasmid profile' of individual strains may be a useful identification guide (Fig. 7)[187] but makes analysis of particular plasmids difficult. Strain MG1363 was constructed by the sequential removal, or 'curing', of the original plasmid complement of *Lc. lactis* subsp. *lactis* NCDO 712 by protoplast-regeneration,[188] and is probably the strain of choice for most researchers. This curing technique is felt to cause less indiscriminate DNA damage than the alternative mutagenic method used to create the closely related plasmid-free derivative LM2301.[189] Recent chromosomal analysis indicates that LM2301 has a smaller genome than that of MG1363, which supports the view that the treatment of LM2301 was more destructive.[190] *Lc. lactis* subsp. *lactis* IL1403 offers an alternative plasmid-free host background.[191] Such plasmid-free hosts offer obvious advantages for the study and isolation of intentionally introduced plasmids.

The development of suitable host backgrounds was only the first step in constructing a good genetic system. A second requirement was to establish useful vector plasmids and the means of introducing them into plasmid-free strains. A plasmid vector can be described as a carrier plasmid which is small, easily detected (usually through carrying antibiotic resistance markers), stably maintained in the strain of interest, and allows the possibility of introducing foreign DNA without influencing these useful characteristics. The first vectors for lactococci were 'shuttle vectors', named for their ability to replicate in a number of different host systems. A particularly useful trait in a shuttle vector is the ability to replicate in *E. coli* in addition to the target strain, since this exposes the plasmid to the highly advanced genetic techniques available in that background. One such plasmid, still very much in use, is the *Streptococcus–E coli* shuttle plasmid, pSA3, developed by Dao & Ferretti.[192] This plasmid possesses both a Gram

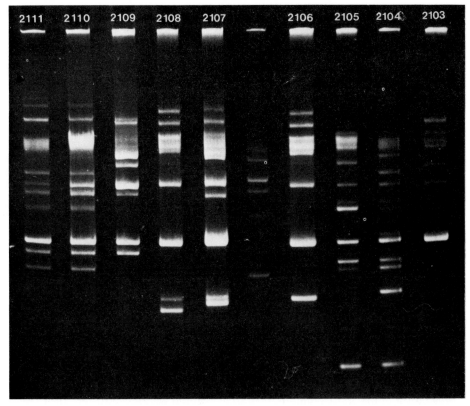

Fig. 7. Plasmid profiles of different strains of *Lactococcus lactis* subsp. *lactis*. The unnumbered track contains marker plasmids (from Gasson, unpublished data).

negative and a Gram positive origin of replication which allows its stable maintenance in both hosts and encodes three antibiotic resistance genes which allow easy detection. A second type of vector was developed by Kok *et al.*[193] They exploited the observation that a small cryptic plasmid, pWV01, from *Lc. lactis* subsp. *cremoris* Wg2, was capable of stable replication in *B. subtilis*.[194] Two antibiotic resistance markers were introduced to this cryptic plasmid to construct the plasmid vector pGK12. One of the unexpected bonuses of this plasmid is that the lactococcal origin of replication is also functional in *E. coli*. It was subsequently shown that pGK12 has an extremely broad host range, including all genera of LAB.[195] A diversity of second generation vectors, capable of performing distinct tasks, has been developed based on both the pWV01 replicon, and the almost identical plasmid pSH71.[196,197] Most strains of *Str. salivarius* subsp. *thermophilus* are plasmid-free, although a few plasmids have been found in individual isolates. Vectors based on these native plasmids are being developed, but currently pSH71-derived vectors are favoured due to their high stability, even at elevated temperatures.[198]

5.3 Gene Transfer in Lactic Acid Bacteria

5.3.1 Transformation and Electroporation

The development of plasmid vectors is redundant without the ability to reintro-
duce the plasmids and recombinant molecules to the organism of interest, a process termed transformation. Initially, this was a significant hurdle to overcome
in LAB, since these strains do not take up DNA naturally. Early transformation
protocols relied upon the formation of protoplasts by enzymatically removing
the cell wall.[199] DNA was subsequently introduced to polyethylene glycol (PEG)-
damaged protoplasts and transformants were recovered by regenerating the cell
wall in osmotically-stabilized media. These methods were a significant break-
through, although only low frequencies of transformation were obtained and the
procedure was tedious. A number of improvements were reported,[200-202] but
transformation remains a somewhat unsatisfactory procedure. In recent years, the
method of choice in almost all laboratories is electro-transformation, or electro-
poration.[195] Briefly, a high voltage is discharged through a cuvette containing both
cells and DNA. Cells are plated immediately on selective media, and transformants are recovered at reproducibly high frequencies. The method is extremely
simple, but the mechanism of DNA uptake is not clear.

5.3.2 Natural Gene Transfer

A number of other techniques are available for the transfer of DNA between
members of the LAB (reviewed by Fitzgerald & Gasson,[203] Steele & McKay[204]).
Conjugation is the most commonly utilized method of gene transfer. In a conjugation event, DNA, usually plasmid DNA, is transferred through cell–cell contact between a donor and a recipient strain. Not all plasmids possess the
conjugation genes necessary to carry out plasmid transfer, although in some
instances a conjugative plasmid may lead to the transfer of a non-conjugative
plasmid.[205] Conjugation has been frequently employed to transfer plasmids to
plasmid-free strains.[206] In a number of instances, high frequency conjugation, accompanied by cell aggregation has been observed.[207,208] It has been proposed that
the aggregation phenomenon requires the presence of two independent genes, a
clumping gene (clu) located on the conjugative plasmid, and an aggregation gene
(agg) located in the chromosome of certain strains.[209] Cell aggregation during
conjugation in Enterococcus faecalis requires sex pheromones, but there is no
evidence that this occurs in lactococci.

Once a plasmid has been introduced to a plasmid-free background, plasmid-
linked functions can be readily detected. Phenotypes such as lactose utilization
(Lac) are the most commonly utilized markers to follow conjugation events. In a
number of studies, Lac transfer has been used to indicate that conjugation has
occurred, and transconjugants are subsequently screened for the co-transfer of a
desired phenotype/plasmid. In this way bacteriocin-producing and phage resistance plasmids have been identified.[203] There is evidence for the existence of
unrelated chromosomally-located sex factors in Lc. lactis subsp. lactis 712 and

IL1401, which may excise and promote transfer of otherwise non-conjugative plasmids.[210] This suggests that such elements may be widespread in lactococci.

Transduction is the transfer of DNA by means of phage. In most cases it can be explained as aberrant packaging of bacterial DNA in phage heads. During infection of a second cell, this DNA may be injected and stably maintained. Transduction of plasmids has been described in both lactococcal and *Lactobacillus* strains.[211-213] One of the benefits of transduction is that temperate phage (see Section 6.3) can be used to transfer chromosomal markers.[214-215] One of the few examples of an improved strain resulting from genetic studies was the result of a transduction event in which a lactose/proteinase plasmid was stabilized in the chromosome of transductants. The stabilized strain was less proteolytic than the parental strain, C2, and was reported to produce a higher quality Cheddar cheese.[212] However, transduction as a means of DNA transfer is limited, in that the phage carrier must be able to infect both donor and recipient. In addition, there are size constraints imposed by the size of the phage head.

5.4 Gene Cloning and Analysis in Lactic Acid Bacteria

In this section, a number of examples of gene cloning, and the subsequent analysis of the cloned genes, will be discussed (for reviews see de Vos *et al*;[216] Mercenier[217]). Cloning, in recombinant DNA terminology, describes the physical linkage of a DNA fragment to a plasmid or phage vector. A typical example is shown in Fig. 8. The immediate benefit of such a situation is that the DNA fragment of interest is now linked to the properties of the vector. The cloned fragment may be introduced to a variety of strains using the vector antibiotic resistance markers and the broad host range origins of replication discussed in Section 5.2. The cloned fragments are also more amenable to the use of genetic techniques such as DNA sequence determination, directed mutations, and studies on gene expression and control. Some examples of this detailed genetic analysis are dealt with in the following sections. Examples concerning phage genetics, phage resistance and bacteriocins will be discussion in Section 6.3.

5.4.1 Proteinase and Peptidase Genes

Early biochemical analysis of casein degradation profiles for different lactococcal strains led researchers to propose three independent proteolytic systems; PI, PII and PIII, one of which, PII, was an artefact (see Section 4.3.1). The responsible *prt* genes were located on plasmids in their respective strains, cloned and sequenced (reviewed by Kok[131]). Two genes are involved in the production of the mature proteinase. One (*prt*P) encodes the proteinase, which is secreted through the cell membrane and anchored in the cell wall by a membrane anchor sequence in the carboxy terminus. This form of the enzyme is inactive unless a lipoprotein, PrtM, is present. This maturation protein in encoded by a gene immediately adjacent to the *prt*P gene. Comparison of the genes revealed that the

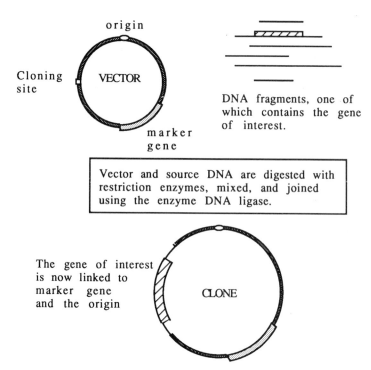

origin

Cloning site

VECTOR

marker gene

DNA fragments, one of which contains the gene of interest.

Vector and source DNA are digested with restriction enzymes, mixed, and joined using the enzyme DNA ligase.

The gene of interest is now linked to marker gene and the origin

CLONE

Fig. 8. Gene cloning strategy, DNA fragments are generated by cutting the source DNA with a restriction enzyme. The vector is cut with the same enzyme to generate compatible ends. The molecules are mixed and the clones selected, usually by screening for plasmids with both the marked gene and the gene of interest.

*prt*P-encoded proteolytic enzymes from Wg2 (PI) and SK11 (PIII) are more than 98% identical. Subtle sequence differences, resulting in amino acid substitutions in the enzymes, are responsible for the diversity of proteolytic specificity. Using genetic techniques, researchers have exchanged segments of the PI and PIII genes, and carried out specific amino acid substitutions, to create enzymes with completely novel specificities. It has also been possible to introduce these plasmid-located genes into the chromosome through directed integration (see Section 5.8). The proteinase is only one component of the proteolytic systems of lactococci. A number of peptidases are also involved and the gene for one of these, encoding XPDAP, has been cloned and sequenced in two laboratories.[218,219]

A different approach to altering the proteolytic specificities of lactococcal strains has been reported by van de Guchte *et al.*[220] In an example of heterologous (foreign) gene cloning, a gene for the *B. subtilis* neutral proteinase was cloned and introduced to a lactococcal strain. The gene product was successfully expressed (produced) in the lactococcal background, leading to a strain with atypical proteinase specificity and much increased activity.

5.4.2 Lactose and Citrate Metabolism

At least three of the genes encoding the enzymes involved in lactose uptake and degradation in lactococci are plasmid-encoded, and all have been cloned and sequenced.[221,222] The genes for Enzyme II (*lac*E), Enzyme III (*lac*F), and pβgal (*lac*G), are arranged in an operon, in which all the genes are controlled simultaneously by the amount and type of carbohydrate in the environment. The *Lb.casei lac*G gene has been cloned and sequenced and shows a high degree of similarity to that of *Lc. lactis* subsp. *lactis*.[223] The genes responsible for the permease (*lac*S) and the βgal (*lac*Z) are also thought to be linked in an operon in *Str. salivarius* subsp. *thermophilus*.[217]

Citrate can be metabolized by a number of LAB (see Section 4.3). In the strains studied so far, the citrate permease gene has been located on a plasmid, which accounts for the ready isolation of Cit⁻ derivatives of these strains. The citrate permease gene from *Lc. lactis* subsp. *lactis* NCDO 176 has been cloned, sequenced and expressed in *E. coli*.[99] The sequence data strongly suggest that the permease is an integral membrane protein, although it shows no similarity to citrate carriers from other species. Transposon mutagenesis (see Section 5.5) of the Cit⁺ *Lc. lactis* subsp. *lactis* 18–16 has demonstrated that the enzyme responsible for the initial degradation of citrate, CL, is chromosomally encoded.[224]

5.4.3 Heterologous Gene Expression

Heterologous (from other genera) gene cloning has also been demonstrated in LAB (reviewed by Kondo[225]). These hosts are GRAS (generally regarded as safe) and as such are suitable organisms for exploitation as bioreactors, i.e. as microorganisms with the potential for the production of valuable products for human use. Van de Guchte *et al.*[226] have reported the cloning and expression of hen egg white lysozyme in *Lc. lactis* subsp. *lactis* IL1403, while the bovine prochymosin B gene has also been fused to expression signals derived from the SK11 *prt*P gene.[227] Depending on the amount of the *prt*P gene in the fusion, the prochymosin product could be detected intracellularly, or in the periplasm and cell-free supernatant. These studies indicate the potential of LAB to produce proteins of non-bacterial origin, provided that the correct signals are fused to the heterologous gene.

5.5 Transposons and Insertion Sequences

Transposons are discreet segments of DNA which can move or 'transpose' between different locations within a cell. The ability to transpose is encoded by a gene or genes carried by the transposon. In certain instances, other easily detected genes, such as drug resistance markers, are also contained within the transposon. A transposon which is so small that it carries only enough information to allow movement is called an insertion sequence (IS). Some transposons and IS elements are thought to cause elevated expression of genes adjacent to their new location. Alternatively, if a transposon 'jumps' into a gene, it will almost certainly inactivate the gene function, and this has been used to determine the location of genes. A number of examples of transposon mutagenesis have also been described in lactococci. Renault & Heslot[228] utilized the conjugative transposon Tn916 to identify and mutate the malolactate pathway, while mutations in maltose and citrate utilization genes have been described using the sister transposon Tn919.[224] However, a generally useful transposition system has not yet been developed in the LAB.

IS elements play an important role in LAB genetics. Initially, it was demonstrated that IS elements mediate cointegrate formation between plasmids in lactococci.[229] This is detected when a non-conjugative plasmid is mobilized by a conjugative plasmid in the same strain through the formation of a plasmid: plasmid cointegrate. IS elements have also been implicated in the mobilization of chromosomal DNA between lactococci.[210] A lactose plasmid bearing an IS element is believed to integrate into the chromosome. After IS-mediated excision, lactose plasmids bearing chromosomal DNA of various lengths have been isolated. The role of an IS in the appearance of a virulent phage of Lb. casei will be dealt with in Section 6.3.5. The cloning and sequence analysis of a number of IS elements from Lactobacillus and lactococcal strains[229-232] suggest that the development of useful transposition systems, based on these small elements, for use in the LAB will be forthcoming.

5.6 Use of Genetically Engineered Strains in Food

Current regulations preclude the use of genetically modified strains within the food industry. Until this restriction is lifted, it might appear that the genetic studies described in the previous sections are of limited practical value. However, this is not the case. Genetic studies in LAB have considerably enhanced our understanding of these bacteria. The prevention of phage problems in Lb. casei fermentations through the elimination of the temperate phage is described in Section 6.3.5. A second example of the practical application of genetic techniques was first described by Klaenhammer's group (see Ref. 233). The conjugative phage resistance plasmid pTR2030 (see Section 6.3.3) was introduced to a number of industrial cheese starter strains. These transconjugants showed an elevated level of phage resistance over the original parent strains. Only conjugation, which is a natural process, was used to transfer the DNA, and no recombinant techniques, or unacceptable genetic markers, were used in the construction of the transconjugants. Therefore, these strains have found acceptance within the dairy industry. While this can be regarded as one of the most exciting advances in the biotechnological area, it should be noted that phage have appeared which are capable of overcoming pTR2030 resistance.[234,235] This is a further example of the genetic variability within the phage population, which will continually lead to the appearance of 'novel' phage strains. It is important to remember that good manufacturing practices will probably never be superseded by biotechnological advances. Other workers, using similar conjugal strategies, have reported the construction of industrially used strains.[236]

Anticipating a relaxation of the prohibition of the use of recombinant strains, some laboratories are developing cloning systems which are composed of acceptable marker genes which would allow the detection of transformants. To be completely acceptable, a 'food grade' vector should ideally be entirely composed of DNA from the target organism, or closely related non-pathogenic strains, and possess a readily selectable marker gene. Candidate marker systems have already been cloned and sequenced and include resistance to nisin,[237] lactose metabolism,[221] phage resistance,[238] and thymidylate synthetase.[239] However, progress in this area is still at a preliminary stage, and no system has found widespread use.

5.7 Chromosomal Integration

One of the early targets of genetic manipulation in the lactococci was to stabilize important phenotypes through the integration of unstable plasmids into the chromosome. This was achieved fortuitously after transduction of Lac plasmids between strains. A more directed approach to achieve integration relies upon homology (similarity) between a plasmid, which does not possess the ability to replicate in the target strain, and the chromosome. In rare instances, the cell recombination system will introduce the plasmid into the chromosome at the point of homology. These instances can be detected by selecting for a suitable marker carried by the plasmid. Successful integration of antibiotic resistance genes has been described in lactococci.[240] An *E. coli* plasmid, which cannot survive in the lactococcal background, was successfully introduced to the lactococcal chromosome at low frequency. Homology between the plasmid and the chromosome was provided by cloning a lactococcal chromosomal fragment in the *E. coli* vector.

6 INHIBITORS OF CULTURE GROWTH

A number of factors may adversely affect the activity of starters, including variations in milk composition due to mastitis or seasonal factors, and cheese manufacturing procedures, e.g. cooking temperature and/or salt level. In addition, a range of inhibitors may be present in milk, e.g. antibiotics, agglutinins, dissolved oxygen, phage, free fatty acids, inhibitory bacteria, the LP system, residual sanitizers and bacteriocins. Adherence to good quality control procedures and good manufacturing practices has helped to minimize the influence of most of these factors in the modern cheese industry. However, problems may still arise from antibiotic residues, bacteriocins and phage.

6.1 Antibiotics

Antibiotic residues are found in milk primarily because of their use to control mastitis in cows. Strict penalty schemes on producers have reduced the impact

TABLE III
Concentrations (μg/ml $\pm$ SD) of Some Antibiotics Causing 50% Inhibition of Growth of Starter Bacteria in Milk [a]

Organism	No. of strains	Antibiotic			
		Penicillin	Cloxacillin	Tetracycline	Strepto-mycin
Lc. lactis subsp. cremoris	4	0·11 ± 0·028	1·69 ± 0·38	0·14 ± 0·02	0·67 ± 0·15
Lc. lactis subsp. lactis	4	0·12 ± 0·025	2·16 ± 0·41	0·15 ± 0·05	0·53 ± 0·18
Str. salivarius subsp. thermophilus	3	0·01 ± 0·002	0·42 ± 0·07	0·19 ± 0·06	10·5 ± 0·29
Lb. delbrueckii subsp. bulgaricus	2	0·03 ± 0·006	0·29 ± 0·04	0·37 ± 0·04	3·0 ± 2·0
Lb. delbrueckii subsp. lactis	1	0·024	0·24	0·6	2·29

[a] From Cogan.[241]

of antibiotics very significantly. Little information is available regarding the levels of antibiotics required to inhibit leuconostocs but among the other starter LAB, considerable inter- and intra-species variations occur in their susceptibility to the antibiotics commonly used in mastitis therapy (Table III). Two important practical points are apparent from this table, viz. mesophilic starters are less susceptible to penicillin and more susceptible to streptomycin than are thermophilic cultures.

6.2 Bacteriocins

Bacteriocins are proteins which exhibit a bacteriostatic or bactericidal action, usually, but not exclusively, against closely related bacteria. Those produced by LAB have been reviewed by Klaenhammer.[242] Several have been purified and their more important properties are shown in Table IV. All are proteins, of varying molecular weight, which can be plasmid or chromosomally encoded, with varying spectra of inhibition/resistance, resistance to heat and pH. Those which inhibit pathogens and spoilage bacteria are potentially useful although to date only nisin, produced by Lc. lactis subsp. lactis, has been used commercially. Because of its broad host range (Clostridium, Staphylococcus, Listeria), it has been questioned whether the term bacteriocin is applicable to this molecule. Nisin has a molecular weight of 3·5 kD and occurs naturally as a dimer. The polypeptide contains a number of unusual amino acids, including lanthionine. Because of this, nisin is sometimes referred to as a lantibiotic. The mechanism of action is thought to involve disruption of the cell membrane which leads to the collapse of the membrane potential and the efflux of low molecular weight compounds.[256]

The nisin gene has been cloned and sequenced, but not expressed so far.[257-259] The gene, variously termed *nis*A or *spa*N, encodes a 57-amino acid (aa) pronisin, which undergoes post-translational processing at the N-terminal 23-aa to give the final 34-aa sequence. Little or nothing is known about the genes and gene products necessary for the processing of pronisin to the active form, although it has been demonstrated that the nisin gene is a part of a larger operon.[260] Nisin production and resistance can be conjugally transferred between cells in the apparent absence of plasmid DNA. Dodd *et al.*[259] have suggested that nisin production and resistance (immunity) genes may be located on a transposable cassette, along with the genes for sucrose metabolism, located in the chromosome. Nisin resistance genes have been cloned from the phage-resistant, nisin-resistant plasmid pNP40,[261] and the plasmid pSF01.[262] Although the nisin resistance gene of pNP40 has been sequenced,[237] the mechanism of action of these resistance determinants has not been elucidated.

The first report of the cloning and sequencing of a bacteriocin gene from LAB was that encoding Helveticin J from the chromosome of *Lb. helveticus* 481.[247] The bacteriocin has a molecular weight of 37·5 kD. The gene for bacteriocin resistance was not confirmed, though it was presumably present on the cloned fragment which encoded production. Another bacteriocin gene has also been cloned and sequenced from *Lactobacillus acidophilus*.[244,245] In this instance, the amino acid sequence of the purified bacteriocin, Lactacin F, was partially determined. This sequence was used as a basis for a 'reverse genetics' approach to design a probe and thereby clone the *laf* gene. The *laf* gene encodes a 75-aa pre-bacteriocin, which is subsequently processed by the removal of 18 aa from the N-terminus to produce the active peptide. It is probable that this cleavage occurs during export of the molecule, and that the N-terminus is a signal sequence. No post-translational processing, or modification of the primary amino acid sequence, was detected in the active Lactacin F polypeptide. These two reports indicate the diversity of size encountered with these inhibitors, ranging from 5 to 37·5 kD. In one instance, at least two bacteriocin genes have also been located on and cloned from a single lactoccocal plasmid, p9B4-6.[263,264] Sequence data revealed that the production and resistance genes are linked in an operon in both cases, and that, in one instance, a third gene is required for the production of active bacteriocin. Both bacteriocin genes have identical amino termini, but differ in their carboxy termini, and in their spectrum of activity. The resistance genes were also identified and shown to encode small polypeptides of 98 and 154 amino acids.[242] An identical bacteriocin and bacteriocin production gene from *L. lactis* subsp. *cremoris* LMG 2130 (termed lactococcin A and *lcn*A, respectively) have also been sequenced.[251]

Domination of mixed-strain cultures is a well documented phenomenon and the possibility that this is due to bacteriocin production cannot be ignored. For this reason, defined strain cultures are screened for compatibility. Two surveys[252,265] have indicated that between 1 and 10% of lactococcal strains produce bacteriocins. The ability to detect a bacteriocin depends upon the indicator strains used, and therefore these figures do not represent the highest possible

TABLE IV

Some Properties of Purified Bacteriocins from Lactic Acid Bacteria

Organism	Name of organism	Sensitive bacteria	Resistant bacteria	Genetic determinant	Mode of action	Heat resistance	pH stability	MW	Other properties	Ref.
Lb. acidophilus N2	Lactacin B	Thermophilic lactobacilli	Mesophilic lactobacilli Clostridia Enterobacteria Ps. aeruginosa S. aureus Lactococci Ec. faecalis	—	Bactericidal (resting cells)	100°C × 1 h	—	6000	Maximum production at pH 6·0	243
Lb. acidophilus 11088	Lactacin F	Thermophilic lactobacilli Enterococci	As above	110 kb plasmid. Gene cloned, sequenced, expressed.	Bactericidal (resting cells) Bacteriostatic (growing cells)	121°C × 15 min	—	2500 56 aas	25 aas sequenced	244, 245
Lb. helveticus 27	Lactocin 27	Thermophilic lactobacilli	—	—	Bacteriostatic (growing cells)	—	—	12 400	Glycoprotein	246
Lb. helveticus 481	Helveticin J	Thermophilic lactobacilli	Mesophilic lactobacilli	Chromosome. Gene cloned, expressed, sequenced.	Bactericidal (resting cells)	<100°C × 0·5 h	—	37 000	Unstable unless produced at pH 5·0	247
Lc. lactis spp. lactis	Nisin	Lactococci Clostridia Staphylococci Listeria	—	Chromosome. Gene cloned, expressed, sequenced.	Bactericidal (resting cells)	121°C × 15 min	Inactivated at pH > 7	3500 34 aas	Unusual aas	248
Lc. lactis spp. lactis biov. diacetylactis S50	Bacteriocin S50	Other lactococci	Lactobacilli S. aureus B. subtilis A Streptococci B Streptococci Enterococci	Chromosome ?	—	100°C × 1 h	2–11	—	—	249

TABLE IV—*contd.*
Some Properties of Purified Bacteriocins from Lactic Acid Bacteria

Producer	Bacteriocin	Spectrum		Genetics	Mode of action	Heat stability	pH	MW	Comments	Ref.
Lc. lactis spp. *lactis* biov. *diacetylactis* S1-67C	—	*Str. thermophilus* Pseudomonas *S. aureus* Enterococci Bacilli Enterobacteria	Thermophilic lactobacilli	—	—	125°C × 1 h	3–9	0·385	—	250
Lc. lactis spp. *cremoris* LMG 2130	Lactococcin A	Other lactococci	Gram negatives	55 kb plasmid. Gene cloned, sequenced.	Bactericidal (resting cells)	—	—	5778 54 aas	IEP pH 9·2 Protein sequenced	251
Lc. lactis spp. *cremoris* 346	Diplococcin	Other lactococci	*Bacillus* *Micrococcus* Enterobacteria Pseudomonas Flavobacteria	54 Md plasmid	Bactericidal (resting cells)	<100° C × 1 h	—	5300 ornithine as N terminal aa	Very unstable; M17 gives protection	252
Lc. lactis spp. *cremoris* 202	Lactostrepcin 5	—	—	—	Bactericidal (resting cells)	—	—	—	Phospholipid ?	253
Ped. acidilactici H	Pediocin AcH	*Lb. plantarum* Leuconostoc *S. aureus* *Cl. perfringens* *L. mono-cytogenes* *Ps. putida*	Salmonella *E. coli* Yersinia	7·4 Md plasmid	Bacterial (resting cells) Bacteriostatic (growing cells)	121°C × 15 min	2·5–9	2700	Produced only at pH < 5·0	254
Ped. acidilactici PAC 1·0	Bacteriocin PA 1	Other pediococci *L. mono-cytogenes* Leuconostocs Mesophilic lactobacilli	Lactococci Staphylococci Micrococci Thermophilic lactobacilli	6·2 Md plasmid	Bactericidal (growing cells)	—	5–7	16 500	—	255

percentage of bacteriocin-producing strains. Despite the widespread ability of
LAB to produce bateriocins, only nisin has been found to be of commercial
value. In 1969 nisin was accepted by the WHO as a legal food additive and has
been used extensively for this purpose in Europe. It was only in 1988 that the
FDA in the US approved the use of nisin in a limited number of pasteurized
cheese spreads, where high moisture and reduced levels of Na$^+$ pose a risk of
botulinal toxin formation in the product.

6.3 Bacteriophage

6.3.1 Occurrence, Taxonomy and Structure

Bacteriophage (phage) are extremely small and relatively simple particles which
are only visible in the electron microscope, Fig. 9. They are composed of a protein coat surrounding a single DNA molecule, the same size as many plasmids
(20–30 kb). Phage are a major problem in cheese manufacture as they lyse the
starter cells, thus preventing acid production. The term 'disturbing phage' is
used to describe such phage. It is also common for phage to be present in the
cheese factory environment or in the culture itself (own phage), which do not
inhibit the ability of the culture to produce acid. Phage have been described
which attack lactococci (reviewed by Jarvis[266]), leuconostocs,[267,268] *Str. salivarius*
subsp. *thermophilus*,[40,269-271] *Lb. delbrueckii* subsp. *bulgaricus, Lb. delbrueckii*
subsp. *lactis*, and *Lb. helveticus*.[270-272] Morphologically, phage for *Lb. helveticus*
are of Bradley's[273] group A, while phage for *Str. salivarius* subsp. *thermophilus*,
Lb. delbreuckii subsp. *lactis* and *Lb. delbrueckii* subsp. *bulgaricus* are generally of
Bradley's group B; phage for leuconostocs can be either group A or B.

   Since they were first isolated by Whitehead & Cox,[274] a number of approaches
to the characterization and classification of phage for lactococci have been
described. Indeed, many of the problems of phage taxonomy highlighted by
Ackermann[275] are experienced in the case of phage for dairy cultures because
much of the data was accumulated for commercial rather than taxonomic reasons; in addition, few attempts have been made to compare phage studied by
different groups. Substantial host-range data have been accumulated, establishing a high degree of diversity.[276,277] The detection of low efficiency phage-host
interactions has been facilitated by the availability of improved media.[39,278,279]
Morphologically, most lactococcal phage fit into the Styloviridae family[275] or
Group B of Bradley[273] in having hexagonal heads and non-contractile tails.
Within this grouping there are three types: prolate and small isometric, which
are common, and large isometric, which are relatively rare.[280-288] Individual
phage isolates are further differentiated by various ultrastructural features such
as collars, baseplates, fibres and whiskers. Morphological groupings of phage
correlate well with serology and DNA homology studies, but not with host
range data.[281,283,284,287] There is considerable evidence, however, that prolate
phages may have broader host ranges than isometric ones.[282,284] The GC content
varies between 32 and 41%[289] and the genome size from 12 to 27 Mdal.[290-294]

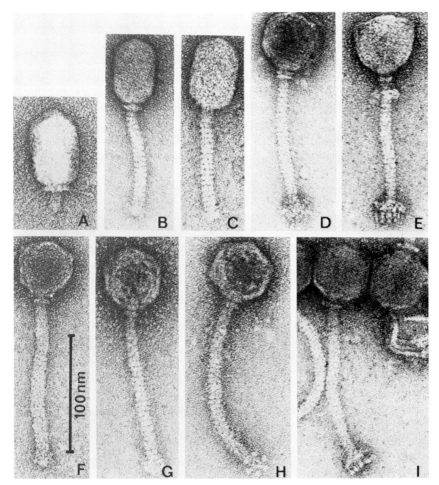

Fig. 9. Electronmicrographs of negatively stained phages active against strains of lacto-cocci. The magnification (see bar) is the same for all pictures (from Lembke *et al.*,[285] reprinted with permission).

6.3.2 Phage-Host Relationships

Phage can enter into two different relationships with their host bacteria—the lytic and lysogenic states. The lytic cycle (Fig. 10) begins with attachment of the phage to specific receptors on the cell surface of the host; this is followed by injection of DNA. Within minutes, phage-directed processes 'hijack' the cell and redirect the host to synthesize additional phage DNA and proteins. Within 45–60 min, approximately 34 (range 2–105) new viable phage particles are produced. The difference in growth rates of phage and cells accounts for the susceptibility of growing cultures to phage attack. Phage-associated lytic enzymes (lysins) may be responsible for inhibition in mixed cultures in which one component strain has been attacked by a lytic phage, leading to the production of significant amounts of lysin which may affect other unrelated strains. Three phage lysins have been purified and shown to have broad specificities.[295-297]

Temperate, or lysogenic, phage have the option of establishing a single copy of their DNA at a specific site in the host chromosome after infection. This lysogenic phage may be dormant within the cell, or, under the correct stimulus (UV irradiation or exposure to mitomycin C), may excise and enter the lytic cycle (reviewed by Davidson et al.[298]). It may be difficult to detect these released particles since a lysogen is generally protected against attack by the released phage by superimmunity. The released phage can occasionally be detected using indicator

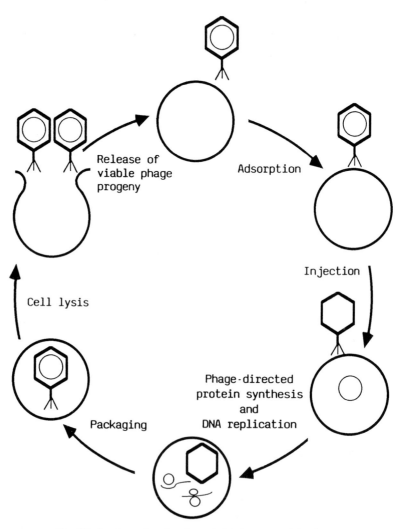

Fig. 10. Lytic cycle of a phage infecting a sensitive host.

strains, or by electron microscopic examination of the supernatant. Several surveys have shown that essentially all LAB are lysogenic[289,299-301] and some are multilysogenic, i.e. release two or more distinguishable phage types;[302,303] however, the number of indicator strains are limited except for the report of Reyrolle et al.[304] who found that 25% of the lactococci tested could act as indicators. Jarvis[288] found considerable homology in the DNAs of three temperate phage, suggesting that the paucity of indicator strains may be due to the use of closely related lysogenic strains which were consequently immune to infection by the induced phage. This conclusion suggests that Reyrolle et al.[304] used a wider range of lysogenically-different strains in their study. The lysogenic nature of most strains has fuelled speculation that such strains are the major source of virulent phage in dairy fermentations. Lawrence et al.[305] found that the removal of strains that released phage spontaneously from cheese starters contributed to a reduction in the levels of phage in the environment. In contrast, Jarvis[288] found

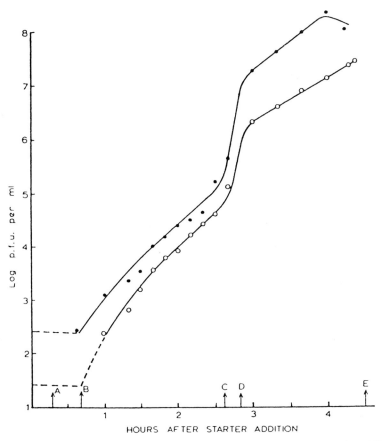

Fig. 11. Phage multiplication during cheese manufacture in two vats. A, addition of rennet; B, cutting; C, running the whey; D, cheddaring; E, salting (from Pearce et al.[307]).

no homology between the DNAs of lysogenic and lytic phages of several strains of lactococci, implying that temperate phage are not the source of lytic phage. Although phage have been shown to be present in some mixed strain starters,[16,306] their ultimate source remains to be determined.

6.3.3 Phage Multiplication During Cheese Manufacture

There is little quantitative information concerning the levels of phage which cause problems in cheesemaking. The kinetics of phage multiplication during cheese manufacture (Fig. 11) have been reported.[307] Contamination of the culture itself is much more important than contamination of the cheese milk. Pearce[308] presented data which showed that between 600 and 6000 phage/ml are necessary to slow down cheese manufacture although in a later paper from the same Institute, Limsowtin and Terzaghi[309] showed that an initial titre of 100 phage/ml and a multiplication factor of 1×10^7 could significantly affect acid production during cheesemaking. Pearce et al.[307] found average phage numbers of 1×10^5/ml and 2.4×10^8/ml in wheys at running and at milling, respectively. These levels inferred an initial level of 5–5000 phage/ml; at the lower level, this is equivalent to about 0.01 ml of an average milling-whey per 6000 litres of milk. Obviously, this is a very low level of contamination and clearly demonstrates the need for adequate control procedures.

6.3.4 Detection and Enumeration of Bacteriophage

Two different methods are used to detect and enumerate phage for LAB–plaque assays and inhibition of acid production. Regardless of the method used, the source of phage must be filter-sterilized (0.45 μm) before being tested. Bulk starters are centrifuged before the supernatant is filter-sterilized, while cheese wheys are usually filtered directly; such wheys may clog the membrane filters but this can be overcome by centrifuging or prefiltering the whey through a coarse filter before filter-sterilization.

Ca^{2+} are considered important for the attachment of phage to the host cells but precipitate in phosphate-buffered media. Ca^{2+} precipitation is overcome in glycerophosphate-buffered media which are used for the lactococci,[39,278] and Str. salivarius subsp. thermophilus[40] but are inhibitory to many thermophilic lactobacilli;[42] for these bacteria, MRS[42] is recommended.[270,272] In the plaque method, dilutions of the filtered-sterilized source, Ca^{2+} and host organism are added to 'sloppy' agar (M17[39] or MRS[42] broth containing 0.7% (w/v) agar) which is quickly poured on solidified agar in a petri dish. Some mesophilic phage multiply much better at 21°C (Daly & Cogan, unpublished data) and others at 35°C.[40] For this reason, plates should be incubated at 21°C, 30°C and 37°C. At 30 and 37°C, plaques can often be seen after a 5 h incubation (Fig. 12) but plates are generally incubated overnight. Information on the effect of temperature on phage for thermophilic cultures is not available. Incubation of plates for thermophilic phage is generally at 37–40°C[40,270,272] aerobically for Str. salivarius subsp. thermophilus,[40,270] and aerobically[270] or anaerobically[272] for the lactobacilli. Mesophilic phage which multiply best at 35°C are known as 'raw milk phage'

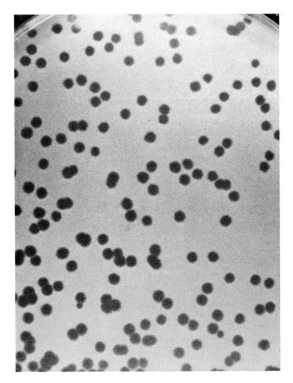

Fig. 12. Plaques formed by phage øc2 on *Lc. lactis* subsp. *lactis* MG1614.

because they multiply better in raw or pasteurized milk than in sterilized milk and only affect acid production during the final stages of cheese manufacture;[40] the reason(s) for the more rapid multiplication in raw milk has not been determined.

In the acid inhibition method, dilutions of the filter-sterilized source are added to samples of pasteurized and sterilized milks, inoculated with the relevant host. Again, it is important to incubate at different temperatures for the reasons outlined above; commonly, these are: 21°C for 12 h, 30°C for 6 h and over the cheese temperature profile for 6 h (90 min at 30°C followed by an increase to 39°C over 45 min; this temperature is held for the remaining 3 h 45 min) for mesophilic cultures. Thermophilic cultures are incubated at 42°C for 4 h.[40] After incubation, the pH is determined and related to that of the control which does not contain whey. A difference of ≥0·3 pH units from the control is presumptive evidence of phage.

It is possible to detect mesophilic phage by the acid inhibition test and yet see no plaques on plates[310] (Daly & Cogan, unpublished data). To determine if such inhibition is due to phage or some other inhibitor, membrane-filtered wheys are prepared from the milks showing inhibition, diluted and retested in the milk inhibition test; persistence of the inhibition on the second growth cycle indicates the presence of phage. Host heterogeneity[311] may be responsible for this

phenomenon but this has not been shown unequivocally. Because acid production by the starter in milk is a key reaction in cheese manufacture, it is probably
better to monitor phage by methods involving inhibition of acid production in milk,
even though, unlike the plaquing method, it does not give absolute evidence of
phage.

6.3.5 Phage Genetics

One of the most elegant studies in phage genetics was carried out by Shimizu-
Kadota et al. on the Lb. casei temperature phage øFSW.[312,313] It was conclusively
demonstrated that the appearance of a disturbing phage (øFSV) in the production of Yakult (a Japanese fermented milk) was a result of a genetic rearrangement in which an insertion sequence (Section 5.6) caused the temperate phage
(øFSW) to become lytic (øFSV) for its own host. Phage øFSV was shown to be
a simple derivative of øFSW. By curing the original host of the lysogenic øFSW,
the problem of phage in Yakult was resolved. This remains the only instance in
which a temperate phage has been confirmed as the source of a lytic phage in
the dairy industry.

Phage DNA can be obtained by extraction from intact phage particles, or
more rapidly by isolation of DNA from infected cells.[314] Compared to other species,
phage in LAB remain underexploited as genetic tools. A number of phage genes
and expression signals have been cloned and sequenced, including a phage lysin
gene,[315] the origin of replication of ø50,[234] and a number of promoters of øBK5-T.[316]
The lactococcal phage, ø50, has been shown to be capable of cloning plasmid
DNA.[317] When cells containing pTR2030 were used in an industrial culture, ø50
was detected in the fermentation vessels. Analysis revealed that ø50 was insensitive to the pTR2030 restriction-modification (R/M) system, but not to other
R/M systems subsequently tested. Cloning and sequence analysis confirmed that
ø50 carried a copy of an active domain of the pTR2030 methylase gene. The
two sequences were 100% identical, confirming that ø50 had 'cloned' the plasmid
gene and rendered itself insensitive to the plasmid restriction system. This example serves to highlight the ability of the apparently simple phage genome to
serve as a template for biological variation. This is not due to any inherent
ability of the phage per se, but most probably to the high numbers of phage produced during an industrial fermentation. These high numbers make probable the
most unlikely events, some of which may be advantageous to the phage.

6.3.6 Bacteriophage Resistance

A large number of plasmids encoding phage resistance have been described in
lactococci (reviewed by Klaenhammer[318]), involving at least three mechanisms:
prevention of phage adsorption, R/M systems, and abortive infection. Prevention of phage adsorption most probably results from the 'blocking' of specific
phage receptors on the cell surface. In the case of the adsorption inhibition plasmid,
pSK112, the blocking agent is believed to be a galactosyl-containing lipoteichoic
acid polymer in the host cell wall.[319] R/M systems encode a restriction enzyme which
recognizes, and cleaves, specific short DNA sequences. Thus, incoming foreign

DNA is degraded. The restriction enzyme is associated with a methylase, which protects the host DNA by specifically modifying those sequences recognized by the restriction enzyme. Two restriction enzymes have been purified from LAB, ScrF1 from *Lc. lactis* subsp. *cremoris* F,[320] and Sth134I from *Str. salivarius* subsp. *thermophilus* 134.[321] Both are type II enzymes, and recognize 5- and 4-bp sequences, respectively. An R/M system has also been described in *Lb. helveticus.*[322] Abortive infection in lactococci applies to any instance in which phage resistance does not result from adsorption inhibition or R/M (reviewed by Sing & Klaenhammer[323]). Strictly, it should be applied only to those cases in which phage development is inhibited, but the phage genome is not degraded. Cell death often results after the abortive response.

Though there are a number of reports of gene cloning, in only a few instances has a phage resistance gene been sequenced. Perhaps the best studied plasmid has been pTR2030 (reviewed by Klaenhammer *et al.*[324]). This plasmid encodes an R/M system designated *Lla*I, and an abortive infection mechanism, Hsp.[325,326] All three genes, restriction (R.*Lla*I), modification (M.*Lla*I), and abortive infection (*hsp*), have been cloned, and the latter two have been sequenced.[238,317] The R/M system was shown to be type II, and most probably recognizes an asymmetric sequence. Interestingly, the gene is composed of two functional units fused into a single gene, and is closely related to the methylase gene, *Fok*I, from *Flavobacterium okeanokoites.* The abortive infection gene *hsp* shows no similarity to other sequenced genes other than the abortive infection gene *abi* from the lactococcal plasmid pCI829, which is identical.[327] The mechanism of action of the *hsp* gene product complies with the definition of abortive infection in that phage DNA replication is prevented,[314] though this is not sufficient to prevent cell death. The R/M system associated with Hsp removes the majority of attacking phage, and thus the drastic response leading to cell death is invoked in only a few cells. The linking of R/M and abortive infection genes will probably emerge as a universal phenomenon, a number of instances of such a linkage have already been described.

Commercially used strains have been shown to possess up to five independent phage resistance mechanisms, including two R/M systems.[233] Recent studies have examined the effect of stacking several resistance plasmids in single strains to attain higher levels of resistance.[234,328,329] An additional source of phage resistance determinants was recently described which employs phage DNA as a means of protecting cells against phage attack.[234] The origin of replication of ø50 was cloned in the shuttle vector pSA3 (Section 5.2). When a strain containing this plasmid is attacked by ø50, the plasmid is replicated instead of the phage, leading to effective resistance. If this method of engineering resistance is generally applicable, it means that each phage may carry its own resistance mechanism which can be exploited through genetic techniques.

6.3.7 Control of Bacteriophage

Rotation of phage-unrelated strains has been a major aspect of the control of phage in the defined cultures (Section 3.2). Mixed cultures contain strains of different phage sensitivities and are inherently less susceptible to phage. When a mixed culture is attacked by phage, it quickly recovers due to growth of the phage-resistance strains (Section 3.1). For this reason, rotations are not used as frequently with mixed cultures. However, mixed strain cultures which have been transferred aseptically in the laboratory (L cultures, Section 3.1) are inherently susceptible to phage. The sensitivity of different cultures to phage is perhaps best exemplified by comparing the performance of P and L cultures in commercial use (Fig. 13).

Phage are normally present in the cheesemaking environment and especially in the whey. Pasteurization (72°C × 15 sec) is an effective way of controlling microbial growth in the dairy industry but the available data (Table V) show that this heat treatment is ineffective in inactivating phage. For this, and other reasons (Section 7.2), milk for bulk culture production is generally heated to 90°C for 20 min.

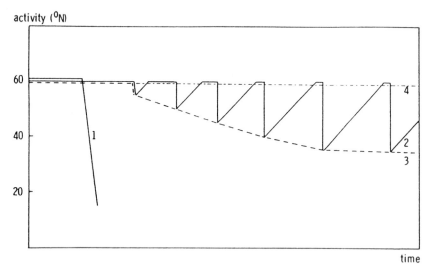

Fig. 13. A schematic picture of the fluctuation in the activities of L and P starters after their introduction to a factory where the starters are propagated daily without (curves 1–3) or with (curve 4) protection against air-borne phage contamination. Curve 1. An L culture introduced once. Starter activity falls-off very rapidly on contamination with disturbing phage because only a few strains are present in the culture. Curve 2. A P culture introduced once. This culture is less sensitive to disturbing phage and its takes more time before the activity is reduced as the recovery is initially 100% within the normal cultivation time (22 h at 20°C). Upon further propagation in the factory, activity is more and more affected, but recovery does occur. This gives rise to fluctuations in activity when new disturbing phage contaminate the culture. Such curves are reminiscent of the predator–prey curves, well known in microbial ecology. Curve 3. A P culture introduced daily. Such a culture is affected by disturbing phage at the same time as a P culture introduced once but there is less recovery. Instead a steady decrease in starter activity occurs. Curve 4. A P culture introduced daily and cultivated with compete protection against phage. Under these conditions contamination of the culture with disturbing phage is prevented and the activity remains high and constant. P cultures are normally used without rotation (from Stadhouders & Leenders[16]).

TABLE V
Comparative Heat Resistance of Phage for Lactococci

Organism	Component	Temp. (°C)	D value (min)	Z value (°C)	Ref.
Lc. lactis subsp. *cremoris* R1	Fast	65·0	16·0	3·3	330
	Slow	65·0	64·0	3·6	—
Lc. lactis subsp. *lactis* 240	—	70·0	14·0	4·9	331
Lc. lactis subsp. *lactis* C10	—	65·0	9·9	5·8	332
Lc. lactis subsp. *cremoris* C1	—	65·0	1·1	5·8	330
Lc. lactis subsp. *lactis* 144F	—	70·0	20·0	5·8	331
Lc. lactis subsp. *cremoris* E8	Fast	65·0	0·6	5·8	333
	Slow	65·0	1·9	10·2	—
Lc. lactis subsp. *lactis* C2	Fast	65·0	202·0	—[a]	332
	Slow	65·0	102·0	31·0	—
Strain 77	Fast	75·0	0·5	5·3	334
	Slow	75·0	1·0	4·8	—
Strain 20	Fast	75·0	0·5	5·6	334
	Slow	75·0	0·9	5·9	—
Strain 8	Fast	75·0	0·9	5·6	334
	Slow	75·0	2·6	5·9	—
Strain 170	Fast	75·0	0·7	6·3	334
	Slow	75·0	2·6	4·6	—
Strain 9	Fast	75·0	1·9	6·3	334
	Slow	75·0	15·9	4·6	—

[a] Could not be calculated since only 1 D value was reported.

Phage infection of bulk cultures is obviously much more serious than phage infection of cheese milk. For this reason, semi-aseptic and aseptic methods of inoculating the bulk starter medium have also received attention. In the Lewis system,[335] inoculation of the starter milk is performed through a rubber septum using a specially designed needle assembly. Inoculation through a steam 'converger' or flame or through an inverted spring-loaded flap valve has also been used.[336] In the latter system, a chlorine solution is placed above the valve; the inoculation flask is pushed quickly through the liquid onto the seat of the flap valve which opens and allows the inoculum to enter the tank; on withdrawal, the valve closes automatically. An aseptic inoculation device (Fig. 14) has been developed in the Netherlands and is commonly used in Dutch cheese factories.[337] These inoculation methods are usually used in combination with High Efficiency Particulate Air (HEPA) filters, with penetration levels of <1 in 1×10^8 phage. The HEPA filtered air enters the tank when cooling begins. Use of HEPA filtered air is also common in bulk starter tanks which are inoculated manually.

Phage inhibitory media (PIM) have been developed which include increased nutrients (e.g. corn steep liquor and yeast extract) and phosphates and/or citrates to chelate the Ca^{2+}, which are essential for phage adsorption. The use of PIM is more common in the US than in Europe. Their use has been questioned on the

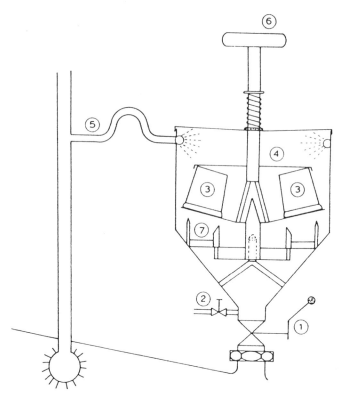

Fig. 14. An aseptic inoculation device. The device is opened using the screw (6). Cartons containing the frozen inoculum are inverted with their foil caps in place and placed on the holder (3). The device is assembled and chlorinated water is allowed to fill the device through the pipe (5). The chlorination also helps to thaw the culture. Residual chlorination fluid is run to waste through the valve (2). The screw (6) is turned to bring the prongs (7) into contact with the foil lid and puncture it. The cartons empty and their contents are added to the starter tank through the valve (1). Any residual chlorine has no effect on the starter as it will be inactivated on contact with the fluid in which the starter cells are suspended (from Leenders & Stadhouders).[337]

basis of cost, effectiveness, cell damage and inhibition of the *Leuconostoc* component of mixed strain cultures.[338–341] This led Richardson and his co-workers[342,343] to develop PIM based on pH control of growth, lactose limitation, low concentrations of phosphate and whey. Cultures are incubated until lactose is exhausted and much greater cell numbers are obtained than in conventional cultures. Such media are called externally-buffered media. Another approach is the use of media containing insoluble buffering agents which solubilize as the pH decreases and help to maintain the pH above 5.3.[344,345] Such media are called internally buffered media and do not require pH electrodes, injectors for the neutralizer or recorders. These media have greater buffering capacities than milk due to the higher concentrations of phosphate and citrates in them and are effective against phage for both mesophilic and thermophilic cultures. Other advantages of internally and externally buffered media include reduced rates of inoculation, reduction in bulk culture requirements, extension of the storage time during which the activity of the culture is retained, and reduced labour costs.

Elimination of the ripening period (i.e. the period between starter and rennet additions) is also useful in controlling phage, since once coagulation has occurred, the phage will be unable to penetrate the curd particles. In factories where cheese vats are filled several times daily, phage levels can build up. These are very effectively controlled by chlorination between fills; a contact time of 10 min with 200 ppm *available* chlorine is usually recommended; it is important not to rinse the residual chlorine from the vat. The use of phage insensitive cultures, limited numbers of strains, physical separation of the starter producing and cheese manufacturing areas, also reduce phage levels in plants.

7 COMMERCIAL PRODUCTION OF CULTURES

7.1 Supply of Inocula

Until the beginning of this century, cheese was still mainly a farm enterprise and the cultures were produced by allowing the adventitious LAB present in the milk to multiply and produce sufficient acid to sour it. As production of cheese became more industrialized, companies were formed to supply cultures, enzymes and other cheesemaking necessities. Initially, cultures were supplied in milk containing added $CaCO_3$ to partially neutralize the acid. These liquid cultures had a relatively short shelf life (< 1 week) and were gradually replaced by vacuum dried cultures which had longer shelf lives but which required several transfers to activate them and build up the necessary volumes required for bulk culture production. Cultures are normally used at a rate of 1–2% (v/v), so a factory processing 500 000 litres of milk/day into cheese requires 5000 litres of bulk culture/day, i.e. 50 litres of inoculum. Thus, starting with a liquid culture of 5 ml, two consecutive 1% subcultures would be required to obtain the 50 litres of inoculum required for bulk culture inoculation.

In the mid 1960s vacuum dried cultures were replaced by frozen concentrated cultures.[346] These cultures retain activity for several months during frozen storage, require no subculturing and are used directly to inoculate the bulk culture medium. In the past decade, freeze drying has been shown to be as effective as frozen storage in retaining high cell numbers with good acid producing activities but, unfortunately not all cultures are resistant to freeze drying. Cryoprotective agents are normally added to cultures prior to freeze drying. Freeze dried cultures have distinct advantages for storage and transport. The parameters for the production of frozen/freeze dried cultures are now well established. Milk- or whey-based media containing other sources of nutrients (e.g. yeast extract or corn steep liquor) are used for growth. Temperature and pH are controlled at 30°C and pH 6·0 for mesophilic cultures, at 42°C and pH 6·0 for *Str. salivarius* subsp.

thermophilus and at 42°C and pH 5·5 for thermophilic lactobacilli. The neutralizer can be NaOH or NH_4OH and cultures are harvested by centrifugation or membrane filtration. Batch fermentations are used with gentle agitation to mix the alkali. This can lead to inhibition by H_2O_2 production and a 'head' of N_2 may be maintained above the culture to prevent O_2 entering the medium. Care must be taken in growing cultures at constant pH since, at pH values above 6·0, Prt⁻ cultures accumulate at high frequencies and result in strain imbalance.[180]

Super-concentrated cultures for direct inoculation of the cheesemilk are produced in a similar manner. These cultures are expensive but are often used by smaller manufacturers since they do not require a bulk culture production facility. A bulk culture contains about $\sim 10^9$ organisms/ml whereas concentrates and super-concentrates contain about 10^{10} cells/g and 10^{13} cells/g, respectively.

7.2 Bulk Culture Preparation

A supply of bulk cultures, free of disturbing phage, is probably the key factor in producing good quality cheese. A variety of media may be used in preparing bulk cultures, e.g. PIM, whole milk, skim milk, or skim milk fortified with extra solids.[335] Reconstituted antibiotic-free skim milk powder, containing 10–14% solids, is commonly used; the medium (and the head-space of the tank) are heated at $\geq 90°C$ for at least 20 min. Some producers prefer to heat-treat the medium in a plate heater (90°C × 20 s) before addition to a clean, pre-sterilized bulk starter tank.[347] This method is not to be recommended since the cleaning and sanitizing procedures may not inactivate all phage (Table V). Heating has several effects. It destroys contaminating bacteria whose growth could compete with culture growth and subsequently cause problems during cheese manufacture. It increases the rate of growth of the culture through increased availability of nitrogen by partial hydrolysis of the milk protein. It destroys the natural inhibitors present in the milk and any phage which might be present in the medium. To further improve the efficacy of phage destruction, the head space may be steamed during the heating period. Phage may enter the starter tank during cooling of the bulk culture medium and therefore either a pressurized tank is used or the air, which is sucked into the tank as it is cooled, is filtered through a HEPA filter.[348] Inoculation is another critical step in bulk culture preparation and several aseptic inoculation devices have been described (Section 6.3.7).

pH controlled growth of cultures has been advocated to control phage (Section 6.3.7) and increase cell numbers. Another way of increasing cell numbers has been suggested in New Zealand where a one or two shot neutralization of the culture after growth followed by further incubation doubles the yield of cells and halves the inoculum required in the cheese vat.

Mesophilic cultures are normally incubated at 21°C even though their optimum is $\sim 30°C$. Most of them have similar growth rates at 27°C but at other temperatures, especially 21°C, considerable differences occur.[18] This has important practical implications for growth because it means that in mixed cultures, domination by the faster-growing components can occur at the traditional incubation temperature. Three parameters: rate of inoculation, temperature of incubation and time of incubation, determine how rapidly a culture grows. For mesophilic cultures, these are 0·5–1% (v/v) inoculum, 21°C and 16 h, respectively, and, under these conditions, mesophilic cultures will produce 0·85% lactic acid in milk and reduce the pH from 6·5 to 4·5. The use of 21°C as the incubation temperature is significant since it means that a culture could be inoculated at 4 pm, incubated at room temperature (21°C) and be ready for cheesemaking 16 h later, i.e. 8 am the following morning. At 21°C, the generation time (GT) of both mixed and defined strain starters in milk is 2·2 h.[185] A mathematical basis can also be found for the choice of 21°C as the incubation temperature: if the GT is 2·2 h, 7·2 generations occur in 16 h. If a 0·5% inoculum is used, a fully grown culture is equivalent to 7·5 generations. Both calculations give approximately the same value for the number of generations.

Thermophilic cultures are normally incubated at 42°C for 8–12 h, although the optimum temperature for different strains of *Str. salivarius* subsp. *thermophilus* and thermophilic lactobacilli shows some variation: 39–46°C and 42–48°C, respectively.[349,350] During incubation, the pH will decrease to about 3·5 and 4·5 for the lactobacilli and streptococci, respectively, resulting in production of 1·2 and 0·6% lactic acid. Continued incubation of thermophilic lactobacilli will result in the production of 2% lactic acid. Such cultures are normally cooled when 1·2% lactic acid has been produced. Both mesophilic and thermophilic cultures can be stored for several days at 4°C without showing significant losses in their activity or ability to produce lactic acid.

7.3 Practical Use of Cultures

The different traditions observed in cheesemaking have been accompanied by a variety of approaches to the use of cultures. In many countries, especially Italy, France and Switzerland, natural whey cultures are still used. Current developments in the use of cultures aim at maximizing process control so as to ensure products of consistent composition and quality. In the Netherlands, this has been achieved by maintaining the proper strain balance in mixed-strain 'P' cultures by careful attention to the conditions of storage and propagation. These starters can be used in large-scale cheesemaking without rotation, provided they are protected from disturbing phage.[16] While this report[16] is a fascinating study of the interactions of phage and cells, it also highlights the complexity of using mixed cultures of unknown composition.

In contrast, many mixed mesophilic cultures used in other countries are sensitive to phage attack and are used in rotations.[26,348] The basis of this practice is that phage remaining from the use of one culture should have cleared by the time a phage-related starter is used again. It is difficult to anticipate problems with this system due to the lack of precise information on the phage-host relationships of the different strains in the mixed cultures. This makes process control difficult and leads to variations in manufacturing schedules. Levels of disturbing phage in cheese factories can be greatly reduced if a small number of strains are used.[16,26]

The use of phage-insensitive defined-strain mesophilic cultures was pioneered in New Zealand. The demands of large-scale cheese production have necessitated the fine-tuning of this system so that only a small number (often a pair) of phage-insensitive strains are used daily without rotation. Indeed, the strains are remarkably phage resistant and have been used for several years without replacement.[351,352] The known identity of the strains facilitates phage monitoring and trouble-shooting in the factory and ensures maximum information on culture performance. It is noteworthy that it is not necessary to have a phage-free system. Indeed, it is very likely that phage are always present but do not interfere with manufacturing schedules. These are slow-replicating phage and were observed as early as 1951.[353]

Carefully selected defined strains are very stable when used in combination with good manufacturing practices (see Section 6.3.7). The occasional appearance of fast-replicating virulent phage can necessitate the use of a replacement strain. This may be a phage-unrelated strain with established cheesemaking characteristics[305] or a BIM of the strain which succumbed to phage in the factory,[23-26] or a transconjugant containing one or more phage resistant plasmids.[233]

8 CONCLUDING COMMENTS

It has been claimed that 10^{23} individual LAB cells are used in the dairy industry each year, confirming them to be the most important industrial bacteria worldwide and underlining their enormous economic impact.[354] There is a general movement away from undefined mixed starters towards the use of single or limited numbers of defined strains with proven performance, capable of producing cheese of consistent quality within the confines of automated systems. This is particularly true for Cheddar cheese. A large part of this chapter was devoted to a review of developments in the genetics of starter cultures and their bacteriophage and this accurately reflects a major effort in many countries to develop our understanding of these bacteria on a molecular level. It is our hope that in documenting this shift in emphasis in a number of laboratories we have not neglected the excellent research on metabolism and taxonomy, nor indeed the lessons learned from our industrial colleagues.

One of the effects of recent research on the taxonomy of LAB has been the reclassification of some of the more common genera. This will make subsequent reviews of the literature more difficult, since not all workers have made the recommended changes. Perhaps the most significant changes include those of *Streptococcus lactis* and *Str. cremoris* to *Lactococcus lactis* subsp. *lactis* and *Lc. lactis* subsp. *cremoris*. It is important to note that *Streptococcus thermophilus* remains within the streptococcal family, though as a subspecies, *Str. salivarius* subsp. *thermophilus*. However, these changes in nomenclature are based on sound scientific principles and should withstand the test of time. The most recent recommendations are included in this chapter.

Phage remain a significant problem within the cheese industry. Improvements in plant hygiene and a more scientific approach to the design and monitoring of starter systems have reduced the problem, but much remains to be done. It is surprising, in view of the amount of effort in this area, that the source of disrupting phage has not been equivocally identified. In general, rather than search for the source, attention has focused on means of protecting strains from these apparently ubiquitous particles. In this respect the development of conjugal strategies for the construction of phage resistant strains acceptable to the food industry has been highlighted. A considerable arsenal of phage resistance mechanisms has been, and continues to be, described in the literature. Many are at the forefront of the genetic technologies developed in other bacteria, and, as such, are not ready for immediate application to food systems. Nonetheless, they point to the strategies by which phage may be controlled in the future. If phage problems diminish as a result of the techniques outlined in Section 6.3, then it can be expected that more subtle differences in flavour, aroma and texture will become uppermost in cheesemakers' priorities. We may expect a return to more complex starters, though composed of defined strains, in an attempt to achieve cheese of the desired high quality consistently. Alternatively, one could envisage 'super strains' in which all necessary traits are combined by genetic manipulation. However, while phage problems remain, consistent acid production will continue to be the most important attribute of any starter system.

ACKNOWLEDGEMENTS

Grateful thanks are extended to those colleagues who allowed us use of their figures in this chapter.

REFERENCES

1. Van Niel, C.B., Kluyver, A.J. & Deux, H.G., 1929. *Biochem. Z.*, **210**, 234.
2. Michaelian, M.B., Hoecker, W.H. & Hammer, B.W., 1938. *J. Dairy Sci.*, **21**, 213.
3. Schleifer, K.H. & Kilpper-Balz, R., 1987. *System. Appl. Microbiol.*, **10**, 1.
4. Garvie, E.I., 1986. *Bergey's Manual of Systematic Bacteriology*, **2**, 1071.
5. Farrow, J.A.E. & Collins, M.D., 1984. *J. Gen. Microbiol.*, **130**, 357.
6. Kandler, O. & Weiss, N., 1986. *Bergey's Manual of Systematic Bacteriology*, **2**, 1209.
7. Thomas, T.D. & Crow, V.L., 1984. *Appl. Environ. Microbiol.*, **48**, 186.
8. Hickey, M.W., Hillier, A.J. & Jago, G.R., 1986. *Appl. Environ. Microbiol.*, **51**, 825.
9. Tinson, W., Hillier, A.J. & Jago, G.R., 1982. *Aust. J. Dairy Sci. Technol.*, **37**, 8.
10. Garvie, E.I., 1960. *J. Dairy Res.*, **27**, 283.
11. Galesloot, T.E. & Hassing, F., 1961. *Neth. Milk Dairy J.*, **15**, 225.
12. Stadhouders, J., 1974. *Milchwissenschaft*, **29**, 329.
13. Andresen, A., Geis, A., Krusch, U. & Teuber, M. 1984, *Milchwissenschaft*, **39**, 140.
14. Lee, D.A. & Collins, E.B., 1976. *J. Dairy Sci.*, **59**, 405.
15. Hugenholtz, J. & Veldkamp, H., 1985. *FEMS Microbiol. Ecology*, **31**, 57.
16. Stadhouders, J. & Leenders, G.J.M., 1984. *Neth Milk Dairy J.*, **38**, 157.

17. Geis, A., Kiefer, B. & Teuber, M., 1986. *Chem. Microbiol. Technol. Lebensm.*, **10**, 9.
18. Hugenholtz, J., Splint, R., Konings, W.M. & Veldkamp, H., 1987. *Appl. Environ. Microbiol.*, **53**, 309.
19. Heap, H.A. & Lawrence, R.C., 1988. In *Developments in Food Microbiology*, ed. R.K. Robinson. Elsevier Applied Science Publishers, London, pp. 149–85.
20. Whitehead, H.R., 1953. *Bacteriol. Rev.*, **17**, 109.
21. Lawrence, R.C & Pearce, L.E., 1972, *Dairy Industries Intern.*, **37**(2), 73.
22. Limsowtin, G.K.Y., Heap, H.A. & Lawrence, R.C., 1977. *N.Z. J. Dairy Sci. Technol.*, **12**, 101.
23. Czulak, J., Bant, D.J., Blyth, S.C. & Crace, J.B., 1979. *Dairy Industries Intern.*, **44**(2), 17.
24. Hull, R.R., 1983. *Aust. J. Dairy Technol.*, **38**, 149.
25. Richardson, G.H., Hong, G.L. & Ernstrom, C.A., 1980. *J. Dairy Sci.*, **63**, 1981.
26. Thunell, R.K., Sandine, W.E. & Bodyfelt, F.W., 1981. *J. Dairy Sci.*, **60**, 2270.
27. Timmons, P., Hurley, M., Drinan, F., Daly, C. & Cogan, T.M., 1988. *J. Soc. Dairy Technol.*, **41**, 49.
28. Heap, H.A. & Lawrence, R.C., 1976. *N.Z. J. Dairy Sci. Technol.*, **11**, 16.
29. Coventry, M.J., Hillier, A.J. & Jago, G.R., 1984. *Aust. J. Dairy Technol.*, **39**, 154.
30. Auclair, J. & Accolas, J.P., 1983. *Ir. J. Food Sci. Technol.*, **7**, 27.
31. Bottazzi, V., Bersani, C., Sarra, P.G. & Magri, M., 1977. *Sci. Tec. Latt. Casearia*, **28**, 430.
32. Coppola, S., Parente, E., Dumontet, S. & La Peccerella, A., 1988. *Le Lait*, **68**, 295.
33. Wood, N.J., 1981. *Dairy Industries Intern.*, **46**(12), 14.
34. Steffen, C., 1980. *Intern. Dairy Fed. Doc.*, **126**, 16.
35. Turner, K.W. & Martley, F.G., 1983. *Appl. Environ. Microbiol.*, **45**, 1932.
36. Valles, E. & Mocquot, G., 1968. *Le Lait*, **48**, 631.
37. Rousseaux, P., Vassal, L., Valles, E., Auclair, J. & Mocquot, G., 1968. *Le Lait*, **48**, 241.
38. Elliker, P.R., Anderson, A.W. & Hannesson, G., 1956. *J. Dairy Sci.*, **39**, 1611.
39. Terzaghi, B.E & Sandine, W.E., 1975. *Appl. Environ. Microbiol.*, **29**, 807.
40. Accolas, J.-P. & Spillmann, H., 1979. *J. Appl. Bacteriol.* **47**, 135.
41. Shankar, P.A. & Davies, F.L., 1977. *J. Soc. Dairy Tech.*, **30**, 28.
42. De Man, J.C., Rogosa, M. & Sharpe, M.E., 1960. *J. Appl. Bacteriol.*, **23**, 130.
43. Radke-Mitchell, L. & Sandine, W.E., 1984. *J. Food Protect.*, **47**, 245.
44. IDF Standard. 117A, 1988. Yogurt—Enumeration of characteristic microorganisms —Colony Count Technique at 37°C.
45. Waes, G., 1968. *Neth. Milk Dairy J.*, **22**, 29.
46. Nickels, C. & Leesment, H., 1964. *Milchwissenschaft*, **19**, 374.
47. Vogensen, F.K., Karst, T., Larsen, J.J., Kringelum, B., Ellekjaer, D. & Waagner-Nielsen, E., 1987. *Milchwissenschaft*, **42**, 646.
48. McDonough, F.E., Hargrove, R.E & Tittsler, R.P., 1963. *J. Dairy Sci.*, **46**, 386.
49. Reddy, M.S., Vedmuthu, E.R., Washam, C.J. & Reinbold, G.W., 1972. *Appl. Microbiol.*, **24**, 947.
50. Kempler, G.M. & McKay, L.L., 1980. *Appl. Environ. Microbiol.*, **39**, 926.
51. Poolman, B., 1990. *Molec. Microbiol.*, **4**, 1629.
52. Hutkins, R.W. & Ponne, C., 1991. *Appl. Environ. Microbiol.*, **57**, 941.
53. Romano, A.H., Trifone, J.D. & Brustolon, M., 1979. *J. Bacteriol.*, **139**, 93.
54. Poolman, B., Driessen, A.J.M. & Konings, W.N., 1987. *Microbiol. Rev.*, **51**, 498.
55. McKay, L.L., Miller, A., Sandine, W.E. & Elliker, P.R., 1970. *J. Bacteriol.*, **102**, 804.
56. Dills, S.S., Apperson, A., Schmidt, M.R. & Saier, Jr., M.H., 1980. *Microbiol. Rev.*, **44**, 385.
57. Lawrence, R.C. & Thomas, T.D., 1979. In *Microbial Technology*, ed. A.T. Bull, D.C. Ellwood & C. Radledge, *Society of General Microbiology, Symposium*, **29**, 187.
58. Farrow, J.A.E., 1980. *J. Appl. Bacteriol.* **49**, 49.

59. Crow, V.L. & Thomas, T.D., 1984. *J. Bacteriol.*, **157**, 28.
60. Hutkins, R.W. & Morris, H.A., 1987. *J. Food Prot.*, **50**, 876.
61. Hemme, D., Nardi, M. & Jette, D., 1980. *Le Lait*, **60**, 595.
62. Hutkins, R., Morris, H.A. & McKay, L.L., 1985. *Appl. Environ. Microbiol.*, **50**, 772.
63. Thompson, J., 1980. *J. Bacteriol.*, **144**, 683.
64. Park, Y.H. & McKay, L.L., 1982. *J. Bacteriol.*, **149**, 420.
65. Thomas, T.D., Turner, K.W. & Crow, V.L., 1980. *J. Bacteriol.*, **144**, 672.
66. Bisset, D.L. & Anderson, R.L., 1974. *J. Bacteriol.*, **117**, 318.
67. Crow, V.L., Davey, G.P., Pearce, L.E & Thomas, T.D., 1983. *J. Bacteriol.*, **153**, 76.
68. Crow, V.L. & Thomas, T.D., 1982. *J. Bacteriol.*, **151**, 600.
69. Thomas, T.D., 1976. *Appl. Environ. Microbiol.*, **32**, 474.
70. Thomas, T.D., Ellwood, D.C. & Longyear, V.M.C., 1979. *J. Bacteriol.*, **138**, 109.
70a. Fordyce, A.M., Crow, V.L. & Thomas, T.D., 1984. *Appl. Environ. Microbiol.*, **48**, 332.
71. Collins, E.B. & Thomas, T.D., 1974. *J. Bacteriol.*, **120**, 52.
72. Mason, P.W., Carbone, D.P., Cushman, R.A. & Waggoner, A.S., 1981. *J. Biol. Chem.*, **256**, 1861.
73. Cogan, T.M., 1987. *J. Appl. Bacteriol.*, **63**, 551.
74. Schmitt, P. & Divies, C., 1991. *J. Ferm. Bioeng.*, **71**, 72.
75. Somkuti, G.A. & Steinberg, D.H., 1979. *J. Food Protect.*, **42**, 885.
76. Ratliff, T.L., Stinson, R.S. & Talburt, D.E., 1980. *Can. J. Microbiol.*, **26**, 58.
77. Garvie, E.I., 1980. *Microbiol. Revs.*, **44**, 106.
78. Smart, J.B. & Thomas, T.D., 1987. *Appl. Environ. Microbiol.*, **53**, 533.
79. Condon, S., 1987. *FEMS Microbiol. Revs.*, **46**, 269.
80. Lucey, C.A. & Condon, S., 1986. *J. Gen. Microbiol.*, **132**, 1789.
81. Cogan, J.F., Walsh, D. & Condon, S., 1989. *J. Appl. Bacteriol.*, **66**, 77.
82. Kaneko, T., Takahashi, M. & Suzuki, H., 1990. *Appl. Environ. Microbiol.*, **56**, 2644.
83. Reiter, B., 1985. In *Developments in Dairy Chemistry—3*, ed. P.F. Fox. Elsevier Applied Science Publishers, London. p. 281.
84. Karahadian, C. & Lindsay, R.C., 1987. *J. Dairy Sci.*, **70**, 909.
85. Turner, K.W., Morris, H.A. & Martley, F.G., 1983. *N.Z. J. Dairy Sci. Technol.*, **18**, 117.
86. Turner, K.W. & Thomas, T.D., 1980. *N.Z. J. Dairy Sci. Technol.*, **15**, 265.
87. Mora, R., Nanni, M. & Panari, G., 1984. *Scienza Tecnica Latt-Cas.*, **35**, 20.
88. Raadsveld, C.W., 1957. *Neth. Milk Dairy J.*, **11**, 313.
89. Fox, P.F., Lucey, J.A & Cogan, T.M., 1990. *Crit. Revs. Food Sci. Nutr.*, **29**, 237.
90. Manning, D.J. & Robinson, H.M., 1973. *J. Dairy Res.*, **40**, 63.
91. Sherwood, I.R., 1939. *J. Dairy Res.*, **10**, 326.
92. Sandine, W.E., Elliker, P.R. & Anderson, A.W., 1959. *J. Dairy Sci.*, **42**, 799.
93. Speckman, R.A. & Collins, E.B., 1968. *J. Bacteriol.*, **95**, 174.
94. Speckman, R.A. & Collins, E.B., 1973. *Appl. Microbiol.*, **26**, 744.
95. Seitz, E.W., Sandine, W.E., Elliker, P.R. & Day, E.A., 1963. *Canad. J. Microbiol.*, **9**, 431.
96. Kempler, G.M. & McKay, L.L., 1981. *J. Dairy Sci.*, **64**, 1527.
97. Harvey, R.J. & Collins, E.B., 1963. *J. Biol. Chem.*, **238**, 2648.
98. Kummel, A., Behrens, G. & Gottschalk, G., 1975. *Arch. Microbiol.*, **102**, 111.
99. David, S., van der Rest, M.E., Driessen, A.J.M., Simons, G. & deVos, W.M., 1990. *J. Bacteriol.*, **172**, 5789.
100. Sesma, F., Gardiol. D., de Ruiz Holgado, A.P. & de Mendoza, D., 1990. *Appl. Environ. Microbiol.*, **56**, 2099.
101. Cogan, T.M., 1981. *J. Dairy Res.*, **48**, 489.
102. Mellerick, D. & Cogan, T.M., 1981. *J. Dairy Res.*, **48**, 497.

103. Bowien, S. & Gottschalk, G., 1977. *Eur. J. Biochem.*, **80**, 305.
104. Malthe-Sorenssen, D. & Stormer, F.C., 1970. *Eur. J. Biochem.*, **14**, 127.
105. Cogan, T.M., Fitzgerald, R.J. & Doonan, S., 1984. *J. Dairy Res.*, **51**, 597.
106. Goelling, D. & Stahl, U., 1988. *Appl. Environ. Microbiol.*, **54**, 1889.
107. De Man, J.C. & Pette, J.W., 1956. *Proc. 14th Intern. Dairy Congr. Rome*, **2**(1) 89.
108. De Man, J.C., 1959. *Recueil Trav. Chem.*, **78**, 480.
109. Veringa, H.A., Verburg, E.H. & Stadhouders, J., 1984. *Neth. Milk Dairy J.*, **38**, 251.
110. Jonsson, H. & Pettersson, H.E., 1977. *Milchwissenschaft*, **32**, 587.
111. Jordan, K.N. & Cogan, T.M., 1988. *J. Dairy Res.*, **55**, 227.
112. Louis-Eugene, S., Ratomahenina, R. & Galzy, P., 1984. *Z. Allg. Mikrobiol.*, **24**, 151.
113. Silber, P., Chung, H., Gargiulo, P. & Schultz, H., 1974. *J. Bacteriol.*, **118**, 919.
114. Bryn, K., Hetland, O. & Stormer, F.C., 1971. *Eur. J. Biochem.*, **18**, 116.
115. Crow, V.L., 1990. *Appl. Environ. Microbiol.*, **56**, 1656.
116. Drinan, D.F., Tobin, S. & Cogan, T.M., 1976. *Appl. Environ. Microbiol.*, **31**, 481.
117. Kaneko, T., Suzuki, H. & Takahashi, T., 1987. *Agr. Biol. Chem.*, **51**, 2315.
118. Kaneko, T., Watanabe, Y. & Suzuki, H., 1990. *J. Dairy Sci.*, **73**, 291.
119. Cogan, T.M., O'Dowd, M. & Mellerick, D., 1981. *Appl. Environ. Microbiol.*, **41**, 1.
120. Keen, A.R. & Walker, N.J., 1974. *J. Dairy Res.*, **41**, 65.
121. Fryer, T.F., 1970. *J. Dairy Res.*, **37**, 17.
122. Thomas, T.D., 1987. *N.Z. J. Dairy Sci. Technol.*, **22**, 25.
123. Thomas, T.D., McKay, L.L. & Morris, H.A., 1985. *Appl. Environ. Microbiol.*, **49**, 908.
124. Kristofferson, T., Gould, I.A. & Harper, W.J., 1959. *The Milk Products J.*, **50**, 14.
125. Reiter, B. & Oram, J.D., 1962. *J. Dairy Res.*, **29**, 63.
126. Bracquart, P. & Lorient, D., 1979. *Milchwissenschaft*, **34**, 676.
127. Garvie, E.I., 1967. *J. Gen. Microbiol.*, **48**, 439.
128. Thomas, T.D & Mills, O.E., 1981. *N.Z. J. Dairy Sci. Technol.*, **16**, 43.
129. Smid, E.J., Plapp, R. & Konings, W.N., 1989. *J. Bacteriol.*, **171**, 6135.
130. Thomas, T.D. & Pritchard, G.G., 1987. *FEMS Microbiol. Revs.*, **46**, 245.
131. Kok, J., 1990. *FEMS Microbiol. Revs.*, **87**, 15.
132. Thomas, T.D., Jarvis, B.D.W. & Skipper, N.A., 1974. *J. Bacteriol.*, **118**, 329.
133. Pearce, L.E., Skipper, N.A. & Jarvis, B.D.W., 1974. *Appl. Microbiol.*, **27**, 933.
134. Ohmiya, K. & Sato, Y., 1975. *Appl. Microbiol.*, **30**, 738.
135. Law, B.A., 1979. *J. Appl. Bacteriol.*, **46**, 455.
136. Exterkate, F.A., 1984. *Appl. Environ. Microbiol.*, **47**, 177.
137. Geis, A., Bockelmann, W. & Teuber, M., 1985. *Appl. Microbiol. Biotech.*, **23**, 79.
138. Argyle, P.J., Mathison, G.E. & Chandan, R.C., 1976. *J. Appl. Bacteriol.*, **41**, 175.
139. Eggimann, B. & Bachmann, M., 1980. *Appl. Environ. Microbiol.*, **40**, 876.
140. Monnet, V., Le Bars, D. & Gripon, J.C., 1987. *J. Dairy Res.*, **54**, 247.
141. Exterkate, F.A. & de Veer, G.J.C.M., 1987. *Syst. Appl. Microbiol.*, **9**, 183.
142. Hugenholtz, J., van Sinderen, D., Kok, J. & Konings, W.N., 1987. *Appl. Environ. Microbiol.*, **53**, 853.
143. Visser, S., Exterkate, F.A., Slangen, C.J. & de Veer, G.J.C.M., 1986. *Appl. Environ. Microbiol.*, **52**, 1162.
144. Kok, J., Leenhouts, K.J., Haandrikman, A.J., Ledeboer, A.M. & Venema, G., 1988. *Appl. Environ. Microbiol.*, **54**, 231.
145. Vos, P., Simons, G., Siezen, R.J. & deVos, W.M., 1989. *J. Biol Chem.*, **264**, 13579.
146. Kiwaki, M., Ilemura, H., Shimizu-Kudota, M. & Hirashima, A., 1989. *Mol. Microbiol.*, **3**, 359.
147. Haandrikman, A.J., Kok, J., Laan, H., Soemitro, S., Ledeboer, A.M., Konings, W.N. & Venema, G., 1989. *J. Bacteriol.*, **171**, 2789.
148. Vos, P., van Asseldonk, M., van Jeveren, F., Siezen, R., Simons, G. & de Vos, W.M., 1989. *J. Bacteriol.*, **171**, 2795.

149. Chandan, R.C., Argyle, P.J. & Mathison, G.E., 1982. *J. Dairy Sci.*, **65**, 1408.
150. El-Soda, M. & Desmazeaud, M.J., 1982. *Canad. J. Microbiol.*, **28**, 118.
151. Ezzat, N., El-Soda, M., Bouillanne, C., Zevaco, C. & Blanchard, P., 1985. *Milchwissenschaft*, **40**, 140.
152. Ezzat, N., Zevaco, C., El-Soda, M. & Gripon, J.C., 1987. *Milchwissenschaft*, **42**, 95.
153. Galesloot, T.E., Hassing, F. & Veringa, H.A., 1968. *Neth. Milk Dairy J.*, **22**, 50.
154. Veringa, H.A., Galesloot, T.E. & Davelaar, H., 1968. *Neth. Milk Dairy J.*, **22**, 114.
155. Bautista, E., Dahiya, R.S. & Speck, M.L., 1966. *J. Dairy Res.*, **333**, 299.
156. Driessen, F.M., Kingma, F. & Stadhouders, J., 1982. *Neth. Milk Dairy J.*, **36**, 135.
157. Thomas, T.D. & Mills, O.E., 1981. *Neth. Milk Dairy J.*, **35**, 255.
158. Mills, O.E. & Thomas, T.D., 1975. *N.Z. J. Dairy Sci. Technol.*, **10**, 162.
159. Stoddard, G.W. & Richardson, G.H., 1986. *J. Dairy Sci.*, **69**, 9.
160. Oberg, C.J., Davis, L.H., Richardson, G.H. & Ernstrom, C.A., 1986. *J. Dairy Sci.*, **69**, 2975.
161. Stadhouders, J., Toepoel, L. & Wouters, J.T.M., 1988. *Neth. Milk Dairy J.*, **42**, 183.
162. Faryke, N.Y., Fox, P.F., Fitzgerald, G.F. & Daly, C., 1990. *J. Dairy Sci.*, **73**, 874.
163. Meyer, J. & Jordi, R., 1987. *J. Dairy Sci.*, **70**, 738.
164. Kiefer-Partsch, B., Bockelmann, W., Geis, A. & Teuber, M., 1989. *Appl. Microbiol. Biotechnol.*, **31**, 75.
165. Zevaco, C., Monnet, V. & Gripon, J.C., 1990. *J. Appl. Bacteriol.*, **68**, 357.
166. Atlan, D., Laloi, P. & Portalier, R., 1990. *Appl. Environ. Microbiol.*, **56**, 2174.
167. Khalid, N.M. & Marth, E.H., 1990. *Appl. Environ. Microbiol.*, **56**, 381.
168. Lloyd, R.J. & Pritchard, G.C., 1991. *J. Gen. Microbiol.*, **137**, 49.
169. Booth, M., Ni Fhaolain, I., Jennings, P.V. & O'Cuinn, G., 1990, *J. Dairy Res.*, **57**, 89.
170. Crow, V.L. & Thomas, T.D., 1982. *J. Bacteriol.*, **150**, 1024.
171. Lowrie, R.J., Lawrence, R.C., Pearce, L.E. & Richards, E.L., 1972. *N.Z. J. Dairy Sci. Technol.*, **7**, 44.
172. Lowrie, R.J., Lawrence, R.C. & Peberdy, M.F., 1974. *N.Z. J. Dairy Sci. Technol.*, **9**, 116.
173. Sullivan, J.J., Mou, L., Rood, J.I. & Jago, G.R., 1973. *Aust. J. Dairy Technol.*, **28**, 20.
174. Mills, O.E. & Thomas, T.D., 1980. *N.Z. J. Dairy Sci. Technol.*, **15**, 131.
175. Stadhouders, J., Hup. G., Exterkate, F.A. & Visser, S., 1983. *Neth. Milk Dairy J.*, **37**, 157.
176. Visser, S., Hup, G., Exterkate, F. A. & Stadhouders, J., 1983. *Neth. Milk Dairy J.*, **37**, 169.
177. Visser, S., Slangen, K.J., Hup, G. & Stadhouders, J., 1983. *Neth. Milk Dairy J.*, **37**, 181.
178. Lindsay, R.C., Day, E.A. & Sandine, W.E., 1965. *J. Dairy Sci.*, **48**, 863.
179. Lees, G.J. & Jago, G.R., 1976. *J. Dairy Res.*, **43**, 75.
180. Hamdan, I.Y., Kunsman, Jr., J.E. & Deane, D.D., 1971. *J. Dairy Sci.*, **54**, 1080.
181. Marshall, V.M. & Cole, W.M., 1983. *J. Dairy Res.*, **50**, 375.
182. Wilkens, D.W., Schmidt, R.H., Shireman, R.B., Smith, K.L. & Jezeski, J.J., 1986. *J. Dairy Sci.*, **69**, 1219.
183. Kanasaki, M., Brehony, S., Hillier, A.J. & Jago, G.R., 1975. *Aust. J. Dairy Technol.*, **30**, 142.
184. Turner, K.W. & Thomas, T.D., 1975. *N.Z. J. Dairy Sci. Technol.*, **10**, 162.
185. Cogan, T.M., 1978. *Ir. J. Food Sci. Technol.*, **2**, 105.
186. Cords, B.R., McKay, L.L. & Guerry, P., 1974. *J. Bacteriol.*, **117**, 1149.
187. Davies, F.L., Underwood, H.M. & Gasson, M.J., 1981. *J. Appl. Bacteriol.*, **51**, 325.
188. Gasson, M.J., 1983. *J. Bacteriol.*, **154**, 1.
189. Efstathiou, J.D. & McKay, L.L., 1977. *J. Bacteriol.*, **130**, 257.
190. Le Bourgeois, P., Mata, M. & Ritzenthaler, P., 1989. *FEMS Microbiol. Lett.*, **59**, 65.

191. Chopin, A., Chopin, M.-C., Moillo-Batt, A. & Langella, P., 1984, *Plasmid*, **11**, 260.
192. Dao, M.L. & Ferretti, J.J., 1985. *Appl. Environ. Microbiol.*, **49**, 115.
193. Kok, J., van der Vossen, J.M.B.M. & Venema, G., 1984. *Appl. Environ. Microbiol.*, **48**, 726.
194. Vosman, B. & Venema, G., 1983. *J. Bacteriol.*, **156**, 920.
195. Luchansky, J.B., Muriana, P.M. & Klaenhammer, T.R., 1988. *Molec. Microbiol.*, **2**, 637.
196. Gasson, M.J. & Anderson, P.H., 1985. *FEMS Microbiol. Lett.*, **30**, 193.
197. de Vos, W.M. 1987. *FEMS Microbiol. Rev.*, **46**, 281.
198. Mercenier, A. & Lemione, Y., 1989. *J. Dairy Sci.*, **72**, 3444.
199. Kondo, J.K. & McKay, L.L., 1982. *Appl. Environ. Microbiol.*, **43**, 1213.
200. Kondo, J.K. & McKay, L.L., 1984. *Appl. Environ. Microbiol.*, **48**, 252.
201. Von Wright, A., Taimisto, A.-M. & Sivela, S., 1985. *Appl. Environ. Microbiol.*, **50**, 1100.
202. Simon, D., Rouault, A. & Chopin, M.-C., 1986. *Appl. Environ. Microbiol.*, **52**, 394.
203. Fitzgerald, G.F. & Gasson, M.J., 1988. *Biochimie*, **70**, 489.
204. Steele, J.L. & McKay, L.L., 1989. *J. Dairy Sci.*, **72**, 3388.
205. Hayes, F., Fitzgerald, G.F. & Daly, C., 1985. *Ir. J. Food Sci. Technol.*, **9**, 77.
206. Kempler, G.M. & McKay, L.L., 1979. *Appl. Environ. Microbiol.*, **37**, 1041.
207. Gasson, M.J. & Davies, F.L., 1980. *J. Bacteriol.*, **143**, 1260.
208. Walsh, P.M. & McKay, L.L., 1981. *J. Bacteriol.*, **146**, 937.
209. van der Lelie, D., Chavarri, F., Venema, G. & Gasson, M.J., 1991. *Appl. Environ. Microbiol.*, **57**, 201.
210. Gasson, M.J., 1990. *FEMS Microbiol. Rev.*, **87**, 43.
211. Gasson, M.J., 1983. *Antoine v. Leeuwenhoek*, **49**, 275.
212. McKay, L.L., Baldwin, K.A. & Efstathiou, J.D., 1976. *Appl. Environ. Microbiol.*, **32**, 45.
213. Raya, R.R., Kleeman, E.G., Luchansky, J.B. & Klaenhammer, T.R., 1989. *Appl. Environ. Microbiol.*, **55**, 2206.
214. Kondo, J.K. & McKay, L.L., 1985. *J. Dairy Sci.*, **68**, 2143.
215. Gasson, M.J. & Davies, F.L., 1984. In *Advances in the Microbiology and Biochemistry of Cheese and Fermented Milk,* ed. F.L. Davies & B.A. Law. Elsevier Applied Science Publishers, London, p. 99.
216. de Vos, W.M., Vos, P., Simons, G. & David, S., 1989. *J. Dairy Sci.*, **72**, 3398.
217. Mercenier, A., 1990. *FEMS Microbiol. Rev.*, **87**, 61.
218. Nardi, M., Chopin, M.-C., Chopin, A., Cals, M.-M. & Gripon, J.C., 1991. *Appl. Environ. Microbiol.*, **57**, 45.
219. Mayo, B., Kok, J., Venema, K., Bockelman, W., Teuber, M., Reinke, H. & Venema, G., 1991. *Appl. Environ. Microbiol.*, **57**, 38.
220. van de Guchte, M., Kodde, J., van der Vossen, J.M.B.M., Kok, J. & Venema, G., 1990. *Appl. Environ. Microbiol.*, **56**, 2606.
221. de Vos, W.M. & Simon, G., 1988. *Biochimie.*, **70**, 461.
222. de Vos, W.M. & Gasson, M.J., 1989. *J. Gen. Microbiol.*, **135**, 1833.
223. Porter, E.V. & Chassy, B.M., 1988. *Gene*, **62**, 263.
224. Hill, C., Daly, C. & Fitzgerald, G.F., 1991. *FEMS Microbiol. Lett.*, **81**, 135.
225. Kondo, J.K., 1989. *J. Dairy Sci.*, **72**, 338.
226. van de Guchte, M., van der Vossen, J.M.B.M., Kok, J. & Venema, G., 1989. *Appl. Environ. Microbiol.*, **1989**, **55**, 224.
227. de Vos, W.M. & Simons, G., 1984. In *Proc. 4th Eur. Congr. on Biotechnology.* Elsevier Scientific Publishers, Amsterdam, p. 458.
228. Renault, P. & Heslot, H., 1987. *Appl. Environ. Microbiol.*, **53**, 320.
229. Polzin, K.M. & Shimizu-Kadota, M., 1987. *J. Bacteriol.*, **169**, 5481.
230. Romero, D.A. & Klaenhammer, T.R., 1990. *J. Bacteriol.*, **172**, 4151.
231. Polzin, K.M. & McKay, L.L., 1991., *Appl. Environ. Microbiol.*, **57**, 734.

232. Haandrikman, A.J., van Leeuwen, C., Kok, J., Vos, P., de Vos, W.M. & Venema, G., 1990. *Appl. Environ. Microbiol.*, **56**, 1890.
233. Klaenhammer, T.R., 1988. *J. Dairy Sci.*, **72**, 3429.
234. Hill, C., Miller, L.A. & Klaenhammer, T.R., 1990. *J. Bacteriol.*, **172**, 6419.
235. Allatossova, T.A. & Klaenhammer, T.R., 1991. *Appl. Environ. Microbiol.*, **57**, 1346.
236. Jarvis, A.W., Heap, H.A. & Limsowtin, G.K.Y., 1989. *Appl. Environ. Microbiol.*, **55**, 1537.
237. Froseth, B.R. & McKay, L.L., 1991. *Appl. Environ. Microbiol.*, **57**, 804.
238. Hill, C., Miller, L.A. & Klaenhammer, T.R., 1990. *Appl. Environ. Microbiol.*, **56**, 2255.
239. Ross, P., O'Gara, F. & Condon, S., 1990. *Appl. Environ. Microbiol.*, **56**, 2156.
240. Leenhouts, K.J., Kok, J. & Venema, G., 1990. *Appl. Environ. Microbiol.*, **56**, 2726.
241. Cogan, T.M., 1972. *Appl. Microbiol.*, **23**, 960.
242. Klaenhammer, T.R., 1988. *Biochimie.*, **70**, 337.
243. Barefoot, S.F. & Klaenhammer, T.R., 1984. *Antimicrob. Agents Chemother.*, **26**, 328.
244. Muriana, P.M. & Klaenhammer, T.R., 1990. *Appl. Environ. Microbiol.*, **57**, 114.
245. Muriana, P.M. & Klaenhammer, T.R., 1990. *J. Bacteriol.*, **173**, 1779.
246. Upretti, G.C. & Hinsdill, R.D., 1975. *Antimicrob. Agents Chemother.*, **7**, 139.
247. Joerger, M.J. & Klaenhammer, T.R., 1990. *J. Bacteriol.*, **172**, 6339.
248. Hurst, A., 1981. *Adv. Appl. Microbiol.*, **27**, 85.
249. Kojic, M., Svircevic, J., Banina, A. & Topisirovic, L., 1991. *Appl. Environ. Microbiol.*, **57**, 1835.
250. Reddy, N.S. & Ranganathan, B., 1983. *Milchwissenschaft*, **38**, 726.
251. Holo, H., Nilssen, O. & Nes, I.F., 1991. *J. Bacteriol.*, **173**, 3879.
252. Davey, G.P. & Richardson, B.C., 1981. *Appl. Environ. Microbiol.*, **41**, 84.
253. Zajdel, J.K., Ceglowski, P. & Dobrzanski, W.T., 1985. *Appl. Environ. Microbiol.*, **49**, 969.
254. Bhunia, A.K., Johnson, M.C. & Ray, B., 1988. *J. Appl. Bacteriol.*, **65**, 261.
255. Pucci, M.J., Vedamuthu, E.R., Kunka, E.S. & Vandenburgh, P.A., 1988. *Appl. Environ. Microbiol.*, **54**, 2349.
256. Ruhr, E. & Sahl, H.-G., 1985. *Antimicrob. Agents Chemother.*, **27**, 841.
257. Kaletta, C. & Entian, K.-D., 1989. *J. Bacteriol.*, **171**, 1597.
258. Buchman, G.W., Banerjee, S. & Hansen, J.N., 1988. *J. Biol. Chem.*, **263**, 16260.
259. Dodd, H.M., Horn, N. & Gasson, M. J., 1990. *J. Gen. Microbiol.*, **136**, 555.
260. Steen, M.T., Chung, Y.J. & Hansen, J.N., 1991. *Appl. Environ. Microbiol.*, **57**, 1181.
261. Froseth, B.R., Herman, R.E. & McKay, L.L., 1988. *Appl. Environ. Microbiol.*, **54**, 2136.
262. von Wright, A., Wessels, S., Tynkkynen, S. & Saarela, M., 1990. *Appl. Environ. Microbiol.*, **56**, 2029.
263. van Belkum, M.J., Hayema, B.J., Geis, A., Kok, J. & Venema, G., 1989. *Appl. Environ. Microbiol.*, **55**, 1187.
264. van Belkum, M.J., Hayema, B.J., Jeeninga, R.E., Kok, J. & Venema, G., 1991. *Appl. Environ. Microbiol.*, **57**, 492.
265. Geis, A., Singh, J. & Teuber, M., 1983. *Appl. Environ. Microbiol.*, **45**, 205.
266. Jarvis, A.W., 1989. *J. Dairy Sci.*, **72**, 3406.
267. Sozzi, T., Poulin, J.M., Meret, R. & Pousaz, R., 1978. *J. Appl. Bacteriol.*, **44**, 159.
268. Shin, C. & Sato, Y., 1981. *Jap. J. Zootech. Sci.*, **52**, 639.
269. Sozzi, T. & Maret, R., 1975. *Le Lait*, **55**, 269.
270. Sozzi, T., Maret, T. & Poulin, J.M., 1976. *Appl. Environ. Microbiol.*, **32**, 131.
271. Reinbold, G.W., Reddy, M.S. & Hammond, E.G., 1982. *J. Food Prot.*, **45**, 119.
272. Accolas, J.-P. & Spillman, H., 1979. *J. Appl. Bacteriol.*, **47**, 309.
273. Bradley, D.E., 1967. *Bacteriol. Rev.*, **31**, 230.
274. Whitehead, H.R. & Cox, G.A., 1935. *N.Z. J. Sci. Technol.*, **16**, 319.

275. Ackermann, H.-W., 1983. In *A Critical Appraisal of Viral Taxonomy*, ed. R.E.F. Mathews. CRC Press, Florida, p. 105.
276. Henning, D.R., Black, C.H., Sandine, W.E. & Elliker, P.R., 1966. *J. Dairy Sci.*, **51**, 16.
277. Chopin, M.-C., Chopin, A. & Roux, C., 1976. *Appl. Environ. Microbiol.*, **32**, 741.
278. Douglas, J., Qanbar-Agha, A. & Philips, V., 1974. *Lab. Pract.*, **23**, 3.
279. Keogh, B.P., 1980. *Appl. Environ. Microbiol.*, **40**, 798.
280. Teuber, N. & Lembke, J., 1974. *Antoine v. Leeuwenhoek*, **27**, 411.
281. Keogh, B.P. & Shimmin, P.D., 1974. *Appl. Microbiol.*, **27**, 411.
282. Terzaghi, B.E., 1976. *N.Z. J. Dairy Sci. Technol.*, **11**, 155.
283. Tsaneva, K.P., 1976. *Appl. Environ. Microbiol.*, **31**, 590.
284. Heap, H.A. & Jarvis, A.W., 1980. *N.Z. J. Dairy Sci. Technol.*, **15**, 75.
285. Lembke, J., Krusch, U., Lompe, A. & Teuber, M., 1980. *Zbl. Bakt. I. Abt. Orig.*, **CI**, 79.
286. Budde-Niekiel, A., Muller, V., Lembke, J. & Teuber, M., 1985. *Milchwissenschaft*, **40**, 477.
287. Jarvis, A.W., 1984. *Appl. Environ. Microbiol.*, **47**, 343.
288. Jarvis, A.W., 1984. *Appl. Environ. Microbiol.*, **47**, 1031.
289. Davies, F.L. & Gasson, M. J., 1984. In *Advances in the Microbiology and Biochemistry of Cheese and Fermented Milk*, ed. F.L. Davies & B.A. Law. Elsevier Applied Science Publishers, London, pp. 127–151.
290. Daly, C. & Fitzgerald, G.F., 1982. In *Microbiology-1982*, ed. D. Schlessinger. American Society for Microbiology, ASM, Washington, DC, p. 213.
291. Loof, M., Lembke, J. & Teuber, M., 1983. *System Appl. Microbiol.*, **4**, 413.
292. Lyttle, D.J. & Peterson, G.B., 1984. *Appl. Environ. Microbiol.*, **48**, 242.
293. Powell, I.B. & Davidson, B.E., 1985. *J. Gen. Virol.*, **66**, 2737.
294. Jarvis, A.W. & Meyer, J., 1986. *Appl. Environ. Microbiol.*, **51**, 556.
295. Oram, J. & Reiter, B., 1965. *J. Gen. Microbiol.*, **40**, 57.
296. Tourville, D.R. & Tokuda, S., 1967. *J. Dairy Sci.*, **50**, 1019.
297. Mullan, W.M.A. & Crawford, R.J.M., 1985. *J. Dairy Res.*, **52**, 123.
298. Davidson, B.E., Powell, I.B. & Hillier, A.J., 1990. *FEMS Microbiol. Rev.*, **87**, 79.
299. Klaenhammer, T.R., 1984. In *Advances in Applied Microbiology*, ed. A.I. Laskin. Academic Press, New York, pp. 1–29.
300. Kurmann, J.L., 1979. *Schweiz. Milchwirtsch. Forsch.*, **47**, 309.
301. Barefoot, S., McArthur, J.L., Kidd, J.K. & Grinstead, D.A., 1990. *J. Dairy Sci.*, **73**, 2269.
302. Jarvis, A.W., 1982. *Proc. 21st Intern. Dairy Congr., Vol. 1, Moscow*, **I**, 2, 314.
303. Chopin, M.-C. & Rousseau, M., 1983. *Le Lait*, **63**, 102.
304. Reyrolle, J., Chopin, M.-C., Letellier, F. & Novel, G., 1982. *Appl. Environ. Microbiol.*, **43**, 349.
305. Lawrence, R.C., Heap, H.A., Limsowtin, G.K.Y. & Jarvis, A.W., 1976. *J. Dairy Sci.*, **61**, 1181.
306. Lodics, T.A. & Steenson, L.R., 1990. *J. Dairy Sci.*, **73**, 2685.
307. Pearce, L.E., Limsowtin, G.K.Y. & Crawford, A.M., 1970. *N.Z. J. Dairy Sci. Technol.*, **5**, 145.
308. Pearce, L.E., 1974. *Proc. 19th Intern. Dairy Congr., New Delhi*, **IE**, 411.
309. Limsowtin, G.K.Y. & Terzhagi, B.E., 1976. *N.Z. J. Dairy Sci. Technol.*, **11**, 251.
310. Hull, R.R. & Brooke, A.R., 1982. *Aust. J. Dairy Technol.*, **37**, 143.
311. Limsowtin, G.K.Y., Heap, H.A. & Lawrence, R.C., 1978. *N.Z. J. Dairy Sci. Technol.*, **13**, 1.
312. Shimizu-Kadota, M. & Sakurai, T., 1982. *Appl. Environ. Microbiol.*, **43**, 1284.
313. Shimizu-Kadota, M., Sakurai, T. & Tsuehida, N., 1983. *Appl. Environ. Microbiol.*, **45**, 669.
314. Hill, C., Massey, I.J. & Klaenhammer, T.R., 1991. *Appl. Environ. Microbiol.*, **57**, 283.

315. Shearman, C.A. & Gasson, M.J., 1989. *Mol. Gen. Genet.*, **218**, 214.
316. Lakshmidevi, G., Davidson, B. & Hillier, A.J., 1990. *Appl. Environ. Microbiol.*, **56**, 934.
317. Hill, C., Miller, L.A. & Klaenhammer, T.R., 1991. *J. Bacteriol.*, **173**, 4363.
318. Klaenhammer, T.R., 1987. *FEMS Microbiol. Rev.*, **46**, 313.
319. Sijtsma, M., Wouters, J.T.M. & Hillingwert, K.J., 1990. *J. Bacteriol.*, **172**, 7126.
320. Fitzgerald, G.F., Daly, C., Browne, L.R. & Gingeras, T.R., 1982. *Nucleic Acids Res.*, **10**, 8171.
321. Solaiman, D.K.Y. & Somkuti, G.A., 1990. *FEMS Microbiol. Lett.*, **67**, 261.
322. de los Reyes-Gavilan, C.G., Limsowtin, G.K.Y., Sechaud, L., Veaux, M. & Accolas, J.-P., 1990. *Appl. Environ. Microbiol.*, **56**, 3412.
323. Sing, W.D. & Klaenhammer, T.R., 1990. *J. Dairy Sci.*, **73**, 2239.
324. Klaenhammer, T.R., Romero, D.A., Sing, W.D. & Hill, C., 1991. In *Genetics and Molecular Biology of Streptococci, Lactococci and Enterococci*, ed. G.M. Dunny, P.P. Cleary & L.L. McKay. American Society for Microbiology, ASM, Washington, DC.
325. Hill, C., Romero, D.A., McKenney, D.S., Finer, K.R. & Klaenhammer, T.R., 1989. *Appl. Environ. Microbiol.*, **55**, 1684.
326. Hill, C., Pierce, K. & Klaenhammer, T.R., 1989. *Appl. Environ. Microbiol.*, **55**, 2416.
327. Coffey A.G., Fitzgerald, G.F. & Daly, C., 1991. *J. Gen. Microbiol.*, **137**, 1355.
328. Josephson, J. & Klaenhammer, T.R., 1990. *FEMS Microbiol. Lett.*, **23**, 71.
329. Coffey, A.G., Fitzgerald, G.F. & Daly, C., 1989. *Neth. Milk Dairy J.*, **43**, 229.
330. Koka, M. & Mikolajcik, E.M., 1967. *J. Dairy Sci.*, **50**, 1025.
331. Zottola, E.A. & Marth, E.H., 1966. *J. Dairy Sci.*, **49**, 1338.
332. Koka, M. & Mikolajcik, E.M., 1970. *J. Dairy Sci.*, **53**, 853.
333. Daoust, D.R., El-Bisi, H.M. & Litsky, W., 1965. *Appl. Microbiol.*, **13**, 478.
334. Chopin, M.-C., 1980. *J. Dairy Res.*, **47**, 131.
335. Lewis, J.E., 1956. *J. Soc. Dairy Technol.*, **9**, 123.
336. Robertson, P.S., 1966. *Dairy Industries Int.*, **31**, 805.
337. Leenders, G.J.M. & Stadhouders, J., 1981. *Report NOV-767*. Nederlands Instituut voor Zuivelonderzoek.
338. La Grange, W.S. & Reinbold, G.W., 1968. *J. Dairy Sci.*, **51**, 1985.
339. Gulstrum, T.J., Pearce, L.E., Sandine, W.E. & Elliker, P.R., 1979. *J. Dairy Sci.*, **62**, 208.
340. Ledford, R.A. & Speck, M.L., 1979. *J. Dairy Sci.*, **62**, 781.
341. Henning, D.R., Sandine, W.E., Elliker, P.R. & Hays, H.A., 1965. *J. Milk Food Technol.*, **28**, 273.
342. Ausavanodom, N., White, R.S., Young, G. & Richardson, G.H., 1977. *J. Dairy Sci.*, **60**, 1245.
343. Wright, S.L. & Richardson, G.H., 1982. *J. Dairy Sci.*, **65**, 1882.
344. Mermelstein, N.H., 1982. *Food Technol.*, **36**(8), 69.
345. Willrett, D.L., Sandine, W.E. & Ayres, J.W., 1982. *Cult. Dairy Products J.*, **17**(3), 5.
346. Gilliland, S.E., 1985. In *Bacterial Starter Cultures for Foods*, ed. S.E Gilliland. CRC Press, Boca Raton, FL, p. 145.
347. Walker, A.L., Mullan, W.M.A. & Muir, M.E., 1981. *J. Soc. Dairy Technol.*, **34**, 78.
348. Bolle, A.C., Leenders, G.J.M. & Stadhouders, J. 1985. *Rapport 121 Nederlands Instituut voor Zuivelonderzoek*.
349. Accolas, J.P., Bloquel, R., Didienne, R. & Regnier, J., 1977. *Le Lait*, **57**, 1.
350. Martley, F.G., 1983. *N.Z. J. Dairy Sci. Technol.*, **18**, 191.
351. Lawrence, R.C. & Heap, H.A., 1986. *IDF Bull.*, **199**, 14.
352. Lawrence, R.C., Heap, H.A. & Gilles, J., 1984. *J. Dairy Sci.*, **67**, 1632.
353. Collins, E.B., 1951. *J. Dairy Sci.*, **34**, 894.
354. Teuber, M. & Loof, M., 1987. In *Streptococcal Genetics*, ed. J.J. Ferretti & R. Curtiss III. American Society for Microbiology, Washington, DC, pp. 250–58.

7

Salt in Cheese: Physical, Chemical and Biological Aspects

T.P. GUINEE

The National Dairy Products Research Centre, Moorepark, Fermoy, Co. Cork, Republic of Ireland

&

P.F. FOX

Department of Food Chemistry, University College, Cork, Republic of Ireland

1 INTRODUCTION

The use of salt (NaCl) as a food preservative dates from pre-historic times and, together with fermentation and dehydration (air/sun), is one of the classical methods of food preservation. So useful and widespread was the use of salt as a food preservative in Classical and Medieval times that it was a major item of trade and was used as a form of currency in exchange for goods and labour. It is perhaps a little surprising that Man discovered the application of salt in food preservation so early in civilization since, in contrast to fermentation and de-hydration, salting is not a 'natural event' in foods but requires a conscious act. It is interesting that the three classical methods of food preservation, i.e. fermentation, dehydration and salting, are all exploited in cheese manufacture and in fact are interdependent. The fourth common method of food preservation, i.e. use of high and/or low temperatures, was less widespread than the others because the exploitation of low temperatures was confined to relatively few areas until the development of mechanical refrigeration about 1870 and, although heating was probably used to extend the shelf-life of foods throughout civilization, its controlled use dates from the work of Nicolas Appert (1794) and Louis Pasteur (c. 1840). In modern cheese technology, temperature control complements the other three methods of food preservation.

In addition to its preservative effect, NaCl plays two other important roles in foods. Man requires ~3·5 g Na per day and although this requirement can be met through the indigenous Na content in foods, added NaCl is a major source in modern western diets. In fact, western diets contain, on average, three to five times more Na than is necessary and excessive intakes of Na have toxic, or at least undesirable, physiological effects, the most significant of which are hypertension and increased calcium excretion which may lead to osteoporosis (for reviews on the dietary significance of Na, see Refs 1–7).

Cheese, even when consumed in large amounts, as in France and Switzerland, makes a relatively small contribution to dietary Na intake (see Chapter 15, Vol. 1) although it may be a major contributor in individual cases where large amounts of high-salt cheese, e.g. Blue, Feta, Domiati, are consumed. Nevertheless, there is interest in many western countries in the production of low-Na cheese, for at least certain sectors in the population, but, as discussed below, this has significant repercussions in cheese manufacture. The most common approach at present is to replace some or all of the NaCl by KCl, but apart from cost, this practice affects the flavour of cheese since the flavour of KCl is distinctly different from that of NaCl and a bitter flavour (not due to abnormal proteolysis) is detectable in cheese containing >1% KCl (see Refs 8–12 for some recent work on the production of low-Na cheese).

The third major feature of the use of NaCl in foods is its direct contribution to flavour. The taste of salt is highly appreciated by many and saltiness is regarded as one of the four basic flavours. Presumably, the characteristic flavour of NaCl resides in the Na moiety since KCl has a distinctly different flavour sensation. At least part of the desirability of salt flavour is acquired but while one can easily adjust to the flavour of foods without added salt, the flavour of salt-free cheese is insipid and 'watery', even to somebody not 'addicted' to salt; the use of 0·8% NaCl is probably sufficient to overcome the insipid taste.[11]

In this chapter, we will concentrate on the role of NaCl in controlling cheese ripening rather than on its dietary and direct flavour effects. NaCl influences cheese ripening principally through its effects on water activity but it probably has some more specific effects also which appear to be only partly due to water activity. Among the principal effects of salt are:

1. control of microbial growth and activity;
2. control of the various enzyme activities in cheese;
3. syneresis of the curd resulting in whey expulsion and thus in a reduction of cheese moisture, which also influence 1 and 2 above;
4. physical changes in cheese proteins which influence cheese texture, protein solubility and probably protein conformation.

2 CONTROL OF MICROBIAL GROWTH

Probably the most extreme example of the use of NaCl for this purpose is in the manufacture of Domiati-type cheeses where 12–15% of NaCl is added to cheese-milk to inhibit bacterial growth and thus maintain milk quality (see Chapter 11, Volume 2). In all other major varieties, NaCl is added after curd formation but nevertheless it plays a major role in regulating and controlling cheese microflora.

The simplest example of this is the regulation of the pH of cheese, which in turn influences cheese ripening and texture.

The pH of cheese may be regulated by:

1. reducing the amount of residual lactose in the curd which is accomplished by washing the curd with water, as practised in Dutch-type cheeses, Tallegio and Cottage;
2. the natural buffering capacity of the cheese and the toxic effect of the lactate anion which establishes a natural lower limit to pH ($\sim$4·5), e.g. Blue, hard Italian varieties;
3. salt addition.

The use of salt to regulate the final pH appears to be almost exclusively confined to British-type cheeses. The curd for most, if not all, non-British cheese is placed in moulds while the pH is still high ($>6·0$) and acid development continues during pressing. Since levels of NaCl $> \sim1·5\%$ inhibit starter activity, such cheeses are salted by immersion in brine or by surface application of dry salt. In British cheeses, e.g. Cheddar-types and Stilton, the pH has almost reached its final, desired level at hooping and salt is added to maintain the pH at that (desired) value. One could probably argue that the method of salting cheese that predominates in a certain region reflects the form of salt available locally: in regions where salt deposits occur, dry salt was readily available and thus permitted the manufacture of cheeses in which dry salt was added to the curd or to the surface of the cheese; in regions where salt was prepared by evaporation of sea water, it would have been more convenient to salt the cheese by immersion in concentrated brine rather than wait for crystallization.

Curd for Cheddar, and similar varieties, contains $\sim0·6$–$1·0\%$ lactose at hooping; this is fermented during the early stages of ripening by continued starter activity but this depends strongly on the salt-in-moisture (S/M) level in the curd and the salt tolerance of the starter. Commercial lactic acid cultures are stimulated by low levels of NaCl but are very strongly inhibited $>2·5\%$ NaCl.[13] Thus, the activity of the starter and its ability to ferment residual lactose is strongly dependent on the S/M level in the curd. This is clearly evident from the data of O'Connor[14] shown in Fig. 1: the pH decreased after salting, presumably due to the action of starter, at S/M levels $<5\%$ but at higher values of S/M, starter activity decreased abruptly and the pH remained high. The grade assigned to the cheese also decreased sharply at S/M levels $>5\%$.

Inhibition of starter occurs within quite a narrow pH range (Fig. 1), emphasizing the importance of precise control of S/M level. However, since the sensitivity of starter cultures to salt varies, the influence of NaCl concentration on post-salting acid production in cheese obviously depends on the starter used and a general value for S/M cannot be definitely stated. *Lactococcus lactis* subsp. *lactis* starters are generally more salt-tolerant than strains of *Lc. lactis* subsp. *cremoris*[15] but there is also considerable variation in salt sensitivity between strains of *Lc. lactis* subsp. *cremoris*.[16,17] If starter activity is inhibited after manufacture, residual lactose will be metabolized by non-starter lactic acid bacteria

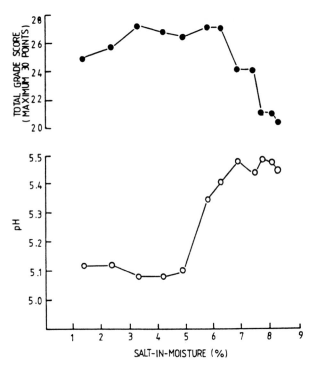

Fig. 1. The correlation between the salt-in-moisture (S/M) levels and the pH (○) at eight weeks, and between the S/M and the total grade score (maximum 30) (●) of batches of curd from the same vats, salted at different rates (from Ref. 18).

but the number of these present, which is influenced by level of contamination at salting, level of S/M, rapidity with which pressed curd is cooled and ripening temperature[19] is usually insufficient to cause significant lactose metabolism for several days and consequently the pH falls slowly.

In the study by Turner & Thomas[17], non-starter bacteria, mainly *Pediococcus*, were more salt-tolerant than starter bacteria and metabolized the lactose with the production of DL-lactate and the racemization of L-lactate. Non-starter bacteria grew in all cheeses but their growth was markedly dependent on temperature and they had little influence on lactose or lactate concentration until numbers exceeded 10^6–10^7 cfu/ml.

The control of lactose metabolism by S/M concentration within a single cheese was clearly demonstrated by Thomas & Pearce[20] (Fig. 2). In this study also, the greater salt-tolerance of non-starter bacteria was clearly apparent, with the fermentation of lactose of D-lactate relatively late in ripening and the racemization of L-lactate to D-lactate.

Although acid production can be uncoupled from cell growth, it is likely that acid production at low salt levels will be accompanied by high cell numbers which tend to lead to bitterness.[21] Not surprisingly, bitterness in Cheddar cheese

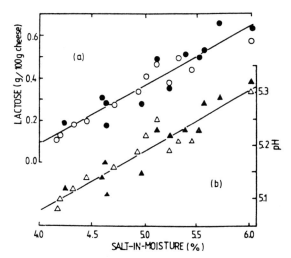

Fig. 2. Effect of S/M concentration on (a) lactose concentration and (b) pH, within single Cheddar cheese blocks. From each of two cheese blocks (open and closed symbols), manufactured at the same time, 12 plugs were removed 14 days after manufacture. Each plug was analysed for S/M, lactose (○,●) and pH (△,▲) (from Ref. 20).

is markedly influenced by S/M level over a very narrow range: *Lc. lactis* subsp. *cremoris* HP generally yielded bitter cheese at S/M levels <4·3% but rarely >4·9%.[22]

In the foregoing discussion on the influence of NaCl on the fermentation of residual lactose in cheese curd by starter microorganisms, it has been assumed that the NaCl is distributed throughout the cheese within a very short period after salting. However, this is not so. Cheddar cheese is usually milled into quite large particles of cross section 2 cm × 2 cm or larger. Obviously, dry salt applied to the surface of such particles requires a considerable period of time to diffuse to the centre of the curd chips and to attain an inhibitory level throughout. Consequently, starter will continue to grow and produce acid at the centre of a chip for a considerable period after growth at the surface has ceased.

Experimental support for this is provided by the experiments of Hoecker & Hammer[23] who measured the salt and moisture contents and pH at the surface and centre of individual chips, prised from a block of Cheddar cheese, over a 72 h period after pressing. Their data showed that the pH fell faster and to a greater extent at the centre, where NaCl concentration was lower, than at the surface. In one experiment, the difference in pH persisted for 72 h but in a duplicate experiment the difference in pH had essentially disappeared after 48 h. Turner & Thomas[17] showed that a higher level of salt addition is required to inhibit lactose metabolism when the curd is milled into large chips than for smaller ones.

In surface-salted Meshanger cheese, Noomen[24] showed considerable zonal variations in the changes in carbohydrate, lactose and pH throughout the cheese in response to variations in S/M concentration. *Streptococcus salivarius* subsp.

thermophilus is considerably less salt-tolerant than *Lc. lactis* subsp. *lactis*;[25] its critical NaCl concentration is 0·4 M (2·34%), corresponding to an A_w of 0·984, compared with 1·1 M NaCl ($A_w = 0·965$) for *Lc. lactis* subsp. *lactis*. *Lactobacillus delbrueckii* subsp. *helveticus* and *Lb. lactis* subsp. *lactis* were also less salt tolerant, being inhibited by 0·95 M and 0·90 M NaCl, respectively.

Data on the sensitivity of *Propionibacteria* to NaCl appear to be variable: Orla-Jensen[26] reported that concentrations of NaCl as low as 0·5% are sufficient to reduce the growth of *Propionibacteria* in a medium containing calcium lactate. However, Antila[27] reported that 3% NaCl is necessary to reduce growth. In fact, salt tolerance appears to be strain and pH-dependent:[28] in a lactate medium, 6% NaCl was required to inhibit the growth of a fast-growing strain of *Propionibacteria* at pH 7·0 and 3% at pH 5·2 whereas a slow-growing strain was more salt-tolerant at pH 5·2 than at pH 7·0. The data reported by Ruegg & Blanc[25] show that *P. shermanii* was the most salt tolerant of the starter species investigated: its critical NaCl concentration was 1·15 M (~6·7%; $A_w = 0·955$). However, Emmental cheese, with NaCl concentrations of ~0·7%, is the least heavily salted among major cheese varieties.

Blue cheeses are among the most heavily salted varieties with 3–5% NaCl (Stilton <3%). Ripening in these varieties is dominated by the enzymes of *Peni­cillium roqueforti* and consequently good growth of this mould is paramount. Germination of *P. roqueforti* spores is stimulated by 1% NaCl but inhibited by >3–6% NaCl, depending on strain; however, the growth of germinated spores on malt extract agar or cheese curd is less dependent on NaCl concentration than is germination and some strains grow in cheese curd containing 10% NaCl, although growth is retarded compared to that in curd containing lower levels of NaCl.[29,30] Morris[31] reported that it is fairly common commercial practice to add 1% NaCl directly to Blue cheese curd before hooping, possibly to stimulate spore germination, although it also serves to give the cheese a more open structure which facilitates mould growth. Since most Blue cheeses are surface-salted, a salt gradient from the surface to the centre exists for a considerable period after manufacture; a high initial level of salt in the outside zone of the cheese will inhibit spore germination at a critical time and a mould-free zone at the outside is a common defect in Blue cheeses.[30]

Growth of *P. camemberti* is also stimulated by low levels of NaCl; <0·8 NaCl, mould growth on Camembert cheese is poor and patchy.[32]

3 INFLUENCE OF NaCl ON ENZYME ACTIVITY IN CHEESE

3.1 Coagulant

With the exception of Emmental and similar high-cooked cheeses, the initial proteolysis in cheese is catalysed by residual coagulant. Application of polyacrylamide gel electrophoresis to cheese ripening[33–36] has shown that in hard and semi-hard, bacterially ripened cheeses, α_{s1}-casein undergoes considerable proteolysis during ripening but β-casein remains unchanged until an advanced stage of ripening. A similar pattern is evident during the early phases of mould-ripening cheeses, when the coagulant is the principal ripening agent[37,38] but fungal proteinases dominate in these cheeses during the later phases of ripening (see Chapter 4, Volume 2).

The hydrolysis of α_{s1}-casein by milk clotting enzymes is greatly influenced by the concentration of NaCl. The proteolytic activities of chymosin, pepsins, *Mucor miehei* and *Endothia parasitica* rennets are stimulated by increasing NaCl concentrations to an optimum of ~6%.[39,40] While activities are inhibited at somewhat higher NaCl levels, proteolysis of α_{s1}-casein occurs up to 20% NaCl.[39,40] In contrast, proteolysis of β-casein by chymosin and pepsins is strongly inhibited by 5% and completely inhibited by 10% NaCl.[39] Sucrose[41] and glycerol[42] selectively inhibit proteolysis of β-casein. KCl, LiCl, NH_4Cl and $CaCl_2$ are as effective as NaCl in inhibiting the proteolysis of β-casein.[32] Since the inhibitory effect of solutes is substrate- rather than enzyme-specific, it appears that NaCl and similar solutes cause some conformational changes in β-casein[43] which renders its chymosin (pepsin)-susceptible bonds less accessible to the enzyme. The nature of these conformational changes does not appear to have been investigated but may arise from the strongly hydrophobic nature of β-casein. The resistance of β-casein in cheese to proteolysis is not dependent solely on the salt concentration since it is also very resistant to proteolysis in salt-free cheeses,[36] suggesting that a high protein concentration is sufficient to induce the necessary conformational change(s). However, a certain level of NaCl (>4·9% S/M) is necessary to prevent the development of bitterness in cheese.[22]

The inhibitory effect of NaCl on proteolysis is pH-dependent, and at low pH, NaCl also alters the proteolytic specificity of chymosin and pepsins: NaCl (2·5%) inhibits the formation of β-III but promotes the formation of β-IV and β-V.[44] The formation α_{s1}-casein peptides, α_{s1}-VII and α_{s1}-VIII, in solution is at least stimulated by and perhaps dependent on, the presence of NaCl (5%) and these peptides are also formed in cheese.[45]

The proteolytic activities of *Mucor miehei* and *Endothia parasitica* rennets of β-casein are less strongly inhibited by NaCl than are those of chymosin or pepsins.[40,46]

3.2 Milk Proteinase

Milk contains several indigenous proteinases[47-51] the most significant of which, alkaline milk proteinase (plasmin), is almost exclusively associated with the casein micelles at the normal pH of milk,[47,52] but dissociates from the micelles as the pH is reduced.[53] Richardson and Elston[53] indicate that the dissociation of plasmin from the casein micelles is pH- and time-dependent and that it occurs at pH 5·7 and possibly higher. However, Grufferty & Fox[52] found no dissociation on holding at pH >4·9 for 4 h. This implies that all the plasmin in milk should accompany the curd in most cheese varieties. However, the concentration of plasmin in Swiss-type cheese is two to three times that in Cheddar[54,55] while the activity in Cheshire cheese is very low[55] suggesting that the plasmin content of cheese may be influenced by the pH at hooping.[55] The difference in the plasmin levels between Cheddar and Swiss cheeses is considered unlikely to be due to pH dependent dissociation of the enzyme as the pH of both cheeses is 6·1–6·4 at whey drainage.[52] It is possible that the reported differences may to be due to different rates of plasminogen activation in the two cheeses due to different processing conditions, especially cooking temperature,[56,57] and possibly the higher pH in Swiss cheese during ripening.[52,56] The increase in pH in Swiss-type cheese during ripening is paralleled by a large increase in plasmin activity.[56] Owing to the relatively high buffering capacity of Swiss-type cheese [as affected by the retention of colloidal calcium phosphate due to the relatively high pitching pH (i.e. ~6·4 compared to ~6·1 for Cheddar)] and its relatively high protein level (i.e. ~29 compared to 24% for Cheddar) and the propionic acid fermentation, the pH of Swiss does not fall as low as, and rises more rapidly than that of Cheddar.

The role of plasmin in cheese ripening has not been extensively studied but the presence of γ-caseins in most cheese suggests at least some activity; it appears to make a significant contribution to the maturation of Gouda,[34,35] Romano-type cheese[58] and of Swiss[54,56,59–60a] in which the coagulant is extensively denatured by the high cooking temperature[60] but it has only a limited role in the ripening of Cheddar[35,61] and soft Meshanger-type cheese.[62] Noomen[62] suggested that plasmin may make a significant contribution to proteolysis in soft cheeses with a surface flora, in which the pH rises markedly during ripening to a value more favourable to the activity of plasmin.

Milk also contains an acid proteinase which apparently has a specificity similar to milk coagulants[63,64] (which are also acid proteinases) and consequently its significance to cheese ripening may be underestimated.

Noomen[62] showed that the activity of alkaline milk proteinase in simulated cheese was stimulated by low concentrations of NaCl up to a maximum at 2% but was inhibited by higher concentrations of NaCl although some activity remained at 8% NaCl. To our knowledge, the influence of NaCl on the activity of acid milk proteinase has not been investigated.

3.3 Microbial Enzymes

There appears to be relatively little information on the influence of NaCl on microbial enzymes in cheese; indirect evidence, e.g. in relation to bitterness in cheese[20,22,65,66] suggests that the activity of starter proteinase is inhibited by moderately high levels of NaCl. *P. roqueforti* lipases[67] and proteinases[68] are inhibited by NaCl concentrations >6%.

4 INFLUENCE OF NaCl ON THE WATER ACTIVITY (A_w) OF CHEESE

In addition to reducing the moisture content of cheese via syneresis, NaCl also reduces the water activity, A_w, of cheese[12] which in the case of young cheese, especially those containing >40% H_2O, is determined almost entirely by NaCl content according to the equation, $A_w = 1-0.033\ m$, where m = molality of NaCl in cheese moisture.[69] In hard and semi-hard cheese, the contributions of lactate and ash (other than NaCl) to reducing A_w must be considered and as cheese matures, the formation of low molecular compounds, notably peptides and amino acids, also influence A_w; these compositional factors have been included in the formulae developed by Ruegg & Blanc[25,70,71] and Fernandez-Salguero et al.[72] for the calculation of A_w. The inhibitory effect of NaCl on the activity of starter is undoubtedly due to its influence on A_w, the inhibitory effect of which is species- and strain-specific.[25,73–77] Presumably, the activity of the various enzyme systems in cheese is also inhibited at reduced values of A_w[73,78–80] but detailed studies are lacking. Various aspects of the water activity of cheese are considered in depth in Chapter 11 of this volume.

5   OVERALL INFLUENCE OF NaCl ON CHEESE RIPENING AND QUALITY

5.1 Cheddar Cheese

The influence of % salt-in-cheese moisture (% S/M) on lactose metabolism in young Cheddar cheese has already been discussed. There appears to be little information available on the influence of % S/M on lipolysis in Cheddar cheese. However, Thakur et al.[8] compared lipolysis in salted (three lots, 1·48–1·79% NaCl) and unsalted Cheddar: the concentration of volatile acids was significantly higher in the unsalted than in the salted cheese due mainly to acetic acid which is presumably a product of lactose metabolism. The concentrations of all individual fatty acids, except linoleic and linolenic (at certain ages), were also higher in the unsalted cheese compared to the control; the authors did not comment on the markedly lower levels of linoleic acid in the unsalted cheese. However, Lindsay et al.[10] found little difference between the levels of free fatty acids in cheeses with low (3·5%) or intermediate (4·2%) S/M levels except for myristic and palmitic acids which were considerably higher in the higher salt cheese. Reduced-sodium cheeses will be discussed in more detail in Section 6.

Proteolysis is considerably more extensive in unsalted than in salted cheese and consequently the body of the former is less firm.[8,12,80] A linear relationship between the extent of degradation of both α_{s1}- and β-caseins in young (1 month) cheese and % S/M is apparent from the data of Thomas & Pearce.[20] During the normal ripening of Cheddar cheese, α_{s1}-casein is the principal substrate for proteolysis with little degradation of β-casein;[33] proteolysis of β-casein is more extensive at low salt levels.[36] However, Thomas & Pearce[20] noted that while the normal products of β-casein degradation (β-I, B-II produced by rennets, and γ-casein by milk proteinase) were not apparent in their studies, the concentration of unhydrolysed β-casein decreased, suggesting that proteolysis of β-casein in low-salt cheese may be due to bacterial proteinases.

At least five studies[81-85] have attempted to relate the quality of Cheddar cheese to its composition. While these authors agree that the moisture content, % S/M and pH are the key determinants of cheese quality, they disagree as to the relative importance of these three parameters.

In a study of 300 Scottish Cheddar cheeses, O'Connor[81] found that flavour and aroma, texture and total score were not correlated with moisture content but were significantly correlated with % NaCl and particularly with pH. Salt content and pH were themselves strongly correlated, as were salt and moisture; a very wide variation in composition was noted. Gilles and Lawrence[82] proposed a grading scheme for young (14 day) Cheddar cheese, based on analysis of cheese made at the New Zealand Dairy Research Institute over many years and also by commercial cheese factories in New Zealand. The influence of cheese composition on quality and compositional grading of Cheddar cheese will be discussed in Chapter 1, Volume 2; suffice it to record here that the S/M specified for premium and First Grade Cheddar in New Zealand are 4·0–6·0 and 4·7–5·7, respectively.

Fox[83] assessed the influence of moisture, salt and pH on the grade of 123, 10-week-old Irish Cheddar cheeses (70 high Quality and 53 'rejects') from six factories and 27 extra-mature, high quality Cheddars. The composition of the cheeses varied widely and while the correlations between grade and any of the compositional factors were poor, a high percentage of cheeses with compositional extremes were downgraded, especially those with salt (<1·4%), high moisture (>39%) or high pH (>pH 5·4). In the samples studied, salt concentration seemed to exercise the strongest influence on cheese quality and the lowest percentage of downgraded cheeses can be expected in the salt range 1·6–1·8% or in the S/M range, 4·0–4·9%. The composition of high quality extra-mature cheeses also varied widely but less than that of the young cheeses. Although the mean salt levels were identical for both groups of cheeses, the spread was much narrower for the mature cheeses and only three had <1·7% NaCl. The mean moisture content of the mature cheeses was 1% lower than that of the regular cheeses.

The grading ratio (ratio of high to low grading cheeses) for 486 14-day-old cheeses produced at the New Zealand Dairy Research Institute was most highly correlated with the % moisture in fat-free substances and second best with % salt.[84] The optimum compositional ranges were: moisture-in-non-fat substances: 52–54%; % S/M: 4·2–5·2%; pH: 4·95–5·15. Cheese with a S/M of 3·1% received the highest grade in a study by Knox[86] although there was little difference in grade in the S/M range 3·1–5·2; quality declined markedly at % S/M >6·4.

A very extensive study of the relationship of the grade and composition of nearly 10 000 cheeses produced in five commercial New Zealand factories was made by Lelievre & Gilles.[85] As in previous studies, considerable compositional variation was evident but the variation was considerably less for some factories than others. While the precise relationship between grade and composition varied from plant to plant, certain generalizations emerged: (i) within the compositional range suggested by Gilles & Lawrence[82] for 'premium' quality cheese, composition does not have a decisive influence on grade, which falls off outside this range;

(ii) composition alone does not provide a basis for grading as currently accept-able to the dairy industry (New Zealand); (iii) moisture-in-non-fat substances was again found to be the dominant factor influencing quality; (iv) within the recom-mended compositional bands, grades declined marginally as moisture-in-non-fat substance (MNFS) increased from 51 to 55%, increased slightly as S/M de-creased from 6 to 4% while pH had no consistent effect within the range 4·9–5·2 and FDM had no influence in the range 50–57%. The authors stress that since specific inter-plant relationships exist between grade and composition, each plant should determine the optimum compositional parameters pertinent to that plant.

Apart from the acid flavour associated with low-salt cheese, bitterness has been reported consistently as a flavour defect in such cheeses. A complex corre-lation exists between the propensity of a cheese to develop bitterness and starter culture, pH, rate of acid development and % salt-in-moisture. There is still some controversy on the development of bitterness (cf. Refs 21, 87–89) but the subject will not be reviewed here.

From the compositional viewpoint, % S/M appears to be the most important factor influencing bitterness.[22] The probability of bitterness developing is greatly increased at % S/M <4·9 and pH, in the normal range encountered for Cheddar, i.e. 4·9–5·3 (where paracasein is most soluble[90] and therefore most susceptible to proteolysis), has little effect except at low S/M values (i.e. <4·9%). Rennet has maximum activity towards paracasein in salt solutions between 2·5 and 4%.[91] The bitterness of peptides is strongly correlated with hydrophobicity.[92,93] The bitter peptides in cheese appear to arise primarily from β-casein[94–95] which might be expected since β-casein is the most hydrophobic casein.[96] The effectiveness of NaCl in preventing bitterness is very likely due to the selective inhibition by NaCl of β-casein hydrolysis.[36,39,40,89]

The protein matrix in young cheese appears to consist of α_{s1}-casein molecules linked through hydrophobic interactions between their amino terminal regions; the primary site for rennet action on α_{s1}-casein is Phe_{23}–Phe_{24}[95] or Phe_{24}–Val_{25},[97] hydrolysis of which leads to the formation of α_{s1}-I-casein and destruction of the matrix. This specific cleavage is considered to be primarily responsible for the loss of firmness (yield value) of cheese during the early stages of ripening.[98,99] However, Luyten[100] found that increased α_{s1}-casein breakdown (as effected by an increased rennet concentration and varied independent of age) in Gouda cheese had little effect on the shortness (which may be best described as the inverse of the yield stress).[101] Indeed the increase in shortness in Gouda cheese on ripening was attributed more to in-depth proteolysis (e.g. NPN formation) than to gross proteolysis.[100] The increase in shortness with maturation may be considered as a result of an upward shift in pH[102] away from the point of maximum yield value, i.e. 5·2–5·35.[101] Indeed, this seems highly probable when one considers the pro-duction of pasta filata-type cheeses such as Mozzarella and Kashkaval: the cheeses flow and stretch over a narrow pH region, 5·2–5·35, outside which flow is very much restricted unless some processing changes, such as the addition of brine to hot water (so as to partially solubilize the casein) section of the kneading machine, are implemented. During the time required for the pH to fall from ~6·1

TABLE I
Influence of NaCl on Proteolysis on Camembert Cheese (4 Weeks Old)

NaCl(%)	Zone	pH	WSN as % TN	pH 4·6 soluble N as % TN	70% ethanol soluble N	5% PTA soluble N
0·20	I	5·5	36·1	43·4	23·7	16·8
	O	6·4	100	54·4	35·7	18·9
0·70	I	5·3	28·7	29·1	15·8	10·4
	O	6·1	100	39·2	28·7	15·5
0·93	I	5·2	17·9	17·3	13·3	12·1
	O	6·0	100	49·5	32·8	15·2
1·14	I	5·2	22·5	23·8	15·8	8·1
	O	6·2	93·7	43·4	28·4	10·4
1·73	I	5·1	26·6	28·3	15·8	8·8
	O	6·45	85·3	37·1	22·7	10·1
2·49	I	5·15	22·2	23·1	18·0	8·3
	O	6·3	63·2	29·8	26·1	9·3

I = inner portions of cheese; TN = total nitrogen; PTA = phosphotungstic acid;
O = outer portions of cheese; WSN = water soluble N.[32]

at pitching to ~5·2 at stretching, little or no degradation of α_{s1}-casein occurs. From the observations of Noomen[103] on Camembert cheese, it is probable that both mechanisms (i.e. NPN formation with consequent movement of pH from the point of maximum yield and hydrolysis of α_{s1}-casein) contribute to cheese softening on maturation to different extents depending on the variety. (In Cheddar cheese the pH remains more or less constant during ripening and yet the cheese softens.) Irrespective of the proteolytic system responsible for the decrease in yield value on ripening, the influence of NaCl on the proteolysis of α_{s1}-casein (see above) and on NPN formation (Table I) partly explains its influence on cheese texture: a weak, pasty body at low salt concentrations and an excessively firm body at high salt levels. The influence of NaCl on cheese moisture levels and possibly on protein conformation/solubility probably also affects cheese texture.

5.2 Blue Cheese

The influence of NaCl concentration on the principal ripening events in Blue cheese was studied by Godinho & Fox.[29,30,37,104] Proteolysis as measured by polyacrylamide gel electrophoresis and the formation of 12% TCA-soluble N was invariably lower in the outer (high salt) region than in the middle or centre (lower salt) zones; the differences were apparent both before visible mould growth (during the first two weeks when coagulant is the principal proteolytic agent) and during the mould phase (after two weeks).[37,38] There was a strong negative correlation between salt concentration and TCA-soluble N. Unfortunately, formation of amino acid N (e.g. PTA soluble N) or other more detailed characterizations of proteolysis were not investigated. With a few exceptions, the pH increased faster at the centre than in the outer region of the cheese, indicating that amino acid catabolism is also influenced by NaCl concentration.

Lipolysis in Blue cheese is also influenced by salt concentration with maximum activity occurring at 4–6% NaCl.[104] However, the concentrations of methyl ketones was relatively independent of salt concentration.

5.3 Camembert Cheese

The ripening of the surface mould-ripened cheeses, Camembert and Brie, is characterized by a very marked softening, almost liquefaction, of the body from the surface to the centre due mainly to the combination of α_{s1}-casein hydrolysis and the decreasing pH gradient from the surface to the centre (due to the production of ammonia by the surface mould, *P. camemberti* and its inward diffusion[103]). Proteolysis by coagulant and starter proteinases is also important and although the proteinases excreted by *P. camemberti* undergo only very limited diffusion in the cheese,[103] peptides produced by them do, apparently, diffuse into the cheese (see Chapter 4, Volume 2).

In this variety also, NaCl concentration has a major influence on proteolysis and pH changes, as well as on surface mould growth (Table I).

6 REDUCED SODIUM CHEESES

While the physiological requirement of Na as a dietary constituent is universally accepted, there is growing concern that the excess ($>\sim3.5$ g/day for healthy adults) induces physiological defects including hypertension (c.f. Refs 4, 6 and 7). Such concern has led to a call for reduced dietary intake of Na, classification of foods (high, medium, low) according to sodium level, declaration of sodium levels on the food labels and an increased demand for reduced-sodium foods, including cheese.[105,106]

In addition to its preservative effect, salt in cheese exerts a major influence on cheese composition, microflora, ripening rates, texture, flavour and quality (c.f. Sections 2–5). Salt levels (%, w/w) in cheese range from ~0.65 in Swiss, 1.6 for Cheddar to ~6.5 for Domiati; equivalent sodium levels (%, w/w) are ~0.26, 0.62 and 2.6, respectively (See Volume 2; Refs 107, 108). Approaches to reduce the Na levels of cheese include reduction in the level of added salt *per se*,[10,109,110] partial or complete substitution of NaCl by other salts such as KCl, $MgCl_2$, $CaCl_2$,[111–113] reduced salt levels in combination with flavour-enhancing substances such as autolysed yeast extracts[112,114] and the use of ultrafiltration—and reverse osmosis retentate-supplemented milks to alter the mineral levels of the cheese.[9,110,115,116] Attempts to produce low sodium processed cheese products include the partial substitution of sodium phosphates with the corresponding potassium phosphates, the use of flavour enhancers (e.g. monosodium glutamate, glucono-delta-lactone, enzyme-modified cheese and cheese pastes)[114] and the use of selected cheese blends and dairy ingredients in the production of emulsifying salt-free processed cheese foods and spreads.

6.1 Cheddar Cheese

In Cheddar cheeses (ripened at 4·4°C for 7 months) with salt levels (%, w/w) ranging from 1·44 to 0·07 (i.e. 4·1–0·2% S/M) a reduction of salt level by 25% to 1·12% (3·1% S/M), while giving a decreased saltiness, gave cheeses which were very comparable to the controls in relation to flavour, texture and overall acceptability.[11] Further reduction of salt level resulted in altered composition (increased moisture and reduced fat levels) and higher water activities which promoted excessive proteolysis and lipolysis and in turn led to defective body and texture (open, soft, greasy) and flavour (unclean, bitter) and consumer unacceptability.[8,11,80] Fitzgerald & Buckley[111] studied the influence of different salts (i.e. KCl, $MgCl_2$, $CaCl_2$) and 1 : 1 mixtures of these salts with NaCl on the quality of Cheddar cheese ripened at 4°C over a 4 month period; salts and NaCl/salt mixtures were added at levels which gave ionic strengths equivalent to the control (i.e. 2·5% added NaCl; 1·44% salt in the cheese). A KCl/NaCl (1 : 1) salt combination gave cheese at 16 weeks which was not significantly different from the controls in terms of flavour, texture and acceptability. The use of KCl, $MgCl_2$ or $CaCl_2$ alone resulted in oversoft cheeses with very bitter and unacceptable flavours. These defects may be attributed to the higher moisture levels and greater proteolysis in the case of $MgCl_2$ and $CaCl_2$ but not in the case of KCl, where the moisture and water-soluble N levels were similar to the controls. Both flavour and texture scores for the $CaCl_2$/NaCl- and Mg Cl_2/NaCl-salted cheeses were significantly lower than the controls. Similar results with NaCl/KCl mixtures were observed by Lindsay et al.[10] who also found that reduction of salt (i.e. NaCl or NaCl/KCl—1:1 mixtures) by ~30% from 1·75 to 1·25 (i.e. 5·14 to 3·6% S/M) resulted in no major differences in flavour, texture or acceptability scores. In Cheddar cheeses (~34% moisture) made with NaCl/KCl mixtures, free fatty acid levels were higher and grading scores were somewhat lower due to a slight bitterness.[10]

Kosikowski[9,115] found that in Cheddar cheeses with a reduced level of NaCl (i.e. 1·05%; ~3·0% S/M), increasing the protein content of the milk, from ~3·36 to 6·26%, prior to renneting, by supplementing the cheese milk with ultrafiltered milk (4·5:1 retentate) in increasing amounts (1·1 to 1·9:1), was paralleled by a decrease in moisture and increases in the Ca (from ~590 to 730 mg/100 g) and P (from ~470 to 556 mg/100 g) levels and in the scores for flavour, body and texture during ripening at 10°C over 4 months. Grading scores for flavour and texture increased to an optimum at milk protein concentrations of 4·97–6·26% in the supplemented milks; at the lower milk protein levels the cheeses become progressively more acidic, bitter, pasty and devoid of cheese flavour. The enhancing effect of increased milk protein level on grading score was attributed to the increased buffering capacity which prevented a rapid pH decline (in the absence of normal salt levels) during moulding and pressing and hence excessive loss of calcium and phosphorus which influence cheese structure and rate of proteolysis.[9,116–118] However, the results of Kosikowski[9] could not be confirmed under practically identical conditions by Lindsay et al.[110] Contrary to the results of Kosikowski,[9] the latter group found that:

(i) the calcium level in low-sodium Cheddar (1% NaCl) made from control milk was not significantly lower than that made from milk supplemented with a 4·5:1 retentate (added at a ratio of 1·9:1, retentate:milk);

(ii) the grading scores of cheeses made from supplemented milk were of the same magnitude as, or slightly lower than, those of the control 'non-supplemented' low-salt Cheddar. Moreover, the former cheeses were generally softer and had a less intense cheese flavour.

The inclusion of reverse osmosis (RO) retentate in UF retentate-supplemented milk gave low-salt Cheddar (1% NaCl) with grading scores similar to the control.[110] Cheese made using RO retentate-supplemented milk had a unique sharpness which could be used to enhance the flavour of other cheese products such as processed cheese.[110]

Undoubtedly, the quality of commercial reduced-sodium cheeses depends on many factors, including pitching pH, the type and amount of residual coagulant in the cheese, types and counts of starter and non-starter bacteria, composition and ripening temperature. Ranges of compositional parameters for good quality (New Zealand) Cheddar as proposed by Lelievre & Gilles[85] are: 4·0–6·0% S/M; 50–57% FDM; 50–56% MNFS and pH 5·0–5·4; outside these ranges quality deteriorates rapidly. With the modern continuous production methods for Cheddar, in combination with blast cooling of blocks, it may be possible (though somewhat more expensive) to produce consistently high quality Cheddar by reducing the MNFS, keeping the pH close to 5·1, avoiding the use of bitter starters and microbial rennets, and ripening at low temperatures (i.e. <5°C).

6.2 Cottage Cheese

Because of its relatively large serving size (~112 g compared to ~66 g for other cheeses), Cottage cheese has been viewed as a potentially high source of dietary sodium.[119] Hence, much interest has focused on various ways of reducing the salt level of Cottage cheese. Wyatt[109] evaluated preference scores for Cottage cheeses in which the NaCl content was reduced stepwise from 1% (control commercial cheese) to 0·25% NaCl. It was concluded that a 35% reduction in salt did not influence consumer response to the cheese compared to the control; however, reduction by 50% or greater resulted in significantly lower scores. Demott et al.[112] also evaluated consumer reactions to low-sodium Cottage cheeses salted with various mixtures of KCl and NaCl and found that sodium levels could be reduced by 50% (by using a 1·26% addition of a NaCl/KCl mixture instead of 1·26% NaCl) without affecting grading scores. Reducing the sodium level by more than 50% resulted in a significant reduction in score.[112,120] Lindsay et al.[110] also found that 50% reduction of salt (by lowering added NaCl) gave no significant changes in consumer acceptability. However, the use of substitutes, i.e. KCl or KCl/NaCl, to reduce sodium levels by 50% gave a significant reduction in quality.

6.3 Other Cheeses

Martens *et al.*[121] reported the successful manufacture of low-sodium Gouda cheese using mixtures of NaCl and KCl in the curd manufacturing and brining processes. While the Na and K levels (mg%) in the control cheese were ~830–650 and 120, those of the reduced-sodium cheese were 200 and ≥200, respectively. However, reduction of salt in dry matter (SDM) in Gouda cheese by ~20% (i.e. from 3·8–4·0 to 3%; 2·3% to 1·75% S/M) is claimed significantly to increase the susceptibility to butyric acid fermentation;[122] to prevent such undesirable fermentation in cheeses at salt levels of <3·8% SDM requires process modifications such as bactofugation of milk and reduction of cheese moisture levels. Lefier *et al.*[113] reported the production of low sodium-Gruyère (~45 mg Na/100 g compared to 272 mg/100 g in the control) by replacing of NaCl with $MgCl_2$. While the degree of proteolysis and the concentrations of free fatty acids were similar in both cheeses, the cheese containing $MgCl_2$ had a more bitter taste and softer body than the control, but was found to be organoleptically acceptable.

Processed cheese products contain relatively high levels of Na (i.e. 1·0–1·5%, w/w; c.f. Ref. 107) because of the inclusion of sodium phosphate emulsifying salts in their formulation. Karahadian & Lindsay[114] produced acceptable low-sodium (75% reduction, 0·34% Na in product) by using reduced-sodium Cheddar cheese and/or various combinations of potassium-based emulsifying salts (citrates, phosphates). The most efficient means of reducing Na in processed cheese products is by eliminating the emulsifying salts, i.e. as in emulsifying salt-free processed cheese foods and spreads which have been available on the Irish market since 1988. The production of such products requires careful blending of cheeses (i.e. high and low calcium cheeses, cheeses with varying pH and degree of proteolysis) and alteration of processing conditions so as to obtain a stable emulsion.[123,124] In the latter products, lack of saltiness is easily overcome by the addition of ingredients such as monosodium glutamate, autolysed yeast extracts, 'high cured cheese', cheese powders, enzyme-modified cheese, cheese pastes and/or acidulants.

7 SALT ABSORPTION AND DIFFUSION INTO CHEESE

7.1 Methods of Salting

There are three principal methods of salting cheese curd:

1. direct addition and mixing of dry salt crystals to broken or milled curd pieces at the end of manufacture, e.g. Cheddar and Cottage;
2. rubbing of dry salt or a salt slurry to the surface of the moulded curds, e.g. Blue-type cheeses;
3. immersion of moulded cheese in brine solution, e.g. Edam, Gouda, Saint Paulin, Provolone etc. Sometimes, a combination of the above methods is used, e.g. Emmental, Parmesan, Romano and Brick.

7.2 Mechanism of Salt Absorption and Diffusion in Cheese

7.2.1 Brine-Salted Cheeses

When cheese is placed in brine there is a net movement of NaCl molecules, as Na$^+$ and Cl$^-$, from the brine into the cheese as a consequence of the osmotic pressure difference between the cheese moisture and the brine. Consequently, the water in the cheese diffuses out through the cheese matrix so as to restore osmotic pressure equilibrium. Gels, including cheese, consist of a sponge-like network consisting of strands of fused paracasein micelles, which gives the mass its structure and a certain degree of rigidity and elasticity; the properties of the interpenetrating fluid are generally not appreciably different from those of corresponding solutions. It would appear, therefore, that NaCl molecules diffusing in cheese moisture, while having a longer distance to travel than in solution (diffusing molecules/ions must travel a circuitous route to by-pass obstructing protein strands and fat globules through which they cannot penetrate) would not be appreciably affected otherwise. However, based on the mobilities of NaCl and H$_2$O in Gouda type cheeses brine-salted under model conditions to obey Fick's law for undimensional brine flow, Geurts et al.[125] concluded that the penetration of salt into cheese and the concomitant outward migration of water could be described as an impeded diffusion process, i.e. NaCl and H$_2$O molecules move in response to their respective concentration gradients but their diffusion rates are much lower than those in pure solution due to a variety of impeding factors. The diffusion coefficient (D) for NaCl in cheese moisture is ~0·2 cm^2/d, though it varies from ~0·1 to 0·3 cm^2/d with cheese composition and brining conditions,[125,126] compared to 1·0 cm^2/d for NaCl in pure H$_2$O at 12·5°C.

Geurts et al.[125] used the term 'pseudo-diffusion coefficient' in relation to the movement of NaCl in cheese moisture since the value of the observed coefficient depended on the net effect of many interfering factors on true diffusion. The discrepancies between the true- and pseudo-diffusion coefficients, i.e. D and D*, respectively, were explained by a simplified model of cheese structure consisting of moisture and discrete spherical fat globules dispersed in a protein matrix comprised of discrete spherical protein particles and 15% (w/v) bound water. Based on theoretical considerations, the impedance of various compositional and structural features intrinsic to the model structure were formulated and their effects on D quantified. The principal factors responsible for impedance of NaCl diffusion in cheese, as postulated by Geurts et al.[125] are:

1. The effect of the protein matrix on the mass ratios of salt and water migrating in opposite directions. The pores (estimated to be ~2·5 nm wide) of the protein matrix exert a sieving effect on both the inward-diffusing NaCl molecules and outward-moving H$_2$O molecules but the effect is more pronounced on the former because of their greater effective diffusion radii, which are approximately twice that of the H$_2$O molecules. Hence, during brining, the H$_2$O flux is approximately twice the NaCl flux. The net outflow of H$_2$O during brining causes the plane of zero mass transfer (a plane

where the average flux of all diffusing species is zero) to recede from the cheese/brine interface into the brine and hence reduces the apparent rate of NaCl diffusion due to the additional path length through which the NaCl molecules must migrate. The interference, which is most pronounced when moisture loss is high, e.g. when using concentrated brines or high brining temperatures, was estimated to reduce the pseudo-diffusion coefficient by a factor of ~0.2, hence $D^* = 0.8\ D$.

2. When the NaCl molecules do enter the cheese, the relatively narrow pore width of the protein matrix exerts a frictional effect on the diffusing NaCl and H_2O molecules and reduces their relative diffusion rates from one in true solution to ~0.5 and 0.75, respectively, in cheese moisture. As the effective pore restriction on the diffusion of NaCl in cheese moisture is determined by its effect on the larger-sized molecule, i.e. NaCl, the pseudo-diffusion coefficient was estimated to be reduced by a factor of 0.5.

3. Frictional effects of protein-bound water. Water binding in cheese (0.1 to 0.15 g H_2O/g para-casein)[127] makes ~10% of the total cheese moisture unavailable for salt uptake and hence reduces the apparent diffusion coefficient. Furthermore, the protein-bound water reduces the relative pore width of the protein matrix, thus retarding further the movement of NaCl and H_2O molecules.

4. The high relative viscosity of cheese moisture. The viscosity of cheese moisture is about 1.27 times that of pure water at 12.5°C due to the presence of dissolved materials, e.g. acids, salts and nitrogenous compounds. NaCl molecules diffusing through the cheese moisture encounter an increased collision frequency with the dissolved substances, and are also affected by the charge fields of these substances; both these factors reduce NaCl mobility and the pseudo-diffusion coefficient is thus reduced by a factor of $1/\eta_{rel}$.

5. Obstructions of fat globules and globular protein particles. On proceeding from one parallel plane to another within the cheese, the diffusing molecules must travel by a circuitous route to bypass obstructing particles. The ratio of the real to the apparent distance travelled is a measure of the obstructions caused by fat globules (λ_f) and protein particles (λ_p). Theoretically, λ_f can vary from $\pi/2$ for a close-pack arrangement to one for a very low-fat system, e.g. skim milk cheese. In experimental Gouda cheeses, λ_f and λ_p were found to vary with composition which altered the volume fractions of the fat and protein phases and reduced the pseudo-diffusion coefficient by a factor of $1/(\lambda_f\lambda_p)$. Typical values of λ_f and λ_p were 1.32 and 1.35, respectively, for a Gouda cheese containing 29% fat and 43% moisture.[125]

Beginning from a simplified model of cheese structure and considering the relative effects of the interfering factors discussed, Geurts et al.[125] postulated a theoretical 'pseudo-diffusion' coefficient,

$$D^* = (0.8 \times 0.5\ D)/(\lambda_f\lambda_p\eta_{rel})$$

While the model cheese structure adopted by Geurts et al.[125] may appear over-simplified in view of the results of electronmicroscopic examinations of cheese structure,[128,129] the calculated impedance derived from it was sufficient to explain the very low diffusion coefficient of NaCl in cheese moisture and the variations of D* with variations in cheese composition and brining conditions.

7.2.2 Direct Mixing of Salt with Milled Curd

When dry salt is distributed over the surface of milled curd or curd granules, some NaCl dissolves in the surface moisture and diffuses slowly inwards a short distance.[130,131] This causes a counterflow of whey from the curd to the surface which dissolves the remaining salt crystals and, in effect, creates a supersaturated brine solution around each particle, provided mixing of curd and salt is adequate. However, because of the relatively large surface area to volume ratio of the curd as a whole, salt uptake occurs from many surfaces simultaneously and less time is required for uptake of an adequate amount of salt in dry-salting milled curd (10–20 min) than in brining whole cheeses (0·5–5 days depending on the dimensions). Some of the 'brine' on the surface of curd particles drains away through the curd mass while more is physically expelled from the curd particles during pressing and is lost in the 'press whey'. As the salt/surface area ratio is usually low, and the period of contact of the curd surface with the concentrated brine layer is relatively short, little localized surface protein contraction occurs compared to that in dry-salted, moulded curds.[131]

7.2.3 Dry Surface Salting of Moulded Pressed Cheese Curd

A block of curd can be regarded as a very large particle and solution of dry salt in the surface moisture layer is a prerequisite for salt absorption in this method also. The counterflow of moisture from the cheese creates a concentrated brine layer on the cheese surface and salt uptake then occurs by an impeded diffusion process. Because the surface is in contact with a concentrated brine for a long time (several days), there is considerable contraction of the curd surface (salting out of protein) and this probably leads to relatively high moisture losses from the surface region and hence a reduction in the inward mobility of NaCl which accounts for the lower rate of salt uptake in this method than in brining.[30]

7.3 Factors Influencing Salt Absorption by Cheese

The only prerequisite for salt absorption by cheese is the existence of a salt-in-moisture gradient between the cheese and the salting medium. However, the quantity of salt absorbed depends on the intrinsic properties of the cheese, the conditions of salting and the duration of salting. As the different procedures of salting all involve salt absorption via an impeded diffusion process, the general factors affecting salt uptake by cheese apply equally to granules or milled curd pieces on mixing with dry salt and moulded cheeses which are brined and/or dry salted. Certain peculiarities of the salting of milled curd pieces, as in Cheddar, which affect salt absorption will be discussed separately.

7.3.1 Concentration Gradient

It is generally accepted that an increase in brine concentration results in higher rates of salt absorption and increased salt-in-moisture levels in the cheese.[125,130,132] However, while the rate of NaCl diffusion is scarcely affected by brine concentration in the range 5–20%,[125,126] the rate of uptake increases at a diminishing rate with increasing brine concentration.[130–133] In model brining experiments, in which cheese slices of different thickness were completely submerged in brine, there was a sharp decrease in the rate of salt absorption as the difference between the NaCl concentration in the cheese moisture and the brine decreased, especially when the initial difference was large.[132] A somewhat similar situation applies to dry-salted cheese: the increase in salt-in-moisture level in Cheddar curd is not proportional to the increase in the level of dry salt added to the milled curd.[14,134] This is attributed to increased salt losses with increased salting rates, which reflects the decreasing effect of the driving force (concentration gradient) in raising the quantity of salt absorbed as the salt-in-moisture level in the cheese approaches that of the brine.

7.3.2 Cheese Geometry

It is generally agreed that the rate of salt absorption increases with increasing surface area to volume ratio of the cheese.[130,132,134] This is most readily observed on comparing the rate of salt uptake by milled curd (e.g. Cheddar) and whole moulded cheeses (Brick, Emmental, Romano and Blue-type cheeses) in brine: in the former, salt absorption occurs from many surfaces simultaneously, and the time required to attain a fixed level of salt is very much less than in brined moulded cheeses.[125,130,132,135–140] While at first sight it may appear that smaller cheeses would have a higher mean salt content that larger ones after brining for equal intervals, this applies only to cheeses of the same shape and relative dimensions as salt uptake is linearly related to the surface area to volume ratio of the cheese.[125,126,132]

In addition to its influence on the surface area to volume ratio, cheese shape also affects the rate of salt absorption via its effect on: (i) the number of directions of salt penetration from the salting medium into the cheese[126] and (ii) the ratio of planar to curved surface area of the cheese.[132,137] Geurts et al.[137] found that on brining Edam-type cheese, the quantity of NaCl absorbed per cm^2 cheese surface was greater for an infinite slab than for a sphere and the relative reduction in salt uptake through curved surfaces increased with brining time and with degree of curvature. In Romano-type cheeses with approximately equal surface area to volume ratios, the rate of salt absorption by rectangular-shaped cheeses (volume: 4000 cm^3; three effective directions of salt penetration) was higher than that by cylindrical cheese (volume: 3400 cm^3; two effective directions of salt penetration at any time during a 9-day brining period (Fig. 3).[141]

7.3.3 Salting Time

It is well established that the quantity of salt absorbed increases with salting time[23,30,130,142,143] but that the rate of salt absorption decreases with time due to a

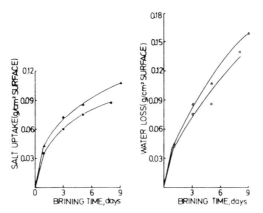

Fig. 3. Influence of cheese shape on salt uptake (▲,●) and moisture loss (△,○) by Romano-type cheese during salting in 19·5% NaCl brine at 23°C. Rectangular cheeses (▲, △), cylindrical cheeses (●,○) (from Ref. 141).

decrease in the NaCl concentration gradient between the cheese moisture and the brine.[132,136] Indeed, the quantity of salt taken up by a cheese is proportional to the square root of brining time [$\sqrt{t}$].[132,137] However, as the curvature of the cheese surface increases, the proportionality of salt uptake with $\sqrt{t}$ is lost and the relative reduction of salt uptake per unit area of cheese surface increases with increasing degree of curvature, and with time.[137] This implies that for cheeses with equal surface area to volume ratios, volumes and compositions, brined under the same conditions, the rate of salt absorption per unit surface area (and hence the cheese as a whole) would be in the order: rectangular > cylindrical > spherical;[139] however, other aspects of cheese geometry affect the mean salt level, as discussed above. Geurts et al.[135] derived a theoretical relationship for the quantity of salt absorbed from a flat surface as a function of brining time:

$$M_t = 2(C - C_o) (D^* \, t/\pi)^{1/2} V_w$$

where M_t = quantity of salt absorbed over time, g NaCl/cm^2
C = salt content of brine, g NaCl/ml
C_o = original salt content of the cheese, g/ml
t = duration of the salting period, days
D^* = pseudo-diffusion coefficient, cm^2/day
V_w = average water content throughout the cheese at time t, g/g.

Applying this theoretical relationship to their model brining experiments on cylindrical Gouda cheeses brine-salted by unidimensional diffusion through one of the planar surfaces in contact with the brine,[125] Geurts et al.[137] found that the predicted values for the quantity of salt absorbed per cm^2 planar surface (M_p) were in close agreement with the experimental values (M_t) over a three-day brining period: $M_t = 0.98 \, M_p$.

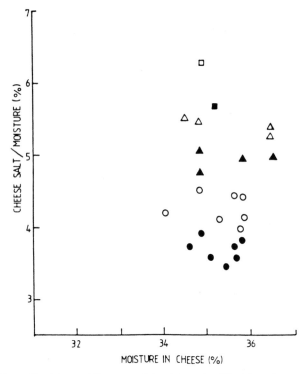

Fig. 4. The effect of salting acidity on salt/moisture in Cheddar cheese: open symbols, curd TA at salting, 0·50%; closed symbols, curd TA at salting, 0·75%. Salting rate 1·8% (w/w): ○ ●; salting rate 2·5%: △ ▲; salting rate 2·75%: □ ■ (from Ref. 134).

7.3.4 Temperature of Curd and Brine

In brining experiments with milled Cheddar chips, Breene *et al.*[130] found that for curd tempered to any temperature in the range 26·7–43·3°C, salt uptake increased with increasing brine temperature in the same range. However, curd tempered to 32°C absorbed salt less readily than curd tempered at lower or higher temperatures before brining. This was attributed to a layer of exuded fat on the surfaces of the curd particles at 32°C which impeded salt uptake; less fat was exuded at lower temperatures while at higher temperatures exuded fat was liquid and dispersed in the brine. Increasing brine temperatures have been found to result in higher mobility of NaCl and higher salt absorption in Gouda,[125,127] Emmental[133] and Romano-type[126,132] cheeses due partly to an increase in true diffusion and partly to an increase in the effective pore width of the protein matrix as non-solvent water decreases with increasing temperature.[127]

7.3.5 Curd pH

While the effect of pH on the rate of salt absorption by whole cheeses has not been investigated, a number of studies have examined the effect of titratable acidity at salting on salt retention by Cheddar cheese curd. Curd salted at low acidity retains more salt than more acidic cheeses (Fig. 4).[18,22,134] Since low acid curd normally contains more moisture than high acid curd one might expect more syneresis and higher salt losses in the former; however, the rate of salt diffusion and salt uptake, all conditions being equal, would be higher in the higher moisture curd chips. Lawrence & Gilles[22] suggest that the observed difference in salt retention may be due to the higher solubility of the curd at the higher pH values, which may effect a higher retention of salt by the curd structure *per se*.[144]

7.3.6 Moisture Content of the Curd

Geurts *et al.*[125,127] showed that the diffusion coefficient and the quantity of salt absorbed by experimental Edam and Gouda-type cheeses during brine-salting generally increased as the moisture content of the curd increased. Similar results were obtained by Byers & Price[142] for brine-salted Brick cheese. Undoubtedly, the higher salt uptake which accompanies increased moisture levels is a consequence of the concomitant increase in the rate (and depth) of penetration into the cheese which has been attributed[125] to an increase in the relative pore width of the protein matrix (volume fraction of protein phase decreases as moisture content increases), which reduces the frictional effect on the inward-diffusing NaCl molecules.

On dry-salting milled Cheddar curd, the reverse situation occurs: as the initial moisture level increases, the rate of salt absorption decreases giving lower salt and salt-in-moisture values in the cheese for a fixed salting rate.[131,134] Such decreases were attributed to greater whey and salt losses from the high-moisture curds; an increase in curd moisture content from 39·1 to 43·4% caused a 30% increase in the amount of whey drainage and a decrease in salt retention from 59 to 43% of the amount applied.[131] Thus, while the extent of salt penetration within each granule increases, there is less salt available for uptake as the initial curd moisture increases (salt causes loss of moisture from the curd and at the same time is itself removed).

7.4 Factors Affecting Salt Uptake in Cheddar Curd

7.4.1 Method of Salting

Breene *et al.*[130] showed that salting of milled Cheddar curd by brining gives a higher rate of salt absorption and a higher level of salt-in-moisture in the pressed curd than dry salting. Differences in absorption rates were explained on the basis of availability of salt at the surfaces of the curd. When dry salt is placed on freshly milled curd, a portion dissolves in the surface moisture, creating a very thin layer of super-saturated brine. The salt-in-moisture gradient between the brine and the cheese moisture results in mutual movements of salt and water in opposite directions in response to their respective concentration gradients. Some water is also 'squeezed out' of the curd due to localized surface contraction (salting-out of the protein matrix) as a result of contact with the super-saturated brine. The moisture level in the curd, which influences whey release, affects the rate of solution

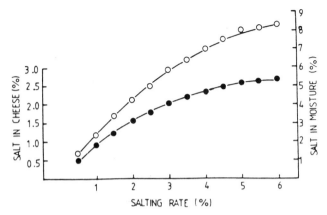

Fig. 5. The relationship between the salt contents (●) and S/M levels (○) of batches of
curd from the same vats, salted at different rates (from Ref. 18).

of surface salt. When curd is placed in brine, salt absorption begins immediately
through all surfaces. Release of whey occurs, as in dry salting, but its extent is
not a limiting factor.[131]

7.4.2 Salting Rate

As expected, an increase in salting rate (especially when the rate is low) increases
the rate of salt absorption by, and whey drainage from, cheese thus giving higher
levels of salt and salt-in-moisture and lower levels of moisture in the cheese after
salting for a fixed time.[14,81,126,130,134,145,146] However, the relationship is curvilinear
(Fig. 5), i.e. the increase in the salt and salt-moisture levels in the cheese is not
proportional to the level of salt added, especially at the higher salting rates, because of higher salt losses at increased salting rates. Although these principles
are probably generally applicable, the precise relationship between salt loss and
retention depends on the pH and moisture content of the curd and the period of
time allowed for salt diffusion into the curd. These interrelationships have been
studied by Sutherland[131] and Gilles.[134]

Sutherland[131] showed that the volume of whey released from the curd and the
percentage of added salt lost increased linearly with the level of salt added (over a
narrower range than that used by O'Connor)[14] while the % moisture decreased and
the % salt, % salt-in-moisture and pH of the cheeses increased in a curvilinear
fashion as the level of added salt was increased. The level of salt addition had no
significant effect on fat losses (~0·25 kg/100 kg curd). The percentage of added
salt lost increased slightly with increasing salting temperature but the proportion
of salt lost during the holding, pre-pressing, period increased markedly. The pH
and % moisture in the finished cheese were essentially unaffected by salting temperature but % salt and % S/M decreased and % fat lost increased markedly. Increasing the duration of mixing salt into the curd had little effect on the volume
of whey released but reduced the percentage of salt lost and hence increased the
salt and S/M contents of the cheese; fat losses were markedly increased. Salt losses

were substantially reduced, and consequently % salt and % S/M were substantially increased, by extending the pre-pressing holding period. Not surprisingly, salt losses increased with increasing moisture content in the curd.

As well as confirming the work of Sutherland,[131] Gilles[134] confirmed that greater salt losses occurred at a higher than at a lower acidity, that salt particle size has little effect on salt retention but milling the curd to smaller particles increases salt retention. Indeed, Gilles[134] maintains that while the best way to regulate the salt content of cheese is to control its moisture content (which can be best done by dry stirring), it is also possible to do so by varying the level of added salt though this is less desirable because of the influence of several factors on salt retention.

7.4.3 Degree of Mixing of Salt and Curd
Extending the stirring time of salted Cheddar curd from 20 s to 6 min caused a significant increase in salt and S/M levels, i.e. from 1·53 to 1·97%, and 4·41 and 5·71%, respectively.[131] Undoubtedly, better mixing leads to salt absorption from more faces and hence there is less 'free' salt to be lost in the press whey. In relation to this, mechanical salting procedures give more uniform distribution of salt in Cheddar cheese than hand or semi-automated salting systems.[86,145–148] Because of the significance of salt level and distribution in relation to cheese quality, salting of Cheddar curd is performed commercially on enclosed inclined, moving, perforated (to allow whey drainage) belts where overhead stirrers continuously turn the curds onto which salt is distributed from an overhead metering device. Improved means of salting cheese curds are being continuously devised.[149–151]

7.4.4 Time Between Salting and Pressing
Extending the holding time between salting and pressing increases the salt and S/M levels in the pressed Cheddar cheese.[130,131,134] The increase is attributed to a higher total absorption and hence a reduction in the physical loss of salt.

7.4.5 Curd Depth During Holding
When the depth of salted Cheddar curd during holding was increased from 12·7 to 68·0 cm, the moisture, salt and S/M levels decreased from 35·1 to 34·9%, 1·81 to 1·68%, and 5·1 to 4·8%, respectively.[131]

The interaction of some factors influencing salt uptake in Cheddar-type curd and brine- or dry-salted cheeses are summarized in Figs 6 and 7, respectively.

7.5 Factors Influencing Salt Diffusion is Cheese During Salting

There is relatively little information on the factors which influence the movement of NaCl in cheese during salting. The first such study was made by Georgakis[152] who related the diffusion of NaCl in Greek Feta to cheese surface area, duration of salting, brine concentration and the fat and moisture contents of the cheese.

More recent studies include those of Minarik[153] and Stêtina et al.[154] In model brining experiments, Geurts et al.[125] quantified the influence of variations in cheese composition and brining conditions on the pseudo-diffusion coefficient

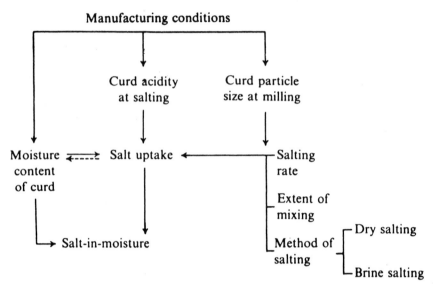

Fig. 6. Principal factors that affect the uptake of salt by Cheddar curd.

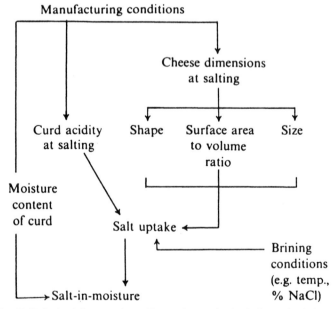

Fig. 7. Principal factors that affect salt uptake by brine-salted cheeses.

(D*) of NaCl in the moisture phase of Gouda cheese. Since the pseudo-diffusion coefficient is an intrinsic property of a given cheese, obviously the factors which affect the movement of salt in cheese during brining presumably also apply to the cheese after brining and hence have a decisive effect on the rate of attainment of salt-in-moisture equilibrium, and hence moisture equilibrium since moisture transport is a direct consequence of salt transport.[125] Although continuing physico-chemical changes during ripening may alter the situation somewhat, it is worth noting that the diffusion coefficient for NaCl in the moisture phase of a dry-salted, 12-week old Cheddar (50% FDM, 37·9% H_2O) corresponded well[155] with that found by Geurts et al.[125] for brine-salted Gouda cheese of similar composition.

The influence of the various factors on NaCl diffusion in Gouda cheese has been studied by Geurts et al.[125] and Guinee.[126]

7.5.1 Moisture

It is generally accepted that the moisture content of cheese affects the rates of salt absorption and/or diffusion.[135,142,152] However, from calculations of diffusion coefficients it has been shown[123,124] that for two cheeses of the same variety, the rate of diffusion is not necessarily higher in the higher moisture cheese; the diffusion coefficient depends on the ratios of fat to solids-not-fat (SNF) and moisture to SNF (the structure of the cheese, which determines the impedance to the diffusing molecules, as discussed above, is dependent on its composition). The results in Table II indicate the importance of cheese composition, and hence structure, in salt diffusion. The diffusion coefficient for NaCl in the cheese moisture (D*) increased with increasing FDM when the % SNF (and hence protein tortuosity, λ_p) decreased and the relative pore width of the protein matrix (λ/d_p) increased (e.g. cheeses 1–8, Table II) but decreased when both the FDM and SNF levels increased (cheeses 9 and 10). In some instances (e.g. cheeses 2, 6 and 9 had ~49% H_2O and cheeses 12 and 14 had ~44% H_2O) the moisture contents were approximately equal but the D*-values differed considerably due to differences in fat (and hence fat tortuosity, λ_f) and protein (SNF), while in other cheeses (e.g. 11 and 13), D* was almost equal while the moisture levels differed appreciably. Similar results, shown in Table III, were obtained by Geurts et al.[125] Therefore, while it is not feasible to study the effect of moisture, or indeed any one compositional parameter separately, on salt flux, it will be attempted to deduce the quantitative effect of moisture separately.

Within a series of cheeses of the same variety with equal FDM, D* increases curvilinearly with moisture content (c.f. Fig. 8).[125] Considering cheeses 9–12, Table II, it is apparent that the contribution of the decreasing fat tortuosity (λ_f) (with increasing moisture content) to the increase in D* was small (D* and D*λ_f increased by a factor of 1·7 and 1·6, respectively, when the moisture content increased from 44·5 to 49·2%). The principal factor affecting the increase in salt flux was the reduced frictional effect on the diffusing molecules as the volume fraction of the protein matrix decreased; hence the relative pore width increased concomitantly with increasing moisture content (c.f. Tables II and III).

TABLE II

Influence of Cheese Composition on Salt Diffusion in Cheese Moisture (Ref. 126)

Cheese code	Properties of unsalted cheese								Obstructing factors		Diffusion coefficients	
	Fat (%)	Moisture (%)	Solids not-fat (%)	Fat-in-dry matter (%)	Moisture in-fat-free cheese (%)	Volume fraction of fat phase (ϕ_f)	Volume fraction of protein matrix (ϕ_p)	Relative pore width of protein matrix (y/d_p)	Fat tortuosity (λ_f)	Protein tortuosity (λ_p)	D^* (cm²/day)	$D^*\lambda_f$ (cm²/day)
1	0·00	53·00	47·00	0·00	53·00	0·00	0·466	0·132	1·00	1·425	0·136	0·136
2	10·88	49·00	40·12	21·33	54·98	0·127	0·442	0·152	1·117	1·409	0·153	0·171
3	19·88	45·93	34·19	36·77	57·33	0·227	0·413	0·178	1·229	1·389	0·203	0·249
4	26·25	43·48	30·27	46·44	58·96	0·297	0·394	0·198	1·296	1·378	0·227	0·294
5	0·00	52·90	47·10	0·00	52·90	0·00	0·467	0·132	1·00	1·425	0·140	0·140
6	10·00	48·93	41·07	19·58	54·37	0·117	0·449	0·146	1·105	1·414	0·202	0·223
7	18·25	44·90	36·85	33·12	54·92	0·207	0·442	0·152	1·208	1·409	0·205	0·248
8	30·28	40·50	29·22	50·89	58·09	0·340	0·404	0·187	1·333	1·383	0·236	0·315
9	18·18	49·20	32·62	37·59	60·13	0·207	0·380	0·213	1·208	1·365	0·295	0·356
10	20·00	47·94	32·06	38·42	59·93	0·227	0·382	0·211	1·229	1·366	0·263	0·323
11	21·00	45·02	37·98	38·20	56·99	0·240	0·417	0·174	1·242	1·392	0·207	0·257
12	21·00	44·44	34·60	37·77	56·25	0·240	0·427	0·165	1·242	1·400	0·176	0·218
13	29·00	41·11	28·89	49·24	57·90	0·326	0·406	0·185	1·322	1·384	0·216	0·285
14	27·66	44·02	28·32	49·41	60·85	0·310	0·371	0·223	1·308	1·358	0·247	0·324

For calculation of ϕ_f and ϕ_p it was assumed that: (a) cheese moisture contained 5% dissolved solids, density = 1 g/ml; (b) the protein matrix consisted of protein + 15% water bound and had a specific gravity of 1·25; (c) the specific gravity of fat = 0·93. Cheeses were from four trials, i.e cheeses 1–4, from trial 1; 5–8, from trial 2; 9–12, from trial 3; and 13, 14 from trial 4. All cheeses were salted in 18·5% NaCl at 20°C for 3–4 days.

TABLE III

Experimentally Determined Diffusion Coefficient (D*) of Salt in Moisture in Cheeses Varying in Properties and Brined Under Different Conditions. Calculation of Diffusion Coefficient in Moisture (D*λv) and of Relative Pore Width of the Protein Matrix (y/d)e in Fat-free Cheese[125]

Brine (g NaCl/ 100g H₂O)	Properties of non-salted cheese				Calculation of factors				Diffusion coefficient (cm²/day)	
	g fat in 100 g DM	pH	Fat content (%)	Moisture content (%)	ϕ_v	λ_v	ϕ_e	$(y/d)_e$	D^*	$D^*\lambda_v$
19·7	12	5·00	5·3	53·0	0·06	1·04	0·42	0·171	0·164	0·170
19·0	12	5·01	4·9	54·0	0·06	1·04	0·41	0·181	0·185	0·192
19·5	12	4·99	5·0	55·0	0·06	1·04	0·40	0·191	0·162	0·168
20·2	22	5·00	10·9	50·2	0·13	1·12	0·42	0·171	0·152	0·171
19·8	50	5·10	33·0	36·2	0·37	1·36	0·44	0·153	0·100	0·136
20·5	50	4·79	31·2	41·1	0·35	1·34	0·37	0·225	0·160	0·215
20·0	50	5·42	30·9	41·5	0·34	1·33	0·37	0·225	0·160	0·213
19·6	50	5·09	30·4	42·2	0·34	1·33	0·37	0·225	0·172	0·229
20·4	50	5·02	29·9	42·5	0·33	1·32	0·36	0·238	0·185	0·245
34·8	50	5·09	29·9	42·5	0·33	1·32	0·36	0·238	0·148	0·196
14·0	50	5·10	30·2	42·5	0·34	1·33	0·36	0·238	0·194	0·258
19·7	50	5·07	29·2	42·9	0·33	1·32	0·37	0·225	0·187	0·248
13·8	50	5·18	29·5	43·0	0·33	1·32	0·36	0·238	0·177	0·234
20·0	50	5·64	28·9	43·4	0·32	1·32	0·36	0·238	0·168	0·221
13·8	50	4·98	26·9	48·0	0·30	1·30	0·31	0·312	0·235	0·305
20·1	50	4·92	26·6	49·0	0·29	1·29	0·30	0·330	0·258	0·333
19·4	50	5·09	25·8	50·1	0·29	1·29	0·29	0·349	0·239	0·309
20·0	62	5·00	39·1	38·5	0·43	1·40	0·34	0·265	0·179	0·251
20·1	62	4·98	37·1	40·8	0·41	1·39	0·32	0·295	0·224	0·311

ϕ_v = volume fraction of fat in cheese, calculated from fat content.
λ_v = tortuosity factor of fat, a function of ϕ_v.
ϕ_e = volume fraction of protein matrix in fat-free cheese.
$(y/d)_e$ = relative pore width of this matrix.
D^* = pseudo-diffusion coefficient in moisture in cheese.
In calculating ϕ_v and ϕ_e it is assumed that cheese moisture contains 5% dissolved substances (density of solution = 1), that the protein matrix consists of protein + 15% water (density 1·25), and that the density of the fat is 0·93. All experiments were carried out at about 12·5°C.

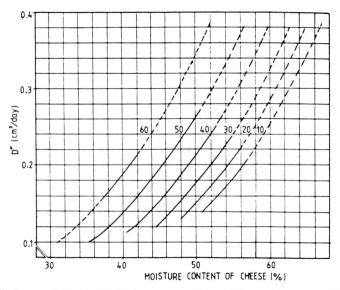

Fig. 8. Diffusion coefficient of NaCl in cheese moisture (D^*) as a function of initial moisture content of the cheese. Parameter is g fat/100 g DM in unsalted cheese. The solid lines are experimental values, broken lines are extrapolations (from Ref. 125).

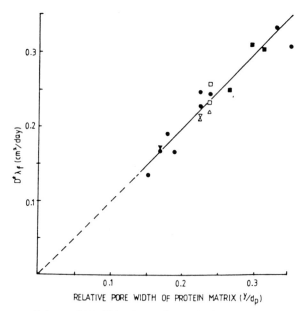

Fig. 9. Diffusion coefficient of NaCl in the moisture in fat-free cheese ($D^* \lambda_f$) as a function of the relative pore width of the protein matrix (y/d_p). Brine concentration, 19–20 g NaCl/100 g H_2O; temperature, 12·5°C; 50% fat in DM; pH 5, unless stated otherwise. g fat/100 g DM: ● 12, ▼ 22, ■ 62; ▽ pH 4·79, △ pH 5·50; brine concentration, 14 g NaCl/100 g H_2O (from Ref. 125).

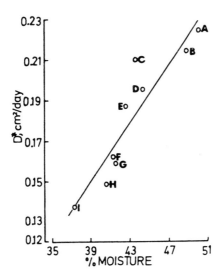

Fig. 10. Dependence of the pseudo-diffusion coefficient of salt in cheese moisture (D*) on the initial moisture content of cheese salted in ~20% NaCl brine at 15–16°C. Blue cheese (A,B), Gouda (C,D), Romano (E), Jarlsberg (F), Emmental (G,I), unsalted milled Cheddar (H) (from Ref. 156).

The relationship between the diffusion coefficient in the fat-free cheese, $D^*\lambda_f$, ($D^*\lambda_f$ can be considered as the 'theoretical' value of D* for a system with the same structural features of cheese but from which the impedance to salt diffusion, due to the physical presence of fat globules, has been eliminated) and the relative pore width of the protein matrix is seen in Fig. 9.[125] While the decrease in the protein tortuosity (λ_p) contributes to the increase in D* associated with increasing moisture content (cheeses 9–12, Table II), its effect is small as it varies little within the range of values for protein volume fraction, ϕ_p encountered.

The role of moisture as the preponderant compositional factor affecting salt flux has been confirmed by Morris *et al.*[156] who found that the values of D* for different commercial cheese varieties (~37·3–49% H_2O; ~23·5–27·5% fat; ~40·5–49·5% FDM; ~28–35% SNF) were directly related to the moisture content of the unsalted cheeses (Fig. 10). Of the variation in D* which could be attributed to compositional parameters (~70% of total variation), ~49, 29 and 22% could be attributed to variations in the relative pore width of the protein matrix, the protein tortuosity and fat tortuosity, respectively.[126]

Perhaps rather surprisingly, it was found that while D* was strongly dependent on the composition and structure of the unsalted cheese, especially the moisture content, it was scarcely affected by variations in composition along the different planes of a cheese resulting from salt uptake as reflected by: (i) the consistency of D* over the region of salt and water movement and with time[125,126,139] and (ii) the almost-constant D* values for brine concentrations in the range of 5–20%

NaCl.[125,126] Indeed, pre-salting cheese to different levels (by mixing dry salt with the curd at the end of manufacture) scarcely affected the penetration depth of the salt over a given brining period during subsequent brine-salting, confirming that compositional changes accompanying salt uptake do not influence D^* significantly.[126] Consideration of the physico-chemical changes in cheese associated with the physical presence of salt *per se* and ageing may provide a tentative explanation.[125] Salting and ageing of cheese are paralleled by a considerable reduction in the amount of protein-bound water and a decrease in the mean diameter of the protein particles[125] and hence an increase in the effective moisture concentration and relative pore width of the protein network. Such changes possibly offset the impeding effects of moisture loss during brining on salt flux and hence D^* remains constant.

7.5.2 Fat Content

As discussed previously, the diffusion coefficient for NaCl in cheese moisture is much lower than that in pure solution (i.e. $\sim 0 \cdot 2$ cm²/day as compared to $\sim 1 \cdot 0$ cm²/day at $12 \cdot 5°C$). This is because salt diffusion in cheese takes place in moisture held in a protein matrix which also occludes fat. Hence, both the sieving effect of the protein matrix and the obstructions of the fat globules and protein strands (through which NaCl cannot penetrate and which increase the real distance travelled by a salt molecule on proceeding from one parallel plane to another), reduce the apparent diffusion rate relative to that in pure H_2O. Therefore, the physical presence of fat *per se* reduces the apparent D-value due to its tortuosity factor, λ_f.

However, D^* increases with fat content in cheeses with equal moisture content (Fig. 8). In unidimensional brine salting experiments, Geurts *et al.*[125] observed that for Gouda cheese with 50% moisture, but with 11 or 26% fat, the D^* values were $0 \cdot 15$ and $0 \cdot 25$ cm²/day, respectively. While the fat tortuosity factor increased with fat level, i.e. $1 \cdot 12$ and $1 \cdot 29$ at 11 and 26% fat, respectively, the relative pore width of the protein matrix also increased (i.e. $0 \cdot 17$ and $0 \cdot 35$ at 11 and 26% fat, respectively). Hence, the increase in D^* with fat content is not due to fat *per se* (which actually reduces D^* by a factor of λ_f) but rather to the concomitant decrease in the protein volume fraction and hence the increase in the relative pore width of the protein matrix; the reduction in the sieve-effect of the protein matrix on the salt molecules overrides the increased obstruction caused by increasing fat levels and hence D^* increases. Indeed, for cheeses of equal moisture content in the fat-free cheese (i.e. cheeses with equal protein volume fractions), D^* is always higher in the cheese with the lower fat content.[125]

From the foregoing it is apparent that the effect of varying any cheese compositional parameter on salt mobility depends on the concomitant changes it causes in the cheese structure (i.e. the ratios of fat to solids-not-fat, and solids-to-moisture). Since increasing fat levels in cheese reduce syneresis,[157–160] D^* should generally increase with increasing fat levels due to the concomitant decrease in the protein volume fraction.

7.5.3 Temperature

Increasing brine (and curd) temperatures are paralleled by increasing diffusion mobilities of NaCl and H_2O in cheese; an increase of ~0·008 cm²/day/°C was found for commercial Gouda and Romano-type cheeses in the brine temperature range 5–25°C.[126] Part (~50%) of this increase was attributed[125] to an increase in true diffusion and the remainder to some effect on diffusion-interfering factors, i.e. possible decreases in the relative viscosity of cheese moisture and the amount of protein-bound water which effects an increase in the relative pore width of the protein matrix (in cheese, water non-solvent for sugars decreases with increasing temperature);[127] both of the latter effects would contribute to decreases in the frictional effect on the diffusing species and the protein tortuosity and thus increases in the relative diffusion rate. Extrapolating the effect of temperature on D* to salted-cheese in which there is large zonal variation of salt and moisture, the higher the storage temperature the shorter should be the time required for equilibration of salt-in-moisture within the cheese mass after salting.

7.5.4 Concentration Gradient

While the concentration gradient is a major determinant of the rate of salt absorption by a cheese during salting, it scarcely affects the mobilities of the diffusing species except during brining in supersaturated salt solutions.[125,126] Although the value of the apparent diffusion coefficient decreases on using saturated brines [~18% lower than that with 5–20% (w/w) NaCl brines at 20°C for Gouda-type cheese][126] the true value would be somewhat higher on allowing for the relatively high water loss which in effect causes the plane of zero mass transfer of all diffusing species to recede further into the brine. However, since the salt-in-moisture level in salted cheese scarcely ever reaches saturation point, the interzonal variations in salt-in-moisture (S/M) levels do not significantly alter the rate of attainment of S/M equilibrium between different cheeses of the same variety.

7.5.5 Cheese Geometry

Cheese geometry influences the rate of attainment of salt-in-moisture equilibrium via its effect on the relative dimensions of the cheese; Guinee & Fox[161] working with commercial Romano-type cheeses of different shapes, showed that at any time during storage, the net difference in S/M concentration along layers of the cheeses increased with layer length. It is worth noting that the depth of salt penetration during brining is proportional to the square root of brining time.[125,126] Using differently shaped cheeses it was found[161] that the rate of attainment of S/M equilibrium is not necessarily directly proportional to the volumes when comparing cheeses of the same variety.

Though not investigated to date, conditions of relative humidity and rates of air circulation during storage possibly alter the rate of attainment of salt and moisture equilibria as a result of alterations in cheese moisture.

7.6 Attainment of Salt and Moisture Equilibria After Salting

While salt absorption is a relatively rapid event [~15–30 min for salt uptake by Cheddar-type curd,[155] and ~7·5 h (Camembert) to ~15 days (e.g. Parmesan) for brine salting], diffusion of salt post-salting, and hence the rate of attainment of S/M equilibrium throughout the cheese mass, is a slow process, e.g. 10–12 days for Limburger,[135,143] 8–12 weeks for Gouda,[162] Brick,[142] Blue[30] and Romano-type cheeses,[140,161] ~40 days for Feta[152] and ~10 months for Parmesan.[138] Salt is fairly uniformly distributed in Cheddar-type cheeses initially, as salt is mixed with the milled curd; however, complete equilibrium is slow and rarely, if ever, reached, giving rise to significant intra- and inter-block variations in the mature cheese.[20,83,135,145–147,155,156,162] In contrast, in cheeses which are salted by immersion in brine and/or by surface application of dry salt there is a large decreasing salt gradient from the surface to the centre and a decreasing moisture gradient in the opposite direction at the completion of salting (c.f. Fig. 11)[126] Due to the slow diffusion of salt from the rind inwards, these gradients disappear slowly and equilibrium of S/M is practically reached at some stage of ripening, depending on size of cheese and curing conditions (c.f. Fig. 12).[126,141]

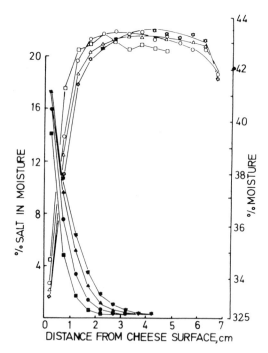

Fig. 11. Moisture content (open symbols) and salt-in-moisture (g NaCl/100 g H$_2$O) (closed symbols) in Gouda cheese (fat-in-dry matter 49·1%; moisture, 43·64%; pH, 5·26) as a function of distance from the salting surface after unidimensional brine salting (20·3% NaCl) for 1 (□,■), 2 (○,●), 3 (△,▲) and 4 (☆,★) days at 15°C (from Ref. 126).

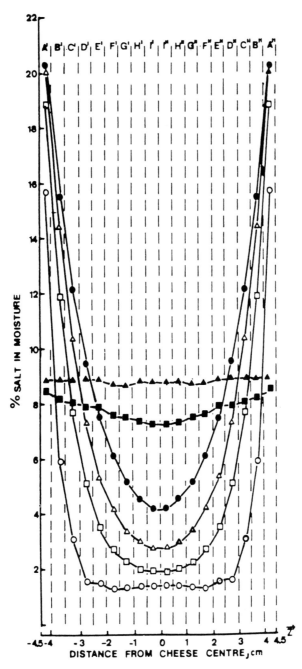

Fig. 12. The mean salt-in-moisture levels in discs A'A", B'/B", C'/C" etc. (as indicated) of cylindrical Romano-type cheese salted in 19·5% NaCl brine at 23°C for 1 (○), 3 (□), 5 (Δ) or 8 (●) days or salted for 5 days and stored wrapped at 10°C for 30 (■) and 83 (▲) days (from Refs 126, 141).

Though the importance of the mean, and the uniformity of, S/M levels in cheese in relation to quality have received much attention,[14,20,36,83,85,163] the factors that affect the diffusion of NaCl and moisture in cheese after salting and hence the rate of attainment of equilibrium have received little study.

7.7 Diffusion of NaCl in Cheddar Cheese

As cited earlier, significant intra- and inter-block variations in salt concentration occur in mature commercial Cheddar cheese, giving rise to considerable variation in the rate of ripening and quality.[8,14,17,20,81,83–85,156,164] The method of salt application used in Cheddar cheese manufacture would appear to be particularly amenable to ensuring accurate control of salt concentration with respect to both level and uniformity. However, in commercial practice, this does not appear to be so,[135,148,162] possibly because of the many factors which influence salt uptake (Fig. 6) and the difficulty of obtaining equal salt distribution in all regions of the vat. A preliminary investigation of brine-salting Cheddar curd chips[130] showed that adequate uptake of salt could be accomplished by holding chips (⅜ in × ⅜ in × 2 in) in 25% brine for 5 min and holding for 15 min after removal from the brine before pressing. Considering the problems encountered in controlling salt uniformity using the present mixing procedures, brine-salting of Cheddar appears to warrant further investigation.

O'Connor[147] assessed the uniformity of salt distribution in cheeses salted by hand, semi-automatic or fully-automatic systems; one plug from every 10th cheese per vat was analysed. While the range of mean salt contents was relatively narrow, there was very considerable intra-vat and inter-vat variation in salt content with greatest variation in the hand-salted cheese and least with the fully automated system; an inverse correlation between salt and moisture contents was apparent. Further evidence of high within-vat variation in salt distribution is provided by O'Connor.[165]

The findings of O'Connor[147,165] were confirmed and extended by Fox[148] who showed that, in general, mechanical salting systems gave more uniform salt distribution than hand-salting systems or a semi-automatic system. Considerable within-block variability (12 samples per 20 kg block) in salt concentration was also demonstrated. Morris[162] also found very large differences in the salt content of blocks from the same vat (spread of 0·6% on a mean of 1·38%).

All the foregoing investigators stress the importance of inter- and intra-block variations in salt content, which is inversely related to moisture content; since it is generally agreed that the quality of cheese is strongly dependent on moisture, salt-in-moisture and pH (which was not reported in any of the above studies), it might reasonably be expected that the quality of cheese also varies between blocks from the same vat and even within the same block. It is normal cheese-grading practice to grade a vat of cheese on the basis of a single plug taken from a single cheese per vat; obviously, the quality of this sample may not be representative of the entire vat. For similar reasons, calculation of mass balances in cheese factories on the basis of a single plug sample per vat may be very inaccurate.

Thus, although salt is fairly well distributed in Cheddar cheese during the initial salting, in contrast to brine and/or dry salted cheese, full equilibrium is approached slowly. Sutherland,[155] who prepared Cheddar cheeses (9·5 kg) with regions of high and low salt, found that equilibria of salt, moisture, and hence, S/M, were not established after 25 weeks ripening at 13°C. Samples situated 7·6 cm apart, which showed an initial difference in S/M concentrations of 4·27%, still showed a difference of 1·56% at the end of the 25-week ripening period. As zones of high and low salt within commercial cheese blocks (~20 kg, ripened at 4–7°C) are likely to be more widely separated, it was concluded that equilibrium of S/M within such cheeses is unlikely. A similar study by Thomas & Pearce[20] showed that there was only a very slight change towards equilibrium of S/M during a 6-month ripening period in Cheddar cheeses prepared with an approximately linear S/M gradient diagonally across the blocks. Equilibration of NaCl in Cheddar cheese intentionally prepared with poor salt distribution was studied by Morris et al.:[156] salt and moisture analyses were performed on samples taken from 32 selected locations in 20 kg blocks (stored at 10°C) over a 24-week ripening period (a similar sampling pattern was used on each of six occasions); the results indicated that there was only a slight equilibration of salt over the 24-week period.

Hoecker & Hammer,[23] who measured the salt and moisture levels at the surface and centre of individual chips, prised from a block of Cheddar, over a 72 h period after pressing, found that salt and moisture equilibria were established within individual chips 48 and 24 h after hooping, respectively (a comparable study by Morris et al.[156] gave almost identical results). However, analysis of two 4-month-old cheeses showed significant intra- and inter-block variation in both variables. Hence, while salt and moisture equilibria are attained relatively rapidly within chips because of the short distance over which NaCl molecules have to diffuse from the surface to the centre, the variations throughout the block, as a result of the different quantities of salt absorbed by individual chips, do not disappear during normal ripening. The foregoing observations suggest that the contracted protein layers (salting-out of protein at chip surfaces possibly occurs because of the high initial S/M concentration before equilibrium is established) at the surface of individual chips and/or microspaces between chips which break the continuity of the interpenetrating gel fluid/moisture (in which salt is dissolved), inhibit movement of salt and water across the chip boundaries, and hence the cheese mass as a whole, even where a concentration gradient exists. (Indeed, milling per se results in the development of a 'skin' which has fewer fat globules and hence a denser protein matrix than the enclosed curd;[166] moreover, light microscopical studies[167–169] show that the 'skin' at milled curd junctions appears much thicker than those of the enclosed granules.) Observations by Morris[162] on salt diffusion in Cheddar cheese lend support to the view that the milled curd pieces 'trap' absorbed salt: the spread in salt levels within individual cheeses at 3 weeks equalled that observed immediately after hooping.

Morris et al.,[156] who also studied salt diffusion in model Cheddar cheese systems, found that equilibrium was established rapidly in cheeses prepared from alternate layers (2 cm thick) of salted and unsalted curd (unmilled) but not in model cheeses prepared from alternate layers (2 cm thick) of salted and unsalted chips. NaCl diffusion across the interface formed between the salted and non-salted layers (from chips) was very slow (Fig. 13) which is in agreement with the results of a similar experiment by McDowall & Whelan.[135] Morris et al.[156] suggested that the fragmented structure of Cheddar cheese (due to its construction from chips) may retard salt diffusion but a further experiment, the results of which showed that the diffusion coefficient for NaCl in the moisture phase of a brine-salted block of Cheddar prepared from unsalted chips at 0·15 cm²/day was as expected from its moisture content (c.f. Fig. 10), could not verify this. Thus, it appears that the contracted protein layers between salted chips, which possibly offer a very tight screening effect on the diffusing molecules, override the effect of low discontinuous gradients in various directions in commercial Cheddar or even at interfaces between salted and unsalted regions where the concentration

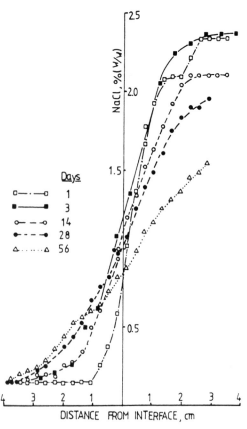

Fig. 13. Distribution of salt (NaCl) throughout a 7·5-cm 'cheese' prepared from half-salted curd and half-unsalted curd at 1, 3, 14, 28 and 56 days after manufacture (from Ref. 156).

gradient is high. In unsalted, milled Cheddar the surfaces of chips would not be as dense as those in dry-salted milled Cheddar due to their higher moisture and fat contents and hence the sieve-effect of the matrix on the diffusing molecules would be much lower than in the latter; indeed the impedance on the salt molecules penetrating the surface layers of milled Cheddar chips during brining is possibly similar to that encountered on penetrating the surface of curd granules *per se* (no light microscopic studies have been reported on unsalted, milled Cheddar curd).

8   EFFECT OF SALT ON CHEESE COMPOSITION

In the light of the findings of O'Connor,[14,81,147] Sutherland[131] and Morris *et al.*[156] which showed that varying salting rates in Cheddar cheese manufacture were associated with large compositional variations in the cheese, the effect of salt on the gross composition of cheese merits brief discussion.

The moisture content of cheese curd is influenced primarily by syneresis of the cheese curd during manufacture which is, in turn, influenced by the composition of the cheese milk, i.e. fat, protein and calcium levels, the level of rennet used, the curd tension at cutting, and curd treatments during manufacture, i.e. size of curd cut, degree of curd agitation, cooking temperature, rate of acid development, extent of dry-stirring of curd and depth of curd during cheddaring and size of pressed cheese.[157,158–160,170–175] Further syneresis occurs on addition of salt after milling (e.g. for Cheddar and Cheshire), during pressing and brine and/or dry salting.

It is generally accepted that there is an inverse relationship between the levels of salt and moisture in cheese. This is most readily observed in brine and/or dry salted moulded cheeses during, or immediately after, salting, where a decreasing salt gradient from surface to the centre is accompanied by a decreasing moisture gradient in the opposite direction (c.f. Fig. 11).[125,135,136,139–142] O'Connor[81] found that a negative correlation exists between the salt and moisture concentrations in commercial Scottish Cheddar cheeses. Although there was considerable scatter, the data of Fox[83] show an inverse correlation between the % moisture and % NaCl in 123 commercial Irish Cheddar cheeses. Direct evidence of this relationship is also apparent from the work of O'Connor[145,146] for cheeses from the same batch of curd salted at different rates.

An inverse correlation between % moisture and % NaCl in Cheddar cheese is not surprising since a considerable volume of whey is released from Cheddar curd following salting and during pressing.[131] The amount of whey released is directly related to the amount of salt added to the curd; roughly half of the whey is released during holding following salting and the other half on pressing. Although other factors, e.g. curd temperature, stirring time after salting, depth of curd and duration of holding time after salting and before pressing, influence the ratio of whey released during holding after salting to that released on pressing, the overall release of whey was not significantly influenced by these factors.[131] The moisture content of the cheese was inversely related to salting rate.[131]

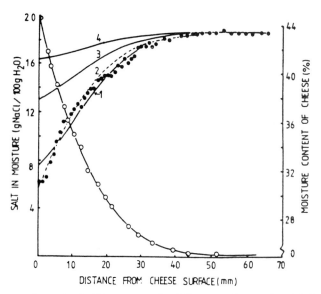

Fig. 14. Moisture (●) and salt (○) content of a full-cream cheese after 8·1 days of brining, as a function of penetration depth. pH 5·64, brine concentration 20·5 g NaCl/100 g H_2O, temperature 12·6°C. Salt contents were calculated from chloride estimations, experimental results (●); (1) moisture content calculated from salt content and a flux ratio (g water : g salt) $p = 2·5$; (2) the same, but p varies from 1·7 at the 'salt front' to 2·9 in the cheese surface; (3) the same, but $p = 1$; (4) the same, but $p = 0$ (from Ref. 125).

Geurts et al.[123] expressed the relative fluxes of NaCl and H_2O during the unidimensional brine-salting of Gouda-type cheese in terms of the flux ratio, p:

$$-\Delta W_x \sim p\Delta S_x$$

where ΔW and ΔS are the changes (from the non-salted cheese) in the g H_2O and g NaCl, respectively, per 100 g cheese solids-not-salt in planes of cheese x cm from the cheese/brine interface. Experimental values for W and S are shown in Fig. 14 together with theoretical curves calculated for various values of p. The experimental curve for W approximated the theoretical curve for $p = 2$ (i.e. when the amount of H_2O leaving the cheese is twice that of the NaCl entering) but varied from 1·5 at the salt front to 2·34 at the brine/cheese interface and was always >1. While a similar trend in p values was observed by Guinee & Fox[139] for commercial Romano-type cheese (salted for 9 days in 19·3% NaCl brine), the value of p varied more, i.e. from 3·75 at the rind to <1 in a region between the rind and the salt front. Guinee[126] concluded that the value of p at a particular location within the region of salt and water movement depends on the concentrations of NaCl and H_2O at the location—indeed this is possibly the reason why p decreases in the direction from the rind inwards,[125,139] along which significant variations of salt and moisture occur as a result of salt uptake *per se*. Indeed, changes in cheese texture and appearance corresponding to the changes in

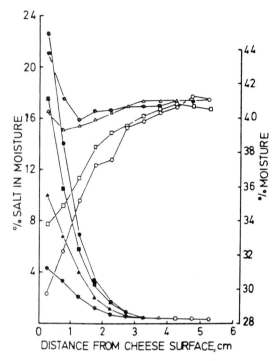

Fig. 15. Moisture content (open symbols) and salt-in-moisture concentration (closed symbols) in Gouda cheese as a function of distance from the salting surface after brine salting for 4 days at 20°C in 5 (☆,★), 12 (△,▲), 20 (□,■) and 24·8 (○,●) % NaCl solution (without calcium) (from Ref. 126).

p which occur in the region of high salt and moisture movement, are visible when a brined cheese is cut parallel to the direction of brine movement, during or shortly after brining.[125,126,176] In the outermost region (0·3–1·3 cm depending on the duration of the brining period) bordering upon the brine, where the S/M levels were high, i.e. ∼>12%, the cheese was hard, brittle, dry and white (indicative of salting-out), whereas further removed from the interface where % S/M >3% and <10%, the cheese was soft, yellowish and somewhat waxy (indicative of swelling); between the 'waxy' layer and the salt front, the cheese had a uniform appearance and resembled the unsalted cheese.[126] The extent of the outer dry white layer is augmented by a low cheese moisture and a low brine pH, i.e. 4·6;[176] such factors contribute further to protein insolubility.

Since the average flux ratio over the region of salt and water movement is >1, there is a net outflow of water which accounts for the commonly-observed volume reduction in cheese during brining and/or dry salting.

Perhaps unexpectedly, salt uptake during brining is sometimes accompanied by an increase in moisture content in the vicinity of the cheese-brine interface,[132,135,136,177] especially in weak brines (<10%, w/v, NaCl solution) without calcium (c.f. Fig. 15).[126,132] Such an effect is associated with the 'soft rind' defect and swelling in cheese and is attributed to a salting-in of the protein matrix in low % NaCl solutions which results in increased protein solubility.[136] Several reports exist on the composition and maintenance of brines and brining times in relation to cheese quality.[133,136,178-183]

8.1 Fat Content

O'Connor[14,81] showed that higher salt levels (especially >2%) in Cheddar cheese are associated with increased fat content, probably due entirely to the decrease in cheese weight as a result of the preponderance of water over salt flux during salting; it is worth noting that O'Connor[81] found a significant inverse relationship between fat and moisture levels ($r = -0.768$) in mature Cheddar cheese. However, on considering the findings of Breene et al.,[130] the fat content may decrease, especially at high salting rates if the curd temperature at salting exceeds 32°C.

8.2 Lactose Content and pH

As discussed in earlier sections, the lactose content and pH of cheese are strongly influenced by the level and time of salt application.[184]

9   CONCLUSION

Clearly, salt plays a multi-faceted role in cheese ripening with an influence on the physical, chemical and biological attributes of the mature cheese. While a considerable amount of information is currently available on many aspects of the significance of salt in cheese and on salt diffusion in cheese curd, many gaps persist.

REFERENCES

1. Anon, 1980. *Food Technol.*, **34**, 85.
2. Joossens, J.V. & Geboers, J., 1983. *Acta Cardiologia*, **1**, 1.
3. Anon, 1984. *Dairy Council Digest* (National Dairy Council, III, USA), **55**(6), 33.
4. Moses, C. (ed), 1980. *Sodium in Medicine and Health: A Monograph*, Salt Institute, USA, pp. 1–126.
5. Abernathy, J.D., 1979. *Food Technol.*, **33**, 57.
6. Tobian, L., 1979. *Amer. J. Clin. Nutr.*, **32**, 2379.
7. McCarron, D.A., Morris, C.D., Henry, H.J. & Stanton, J.L., 1984, *Science*, **224**, 1392.
8. Thakur, M.K., Kirk, J.R. & Hedrick, T.I., 1975. *J. Dairy Sci.*, **58**, 175.
9. Kosikowski, F.V., 1983. *J. Dairy Sci.*, **66**, 2494.
10. Lindsay, R.C., Hargett, S.M. & Bush, S.C., 1982. *J. Dairy Sci.*, **65**, 360.
11. Schroeder, C.L., Bodyfelt, F.W., Wyatt, C.J. & McDaniel, M.R., 1985. *J. Dairy Sci.*, **68** (Suppl. 1), 66.
12. Schroeder, C.L., Bodyfelt, F.W., Wyatt, C.J. & McDaniel, M.R., 1988. *J. Dairy Sci.*, **71**, 2011.

13. Irvine, D.M. & Price, W.V., 1961. *J. Dairy Sci.*, **44**, 242.
14. O'Connor, C.B., 1974. *Irish Agric. Creamery Rev.*, **27**(1), 11.
15. Dawson, D.J. & Feagen, J.T., 1957. *J. Dairy Res.*, **24**, 210.
16. Martley, F.G. & Lawrence, R.C., 1972. *N.Z. J. Dairy Sci. Technol.*, **7**, 38.
17. Turner, K.W. & Thomas, D.T., 1980. *N.Z. J. Dairy Sci. Technol.*, **15**, 265.
18. Lawrence, R.C. & Gilles, J., 1982. *N.Z. J. Dairy Sci. Technol.*, **17**, 1.
19. Fryer, R.F., *Proc. 21st Intern. Dairy Congr. (Moscow)*, **I**, 1, 485.
20. Thomas, T.D. & Pearce, K.N., 1981. *N.Z. J. Dairy Sci. Technol.*, **16**, 253.
21. Lowrie, R.J. & Lawrence, R.C., 1972. *N.Z. J. Dairy Sci. Technol.*, **7**, 51.
22. Lawrence, R.C. & Gilles, J., 1969. *N.Z. J. Dairy Sci. Technol.*, **4**, 189.
23. Hoecker, W.H. & Hammer, B.W., 1943. *Food Res.*, **9**, 278.
24. Noomen, A., 1977. *Neth. Milk Dairy J.*, **31**, 75.
25. Ruegg, M. & Blanc, B., 1981. In *Water Activity: Influence on Food Quality;* ed. L.B. Rockland & G. F. Stewart. Academic Press, New York, p. 791.
26. Orla-Jensen, S., 1926. *Landw. J., Ahnb.*, **20**, 437. (Cited from Langsrud, T. & Reinbold, G.W., 1974. *Milk Food Technol.*, **37**, 26).
27. Antila, M., 1955. *Meijerit. Aikausk.*, **16**, 7. (Cited from Langsrud, T. & Reinbold, G.W., 1974. *Milk Food Technol.*, **37**, 26).
28. Rollman, N.O. & Sjostrom, G., 1946. *Svenska Mejeritidningen*, **38**, 199, 209. (Cited from *Dairy Sci. Abstr.*, 1948–50, **11**, 33).
29. Godinho, M. & Fox, P.F., 1981. *Milchwissenschaft*, **36**, 205.
30. Godinho, M. & Fox, P.F., 1981. *Milchwissenschaft*, **36**, 329.
31. Morris, H.A., 1981. *Blue-Veined Cheeses*, Pfizer Cheese Monographs, Vol. 7. Pfizer Inc., New York.
32. O'Nulain, M., 1986. MSc thesis, National University of Ireland.
33. Ledford, R.A., O'Sullivan, A.C. & Nath, K.R., 1966. *J. Dairy Sci.*, **49**, 1098.
34. Visser, F.M.W. & de Groot-Mostert, A.E.A., 1977. *Neth. Milk Dairy J.*, **31**, 247.
35. Creamer, L.K., 1975. *J. Dairy Sci.*, **58**, 287.
36. Phelan, J.A., Guiney, J. & Fox, P.F., 1973. *J. Dairy Res.*, **40**, 105.
37. Godinho, M. & Fox, P.F., 1982. *Milchwissenschaft*, **37**, 72.
38. Hewedi, M. & Fox, P.F., 1984. *Milchwissenschaft*, **39**, 198.
39. Fox, P.F. & Walley, B.F., 1971., *J. Dairy Res.*, **38**, 165.
40. Gouda, A., 1987. *Egyptian J. Dairy Sci.*, **15**, 15.
41. Creamer, L.K., 1971. *N.Z. J. Dairy Sci. Technol.*, **6**, 91.
42. Al-Mzaien, K., 1985. PhD thesis, National University of Ireland.
43. Barford, R.A., Kumosinki, T.F., Parris, N. & White, A.E., 1988. *J. Chromat.*, **458**, 57.
44. Mulvihill, D.M. & Fox, P.F., 1978. *Ir. J. Food Sci. Technol.*, **2**, 135.
45. Mulvihill, D.M. & Fox, P.F., 1980. *Ir. J. Food Sci. Technol.*, **4**, 13.
46. Phelan, J.A., 1985. PhD thesis, National University of Ireland.
47. Humbert, G. & Alais, C., 1979. *J. Dairy Res.*, **46**, 559.
48. Fox, P.F., 1981. *Neth. Milk Dairy J.*, **35**, 233.
49. Visser, S., 1981. *Neth. Milk Dairy J.*, **35**, 65.
50. Reimerdes, E.H., 1982. In *Developments in Dairy Chemistry—1: Proteins,* ed. P.F. Fox. Elsevier Applied Science Publishers, London, p. 271.
51. Grufferty, M.B. & Fox, P.F., 1988. *J. Dairy Res.*, **55**, 609.
52. Grufferty, M.B. & Fox, P.F., 1988. *N.Z. J. Dairy Sci. Technol.*, **23**, 153.
53. Richardson, B.C. & Elston, P.D., 1984. *N.Z. J Dairy Sci. Technol.*, **19**, 63.
54. Richardson, B.C. & Pearce, K.N., 1981. *N.J. J. Dairy Sci. Technol.*, **16**, 209.
55. Lawrence, R.C., Gilles, J. & Creamer, L.K., 1983. *N.Z. J. Dairy Sci. Technol.*, **18**, 175.
56. Ollikainen, P. & Nyberg, K., 1988. *Milchwissenschaft*, **43**, 497.
57. Farkye, N.Y. & Fox, P.F., 1990. *J. Dairy Res.*, **57**, 413.
58. Guinee, T.P. & Fox, P.F., 1984. *Ir. J. Food Sci. Technol.*, **8**, 105.
59. Sweeney, K., 1984. MSc thesis, National University of Ireland.
60. Ollikainen, P. & Kivelä, T., 1989. *Milchwissenschaft*, **44**, 204.

60a. Matheson, A.R., 1981. *N.Z. J. Dairy Sci. Technol.*, **16**, 33.
61. Green, M.L. & Foster, P.M.D., 1974. *J. Dairy Res.*, **41**, 269.
62. Noomen, A., 1978. *Neth. Milk Dairy J.*, **32**, 26.
63. Kaminogawa, S. & Yamauchi, K., 1972. *Agric. Biol. Chem.*, **36**, 2351.
64. Kaminogawa, S., Yamauchi, K., Miyazawa, S. & Koga, Y., 1980. *J. Dairy Sci.*, **63**, 701.
65. Sullivan, J.J. & Jago, G.R., 1972. *Aust. J. Dairy Technol.*, **27**, 98.
66. Stadhouders, J. & Hup, G., 1975. *Neth. Milk Dairy J.*, **29**, 335.
67. Morris, H.A. & Jezeski, J.J., 1953. *J. Dairy Sci.*, **36**, 1285.
68. Madkor, S., 1985. PhD thesis, Mania University, Egypt.
69. Marcos, A., Alcala, M., Leon, F., Fernandez-Salguero, J. & Esteban, M.A., 1981. *J. Dairy Sci.*, **64**, 622.
70. Ruegg, M. & Blanc, B., 1977. *Milchwissenschaft*, **32**, 193.
71. Ruegg, M., 1985. In *Properties of Water in Foods*, ed. D. Simatos & J.L. Multon. Martinus Nijhoff Publishers, Dordrecht, pp. 603–25.
72. Fernandez-Salguero, J., Alcala, M., Marcos, A. & Esteban, M.A., 1986. *J. Dairy Res.*, **53**, 639.
73. Streit, K., Ruegg, M. & Blanc, B., 1979. *Milchwissenschaft*, **34**, 459.
74. Larsen, R.F. & Anon, M.C., 1989. *J. Food Sci.*, **54**, 917.
75. Larsen, R.F. & Anon, M.C., 1989. *J. Food Sci.*, **54**, 922.
76. Larsen, R.F. & Anon, M.C., 1990. *J. Food Sci.*, **55**, 708.
77. Lacroix, C. & Lachance, O., 1988. *Canadian Institute Food Sci. Technol. J.*, **21**, 501.
78. Acker, L., 1969. *Food Technol.*, **23**, 1257.
79. Rockland, L.B. & Nishi, S.K., 1980. *Food Technol*, **34**(4), 42.
80. Pagana, M.M. & Hardy, J., 1986. *Milchwissenschaft*, **41**, 210.
81. O'Connor, C.B., 1971. *Irish Agric. Creamery Rev.*, **24**(6), 5.
82. Gilles, J. & Lawrence, R.C., 1973. *N.Z. J. Dairy Sci. Technol.*, **8**, 148.
83. Fox, P.F., 1975. *Ir. J. Agric. Res.*, **14**, 33.
84. Pearce, K.N. & Gilles, J., 1979. *N.Z. J. Dairy Sci. Technol.*, **14**, 63.
85. Lelievre, J. & Gilles, J., 1982. *N.Z. J. Dairy Sci. Technol.*, **17**, 69.
86. Knox, J., 1978. *Dairy Ind. Intern.*, **43**(4), 31, 34.
87. Mills, O.E. & Thomas, T.D., 1980. *N.Z. J. Dairy Sci. Technol.*, **15**, 131.
88. Stadhouders, J., 1978. *Proc. 20th Intern. Dairy Congr. (Paris)*, 39ST.
89. Stadhouders, J., Hup, G., Exterkate, F.A. & Visser, S., 1983. *Neth. Milk Dairy J.*, **37**, 157.
90. Creamer, L.K., 1985. *Milchwissenschaft*, **40**, 589.
91. Stadhouders, J., 1962. *Proc. 16th Intern. Dairy Congr.*, **B353**.
92. Ney, K.H., 1971. *Lebensmittel u. Forsch.*, **147**, 64.
93. Guigoz, Y. & Solms, J., 1976. *Chemical Senses and Flavour*, **2**, 71.
94. Visser, S., Slangen, K.J., Hup, G. & Stadhouders, J., 1983. *Neth. Milk Dairy J.*, **37**, 181.
94a. Visser, S., Hup, G., Exterkate, F.A. & Stadhouders, J., 1983. *Neth. Milk Dairy J.*, **37**, 169.
95. Hill, R.D., Lahav, E. & Givol, D., 1974. *J. Dairy Res.*, **41**, 147.
96. Walstra, P. & Jenness, R., 1984. *Dairy Chemistry and Physics*. John Wiley, New York, Chapter 6.
97. Creamer, L.K. & Richardson, B.C., 1974, *N.Z. J. Dairy Sci. Technol.*, **9**, 9.
98. de Jong, L., 1976. *Neth. Milk Dairy J.*, **30**, 242.
99. Creamer, L.K. & Olson, N.F., 1982. *J. Food Sci.*, **47**, 631.
100. Luyten, H., ICODRL Experts Meeting: New Insights in Cheese Ripening, NIZO, June 19–20, 1990.
101. Luyten, H., van Vliet, T. & Walstra, P., 1987. *Neth. Milk Dairy J.*, **41**, 285.
102. Lawrence, R.C., Creamer, L.K. & Gilles, J., 1987. *J. Dairy Sci.*, **70**, 1748.
103. Noomen, A., 1983, *Neth. Milk Dairy J.*, **37**, 229.

104. Godinho, M. & Fox, P.F., 1981. *Milchwissenschaft*, **36**, 476.
105. Demott, B., 1985. *Cultured Dairy Products J.*, **20**(4), 6.
106. Petik, S., 1987. *Cultured Dairy Products J.*, **22**(1), 12.
107. United States Department of Agriculture, Agriculture Handbook, No. 8–1: *Composition of food and egg products, raw, processed*, prepared by Consumer and Food Economics Institute (L.P. Posati & M.L. Orr), 1976.
108. Kindstedt, P.S. & Kosikowski, F.V., 1988. *J. Dairy Sci.*, **71**, 285.
109. Wyatt, C.J., 1983. *J. Food Sci.*, **48**, 1300.
110. Lindsay, R.C., Karahadian, C. & Amudson, C.H., Low sodium cheese: an overview and properties of Cheddar cheese made with UF and RO retentate supplemented milk. *Proc. IDF Seminar, Atlanta, Ga, USA*, 8–9 October 1985, p. 55.
111. Fitzgerald, E. & Buckley, J., 1985. *J. Dairy Sci.*, **68**, 3127.
112. Demott, B.J., Hitchcock, J.P. & Davidson, P.M., 1986. *J. Food Protec.*, **49**(2), 117.
113. Lefier, D., Grappin, R., Grosclaude, G. & Curtat, G., 1987. *Lait*, **67**, 451.
114. Karahadian, C. & Lindsay, R., 1984. *J. Dairy Sci.*, **67**, 1892.
115. Kosikowski, F.V., 1985. Low sodium chloride ripened rennet cheese by ultrafiltration. US Patent No. 4515815.
116. Kindstedt, P.S. & Kosikowski, F.V., 1984. *J. Dairy Sci.*, **67** (Suppl. 1), 58.
117. Kindstedt, P.S. & Kosikowski, F.V., 1986. *J. Dairy Sci.*, **69** (Suppl. 2), 78.
118. Kindstedt, P.S. & Kosikowski, F.V., 1984. *J. Dairy Sci.*, **69** (Suppl. 1), 78.
119. Marsh, A.C., Klippstein, R.N. & Kaplan, S.D., 1980. *The Sodium Content of your Food*, USDA Home Garden Bull., No. 233.
120. Demott, B.J., Hitchcock, J.J. & Sanders, O.G., 1984. *J. Dairy Sci.*, **67**, 1539.
121. Martens, R., van den Poorten, R. & Naudts, M., 1976. *Revue de l'Agriculture*, **29**, 681. (Cited from *Dairy Sci. Abstr.*, 1977, **39**(1), 70.)
122. van den Berg, G., de Vries, A.E. & Stadhouders, J., 1986. *Voedingsmiddelentechnologie*, **19**(7), 37. (Cited from *Dairy Sci. Abstr.*, 1988, **50**(1), 210.)
123. Guinee, T.P., 1991. In *First Food Ingredients Symposium*, ed. M.K. Keogh. National Dairy Products Research Centre, Ireland, pp. 74–87.
124. McAuliffe, J.P. & O'Mullane, T.A., 1989. Preparation of cheese from natural ingredients. UK Patent application GB2237178A.
125. Geurts, T.J., Walstra, P. & Mulder, P., 1974. *Neth. Milk Dairy J.*, **28**, 102.
126. Guinee, T.P., 1985. PhD thesis, National University of Ireland.
127. Geurts, T.J., Walstra, P. & Mulder, P., 1974. *Neth. Milk Dairy J.*, **28**, 46.
128. Kimber, A.M., Brooker, B.E., Hobbs, D.G. & Prentice, J.H., 1974. *J. Dairy Res.*, **41**, 389.
129. Kalab, M., 1979. *Scanning Electron Microscopy*, **III**, 261.
130. Breene, W.M., Olson, N.F. & Price, W.V., 1965. *J. Dairy Sci.*, **48**, 621.
131. Sutherland, B.J., 1974. *Aust. J. Dairy Technol.*, **29**, 86.
132. Guinee, T.P. & Fox, P.F., 1985. *Food Chem.*, **19**, 49.
133. Chamba, J.F., 1988. *Lait*, **68**, 121.
134. Gilles, J., 1976. *N.Z. J. Dairy Sci. Technol.*, **11**, 219.
135. McDowall, F.H. & Whelan, L.A., 1933. *J. Dairy Res.*, **4**, 147.
136. Geurts, T.J., Walstra, P. & Mulder, H., 1972. *Neth. Milk Dairy J.*, **26**, 168.
137. Geurts, T.J., Walstra, P. & Mulder, H., 1980. *Neth. Milk Dairy J.*, **34**, 229.
138. Resmini, P., Volonterio, G., Annibaldi, S. & Ferri, G., 1974. *Scienza e Tecnica Lattiero-Casearia*, **25**, 149.
139. Guinee, T.P. & Fox, P.F., 1983. *J. Dairy Res.*, **50**, 511.
140. Guinee, T.P. & Fox, P.F., 1983. *Ir. J. Food Sci. Technol.*, **7**, 119.
141. Guinee, T.P. & Fox, P.F., 1986. *Ir. J. Food Sci. Technol.*, **10**, 73.
142. Byers, E.L. & Price, W.V., 1937. *J. Dairy Sci.*, **20**, 307.
143. Kelly, C.D. & Marqurdt, J.C., 1939. *J. Dairy Sci.*, **22**, 309.
144. Dolby, R.M., 1941. *N.Z. J. Dairy Sci. Technol.*, **22**, 289A.
145. O'Connor, C.B., 1970. *IFST Proc.*, **3**(3), 116.

146. O'Connor, C.B., 1973. *Irish Agric. Creamery Rev.*, **26**(11), 19.
147. O'Connor, C.B., 1968. *Dairy Ind. Intern.*, **33**, 625.
148. Fox, P.F., 1974. *Irish J. Agric. Res.*, **13**, 129.
149. Zahlus, A., 1986. European Patent application EP 0181771A1—Improvements in processing of curd.
150. Ryskowski, J., Kobiela, A., Rozumowicz, K., Klepacki, J. & Soltys, W., 1989. European Patent application EP 0333898A1—Method and device for salting cheese.
151. Cosentino, R., Bonapace, F. & Cocchi, S., 1987. European Patent application EP 0231984A2—Salting process and apparatus for electrostatic deposition of particulate salt on cheese.
152. Georgakis, S.A., 1973. *Milchwissenschaft*, **28**, 500.
153. Minarik, R., 1985. *Prumysl Potravin*, **36**, 475. (Cited from *Dairy Sci. Abstr.*, 1986, **48**(9), 587.)
154. Stětina, R., Brezina, P., Jelinek, J. & Nekovár, P., 1989. *Prumysl Potravin*, **40**, 260. (Cited from *Dairy Sci. Abstr.*, 1990, **52**, 230.)
155. Sutherland, B.J., 1977. *Aust. J. Dairy Technol.*, **32**, 17.
156. Morris, H.A., Guinee, T.P. & Fox, P.F., 1985. *J. Dairy Sci.*, **68**, 1851.
157. Whitehead, H.R., 1948. *J. Dairy Res.*, **15**, 387.
158. Marshall, R.J., 1982. *J. Dairy Res.*, **49**, 329.
159. Walstra, P., van Dijk, H.J.M. & Geurts, T.J., 1985. *Neth. Milk Dairy J.*, **39**, 209.
160. van Dijk, H.J.M. & Walstra, P., 1986. *Neth. Milk Dairy J.*, **40**, 3.
161. Guinee, T.P. & Fox, P.F., 1986. *Ir. J. Food Sci. Technol.*, **10**, 97.
162. Morris, T.A., 1961. *Aust. J. Dairy Technol.*, **16**, 31.
163. Lawrence, R.C. & Gilles, J., 1980. *N.Z. J. Dairy Sci. Technol.*, **15**, 1.
164. Lawrence, R.C., Heap, H.A. & Gilles, J., 1984. *J. Dairy Sci.*, **67**, 1632.
165. O'Connor, C.B., 1973. *Irish Agric. Creamery Rev.*, **26**(10), 5.
166. Brooker, B.E., 1979. In *Food Microscopy*, ed. J.G. Vaughan. Academic Press, London, p. 273.
167. Rammell, C.G., 1960. *J. Dairy Res.*, **27**, 341.
168. Kalab, M., Lowrie, R.J. & Nichols, D., 1982. *J. Dairy Sci.*, **65**, 1117.
169. Lowrie, R.J., Kalab, M. & Nichols, D., 1982. *J. Dairy Sci.*, **65**, 1122.
170. Emmons, D.B., Price, W.V. & Swanson, A.M., 1959. *J. Dairy Sci.*, **42**, 966.
171. Lawrence, A.J., 1959. *Aust. J. Dairy Technol.*, **14**, 166.
172. Lawrence, A.J., 1959. *Aust. J. Dairy Technol.*, **14**, 169.
173. Ayar, J. & Wallace, G.M., 1970. *Proc. 18th Intern. Dairy Congr. (Sydney)*, **1E**, 47.
174. Lelievre, J., 1977. *J. Dairy Res.*, **44**, 611.
175. Geurts, T.J., 1978. *Neth. Milk Dairy J.*, **32**, 112.
176. Bochtler, K., 1987. *Deutsche Milchwirtschaft.*, **38**, 1566.
177. van den Berg, G., Stadhouders, J., Smale, E.J.W.L., de Vries, E. & Hup, G., 1976. *Nordeuropaeisk Mejeri-Tidsskrift*, **43**, 363.
178. Schaegis, P., 1988. Procédé de régénération des saumures de fromagerie et installation pour sa mise en oeuvre. French Patent Application, FR2604339A1.
179. Blanchard, M., 1987. *Technique Laitière & Marketing* (1024), 42. (Cited from *Dairy Sci Abstr.*, **50**(4), 189.)
180. Brazhnikov, A.M., Shcherbina, B.V., Polishchuk, P.K., Karpichev, S.V. & Dunchenko, N.I., 1986. *Izvestiya Vysshikh Uchebnyk Zavedenii, Pishchevaya Tekhnologiya*, **2**, 46. (Cited from *Dairy Sci., Abstr.*, **50**(1), 23.)
181. Cohen-Maurel, E., 1987. *Revista Española de Lecheria*, **14**, 5–10. (Cited from *Dairy Sci. Abstr.*, **50**(5), 239.)
182. de Vries, A.E., 1979. *North European Dairy J. (Nordeuropaeisk Mejeri-Tidsskrift)*, **45**(7/8), 193.
183. Jakubowski, J., 1968. *Milchwissenschaft*, **23**, 282.
184. Fox, P.F., Lucey, J.A & Cogan, T.M., 1990. *Crit. Rev. Sci. Nutr.*, **29**, 237.

8

Cheese Rheology

J.H. Prentice

Formerly of the National Institute for Research in Dairying, Shinfield, Reading, RG7 9AT, UK*

K.R. Langley

Institute of Food Research (Reading Laboratory), Earley Gate, Whiteknights, Reading, RG6 2EF, UK

&

R.J. Marshall

University of North London, Holloway Road, London, N7 6DB, UK

1 INTRODUCTION

Rheology is formally defined as the study of the flow and deformation of matter. In everyday experience, not all cheeses appear to flow, though some of the softer ones, such as Brie or Camembert, obviously do. However, it will be shown later that under many conditions even the harder cheeses may be caused to flow. The second part of the definition, the study of deformation, is more immediately applicable to describing the properties of any cheese since deformation may embrace any aspect of the change of shape of a sample.

Before proceeding to any detailed discussion of the application of rheological methods to the examination of a cheese, it is pertinent to consider a few of the basic principles. As rheology is a branch of physics, these will be familiar to those who have been brought up in the ways of classical physics, but may appear less familiar to others. The layman would probably describe cheese as a solid, certainly the hard cheeses would be so classified and many of the softer ones partake rather more of the nature of a solid than a liquid. The characteristic by which a layman distinguishes a solid is its rigidity, that is, its ability to maintain indefinitely its particular shape. In fact, the precise physical property which describes a solid is known as its rigidity. Without defining it formally at this stage, rigidity is a measure of the relation between the effort which has to be applied to a sample and the deformation which results. It will be understood that if the material is a true solid in the strict physical sense, then this relation

* Present address: 3, Millbrook Dale, Axminster, Devon.

will be invariant for that material. Furthermore, once the source of the effort is withdrawn, the sample will recover its original shape spontaneously. The physicist will describe such a sample as an elastic solid.

By contrast, the characteristic by which a layman distinguishes a liquid is its fluidity, that is, its ability to flow into and take up the shape of any container which may hold it. With a certain perverseness, actually more apparent than real, the physicist uses the inverse concept for his characteristic property of a liquid, which is called the viscosity. This is, when defined in the same way as before, the relation between the effort applied to the sample and the rate at which flow ensues. Again, if the sample is a true liquid, the viscosity will be invariant for the material and upon cessation of the effort the *rate* of flow will spontaneously return to the status quo, i.e. flow will cease. The physicist describes this material as a viscous liquid. If we now replace the word 'flow' in the previous sentences by 'deformation', it can be seen that an elastic solid is characterized by the amount of effort required to produce a certain *extent* of deformation whilst the viscous liquid is characterized by the effort required to produce a given *rate* of deformation. The dimension of time does not enter into the measurement of the characteristic property of a solid whereas it is equally important with the spatial dimensions in the measurement of the characteristic physical properties of a liquid.

Solid and liquid may be regarded as the 'black' and 'white' of classical material physics. The rheologist is seldom concerned with these except as reference points but is concerned with the whole spread of that uncertain grey area between, where the material exhibits some of the characteristics of a solid and simultaneously some of the characteristics of a liquid.

It is now possible to define a third category of material. Any material which falls within the grey area, exhibiting both elastic and viscous properties, is known as a viscoelastic material. Cheese then fits neatly into this category, as indeed do almost all foodstuffs which are destined to be eaten.

2   DEFINITIONS

Before proceeding further to discuss the rheological properties of cheese or their measurement it is pertinent to give more precise definitions of the terms commonly used by rheologists. What has loosely been described in the preceding paragraphs as the 'effort', using a vernacular term, is more precisely known as the *stress*. This is denoted by the Greek letter sigma (σ). If the stress is applied, as an example, by means of a weight placed on the top of a sample, as is shown in Fig. 1a, the downward force, using SI units, is given by the weight in kg multiplied by the acceleration due to gravity (about 9.81 m/s^2). This is then in Newtons (N). If the force in Newtons is divided by the area of the surface to which it is applied, this gives the stress on the sample measured in pascals (Pa). The force could equally well be applied tangentially, as in Fig. 1b. The stress is still the force per unit area of the surface to which it is applied. If, as a result of the application of the stress as in Fig. 1a, the height of the sample, which may originally be denoted by h_0, is

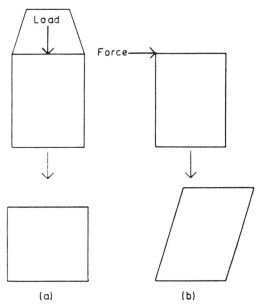

Fig. 1. Application of stress to a sample: (a) compressive strain, (b) shear.

changed by an amount δh, then the fractional change $\delta h/h_0$ defines the *strain*. This is denoted by the Greek letter ϵ. As this is the result of dividing a length by another length, it is a dimensionless number. Whilst the ratio $\delta h/h_0$ is the formal definition of the strain, it will be obvious that, except when δh is infinitesimal, the height of the sample changes as the deformation proceeds. It is arguable that for finite, and particularly for large values of strain, this change should be allowed for. This leads to a new definition of the strain as:

$$\epsilon = \ln(h_0/h_1) \tag{1}$$

where h_0 and h_1 are the original and final height of the sample, respectively. Again, this is a dimensionless number. Which definition of strain an author uses depends largely upon his own individual preferences. In the present work the second (logarithmic) form will be used and the fractional change in a situation such as in Fig. 1a will be described as the compression. In accordance with common usage, this will be given as a percentage to distinguish it from the strain given in decimal form. When the stress is applied tangentially, as in Fig. 1b, the strain which results is described as a *shear*. This is defined as the relative distance through which two parallel planes in the sample travel, divided by the distance separating them and is given the Greek symbol gamma (γ). As before, it is a length divided by a length and is a dimensionless number.

The physical property *rigidity* can now be defined more precisely. If the layman's terms, effort and deformation are replaced by the precise physical entities stress and shear, the *modulus of rigidity* is now defined as the stress divided by the shear and this is given the symbol, n:

$$n = \sigma/\tau. \tag{2}$$

The dimensions of rigidity are the same as those of the stress and in SI units rigidity will be measured in N/m^2. Although this is dimensionally the same as pascals, it is convenient to preserve a distinction and reserve the use of pascals for stresses only.

Returning to the compression of a sample as in Fig. 1a, there is another ratio to be calculated. This is the ratio of the stress to the strain and is termed the *modulus of elasticity* (sometimes known as Young's modulus), and is denoted by the capital letter *E*.

$$E = \sigma/\epsilon \tag{3}$$

It will be obvious that if the material composing the sample is incompressible, i.e. its volume is unchanged as a result of any external stresses acting upon it, as the sample is compressed in the vertical direction it will expand in the two other orthogonal directions. Part of the stress will be used in producing the vertical strain and part in producing horizontal strains. If the material is isotropic, i.e. if its properties are the same in any direction, it will follow that the stress is divided equally between the three orthogonal directions. The modulus of elasticity is therefore three times the modulus of rigidity.

If, on the other hand, the material is compressible, some of the stress will be expended in compressing the sample and the factor 3 falls to 2 in the extreme case in which no lateral expansion accompanies the vertical compression. The significance of this in the present context is that some research workers quote the results of their measurements on cheese in terms of a rigidity modulus and some in terms of an elasticity modulus. As a rough approximation, cheese may be taken as only slightly compressible, so that the factor relating the two is generally nearer to 3 than 2, and the use of the factor 3 may be sufficiently accurate to compare most workers' results, but an exact comparison is only possible if the compressibility is precisely known.

Turning now to liquid properties, the *rate of deformation* is given by the strain divided by the time taken to reach it, if this rate is constant. In the more general case, it is given by the time derivative of the strain, $(\dot{\gamma})$. This being a pure number divided by a time, it has the dimensions of the inverse of time (T^{-1}) and is usually measured in reciprocal seconds (s^{-1}). In the case of a true liquid it is the tangential forces which give rise to flow, whence the *shear rate* is the relevant rate of deformation. The viscosity, denoted by η, is defined as the ratio of the applied stress to the induced shear rate, $\eta = \sigma/\dot{\gamma}$, and this is measured in pascal-seconds in the SI system. In passing, it may be remarked that the logic of using viscosity instead of fluidity for the characteristic property of a liquid now becomes evident. Both rigidity and viscosity vary in the same sense; in each case a greater numerical value of the property denotes a greater stress being required to achieve the deformation.

The rheological properties of cheese may be studied in two very different ways. The quality of a cheese has long been assessed by graders by feel and by mouth.

Apart from those aspects associated with flavour, these graders' judgements are largely rheological in character. In the early days they were the only rheological tests known and much of the early literature attempted to describe rheological properties in terms such as those used by graders. There are severe semantic difficulties in understanding completely some of the earliest research. Graders, whose skill is largely a craft skill, often found it difficult to define precisely the terms they used and other research workers tended to give them their own interpretation, so that confusion arose when trying to distinguish between terms such as consistency and body, hardness and firmness, chewiness and meatiness. The difficulty was compounded on the international scene as these words do not translate precisely into other languages. Even American and English usage may differ. In more recent years, this subjective approach has been much more highly organized,[1-3] and has become a distinct branch of rheology under the name Texture Studies, the work 'texture' being used in this context with a rather different connotation from its OED definition. 'Texture' in this context is defined as the rheological properties as appraised by the senses and in the case of foodstuffs, largely by mouth.

Alongside texture studies, rheological properties of cheese may be measured instrumentally with one of two principal objectives in mind. The more obvious one for the practical cheesemaker is to seek physical measurements which may assist him by providing him with a means of quality control, or to assist the grader in carrying out his assessment, or even to displace him ultimately with a fully automated system that works independently of the human factor. The other aim is the more fundamental one of studying the structure of the cheese itself. It was the first of these two aims which motivated almost all the early work. It is a measure of the complexity of the problem that, 60 years later, solutions to the problem are still being sought. It may well be that in seeking parity between instrumental measurements and subjective judgements of rheological properties one is asking the wrong question and one to which there is no easy answer. The real question which should then be asked is: what can instrumental measurements tell us about the properties and in particular the structure of the material on which those properties depend and how can this information be useful to the manufacturers and the consumers? It is not the purpose of this chapter to answer the philosophical question. Nor is it possible to give a categorical answer to the alternative question. However, since rheological behaviour of any material is largely dependent upon its internal architecture, rheological measurements are particularly suited to investigating structures and much of the remainder of this chapter will be devoted to a discussion of the types of measurements which can be made and to some of the results which have so far been achieved.

3 EMPIRICAL MEASUREMENTS

The instrumentation of cheese rheology may be sub-divided into two more or less separate categories. Historically, the first of these is that of instruments of an *ad hoc* nature, which were designed to give some indication of firmness or springiness or similar qualities which the grader or the consumer attributes to the cheese. These were generally unpretentious and hence usually inexpensive and empirical in their mode of action. Because of this, measurements made by them cannot be directly analysed and expressed in terms of basic rheological parameters. Notwithstanding this limitation, they may have a useful place in the cheese scientist's repertory and much useful early work on the rheology of cheese was carried out using them.

The grader, in the course of his examination of a cheese, presses into the surface with the ball of his thumb or finger. One of the more successful earlier instruments attempted to simulate this thumb action by driving a hemisphere slowly into the surface of a cheese under the action of a load. This became known as the Ball Compressor.[4] It is possible, making a number of simplifying assumptions, to convert the reading of the depth of the indentation into a modulus[5,6] analogous to the modulus of rigidity, using the formula

$$G = 3M/[16(RD^3)^{1/2}] \tag{4}$$

where M is the applied load, R the radius of the indentation and D its depth. The test was slow but had the merit of being non-destructive and could be carried out on the whole cheese in the store. Using it, it has been possible to demonstrate the considerable variation in firmness over the surface of a whole Cheddar or Cheshire cheese, that the firmness differs between the upper and lower sides and that this is influenced by the frequency with which the cheese is turned in the store during the maturation period and the time which had elapsed since it was last turned.[7,8] The implication of this is that, whatever instrumental measurement is made, it is not possible to assign a single number to any property of the cheese nor is it possible to assess the properties of the whole cheese by means of a measurement made at a single point in that cheese.

Clearly, there is a need for some form of non-destructive testing and for many purposes there is no reason why it should not be empirical provided that the implications are understood. The Ball Compressor had the merit of cheapness and simplicity but the time taken to obtain a representative reading militated against its use outside the research laboratory. The problem of devising a suitable test is still unsolved. Unfortunately, the application of ultrasonic techniques, which has proved so useful in testing many engineering materials, has proved unrewarding with cheese.[9,10] This is because the dimensions of the cracks and other inhomogeneities in cheese are commensurate with or sometimes larger than the wavelength of the ultrasound and large-scale scattering takes place. It is difficult for a pulse to penetrate the body of the cheese and both velocity of propagation and attenuation are more influenced by the scattering than by the properties of the bulk.

One other empirical test deserves to be mentioned. This is the penetrometer.[11-13] It is not quite non-destructive but almost so, as it only requires a needle to be driven into the body of the cheese and no separate sampling is required. The penetrometer test may take several forms. As an example, a needle may be allowed to penetrate under the action of a fixed load,[11,12] or it may be forced into the cheese at a predetermined rate[13] and the force required measured. Whichever mode of operation is used, let the actual motion be considered. As the needle penetrates the cheese, that part of the cheese immediately ahead of it is ruptured and forced apart. If the needle is thin, the actual deformation normal to the axis is small, so that the force required to accomplish this may be neglected. On the other hand, the progress of the needle is retarded by the adhesion of its surface to the cheese through which it passes. This may be expected to increase with the progress of the penetration until a point is reached at which the restraining force matches the applied load and the penetration ceases. If a suitable diameter of needle and a suitable weight have been chosen, this test may be completed in a few seconds. This test is more useful for cheeses whose body is reasonably homogeneous on the macroscopic scale, such as some Dutch and Swiss cheeses. With cheeses such as Cheddar or Cheshire, the heterogeneities are generally too large and the penetration becomes irregular. The needle may pass through weaknesses in the structure, or even cracks, and so gives rise to the impression of a cheese less firm than it really is, or the point of the needle may attempt to follow a line of weakness, not necessarily vertical. As a result, there will be additional lateral forces acting on the needle and its penetration will be arrested prematurely.

The measurements made with a penetrometer cannot be converted to any well defined physical constant. Both the cohesive forces within the cheese and the adhesive forces between the cheese and the surface of the needle are a consequence of the forces binding the structure together. By inference, these are related to its viscoelastic properties, but there is no simple theory which attempts to establish these relations. It has been shown experimentally[14] that there is a statistically significant correlation between firmness as measured by the Ball Compressor and by penetration, but also that this differs between types of cheese.[15] It has also been shown experimentally[12] that a curvilinear relation exists between the resistance to penetration and an elastic modulus of some Swiss cheese, as calculated from a compression experiment. This is shown in Fig. 2. A curve such as this has only a limited usefulness since it only relates to one type of cheese, but it does give some idea of the magnitude of the forces involved. Except where such an experimental relation is available, penetrometer measurements can only be regarded as entirely empirical.

4 PHYSICAL MEASUREMENT—INTRODUCTORY

The alternative to making purely empirical measurements is to attempt to make objective measurements of recognized physical properties. It has already been remarked that the rheological properties of any material involve both deformation and flow. A complete description of the behaviour of any sample may be said to be given by the dependence of the strain on both the applied stress and the time, i.e.

$$\epsilon = f(\sigma, t) \tag{5}$$

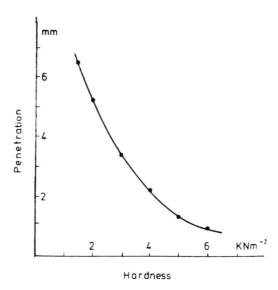

Fig. 2. Penetrometer readings compared with compression modulus (based on data in Ref. 12).

Graphically, this may be represented by a surface in a three-dimensional diagram. If, for any value of t, the strain is linearly related to the stress, the material is said to be linear viscoelastic. Many substances come within this category for small values of strain and stress; cheeses vary in this respect. For a linear viscoelastic substance, a two-dimensional graph of strain vs time gives a sufficient description; even when there is some departure from this linearity the graph for any fixed stress gives a useful insight into the behaviour.

The simplest rheological experiment which can be conceived[16] is to apply a stress to a sample of known dimensions and to observe the progress of the strain with time. When drawn as a graph this is known as the creep curve. This may be continued indefinitely and eventually some equilibrium will be established. Either the strain will become constant, which occurs when the sample is more akin to a solid, or a constant strain rate will be established if the sample is more akin to a liquid. Instead of allowing the experiment to continue indefinitely, the stress may be removed when a predetermined strain has been reached, or alternately after a predetermined time. Should the sample possess any elastic characteristic, some energy will have been stored up within it and the strain will decrease again. This part of the graphical representation is known as the recovery curve. The creep and recovery curves between them contain all the information about the rheological behaviour of the sample under that particular stress. The experiment could also be performed in a different way. Instead of applying a known fixed stress and observing the strain, the sample could be constrained to deform at a known rate and the stress required to maintain this rate of deformation could be measured. Again, the action may be stopped at any point. If, at this point, the strain is held constant the stress may be followed as it relaxes. Again, the complete cycle contains all the available information about the rheological properties of the sample.

The foregoing is admittedly somewhat of an over-simplification. It applies strictly only to those materials which are linear viscoelastic and only to those whose properties are not altered by the action of the strain itself. In the case of cheese, subjecting it to a strain, other than a very small one, certainly modifies its properties and a repetition of the simple rheological experiment will give a different response curve. Indeed, both the extent and the duration of the strain need to be taken into account. A complete rheological description of any sample will need, therefore, not only a creep and recovery curve or a compression and relaxation curve but also a statement of the history of the straining of the sample previous to commencing the experiment. Notwithstanding these caveats, a single rheological experiment on a sample of cheese will yield much useful information. It is pertinent to consider the shape of these response curves and how they may be analysed in terms of easily recognizable physical parameters.

5 MODELS

A convenient way sometimes used by rheologists to describe viscoelastic behaviour is by postulating simple models[17] which assume that the material possesses simultaneously both solid and fluid properties. It must be emphasized that these are purely mathematical models, not physical ones, and should not be interpreted as implying any particular structure within the material. Nevertheless, it will be seen later that in some favourable cases it is possible to relate these models to the actual structure of the cheese. The models assume that the structure is built up of solid and fluid elements. The solid element is represented by an ideal spring in which the deformation is always strictly proportional to the force applied to it and the fluid element by an ideal dashpot, the rate of deformation of which is strictly proportional to the applied force. A viscoelastic model is then made up by combining these elements. Starting with just these two units it is possible to combine them in two different ways to make two new models. If they are placed in tandem, as in Fig. 3a, the total displacement (i.e. the strain) is the sum of the displacement of the two units. The mathematical expression of this is:

$$\epsilon = \sigma/G + \sigma \cdot t/\eta \qquad (6)$$

The first term on the right-hand side expresses the elastic component of the strain and this is independent of the time during which the stress is applied, whilst the second term expresses the viscous component, which is proportional to the duration of the stress. If the stress is removed after some given time, t, the spring, no longer constrained, returns to its original length immediately, but that part of the displacement due to the dashpot remains. A complete creep and recovery curve for this type of behaviour is given in Fig. 4a. This model is

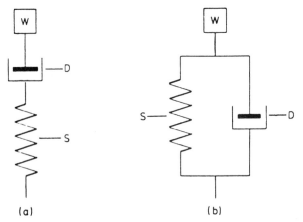

Fig. 3. Viscoelastic models: (a) Maxwell body, (b) Kelvin body.

commonly known as the Maxwell Body. Already it may be seen that the Maxwell body gives a better representation of the rheological behaviour of cheese than one gets by assuming that it is a simple solid or liquid.

If the two units are combined in the other possible way, i.e. in parallel, so that they each have the same compression but they share the stress between them as in Fig. 3b, the equation expressing this is:

$$\sigma = G \cdot \epsilon + \eta \epsilon \tag{7}$$

Solving this equation so that it may be compared with eqn (6) above gives:

$$\epsilon = [\sigma/\eta] [1 - \exp(-G \cdot t/\eta)] \tag{8}$$

This is the equation for the Kelvin body. The characteristic curve for the complete rheological experiment now becomes that shown in Fig. 4b. Once again, there may be some resemblance between this curve and the behaviour experienced with cheese. In this case, the reaction to the application of a load or its

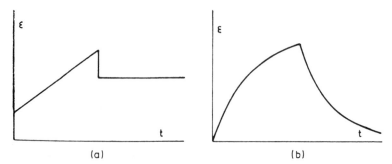

Fig. 4. Creep and recovery curve: (a) Maxwell body, (b) Kelvin body.

removal is not instantaneous, but takes place gradually and only eventually reaches an equilibrium situation.

More sophisticated models may be constructed using the Maxwell and Kelvin bodies themselves as building blocks and combining them in tandem or in parallel. The mathematical expressions for their responses are readily obtainable. This process may be repeated indefinitely, but the expressions become more cumbersome as the complexity increases. From the practical point of view, little purpose is served by making very complex models: the main use of a model is to give a simple representation, albeit not necessarily exact, of any particular pattern of behaviour. Furthermore, the smaller the number of parameters that are introduced the more likely it is to be useful and comprehensible. Moreover, there is a very practical consideration that, as the complexity of the model increases, so it rapidly becomes more difficult to analyse the experimental data. Indeed, the precision of the data may well be insufficient to allow more than a small number of parameters to be identified with any pretensions of accuracy. A simple model, probably with only two or three elements, inexact though it may be, is often the most generally useful in describing the behaviour of cheese.

Another model frequently of use is that of the plastic material. This is a material which behaves as an ordinary solid under the action of low stresses, i.e. it deforms and recovers elastically, but once a critical stress is exceeded, it flows as a fluid, the rate of flow being proportional to the excess of the stress over the critical stress which initiates the flow. This is known as the Bingham body and may be conceived as a Maxwell body, Fig. 3a, in which the dashpot element is locked until the critical stress is reached. This stress is known as the yield stress. For the Bingham body, the equation relating strain and stress becomes:

$$\epsilon = \sigma/n, \ (\sigma \leq \sigma_y); \ \epsilon = \sigma_y/n + (\sigma - \sigma_y) \cdot t/\eta, \ (\sigma > \sigma_y) \tag{9}$$

More commonly in food materials, the situation is not so clear cut. The yield does not occur suddenly at a definite stress but rather over quite a range of stresses and the subsequent flow is not necessarily Newtonian. In this case, where the ratio of the strain to the stress increases progressively with increasing stress, the material is known as pseudoplastic.

6 PRACTICAL INSTRUMENTATION

Before discussing more sophisticated quantitative tests, two popular semi-empirical tests may be mentioned. The first of these is in effect a cross between the Ball Compressor and the penetrometer. It is the Cone penetrometer, originally developed in the hydrocarbon industry for the evaluation of stiff greases. This depends on a different concept. It assumes that the sample material is plastic, i.e. it behaves as a solid when acted upon by small stresses, but once a critical stress is exceeded it flows as a viscous liquid. This critical stress is known as the yield stress. In action, a loaded cone with its apex downwards is held initially just in contact with the upper surface of the cheese and then released. Equilibrium is soon reached, normally within a few seconds. The stress, which is equal to the load divided by the area of contact, is infinite at the commencement, since only the apex touches the surface. As penetration proceeds the cross-sectional area increases in proportion to the square of the depth of penetration, whilst the cheese in the vicinity of the cone is caused to flow away from the point of contact in a lateral direction. Equilibrium is rapidly approached when the applied stress no longer exceeds the yield stress. The vertical stress may then be calculated from the penetration and the angle of the cone:

$$Y = Mg \cdot (\cot^2 \alpha/2)/\pi h^2 \tag{10}$$

where α is the apical angle of the cone, h the penetration and M the applied load. This stress is greater than the yield stress of the sample since it does not take into account the stresses involved in causing the sample to flow in a lateral direction. An estimate of the yield stress, σ_y, may be obtained by multiplying the stress Y by $\frac{1}{2} \sin \alpha$[18] so that

$$\sigma_y = Mg (\sin \alpha \cdot \cot^2 \alpha/2)/2\,\pi h^2 \tag{11}$$

The other test is the 'puncture' test,[19] in which a rod is driven into the sample and the resistance to its motion measured. There are at least three principal factors affecting this. One is the compression of the sample ahead of the rod, the second is the force required to cut through the sample at the leading edge of the rod and the third is the frictional resistance between the surface of the rod and the surrounding sample. If the sample is semi-infinite so that any compression in a direction perpendicular to the motion may be ignored, the first two factors will be constant once an equilibrium has been established whilst the third will increase linearly with the penetration. By using rods of different cross-sectional shape[19,20] such that the perimeter remains constant but the cross-sectional area differs, it is possible to separate the effects, and theory exists for this mode of operation.

More usually, the test has been used on small samples of cheeses contained within a rigid box.[21-24] In this manner of use, further forces come into play, of which the most important is the lateral force on the rod. At the same time the compressive forces ahead of the rod may no longer be considered constant as the distance between the leading face of the rod and the bottom of the sample diminishes. Some workers[21,25,26] have treated this as the principal force and have calculated a quasi-modulus by dividing the measured stress by the compressive strain. It will be evident that the quasi-modulus so calculated will be greater, by an arbitrary amount, than any true modulus which could be obtained from a simple unrestrained compression test. Nevertheless, it has proved to be a useful semi-empirical test.

In terms of practical instrument engineering, it is easier to apply a deformation and to observe the resulting stress in the sample than to apply a stress smoothly without any jerk at the moment of application. The currently most popular measurements use an instrument in which a sample cut from the bulk of the cheese is compressed between two parallel plates, one of which is fixed and the other driven at a constant rate. The total thrust resulting from this is measured or recorded. The plotted curve or the recorder trace now has, as one axis, time, which is proportional to the deformation, since the motion is linear and as the other, the total thrust. It may be used directly in this form when cheeses give a characteristic pattern in which certain prominent features may be identified. Alternatively, the curve may be redrawn, converting the instrumental parameters to true stress and strain.

7 FORCE-COMPRESSION TESTS—THEORY

In order to relate instrumental measurements to the models described above and hence to obtain elastic and viscous constants relating to the samples, it is necessary to consider the way in the which operation of an instrument differs from the basic rheological experiment proposed.[27] In a compression instrument it is usually the rate at which the plates approach each other which is constant and not the rate of strain. Also, it is the total force which is measured and not the stress, which is the force per unit area. Variations in cross-section were ignored in the simple theory. As a result, one cannot obtain true stress–strain curves directly from the instrument; the force–time curves distort the relation between stress and strain.

On the other hand, if cheese were considered to be viscoelastic it would not deform so simply,[28-30] but in such a way that the lateral movement near the ends is greater than near the middle, resulting in a concave shape, as in Fig. 5b. In this case the stresses at any instant are not uniform throughout the sample and any simple correction gives only an average value. When the sample is linear viscoelastic, this is relatively unimportant, but when there are serious departures from

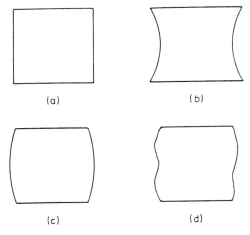

Fig. 5. Compression of a cylinder: (a) ideal, (b) concave distortion, (c) barrel distortion, (d) complex.

linearity an average tends to 'blur' the result by distorting the shape of the stress–strain curve. Further complications arise when there is friction between the plates and the sample.[28] The lateral movement of the sample layers near the plates is restricted and further internal stresses develop within the sample. The result is barrel-shaped distortion as in Fig. 5c. This may be sufficient to obscure any tendency to concavity; sometimes an even more complex shape develops as in Fig. 5d.

The effect of non-linearity in the strain rate is rather more serious. Writing as before

$$\epsilon = 1n(h_0/h_1) \tag{11}$$

and

$$h = h_0(1 - at) \tag{12}$$

where a is a constant defining the rate at which the plates approach gives

$$\dot{\epsilon} = ah_0/h = ae^\epsilon \tag{13}$$

The strain rate increases exponentially as the compression proceeds, becoming infinite at the point where the plates would come into contact were the experiment to be carried that far. As a result, the stress–strain curve becomes progressively distorted as the sample becomes more compressed. If ϵ and $\dot{\epsilon}$ are inserted into the constitutive equations for the basic viscoelastic models and the new equations solved the relations between stress and strain for the force–compression test are obtained.

Taking the Kelvin body first, the solution is straightforward:

$$\sigma = G\epsilon + a\eta\dot{\epsilon} \tag{14}$$

This is the equation of a curve with a finite intercept on the stress-axis proportional to the rate of deformation and to the viscous component and an initial slope proportional to the elastic term. It is concave upwards throughout.

The solution for the Maxwell body is less simple. Making the same substitutions leads to a solution in the form of an infinite series:

$$\sigma = G \exp\left(\frac{G}{a\eta} e^{-\epsilon}\right) \left[\epsilon + \sum_1^\infty \frac{\left(-\dfrac{G}{a\eta}\right)^i (1 - e^{-ei})}{ii!}\right] \tag{15}$$

At the commencement of the compression the stress is zero (there is no intercept) and the initial slope is proportional to the elastic component. At first the curve is convex upwards, but as the compression proceeds a point of inflexion is reached depending upon the ratio of the viscous to the elastic component and on the rate of compression. Ultimately, the curve becomes asymptotic to a straight line through the origin and with a slope given by the elastic constant.

If the rheological behaviour of the sample conforms to that of a simple viscoelastic model, an inspection of stress–strain curve obtained in a force–compression test will suffice to decide which model is appropriate. In general, it is unlikely that any simple two-element model would actually fit the behaviour of a material so heterogeneous as cheese. More sophisticated models may give a better representation but, as previously pointed out, the theory becomes more complex, the analysis becomes more difficult, and the greater number of constants required frustrates the aim of simplicity.

As an alternative hypothesis, consider what happens if the cheese is treated as if it has only a single modulus to describe the relation between stress and strain at any point on the compression curve. Taking first the observed stress. As the sample is compressed in a vertical direction its lateral dimensions change. In an ideal situation, if the material is incompressible and the plates are perfectly smooth so that there is no friction between them and the sample, i.e. no lateral forces act to restrain the free movement of the sample end surfaces, the cross-sectional area increases inversely as the height of the sample decreases and the sample deforms as shown diagrammatically in Fig. 5a. The true stress at any compression may be obtained directly by multiplying the force per original unit area by the fractional height of the sample. If the sample is compressible, its volume changes and the relation between the sample radius (a) and height (h) is given by:

$$r^2/r_o^2 = 1 + 2\mu(h_o/h)(1 - h/h_o) \tag{16}$$

where r_o, h_o are the original radius and height and μ is the Poisson's ratio. It is convenient to write λ for the compression ratio; then

$$r^2/r_o^2 = 1 + 2\mu(1 - \lambda)/\lambda \tag{17}$$

which reduces to

$$r^2/r_o^2 = 1/\lambda \tag{18}$$

when Poisson's ratio, μ, $= 1/2$, i.e. the material is incompressible.

If it were assumed that the sample deforms as an incompressible elastic solid, the modulus is defined as:

$$E = \sigma/\epsilon \tag{3}$$

whence the compressive force is

$$F = E \cdot \pi r_o^2 \, (1/\lambda)\ln(1/\lambda) \tag{19}$$

Hence, a plot of the measured force vs $(1/\lambda)\ln(1/\lambda)$ should yield the modulus, E, as the slope.[31]

At small deformations, $\ln(1/\lambda)$ may be replaced by $(1 - \lambda)$, whence:

$$F \approx E \cdot \pi r_o^2 \, (1 - \lambda)/\lambda \tag{20}$$

which reduces to the simple form[32]

$$F \approx A \cdot \delta h/(h_o - \delta h) \tag{21}$$

where A is the original cross-sectional area of the sample.

Another approach[33] draws on the similarity of the force-deformation curve given by eqn (19) to that of a pseudoplastic material. Many pseudoplastic materials approximately obey a power-law relationship. By analogy we may then write:

$$F \approx \pi r_o^2 \, K(h/h_o)^N \tag{22}$$

This equation has been derived empirically for both non-yielding materials, e.g. agar gel, apple and potato flesh and yielding materials, e.g. Cheddar cheese. The constant N is a measure of the deviation from linearity for the force-deformation plot. When $N > 1$ the curve is concave upward and when $N < 1$ it is convex upward. Values of $N > 1$ are typical of non-yielding materials and $N < 1$ is associated with yielding materials. The value of k is a measure of the 'stiffness'. At low strains, typically $<10\%$, eqns (19) and (22) give very similarly shaped curves when $N \approx 1\cdot05$. This value is higher than that obtained for cheese, which is typically in the range 0·6 to 1·0, suggesting that even under moderate loading cheese will yield, i.e. the structure becomes less stiff.

The deformation when yielding occurs may be very easily shown if a 'normalized' stress function is used. The 'normalized' stress function is defined[34] as:

$$F(\sigma) = 2 \, F/\pi r_o^2 \, (\lambda - \lambda^{-2}) \tag{23}$$

This can be plotted[34] against $1/\lambda$. Equation (23) has been derived by considering the energy stored during deformation. Application of mechanical stress causes the material to change shape, thus resulting in a change of stored energy. The strain energy is defined by the strain energy function, W,

$$W = C_1 \cdot (I_1 - 3) + C_2 \cdot (I_2 - 3) \tag{24}$$

where C_1 and C_2 are constants and I_1 and I_2 are the determinants of the material's dimensions, width, height and thickness.

Combining eqns (20) and (23) leads to

$$F(\sigma) = E \cdot /2(\lambda^2 + \lambda + 1) \tag{25}$$

for small strains. As λ tends to 1, i.e. no compression, the value of $F(\sigma)$ tends to the value $E/3$, so that the initial modulus can be determined from the intercept of this plot.

8   COMPRESSION TESTS—EXPERIMENTAL

Compression tests on cheese are usually carried on until the strain reached is far greater than that at which any simple viscoelastic theory might reasonably be expected to apply, often reaching a compression to as little as 20% of the original height. Such drastic treatment may result in destruction of any structure which may have been present in the original. A typical compression curve is shown in Fig. 6. At the outset it is sometimes difficult to decide whether there is an intercept on the stress axis or whether the stress rises smoothly and steeply from zero as the sample is strained. This is particularly so if the rate of response of the

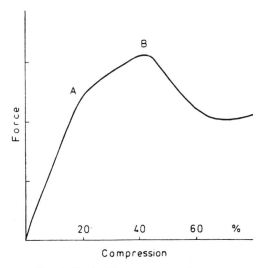

Fig. 6. Typical force-compression curve.

recorder is commensurate with the rate of build-up of the stress.[35] There is also another practical difficulty. It is not easy to prepare a sample of cheese in which the opposing end faces are exactly parallel. If there is a slight angle between the two faces, in the initial stages the take-up of strain on the sample as a whole may be gradual rather than instantaneous.

As the compression proceeds there is a smooth increase of stress with strain and this portion of the curve may be analysed along the lines suggested above. If the compression is stopped during this stage, the cheese would recover, completely or partially, and a repeat of the curve could be obtained having the same general shape, showing that the structure still remained intact, though perhaps somewhat modified. Eventually a point is reached, point A in Fig. 6, at which the slope becomes noticeably less as the structure begins to break down. This change of slope has been attributed to the onset of the development of cracks within the structure which then spread spontaneously.[12,36] The more homogeneous the cheese within the sample, the more likely this is to occur at all points within the structure at the same time and the change of slope will be clearly defined. If the cheese is very heterogeneous, the change in slope may be very diffuse. Once this point has been passed the cracks continue to spread at an increasing rate until point B is reached at which the breakdown of the structure overtakes the rate of build-up of stress through further compression and a peak in the curve is reached. This peak value, though it is obviously dependent on a nice balance between spontaneous disintegration of the structure and the rate of deformation applied by means of the instrument, is the most easily determined parameter, and has been used as a measure of firmness (or hardness).[12]

After this peak, there may be a fall in the stress as the breakdown becomes catastrophic until eventually the fragmented particles become reorganized in a

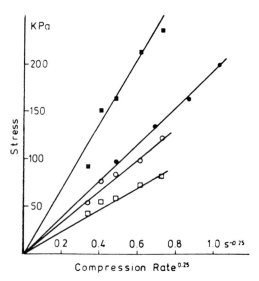

Fig. 7. Effect of rate of compression on stress at yield point: ● Double Gloucester, ■ Cheddar, ○ Leicester, □ Cheshire (based on data from Refs 36 and 37).

new, compacted arrangement and once again take up stress which then rises further. Using this explanation, the test is more or less empirical. The two parameters most commonly used are the peak (point B in Fig. 6) and the stress required to reach a given degree of compression, usually 80%, i.e. the stress required to reduce the height to 20% of the original. This is sometimes taken to be representative of the firmness as judged in the first bite during mastication. The stress at point B is often referred to as the 'yield value'. It should not be confused with the yield stress as determined, for instance, by a cone penetrometer, which is a constant associated with the model for a plastic material. The yield value at point B is not a material constant defining the breakdown of the sample: breakdown has already been occurring at least since the inflexion point A and possibly earlier. The peak value at B only indicates that at that point the rate of collapse of the stress-supporting structure overtakes the build-up of stress due to compression.

It is to be expected that the yield value obtained in a compression test is dependent upon the rate at which the compression is carried out. Measurements which have been carried out on four hard cheeses, Cheddar, Cheshire, Leicester[37] and Double Gloucester[36] all confirm this. The results are shown in Fig. 7 in which the yield stress has been plotted against the fourth root of a (eqn 11). Straight lines through the origin have been drawn through each set of points. Bearing in mind that this is a destructive test wherein each measurement has to be made on a different sample, the fit of the lines to the experimental points is acceptable. There is no theoretical significance in this fit: the result is quite empirical but there is a fortuitous practical benefit. It is possible, using this finding, to make plausible reductions to a common rate of compression of measurements

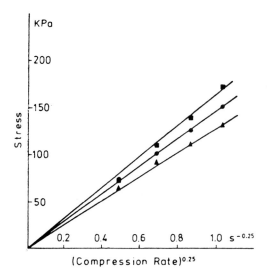

Fig. 8. Effect of rate of compression on stress at various compressions: ● 10%, ■ 40%, ▲ 70% (based on data from Ref. 36).

made at different rates of compression, so that results obtained by workers in different laboratories may be compared.

Pursuing this variation a little further, the measurements on Double Gloucester cheese[36] showed that this relation between stress and rate of deformation held good, not only at the peak but throughout the compression. The results for three compressions, 10%, 40% and 70%, are shown in Fig. 8. The fact that the relation between rate of deformation and stress is more or less constant before, at and beyond the yield point lends some support to the idea that there is no significant change in the processes taking place within the cheese. The strain at any point results from a balance between the rate of collapse of the structure, i.e. the spread of the cracks, and the build-up of stress due to compression and not to any pronounced viscoelastic flow followed by a sudden breakdown.

9 RELAXATION—THEORY

When the simple model is evidently insufficient to describe a relaxation or recovery curve, it may be difficult to carry out an analysis of the experimental curve. This is particularly the case if there is any doubt about the precision of the data. In this case, an empirical treatment may sometimes be useful.[38,39] If one writes Y for the decaying parameter, where

$$Y = (F_o - F_t)/F_o \tag{26}$$

F_t being the measured value of the stress or strain at time t, Y is the fractional recovery of that stress or strain. It has been found for many complex viscoelastic materials that the decay may be represented, to a fair degree of approximation, by an expression of the form:

$$1/Y = k_1/t + k_2 \tag{27}$$

The constants k_1 and k_2 can readily be found by rearranging this equation in the form:

$$t/Y = k_1 + k_2 \cdot t \tag{28}$$

$1/k_2$ is the extent to which the parameter ultimately decays. It is zero for an elastic or a Kelvin body and unity for a liquid or a Maxwell body. For more complex models it lies somewhere in between. The ratio, k_2/k_1, is a measure of the rate of decay.

10   RHEOLOGY AND STRUCTURE

In general terms, cheese may be considered as a composite material. The properties of the casein matrix are modified by the presence of fat particles (the filler), brine, small holes and cracks and the boundaries between curd granules. Treatment of the curd during manufacture may also add particular structure to the system. The rheological properties of the fat are added to those of the casein matrix so that the cheese as a whole is viscoelastic.

The casein is the main structural component and forms a network which may be divided by curd granule boundaries, fat particles, pockets of water (brine) and gas bubbles. Generally, the casein network extends in all directions, forming a cage, the rigidity of which depends upon the degree of openness, the amount of water bound to the casein and the presence of fat and free water. Water in cheese acts as a plasticizer,[40] so that more water will make the casein more plastic and vice versa. The amount of water so bound is quite small, most of the water existing as free water with dissolved salts. The structure of the casein network is laid down at the start of cheesemaking[41] and is only modified later if the curd undergoes special processing such as in the manufacture of Pasta Filata-type cheese where it is stretched in hot water. The initial conditions of cheesemaking determine the degree of aggregation of the casein, so that the formation of an open network such as when cheese is made from milk concentrated by ultrafiltration give an open cheese that has an atypical texture.[41] The basic structure of the casein network is modified during cheesemaking by the amount of acid produced by the starter organisms which alters the amount of Ca in the casein[42] and it is possible to classify cheeses according to their characteristic ranges of Ca to solids-not-fat. Thus, many of the differences between cheese varieties arise from the initial coagulation conditions and the amount of acid produced during cheesemaking.

As has been said, it is the casein that gives cheese its solidity. There is a minimum amount of casein below which any continuous network cannot exist because the casein chains form in the spaces between the fat globules. The exact amount of casein necessary will depend, therefore, not only on the size and size distribution of the casein micelles but also on the size and size distribution of the fat globules. Once that minimum has been exceeded, additional casein will only serve to strengthen the branches and the junctions. For a hard cheese, around

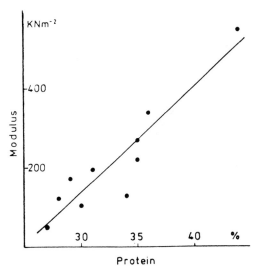

Fig. 9. Relation between firmness of cheeses and total protein content (based on data from Ref. 22).

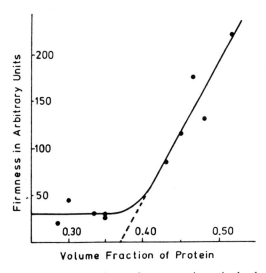

Fig. 10. Relation between firmness and protein content in a single cheese type (based on data from Ref. 43).

25% by weight must be protein in order to provide a rigid matrix throughout and as the quantity of casein is increased further there is an approximately linear relation between protein content and firmness. Figure 9 shows such a relation for ten different types of hard cheese. As an example of the effect of the protein content within a single cheese variety it is necessary to look at a soft cheese.[43] Figure 10 shows the firmness, measured on an arbitrary scale, of some Meshanger cheeses plotted against the amount of protein, expressed as the volume fraction of protein in fat-free cheese. It is clear that in this case unless sufficient protein is present to occupy almost 40% of the volume, excluding fat, there is little rigidity and the cheese is effectively a soft paste. Above this critical value the rigidity builds up rapidly. Although it has been claimed that the casein matrix gives rigidity to the cheese there is still a theoretical question to be answered. Is the matrix continuous, so that it may be treated as a solid body, or does cheese flow, albeit imperceptibly, however small the stress because of discontinuities in the structure? In theory, an examination of the stress–strain curves (page 316) at very low strains should reveal this information. If there is a continuous structure there will be a finite stress before any flow is initiated, though there may be some elastic deformation. Most of the force-compression curves which have been obtained on commercially available instruments have not been precise enough in this region, for the instruments were not designed for this purpose. By replacing the conventional pen recorder with a computer interface, thereby reducing the response time to about 1/1000 sec, it was possible to show that for one particular homogeneous cheese (Galbanino), the early part of the curve bore a very close resemblance to that for a Maxwell body[44] (Fig. 11) which would suggest that there is no continuous rigid structure. Some measurements have also been made on other cheeses, at rather higher than room temperatures and hence requiring rather lower overall stresses, in which the cheeses were compressed to various

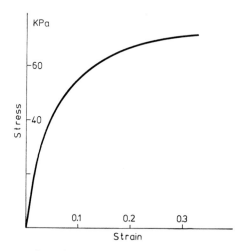

Fig. 11. Compression of Galbanino cheese (based on data from Ref. 44).

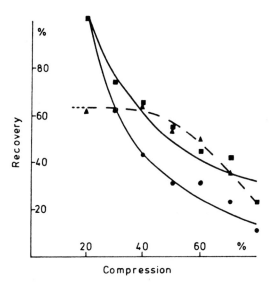

Fig. 12. Relation between recovery of cheese and previous compression: ● Cheddar, ■ Münster, ▲ Mozzarella (based on data from Ref. 45).

strains and then allowed to recover (Fig. 12). These showed that one Cheddar and one Münster cheese recovered completely, i.e. there had been no permanent flow, while a Mozzarella certainly did not recover. Cheddar cheese has also been studied by means of relaxation experiments.[39] Samples were compressed to a constant deformation, though under different stresses and the subsequent relaxation of stress followed. Recalling the empirical theory on page 322 (eqn 27)

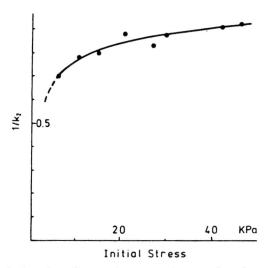

Fig. 13. Amount of relaxation after previous stress (see text; based on data from Ref. 39).

it was found that the value of $1/k_2$ was not independent of the stress applied. The greater the stress, and hence the faster the deformation to reach the fixed strain, the less the internal stresses developed within the cheese were able to relax during the compression and the structure was broken down to a greater extent. Figure 13 shows the relation between $1/k_2$ and stress was curvilinear, particularly at the low stress end of the experimental range. It is unfortunate that these experiments were not extended to much lower stresses so that it might have been possible to extrapolate the curve with greater confidence to zero stress. Nevertheless it appears likely that an extrapolation would lead to a finite value of $1/k_2$. If this were so it would indicate that there was no continuous structure throughout the cheese. To establish this with confidence requires measurements to be carried out at very small strains. These have been carried out using the force-compression test on cheese[46] and on cheese analogues[47] and it appears that breakdown of the structure does indeed occur at a very early stage, possibly below 1% compression in the analogues. Another way of carrying these out is to apply a small sinusoidal vibration to one surface of a sample and observe the stress transmitted through it. It is customary to analyse the results of such an experiment in the first instance as if it were a Kelvin body. A total modulus is calculated which is the resultant of the out-of-phase elastic and viscous components. Measurements made at a single frequency cannot serve to distinguish one viscoelastic model from another.[48] However, if the sample is truly a Kelvin body, the moduli determined are independent of frequency. Neither Cheddar nor Gouda cheese has constant moduli even when the strain is as low as $\epsilon = 0.04$. One must infer that this small strain is sufficient to cause internal cracks or slip planes to develop within the cheese.

Summarizing the foregoing paragraphs it appears that the rheological role of the casein in cheese is to provide a continuous elastic framework for the individual granules. Where casein chains lying on the surface of neighbouring granules are contiguous they may be bonded together either by physical bonds or by chemical bonds which develop during the ripening of the cheese, giving some rigidity to the agglomeration of granules. If these bonds exist, some at least must be very weak, since the whole cheese mass may be readily constrained to deform inelastically at quite low stresses. There is a clear need for still further definitive research into the relation between the casein structure and the rheological properties of the finished cheese, embracing microscopic examination, chemical analyses and the whole range of rheological techniques available.

The fat in cheese is normally held there by entrapment within the casein network and between the curd particles, the only interaction between it and the casein being by friction. In cheese made from homogenized milk the fat globules are smaller and of more uniform size. During homogenization the natural fat globule membrane is partly replaced by casein micelles[49] and this casein will probably bind to the casein network formed by the rennet action. Thus, the fat can interact strongly with the matrix. Cheese made from homogenized milk is smoother and has a finer texture than normal but is also firmer and more elastic because the fat globules contribute to the overall rheological properties.[50] The fat in normal cheese has a complex role different from that of most fillers in composite materials. If the fat is considered as a filler in a composite material, then the influence of that filler will depend upon: (1) the rigidity of the filler, with respect to the rigidity of the matrix, (2) the volume fraction of the filler and (3) the mechanical interaction between the filler and the matrix.[51] At low temperatures the fat globules will be mainly solid and add to the rigidity of the casein matrix. At intermediate temperatures, such as found in storage, the fat is plastic and adds to the rheological properties in a complex way. At the initial stages of deformation, the fat deforms with the casein matrix and energy is required for this but as the deformation increases the fat flows more easily and tends to lubricate the deformation. If sufficient energy is applied the cheese will fracture and the fat will flow and spread out over the fracture surfaces.[47] With Gouda cheese at 20°C, the compression modulus is independent of the fat content[51] indicating that the modulus of the casein matrix and the fat are the same. At higher temperatures (above about 20°C), the modulus of the fat globules depends on their surface tension and size.[51] Most of the fat is liquid and an increase in fat content will decrease the strength of the cheese.[51] If the cheese is strained sufficiently, fracturing will commence and the fat forms weaknesses through which fractures propagate.[52] Recent evidence from model cheese systems[47] and from experiments on Gouda type cheese (Marshall, unpublished data) shows that fractures start after less than 2% deformation in compression tests and are highly dependent upon the fat content of the cheeses. Once the fractures have started, the fat, which is mostly liquid, spreads out over the fracture surfaces and lubricates them so reducing the energy required for further fracture.[47] During eating, the fat coats not only the surface of the cheese but also parts of the mouth having a major influence on mouth feel and flavour perception. In addition to the casein matrix and the included fat, there is the interstitial brine that will also act as a lubricant between the curd particles as the cheese is deformed. This brine tends to be squeezed out from between the particles in a manner similar to that of the fat. During eating, once outside the cheese, the brine mixes with saliva to give 'moistness' to the cheese and carries flavours to the taste buds.

The cheese consists of more than just casein, fat and brine. During manufacture the curd is cut into pieces several times, then pressed together as in cheddaring and finally is probably milled before being pressed in hoops. The curd particles become more or less distorted, depending on the exact procedure used so that this coarser level of structure imposes a further set of rheological properties on those of the casein, fat and brine. Therefore, there is the possibility of a whole range of sizes of weaknesses and flaws which will affect the behaviour of the cheese. Finally, the casein matrix is modified by proteolysis that occurs during ripening. The residual rennet in the cheese starts the protein breakdown and enzymes from the starter and from moulds continue it during ripening. The rheological effects of the components of cheeses are often overshadowed or added to by the effects of the gross structure of the cheese itself. In Cheddar and Cheshire cheeses, the boundaries between curd grains form weaknesses along and around which the structure fails (Fig. 14).[52] With processed and similar cheese, there are

Fig. 14. Scanning electron micrograph of Cheddar cheese compressed to failure. Major fractures (a) occur around curd grains (b).

Fig. 15. Scanning electron micrograph of processed cheese compressed to failure. Major fractures occur at weaknesses such as folds (a) or across the matrix (b).

Fig. 16. Scanning electron micrograph of Mozzarella cheese compressed to failure. Major fractures (a) occur between the fibres of the cheese (b).

no such boundaries so that the structural failure will start at other flaws such as bubbles or fat globules (Fig. 15).[47,51] In Mozzarella, the stretching process during manufacturing produces a cheese with fibrous structure (Fig. 16). When cylindrical samples of this type of cheese are tested by uniaxial compression, the values of Young's modulus and maximum stress are virtually the same whether it is compressed along or across the direction of the internal fibres.[53] This, of course, makes sense in view of what is happening during compression. With cylinders of cheese compressed uniaxially the stresses are distributed radially throughout the sample so that stresses will be both along and across the direction of the internal fibres. If such a cheese were subjected to tensile stress, a difference in strength depending upon orientation would be expected. Such differences have not been shown for cheese, probably because of the difficulties of setting up tensile testing. They have, however, been found with meat tested with different orientations with respect to the muscle fibres.[54]

Within an individual cheese there may be considerable variation in rheological properties depending upon whether the cheese has a rind, how it is turned during ripening, the moisture content and distribution and the proteolytic activity within the cheese. Cheeses with a rind may lose water rapidly from their surface so that the composition of the core of the cheese will be quite different from the outer regions. Turning of cheeses subjects the top and bottom regions to alternating high and low compressive stresses under their own weight, while the central core is subject to much less variation. This gives, in an otherwise uniform cheese, a distribution of firmness that varies with distance from the surface (Fig. 17). The variation between cheeses within a single batch from a single variety is quite great and depends partly upon the method of measurement. In compression tests, variations of the order of 10–20% are often found[55] but with penetrometer methods the variation may be as much as 25–30% because the penetrometer only acts on a limited quantity of the sample. Variations between cheese batches may be considerably larger but rheological measurements do not always distinguish the differences as well as sensory assessment.[55]

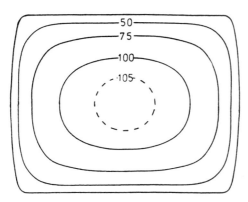

Fig. 17. Typical distribution of firmness throughout a mature cheese (based on data from Ref. 56).

11 SOME EXPERIMENTAL RESULTS

Most of the rheological methods described above have been used at one time or another for the study of cheese. Although the most satisfactory of these for routine measurements is the force-compression test, there has been no consensus of opinion regarding the most suitable operating conditions. Sample size, rate of compression and temperature have all varied. Yet, perhaps the most interesting feature is that, in spite of the differences and the different varieties of cheese which have been examined, the general shapes of the curves are so similar. This underlines its general usefulness.

There are several important considerations to be taken into account when deciding upon the operating conditions. The sample should be as large as possible[57] to be representative and to avoid local inhomogeneities. On the other hand, it should be small enough so that it can be seen that it does not include hidden

TABLE I
Rheological Measurements on Cheese by Force-Compression Tests

Variety	Initial Slope (kN/m^2)	Yield Value (kPa)	Yield Strain	Stress at 80% (kPa)	Ref.
Appenzell	22	7	0·63	6·1	58
Caerphilly	7·5	64	0·17	46	36
Cheddar	—	—	—	12	59
	—	—	—	14	59
	48	8	0·2	11	43
	—	44	0·21	—	37
	180	23	0·21	22	60
	—	108	—	—	37
Cheshire	—	44	0·33	—	37
Double Gloucester	1000	94	0·24	54	36
Edam	*c.* 500	146	0·63	71	36
Emmental	—	—	—	19	59
	18	12	1·05	9	58
Galbanino	900	—	—	—	44
Gouda	405	69	0·37	69	61
	390	68	0·72	36	36
Gruyère	77	15	0·51	9	58
Lancashire	*c.* 1250	87	0·2	95	36
Leicester	—	50	—	—	37
	—	48	0·3	37	28
Montasio	470	125	0·58	—	62
Mozzarella	15	2·5	0·55	10	45
Münster	6	2·8	0·1	10	45
Parmesan	1980	112	0·14	—	62
Pecorino Romano	2840	187	0·25	—	62
Provolone	240	57	0·53	—	62
Sbrinz	195	22	0·41	16	58
Tilsit	30	7	0·78	—	58

cracks. As a compromise, most workers have used samples with dimensions between 10 and 25 mm. Cylindrical samples have usually been preferred[56] because their symmetry tends to minimize irregular crack development. On the other hand, it is easier to prepare rectangular samples having exactly known dimensions.

A wide range of compression rates have been used, from 2 mm/min up to 100 mm/min; the strain rate, of course, depends upon both this and the sample height. The instrument produces curves of total force against linear travel. As already indicated on page 316, the conversion is readily effected if it is assumed that the volume remains constant; the cross-sectional area then varies inversely to the height. This is not a justifiable assumption if severe distortion of the cylindrical shape occurs, but it must be accepted in the absence of a better one.

Table I summarizes the principal measurements which have been made on hard cheese using the force-compression test. In order to produce this table, original data from the references cited were used to draw new stress–strain curves and the appropriate values were read from these curves. To make it possible to compare various results these values were then 'corrected' for the rate of compression and temperature, so that the table refers to an initial strain rate of 0·05 and a temperature of 20°C. Recalling the treatment on page 316, all the curves showed a tendency to be convex-upward at first and showed no intercept on the stress-axis. This is consistent with the use of a Maxwell body-type of model as a first approximation of cheese behaviour. In this (eqn 14), the initial slope is a measure of the elastic modulus.

The most striking feature of the results tabulated is the wide variation which occurs within one single variety of cheese (Cheddar). The origin and history of the different samples was not generally documented; it is not possible therefore to draw any conclusions as to the origin of this variation.

Table II summarizes a number of results which have been obtained by other methods, where it has been possible to calculate a modulus from the available data. As far as a comparison of the figures in the two tables is possible, it may be recalled that the initial slope of the force-compression curve is determined by the modulus of the elastic element, so that this must be most nearly comparable with any quasi-modulus determined by the punch test or by simple compression. The yield stress measured by a cone penetrometer is more likely to be comparable with the yield value in the force-compression test since both are influenced by the breakdown of structure. As before, the lack of adequate documentation makes if difficult to make any meaningful comparisons between the results obtained in different laboratories at different times.

One aspect of the rheological properties of cheese which has been fairly extensively researched is the development of firmness during ripening. As has been mentioned earlier, during this process all three principal constituents undergo change. Moisture evaporates from the surface of the cheese so that during the course of time a moisture gradient is set up and even the centre is considerably drier than it was originally. The protein in its matrix, as the available water is reduced and as residual bacteria and enzymes continue to act, undergoes a progressive change.[60,63-65] Some of the glycerides in the fat slowly crystallize, result-

TABLE II
Rheological Measurements on Cheese by Various Methods

Variety	Modulus (kN/m²)	Method	Ref.
Brie	1.33	Extruder	66
Sbrinz	41	Extruder	66
Chanakh	58	Cone	67
Emmental	40	Cone	67
	3·5	Cone	68
Gouda	44	Cone	67
Kostroma	54	Cone	67
Lori	40	Cone	67
Cheddar	195	Punch	22
Edam	340	Punch	22
Emmental	220	Punch	22
Gouda	270	Punch	22
Kachkaval	120	Punch	65
Mozzarella	170	Punch	22
	22	Punch	69
Münster	120	Punch	22
Parmesan	550	Punch	22
Provolone	130	Punch	22
Cheddar	270–400	Compression	70
Mozzarella	80	Compression	71
Kachkaval	60–100	Ball Compressor	72

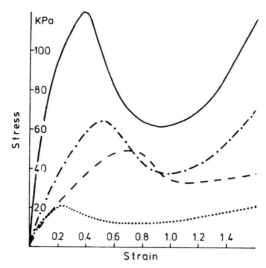

Fig. 18. Force-compression curves for young and mature cheeses: —— Gouda (mature), — ∎ — Gouda (young), ∎∎∎∎ Cheddar (mature), – – – Cheddar (young) (adapted from Refs 61 and 63).

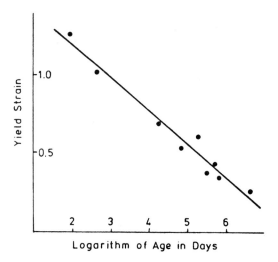

Fig. 19. Relation between age and strain at yield point for Cheddar cheese (based on data from Ref. 63).

ing in a more solid mass of fat. These changes are reflected in the change which takes place in the cheese.

The force-compression curves for mature cheese may be considerably different from those for green cheese, as may be seen from Fig. 18, which shows some curves for Cheddar[63] and Gouda[61] cheeses. During ageing the strain which the cheese could sustain before breaking down (at the yield point) progressively

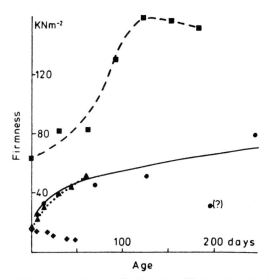

Fig. 20. Variation of firmness with age: ● Cheddar, ■ Kachkaval, ▲ Russian, ◆ Unspecified French cheese (based on data from Refs 63, 65, 73, 74).

decreased more or less exponentially. Figure 19 shows this for a batch of Cheddar cheese. On the other hand, whilst the stress at this point decreased somewhat in those Cheddar cheeses, in the Gouda cheese the ultimate strength of the matrix increased. As the actual balance between the different mechanisms varies from one cheese type to another, so will the paths of the change in firmness vary. Figure 20 shows some of the results which have been reported.[65,70,73,74]

In addition to changes in firmness, changes during ripening have also been reported to take place in springiness. Degradation of the protein, reduction of the free water and the firming up of the fat all tend to reduce the springiness. This does not show directly in single force-compression curves but it has been consistently observed empirically[41,65,74–76] on a number of varieties of cheese.

12   SUBJECTIVE JUDGEMENTS

Finally, it remains to consider briefly how the rheological properties measured by means of instruments,[77] as just described, relate to the grader's or the consumer's assessment of the cheese. Throughout this chapter the emphasis has been on the measurement of firmness of cheese which arises from its structure. It should be stressed that for any comparisons to be made between subjective and instrumental methods it is just as important that the subjective terminology is unambiguously defined as it is that the instrumental measurements are precise. Early experiments on single varieties of cheese showed that the subjective assessment of firmness, which is a fairly simple concept, correlated very highly with measurements made on simple instruments such as the Ball Compressor.[15,68] The simple correlation by itself does not, however, indicate to what extent the instrumental measurements could be used to predict a typical user's appraisal of firmness. Nor can the result on a single variety be extrapolated to all types of cheese without reservation. Nevertheless, the success was encouraging.

Some authors have, on the basis of their own experience, arbitrarily assigned a specific significance in terms of user appraisal to a particular instrumental reading. For instance, a quasi-modulus obtained in a simple compression test,[25] a yield value obtained by a cone penetrometer[78] and the stress at 80% compression[59] in a force-compression test have all been proposed as measurements of firmness. Recovery after a limited compression has been used as an indication of springiness.[22,25,59,74,79] These intuitive opinions were probably quite adequate for the purpose of making broad comparisons within a particular course of investigation, but need to be read with caution since they lack generality.

The force-compression test provides a number of possibly useful parameters and two of these merit a little consideration. Figure 21 shows an example in which the assessment of the hardness of a number of different cheeses has been plotted against the stress at 80% compression,[59] and a logarithmic regression line has been drawn through the points. The standard deviation about the regression line is less than the scatter of any of the individual subjective assessments arising from differences among the panel. It is clear that within this laboratory this

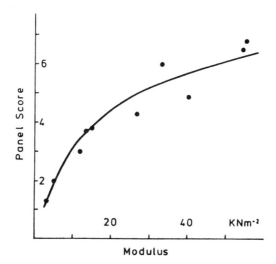

Fig. 21. Comparison of subjective and objective measurements of firmness (based on data from Ref. 59).

instrumental parameter could be used to predict that panel's assessment. In the same series of experiments the instrumental measurement was carried further: the compression was reversed and the recovered height after the first compression observed. At the same time the springiness was assessed subjectively. This is a less well defined concept and its transpired that there was much poorer agreement between panel members. Nevertheless, a highly significant correlation was obtained between the measured recovery and the springiness.

These examples show that rheological measurements can successfully predict users' assessments. What remains is to establish a consensus on instrumental practice and on the definitions of the subjective terms. International cooperation is already fostering the first. The semantic problem is not too difficult.[1-3]

13 SOME PRACTICAL RECOMMENDATIONS FOR RHEOLOGICAL TESTING

All cheeses are viscoelastic and therefore their rheological behaviour will depend upon the rate and the extent of strain.[62] As the cheese is deformed, stress relaxation occurs through rearrangement of the components and the breakage and reformation of bonds within the matrix. As the rate of deformation increases, fewer bonds are reformed and eventually catastrophic failure of the structure may occur.[62] When measuring the rheological properties of cheese it is essential to bear in mind the purpose of the measurement; this will dictate the conditions under which testing should be carried out. If the purpose is to determine the basic material properties of the cheese to understand how the components interact or to measure properties relevant to storage and handling where the rate of deformation is quite slow, then the test conditions should reflect this. For example, deformation should be only a few percent and the rate should be suitably slow to avoid fracture within the samples. On the other hand, if the rheological properties are to be compared with sensory properties, different conditions must be used. The closing speed of the jaws in mastication is about 15–30 mm/sec[80] and solid foods are chewed so that multiple fracturing occurs. Therefore, the rheological test conditions should be equivalent, say 1000 mm/min deformation rate and deformation carried on beyond the point of failure if possible. However, it is important to bear in mind that many instruments used in rheological testing have recorders that are not capable of responding fast enough to record the complete curve. It is better to attach a fast response recorder[47] in place of the one normally used if deformation rates greater than about 200 mm/min are envisaged or to collect the data via a computerized system that can cope with the high rate. When making comparisons between rheological measurements and sensory assessment of texture, even if the conditions are as close as practicable, the relationships may not be very good because of the relative simplicity of the rheological test compared with the process of mastication. Therefore, it may only be possible to make sensible comparisons between fracture properties and certain stages in mastication such as the first bite.

The aspect ratio of the sample under test is very important. With cylindrical samples the length/diameter ratio should be >1 to avoid excessive effects due to friction between the sample and the compression plates.[62] As the aspect ratio increases, the observed rheological properties tend to become independent of sample size.[62]

Also important is care in the preparation of samples. As has been shown above (page 326), some cheese samples appear to begin to yield under fracture at very low strains. It is not unlikely that the act of extraction of core samples from the cheese itself may introduce damage into a visually 'fracture free' material. Unless extreme care has been exercised it is not possible to assess whether the observed yield is real or an artefact.

REFERENCES

1. Szczesniak, A.S., 1963. *J. Food Sci.*, **28**, 385.
2. Szczesniak, A.S. & Bourne, M.C., 1969. *J. Texture Studies*, **1**, 52.
3. Jowitt, R., 1974. *J. Texture Studies*, **5**, 351.
4. Caffyn, J.E. & Baron, M., 1947. *The Dairyman*, **64**, 345.
5. Mohsenin, N.N., 1970. *Physical Properties of Plant and Animal Materials.* Gordon and Breach, New York, p. 288.
6. Mohsenin, N.N., Morrow, C.T. & Young, Y.M. 1970. *Proc. 5th Int. Congr. on Rheology*, **2**, 647.
7. Wearmouth, W.G., 1953. *Dairy Ind.*, **20**, 726.
8. Cox, C.P. & Baron, M., 1955. *J. Dairy Res.*, **22**, 386.
9. Konoplev, A.D., Krashenin, P.F. & Tabachnikov, V.P., 1974. *Trudy vses. nauchno-issled. Inst. masl. i syr. Prom.*, **17**, 40, 101, 106.

10. Poulard, S., Roucou, J., Durrange, G. & Manry, J., 1974. *Rheol. Acta*, **13**, 761.
11. Baron, M., 1949. *Dairy Ind.*, **14**, 146.
12. Eberhard, P. & Flückiger, E., 1978. *Schw. Milch Ztg.*, **104**, 24.
13. Tabachnikov, V.P., Borkov, V. Ya., Ilyushkin, V.B. & Tetereva, I.I., 1979. *Trudy vses. nauchno-issled. Inst. masl i syr. Prom.*, **27**, 61, 118.
14. Baron, M. & Harper, R., 1951. *Dairy Ind.*, **16**, 45.
15. Wearmouth, W.G., 1954. *Dairy Ind.*, **19**, 213.
16. Prentice, J.H., 1984. *Measurements in the Rheology of Foodstuffs*. Elsevier Applied Science Publishers, London, p. 9.
17. Prentice, J.H., 1984. *Measurements in the Rheology of Foodstuffs*. Elsevier Applied Science Publishers, London, p. 12.
18. Mottram, F.J. 1961. *Lab. Practice*, **10**, 767.
19. Bourne, M.C., 1979. In *Food Texture and Rheology*, ed. P. Sherman. Academic Press, London, p. 95.
20. deMan J.M., 1969. *J. Texture Studies*, **1**, 114.
21. Kamel, B.S. & deMan, J.M., 1975. *Lebensm. Wiss. u. Technol.*, **8**, 123.
22. Chen, A.H., Larkin, J.W., Clark, C.J. & Irvine, W.E., 1979. *J. Dairy Sci.*, **62**, 901.
23. Tabachnikov, V.P., 1974. *Proc. 19th Int. Dairy Congr.*, **1E**, 511.
24. Tabachnikov, V.P., 1974. *Trudy vses. nauchno-issled. Inst. masl. i syr. Prom.*, **17**, 84, 104, 109.
25. Davidov, R. & Barabanshchikov, N., 1950. *Moloch. Prom.*, **9**(4), 27.
26. Ramanauskas, R., Urbene, S., Galginaitye, L. & Matulis, P., 1979. *Trudy Lit. Fil. vses. nauchno-issled. Inst. masl. i syr. Prom.*, **13**, 64.
27. Sheth, B.B., 1976. *J. Texture Studies*, **7**, 157.
28. Carter, E.J.V. & Sherman, P., 1978. *J. Texture Studies*, **9**, 311.
29. Boyd, J. & Sherman, P., 1975. *J. Texture Studies*, **6**, 507.
30. Hammerle, J.R. & McClure, W.F., 1971. *J. Texture Studies*, **2**, 31.
31. Langley, K.R. & Green, M.L., 1989. *J. Dairy Res.*, **56**, 275.
32. Swyngedau, S., Nussinovitch, A., Roy, I., Peleg., M. & Huang, V., 1991. *J. Texture Studies*, **56**, 756.
33. Roy, I. & Peleg, M., 1989. *J. Texture Studies*, **19**, 453.
34. Bagley, E.B., 1987. In *Physical Properties of Foods—2*, ed. R. Jowitt, F. Escher, M. Kent, B. McKenna & M. Roques. Elsevier Applied Science Publishers, London, p. 345.
35. Voisey, P.W. & Kloek, M., 1975. *J. Texture Studies*, **6**, 489.
36. Shama, F. & Sherman, P., 1973. *J. Texture Studies*, **4**, 344.
37. Dickinson, E. & Goulding, I.C., 1980. *J. Texture Studies*, **11**, 51.
38. Peleg, M., 1979. *J. Food Sci.*, **44**, 277.
39. Peleg, M., 1980. *J. Rheol.*, **24**, 451.
40. Taneya, S., Izutsu, T. & Sone, T., 1979. In *Food Texture and Rheology*, ed. P. Sherman. Academic Press, London, p. 369.
41. Green, M.L., Turvey, A. & Hobbs, D.G., 1981. *J. Dairy Res.*, **48**, 343.
42. Lawrence, R.C., Creamer, L.K. & Gilles, J., 1987. *J. Dairy Sci.*, **70**, 1748.
43. De Jong, L., 1978. *Neth. Milk Dairy J.*, **32**, 1.
44. Masi, P., 1989. *J. Texture Studies*, **19**, 373.
45. Imoto, E.M., Lee, C.-H. & Rha, C.K., 1979. *J. Food Sci.*, **44**, 343.
46. de Mariscal, P., 1987. MSc thesis, Reading University, England.
47. Marshall, R., 1990. *J. Sci. Food Agric.*, **50**, 237.
48. Prentice, J.H., 1984. *Measurements in the Rheology of Foodstuffs*. Elsevier Applied Science Publishers, London, p. 177.
49. Mulder, H. & Walstra, P., 1974. *The Milk Fat Globule. Emulsion Science as applied to Milk Products and Comparable Foods*. Commonwealth Agricultural Bureaux, Farnham Royal.
50. Emmons, D.B., Kalab, M., Larmond, E. & Lowrie, R.J., 1980. *J. Texture Studies*, **11**, 15.

51. Luyten, H., 1988. PhD thesis, Wageningen Agricultural University, Wageningen.
52. Green, M.L., Marshall, R.J. & Brooker, B.E., 1985. *J. Texture Studies.*, **16**, 351.
53. Apostolopoulos, C., 1991. Pers. Comm.
54. Purslow, P.P., 1989. *J. Biomechanics*, **22**, 21.
55. Voisey, P.W., 1975. *J. Texture Studies*, **6**, 253.
56. Steffen, C., 1976. *Schw. Milch. Forsch.*, **7**, 353.
57. Peleg, M., 1977. *J. Food Sci.*, **42**, 649.
58. Eberhard, P. & Flückiger, E., 1981. *Schweiz. Milch. Ztg.*, **107**, 23.
59. Lee, C.-H., Imoto, E.M. & Rha, C.K., 1978. *J. Food Sci.*, **43**, 1600.
60. Creamer, L.K., Zoerb, H.F., Olson, N.F. & Richardson, T., 1982. *J. Dairy Sci.*, **65**, 902.
61. Culioli, J. & Sherman, P., 1976. *J. Texture Studies*, **7**, 353.
62. Masi, P., 1987. In *Physical Properties of Foods—2*, ed. R. Jowitt, F. Escher, M. Kent, B. McKenna & M. Roques. Elsevier Applied Science Publishers, London, p. 383.
63. Creamer, L.K. & Olson, N.F., 1982. *Proc. 21st Dairy Congr.*, **1**(1), 474.
64. Kimber, A.M., Brooker, B.E., Hobbs, D.G. & Prentice, J.H., 1974. *J. Dairy Res.*, **41**, 389.
65. Stefanovich, R., 1973. *Dechema Monograph*, **77**, 211.
66. Ruegg, M., Eberhard, P., Moor, U., Flückiger, E. & Blanc, B., 1980. *Schw. Milch. Forsch.*, **9**, 3.
67. Khachatryan, G.G., Dilanyan, K.Zh., Tabachnikov, V.P. & Tetereva, I.I., 1974. *Proc. 19th Int. Dairy Congr.*, **1E**, 717.
68. Flückiger, E. & Siegenthaler, E., 1963. *Schw. Milch Ztg. (Wiss. Beil.)*, **89**, 707.
69. Yang, C.S.T. & Taranto, M.V., 1982. *J. Food Sci.*, **42**, 906.
70. Weaver, J.C. & Kroger, M., 1978. *J. Food Sci.*, **43**, 579.
71. Cervantes, M.A., Lund, D.B. & Olson, N.F., 1983. *J. Dairy Sci.*, **66**, 204.
72. Szabo, G., 1974. *Proc. 19th Int. Dairy Congr.*, **1E**, 505.
73. Ostojic, M., Miocinovic, D. & Niketic, G., 1982. *Mljekarstvo*, **32**, 139.
74. Le Bars, D. & Bergère, J.L., 1976. *Lait*, **56**, 485.
75. Nikolaev, B.A. & Abdullina, R.M., 1969. *Moloch. Prom.*, **30**(7), 27.
76. Kunakhov, I.M., 1967. *Trudy Vologod. Moloch. Inst.*, **55**, 70.
77. Bourne, M.C., 1982. In *Food Texture and Viscosity. Concept and Measurement.* ed. M.C. Bourne. Academic Press, New York, p. 24.
78. Baron, M. & Harper, R., 1950. *Dairy Ind.*, **15**, 407.
79. Guinee, T.P. & Fox, P.F., 1983. *J. Dairy Res.*, **50**, 511.
80. Langley, K.R. & Marshall, R.J., 1993. *J. Texture Studies* (in press).

9

Cheese: Methods of Chemical Analysis

P.L.H. McSweeney & P.F. Fox

Department of Food Chemistry, University College, Cork, Republic of Ireland

1 Introduction

Cheese is subjected to chemical analysis for a variety of reasons, such as to ascertain its composition for nutritional purposes, to ensure its compliance with standards of identity, to assess the efficiency of production or as an index of quality (see Ref. 1). Chemical analyses are of critical importance to the dairy scientist involved in cheese research, to analysts working on quality assurance and for regulation of the production process. This chapter will review the principal methods available for the chemical analysis of cheese, with particular reference to the ripening process and to techniques used in research. As far as we are aware, the methodology used to monitor the biochemistry of cheese ripening has not been comprehensively reviewed, although some aspects have been, e.g. proteolysis.[2-5]

2 Compositional Analysis

Gross compositional analysis of cheese is conducted in accordance with standard methods published by the International Dairy Federation and the Association of Official Analytical Chemists. Standard methods for moisture, ash, protein, fat, acidity and anion analysis are listed in Table I and will not be discussed further. Enzyme assays for lactose, lactic acid and citrate have been introduced recently.[6-9] The enzymatic assay for lactate is particularly useful since it permits determination of the D- and L-isomers, the proportions of which provide useful information on the activity of starter and non-starter lactic acid bacteria in cheese during ripening. There does not appear to be a standard method for determining the pH of cheese. The method used in our laboratory is as follows; grated cheese (10 g) is thoroughly blended with 10 ml H_2O using a mortar and pestle and the pH of the resulting slurry is measured potentiometrically. However, it may be preferable to measure the pH of the cheese directly (e.g. Ref. 10) to minimize changes caused by alteration of the salt balance of the cheese.

TABLE I
Standard Methods for Compositional Analysis of Cheese

Constituent	Method
Total solids (moisture)	IDF 4A:1982[15] AOAC 926.08, 948.12, 969.19, 977.11 (1990)[16]
Ash	AOAC 935.42 (1990)[16]
Fat	IDF 5B:1986[15] AOAC 933.05 (1990)[16]
Protein (total nitrogen)	IDF 25:1964[15] AOAC 920.123 (1990)[16]
Chloride	IDF 88:1979[15] AOAC 935.43, 983.14 (1990)[16]
Salt (NaCl)	AOAC 975.20 (1990)[16]
Citrate	IDF 34B:1971[15] AOAC 976.15 (1990)[16]
Phosphorus	IDF 33C:1987[15]
Nitrate/nitrite	IDF 84A:1984[15] AOAC 976.14 (1990)[16]
Acidity	AOAC 920.124 (1990)[16]

Calcium can be quantified by titration with ethylenediamine tetra-acetic acid (EDTA), using ammonium purpurate (murexide) as indicator or by means of oxalate precipitation, atomic absorption spectrophotometry or ion-specific electrodes (it should be noted that the later method measures only free calcium ions and not total concentration).

The concentration of Na^+ can be quantified specifically by ion-selective electrodes, atomic absorption or flame spectrophotometry.

The water activity (a_w) of cheese can be determined by a variety of methods, including psychrometry, cryoscopy, dew-point hygrometry and isopiestic equilibration. A number of regression equations have been developed to predict a_w from chemical composition for various cheeses.[11,12] The subject of the a_w of cheese is discussed more fully in Chapter 11.

Other methods of gross compositional analysis have been proposed, e.g. Frank & Birth[13] applied near infra-red reflectance spectroscopy to measure fat, protein, moisture and moisture in non-fatty substances in a range of cheese varieties. Cronin & McKenzie[14] quantified fat in a variety of foodstuffs, including Cottage and Cheddar cheese, by infrared transmittance spectrophotometry of a solvent extract of cheese, measuring the carbonyl ester stretch of the triglyceride at 5·72 μm relative to a methyl silicone internal standard.

3 Methods Used to Monitor Cheese Ripening

3.1 Assessment of Proteolysis

3.1.1 Extraction and Fractionation of Cheese Nitrogen

Introduction. Proteolysis in cheese involves a complex and dynamic series of events. It is desirable and sometimes necessary to separate the various proteins, peptides and amino acids which result from the action of the proteolytic enzymes from the milk, rennet, starter and non-starter bacteria. Various approaches have been used to extract and fractionate cheese nitrogen, as an aspect of the study of proteolysis in general and to prepare extracts for the identification and quantification of peptides and free amino acids.

Water soluble extracts. Quantitation and characterization of nitrogen in a water extract of cheese is a commonly used index of cheese ripening (see Ref. 3) and several procedures have been developed (see Ref. 17). Water-soluble extracts are also frequently used for the isolation of peptides and amino acids.[3]

Mabbitt[18] quantified the free amino acids in the aqueous phase of Cheddar cheese, prepared from cheese press juice by the removal of fat and proteinaceous material, by ion-exchange chromatography. In preparation for another chromatographic study of free amino acids in Cheddar, Bullock & Irvine[19] homogenized cheese samples in a Waring blender, followed by adjustment to pH 6·2 and filtration; the filtrate was further fractionated using ethanol. Stadhouders[20] macerated Edam cheese with water in a mortar at 50°C and diluted the paste to volume, taking into consideration the moisture content of the cheese. The homogenates were held at 25°C for 16 h and centrifuged; the supernatants were filtered and analysed for nitrogen by the Kjeldahl method. Water-soluble nitrogen was used as an index of proteolysis in aseptic cheese by Reiter *et al.*[21]

McGugan *et al.*[22] prepared a water-soluble extract of cheese by centrifugally defatting the cheese, extraction of the defatted residue with methanol/water/ methylene chloride, followed by repeated water extractions of the residue. A water-soluble extract prepared according to this method was analysed by HPLC by Pham & Nakai[23] and the procedure was also adopted by Aston & Creamer[24] who omitted extraction with methanol/water/methylene chloride.

Kuchroo & Fox[25] compared various extraction procedures for Cheddar cheese. All but one of several homogenization techniques yielded essentially similar results; a stomacher was used for routine work. Homogenization temperature had little effect on the level of extractable N, which increased with the ratio of water to cheese; a ratio of 2:1 was recommended and 90% of the potentially water-extractable N was obtained in two extractions. The final procedure recommended is: grated cheese is homogenized in a stomacher at 20°C for 10 min with twice its weight of water; the slurry is held at 40°C for 1 h, centrifuged and filtered. The residue can be re-extracted to increase the yield of extract. This procedure has been used by a number of workers.[26-31] O'Sullivan & Fox[30] showed that the water-soluble fraction of Cheddar cheese is very heterogeneous.

The preparation of a water-soluble extract is an efficient procedure for separating the small peptides from proteins and larger peptides in cheese.[17]

Extraction at pH 4·6.   pH 4·6-soluble N is also widely used as an index of cheese ripening. The pH of a water extract of internal bacterially ripened cheese (e.g. Cheddar, Swiss, Dutch) is approximately 5·2.[4] Thus, for these cheeses there is little difference between levels of N soluble in water or in buffers at pH 4·6.[25] However, for cheeses which are characterized by a significant rise in pH during ripening, water soluble N may be far higher than the levels of N soluble at pH 4·6.[4]

Dahlberg & Kosikowsky[32] extracted Cheddar with a buffer intended to maintain the pH of the cheese and to mimic the ionic composition of the aqueous phase. Kuchroo & Fox[25] reported that the level of N extracted by this method was only 30% of that in a water-soluble extract. It was suggested that this low extraction rate was due to the presence of $CaCl_2$ in the buffer, which might precipitate otherwise soluble casein-derived peptides. An extraction method using sodium citrate, in which the extract had a pH of $4·40 \pm 0·05$, was proposed by Mogensen.[33] This procedure was adopted by Vakaleris & Price[34] and Vakaleris et al.:[35] cheese was dispersed in a 0·5 M sodium citrate solution, and the pH of the dispersion adjusted to 4·4 with HCl. This approach has also been adopted by a number of other workers (e.g. Refs 36–38). Kuchroo & Fox[25] found that this method gave slightly lower values for soluble N than extraction with water. The extraction is more difficult to perform but may be easier to standardize.[4] Christensen et al.[17] favoured this approach over that of Dahlberg & Kosikowsky[32] because of the better dispersion of the cheese at pH 7.

O'Keeffe et al.[39] adjusted the pH of a water extract of cheese to 9·0 to inactivate rennet prior to adjustment of the pH to 4·6; some dissociation of the peptides was also achieved. This approach was also used by O'Keeffe et al.[40]

Extraction at pH values about the isoelectric point of casein is used to isolate small and medium sized peptides.[3] O'Keeffe et al.[39] found that pH 4·6-soluble N was produced mainly by the activity of rennet. Proteose peptones and whey proteins are also soluble at pH 4·6, but their contribution to pH 4·6-soluble nitrogen is small. γ-caseins are precipitated at pH 4·6 (see Ref. 3).

The pH 4·6-soluble fraction of cheese is very heterogeneous.[25] O'Keeffe et al.[40] found whey proteins, proteose peptones and a variety of peptides in a pH 4·6-soluble extract. Reville & Fox[41] reported that about 20% of the total N was soluble at pH 4·6 in a 6 months-old Cheddar and they concluded that fractionation at pH 4·6 was the most suitable extraction method for young cheeses.

Fractionation with CaCl₂. $CaCl_2$ has been used to fractionate cheese nitrogen by a number of workers. Dahlberg & Kosikowsky[32] included $CaCl_2$ at 0·02 M in their extraction buffer. This results in the recovery of only 30% of water-soluble N, presumably due to precipitation of casein-derived peptides by $CaCl_2$.[25]

Noomen[42] and Venema et al.[43] homogenized cheese in a $CaCl_2(0.137$ M$)$/ NaCl(0.684 M) solution; the homogenate was adjusted to pH 5.1, acidified with 0.25 M HCl (final pH ca. 1.6), held at 55°C for 30 min, then overnight at room temperature and filtered. Visser[44] homogenized cheese in a 0.55% $CaCl_2$-4% NaCl solution; the homogenate was defatted and adjusted to pH 7.5, centrifuged and the supernatant clarified by filtration.

Kuchroo & Fox[25] found that only 40% of the water soluble nitrogen was soluble in 0.1 M $CaCl_2$. Increasing the concentration of $CaCl_2$ above 0.05% at or above pH 7.0 had little influence on extraction.[42] The increase in $CaCl_2$-soluble N correlates with the age of the cheese and the extracts contain whey proteins, peptides and amino acids, while the $CaCl_2$-insoluble fraction contains caseins and high molecular weight peptides, similar to those in the water-insoluble fraction.[17]

Fractionation with NaCl.   Fractionation of cheese N with a 5% NaCl brine was employed by Chakravorty et al.[45] and Gupta et al.[46] Reville & Fox[41] found that 93% of the total N of a 12 months-old Cheddar cheese was soluble in 5% NaCl and concluded that fractionation with NaCl is suitable only for very young cheeses. NaCl (5%)-soluble N and unfractionated cheese are indistinguishable electrophoretically;[41] thus α-, β- and γ-caseins are extracted as well as peptides.[3]

Inclusion of $CaCl_2$ in the NaCl extraction solution reduces the percentage N extracted (see Ref. 4). Addition of $CaCl_2$ to 0.1 M to a water extract of cheese and adjusting the pH to 4.6 precipitates 60% of the water soluble N.[25]

Christensen et al.[17] concluded that fractionation with 5% NaCl is not as discriminating for the fractionation of cheese N as water.

Fractionation with chloroform/methanol.   Harwalkar & Elliott[47] extracted the bitter and astringent peptides in Cheddar with chloroform/methanol. Freeze-dried cheese was blended with chloroform/methanol (2:1, v/v) and filtered. Water was added to the filtrate and the mixture allowed to separate. The chloroform layer, containing lipids, was discarded. The methanol was evaporated from the aqueous methanolic layer, resulting in the formation of a precipitate; the supernatant was very bitter and astringent. This approach was also used by Visser[48] and Visser et al.[49] The latter used a freeze dried water- or dilute NaCl-soluble extract as starting material. The chloroform phase was extracted with methanol/water (50:30, v/v); the extract was freed of methanol and the aqueous residue diafiltered using 500 D membranes; the retentate contained the bitter peptides.

The chloroform/methanol procedure consistently extracts more N than water, which is understandable considering the relatively hydrophobic nature of many casein-derived peptides, but both extracts yielded identical chromatograms on Sephadex G-25.[3] Extraction with chloroform/methanol is useful for isolating the more hydrophobic peptides in cheese.

Butan-1-ol.   Butanol has been used to extract bitter peptides from casein hydrolysates but does not appear to have been applied to the fractionation of cheese.[4]

Fractionation with trichloroacetic acid (TCA).   The protein precipitant, TCA, has been applied by a number of workers to fractionate cheese nitrogen. The concentrations used have varied from 2% or 2·5%[25,26,30,41] to 12%.[25,26,40,41,43] Most authors have used TCA to fractionate water-soluble N, but O'Sullivan & Fox[30] used 2% TCA to fractionate the UF retentate of a water-soluble fraction.

Reville & Fox[41] found that about 14% of the total N of a 6 months-old cheese was soluble in 2·5% TCA. It was noted that 2·5% TCA was the best extractant for mature cheese, but other methods are recommended if further character-ization of peptides is to be performed. Kuchroo & Fox[26] found that 90% of the water soluble N in Cheddar was soluble in 2% TCA while O'Sullivan & Fox[30] found that 50–60% of a UF retentate (10 000 Da membranes) was soluble in this solvent. Ion-exchange chromatography confirmed the heterogeneity of the TCA soluble and insoluble fractions.[30]

Rennet is responsible for the production of some of the 12% TCA-soluble N but the N levels in this fraction are higher in cheeses acidified by starter than in chemically acidified cheeses,[21,39] indicating that starter peptidases are responsible for the formation of some of the 12% TCA soluble N. Reville & Fox[41] found that about 13% of total N in mature Cheddar was soluble in 12% TCA; this fraction, which is very heterogeneous,[26] increases with age and Venema *et al.*[43] reported that the ratio 12% TCA soluble N : total N is a better index of matu-rity than water soluble N : total N.

There is no 'precipitation threshold' relating peptide size to solubility in TCA but all peptides studies by Yvon *et al.*[50] containing fewer than seven amino acid residues were soluble in 12% TCA; peptide solubility is related to hydrophobicity.

A disadvantage of TCA for peptide fractionation is the necessity to remove it prior to further analysis of the fractions, for example by chromatography or electrophoresis. Small peptides and free amino acids will be lost on dialysis. Other methods for removal of TCA include repeated ether extraction, gel perme-ation and ion-exchange chromatography.[4] The use of 70% ethanol, which gives similar precipitation levels, is preferable because ethanol can be readily removed by evaporation.[4]

Ethanol.   Precipitation of proteins and peptides by ethanol is a classical pro-tein fractionation method and has been used extensively to fractionate cheese (see Ref. 17). However, various ethanol concentrations have been used, presum-ably with different results. Edwards & Kosikowski[51] used about 50% while Ismail & Hansen[36] extracted amino acids from a sodium citrate/HCl-soluble fraction using ethanol at 80%. Peptide fractions soluble in 2% or 12% TCA were precipitated by 80% ethanol and 75% acetone acidified to pH 4·6 with HCl.[52] Gonashvily[53] used 65% ethanol while Aston & Creamer[24] precipitated peptides from the water-soluble fraction of Cheddar with methanol in the ratio of 2 : 1, extract : methanol.

However, 70% has been used most widely.[25–27,41] Reville & Fox[41] found that 12% TCA and 70% ethanol had approximately similar extraction rates. The ethanol-soluble extracts contained a considerable amount of nitrogenous material in addition to free amino acids, but no protein bands could be detected in 70% ethanol- or 12% TCA-soluble fractions by gel electrophoresis. Material soluble in 12% TCA or 70% ethanol gave similar patterns on high voltage electrophoresis, but differences were detected in the insoluble fractions.[26] The 70% ethanol-insoluble fraction contains proteins, large and medium peptides.[27]

Fractionation of a UF retentate (10 000 Da) of a water-soluble extract of Cheddar using increasing concentrations of ethanol (30–80%) showed that most of the precipitable peptides were precipitated by 30% ethanol at pH 5·2; fractionation of UF retentate with 30% ethanol at pH 6·5, followed by adjustment of the filtrate to pH 5·5 is quite effective.[53a]

Ethanol precipitation may be used to fractionate cheese or to sub-fractionate water-soluble extracts and should be preferred to the largely equivalent 12% TCA owing to the ease of removal of ethanol by evaporation.[17]

Phosphotungstic acid (PTA). PTA-H_2SO_4 is a very discriminating protein precipitant. Free amino acids, except dibasic amino acids, are soluble in 5% PTA but peptides greater than about 600 Da are precipitated.[54] PTA (2·5% or 5%)-soluble nitrogen has been widely used as an index of free amino acids in cheese.[20,37,54–60] PTA-soluble nitrogen increases with age[56] and is produced primarily by the action of microbial peptidases. The free amino acid profile of the PTA soluble fraction has been determined.[54,60a,60b] The peptides in the PTA soluble fraction of blue cheese (Cabrales) have been fractionated by HPLC and gel filtration and characterized.[60a,60b] Reiter *et al.*[21] reported that 5% PTA soluble-N constituted 3·0% of total N of Edam at 3 months.

Sulphosalicylic acid (SSA). SSA is a discriminating protein precipitant.[4] Water-soluble extracts of cheese have been prepared for amino acid analysis by treatment with 3% SSA.[21] SSA has also been used to prepare samples for amino acid analysis.[61] However, Kuchroo & Fox[25] found that 2·5% SSA precipitated only 10% of the water-soluble N. The water-soluble protein and peptides in the fractions obtained with SSA have not been characterized (see Ref. 17).

Picric acid. Picric acid is supposed to be the most discriminating protein precipitant[4] and Reville & Fox[41] found that 0·85% picric acid-soluble extracts of cheese contained the lowest levels of N of the methods examined. It was considered to be the most suitable extractant for amino acids but small peptides are also soluble in picric acid.[62]

However, picric acid interferes with the determination of N by Kjeldahl or spectrophotometric methods.[4,41]

Ba(OH)$_2$/ZnSO$_4$. Free amino acids were extracted from Cheddar by Hickey *et al.*[63] using Ba(OH)$_2$/ZnSO$_4$. Cheese was macerated in a 0·15 M Ba(OH)$_2$–0·14 M ZnSO$_4$ solution, filtered and residual fat removed by extraction with chloroform/ethanol (1:1). The aqueous phase was filtered and the filtrate freeze dried and suspended in a sodium citrate buffer prior to amino acid analysis. No data were given on extraction levels or whether peptides were soluble. This reagent does not appear to be widely used as an extractant for N from cheese.

Ethylenediaminetetraacetic acid (EDTA).   Kuchroo & Fox[26] fractionated a water-soluble extract of Cheddar with 0·1 M EDTA; approximately 30% of the water-soluble N was precipitated.

Dialysis and ultrafiltration.   While many authors have attempted to fractionate water soluble extracts of cheese by exploiting differences in the relative solubility of its components in various reagents, others have used fractionation methods based on molecular size, using dialysis or ultrafiltration.

Kuchroo & Fox[26] found that exhaustive dialysis (96 h, four changes) was a simple method for partitioning water-soluble peptides and that it was suitable for relatively large samples. It gave clean-cut fractionation and appeared more effective than many other fractionation techniques. About 50% of the water-soluble N in a mature Cheddar was dialysable; the diffusate was completely soluble and the retentate 50% soluble in 70% ethanol. This technique was applied by Kuchroo & Fox[28] as a step in their fractionation scheme.

Diafiltration through membranes with a nominal molecular weight cut-off of 500 Da was used by Visser *et al.*[49,64] to remove very low molecular weight components from a bitter extract of Gouda. Aston & Creamer[24] used ultrafiltration with 1000 Da cut-off membranes to fractionate water-soluble N, while O'Sullivan & Fox[30] used 10 000 Da membranes. The 10 000 Da permeate contained no peptides detectable by PAGE and 40–50% of water soluble N was permeable. All peptides in the permeate are soluble in 2% TCA but some peptides in the UF retentate are insoluble.

Rejection of hydrophobic peptides by UF membranes and aggregation of small peptides are potential limitations in this approach. However, UF allows fractionation of large samples and does not require solvents, both of which facilitate taste panel work (see Ref. 4).

Trifluoroacetic acid/Formic acid.   Following a comparison of various extractants/precipitants for the fractionation of cheese nitrogen, Bican & Spahni[65] recommended the following procedure: homogenize cheese (2 g) dispersed in 25 ml of an extractant consisting of 1% (v/v) trifluoroacetic acid (TFA), 5% (v/v) formic acid, 1% (w/v) NaCl and 1 M HCl in a Polytron-Kinematica homogenizer at 4°C. Centrifuge the homogenate at 9750 g for 30 min at 4°C. Filter the supernatant and load onto a Sep-Pak C_{18} cartridge, wash with 0·1% TFA and elute the peptides with 80% aqueous acetonitrile containing 0·1% TFA. The peptide fraction was analysed by HPLC and fingerprinted by thin layer chromatography and high voltage electrophoresis on silica gel plates.

Fractionation schemes.   Since proteolysis in cheese involves a complex and dynamic series of events, it is desirable to fractionate the heterogeneous mixture of hydrolysis products to facilitate analysis. Various fractionation schemes have been proposed.[4,24,28,30]

A modification of the fractionation scheme proposed by Fox[4] is summarized in Fig. 1. A water-soluble extract is prepared according to the method of Kuchroo & Fox.[25] The extract is diafiltered through 10 000 Da membranes, the retentate sub-fractionated by ethanol (30%) at pH 6·5 and 5·5 and the permeate by 5% PTA. All the fractions prepared by this scheme are heterogeneous and

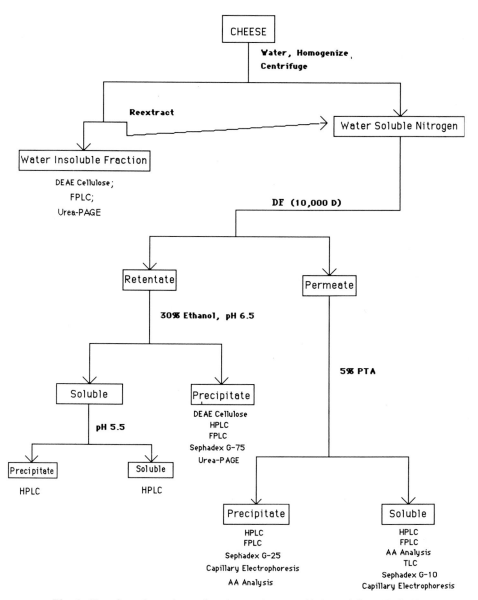

Fig. 1. Fractionation scheme for cheese nitrogen. (Adapted from Ref. 4.)

other reagents/techniques, e.g. fractionation with ethanol, acetone, chloroform, PTA, SSA, gel filtration, HPLC, FPLC, preparative gel electrophoresis or TLC are required to isolate homogeneous peptides.[4]

3.1.2 Methods for the Direct Measurement of Proteolysis in Cheese

Introduction. Many methods used to assess proteolysis in cheese (e.g. Kjeldahl determination of the N content of various fractions) are time-consuming and therefore it is desirable to develop direct methods to estimate proteolysis. Direct approaches adopted include estimation of the tyrosine and tryptophan content of cheese fractions by absorbence at 280 nm, protein dye-binding, liberation of ammoniacal nitrogen and, most importantly, determination of free amino groups by methods such as the formol titration, reaction with TNBS, ninhydrin, OPA or fluorescamine. There is a growing demand for such rapid methods for assessing the degree of cheese maturation.[66]

Formation of ammoniacal nitrogen. This was used as an index of proteolysis in Ulloa cheese by Ordonez & Burgos.[67] Ammoniacal nitrogen levels, measured by direct Nesslerization of a non-protein N preparation, constituted about 4 mg g^{-1} dry matter and increased slightly during ripening.

Measurement of tryptophan and tyrosine. Measurement of the 'soluble' tyrosine and tryptophan content of cheese is a well established method of assessing proteolysis. Hull[68] used the Folin-phenol reagent to assess proteolysis in milk. When compared on 12% TCA filtrates, the Folin reagent was less sensitive than TNBS for the assessment of proteolysis in cheese.[69] Singh & Ganguli[70] also applied the Folin reagent to TCA-soluble fractions of cheese.

Vakaleris & Price[34] measured the A_{280} of a sodium citrate-HCl extract (pH 4·6) of cheese as an index of proteolysis.

Dye-binding methods. At about pH 2, anionic dyes, e.g. amido-black, acid orange 12 and orange G, react with the positively charged lysyl and arginyl residues of proteins, leading to precipitation. The precipitate is removed by centrifugation or filtration and the excess dye measured spectrophotometrically. The amount of dye bound is proportional to the concentration of protein. However, low molecular weight proteins and peptides react only slowly, leading to poor separation and turbid filtrates or centrifugal supernatants.[71]

A decrease in the dye-binding capacity of a dispersion of Cheddar cheese with age was demonstrated by Ashworth.[71] Kroger & Weaver[72] reported that a dye binding method could be used as an index of proteolysis in cheese but Kuchroo et al.[29] concluded that the dye-binding method of McGann et al.,[73] using amido black, is not a suitable method for assessing proteolysis in cheese.

Formol titration. The formal titration is a simple method for estimating amino groups in milk and may be used to measure the extent and rate of proteolysis.

The method involves adding a formaldehyde solution, neutralized with NaOH, to a sample of milk and titrating to the phenolphthalein end-point.[73] This procedure has been used to estimate proteolysis in cheese (e.g. Ref. 35) but it is now considered obsolete owing to difficulties caused by variations in the buffering capacity of cheese.[66]

Buffering capacity. The buffering capacity of cheese increases during ripening owing to the formation of ammonia, amino and imino groups. This principle was used by Ollikainen[74] to assess proteolysis in cheese; it was claimed that the method is rapid and convenient and as accurate as colorimetric methods.

p-*Benzoquinone.* Reaction with p-benzoquinone provides the basis of a colorimetric procedure for assaying peptide bond cleavage. The method of Lorenz[75] was compared by Ollikainen[74] with the above titrimetric procedure.

Trinitrobenzenesulphonic acid. 2,4,6-Trinitrobenzenesulphonic acid (TNBS) reacts quantitively with primary amines, producing a chromophore which absorbs maximally at 420 nm. The chromophore remains attached to the amino acid, peptide or protein (Fig. 2).

The reagent, which was introduced by Satake et al.,[76] was used by Fields[77] to quantify amino groups in proteins and peptides. The reaction is performed at an alkaline pH and stopped by lowering the pH. TNBS also reacts slowly with the hydroxyl ion, a reaction which is catalysed by light. Since ammoniacal nitrogen produces only 20% of the absorbence of amino groups when reacted in equimolar concentrations with TNBS,[78] this reagent might underestimate proteolysis in cheeses which have undergone significant deamination.[79] Clegg et al.[78] concluded that TNBS is not as sensitive as ninhydrin for assessing proteolysis in cheese but is preferable owing to the simple analytical procedure; they proposed a correction factor for ammoniacal nitrogen. A disadvantage of the TNBS method is that the dry powder is explosive and prolonged storage leads to high blank values.[66]

Habeeb[80] and Adler-Nissen[81] measured the absorbence of the TNBS-amino group complex at 340 nm whereas in the procedure of Fields,[77] the absorbence of a sulphite-TNBS-amino group complex is measured at 420 nm. Barlow et al.[82] reported that the latter is preferable since the TNBS-amino group complex itself gave an unstable colour complex and the procedure is liable to operator error. Adler-Nissen[83] developed a procedure to assess the degree of hydrolysis of food proteins using TNBS. A linear relationship was found between the concentration of α-amino groups and A_{420} but the relationship varied between proteins, owing to variations in the concentration of ϵ-amino groups.

The spectroscopic method of Hull[68] was compared with the TNBS method by Samples et al.[69] who found that the latter was a better index of proteolysis in cheese. Jarrett et al.[54] used the TNBS assay to quantify free amino acids in 5% PTA-soluble fractions of cheese. Kuchroo et al.[29] found that the TNBS method was reproducible for monitoring proteolysis in cheese and concluded that while the method could be applied to unfractionated cheese, it is more sensitive if applied to pH 4·6- or 12% TCA-soluble extracts owing to a lower background colour, caused by reaction of TNBS with ϵ-amino groups. A strong correlation was found by Madkor et al.[83] between water soluble N and the A_{420} of the TNBS complex for a water-soluble extract of Stilton cheese.

Barlow et al.[82] considered that application of the TNBS method of Fields[77] to a water soluble extract, prepared according to Kuchroo & Fox,[25] would provide

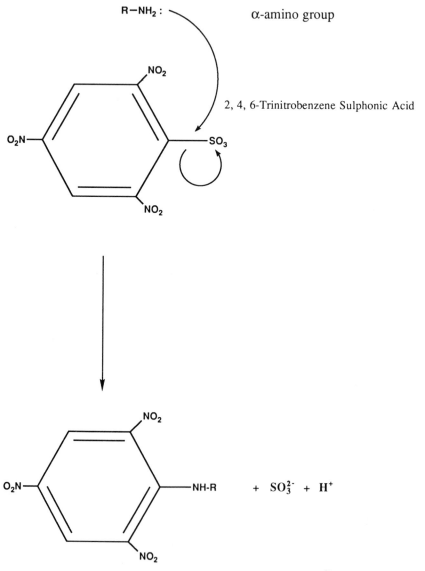

Fig. 2. Reaction of TNBS with α-amino groups.[81]

a simple routine method for the determination of soluble N in cheese. Whole cheese was analysed by Polychroniadou[84] who found a good correlation between the TNBS method and the ratio of water-soluble N to total N, as determined by the Kjeldahl procedure. A method was described for the assessment of proteolysis in whole cheese and fractionation of the cheese prior to analysis was considered to be unnecessary.

Humbert et al.[85] developed a new solvent (a patented mixture of organic acids and detergents) to clarify the sample after the TNBS reaction.

Ninhydrin.   The use of ninhydrin to assay for proteolysis in cheese is based on the spectrophotometric estimation of the chromophore formed when ninhydrin reacts with free amino groups (Fig. 3). The purple chromophore, named Ruhemann's purple after its discoverer,[86] does not remain attached to the protein or peptide and thus is not precipitated during procedures necessary to clarify the sample prior to spectrophotometric analysis at 570 nm.[66] This is considered to be a major advantage of the technique.[81] Ninhydrin reacts with ammonia almost as readily as with amino groups and thus ninhydrin values are consistently higher than those found by the TNBS procedure, the discrepancy being due to ammonia.[78] Ninhydrin is more sensitive than TNBS, but the latter was preferred by Clegg et al.[78] because the analytical procedure is simpler.

Moore & Stein[87] applied the ninhydrin reaction to quantify amino acids in chromatographic eluates. The procedure was modified by Moore & Stein[88] and further modified by Moore[89] who used aqueous dimethyl sulphoxide as solvent.

Doi et al.[90] described a number of ninhydrin-based assays for peptidase activity. Two were modifications of the methods of Moore & Stein[87,88] and two were based on the Cd-ninhydrin assay of Tsarichenko.[91] It was found that the Cd-ninhydrin reagent was more selective for the amino group of free amino acids than the amino groups of peptides or proteins and was the most sensitive of several ninhydrin reagents, including Sn-ninhydrin.

Ninhydrin-reactive lysine in food proteins was measured by Friedman et al.[92] who optimized the production of chromophore. Pearce et al.[79] developed a Li-ninhydrin assay for proteolysis in ripening cheese based on that of Friedman et al.[92] Cheese was dispersed in a citrate buffer and an aliquot of this solution heated with aqueous dimethyl sulphoxide, ninhydrin and lithium acetate solution at pH 5·2. The mixture was diluted and A_{570} determined. The method correlated well with conventional amino acid analyses.

Folkertsma & Fox[93] applied the Cd-ninhydrin assay of Doi et al.[90] to the assay of proteolysis in cheese; the reagent was found to be about five times as sensitive as TNBS for the measurement of amino acid nitrogen and could be performed on citrate-, water- or PTA-soluble fractions but not on TCA-soluble fractions as the latter appeared to interfere with colour development.

Fluorescamine.   Fluorescamine (4-phenylspiro [furan-2 (3H), 1'-phthalan]-3,3'-dione), is used to quantify amino acids and peptides in the picomole range. Fluorescamine, which was introduced by Weigele et al.,[94] reacts with primary

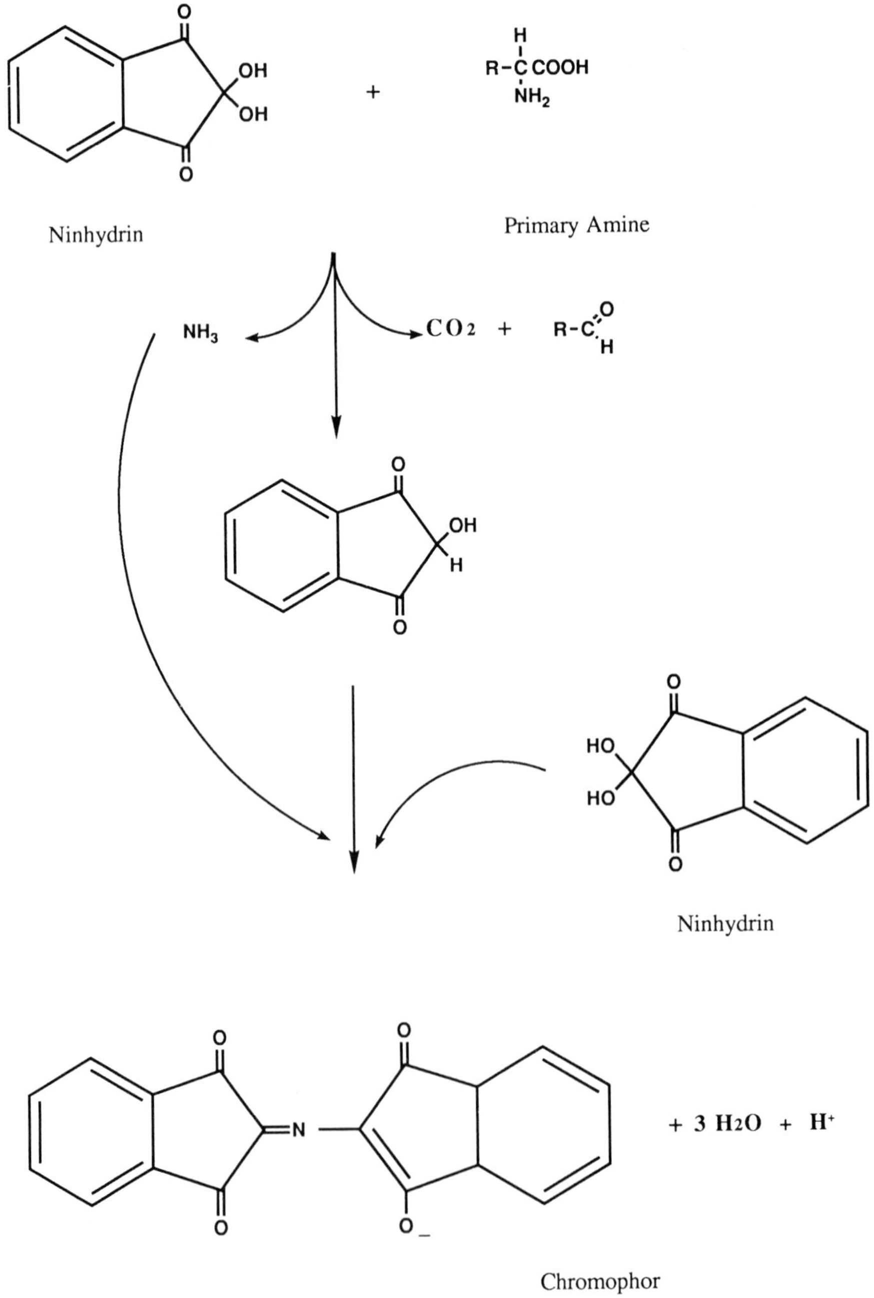

Fig. 3. The ninhydrin reaction.

amino groups to produce a fluorophor which is assayed at 390 nm excitation and 475 nm emission (Fig. 4). It reacts at room temperature with water or a primary amine, the latter reaction being far faster.[95] Fluorometric methods of analysis are desirable because of their high sensitivity.

Pearce[96] used this reagent to measure the release of glycomacropeptide from κ-casein by rennet action at an excitation wavelength of 395 nm and emission wavelength of 480 nm. Aliquots of TCA filtrate were mixed with 4% borax solution and an aliquot of fluorescamine solution in acetone added rapidly. The reaction product was protected from light to prevent decomposition prior to measurement of the fluorescence intensity. The hydrolysis of κ-casein was also

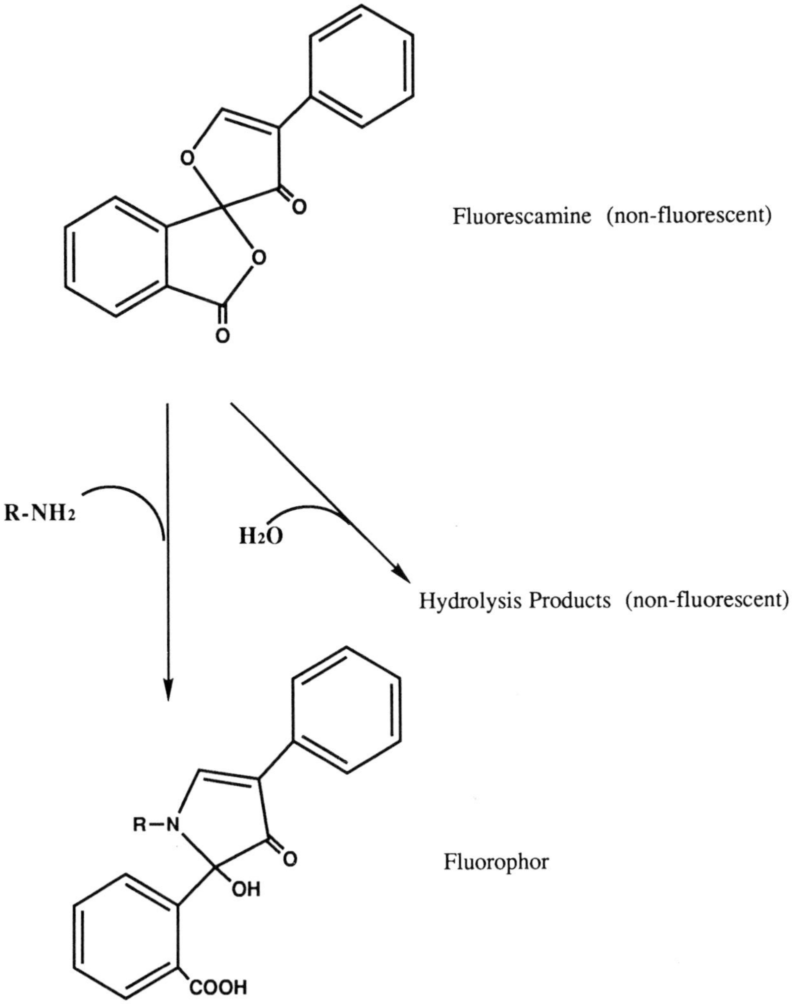

Fluorescamine (non-fluorescent)

R-NH₂

H₂O

Hydrolysis Products (non-fluorescent)

Fluorophor

Fig. 4. Reaction of fluorescamine with water and primary amino groups.[95]

studied using fluorescamine by Beeby,[97] who found that the procedure could be used to estimate the concentration of this protein in milk.

This reagent was used by Creamer *et al.*[98] to quantify acid-soluble proteins, peptides and amino acids from cheese; they noted that the fluorescamine procedure appeared to give more consistent results than did the TNBS method.

o-Phthaldialdehyde.   *o*-Phthaldialdehyde (OPA) reacts with β-mercaptoethanol and primary amines to form a fluorescent complex (1-alkylthio-2-alkylisoindole), which also absorbs strongly at 340 nm (Fig. 5).[99] Church *et al.*,[99] who used OPA to quantify proteolysis in milk protein systems, noted that the method is more accurate than the measurement of A_{280} and is more convenient than the ninhydrin, TNBS or fluorescamine procedures. OPA has been applied by Frister *et al.*[100] to monitor proteolysis in cheese.

3.1.3 Electrophoretic Analysis of Cheese

Electrophoresis has been applied widely to study primary proteolysis in cheese.[4] Since only proteins and large peptides can be visualized by staining, the

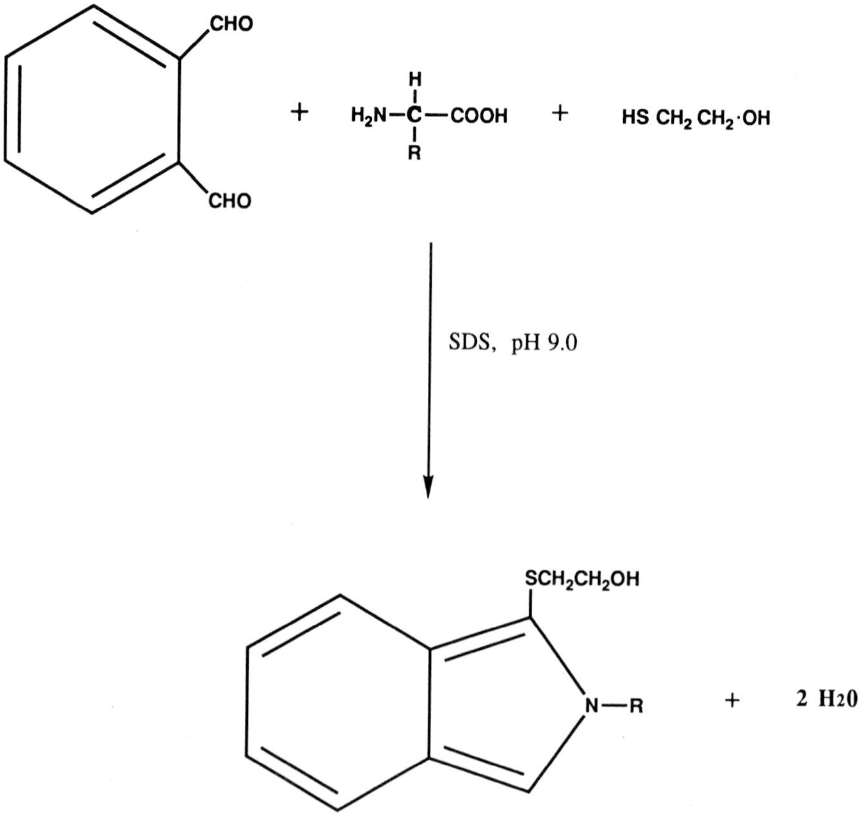

Fig. 5. Reaction of *o*-Phthaldialdehyde with a primary amine.[99]

technique is limited to the assessment of casein loss and the formation and subsequent hydrolysis of the primary products of casein proteolysis. However, it is a powerful technique for studying proteolysis during the early stages of cheese ripening.[101] The peptides in a 10 000 Da UF permeate do not stain on urea-PAGE, but the retentate of the WSN contains several detectable peptides.[30]

The first application of electrophoresis to the study of cheese ripening appears to be that of Lindqvist et al.[102] who used paper electrophoresis to study the ripening process in a number of cheese varieties. Lindqvist & Storgards[103] used free boundary electrophoresis for this purpose. Isoelectric focusing was used by Trieu-Cuot & Gripon,[104] as was two-dimensional electrophoresis (isoelectric focusing in one dimension, SDS-PAGE in the other). High-voltage paper electrophoresis (e.g. Ref. 26) is considered to be of limited value (see Ref. 4). Fingerprinting by a combination of thin layer chromatography and high voltage electrophoresis on silica gel plates was considered by Bican & Spahne[65] to be very satisfactory for the characterization of peptides but to be laborious.

Although starch gel electrophoresis has been used by some workers, electrophoresis in polyacrylamide gels (PAGE) has found the most widespread application (see Ref. 4). The literature was reviewed by Creamer[101] and Shalabi & Fox.[105] The latter compared several electrophoretic procedures for the analysis of cheese and strongly recommended the stacking gel system of Andrews[106] in gels containing 6 M urea. The direct staining procedure of Blakesley & Boezi[107] with Page Blue G90 gave very satisfactory results. It was concluded that SDS-PAGE were inferior to urea-PAGE for cheese analysis and its use was not recommended.

After staining, the electrophoretograms are usually recorded by photography, although densitometry or excision and elution of the stained bands followed by spectrophotometric quantitation have also been used (see Ref. 101).

The difficulty in obtaining quantitative data is a serious limitation of electrophoresis. Creamer[101] recommends that several control samples be included in each gel and that comparisons should only be made between samples on the same gel. He emphasizes that band dimensions are critical for densitometry and that dye uptake is a function of the protein as well as staining and destaining protocol.

3.1.4 Chromatographic Methods to Assess Proteolysis in Cheese

As discussed in Section 3.1.5, paper chromatography has been used by many workers to quantify free amino acids in cheese.[45,108–111]

Various forms of chromatography have been used to characterize peptides in cheese. O'Keeffe et al.[40] used single dimensional paper chromatography (butanol/acetic acid/water/pyridine, 30:6:24:20) to separate peptides from fractions obtained by gel filtration. Kuchroo & Fox[26] characterized the peptides in a water-soluble extract of Cheddar cheese by paper chromatography with butanol/pyridine/acetic acid/water (35:25:10:30, v/v) as solvent. Kuchroo & Fox[27] also used these chromatographic conditions to characterize peptides in fractions from gel permeation and hydrophobic chromatography. Paper chromatography is cheap and straightforward but suitable only for assessing the complexity of systems containing only small peptides.[111]

Thin layer chromatography on silica gel has been used to characterize peptides in cheese fractions.[26-28,49,51] Different solvent systems have been used, e.g. n-propanol/water (70/30, v/v) and n-propanol/acetic acid-water (5:1:3). Ninhydrin is normally used to develop the plates, although UV fluorescence of the spots was used by Edwards & Kosikowski.[51] Preparative TLC has been used[51] to isolate bitter peptides from Cheddar. TLC is a simple, straightforward method for separating peptides and is normally superior to paper chromatography, although Kuchroo & Fox[27] were able to partially resolve peptides by PC in contrast to TLC.

Column chromatography on silica gel was used by Visser et al.[112] to study the bitter peptides from rennet-treated casein. Peptide fractions were eluted with n-propanol/water (7:3, v/v) and monitored at 206 and 280 nm. The authors point to the potential of the technique but it does not appear to have been widely used.

Metal chelate (ligand exchange) chromatography on Cu-Sephadex was used by Ney[113] to isolate peptides from Cheddar. This technique was developed further by Mojarro-Guerra et al.[114] for the separation of peptides from amino acids; the peptides were further fractionated by chromatography on Aminex A6 or Durrum D6 resins. Kuchroo & Fox[27] separated the peptides in a water-soluble extract of Cheddar by hydrophobic chromatography on phenyl- and octyl-Sepharose CL-4B. Chromatography in Duolite S-571 was unsuccessful since the peptides in the water-soluble fraction of the cheese did not absorb onto this medium. Visser et al.[49] isolated bitter peptides by hydrophobic chromatography on Sephadex LH 20.

Gel permeation chromatography has been used widely to study cheese peptides. When a column has been calibrated, the technique allows the estimation of molecular weights. The majority of workers have used Sephadex gels of various pore sizes (Sephadex G10 will resolve molecules with molecular weights of 700 to 1500 Da; G25, 1000–5000 Da; G50, 1500–30 000 Da and G75, 3000–80 000 Da; Ref. 3). Cheese preparations chromatographed have varied from defatted cheese[115] to fractions of the dialysate of a water-soluble extract.[28] A wide variety of eluents have been used. Chromatograms are generally quantified by spectrophotometry at UV wavelengths; 280 nm is generally used and is suitable for fractions containing large peptides but for fractions containing smaller peptides absorbence of the carbonyl group in the peptide bond at a lower wavelength, e.g. 220 nm[116,117] is preferable, since smaller peptides may not contain aromatic residues, which are relatively scarce in caseins. Measurement of the amino functional group by reaction with ninhydrin (e.g. Ref. 118) or TNBS are alternatives. Studies in which classical gel permeation chromatography was used are summarized in Table II.

A number of early workers quantified free amino acids by classical ion-exchange chromatography and auto-analysers based on this technique are widely used for this purpose (see Section 3.1.5). Although ion-exchange chromatography

TABLE II
Studies Involving Gel Permeation Chromatography of Cheese Peptides

Cheese	Medium[a]	Fraction studied	Eluent	Detection	Reference
Cheddar	G 10 G 25 G 50 G 150	Freeze-dried water-soluble N	water	UV, 280 nm	27
Cheddar	G 10 G 25	Fractions of diffusate of water-soluble N	water	UV, 280 nm	28
Cheddar	G 50	Water-soluble N	pyridine-acetic acid	UV, 280 nm	39, 123, 123a
Cheddar	G 25 G 50	Bitter fraction	water	UV, 280 nm	47
Cheddar	G 25	Bitter extracts	0·1 N NH$_4$OH	UV, 254 nm	51
Limburger	G 100	Defatted cheese dissolved in 0·033 M acetate buffer, pH 7	0·033 M acetate buffer	UV, 280 nm	115
Cheddar	G 100	Cheese dissolved in solvent 0·1 M tris-citrate, pH 8·6 1 mM Na$_2$-EDTA, 1 mM NaN$_3$, 6 M urea, 1 mM dithiothreitol, aqueous fraction prepared centrifugally	solvent	UV, 280 nm, UV, 220 nm	116, 117, 119
Aseptic cheese	G 25	pH 4·6 soluble	1 M acetic acid	UV, 280 nm; ninhydrin	118
Svecia	G 25	Cheese homogenized in water, EDTA, 1 N NaOH; defatted centrifugally	1. NH$_4$OH 2. 0·1 N Sigma '7–9' 3. Glycine buffer, pH 10·5 4. 0·1 N NaCl	UV, 253·7 nm; conductivity	120

[a] Sephadex gels unless otherwise specified.

(continued)

TABLE II—*contd.*
Studies Involving Gel Permeation Chromatography of Cheese Peptides

Cheese	Medium[a]	Fraction studied	Eluent	Detection	Reference
Svecia Adelost Hard cheese (37-yr-old)	G 25 G 200	Na-citrate extract, dialysis	—	UV; conductivity	121
Domiati	G 25	Cheese homogenized in 1 M acetic acid, defatted centrifugally	1 M acetic acid	UV, 280 nm; ninhydrin	122
Butterkase	G 25 G 50	Water-soluble N	water	UV, 280 nm	123
Unspecified	G 100	Cheese homogenized in water, EDTA, 1 N NaOH, defatted centrifugally	phosphate buffer $\Gamma/2 = 0 \cdot 06$, pH 7·1	UV	124
Cheddar	Sepharose 6B	High molecular weight peptides	6 M guanidine-HCl	UV, 280 nm	125
Aseptic cheese	G 25 Biogel A5m	pH 4·6 soluble and insoluble N	1 M acetic acid	UV, 280 nm, UV, 254·5 nm	126
Provolone	G 10 G 25 G 50 G150 Sepharose 6B	Freeze-dried water-soluble N	water	UV, 280 nm	127
Gomonedo	G 10	PTA soluble N	water	UV, 280 nm	60a
Cabrales	G 10	PTA soluble N	water	UV, 280 nm	60b

[a] Sephadex gels unless otherwise specified.

TABLE III
Studies Involving Ion-Exchange Chromatography of Cheese Peptides

Cheese	Medium	Fraction studied	Elution conditions	Detection	Reference
Cheddar	DEAE-cellulose	70% ethanol-soluble and insoluble fractions of water-soluble N retentate	0·02 N Na phosphate buffer, pH 6·5, NaCl gradient: 0 to 0·5 M	UV, 280 nm	28
Cheddar	DEAE-cellulose	2% TCA-soluble fractions of 10 000 Da UF retentate of water-soluble extract	0·02 M Na phosphate buffer, pH 6·4; gradient: 0 to 0·5 M NaCl	UV, 280 nm	30
Butterkase	Dowex 50 W x2	Fractions from gel permeation chromatography	linear gradient: 2·5 l 0·2 M pyridine acetate, pH 3·1 2·5 l 2 M pyridine acetate, pH 5	ninhydrin	123
Cheddar	DEAE-cellulose	Cheese dispersed in Na-citrate solution, defatted, urea mercaptoethanol added	4·5 M urea, pH 5·5, 0 to 0·5 M NaCl gradient	UV, 280 nm	125
Provolone	Cationic resin on auto-analyser	Fractions from gel-permeation chromatography	linear pyridine gradient, 0·2 to 2 M, pH 2·8 to 6·5	ninhydrin	127
Cheddar	DE 52	Cheese dispersed in 6 M urea	6 M urea, pH 7·0 linear gradient: 0–0·05 M NaCl	UV, 280 nm	128
Gouda		Fractions from ion-exchange chromatogram	6 M urea, pH 8·5 linear gradient 0–0·3 M NaCl	UV, 280 nm	

TABLE IV
Studies Involving Reversed-Phase HPLC of Cheese Peptides

Cheese	Extract	Column	Elution conditions	Detection	Reference
Cheddar	Water extract	Adsorbosphere C_8 Spherisorb S_5 C_8 Vydac C_{18}	Isocratic: 0·1 M phosphate buffer, pH 6·0	UV, 220 nm	23
Cheddar Cheshire Gouda	Water extract	Spheri-5 RP-18	Gradient: TFA: H_2O: CH_3CN	UV, 214 nm, UV, 280 nm	24, 129
Cheddar	Fractions from gel-permeation chromatography	Waters RP-18	Isocratic: 0·01 M KH_2PO_4, pH 4·3	UV, 200 nm	28
Edam Gouda Swiss Cheddar Parmesan	Extract according to Ref. 22	Adsorbosphere C_8	Ternary gradient 0·1% TFA: CH_3CN: CH_3OH	UV, 220 nm	130
Cheddar	Water extract	Adsorbosphere C_8	Gradient: TFA, H_2O, CH_3CN	UV, 220 nm	131
Gouda	pH 4·6 soluble extract	Zorbax ODS	Gradient: TFA, H_2O, CH_3CN	UV, 230 nm	132

Danbo Havarti	pH 4·6 insoluble extract	Nucleosil 10C$_{18}$	Gradient H$_2$O, TFA: H$_2$O, CH$_3$CN	UV, 280 nm	133
Cheddar-type	Water extract treated with methanol, hexane and permeation chromatography on Sephadex G 25	Pharmacia Pep HR 5/5	Gradient, H$_2$O, TFA: CH$_3$CN, TFA	UV, 214 nm	134, 135
Cheddar	Water extract	Pharmacia Pep HR 5/5	Gradient, H$_2$O, TFA: CH$_3$CN, TFA	UV, 214 nm	136
Blue	Water, 5% PTA soluble extracts	Ultrasphere ODS	Gradient, H$_2$O, TFA: CH$_3$CN, TFA	UV, 230 nm	60a
Appenzel	1% TFA-5% formic acid 1% NaCl 1% HCl	Vydec 218 TP54	TFA, CH$_3$CN	UV, 215 nm	65

RP-HPLC   Reversed-Phase High Performance Liquid Chromatography.
TFA   Trifluoroacetic Acid.

on DEAE-cellulose is widely used to fractionate milk proteins, this medium has had limited use for the fractionation of cheese peptides.[4] Creamer and Richardson[125] isolated α_{s1}-I casein from Cheddar by chromatography in DEAE-cellulose. Chromatography on DEAE-cellulose is suitable for fractionating 2% TCA soluble and insoluble fractions of a 10 000 Da UF retentate of a water-soluble extract of Cheddar cheese.[30] Ion-exchange chromatography on DEAE-cellulose has considerable potential for fractionating water-insoluble peptides in cheese.[53a] Mulvihill & Fox[137] used phosphocellulose to fractionate peptides produced from α_{s1}-casein by chymosin, but this medium does not appear to have been used for cheese analysis. A few authors have used ion-exchange chromatography on Dowex 50 resins to fractionate cheese peptides (e.g. Ref. 124). Studies on cheese in which classical ion-exchange chromatography has been used are summarized in Table III.

The introduction of gel-permeation and ion-exchange HPLC and FPLC will reduce considerably the work-load involved in these forms of chromatography.

Reversed-phase high performance liquid chromatography (RP-HPLC) has been used increasingly to characterize peptides in casein hydrolysates (e.g. Ref. 138) and gel-permeation HPLC has been used to characterize caseins, whey proteins and peptides.[139-141] HPLC analysis of cheese peptides is becoming increasingly popular, especially on reversed-phase columns. Water extracts have generally been studied but other peparative techniques have included pH 4·6-soluble and -insoluble fractions and classical gel-permeation chromatography.[28] Gradient elution with water/acetonitrile and trifluoroacetic acid as an ion-pair reagent is most commonly used but isocratic conditions have been used by some workers. Detection is generally by UV spectrophotometry, usually at a wavelength in the region of 200–230 nm (which measures the carbonyl in the peptide bond), although 280 nm has been used in cases where larger peptides, which are more likely to contain aromatic residues, are expected. Fluorescence detection has also found limited use. Studies involving reversed-phase HPLC are summarized in Table IV.

To date, HPLC has been confined to the research laboratory; difficulties remain with data interpretation and the development of a solvent system which will permit tasting of the peptide fractions.[142] However, Smith & Nakai[130] discuss the potential of HPLC for the routine assessment of cheese quality.

Fast protein liquid chromatography (FPLC) on Mono-S or Mono-Q columns gives better resolution of bovine milk proteins than classical chromatography on DEAE-cellulose.[143,144] Wilkinson et al.[145] fractionated press juice from Cheddar by gel permeation FPLC on a Superose-12 column; changes in the peptide profile were evident during ripening. Preliminary results indicate that gel permeation FPLC with a Superose-12 column is valuable for the characterization of peptides in fractions obtained from Cheddar on application of the fractionation scheme shown in Fig. 1 (E.D. Breen & P.L.H. McSweeney, pers. comm.). FPLC is suitable for small-scale preparative work; fractions can be collected, freeze-dried and analysed by PAGE. Preliminary results indicate that preparative ion exchange FPLC is of value in studying larger peptides from Cheddar.[53a] Ion exchange and gel permeation FPLC greatly reduce the work-load involved in these forms of column chromatography, while increasing speed and reproducibility. Thus, it seem likely the FPLC will achieve more widespread application for monitoring proteolysis in cheese.

3.1.5 Free Amino Acids in Cheese

Analysis of free amino acids.   Paper chromatography was used by many early workers to semi-quantify amino acids in cheese.[19,108–110,146,147] Solvent systems used have included phenol-water-collidine-lutidine[108,146] or a combination of various buffered solvents. Chromatograms were generally developed with ninhydrin and amino acids quantified by reference to spot area and colour intensity or by a photoelectric scanner. However, at best, paper chromatography is semi-quantitative and identification of all the protein amino acids is difficult. It has thus been rendered obsolete for quantitative amino acid analysis although it may still be useful for qualitative work.

Mabbitt[18] used ion-exchange chromatography on Dowex-50 columns to quantify a number of amino acids in press juice from Cheddar cheese. Short (15 cm) columns of Dowex-50 were employed by Hintz *et al.*[148] to study free amino acids in a water soluble extract of Swiss cheese. Ali & Mulder[149] also used a method based on that of Moore & Stein[150] to separate amino acids using Dowex-50. They found that the free amino acid pattern of mature Edam did not vary significantly between samples, that amino acids contribute only to the basic taste and that the flavour of Edam cannot be attributed solely to free amino acids.

Classical ion-exchange chromatography is laborious and has been superseded by automated amino acid analysers based on ion-exchange chromatography which has greatly facilitated the analysis of free amino acids in cheese. Amino acid analysis is relatively simple, accurate and quantitative and requires little sample preparation. However, the capital cost of the instrument is higher than those for other methods of amino acid analysis.

Free amino acids were one of the indices of proteolysis considered by Reiter *et al.*[21] who used automated ion-exchange chromatography of a water-soluble extract of cheese, deproteinized by sulphosalicilic acid. Visser[44] used a technique similar to that of Kosikowsky,[146] i.e. the ethanol soluble fraction of a water-soluble extract of cheese. A similar preparation method was used by Polychroniadou & Vlachos.[151] Ion-exchange chromatography was used by Weaver *et al.*[152] to prepare samples for amino acid analysis; cheese samples were homogenized in a picric acid solution and an internal standard added. The samples were then centrifuged and the supernatant passed through a column of Dowex 2X8–200.HCl; the eluate was used for amino acid analysis. An essentially similar approach was adopted by Shindo *et al.*[153] Omar[154] used sulphosalicilic acid to deproteinize a Na-citrate dispersion of Ras cheese for amino acid analysis while Hickey *et al.*[63] used $Ba(OH)_2/ZnSO_4$ to deproteinize cheese.

Polo et al.[155] and Ramos et al.[61] used reverse-phase HPLC to measure free amino acids in cheese as o-phthaldialdehyde derivatives.

Amino acids can be quantified by gas chromatography after derivatization with heptafluorobutyric anhydride (HFBA) to yield n-heptafluorobutyryl isobutyl derivatives. However, packed GC columns do not give adequate resolution (e.g. Ref. 156) and as far as we are aware have not been used for cheese analysis. However, Wood et al.[157] and Laleye et al.[158] applied capillary GC to quantify free amino acids (as HFBA derivatives) in cheese. This technique gave good recovery of added amino acids and only one column was necessary. All protein amino acids could be resolved in a single chromatogram and speed and accuracy were comparable to that of automated amino acid analysers but with greatly reduced equipment costs. However, this technique has not been used as widely as automated amino acid analysers based on ion-exchange chromatography.

Amino acid analyses of various cheese varieties. Studies in which the free amino acids in cheese have been quantified are summarized in Table V. Amino acid concentrations found in selected varieties are given in Table VI; the values are intended only as examples and further data are presented for individual varieties in Volume 2.

Weaver et al.[152] found that the total concentration of free amino acids in an 8 months-old Cheddar was 9967 μg g^{-1}. Val, Tyr, Phe, Glu and Leu accounted for about 80% of the free amino acids at all stages of ripening; Leu showed the largest increase during ripening. The findings of Wood et al.[157] were in general agreement with these results but Hickey et al.[63] found Glu and Asp to be relatively low.

Wood et al.[157] noted that the relative proportions of free amino acids in a Swiss-type cheese were similar to those in Cheddar with the exception of Pro, which is high in Swiss cheese in which it is an important flavour compound.

Kosikowsky & Dahlberg[147] reported that Roquefort-type cheeses have higher concentrations of free amino acids than Camembert and this was confirmed by Ismail & Hansen.[36]

Most of the studies on free amino acid in bacterial surface-ripened cheeses have been performed using paper chromatography. Kosikowsky & Dahlberg[147] found lower levels of free amino acids in Limburger-type cheese than in Roquefort or Emmental. This result was unexpected.

Polychroniadou & Vlachos[151] found that Leu (18% of total free amino acids), Phe, Lys and Val were the principal free amino acids in Teleme cheese, together making up 50% of the total in a 4 months-old cheese; the development of flavour was correlated with amino acid levels. Mahon cheese was studied by Polo et al.[155] who found that Phe, Val, Pro, Glu and Ile were the principal free amino acids, contributing 67 to 80% of the total. The free amino acid profile of Ras cheese was similar to that of some other hard cheese varieties (e.g. Emmental); the young cheese contained mainly Phe, Val, His, Ala, Pro and Lys while a 4 months-old cheese contained more Glu, Leu, Phe, Val and less Asp, Tyr, Gly, His and Ser.[154]

TABLE V
Studies on Free Amino Acids in Cheese

Variety	Analytical method	Reference
Cheddar	Ion-X	18
Cheddar, Edam	AAA	21
Samsoe, Maribo, Danbo, Havarti, Danish Camembert, Danablue	AAA	36
Cheddar	AAA	39
Gouda	AAA	44
Cheddar	PC	45
Cabrales	HPLC	61
Cheddar	AAA	62
Cheddar	AAA	63
Cheddar	PC	108, 146
Cheddar, Emmental, Edam, Gouda, Svecia, Herrgard, Roquefort, Domestic Blue, Gorgonzola, Limburger, Brick, Liederkranz, Asiago, Provolone, Primost	PC	110
Tilsit	PC	110
Mansoura	PC	111
Emmental, Domestic Swiss, Roquefort, Gorgonzola, Blue, Liederkranz, Limburger, Brick, Camembert, Asiago, Provolone, Gouda, Gruyère-type, Primost-type	PC	147
Domestic Swiss	Ion-X	148
Edam	Ion-X	149
Teleme	AAA	151
Cheddar	AAA	152
Gouda	AAA	153
Ras	AAA	154
Mahon	AAA?	155
Cheddar, Swiss-type, Edam, Gouda, Jarlsberg	Cap-GC	157
Cheddar	Cap-GC	158
Rokpol	AAA	159
Bryndza	—	160
Fior de Latte	—	161
Cheddar	—	162
Feta	AAA	163
Kashkawal	—	164
Emmental, Gruyère, Appenzel, Camembert, Brie	—	165
Teleme	—	166
Parmesan	—	167
Cheddar	HPLC	168
Cheddar	—	169
Ras	—	170

AAA   Automated Amino Acid Analyzer.   Ion-X Ion Exchange
Cap-GC Capillary Gas Chromatography.   Chromatography.
HPLC   High Performance Liquid Chromatography. PC   Paper Chromatography.

3.2 Assessment of Lipolysis

3.2.1 Introduction

The degree of lipolysis in cheese depends on the variety and varies from slight to extensive. Extensive lipolysis in internal bacterially-ripened cheeses (e.g. Cheddar, Gouda, Swiss) is undesirable, while in mould-ripened cheese, lipolysis is essential for flavour development. A number of procedures have been developed to quantify lipolysis.

TABLE VI
Free Amino Acid Composition of Selected Cheeses

Amino Acid	Variety				
	Cheddar		Swiss	Gouda	Blue[e]
	mol %[a]	μmol/g[b]	μmol/g[c]	mg/g[d]	
Ala	4·0	4·04	5·66	0·50	1729
Arg	1·5	6·37	n.d.	1·18	445
Asn	—	—	—	—	—
Asp	2·3	11·65	5·40	0·45	860
Cys	—	0·37	n.d.	—	—
Gln	—	—	—	—	—
Glu	19·7	21·62	9·79	3·38	4775
Gly	3·2	4·14	1·74	0·37	466
His	—	—	—	0·22	449
Ile	2·2	3·58	—	0·65	1929
Leu	18·7	21·19	15·72	3·34	3015
Lys	11·8	7·8	1·12	2·02	2231
Met	1·9	2·95	0·42	0·66	1111
Phe	7·7	9·02	6·40	1·67	1785
Pro	3·4	2·95	2·41	—	3076
Ser	13·2	4·00	2·65	0·54	1045
Thr	—	3·78	2·80	1·55	1123
Trp	—	—	—	—	552
Tyr	2·6	3·75	1·03	1·00	1154
Val	7·6	9·47	8·36	1·13	2408
Orn	—	n.d.	n.d.	—	77
Asn, Gln	—	—	—	—	876
Citrulline	—	—	—	—	423
γ-ABA	—	—	—	—	221

n.d. = not detected.

— = not analysed.

γ-ABA = γ-amino butyric acid.

[a] Salji & Kroger;[62] 3 weeks-old cheese analysed by amino acid analyser.

[b] Wood et al.[157] 8 months-old cheese analysed by capillary GC.

[c] Wood et al.[157] 2 months-old Swiss-type cheese analysed by capillary GC.

[d] Visser[44] 6 months-old cheese made with starter Lactococcus lactis spp. cremoris E8 under aseptic conditions and analysed by amino acid analyser.

[e] Ismail & Hansen[36] 248 days-old Danablue cheese analysed by amino acid analyser, results expressed as mg amino acid residue/15·7 g total N.

3.2.2 Copper Soaps Methods

Spectrophotometric methods based on the formation of copper soaps for the estimation of free fatty acids in milk were described by Koops & Klomp[171] and Shipe et al.[172] However, Bynum et al.[173] obtained low recovery of free fatty acids from Cheddar cheese by the above versions of the Cu-soaps method which they modified (below) for application to cheese.

Cheese samples (200 mg) are mixed with 0·4 ml 0·7 M HCl and 7·5 ml copper soaps reagent [100 ml 1·0 M $Cu(NO_3)_2 \cdot 2 \cdot 5H_2O$ plus 50 ml triethanolamine diluted to 1 litre with saturated NaCl and adjusted to pH 8·3 with 1·0 M NaOH], incubated at 60°C for 10 min, cooled and mixed vigorously. Copper soaps of free fatty acids are extracted with 20 ml chloroform/heptane/methanol (49:49:2, v/v/v) by shaking gently for 30 min, followed by centrifugation. An aliquot (5 ml) of the solvent layer is added to 0·2 ml 0·5% (w/v) sodium diethyldithiocarbamate in butan-1-ol. The colour of the resulting complex is measured spectrophotometrically at 440 nm. A standard curve is prepared using palmitic acid, which is one of the principal free fatty acids in cheese.[173]

The copper soaps of free fatty acids partition between the aqueous and apolar solvent phases, and therefore short chain fatty acids ($<C_{10}$) may not be extracted fully. According to Bynum et al.,[173] the recovery of octanoic acid was 60% of the predicted level and the authors suggest that the recovery of shorter chain free fatty acids would be lower. However, Woo & Lindsay[174] showed that short chain fatty acids as a percentage of total free fatty acids were similar for rancid and non-rancid Cheddar cheese and that long chain fatty acids were dominant, although they may not have as significant an impact on flavour as short chain acids. Therefore, Bynum et al.[173] considered that the poor recovery of short chain fatty acids was not a serious limitation of the copper soaps procedure. Inclusion of methanol (2%) in the extraction solvent minimizes interference from copper complexes of phospholipids.[171]

3.2.3 Acid Degree Value

The acid degree value (ADV) has been used for many years as an index of lipolysis in dairy products. Free fat is obtained by the combined action of detergent, ion-exchange and heat and is separated from the aqueous phase by centrifugation. Aliquots of the free fat are weighed, dissolved in solvent and the free fatty acids titrated with alcoholic KOH (~0·02 M) using methanolic phenolphthalein as indicator.[175]

The Bureau of Dairy Industry (BDI) method, developed by Thomas et al.,[176] was applied to Cheddar cheese by Dulley & Grieve.[177] Cheese samples were dispersed in 2% sodium citrate prior to de-emulsification and titration. Salji & Kroger[62] used a similar method to measure the fat acidity of Cheddar cheese. A finely ground sample of cheese (10 g) was placed in a Babcock cream butyrometer and 20 ml BDI reagent (30 g Triton X-100, 70 g Na tetraphosphate, H_2O to 1 litre) added. The mixture was then heated at 100°C for 20 min and centrifuged to liberate the fat. Aqueous methanol was added to bring the fat into the neck of the butyrometer. An aliquot (1 ml) of free fat was titrated with alcoholic KOH.

Deeth & Fitzgerald[175] consider the ADV method to be tedious, but to be reliable in the hands of a competent operator for foods containing >2% fat; they report that Cheddar cheese with an ADV > 3·0 will have a rancid off-flavour.

3.2.4 Gas Chromatography

Analysis of free fatty acids by gas chromatography has been applied widely to cheese. A preparative procedure is generally necessary and in many cases the esters of free fatty acids are chromatographed.

Aqueous and solvent extractions have been used (e.g. Refs 178, 179) but this approach has been criticized owing to partitioning effects[180] and the extraction of compounds which interfere with analysis.[178,179] Martin-Hernandez et al.[181] described a simple extraction-derivitization procedure for free fatty acids in cheese: cheese is acidified and homogenized with diethyl ether; the organic phase is dried with anhydrous Na_2SO_4 and the fatty acids methylated by treatment with 20% tetraethylammonium hydroxide in methanol prior to GC analysis.[181]

Anion-exchange resins have been used to prepare samples for free fatty acid analysis in cheese (e.g. Ref. 182) but these methods have been criticized because of incomplete recovery, glyceride hydrolysis and they are time-consuming.[174]

In the method of McNeill & Connolly,[183] free fatty acids were extracted from Cheddar and Emmental with diethyl ether/HCl and the extract passed through a column of silicic acid to remove phospholipids. Free fatty acids were absorbed on Amberlyst A-26 ion-exchange resin and methylated with a 5% solution of HCl in methanol. The methyl esters were recovered by salting out from a mixture of the ion-exchange resin and diethyl ether with saturated aqueous NaCl. The organic layer was dried with anhydrous Na_2SO_4 prior to GC analysis.

Alkaline arrestant columns have also been used to prepare samples for GC analysis.[174] McCarthy & Duthie[184] introduced such a column which has been applied to cheese by a number of workers, e.g. Thakur et al.,[185] who chromatographed fatty acids as butyl esters and Harte & Stine[186] but the method causes glyceride hydrolysis.[174] An improved silicic acid–KOH column was developed and applied to butter by Woo & Lindsay[187] but the lactic acid in cheese interferes with the analysis and the relatively large amounts of water-induced glyceride hydrolysis.[174] However, Woo & Lindsay[174] developed a technique for Cheddar cheese involving an ethylene glycol pre-column (to remove lactic acid) and an arrestant column modified to remove water. Ha & Lindsay[188] chromatographed volatile ($<C_{12}$) free and total branched-chain fatty acids from Parmesan cheese as butyl esters and identified them by GC-MS.

Triglycerides and partial glycerides in cheese have been studied by Contarini et al.[189] who extracted fat from Gorgonzola by hexane and diethyl ether; aliquots were subjected to TLC and the chromatograms developed with 2,7-dichlorofluorescin in ethanol. The bands corresponding to mono-, di- and tri-glycerides were recovered and extracted with chloroform and diethyl ether. Mono- and diglycerides were chromatographed by capillary GC as trimethylsilane derivatives and triglycerides were chromatographed directly. Triglyceride analysis gave no useful information on cheese ripening but comparisons between partial glycerides were valuable. Studies involving the analysis of shorter chain acids include Refs 168 and 177–179.

3.3 Assessment of Glycolysis

The metabolism of lactose to lactic acid by starter is fundamental to most, if not all, varieties of cheese. L-Lactate is produced by mesophilic and some thermophilic starters while a mixture of D- and L-isomers is produced by other thermophiles. In certain varieties (e.g. Swiss), the lactate produced serves as a substrate for further microbial metabolism and in many varieties the L-isomer produced by the starter is converted to a racemic mixture by the non-starter flora of the cheese (see Ref. 190). Citrate is an important substrate and its metabolism leads to flavour compounds (diacetyl, acetoin and 2,3-butylene glycol) in some cheeses. Many of the products of glycolysis are quantified by enzyme assay and in these cases the enzymatic procedure has superseded earlier methods because of increased sensitivity and specificity. Only enzymatic methods can distinguish between D- and L-lactate.

Lactose can be measured by an enzyme assay, as can D-galactose and D-glucose. Suppliers of test kits include Boehringer Mannheim GmbH (Mannheim, Germany). Lactose is hydrolysed by β-galactosidase to D-glucose and D-galactose. The galactose produced is then oxidized by treatment with galactose dehydrogenase with the concomitant conversion of NAD^+ to NADH, which is monitored spectrophotometrically at 334, 340 or 365 nm and is stoichiometrically related to the concentration of lactose. Glucose is phosphorylated by ATP and hexokinase to glucose-6-phosphate (G6P) which is then oxidized by $NADP^+$ in the presence of G6P dehydrogenase; the NADPH formed is quantified spectrophotometrically.[6,7]

Enzyme assay test kits for D- and L-lactate are available (e.g. Boehringer Mannheim) which use L- and/or D-lactate dehydrogenase (LDH) and glutamate-pyruvate transaminase (GPT). Lactate is oxidized by NAD^+ in the presence of LDH to pyruvate and NADH. The equilibrium of this reaction normally lies in favour of lactate but the pyruvate formed reacts with L-glutamate in the presence of GPT, shifting the equilibrium in flavour of NADH. The concentration of NADH is measured spectrophotometrically and is stoichiometrically related to the concentration of D- or L-lactate.[8]

Acetic acid in cheese can also be measured by an enzyme assay (e.g. from Boehringer Mannheim). Acetyl-coenzyme A is formed from acetate and ATP in the presence of acetyl-CoA synthetase. Acetyl CoA reacts with oxaloacetate and water in the presence of citrate synthetase to form citrate and free CoA. The oxaloactate necessary for this reaction is formed via the action of malate dehydrogenase from malate and NAD^+ with the concomitant formation of NADH, the concentration of which is related, non-linearly, to the acetate concentration.[191]

Citrate in cheese can be quantified by chemical methods (IDF 34B:1971, Ref. 15; AOAC 920.126, 976.15, 1990, Ref. 16) or by an enzyme assay (e.g. from Boehringer Mannheim). In the latter, citrate is converted to oxaloacetate and acetate by citrate lyase. In the presence of malate dehydrogenase and L-LDH, oxaloacetate and its decarboxylation product, pyruvate, are reduced to L-malate and L-lactate, respectively, with the concomitant oxidation of NADH to NAD^+, which is measured spectrophotometrically.[9]

Diacetyl can be quantified in milk by chemical methods. Walsh & Cogan[192,193] described a quantitative colorimetric method for diacetyl and acetoin in milk. Diacetyl is extracted by steam distillation and treated with hydroxylamine to form dimethylglioxime which is converted to a pink ammono-ferrous dimethyl-glyoximate complex by reaction with $FeSO_4$ in alkaline solution. Essentially all the diacetyl is recovered in the first 10 ml of distillate, thus effectively separating it from acetoin and enabling their separate estimation.[192] Keen & Walker[194] measured diacetyl in a steam distillate of Cheddar colorimetrically by reaction with 3,3'-diaminobenzidine tetrahydrochloride and HCl. Acetoin and 2,3-butylene glycol were quantified by extraction with methylene chloride; the extract was separated from the residue, dried with anhydrous Na_2SO_4 and reduced in volume by rotary evaporation. The extract was then equilibrated with water when acetoin and 2,3-butylene glycol passed into the aqueous phase which was clarified by treatment with $BaCl_2/NaOH/ZnSO_4$. Acetoin was quantified directly in the aqueous phase by treatment with α-naphtol. 2,3-Butylene glycol was quantified in an aliquot of clarified aqueous phase: traces of acetoin and diacetyl were removed by ion-exchange chromatography and the 2,3-butylene glycol oxidized to acetoin with bromine water and quantified as acetoin by reaction with α-naphtol. 2-Butanol in a cheese homogenate was quantified by gas chromatography.[194] Thornhill & Cogan[195] described a gas chromatographic method for quantifying acetic acid, acetoin, racemic and meso-2,3-butylene glycol in broth media.

Keen & Walker[179] described a gas chromatographic method for the determination of propionate in cheese. This was applied to Swiss cheese by Turner et al.[196] Marsili[168] measured pyruvic, lactic, acetic and propionic acids by ion-exchange HPLC in aqueous extracts prepared by $ZnSO_4$ and $Ba(OH)_2$ of Cheddar and monitored acetone, 2-butanone, ethanol, 2-pentanone, 2-butanol and n-propanol by GC of headspace gases.

4.1 Hydroxy Acids and Fatty Acid Lactones in Cheese

Although it is generally agreed that lactones may play a role in the overall flavour of cheese, they have received little attention. They may be quantified by gas chromatography and various extraction and preparation procedures have been developed. O'Keefe et al.[197] separated cheese lipids by centrifugation at 40°C and subjected the lipid fraction to molecular distillation at 70°C and 10^{-5} mm Hg. Free fatty acids in the distillate were then removed by saponification and the lactones quantified by GC-MS. Wong et al.[198] developed an adsorption chromatographic method using a column of aluminium oxide, sodium sulphate and Celite 545 with which a cheese sample was mixed. The column was eluted with acetonitrile and the eluate concentrated prior to GC. This procedure was modified by Wong et al.:[199] the acetonitrile extract was evaporated under nitrogen, sodium sulphate was added and the residue re-extracted with acetonitrile and passed through a small column of alumina before final concentration and analysis by GC.

4.2 Methyl Ketones

Patton[200] prepared a fraction rich in methyl ketones by steam distillation and ether extraction; the methyl ketones in the extract were separated by fractional distillation. Morgan & Anderson[201] reacted a distillate from a cheese/water suspension with 2,4-dinitrophenylhydrazine (DNPH) and identified the 2,4-dinitrophenyl-hydrazones by paper chromatography. Sato et al.[202] also employed a steam distillation step. Kanisawa & Itoh[203] measured the methyl ketones produced by fungi in lipolysed milk-fat by steam distillation and gas chromatography.

Schwartz et al.[204] described a method for the quantitative isolation of mono-carbonyls from fats and oils by reaction with DNPH and adsorption of the derivatives onto activated magnesia; most of the lipid material was removed by elution with hexane and the DNPH derivatives then eluted with nitromethane/chloroform and fractionated on activated alumina into four classes (methyl ketones, saturated aldehydes, 2-enals and 2,4-dienals) which were identified by chromatographic procedures. This method has been applied to Blue, Camembert and Roquefort.[205-208] Godinho & Fox[209] also used this method but measured the DNPHs of total carbonyls spectrophotometrically.

Dartley & Kinsella[210] prepared DNPH derivatives from a hexane extract of Blue cheese and fractionated them into classes on activated Celite-Sea Sorb 43. Methyl ketones were then regenerated from the dinitrophenylhydrazones by treatment with H_2SO_4 and analysed by gas chromatography. Keen & Walker[194] quantified 2-butnone in a cheese/water homogenate by refluxing followed by GC. Manning[211] measured 2-pentanone in Cheddar headspace gas directly by GC.

4.3 Biogenic Amines

Decarboxylation of free amino acids leads to the formation of amines, many of which are biologically active and have physiological importance. The principal amines in cheese are histamine, tyramine, tryptamine, putricine, cadaverine and phenylethylamine.[212]

Spector et al.[213] developed a fluorometric assay for tyramine in tissue, involving homogenization and extraction followed by the formation of a fluorophor by reaction with 1-nitroso-2-naphtol. This assay was applied to Gruyère by Price & Smith.[214] Voigt et al.[215] measured tyramine, tryptamine and histamine in a variety of cheeses by TLC; the plates were visualized using 0·2% 7-chloro-4-nitribenzofuran with which the amines formed fluorescent spots which were scraped off the plate and quantified spectrofluorometrically. Colonna & Adda,[216] using a different extraction procedure, quantified tyramine, histamine and tryptamine in a number of French cheeses by TLC, as described by Voigt et al.[215] Taylor et al.[217] considered the chromatographic separation step to be cumbersome and described a simplified method for the determination of histamine in foods, involving selective extraction, reaction with o-phthalaldehyde and spectrofluorometric assay. They determined the histamine content of Cheddar cheese by this method, which was also used by Antila et al.[218] to determine the histamine content of 10 varieties of cheese consumed in Finland. Kaplan et al.[219] considered such fluorometric assays for amines to be laborious and prone to error owing to long extraction procedures; GC was preferred.

Evans et al.[220] quantified tyramine in a number of cheese varieties by TLC; the chromatograms were visualized by a sulphanilic acid-containing reagent and the spots corresponding to tyramine eluted and quantified spectrophotometrically.

Gas chromatographic methods have been used by a number of workers.[158,213,219,221] Spector et al.[213] measured tyramine in tissue extracts directly by GC. Kaplan et al.,[219] who considered direct GC methods to be unsuitable for the determination of amines, recommended derivitization with trifluoroacetic acid prior to GC analysis. Laleye et al.[158] extracted a number of biogenic amines from Cheddar and formed N-heptafluorobutyryl-isopropyl derivatives which were quantified by capillary GC. Staruszkiewicz & Bond[221] analysed cadaverine, putrescine and histamine in a number of foods, including several cheese varieties, by GC of perfluoropropionyl derivatives.

Staruszkiewicz[222] quantified the o-phthalicdicarboxaldehyde derivative of histamine by HPLC with fluorescent detection at 350 nm excition and 444 nm emission. Chambers & Staruszkiewicz[223] used this method to quantify amines in a number of cheeses. Tyramine, phenylethylamine, histamine and tryptamine were analysed in extracts of a number of cheese varieties by Koehler & Eitenmiller[224] and Chang et al.[225] using HPLC without derivitization and with UV detection. Suhren et al.[226] adapted an autoanalyser to measure histamine and tyramine in dairy products. Antila et al.[227] reacted the amines in cheese extracts with dansyl chloride and quantified the derivatives by HPLC with UV detection. An HPLC procedure with amperometric detection was used by Takeba et al.[228] to determine tyramine in cheese; it was claimed that this detection procedure was 25 times more sensitive than fluorometric detection. Sumner et al.[229] used a variation of the AOAC standard procedure for measuring histamine in seafoods (AOAC 18.067–18.071, 1984, Ref. 230) to study the production of histamine in Swiss cheese; cheese was homogenized with methanol, heated to 60°C × 10min and cooled to room temperature; aliquots were chromatographed on Dowex I-X8 and the histamine quantified fluorimetrically by reaction with o-phthalicdicarboxaldehyde. Another AOAC standard method for histamine determination in seafoods involves a bioassay using guinea pig intestine (AOAC 18.060– 18.063, 1984, Ref. 230) but this does not appear to have been applied to cheese.

4.4 Sulphur Compounds

Volatile sulphur compounds, e.g. H_2S, methanethiol, methional, dimethylsulphide, occur in cheese at levels characteristic of the variety and ripening conditions and have been implicated in cheese flavour.[231,232]

Walker[233] heated an aqueous dispersion of shredded Cheddar cheese under reflux conditions and flushed the headspace with a nitrogen stream. The gas from the reflux was passed through a series of traps to remove specific classes of compounds (crystalline lead acetate to remove H_2S; $HgCl_2$ solution [3%] for sulphides and disulphides and $Hg(CN)_2$ solution [3%] for mercaptans). The odour of the gas was noted after each trap to establish the effect of removing each class of compound from the gas stream. Tsugo & Matsuoka[234] used a similar series of traps, but with 4% $Hg(CN)_2$ solution, to separate and identify various sulphur compounds from a semi-soft white mould-type cheese by chemical means. Lawrence[235] flushed an aqueous slurry of cheese with nitrogen, trapped the H_2S in a 1% Zn-acetate/NaOH solution and quantified it by the development of methylene blue from p-diamino-NN-dimethylaniline hydrochloride and $FeCl_3$ which was measured spectrophotometrically at 678 nm. McGugan *et al.*[236] stripped volatiles from centrifugally defatted Cheddar, made with or without starter, by vacuum distillation and trapping in liquid nitrogen. The distillates were then analysed by GC for a number of volatiles, including dimethylsulphide. Manning & Robinson[237] used headspace trapping, methanol extraction and reduced pressure distillation to prepare samples for GC. Headspace gases were considered to contain the essential components of cheese aroma in the correct proportions and the sampling conditions were such that labile components were not lost or damaged. Methanol extraction was not considered suitable and distillation under low pressure was the preferred method of sample preparation because of drawbacks in the headspace analysis techniques used. Dimethylsulphide and H_2S were analysed by this method. Manning[238] modified the distillation apparatus of Manning & Robinson[237] and quantified H_2S, methanethiol and dimethylsulphide by GC. Manning *et al.*[239] developed an improved technique for headspace analysis: a plug of cheese was removed and the entrance to the hole immediately sealed with a metal cap fitted with a septum; the atmosphere was allowed to equilibrate for 1 h and 5 ml aliquots of the headspace gas were then removed by inserting a syringe needle through the septum and used for GC analysis. This method has been used by a number of workers[211,239-243] to analyse cheese volatiles.

4.5 Detection of Adulteration of Ovine and Caprine Milks and Cheeses

Adulteration of ewes' or goats' milk for cheesemaking with less expensive bovine milk has led to the development of techniques which permit the detection of mixed-species milks. This topic has been reviewed by Ramos & Juarez.[244,245]

Various authors have used differences in fatty acid profiles to detect adulteration (e.g. Refs 246, 247). Such chromatographic methods have limited sensitivity and will not detect adulteration with skim milk, but they can be applied to mature cheese since relatively small changes occur in fat composition during maturation.[244] Ramos & Juarez[245] suggest that differences in triglycerides might be used to identify milks from different species. The absence of β-carotene from sheep's and goats' milks may permit the detection of bovine milk in caprine or ovine milks.[245]

Differences between HPLC profiles of tryptic hydrolysates of caseins from different species have been demonstrated;[248] however, the effect of proteolysis during cheese ripening was not studied. The amino acid compositions of caprine[249] and ovine[250] caseins differ from bovine casein.

Electrophoretic techniques have been developed to detect mixtures of milks from different species. Aschaffenburg & Dance[251] showed that bovine α_{s1}-casein had a faster electrophoretic mobility than caprine α-casein while Ramos et al.[247] demonstrated differences between ovine and bovine α_{s1}-caseins which have been exploited by a number of workers to detect mixed-species milks (see Ref. 245). Proteolysis during cheese ripening interferes with this procedure although Pierre & Portmann[252] achieved satisfactory results up to 15 days of ripening for Saint-Maure and Chabichou cheeses. However, Ramos & Juarez[245] considered that the detection of bovine milk in industrially produced sheep's cheeses by electrophoresis is difficult. Differences between ovine and bovine para-κ-caseins permitted the detection by isoelectric focusing of bovine milk in ewes' milk Percorino cheese after five months of ripening.[253] Addeo et al.[254] also used isoelectric focusing and densitometry to study the composition of cheese made with caprine and ovine milks. Krause et al.[255] applied isoelectric focusing to determine interspecies differences in the γ-caseins.

Interspecies differences in the electrophoretic mobility of whey proteins may be used to identify mixed-species milks but not cheeses.

The higher xanthine oxidase activity of bovine milk was used by Monget et al.[256] to detect adulteration of goats' milk. The test was reported to be simple, rapid and could be applied to raw or pasteurized ($<75°C \times 20$ s) milk or fresh cheese and allowed detection of 2% added bovine milk.

Pollman[257] found differences in the Ca/Mg ratio in cheeses made from cows' or sheep's milk.

Immunological methods are well suited for analysis of mixed-species milks because of their sensitivity, specificity and their availability in kit form. However, when preparing antisera to proteins from a particular milk, the antisera will react not only with that milk but with the milks of many other species also; however, techniques have been developed to overcome this (see Ref. 245). Ramos & Juarez[245] report that Durand et al.[258] detected bovine milk in ewes' milk by immunodiffusion in agar with antisera to cow blood serum and Levieux[259] chose immunoglobulin G_1 as the antigen. Caseins have relatively low antigenicity but to overcome this difficulty, Aranda et al.[260] used an immunodotting technique to detect bovine caseins in ewes' milk and cheese. Rodriguez et al.[261] developed an indirect enzyme-linked immunosorbent assay (ELISA) for bovine caseins and used it to assay for bovine milk in sheep's milk and cheese.

5 OBJECTIVE INDICES OF CHEESE RIPENING AND QUALITY

5.1 Introduction

As discussed in Section 1, chemical analyses of cheese are performed for one of several reasons. Possibly the most ambitious is the development of objective methods for assessing cheese quality. The organoleptic qualities of cheese are determined by the physical and chemical changes which occur during ripening and are traditionally assessed by sensory evaluation of flavour, body and texture, finish and overall quality by experienced judges or trained consumer panellists. However, sensory analyses are subjective, and although at present the best index of consumer acceptability, they provide data that are difficult to evaluate scientifically and to compare between laboratories or studies. Therefore, chemical and physical analyses of cheese are used to monitor ripening objectively and to assess quality, usually to complement sensory evaluation. While it may never be possible to assess cheese quality by chemical criteria alone, they can provide very valuable indices of quality. The following is an attempt to describe how some of the analytical methods described in the proceeding sections may be valuable for assessment of quality. The subject was reviewed by Farkye & Fox.[262]

5.2 Chemical Indices

5.2.1 Proteolysis

For most hard and semi-hard cheeses, proteolysis is the most commonly used index of maturity. Most, if not all, nitrogenous compounds that contribute to cheese flavour are soluble in aqueous solvents. Solubility in water or at pH 4·6 is probably most widely used for initial fractionation of cheese N or as a crude index of proteolysis; although relatively simple to perform, it may not be sufficiently discriminating as an index of quality. Both PTA and 12% TCA-soluble N correlate significantly ($p < 0.001$) with the age and flavour intensity of Cheddar cheese.[56] Kroger & Weaver[72] reported that the dye-binding capacity of cheese correlated strongly ($r = 0.85$) with the concentration of free amino acids and suggested its use in determining the chemical age of cheese; however, Kuchroo et al.[29] found that a dye-binding technique using amido black was not very useful for this purpose. Since the concentration of free amino acids in cheese is reported[56] to correlate strongly with the age and maturity of cheese, chemical indicators of free amino acid concentration, e.g. TNBS[29,54,69,79,83] or ninhydrin-reactive groups[79,93] should be useful indicators of cheese age and quality; while there is a substantial amount of information relating these indices to age, there is much less relating them to quality. Marsili[168] reported that the best predictors of the proteolytic age of Cheddar were leucine ($r^2 = 0.532$), methionine ($r^2 = 0.525$) and glutamic acid ($r^2 = 0.477$).

Pham & Nakai[23] analysed water-soluble N (WSN) from 41 commercial Cheddar cheeses of difference ages by reverse phase-HPLC. They observed 13 chromatographic peaks, the relative areas of which varied with the age of the cheese, suggesting that discriminate HPLC analysis of WSN should be a good method for objective evaluation of cheese maturity. Santa-Maria et al.[263] successfully applied linear discriminant analysis of 18 parameters of proteolysis for the age classification of Manchego cheese.

Polyacrylamide gel electrophoresis (PAGE) is widely used to monitor primary proteolysis in cheese and is mostly related to the activities of residual coagulant and plasmin. α_{s1}-Casein is hydrolysed faster than β-casein in cheese by all commercial coagulants but the plant rennets produce very different products.[264,265] Thus PAGE can provide information on the type of coagulant used, which may be indicative of quality. The level of residual α_{s1}-casein is a good index of the level of general proteolysis in relatively young cheeses (e.g. ≤ 3 months for Cheddar or Gouda) but is probably not sufficiently discriminating to be a useful index of quality. Chymosin, bovine and porcine pepsins hydrolyse only α_{s1}-casein in cheese while proteolysis of β-casein in bacterially-ripened cheeses is due mainly to plasmin activity, which appears to be directly related to cooking temperature.[262] Plasmin is active in most cheeses and is considered to make a significant contribution to proteolysis in Romano-type,[266] Swiss and Dutch-type cheese;[267] its activity is clearly indicated by the level of γ-casein but the relevance of this for assessment of cheese quality is not apparent although it is clearly indicative of age.

5.2.2 Lipolysis

The extent of lipolysis during cheese ripening varies among varieties.[268,269] In Cheddar, Emmental, Gruyère and Gouda-type cheeses, the level of lipolysis is low, while in Blue mould-ripened and hard Italian varieties, extensive lipolysis occurs. Obviously, the concentration of free fatty acids increases during ripening and has a major impact on cheese flavour.[174,267,270-272] A certain concentration of free fatty acids in the correct balance appears to be necessary for optimum flavour, but an excess of fatty acids or the incorrect balance leads to off-flavours, especially in mild cheese varieties. As far as we are aware, the only attempt to correlate the 'lipolytic age' of Cheddar cheese with the concentrations of individual free fatty acids is that of Marsili[168] who found that C_{10} ($r^2 = 0.375$), C_{14} ($r^2 = 0.227$), C_{12} ($r^2 = 0.198$) and C_{16} ($r^2 = 0.129$) were the best indicators although considerably less useful as indicators of age than certain products of glycolysis and proteolysis. Such correlations for the more highly lipolysed cheeses, i.e. Blue and Italian, appear to be lacking although the flavour intensity and quality of Italian cheese is strongly influenced by the concentration and balance of free fatty acids.[269]

Liberated fatty acids may be modified to products which may influence cheese quality. This is particularly true for methyl ketones which are responsible for the characteristic flavour of blue mould cheeses; their formation has been extensively studied (see Ref. 273). The concentration and ratios of methyl ketones are useful indicators of the maturity and quality of blue cheeses. A homologous series of δ-lactones (C_{10}–C_{18}) occurs in Cheddar (and probably other) cheeses and their concentrations increase during ripening up to about two months after which they remain constant or decline slightly.[199] Although the concentration of δ-lactones did not correlate with cheese flavour, they may be a useful index of age up to two months. The concentration of γ-dodecalactone was considerably lower than those of the δ-lactones but increased in parallel with the intensity of cheese flavour and hence may be a useful index of maturity. Further work in this area appears warranted.

5.2.3 *Lactose Fermentation and Changes in Lactate*

Fermentation of lactose to lactic acid during cheesemaking reduces the pH of cheese curd to values which prevent the growth of pathogenic bacteria. The rate and extent of acidification of cheese strongly influence cheese quality (see Refs 190, 267, 274) and hence the pH of cheese is a useful indirect index of quality. However, the pH of cheese increases during ripening (see below) and hence the pH of mature cheese may not reflect that of fresh curd. Although ~98% of the lactose of milk is removed in the whey as lactose or lactic acid, cheese curd contains 0·7–1·5% lactose (see Ref. 190; Chapter 10). The concentration of residual lactose in cheese curd depends on the method of manufacture, the type and activity of the starter and especially on the concentration of salt in the moisture phase. Although considerable changes occur in the concentrations of lactose and lactic acid during ripening (see Ref. 190), there are rarely, or not at all, used as indices of ripening. Except in cases where the concentration of salt is so high (>~6% salt-in-moisture) as to prevent bacterial growth, the lactose concentration in Cheddar essentially decreases to zero within about two weeks. In Dutch type cheeses, lactose is completely fermented to lactic acid within 24 h. In high-cooked varieties in which a thermophilic starter (*Str. thermophilus* and a *Lactobacillus* spp.) is used, the concentration of lactose also essentially decreases to zero within one week. In these cheeses, galactose accumulates initially (due to the inability of *Str. thermophilus* to metabolize galactose) but it, also, rapidly reaches zero (see Ref. 190). In mould-ripened cheeses, lactose is completely metabolized within a short period. Therefore, lactose and its constituent monosaccharides are of little value as indices of cheese maturity.

In most cheese varieties, lactose is metabolized to L-lactic acid. In Cheddar, Dutch and many other varieties, L-lactate is isomerized to D-lactate by non-starter lactic acid bacteria (NSLAB) so that a racemic mixture of lactic acid exists after 3–4 months (see Ref. 190). The rate of racemization of lactate depends on the NSLAB population and hence is an index of their numbers and activity rather than of cheese maturity. Although insufficient studies have been undertaken on the rate of racemization of L-lactate in cheese, it is unlikely to be a reliable index of cheese maturity due to variations in NSLAB populations.

In Swiss-type cheese, lactate is converted to propionate, acetate and CO_2 (3 lactate → 2 propionate + 1 acetate + CO_2) when the growth of propionic acid bacteria is initiated on transfer to the hot room (see Ref. 190). During certain periods, the concentrations of lactate, propionate and acetate are good indices of the activity of propionic acid bacteria and hence of correct maturation; the molar ratio of propionate and acetate should be 2:1 and is a good index of the quality of Swiss-type cheese but is less useful as an index of its age.

In surface mould-ripened cheeses, lactate is metabolized to CO_2 and H_2O by the metabolic activity of *P. camemberti*. Thus, the decrease in lactate concentration may be a useful index of the maturity of these cheeses during a certain period.

5.2.4 Changes in pH

The formation of lactic acid during manufacture and the metabolism of residual lactose during the initial stages of ripening reduces the pH of cheese to $\sim 5 \pm 0.3$, depending on variety, after all the lactose has been metabolized. During ripening, the pH of cheese rises due to the formation of alkaline nitrogenous compounds and/or catabolism of lactic acid. The pH of Cheddar increases by only ~ 0.1 unit after six months of ripening;[267] the pH of Gruyère increases from 5.35 to 5.95[271] and that of Gouda from ~ 5.1 to 5.3–5.9 (the extent of this increase appears to depend on the season of manufacture for unidentified reasons).[267] The pH of Camembert increases from ~ 4.8 to as high as 7.5[104,275] while the pH of Blue cheese increases from ~ 4.8 to ~ 7.[209] Obviously, changes in pH of this magnitude during ripening are useful indicators of the age of cheeses, especially mould-ripened varieties, but pH is of little use as an index of the maturity of Cheddar. The increase in pH is essential for the desirable textural changes in surface-ripened cheeses but we are not aware of studies on the correlation between the quality of other varieties and their final pH.

5.2.5 Volatile Organic Compounds

Volatile organic compounds, e.g. acids, carbonyls, lactones, H_2S, methanethiol, dimethylsulphide, etc., occur in cheese at concentrations depending on the variety and ripening conditions. The contribution of volatile compounds to cheese flavour was reviewed by Aston & Dulley[231] (see Chapter 10). Volatile compounds are responsible for cheese aroma and contribute to both background and characteristic flavour, but their concentrations in cheese are variable.[242] McGugan et al.[22] noted that volatiles from un-deodorized cheese fat contributed to flavour but when the volatiles were reconstituted in fat from mild or aged cheeses, the intensity of cheese flavour did not increase significantly, suggesting that quantitation of volatile compounds as an indicator of ripening may underestimate the chemical age of cheese. Marsili[168] reported that the concentrations of alcohols, aldehydes and ketones were poor indicators of the age of Cheddar. However, he found that acetate ($r^2 = 0.395$) and especially propionate ($r^2 = 0.705$) were good indices of the glycolytic age of Cheddar cheese; the concentration of propionate was the best single index of cheese maturity. The mechanism and pathway for the formation of propionate in Cheddar are not obvious; Keen & Walker[179] found no propionate in Cheddar and Dutch-type cheese.

5.3 Physical Indices

5.3.1 Cheese Rheology

In many respects, the rheological properties (body and texture) of cheese are as important as its flavour and are routinely assessed (organoleptically) in cheese grading. Cheese rheology is reviewed in Chapter 8 and textural changes in cheese during ripening have been discussed by Lawrence et al.[267] There are relatively few data on the rheological changes in cheese during ripening. Creamer & Olson[276] showed that for Cheddar, yield strain decreased linearly with logarithm of age (days). It is thought that the breakdown of α_{s1}-casein to α_{s1}-I is the most significant reaction responsible for the initial softening of cheese body and texture,[276,277] indicating that textural changes during ripening are related to proteolysis, particularly by residual rennet. Indeed the hardness and springiness of Cheddar are correlated with proteolysis.[278] The large and obvious changes in the texture of Brie and Camembert are due in part to the increase in pH but an appropriate level of proteolysis is also required for proper texture development.[279,280]

Obviously, physical methods for assessing the body and texture of cheese, to replace subjective, sensory methods, are desirable. However, Green et al.[281] cautioned against instrumental measurements of cheese texture unless these involve fracture and mechanisms similar to those that occur during normal grading and consumption. One of the principal difficulties in physically measuring the rheological properties of cheese is obtaining representative samples: many varieties, e.g. Cheddar, Cheshire, some Blue cheese, are heterogeneous and fractured, while the eyes in Swiss-type cheese create obvious problems.

5.4 Conclusions

Cheese ripening is a complex phenomenon. Attempts to understand this phenomenon have led to the development of objective measurements which, it is hoped, will be able to predict the final quality of cheese. Sensory evaluation of cheese relies on the ability of a grader to examine a small random sample from a batch of cheese and use his skills to predict its quality at a later date. The results of sensory evaluation are subjective. Various objective chemical and rheological approaches to evaluating cheese ripening have been developed and assessed. However, in order to predict accurately the ripening process by chemical and rheological methods, the manufacturing history of the cheese, including the types and activities of starter and rennet used and ripening conditions, need to be taken into consideration. While chemical and physical parameters may be good indicators of maturity, their ability to assess cheese quality is less certain, at least until much more precise knowledge of the biochemistry of cheese ripening has been accumulated.

REFERENCES

1. Lawrence, R.C. & Gilles, J., 1987. In *Cheese: Chemistry Physics and Microbiology*, 1st edn, Vol. II., ed. P.F. Fox. Elsevier Applied Science Publishers, London, p 1.
2. Grappin, R., Rank, T.C. & Olson, N.F., 1985. *J. Dairy Sci.*, **68**, 531.
3. Rank, T.C., Grappin, R. & Olson, N.F., 1985. *J. Dairy Sci.*, **68**, 801.
4. Fox, P.F., 1989. *J. Dairy Sci.*, **72**, 1379.
5. International Dairy Federation, IDF, F-Doc 173, 1990.
6. Boehringer Mannheim, 1986. In *Methods of Biochemical Analysis and Food Analysis*, Boehringer Mannheim GmbH, Biochemica, Mannheim, Germany, p. 66.
7. Boehringer Mannheim, 1986. In *Methods of Biochemical Analysis and Food Analysis*, Boehringer Mannheim GmbH, Biochemica, Mannheim, Germany, p.70.

8. Boehringer Mannheim, 1986. In *Methods of Biochemical Analysis and Food Analysis*, Boehringer Mannheim GmbH, Biochemica, Mannheim, Germany, p. 62.
9. Boehringer Mannheim, 1986. In *Methods of Biochemical Analysis and Food Analysis*, Boehringer Mannheim GmbH, Biochemica, Mannheim, Germany, p. 20.
10. Marcos, A., Esteban, M.A. & Alcala, M., 1990. *Food Chem.*, **38**, 189.
11. Esteban, M.A. & Marcos, A., 1990. *Food Chem.*, **35**, 179.
12. Esteban, M.A., Marcos, A., Alcala, M. & Gomez, R., 1991. *Food Chem.*, **40**, 147.
13. Frank, J.T. & Birth, G.S., 1982. *J. Dairy Sci*, **65**, 1110.
14. Cronin, D.A. & McKenzie, K., 1990. *Food Chem.*, **35**, 39.
15. International Dairy Federation, Standard Methods, Brussels.
16. Official Methods of Analysis, 15th edn, Vol. 2, Association of Official Analytical Chemists, Washington, DC, 1990.
17. Christensen, T.M.I.E., Bech, A.-M. & Werner, H., 1990. In F-Doc, 173, International Dairy Federation, Brussels, pp. 7–20.
18. Mabbitt, L.A., 1955. *J. Dairy Res.*, **22**, 224.
19. Bullock, D.H. & Irvine, O.R., 1956. *J. Dairy Sci.*, **39**, 1229.
20. Stadhouders, J., 1960. *Neth. Milk Dairy J.*, **14**, 83.
21. Reiter, B., Sorokin, Y., Pickering, A. & Hall, A.J., 1969. *J. Dairy Res.*, **36**, 65.
22. McGugan, A.W., Emmons, D.B. & Larmond, E., 1979. *J. Dairy Sci.*, **62**, 398.
23. Pham, A.M. & Nakai, S., 1984. *J. Dairy Sci.*, **67**, 1390.
24. Aston, J.W. & Creamer, L.K., 1986. *N.Z. J. Dairy Sci. Technol.*, **21**, 229.
25. Kuchroo, C.N. & Fox, P.F., 1982. *Milchwissenschaft*, **37**, 331.
26. Kuchroo, C.N. & Fox, P.F., 1982. *Milchwissenschaft*, **37**, 651.
27. Kuchroo, C.N. & Fox, P.F., 1983. *Milchwissenschaft*, **38**, 76.
28. Kuchroo, C.N. & Fox, P.F., 1983. *Milchwissenschaft*, **38**, 389.
29. Kuchroo, C.N., Rahilly, J. & Fox, P.F., 1983. *Ir. J. Food Sci. Technol.*, **7**, 129.
30. O'Sullivan, M. & Fox, P.F., 1990. *J. Dairy Res.*, **57**, 135.
31. Farkye, N.Y. & Fox, P.F., 1991. *J. Agric. Food Chem.*, **39**, 786.
32. Dahlberg, A.C. & Kosikowsky, F.V., 1947. *J. Dairy Sci.*, **30**, 165.
33. Mogensen, M.T.S., 1947. *Arssk. Alnarps. Lantsbruk., Mejeri-Tradgardsinst*, 279; cited from *Chem. Abstr.*, 1948, **42**, 8987.
34. Vakaleris, D.G. & Price, W.V., 1959. *J. Dairy Sci.*, **42**, 264.
35. Vakaleris, D.G., Olson, N.F., Price, W.V. & Knight, S.G., 1960. *J. Dairy Sci.*, **43**, 1058.
36. Ismail, A.A. & Hansen, K., 1972. *Milchwissenschaft*, **27**, 556.
37. Gripon, J.C., Desmazeaud, M.J., Le Bars, D. & Bergere, J.C., 1975. *Lait*, **55**, 502.
38. Furtado, M.M. & Partridge, J.A., 1988. *J. Dairy Sci.*, **71**, 2877.
39. O'Keeffe, R.B., Fox, P.F. & Daly, C., 1976. *J. Dairy Res.*, **43**, 97.
40. O'Keeffe, A.M., Fox, P.F. & Daly, C., 1978. *J. Dairy Res.*, **45**, 465.
41. Reville, W.J. & Fox, P.F., 1978. *Ir. J. Food Sci. Technol.*, **2**, 67.
42. Noomen, A., 1977. *Neth. Milk Dairy J.*, **31**, 163.
43. Venema, D.P., Herstel, H. & Elenbaas, H.L., 1987. *Neth. Milk Dairy J.,* **41**, 215.
44. Visser, F.M.W., 1977. *Neth. Milk Dairy J.*, **31**, 210.
45. Chakravorty, S.C., Spinivasan, R.A., Babbar, I.J., Dudani, A.T., Burde, S.D. & Iya, K.K., 1966. *Proc. XVII Intern. Dairy Congr. (Munich)*, D, 187.
46. Gupta, S.K., Whitney, R.M. & Tuckey, S.L. 1974. *J. Dairy Sci.*, **57**, 540.
47. Harwalkar, V.R. & Elliott, J.A., 1971. *J. Dairy Sci.*, **54**, 8.
48. Visser, F.M.W., 1977. *Neth. Milk Dairy J.*, **31**, 265.
49. Visser, S., Slangen, K.J., Hup, G. & Stadhouders, J., 1983. *Neth. Milk Dairy J.*, **37**, 181.
50. Yvon, M., Chabanet, C. & Pelisser, J.-P., 1989. *Intern. J. Peptide Protein Res.*, **34**, 166.
51. Edwards, J. & Kosikowski, F.V., 1983. *J. Dairy Sci.*, **66**, 727.
52. Poznanski, S., Habaj, B., Rymaszewski, J. & Rapczynski, T., 1966. *Proc. XVII Intern. Dairy Congr. (Munich)*, D, 555.

53. Gonashvily, Sh., 1966. *Proc. XVII Intern. Dairy Congr. (Munich)*, **D**, 289.
53a. Breen, E.D., 1992. MSc Thesis, National University of Ireland, Cork, Republic of Ireland.
54. Jarrett, W.D., Aston, J.W. & Dulley, J.R., 1982. *Aust. J. Dairy Technol.*, **37**, 55.
55. Kleter, G., 1976. *Neth. Milk Dairy J.*, **30**, 254.
56. Aston, J.W., Durward, I.G. & Dulley, J.R., 1983. *Aust. J. Dairy Technol.*, **38**, 55.
57. Aston, J.W., Grieve, P.A., Durward, I.G. & Dulley, J.R., 1983. *Aust. J. Dairy Technol.*, **38**, 59.
58. Nunez, M., Garcia-Aser, C., Rodriguez-Martin, M.A., Medina, M. & Gaya, P., 1986. *Food Chem.*, **21**, 115.
59. Fernandez del Pozo, B., Gaya, P., Medina, M., Rodriguez-Martin, M.A. & Nunez, M., 1988. *J. Dairy Res.*, **55**, 457.
60. Ardo, Y. & Pettersson, H.-E., 1988. *J. Dairy Res.*, **55**, 239.
60a. Gonzalez de Llano, D., Polo, C. & Ramos, M., 1991. *J. Dairy Res.*, **58**, 363.
60b. Gonzalez de Llano, D., Ramos, M. & Polo, C., 1987. *Chromatographia*, **23**, 764.
61. Ramos, M., Caceres, I., Polo, C., Alonso, L. & Juarez, M., 1987. *Food Chem.*, **24**, 271.
62. Salji, J.P. & Kroger, M., 1981. *J. Food Sci.*, **46**, 1345.
63. Hickey, M.W., van Leeuwen, H., Hiller, A.J. & Jago, G.R., 1983. *Aust. J. Dairy Technol.*, **38**, 110.
64. Visser, S., Hup, G., Exterkate, F.A. & Stadhouders, J., 1983. *Neth. Milk Dairy J.*, **37**, 169.
65. Bican, P. & Spahni, A., 1991. *Lebensm. Wiss. u. Technol.*, **24**, 315.
66. Ardo, Y. & Meisel, H., 1990. In Doc 173, International Dairy Federation, Brussels, pp. 21–26.
67. Ordonez, J.A. & Burgos, J., 1977. *Lait*, **57**, 150.
68. Hull, M.E., 1947. *J. Dairy Sci.*, **30**, 881.
69. Samples, D.R., Richter, R.L. & Dill, C.W., 1984. *J. Dairy Sci.*, **67**, 60.
70. Singh, A. & Ganguli, N.C., 1972. *Milchwissenschaft*, **27**, 412.
71. Ashworth, U.S., 1966. *J. Dairy Sci.*, **49**, 133.
72. Kroger, M. & Weaver, J.C., 1979. *J. Food Sci.*, **44**, 304.
73. McGann, T.C.A., Mathiassen, A. & O'Connell, J.A., 1972. *Lab. Pract.*, **21**, 628.
74. Ollikainen, P., 1990. *J. Dairy Res.*, **57**, 149.
75. Lorenz, K., 1974. *Z. Analytische Chemie*, **269**, 182.
76. Satake, K., Okuyama, T., Ohashi, M. & Shinoda, T., 1960. *J. Biochem.*, **47**, 654.
77. Fields, R., 1971. *Biochem. J.*, **124**, 581.
78. Clegg, K.M., Lee, Y.K. & McGilligan, J.F., 1982. *J. Food Technol.*, **17**, 517.
79. Pearce, K.N., Karahalios, D. & Friedman, M., 1988. *J. Food Sci.*, **53**, 432.
80. Habeeb, A.F.S.A., 1966. *Anal. Biochem.*, **14**, 328.
81. Adler-Nissen, J., 1979. *J. Agric. Food Chem.*, **27**, 1256.
82. Barlow, I.E., Lloyd, G.T. & Ramshaw, E.H., 1986. *Aust. J. Dairy Technol.*, **41**, 79.
83. Madkor, S.A., Fox, P.F. & Metwally, N.H., 1984. *Ir. J. Food Sci. Technol.*, **8**, 84.
84. Polychroniadou, A., 1988. *J. Dairy Res.*, **55**, 585.
85. Humbert, G., Guingamp, M.-F., Kouomegne, R. & Linden, G., 1990. *J. Dairy Res.*, **57**, 143.
86. Ruheman, S., 1910. *J. Chem. Soc.*, **98**, 2025.
87. Moore, S. & Stein, W.H., 1948. *J. Biol. Chem.*, **176**, 367.
88. Moore, S. & Stein, W.H., 1954. *J. Biol. Chem.*, **211**, 907.
89. Moore, S., 1968. *J. Biol. Chem.*, **243**, 6281.
90. Doi, E., Shibata, D. & Matoba, T., 1981. *Anal. Biochem.*, **118**, 173.
91. Tsarichenko, A.P., 1966. *Nauch. Tr. Krasnodar. Gos. Pedagog. Inst.* No. 70, 86; cited from *Chem. Abstr.*, **67**, No. 79479c.
92. Friedman, M., Pang, J. & Smith, G.A., 1984. *J. Food Sci.*, **49**, 10.
93. Folkertsma, B. & Fox, P.F., 1992. *J. Dairy Res.*, **59**, 217.

94. Weigele, M., De Barnardo, S.L., Tengi, J.P. & Leimgruber, W., 1972. *J. Am. Chem. Soc.*, **94**, 5927.
95. Udenfried, S., Stein, S., Bohlen, P., Darman, W., Leimgruber, W. & Weigele, M., 1972. *Science*, **178**, 871.
96. Pearce, K.N., 1979. *N.Z. J. Dairy Sci. Technol.*, **14**, 233.
97. Beeby, R., 1980. *N.Z. J. Dairy Sci. Technol.*, **15**, 99.
98. Creamer, L.K., Lawrence, R.C. & Gilles, J., 1985. *N.Z. J. Dairy Sci. Technol.*, **20**, 185.
99. Church, F.C., Swaisgood, H.E., Porter, D.H. & Catignani, G.L., 1983. *J. Dairy Sci.*, **66**, 1219.
100. Frister, H., Meisel, H. & Schlimme, E., 1989. *Kieler Milchwirtschaftliche Forschungsberichte*, **41**, 237.
101. Creamer, L.K., 1990. In F-Doc 173, International Dairy Federation, Brussels, p. 27.
102. Lindqvist, B., Storgards, T., Goransson, M.-B., 1953. *Proc. XIII Intern. Dairy Congr. (The Hague)*, **3**, 1261.
103. Lindqvist, B. & Storgards, T., 1959. *Proc. XV Intern. Dairy Congr. (London)*, **2**, 679.
104. Trieu-Cuot, P. & Gripon, J.-C., 1982. *J. Dairy Res.*, **49**, 501.
105. Shalabi, S.I. & Fox, P.F., 1987. *Ir. J. Food Sci. Technol.*, **11**, 135.
106. Andrews, A.T., 1983. *J. Dairy Res.*, **50**, 45.
107. Blakesley, R.W. & Boezi, J.A., 1977. *Anal. Biochem.*, **82**, 580.
108. Kosikowsky, F.V., 1951. *J. Dairy Sci.*, **34**, 235.
109. Storgards, T. & Lindqvist, B., 1953. *Milchwissenschaft*, **8**, 5.
110. Clemens, W., 1954. *Milchwissenschaft*, **9**, 195.
111. Mahmoud, S.A.Z., Khader, A.E., Salem, O.M., Moussa, A.M. & Kebary, K.M., 1983. *Minufiya J. Agric. Res.*, **6**, 207, cited from *Food Sci. Technol. Abstr.*, 1984, **16**, 6 P1446.
112. Visser, S., Slangen, K.J. & Hup, G., 1975. *Neth. Milk Dairy J.*, **29**, 319.
113. Ney, K.H., 1985. *Milchwissenschaft*, **40**, 207.
114. Mojarro-Guerre, S.H., Amado, R., Arrigoni, G. & Solms, J., 1991. *J. Food Sci.*, **56**, 943.
115. Tokita, F. & Hosono, A., 1968. *Milchwissenschaft*, **23**, 758.
116. Foster, P.M.D. & Green, M.L., 1974. *J. Dairy Res.*, **41**, 259.
117. Green, M.L. & Foster, P.M.D., 1974. *J. Dairy Res.*, **41**, 269.
118. Gripon, J.C., Desmazeaud, M.J., Le Bars, D. & Bergere, J.L., 1977. *J. Dairy Sci.*, **60**, 1532.
119. Green, M.H. & Stackpoole, A., 1975. *J. Dairy Res.*, **42**, 297.
120. Lindqvist, B., 1962. *Proc. XVI Intern. Dairy Congr. (Copenhagen)*, **B**, 673.
121. Lindqvist, B., Lindberg, I. & Molin, H., 1963. *Milchwissenschaft*, **18**, 12.
122. Abd-El-Salam, M.H. & El-Shibiny, S., 1972. *J. Dairy Res.*, **39**, 219.
123. Huber, L. & Klostermeyer, H., 1974. *Milchwissenschaft*, **29**, 449.
123a. Nath, K. R. & Ledford, R. A., 1973. *J. Dairy Sci.*, **56**, 710.
124. Bachmann, M. & Schaub, W., 1974. *Proc. XIX Intern. Dairy Congr. (New Delhi)*, **1E**, 271.
125. Creamer, L.K. & Richardson, B.C., 1974. *N.Z. J. Dairy Sci. Technol.*, **9**, 9.
126. Gripon, J.C., Desmazeaud, M.J., Le Bars, D. & Bergere, J.L., 1975. *Lait*, **55**, 502.
127. Santoro, M., Leone, A.M., La Notte, E. & Liuzzi, V.A., 1987. *Milchwissenschaft*, **42**, 709.
128. Creamer, L.K., 1975. *J. Dairy Sci.*, **58**, 287.
129. Creamer, L.K., Aston, A. & Knighton, D., 1988. *N.Z. J. Dairy Sci. Technol.*, **23**, 185.
130. Smith, A.M. & Nakai, S., 1990. *Can. Inst. Food Sci. Technol. J.*, **23**, 53.
131. Amantea, G.F., Skura, B.J. & Nakai, S., 1986. *J. Food Sci.*, **51**, 912.
132. Kaminogawa, S., Yan, T.R., Azuma, N. & Yamauchi, K., 1986. *J. Food Sci.*, **51**, 1253.

133. Christensen, T.M.I.E., Kristiansen, K.R. & Madsen, J.S., 1989. *J. Dairy Res.*, **56**, 823.
134. Cliffe, A.J., Revell, D. & Law, B.A., 1989. *Food Chem.*, **34**, 147.
135. Cliffe, A.J. & Law, B.A., 1990. *Food Chem.*, **36**, 73.
136. Cliffe, A.J. & Law, B.A., 1991. *Food Biotechnol.*, **5**, 1.
137. Mulvihill, D.M. & Fox, P.F., 1979. *J. Dairy Res.*, **46**, 641.
138. Bican, P., 1983. *J. Dairy Sci.*, **66**, 2195.
139. Van Hooydonk, A.C.M. & Olieman, C., 1982. *Neth. Milk Dairy J.*, **36**, 153.
140. Bican, P. & Blanc, B., 1982. *Milchwissenschaft*, **37**, 592.
141. Vreeman H.J., Visser, S., Slangen, C.J. & van Riel, J.A.M., 1986. *Biochem. J.*, **240**, 87.
142. Ardo, Y. & Gripon, J.C., 1990. In F-Doc 173, International Dairy Federation, Brussels, pp. 59–68.
143. Barrefors, P., Ekstrand, B., Fagerstam, L., Larson-Raznikiewicz, M., Scharr, J. & Steffner, P., 1985. *Milchwissenschaft*, **40**, 257.
144. St. Martin, M. & Paquin, P., 1990. *J. Dairy Res.*, **57**, 63.
145. Wilkinson, M.G., Guinee, T.P., O'Callaghan, D.M. & Fox, P.F., 1992. *Le Lait*, **72**, 449.
146. Kosikowsky, F.V., 1951. *J. Dairy Sci.*, **34**, 228.
147. Kosikowsky, F.V. & Dahlberg, A.C., 1954. *J. Dairy Sci.*, **37**, 167.
148. Hintz, P.C., Slatter, W.L. & Harper, W.J., 1956. *J. Dairy Sci.*, **39**, 235.
149. Ali, L.A.M. & Mulder, H., 1961. *Neth. Milk Dairy J.*, **15**, 377.
150. Moore, S. & Stein, W.H., 1954. *J. Biol. Chem.*, **211**, 893.
151. Polychroniadou, A. & Vlachos, J., 1979. *Lait*, **59**, 234.
152. Weaver, J.C., Kroger, M. & Thompson, M.P., 1978. *J. Food Sci.*, **43**, 579.
153. Shindo, K., Sakurada, K., Niki, R. & Arima, S., 1980. *Milchwissenschaft*, **35**, 527.
154. Omar, M.M., 1984. *Food Chem.*, **15**, 19.
155. Polo, C., Ramos, M. & Sanches, R., 1985. *Food Chem.*, **16**, 85.
156. March, J.F., 1975. *Anal. Biochem*, **69**, 420.
157. Wood, A.F., Aston, J.W. & Douglas, G.K., 1985. *Aust. J. Dairy Technol.*, **40**, 166.
158. Laleye, L.C., Simard, R.E., Grosselin, C., Lee, B.H. & Giroux, R.N., 1987. *J. Food Sci.*, **52**, 303.
159. Kostyra, H. & Damicz, W., 1979. *Zes. Naukowe. Akadem. Rolniczo Technic. Olsztynie Technol. Zywnosci.*, **15**, 91; cited from *Food Sci. Technol. Abstr.*, 1981, **13**, 6P1024.
160. Desyatnikova, O.I. & Alexaev, N.S., 1979. *Izvestiya Vysshikh Uchebnykh Zavedeii Pishchevaya Tekhnologiya*, **5**, 29; cited from *Food Sci. Technol. Abstr.*, 1981, **13**, 2P309.
161. Notte, E., La Santoro, M., Leone, A.M. & Vitagliano, M., 1980. *Sci. Tec. Lattiero-Casearia*, **31**, 19; cited from *Food Sci. Technol. Abstr.*, 1980, **12**, 21P2047.
162. Ali, M.M., Al-Dahhan, D.H. & Abo-Elnaga, I.G., Zanco, A., 1980. *Pure Appl. Sciences*, **6**, 97; cited from *Food Sci. Technol. Abstr.*, 1984, **16**, 9P2064.
163. Ucuncu, M., 1981. *Molkerei-Zeitung wett der Milch*, **35**, 634; cited from *Food Sci. Technol. Abstr.*, 1982, **14**, 2P298.
164. Buruiana, L.M. & Zeidan, A.N., 1982. *Egypt. J. Dairy Sci.*, **10**, 209; cited from *Food Sci. Technol. Abstr.*, 1984, **16**, 2P334.
165. Lavanchy, P. & Buhlmann, C., 1983. *Schweiz. Milchwirtschaft. Forsch.*, **12**, 3; cited from *Food Sci. Technol. Abstr.*, 1984, **16**, 3P623.
166. Buruiana, L.M. & Farag, S.I., 1983. *Egypt. J. Dairy Sci.*, 1983, **11**, 53; cited from *Food Sci. Technol. Abstr.*, 1984, **16**, 8P1828.
167. Resmini, P., Pellegrino, L., Pazzaglia, C. & Hogenboom, J.A., 1985. *Scienza Tec. Lattiero-Casearia*, **36**, 557; cited from *Food Sci. Technol. Abstr.*, 1988, **20**, 12P126.
168. Marsili, R., 1985. *J. Dairy Sci.*, **68**, 3155.
169. Hwang, J.H., Huh, J.W. & Yu, J.H., 1987. *Korean J. Dairy Sci.*, **9**, 101; cited from *Food Sci. Technol. Abstr.*, 1988, **20**, 4P98.

170. Attia, I.A. & Gooda, E., 1987. *Egypt. J. Dairy Sci.*, **15**, 135; cited from *Food Sci. Technol. Abstr.*, 1987, **19**, 11P82.
171. Koops, J. & Klomp, H., 1977. *Neth. Milk Dairy J.*, **31**, 56.
172. Shipe, W.F., Senyk, G.F. & Fountain, K.B., 1980. *J. Dairy Sci.*, **63**, 193.
173. Bynum, D.G., Senyk, G.F. & Barbano, D.M., 1984. *J. Dairy Sci.*, **67**, 1521.
174. Woo, A.H. & Lindsay, R.C., 1982. *J. Dairy Sci.*, **65**, 1102.
175. Deeth, H.C. & Fitz-Gerald, C.H., 1976. *Aust. J. Dairy Technol.*, **31**, 53.
176. Thomas, E.L., Nielson, A.J. & Olson, J.C. Jr., 1955. *Am. Milk Rev.*, **77**, 50.
177. Dulley, J.R. & Grieve, P.A., 1974. *Aust. J. Dairy Technol.*, **29**, 120.
178. Ledford, R.A., 1969. *J. Dairy Sci.*, **52**, 949.
179. Keen, A.R. & Walker, N.J., 1974. *J. Dairy Res.*, **41**, 397.
180. Humbert, E.S. & Lindsay, R.C., 1969. *J. Dairy Sci.*, **52**, 1862.
181. Martin-Hernandez, M.C., Alonso, L., Juarez, M. & Fontecha, J., 1988. *Chromatographia*, **25**, 87.
182. Bills, D.D., Khatri, L.L. & Day, E.A., 1963. *J. Dairy Sci.*, **46**, 1342.
183. McNeill, G.P. & Connolly, J.F., 1989. *Ir. J. Food Sci. Technol.*, **13**, 119.
184. McCarthy, R.D. & Duthie, A.H., 1962. *J. Lipid Res.*, **3**, 117.
185. Thakur, M.K., Kirk, J.R. & Hedrick, T.I., 1975. *J. Dairy Sci.*, **58**, 175.
186. Harte, B.R. & Stine, C.M., 1977. *J. Dairy Sci.*, **60**, 1266.
187. Woo, A.H. & Lindsay, R.C., 1980. *J. Dairy Sci.*, **63**, 1058.
188. Ha, J.K. & Lindsay, R.C., 1990. *J. Dairy Sci.*, **73**, 1988.
189. Contarini, G., Galliena, C., Toppino, P.M. & Amelotti, G., 1990. *Proc. 11th Intern. Symp. Capillary GC., Monterey, CA, USA*, pp. 407–420.
190. Fox, P.F., Lucey, J.A. & Cogan, T.M., 1990. *CRC Crit. Rev. Food Sci. Technol.*, **29**, 237.
191. Boehringer Mannheim, 1986. In *Methods of Biochemical Analysis and Food Analysis*, Boehringer Mannheim GmbH, Biochemica, Mannheim, Germany, pp. 8–10.
192. Walsh, B. & Cogan, T.M., 1974. *J. Dairy Res.*, **41**, 25.
193. Walsh, B. & Cogan, T.M., 1974. *J. Dairy Res.*, **41**, 31.
194. Keen, A.R. & Walker, N.J., 1974. *J. Dairy Res.*, **41**, 65.
195. Thornhill, P.J. & Cogan, T.M., 1984. *Appl. Environ. Microbiol.*, **47**, 1250.
196. Turner, K.W., Morris, H.A. & Martley, F.G., 1983. *N.Z. J. Dairy Sci.Technol.*, **18**, 117.
197. O'Keefe, P.W., Libbey, L.M. & Lindsay, R.C., 1969. *J. Dairy Sci.*, **52**, 888.
198. Wong, N.P., Ellis, R., La Croix, D.E. & Alford, J.A., 1973. *J. Dairy Sci.*, **56**, 636.
199. Wong, N.P., Ellis, R. & La Croix, D.E., 1975. *J. Dairy Sci.*, **58**, 1437.
200. Patton, S., 1950. *J. Dairy Sci.*, **33**, 680.
201. Morgan, M.E. & Anderson, E.O., 1956. *J. Dairy Sci.*, **39**, 253.
202. Sato, M., Honda, T., Yamada, Y., Takada, A. & Kawanami, T., 1966. *Proc. XVII Intern. Dairy Congr. (Munich)*, **D**, 539.
203. Kanisawa, T. & Itoh, H., 1984. *J. Jap. Soc. Food. Sci. Technol.*, **31**, 477; cited from *Food Sci. Technol. Abstr.*, 1986, **18**, 2P137.
204. Schwartz, D.P., Haller, H.S. & Keeney, M., 1963. *Anal. Chem.*, **35**, 2191.
205. Schwartz, D.P. & Parks, O.W., 1963. *J. Dairy Sci.*, **46**, 989.
206. Schwartz, D.P. & Parks, O.W., 1963. *J. Dairy Sci.*, **46**, 1136.
207. Schwartz, D.P., Parks, O.W. & Boyd, E.N., 1963. *J. Dairy Sci.*, **46**, 1422.
208. Jolly, R.C. & Kosikowski, F.V., 1975. *J. Dairy Sci.*, **58**, 846.
209. Godinho, M. & Fox, P.F., 1981. *Milchwissenschaft*, **36**, 476.
210. Dartley, C.K. & Kinsella, J.E., 1971. *J. Agr. Food. Chem.*, **19**, 771.
211. Manning, D.J., 1978. *J. Dairy Res.*, **45**, 479.
212. Renner, E., 1987. In *Cheese: Chemistry, Physics and Microbiology*, 1st edn, ed. P.F. Fox. Elsevier Applied Science Publishers, London, p. 345.
213. Spector, S., Melmon, K., Lovenberg, W. & Sjoerdsma, A., 1963. *J. Pharmacol. Exper. Ther.*, **140**, 229.

214. Price, K. & Smith, S.E., 1971. *Lancet*, **1**(1), 130.
215. Voigt, M.N., Eitenmiller, R.R., Keohler, P.E. & Hamdy, H.K., 1974. *J. Milk Food Technol.*, **37**, 377.
216. Colonna, P. & Adda, J., 1976. *Lait*, **56**, 143.
217. Taylor, S.L., Lieber, E.R. & Leatherwood, M., 1978. *J. Food Sci.*, **43**, 247.
218. Antila, P., Luomanpera, E. & Antila, V., 1982. *Proc. XXI Intern. Dairy Congr. (Moscow)*, **1**(1), 465.
219. Kaplan, E.R., Sapeika, N. & Moodie, I.M., 1974. *Analyst*, **99**, 565.
220. Evans, C.S., Gray, S. & Kazim, N.O., 1988. *Analyst*, **113**, 1605.
221. Staruszkiewicz, W.F. & Bond, J.F., 1981. *J. Assoc. Off. Anal. Chem.*, **64**, 584.
222. Staruszkiewicz, W.F, 1977. *J. Assoc. Off. Anal. Chem.*, **60**, 1131.
223. Chambers, T.L. & Staruszkiewicz, W.F., 1978. *J. Assoc. Off. Anal. Chem.*, **61**, 1092.
224. Koehler, P.E. & Eitenmiller, R.R., 1978. *J. Food Sci.*, **43**, 1245.
225. Chang, S.F., Ayres, J.W. & Sandine, W.E., 1985. *J. Dairy Sci.*, **68**, 2840.
226. Suhren, G., Heeschen, W. & Tolle, A., 1982. *Milchwissenschaft*, **37**, 143.
227. Antila, P., Antila, V., Mattila, J. & Hakkarainen, H., 1984. *Milchwissenschaft*, **39**, 81.
228. Takeba, K., Maruyama, T., Matsumoto, M. & Nakazawa, H., 1990. *J. Chromatog.*, **504**, 441.
229. Sumner, S.S., Roche, F. & Taylor, S.L., 1990. *J. Dairy Sci.*, **73**, 3050.
230. *Official Methods of Analysis*, 14th edn., 1984. Association of Official Analytical Chemists, Washington, DC, p. 339.
231. Aston, J.W. & Dulley, J.R., 1982. *Aust. J. Dairy Technol.*, **37**, 59.
232. Adda, J., Gripon, J.C. & Vassal, L., 1982. *Food Chem.*, **9**, 115.
233. Walker, J.R.L., 1959. *J. Dairy Res.*, **26**, 273.
234. Tsugo, T. & Matsuoka, H., 1962. *Proc. XVI Intern. Dairy Congr. (Copenhagen)*, **B**, 385.
235. Lawrence, R.C., 1963. *J. Dairy Res.*, **30**, 235.
236. McGugan, W.A., Howsam, S.G., Elliott, J.A., Emmons, D.B., Reiter, B. & Sharpe, M.E., 1968. *J. Dairy Res.*, **35**, 237.
237. Manning, D.J. & Robinson, H.M., 1973. *J. Dairy Res.*, **40**, 63.
238. Manning, D.J., 1974. *J. Dairy Res.*, **41**, 81.
239. Manning, D.J., Chapman, H.R. & Hosking, Z.D., 1976. *J. Dairy Res.*, **43**, 313.
240. Manning, D.J. & Price, J.C., 1977. *J. Dairy Res.*, **44**, 357.
241. Manning, D.J., 1979. *J. Dairy Res.*, **46**, 531.
242. Manning, D.J. & Moore, C., 1979. *J. Dairy Res.*, **46**, 539.
243. Aston, J.W. & Douglas, K., 1983. *Aust. J. Dairy Technol.*, **38**, 66.
244. Ramos, M. & Juarez, M., 1984. Bull. 181, International Dairy Federation, Brussels, p. 3.
245. Ramos, M. & Juarez, M., 1986. Bull. 202, International Dairy Federation, Brussels, p. 175.
246. Palo, V., 1975. A 7-Doc 3, International Dairy Federation, Brussels.
247. Ramos, M., Martinez-Castro, I. & Juarez, M., 1977. *J. Dairy Sci.*, **60**, 870.
248. Kaiser, K.P. & Krause, I., 1985. *Z. Lebensm. Unters. Forsch.*, **180**, 181.
249. Lavoille, B., 1976. *Ann. Fals. Exp. Chim.*, **69**, 535.
250. Sawaya, W.N., Safi, W.J., Al-Shalhat, A.F. & Al-Mohammad, H.M., 1984. *Milchwissenschaft*, **39**, 90.
251. Aschaffenburg, R. & Dance, J.E., 1968. *J. Dairy Res.*, **35**, 383.
252. Pierre, A. & Portmann, A., 1970. *Ann. Technol. Agric.*, **19**, 107.
253. Addeo, F., Anelli, G., Stingo, C. Chianese, L., Petrilli, P. & Scudiero, A., 1984. *Latte*, **9**, 37.
254. Addeo, F., Anelli, G. & Chianese, L., 1986. Bull. 202, International Dairy Federation, Brussels.
255. Krause, I., Belitz, H.D. & Kaiser, K.P., 1982. *Z. Lebensm. Unters. Forsch.*, **174**, 195.

388   P.L.H. McSWEENEY & P.F. FOX

256. Monget, D., Gelin, M. & Laviolette, P., 1979. *Lait*, **59**, 117.
257. Pollman, R.M., 1984. *J. Assoc. Off. Anal. Chem.*, **67**, 1062.
258. Durand, M., Meusnier, M., Delahaye, J. & Prunet, P., 1974. *Boll. Ac. Vet.*, **47**, 247.
259. Levieux, D., 1978. *Proc. XX Intern. Dairy Congr. (Paris)*, Paper 15 ST.
260. Aranda, P., Oria, R. & Calvo, M., 1988. *J. Dairy Res.*, **55**, 121.
261. Rodriguez, E., Martin, R., Garcia, T., Hernandez, P.E. & Sanz, B., 1990. *J. Dairy Res.*, **57**, 197.
262. Farkye, N.Y. & Fox, P.F., 1990. *Trends Food Sci. Technol.*, **1**, 37.
263. Santa-Maria, G., Ramos, M. & Ordonez, J.A., 1986. *Food Chem.*, **19**, 225.
264. Phelan, J.A., 1985. PhD Thesis, National University of Ireland, Cork, Republic of Ireland.
265. Yiadom-Farkye, N.A., 1986. PhD Thesis, Utah State University, Logan, UT, USA.
266. Guinee, T.P. & Fox, P.F., 1984. *Ir. J. Food. Sci. Technol.*, **8**, 105.
267. Lawrence, R.C., Creamer, L.K. & Gilles, J., 1987. *J. Dairy Sci.*, **70**, 1748.
268. Woo, A.H., Kollodge, S. & Lindsay, R.C., 1984. *J. Dairy Sci.*, **67**, 874.
269. Woo, A. H. & Lindsay, R.C., 1984. *J. Dairy Sci.*, **67**, 960.
270. Arnold, A.G., Shahani, K.M. & Dwivedi, B.K., 1975. *J. Dairy Sci.*, **58**, 1127.
271. Zerfiridis, G., Zafopoulou-Mastrogiannaki, A. & Litopoulou-Tzanekati, E., 1984. *J. Dairy Sci.*, **67**, 1397.
272. Barbano, D.M., 1985. In *New Dairy Products via New Technology*, Proc. IDF Seminar, Atlanta, GA, USA, pp. 31–53.
273. Kinsella, J.E. & Hwang, D.H., 1976. *CRC Crit. Rev. Food Sci. Nutr*, **8**, 191.
274. Lawrence, R.C., Heap, H.A. & Gilles, J., 1984. *J. Dairy Sci.*, **67**, 1632.
275. Karahadian, G. & Lindsay, R.C., 1987. *J. Dairy Sci.*, **70**, 909.
276. Creamer, L.K. & Olson, N.F., 1982. *J. Food Sci.*, **47**, 631.
277. de Jong, L., 1976. *Neth. Milk Dairy J.*, **30**, 242.
278. Fedrick, I.A. & Dulley, J.R., 1984. *N.Z. J. Dairy Sci. Technol.*, **19**, 141.
279. Noomen, A., 1983. *Neth. Milk Dairy J.*, **37**, 229.
280. Lenoir, J., 1984. Bull. 171, International Dairy Federation, Brussels. pp. 3–20.
281. Green, M.L., Marshall, R.J. & Brooker, B.E., 1985. *J. Texture Studies*, **16**, 351.

10

Biochemistry of Cheese Ripening

P.F. Fox,[a] J. Law,[b] P.L.H. McSweeney[a] & J. Wallace[a]

[a]Department of Food Chemistry, [b]Department of Microbiology,
University College, Cork, Republic of Ireland

1 INTRODUCTION

As discussed in Chapter 1, cheese manufacture essentially involves concentrating the fat and casein of milk 6–12-fold by coagulating the casein, enzymatically or isoelectrically, and inducing syneresis of the coagulum which can be controlled by various combinations of time, temperature, pH, agitation and pressure. At the end of the manufacturing phase, all the rennet (enzymatically)-coagulated cheeses are essentially very similar, consisting of a matrix of calcium para-caseinate in which various proportions of lipids are dispersed and with moisture contents typically in the range 35–50%. Depending on the cooking temperature used during manufacture and the moisture content, fresh rennet cheeses are more or less 'rubbery' and are essentially flavourless. Although they may be consumed in this state, this is not usually done. Instead, they are matured (ripened) for periods ranging from about three weeks (e.g. Mozzarella) to two or more years, depending on the moisture content of the cheese and the intensity of flavour desired.

The basic composition and structure of cheese are determined by the curd manufacturing operations but it is during ripening that the individuality and unique characteristics of each cheese variety develop, as influenced by the composition of the curd and other factors, e.g. the microflora established during manufacture. Some bacterial growth does occur in cheese during ripening, especially of the non-starter lactic acid bacteria, and of moulds in the case of the mould-ripened varieties. Although the actual growth of these microorganisms does contribute to cheese ripening, perhaps very significantly in some varieties, cheese ripening is essentially an enzymatic process.

The ripening of the principal cheese varieties is reviewed in separate chapters in Volume 2. The objective of this chapter is to overview the general aspects of the biochemistry of cheese ripening, which are more or less common to all varieties, and the agents responsible for ripening.

2 RIPENING AGENTS

Four, and possibly five, agents are involved in the ripening of cheese: (1) rennet or rennet substitute (i.e. chymosin, pepsin or microbial proteinases); (2) indigenous milk enzymes, which are particularly important in raw milk cheeses; (3) starter bacteria and their enzymes, which are released after the cells have died and lysed; (4) enzymes from secondary starters (e.g. propionic acid bacteria, *Brevibacterium linens*, yeasts and moulds, such as *Penicillium roqueforti* and *P. candidum*) are of major importance in some varieties; (5) non-starter bacteria, i.e. organisms that either survive pasteurization of the cheesemilk or gain access to the pasteurized milk or curd during manufacture; after death, these cells lyse and release enzymes. The contribution of enzymes from non-starter bacteria to cheese quality is controversial; there is a commonly held view that in the case of Cheddar and Dutch cheeses, species of *Lactobacillus, Pediococcus* and *Micrococcus* probably have negative effects on cheese quality, although they almost certainly contribute to the intensity of cheese flavour.

There has been interest for about 30 years in devising model systems that would permit quantitation of the contribution of each of these five agents to cheese ripening and to the secondary reactions. The techniques developed eliminate one or more of the above agents, thereby enabling its role to be assessed, directly or indirectly.

Non-starter bacteria may be eliminated by using an aseptic bucket milking technique, developed by Perry & McGillivray;[1] the teat cups and clusters were chemically sterilized and the bucket steam-sterilized. Cows were screened for the bacteriological quality of their milk and animals with counts <100 cfu. ml^{-1} selected; prior to milking, their udders were cleaned with a quaternary ammonium solution. An essentially similar approach was used by O'Keeffe *et al.*[2] who obtained milk with a total bacterial count of <500 cfu. ml^{-1}. Kleter & de Vries[3] included a cooling coil between the cluster and bucket and succeeded in achieving counts averaging 46 cfu. ml^{-1}. This approach was also adopted by Visser.[4] Reiter *et al.*[5] withdrew milk aseptically by means of a teat cannula, but the quantities obtained (1 litre) were sufficient to produce cheeses of only about 100 g.

Having collected low-count milk, a heating step is usually employed to reduce bacterial counts further. Perry & McGillivray[1] used batch pasteurization (68°C × 5 min) in a steam-jacketed cheese vat. Chapman *et al.*,[6] who did not use an aseptic milking technique, used HTST pasteurization (71·6°C × 17 s) to produce low-count milk. Reiter *et al.*,[7] Visser[4,8] and Kleter[9] also used HTST pasteurization. Turner *et al.*[10] investigated the thermal destruction of various strains of bacteria with reference to the production of milk for aseptic cheesemaking. It was noted that a reduction of 10^8 was required to produce cheese with non-starter counts of <100 cfu per 10 kg cheese from milk with an initial count of 10^3 cfu. ml^{-1} It was concluded that a heat treatment of 83°C × 15 s or 72°C × 58 s would be necessary to ensure this death rate but a heat treatment of 72°C × 15 s would be sufficient for milk with an initial non-starter count of 10 cfu. ml^{-1} An LTLT regime (63°C × 30 min) was adopted by Reiter et al.[5] and O'Keeffe et al.[2] Le Bars et al.[11] used a UHT treatment and offset the ill-effects of the high heat treatment on the rennetability of milk by using a higher rennet concentration, a higher setting temperature and supplementing with $CaCl_2$.

If the cheese curd is to be chemically acidified, antibiotics can be added to the cheesemilk to inhibit the growth of any remaining (or contaminating) bacteria. Nisin, penicillin and streptomycin were used by Le Bars et al.[11] and O'Keeffe et al.[2] Addition of antibiotics is probably necessary to achieve aseptic starter-free cheese.

Cheese with controlled microflora must be manufactured under aseptic conditions. Enclosed vats equipped with integral rubber gauntlets were used by Mabbitt et al.[12] and modified by Perry & McGillivray[1] to include pressurized or sterile air. Chapman et al.[6] and Reiter et al.[5,7] used a similar technique. Le Bars et al.[11] made cheese in an aseptic room (5 × 3 m) with a filtered air supply and the cheesemakers were clothed in sterile garments. O'Keeffe et al.[2,13,14] made cheese in 20 litre vats contained in thermostatically controlled water baths in a laminar air-flow unit.

Acidification of cheese curd to ~pH 5, which is an essential element of cheese manufacture, is normally achieved by in-situ production of lactic acid by a starter culture. If the contribution of starter to cheese ripening is to be assessed, the use of starter must be avoided and acidification is then accomplished by pre-formed acid or acidogen. The use of direct chemical acidification in cheese-making has a long history and is now widely used commercially in the manufacture of certain varieties (see Chapter 15, Volume 2).

Early workers encountered difficulties in controlling pH when dilute acid was used for direct acidification. Mabbitt et al.[15] largely overcame this problem by using an acidogen, preferably gluconic acid-δ-lactone (GDL), which hydrolyses to gluconic acid at a predictable rate in aqueous solutions. O'Keeffe et al.[13] found that GDL, used as recommended by Mabbitt et al.,[15] caused excessively rapid acidification which caused extensive demineralization of the casein micelles Demineralization was considered to be responsible for the excessively rapid rate of proteolysis observed in chemically acidified cheese but this may have been due to increased retention of rennet in over-acid curd.[16] O'Keeffe et al.[13] overcame this problem by using incremental addition of lactic acid to mimic the pH drop during cooking, followed by the addition of GDL to the curd.

The manufacture of rennet-free cheese is necessary if the contribution of rennet to ripening is to be assessed. Rennet must be used to form a para-casein curd so the objective is to denature the rennet after it has completed the first stage of the rennet reaction. To date, three techniques have been developed to achieve this. Visser[4] used cheesemilk which was depleted to Ca and Mg by treatment with an ion-exchange resin; at the reduced Ca concentration, the enzymatic phase of renneting could be completed without coagulation. The rennet was subsequently inactivated by heat treatment (72°C × 15 s), the milk cooled to 5°C and $CaCl_2$ added. To induce coagulation, the milk was heated dielectrically to avoid agitation. Cheesemaking was then completed in aseptic vats. This technique was used by Visser[8,17,18] and Visser & de Groot-Mostert.[19]

O'Keeffe *et al.*[20] used porcine pepsin as coagulant; after the gel had formed, it was cut and the pH of the curd-whey mixture raised to ~7·0; this technique has been used subsequently in this laboratory with satisfactory results. Mulvihill *et al.*[21] demonstrated the potential of piglet gastric proteinase for the manufacture of rennet-free cheese; this enzyme hydrolysis bovine κ-casein but appears to be inactive on α_{s1}- or β-caseins or to be inactivated rapidly during the early stages of cheesemaking. Its use in cheesemaking was demonstrated by small-scale experiments only. Immobilized rennets could be a fourth approach but the leaching of enzyme from the solid support makes this technique unsuitable for the production of rennet-free curd.

Eliminating the proteolytic effects of plasmin presents formidable difficulties and its contribution to proteolysis in cheese has been assessed indirectly in most cases, i.e. in cheese from which all other agents have been eliminated. Plasmin is inhibited by soy-bean trypsin inhibitor which should be suitable for the inhibition of plasmin activity in cheese but no studies to evaluate this approach have been reported. The high heat stability of plasmin and the finding that its activity is increased by high cooking temperatures[22] suggest that a model system could be developed in which aseptic curd is produced, the rennet denatured by a suitable cooking temperature and the curd acidified by GDL; such a system would allow plasmin to act in isolation.

6-Aminohexanoic acid (AHA) is a non-competitive inhibitor of plasmin but does not inhibit chymosin or bacterial peptidases. It was used by Farkye & Fox[23] to assess the role of plasmin in Cheddar cheese made without aseptic precautions and with normal lactic acid starter. γ-Casein bands on electrophoretograms were less intense in cheeses containing AHA than in the control, suggesting that plasmin plays a role in Cheddar cheese ripening. It was necessary to use a relatively high concentration of AHA to inhibit the plasmin in cheese curd, which appeared to cause increased syneresis and consequently reduced the moisture content of the cheese. Further, since AHA contains N, the background level of soluble N is greatly increased.

Several specific irreversible inhibitors of serine proteinases were described by Harper *et al.*[24] who recommend dichloroisocoumarin; as far as we are aware, none of these inhibitors have been used in cheese studies.

3   REACTIONS INVOLVED IN RIPENING

Three primary events occur during cheese ripening, i.e. glycolysis, proteolysis and lipolysis. These primary reactions are mainly responsible for the basic textural changes that occur in cheese curd during ripening and are also largely responsible for the basic flavour of cheese. However, numerous secondary changes occur concomitantly and it is these secondary transformations that are mainly responsible for the finer aspects of cheese flavour and also modify cheese texture.

The biochemistry of the primary events in cheese ripening is now fairly well characterized but the secondary events are understood only in general terms.

3.1 Glycolysis

During the manufacturing phase, lactose is converted to lactic acid (mainly the L-isomer) by the starter bacteria. In the case of Cheddar-type cheeses, most of the lactic acid is produced in the vat before salting and moulding whereas for most other varieties, acidification occurs mainly after the curds have been placed in moulds. For most or all varieties, the pH of the curd reaches ~5 within ~12 h of the start of cheesemaking.

The rate and extent of acidification have a major impact on cheese texture via demineralization of the casein micelles (see Refs 16, 25, 26, 27) and also on cheese proteolysis owing to the increased susceptibility of demineralized casein micelles to proteolysis[13] and/or to greater retention of chymosin at low pH.[16] However, such aspects will not be considered here.

Although ~98% of the lactose is removed in the whey as lactose or lactate,[28] cheese curd still contains 0·8–1·5% lactose at the end of manufacture. Under normal circumstances, this residual lactose is metabolized quickly, predominantly to L-lactate, mainly through the activity of the starter. The complete and rapid metabolism of residual lactose and its component monosaccharides is essential for the production of good quality cheese; the subject has been reviewed by Fox et al.[27] The residual lactose is fermented relatively quickly to an extent dependent on the salt-in-moisture (S/M) content of the curd.[29,30] Lc. cremoris is more salt-sensitive than Lc. lactis, which in turn is more sensitive than non-starter lactic acid bacteria (NSLAB)[29] and therefore the % S/M also determines the products of post-manufacture lactose fermentation. At low S/M concentrations and low populations of NSLAB, residual lactose is converted mainly to L-lactate by the starter. At high populations of NSLAB, e.g. at high storage temperatures, considerable amounts of D-lactate are formed, partly by fermentation of residual lactose and partly by isomerization of L-lactate.[29] At high S/M levels (e.g. 6%), lactose concentration falls very slowly at low NSLAB populations and changes in lactate are slight. The quality of cheese is strongly influenced by the fermentation of residual lactose, as is evident from the data of O'Connor.[31] The pH decreases after salting, presumably due primarily to the continued action of the starter at S/M levels <5%, but at higher levels of S/M, starter activity decreases abruptly, as indicated by the high levels of residual lactose and high pH. The quality of the cheeses also decreases sharply at >5% S/M.[31]

Dutch-type cheese contains up to ~1·4% lactose at pressing but this decreases to <0·1% after pressing and to undetectable levels after brining.[32] The lactose is fermented by the starter to L-lactate.

The fermentation of lactose in Swiss-type cheeses is quite complex.[33] Typically, Emmental contains ~1·7% lactose 30 min after moulding which is rapidly metabolized by Streptococcus salivarius subsp. thermophilus to reach very low levels within 12 h with the production of up to 0·8% L-lactate. Only the glucose moiety of lactose is metabolized by *Str. thermophilus* and consequently, galactose accumulates to a maximum of ~0·7% at ~10 h. The lactobacilli then begin to multiply (after the curd has cooled somewhat) and metabolize galactose to a mixture of D- and L-lactate, reaching ~0·35 and 1·2%, respectively, at 14 days, by which time all sugars are normally completely metabolized. Thereafter, the concentrations of L- and D-lactate change little until the cheese is transferred to the hot-room, when the propionic acid bacteria begin to grow.

3.1.1 Changes in Lactate During Ripening

Racemization. The concentrations of lactate in Camembert, Swiss and Cheddar have been reported to be 1·0, 1·4 and 1·5%, respectively.[29,33,34] The fate of lactate in cheese during ripening had received little attention until recent studies[29,30,35] showed that experimental and commercial Cheddar cheeses contain considerable concentrations of D-lactate, which could be formed from residual lactose by lactobacilli or by racemization of L-lactate. The latter involves oxidation of L-lactate by L-LDH to pyruvate which is then reduced to D-lactate by D-LDH. Except in cases where the post-milling activity of the starter is suppressed (e.g. by S/M >6%), racemization is likely to be the principal mechanism.[36] These authors showed that pediococci are probably responsible for racemization: all 27 pediococci isolated from Cheddar cheese and *P. pentosaceus* NCDO 1220 were capable of converting L-lactate to D-lactate, eventually producing a racemic mixture, while only 5 of 16 *Lactobacillus* isolates were capable of racemizing L-lactate, at much slower rates and to a lesser extent than the pediococci. Both lactobacilli and pediococci possess L(+)-LDH and D(−)-LDH, both of which are NAD$^+$-dependent. Racemization of L-lactate by both pediococci and lactobacilli was pH dependent (optima: 4 to 5) and was retarded by NaCl concentrations >2% or >6% for pediococci and lactobacilli, respectively. Racemization of lactate in a Cheddar cheese inoculated with pediococci was complete in ~19 days, while this required ~3 months in a control cheese with much lower levels of NSLAB, especially pediococci.[36]

Racemization of L-lactate appears to occur in several cheese varieties.[36] Commercial Gouda contains a relatively low proportion of D-lactate, probably due to the short ripening time. The level of L- or D-lactate in Camembert is very low due to the metabolism of lactate by the mould, as discussed below. Metabolism of lactate by the mould in blue cheese might also be expected, but this does not appear to occur.

The racemization of L-lactate is probably not significant from the flavour viewpoint, but D-lactate may have undesirable nutritional consequences in infants. Calcium D-lactate is less soluble than Ca-L-lactate and may crystallize in cheese, causing undesirable white specks, especially on cut surfaces.[37-39]

Metabolism of lactate. Oxidation of lactate can also occur in cheese. Pediococci produce 1 mol of acetate and 1 mol of CO_2 and consume 1 mol of O_2 per mole of lactate utilized.[40] The pH optimum is 5 to 6 and depends on the lactate concentration. The concentration of lactate in cheese exceeds that required for optimal oxidation, and lactate is not oxidized until all sugars have been exhausted. The lactate oxidative system remains active in 6 months-old cheese.

The oxidative activity of suspensions of starter, and NSLAB isolated from cheese, on lactose, lactate, citrate, free amino acids and peptides was studied by Thomas.[41] Starter bacteria oxidized lactose only; *L. casei* oxidized citrate, while *L. plantarum, L. brevis* and *P. pentosaceus* oxidized lactose, peptides, and L- and D-lactate, but not citrate. These results suggest that oxidation of lactate to acetate in cheese depends on the NSLAB population and on the availability of O_2, which is determined by the size of the block and the oxygen permeability of the packaging material.[42] Acetate, which may also be produced by starter bacteria from lactose[43] or citrate or from amino acids by starter bacteria and lactobacilli,[44] is usually present at fairly high concentrations in Cheddar cheese and is considered to contribute to cheese flavour, although high concentrations may cause off-flavours.[45] Thus, the oxidation of lactate to acetate must make some contribution to Cheddar cheese flavour.

Presumably, the oxidation of L-lactate to acetate occurs in all hard and semi-hard cheeses.

The metabolism of lactate is very extensive in surface mould-ripened varieties, e.g. Camembert, Brie, Carré de l'Est. The concentration of lactate in these cheeses at 1 d is ~1·0%, produced mainly or exclusively by the mesophilic starter, and hence, presumably, is L-lactate, Secondary organisms quickly colonize and dominate the surface of these cheeses—first *Geotrichum candidum* and yeasts, followed by *Penicillium caseicolum*, and, in traditional manufacture, by *Brevibacterium linens* which does not colonize the cheese surface until the pH has increased to >5·8. *G. candidum* and *P. caseicolum* rapidly metabolize lactate to CO_2 and H_2O, causing an increase in pH. Deacidification occurs initially at the surface, resulting in a pH gradient from the surface to the centre and causing lactate to diffuse outward. When the lactate has been exhausted, *P. caseicolum* metabolizes proteins, producing NH_3, which diffuses inwards, further increasing the pH. The concentration of calcium phosphate at the surface exceeds its solubility at the increased pH and precipitates as a layer of $Ca_3(PO_4)_2$ on the surface, thereby causing a calcium phosphate gradient within the cheese, resulting in its outwards diffusion; reduction of the calcium phosphate concentration in the interior helps to soften the body of the cheese. The elevated pH stimulates the action of plasmin, which, together with residual coagulant, is responsible for proteolysis in this cheese rather than proteinases secreted by the surface microorganisms, which, although very potent, diffuse into the cheese to only a very limited extent, although peptides produced by them at the surface may diffuse into the body of the cheese. The combined action of increased pH, loss of calcium (necessary for the integrity of the protein network) and proteolysis are necessary for the very considerable softening of the body of Brie and Camembert (see Refs 34, 46, 47).

As with other varieties, L-lactate predominates initially in Romano cheese, reaching a maximum of ~1·9% at one day.[48] The concentration begins to decrease at 10 days and reaches 0·2–0·6% at 150–240 days. Some of the decrease is accounted for by racemization to D-lactate which reaches a maximum at

~90 days (up to 0·6% in some batches) and then declines somewhat. Acetate may reach very high levels (1·2%) at ~30 days but decreases to ≥0·2% at 90 days. The agents responsible for acetate metabolism have not been identified but yeasts may be involved.

In Swiss-type cheese, the propionibacteria metabolize lactate to propionate, acetate and CO_2:

$$3CH_3 - \underset{\underset{OH}{|}}{\overset{\overset{H}{|}}{C}} - \overset{O}{\overset{\|}{C}}-OH \rightarrow 2CH_3CH_2\overset{O}{\overset{\|}{C}}OH + CH_3\overset{O}{\overset{\|}{C}}OH + CO_2$$

The CO_2 generated is responsible for eye development, a characteristic feature of these varieties. L-lactate is preferentially metabolized over D-lactate by propionibacteria[49] to reach 0·2% after ~20 days in the hot room.[33] In fact, the concentration of D-lactate continues to increase to ~0·4% during the early days in the warm room, before being metabolized by propionibacteria. Increasing the number of *Lactobacillus* accelerates sugar metabolism and causes higher concentrations of both D- and L-lactate but suppresses the growth of propionibacteria and delays the production of propionate and acetate. In the absence of lactobacilli or with gal⁻ lactobacilli, galactose accumulates and no D-lactate is formed. Therefore, the proportion of lactobacilli in the starter probably influences the production of CO_2 and volatile acids.

3.1.2 Citrate Metabolism

The relatively low concentration of citrate in milk (~8 mM) belies the importance of its metabolism in many cheeses made using mesophilic cultures (for reviews see Refs 50, 51). Citrate is not metabolized by *Lc. lactis* or *Lc. cremoris*, but is metabolized by *Lc. lactis* subsp. *diacetylactis* and *Leuconostoc* spp. with the production of diacetyl and CO_2. It is not metabolized by *S. thermophilus* or by thermophilic lactobacilli,[52] but several species of mesophilic lactobacilli metabolize citrate with the production of diacetyl and formate;[53] the presence of lactose influences the amount of formate formed.

Citrate is not used as an energy source by *Lc. lactis* subsp. *diacetylactis* or *Leuconostoc* spp., but is metabolized very rapidly in the presence of fermentable carbohydrate by the pathway outlined in Fig. 1. Due to CO_2 production, citrate metabolism is responsible for the characteristic eyes of Dutch-type cheese, and for the undesirable openness and floating curd in Cheddar and Cottage cheese, respectively. Due mainly to the formation of diacetyl, citrate metabolism is very significant in aroma/flavour formation in Cottage cheese, Quarg, and many fermented milks. Diacetyl also contributes to the flavour of Dutch-type cheeses and possibly of Cheddar cheese.[45,54-56] Acetate produced from citrate may also contribute to cheese flavour.

Approximately 90% of the citrate in milk is soluble and most of it is lost in the whey; however, the concentration of citrate in the aqueous phase of cheese is

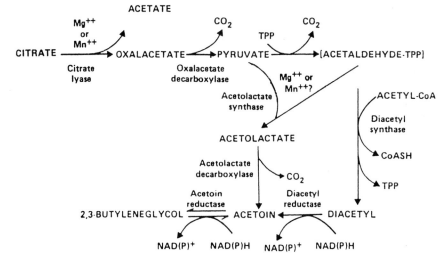

Fig. 1. Pathway for the metabolism of citrate in *Str. lactis* subsp. *diacetylactis* and *Leuconostoc* spp. (from Ref. 51).

~3 times that in whey,[57] presumably reflecting the concentration of colloidal citrate. Cheddar cheese contains 0·2 to 0·5% (w/w) citrate. Fryer *et al.*[57] showed, using cheese with a controlled microflora, that in cheese made using *Lc. cremoris* only, citrate remained constant at 0·2% up to three months, but decreased to 0·1% at six months. Cheese made using *Lc. cremoris* plus *Lc. lactis* subsp. *diacetylactis* contained no citrate at three months. Although *Lb. casei* could ferment citrate in milk, the concentration of citrate in cheese made using *Lc. cremoris* and *Lb. casei* decreased at about the same rate as in cheese made with *Lc. cremoris* alone. Thomas[42] showed that citrate in Cheddar cheese decreased slowly to almost zero at six months, presumably as a result of metabolism by lactobacilli which became the major component of the NSLAB flora. Inoculation of cheesemilk with *Lb. plantarum* accelerated the depletion of citrate. Pediococci did not appear to utilize citrate.

Although the pathway shown in Fig. 1 is probably the major pathway for the metabolism of citrate by LAB, the possibility that lactate may be formed from the pyruvate produced from citrate cannot be overlooked. Three out of four strains of cit+ leuconostocs growing on glucose plus citrate produced no diacetyl or acetoin and more lactate than could be accounted for in terms of the amount of glucose used, suggesting that pyruvate derived from citrate was being reduced to lactate.[58]

3.2 Proteolysis

Proteolysis is the most complex of the three primary events during cheese ripening and is possibly the most important for development of flavour and texture, especially in internal bacterially-ripened cheeses (reviews include Refs 59–62).

Proteolysis contributes to cheese ripening in at least four ways: (1) a direct contribution to flavour via amino acids and peptides, some of which may cause off-flavours, especially bitterness, or indirectly via catabolism of amino acids to amines, acids, thiols, thioesters, etc.; (2) greater release of sapid compounds during mastication; (3) changes in pH via the formation of NH_3; (4) changes in texture arising from breakdown of the protein network, increase in pH and greater water binding by the newly-formed amino and carboxyl groups. Although the ripening of some varieties (e.g. Blue and Romano) is dominated by the consequences of lipolysis, proteolysis is more or less important in all varieties. In the case of Cheddar and Dutch-type cheeses, and probably other varieties, many authors regard proteolysis as the most important biochemical event during ripening. A high correlation exists between the intensity of Cheddar cheese flavour and the concentration of free amino acids.[63,64] Attempts have been made to develop proteolytic indices of cheese maturity (see Refs 65, 66); although such indices correlate well with age and maturity, they fail to detect off-flavours and should therefore be regarded as complementary to the organoleptic assessment of quality.

Proteolysis in cheese is a popular research subject and a considerable amount of information is now available on the level and type of proteolysis in the principal cheese groups. Methods for monitoring and quantifying proteolysis have been reviewed comprehensively (Chapter 9).[62,67]

3.2.1 Proteolytic Agents in Cheese

Four, and in some varieties five, agents contribute to proteolysis in cheese: (1) rennet or rennet substitute; (2) indigenous milk proteinases, especially plasmin; (3) starter proteinases and peptidases released from lysed cells; (4) proteinases and peptidases from secondary microorganisms, e.g. propionic acid bacteria, *Br. linens*, yeasts and moulds (*Penicillium roqueforti* and *P. candidum*) are of major importance in some varieties and (5) enzymes from non-starter bacteria.

3.2.2 Contribution of Coagulant

Only ~3–6% of the rennet added to cheesemilk is retained in the curd but the amount retained is influenced by the type of rennet and the cooking temperature. The proportion of chymosin retained in the curd is strongly influenced by the pH at whey drainage, increasing as the pH decreases.[16,68] According to Green & Foster,[69] little active pepsin remains in Cheddar cheese but O'Keeffe *et al.*[70] found significant activity; presumably the discrepancy can be explained by differences in cheese pH since porcine pepsin is very unstable in the pH range 6·5–7·0. A smaller proportion of microbial rennet is retained in the curd and this is independent of pH.[16,68] Very little, if any, coagulant survives the cooking conditions used for Swiss cheeses[71] although some appears to survive in Mozzarella, as indicated by the formation of α_{s1}-I casein.[72]

All the principal commercial rennets are acid proteinases that show specificity for peptide bonds to which hydrophobic residues supply the carboxyl group; all show generally similar specificity on the B-chain of insulin.[73] The proteolytic specificity of calf chymosin and the principal rennet substitutes on α_{s1}- and β-caseins is fairly well established and these findings can, largely, be extended to cheese.

β-Casein in solution is sequentially hydrolysed at bonds 192–193, 189–190, 163–164 and 139–140 to yield the peptides β-I^1, β-I^{11}, β-II and β-III, respectively; bonds 165–166, 167–168 may also be hydrolysed to yield peptides indistinguishable electrophoretically from β-II and at low pH (2–3), bond 127–128 is also hydrolysed to yield β-IV.[74-76] The hydrolysis of β-casein by chymosin is strongly inhibited by 5%, and completely by 10%, NaCl.[77] The reasons for this inhibition are not clear but a similar effect is produced by sucrose or glycerol[78] or by high protein concentrations. The C-terminal region of β-casein is very hydrophobic and undergoes temperature-dependent hydrophobic interactions. It is likely that such associations occur in cheese and render the chymosin-susceptible bonds, which are located in this region, inaccessible to chymosin. Presumably, the effect is related to water activity, a_w, but the influence of varying a_w directly on the proteolysis of individual caseins has not been studied.

When animal rennets are used, β-casein is quite resistant to proteolysis in bacterially ripened cheeses throughout ripening and in mould-ripened cheeses until fungal proteinases become dominant after mould growth, e.g. ~50% of the β-casein survives after six months,[19,30,79] The β-peptides normally produced by rennet, i.e. β-I, β-II, do not appear, suggesting that plasmin and/or bacterial proteinases are responsible for the hydrolysis of β-casein in these cheeses. NaCl in cheese is undoubtedly an inhibitory factor but even in the absence of NaCl, the extent of β-casein hydrolysis by animal rennets is slight.[80] In contrast, microbial rennets degrade α_{s1}- and β-caseins at about equal rates in cheese.[25,81] Although not fully characterized, the specificity of *M. miehei* proteinase on β-casein appears to differ from that of chymosin and its activity is less sensitive to NaCl.[81]

α_{s1}-Casein in solution has several chymosin-susceptible bonds, hydrolysis of which is dependent on pH and NaCl.[82,83] In contrast to β-casein, NaCl up to 5% stimulates the hydrolysis of α_{s1}-casein and significant proteolysis occurs in the presence of 20% NaCl.[77] Mulvihill & Fox[83] reported that α_{s1}- casein is hydrolysed initially to α_{s1} f1–23 and α_{s1}-I (f24–199) and later to α_{s1}-V (f30–199), α_{s1}-VII (f56–179) and small amounts of α_{s1}-II (f24–169). The specificity of chymosin on α_{s1}-casein has been reinvestigated by McSweeney (unpublished data). At pH 6·5, the cleavage sites, in order of susceptibility appear to be Phe^{23}–Phe^{24}, Trp^{164}–Tyr^{165}, Leu^{156}–Asp^{157}, Ala^{158}–Tyr^{159} and Phe^{153}–Tyr.^{154} At pH 5·2 in the presence of 5% NaCl an additional bond, which has not yet been identified, is hydrolysed. The resulting small peptides formed are α_{s1} f1–23, f165–199, f157–164, f154–158, f154–164 and the corresponding larger peptides are α_{s1} f24–199, f24–164, f24–153 and f24–158 which may correspond to α_{s1}-I, α_{s1}-II, α_{s1}-III, and α_{s1}-IV, respectively. In Cheddar and Dutch-type cheeses, α_{s1}-casein is completely degraded to α_{s1}-I and some further products by the end of ripening.[17,79] In mould-ripened cheese, α_{s1}-casein is completely degraded to at least α_{s1}-I prior to the mould-ripening phase and very extensive degradation occurs thereafter due to the action of fungal proteinases and peptidases.[84]

The specificity of *M. miehei* proteinase on α_{s1}-casein appears to be generally similar to that of chymosin but the relative rates of hydrolysis of the various susceptible bonds by the two enzymes differ and consequently α_{s1}-I casein does not accumulate in cheese made with the microbial rennet.[16,25,81] Apparently, the activity of *M. miehei* proteinase on isolated α_{s1}-casein is very low in the absence of NaCl but is markedly stimulated by the presence of 2% NaCl.[81]

The action of chymosin and other rennets of α_{s2}-casein has received little attention. Chymosin produces several peptides from α_{s2}-casein in solution but the cleavage sites have not yet been identified (McSweeney & Fox, unpublished data). PAGE of Cheddar and similar cheeses indicates the presence of peptides with electrophoretic mobilities similar to α_{s2}-casein, but these could have originated from α_{s1}-casein. Further work in this area is required. Para-κ-casein appears to be resistant to chymosin.[69] Although the γ-caseins contain the chymosin-susceptible bonds of β-casein, γ-caseins accumulate in cheese during ripening; presumably these bonds are inaccessible in γ-caseins, as they are in β-casein. The action of chymosin on γ-caseins in solution does not appear to have been reported.

Although α_{s1} f1–23 and α_{s1}-I (f24–199) casein are probably the principal peptides produced in cheese through the action of chymosin and pepsins, it is very likely that other peptides are also produced. However, this area has been largely unexplored and should make an interesting area of study.

3.2.3 Significance of Secondary Coagulant Proteolysis

Studies on aseptic curd and cheese with controlled microflora have shown that the coagulant is responsible for the level of proteolysis detected by gel electrophoresis and for most of the nitrogen soluble in water or at pH 4·6; however, little TCA- and PTA-soluble N is produced by the coagulant (e.g. Refs 14, 18, 19 and 70). Proteolysis by rennet is believed to be responsible for the softening of cheese texture early during ripening via the hydrolysis of α_{s1}-casein to α_{s1}-I, which is sufficient to break the continuous protein matrix.[85,86] Undoubtedly, further proteolysis by coagulant, plasmin and bacterial proteinases modifies the texture further. Even in surface mould-ripened cheeses, and probably in smear cheeses, coagulant is considered to be essential for the development of proper texture, e.g. in Camembert, although the very marked increase in pH (to 7) caused by the catabolism of organic acids and the production of ammonia (by deamination of amino acids) is also essential.[46,47] The proteinases excreted by the mould diffuse into the cheese to only a slight extent and contribute little to proteolysis within the cheese, although peptides produced by these enzymes in the surface layer may diffuse into the cheese.

Little, if any, coagulant survives the cooking process in high-cooked cheeses, and plasmin, the activity of which is high in these cheeses, probably plays a significant role in primary proteolysis.

The secondary proteolytic action of the coagulant influences flavour in three ways:

1. Some rennet-produced peptides are small enough to influence flavour. Unfortunately, some of these peptides are bitter and excessive proteolysis, e.g. due to too much or excessively proteolytic rennet or unsuitable environmental conditions, e.g. too much moisture or too little NaCl, leads to bitterness.

2. Rennet-produced peptides serve as substrates for microbial proteinases and peptidases which produce small peptides and amino acids. These contribute at least to background flavour, and perhaps, unfortunately, to bitterness if the activity of such enzymes is excessive. Catabolism of amino acids by microbial enzymes, and perhaps alterations via chemical mechanisms, leads to a range of sapid compounds—amines, acids, NH_3, thiols—which are major contributors to characteristic cheese flavours.

3. Alterations in cheese texture appear to influence the release of flavourful and aromatic compounds, arising from proteolysis, lipolysis, glycolysis and secondary metabolic changes, from cheese during mastication. This may be the most significant contribution of proteolysis to cheese flavour.[87]

3.2.4 Plasmin

Milk contains several indigenous proteolytic enzymes, of which plasmin (EC 3.4.21.7) is the principal one (see Ref. 88 for review). Plasmin is a trypsin-like proteinase with a pH optimum of ~7·5 and a high specificity for peptide bonds involving lysine residues. In fact, milk contains the entire plasmin system: plasmin, plasminogen, plasmin inhibitors, plasminogen activators and inhibitors of plasminogen activators. Plasmin and plasminogen are associated with the casein micelles and accompany the micelles into the cheese curd, whereas the other components of the plasmin system are in the serum (whey) phase and are removed in the whey on cheesemaking, Milk contains at least four times as much plasminogen as plasmin and some activation probably occurs during storage, at least of certain products. In milk, β-casein is the most susceptible substrate, hydrolysis of which leads to the formation of the γ-caseins and some of the proteose peptones.

Although β-casein potentially has 15–17 (depending on the genetic variant) plasmin-susceptible bonds, i.e. those containing Lys or Arg residues, only three are hydrolysed at significant rates, i.e. Lys–Lys (28–29), Lys–His (105–106), Lys–Gln (107–108) yielding γ^1-CN (f 29–209), γ^2-CN (f 106–209), γ^3-CN (f 108–209), PP8 fast (f 1–28), PP5 (f 1–105/107) and β-CN (f 29–105/107). α_{s2}-Casein is also a good substrate; the principal cleavage sites have been identified,[89,90] but the actual products have not yet been characterized, at least not in milk. In solution, α_{s1}-casein is also quite susceptible but appears to be resistant in milk; λ-caseins are among the possible products.

Most of the plasmin/plasminogen in milk is retained in cheese and probably contributes at least to some extent to the ripening of most cheese varieties.[22] Swiss-type cheeses contain twice as much plasmin activity as Cheddar (Table I), which Lawrence et al.[26] suggested is due to the difference in the pH of these cheeses at draining. However, Grufferty & Fox[91] and Madkor & Fox[92] found

TABLE I
Plasmin Activities in Various Commercial and
Experimental Cheeses (from Ref. 22)

Cheese type	Plasmin activity AMC units/g
Commercial	
Emmental	2·69 ± 0·30
Cheddar	1·07 ± 0·09
Cheshire	0·76
Gouda	1·85
Blarney	2·59
Leicester	1·40
Wensleydale	1·35
Romano-type	2·91
Experimental	
Emmental	1·71
Cheddar	1·26
Cheshire	0·86

that in the pH range 4·6–6·6, plasmin remains associated with the casein micelles, which suggests that within the range relevant to rennet-coagulated cheeses, the pH at draining does not affect plasmin retention in cheese. Farkye & Fox[22] showed that neither draining pH nor the method of salting affected plasmin activity in cheese. However, plasmin activity in micellar casein dispersions, cheese and rennet curd increased with increasing cooking temperature, apparently due to the activation of plasminogen (Table II). The results suggest that plasmin activity in cheese depends on the cooking temperature used during manufacture, and hence the reported difference between Swiss and Cheddar.

TABLE II
Effect of Cooking Temperature on Plasmin Activity in Cheddar Cheese and
Rennet Curd (from Ref. 22)

Sample	Cooking temperature, °C	Plasmin activity, AMC units/g	
		Initial	Total[a]
Cheddar cheese	31	1·17	—
	39	1·71	—
	52	2·84	—
Rennet curd	30	1·09 ± 0·15	14·58 ± 0·84
	40	1·23 ± 0·09	10·55 ± 0·09
	50	2·42 ± 0·13	9·61 ± 0·34
	60	3·17 ± 0·24	8·66 ± 0·44

[a] After activation of plasminogen with urokinase.

TABLE III

Effect of 6-Amino Hexanoic Acid on the Formation of Water Soluble N in Cheese During Ripening (from Ref. 23)

Cheese	Water Soluble N/Total N (%)			
	1 d	1 m	3 m	6 m
Control	6.00 ± 0.14	12.86 ± 0.19	20.61 ± 0.40	27.58 ± 0.32
Experimental	10.25 ± 0.41	13.90 ± 0.10	20.41 ± 0.58	23.37 ± 0.75
		Increase in Soluble N		
Control	—	6.86	14.61	21.58
Experimental	—	3.65	10.16	13.12

Pasteurization increases plasmin activity in milk, possibly by inactivation of plasmin inhibitors or by increasing the rate of activation of plasminogen (see Ref. 88). It is proposed that a similar situation occurs in cheese (Table II), i.e. more plasminogen is converted to plasmin in high-cook cheeses because the inhibitors of plasminogen activators are inactivated. Since the various inhibitors in the plasmin system are located in the serum phase of milk, they are probably lost in the whey during cheesemaking.

Although it appears to be generally accepted that plasmin contributes to proteolysis in cheese, there is little quantitative information available. Although in internal bacterially ripened cheeses, β-casein is more resistant to proteolysis than α_{s1}-casein, it does undergo proteolysis. Since β-casein-I, the primary chymosin-produced product of β-casein, is not formed, the hydrolysis of β-casein would appear to be due to the action of plasmin and/or starter proteinases. The concentration of γ-caseins increases during the ripening of several cheeses, e.g. in Gouda,[19] Emmental, Romano and Gruyère.[93-95] Farkye & Fox[23] added 6-amino hexanoic acid (AHA) to cheese curd to inhibit plasmin and showed considerable difference in the levels and patterns of proteolysis between experimental and control cheeses (Table III). As discussed previously, other inhibitors, e.g. dichloroisocoumarin, may be preferable to AHA.

Casein micelles in milk are capable of binding at least 10 times the level of plasmin/plasminogen that occurs in milk.[96] When Cheddar cheese was prepared from milk supplemented with up to six times the indigenous level of plasmin, the additional enzyme was incorporated into the cheese curd and significantly accelerated proteolysis and ripening without concomitant defects.[96]

An alternative approach to ascertaining the potential contribution of plasmin to proteolysis in cheese was adapted by Guo & Fox (unpublished) to treated sodium caseinate (2%, w/v) in acetate buffer, pH 5.2, and containing 5% NaCl, i.e. conditions which simulate the aqueous phase of cheese, with plasmin and fractionated the hydrolysate by methods normally used to monitor cheese ripening; it was found that considerable amounts of pH 4.6 and 2% TCA-soluble N but very little ultrafiltrable N (10 000 Da cut-off membranes) were produced.

It appears that plasmin has the potential to make a significant contribution to proteolysis in cheese, but further work on its quantitation is necessary.

3.2.5 Proteolytic System of Lactic Acid Bacteria

Lactic acid bacteria (LAB) are commercially important because of their application in the production of a wide variety of fermented foods, especially cheese, yoghurt and other cultured dairy products. Dairy starter cultures include mesophilic *Lactococcus* and *Leuconostoc* species, thermophilic *Lactobacillus* species and *Streptococcus salivarius* subsp. *thermophilus* These cultures must grow to high cell densities in milk to produce lactic acid at rates appropriate to the particular manufacturing process. Fast acid production, with a concomitant decrease in pH, is a major factor controlling the growth of spoilage and pathogenic bacteria.

The lactic acid bacteria are nutritionally fastidious and have complex amino acid requirements (see Ref. 97). The concentration of free amino acids in milk is insufficient to support optimal starter growth. To overcome this difficulty lactococci possess a proteolytic system which involves the concerted action of proteinases and peptidases which ultimately hydrolyse milk proteins to the free amino acids essential for their nutrition.

3.2.6 Proteinases

The first step in the cascade of reactions leading to the production of amino acids from casein involves proteinases. Lactococcal proteinases have been well characterized biochemically and genetically. It is now generally accepted that they are cell wall-associated except in the case of *L. lactis* subsp. *cremoris* ML1 which secretes proteinase into the culture medium.[98] Proteinase activity can be released from the cell envelope by incubating in a calcium-free buffer, a characteristic which is often exploited as the first step in isolation of these proteinases. Incubation of cells with lysozyme or phage lysin also leads to the release of the cell envelope-anchored proteinase (see Ref. 97). Akuzawa *et al.*[99] identified two intracellular metalloproteinases (MW: 62 000 and 98 000), a serine proteinase (MW: 160 000) and a cysteine proteinase (MW: 12 000) in an intracellular preparation of *L. lactis* IAM 1198, all of which could hydrolyse casein. The methods used to prepare these intracellular fractions were such that contamination with the cell envelope-associated proteinase(s) could have occurred.

3.2.7 Biochemical characteristics

The cell wall-associated proteinases of LAB are serine proteinases (inhibited by PMSF and DFP). Lactococcal proteinases show a high degree of homology with the subtilisins, the serine proteinases produced by *Bacillus* species; proteinases from both sources have the same amino acid triad, i.e. aspartic acid, histidine and serine, at their active sites. Homology is conserved in the substrate-binding region, but lactococcal proteinases contain an additional region at the C-terminus which is also involved in substrate binding and is absent from the subtilisins.[100] Furthermore, subtilisins are excreted by producing cells[101] while the lactococcal proteinases are cell wall-associated.

Lactococcal cell wall proteinases have a molecular weight of ~145 kDa and a pH optimum of 5·5–6·2.[97] On the basis of pH and temperature optima, Exterkate[102] proposed the first classification of lactococcal proteinases. Three types of activity were identified: PI and PIII, which were optimally active at acid pH values and at 30 and 40°C, respectively, and PII which was optimally active at neutral pH and at 30°C. A further classification of lactococcal proteinases, based on substrate specificity, was proposed by Visser et al.,[103] who analysed the activity of proteinases from a number of strains of L. lactis subsp. cremoris on bovine caseins. Two main types of proteinase activity, PI and PIII, were distinguished: PI, present in L. lactis subsp. cremoris HP and Wg2, degrades β-casein with only a very slow hydrolysis of α-casein, while PIII, present in L. lactis subsp. cremoris AM1 and SK11, degrades α_{s1}-casein in addition to β- and κ-caseins. Visser et al.[103] suggested that the PII activity observed by Exterkate[102] could be due to differences in the stability of PI under the reaction conditions used to define PI and PII.

3.2.8 Genetics of Lactococcal Proteinases
It has long been observed that proteinase production is an unstable trait in lactococci. Plasmid curing has provided evidence for plasmid association of proteolytic activity.[100] Further evidence to support the extra-chromosomal location of proteinase genes was obtained when it was shown that the ability to produce proteinase could be transferred by the gene transfer mechanisms of transduction and conjugation.[104] Following identification of the specific plasmids in lactococcal strains Wg2, SK11 and 763 on which the proteinase genes are localized, deletion/subcloning analyses and nucleotide sequencing of these genes revealed the presence of two essential, adjacent, but divergently aligned open reading frames, designated prtP and prtM.[100] prtP encodes a 200 kDa inactive pre-pro-proteinase while prtM encodes a 30 kDa trans-acting lipoprotein involved in the activation of the pre-pro-proteinase to the mature active proteinase.[105] prtP is attached to the cell wall at its C-terminus via a C-terminal membrane-anchoring sequence, deletion of which results in secretion of the enzyme. The N-terminus encodes the catalytic site and the 33 amino acid-long signal peptide sequence. An overall homology of 98% was observed at the amino acid and nucleotide levels for the Wg2, SK11 and 763 proteinases.[106] The most striking difference between SK11 and Wg2 proteinases is an additional 60 amino acids close to the C-terminus of the former.[107] Apart from these 60 amino acids, there are only 43 amino acid differences between Wg2 and SK11 proteinases and only 18 amino acid differences between Wg2 and 763 proteinases. The prtM gene products from Wg2 and SK11 are identical.[100]

3.2.9 Proteinase Activation
Several models for activation of the inactive 200 kDa pre-pro-proteinase to its active 145 kDa form have been proposed but that by Kok[100] seems to be the most probable. The pre-pro-proteinase, which is transcribed from prtP, is transported across the cell membrane and the 33 amino acid signal peptide at the

N-terminus cleaved off by a signal peptidase. The inactive proteinase percursor (pro-proteinase) becomes anchored in the cell membrane via a C-terminal membrane-anchoring sequence. It is postulated that the *prtM* product, which is a lipo-protein located in the membrane, is involved in the removal of the 154 amino acid pro-region at the N-terminus. Removal of 187 residues, i.e. 33 in the signal peptide and 154 as the pro-region, from the N-terminus results in a protein of predicted size, i.e. 187 kDa. Since the observed size of the mature isolated pro-teinase is 145 kDa, further processing must occur to account for the difference between 187 and 145 kDa. Laan & Konings[108] showed that aspartic acid 188 of the pre-pro-proteinase represents the N-terminus of the 145 kDa Wg2 proteinase and several of its degradation products. This finding, which has also been made by others,[109,110] indicates that once the 33 amino acids of the signal peptide and the 154 amino acids of the pro-region have been removed, no further processing occurs at the N-terminus, and therefore additional processing must occur at the C-terminal region. Evidence is now being produced[110] which indicates that in the absence of Ca^{2+}, self digestion sites are exposed in the C-terminal region of the proteinase which leads to autocatalytic degradation, resulting in the 145 kDa proteinase released from the cells in a calcium-free buffer.[108] *In vitro*, e.g. in Ca-free buffers, further processing at the C-terminal region occurs, leading to the formation of active enzymes, of varying size, the smallest isolated to date being 87 kDa.[108]

Peptidases in Lactococcus. As described previously, the lactococcal pro-teinases initiate degradation of the casein substrate to polypeptides which are then further hydrolysed by peptidases to yield peptides and amino acids which are necessary for cellular nutrition. Compared to the proteinases, relatively little is known about the peptidases of lactococci although techniques for the charac-terization of their substrate specificities are well developed. The less developed status of peptidase studies may be explained in part by the fact that, unlike proteinase genes which are plasmin-encoded and relatively easy to locate and isolate, peptidase genes are chromosomally encoded.

Much of the earlier work focused on survey-type studies to identify the differ-ent peptidase specificities in lactococci and to determine their cellular location. However, this work has suffered from the limitations of the experimental tech-niques used for cell fractionation. Despite this, there is now strong evidence for the existence of intracellular, cell membrane and cell wall-bound peptidases of varying molecular weights and pH and temperature optima. Some of these lactococcal peptidases are activated by several metals, e.g. Mg, Co, Zn, Mn, and almost all are inhibited by EDTA. The principal peptidases in the lactococci are exopeptidases which catalyse cleavage of one or two amino acids from the free N-terminal of the peptide chain. Exopeptidase activity in the lactococci is exemplified by amino-, di-, tri- and tetrapeptidases, Endopeptidases can cleave large peptides at some bond within the peptide, distant from the carboxyl or amino terminus. In this respect, a proteinase can be regarded as an endo-peptidase.

3.2.10 Aminopeptidase

Evidence now suggests that aminopeptidases with broad substrate specificities are common in the lactococci. Exterkate and de Veer[111] purified to homogeneity a membrane-bound L-α-glutamyl (aspartyl)-peptide hydrolase from *L. cremoris* HP which had a molecular weight of 130 000 and appeared to be a trimer. The enzyme was completely inhibited by chelating agents and the bivalent metal ions, Cu^{2+} and Hg^{2+}, and is therefore a metalloenzyme. The activity of the apoenzyme could be restored with Co^{2+} or Zn^{2+}. This enzyme resembles mammalian aminopeptidase A. Geis *et al.*[112] isolated a cell wall-associated peptidase of 36 kDa from *L. cremoris* AC1 which preferentially hydrolysed L-lysyl-p-nitroanilides, had an optimum temperature of 40°C and was irreversibly inhibited by EDTA. This aminopeptidase had a broad substrate specificity and hydrolysed large peptides produced from β-casein by lactococcal proteinases. A 95 kDa aminopeptidase from *L. cremoris* Wg2, which has also been purified and characterized,[113] hydrolysed a wide range of substrates, including di-, tri-, tetra- and pentapeptides as well as some bioactive oligopeptides but did not hydrolyse dipeptides containing N-terminal alanine or phenylalanine. This enzyme had a pH optimum of 7, a temperature optimum of 40°C and was completely inactivated by chelating agents. The sequence of the 33 N-terminal residues was determined.

An aminopeptidase with a molecular weight of 50 kDa has been isolated from *L. cremoris* AM2 which, unlike other aminopeptidases, is probably not a metalloenzyme.[114] A broad substrate specificity aminopeptidase, which was capable of hydrolysing aminoacyl-para-nitroanilides, di-, tri- and tetrapeptides, was isolated from *L. cremoris* nTR by Kaminogawa *et al.*.[115] These authors classified 11 strains of lactococci based on their activity profiles on a wide range of substrates, including di-, tri- and tetrapeptides, into three clusters. DEAE chromatography of the cell-free extract of a representative of each cluster indicated three to four different types of peptidase activity in each strain. This study highlighted the broad spectrum of peptidase activity in the lactococci.

3.2.11 Dipeptidases

Hwang *et al.*[116] were the first to describe the purification of a dipeptidase capable of hydrolysing L–Leu–Gly from a *Lactococcus (L. cremoris* H61). The enzyme, which had a molecular weight of 100 000, a pH optimum of 8·0 and was stable at 50°C, was inhibited by EDTA and reactivated by Co^{2+}. It had a wide substrate specificity on dipeptides except those having glycine or proline at the N-terminus. These authors suggested that, due to its broad substrate specificity, the peptidase could be expected to accelerate the production of amino acids in cheese during ripening.

Van Boven *et al.*[117] purified a dipeptidase (MW = 49 000) from *L. cremoris* Wg2. Again, it was a metalloenzyme with a pH optimum of 8·0 and a temperature optimum of 50°C. Like the H61 dipeptidase, it had a broad substrate specificity but some differences between these dipeptidases were apparent, e.g. glutamate dipeptides were hydrolysed by the H61 dipeptidase but not by that from Wg2. Both enzymes were inhibited by reducing agents and required metal ions for activity; however, 1 mM Co^{2+} inhibited the activity of the Wg2 enzyme but enhanced that of the H61 dipeptidase.

3.2.12 Tripeptidases

Kolstad & Law[118] described the peptide specificities of cell-wall bound and intracellular peptidases from *L. lactis* and *L. cremoris*. Cross-contamination of the cell-wall fraction with cytoplasm was less than 3%, as indicated by cytoplasmic markers such as aldolase and lactate dehydrogenase. The cell-wall fraction of both strains contained a tripeptidase active against Leu.Leu.Leu. A second peptidase which hydrolysed a wide range of di- and other tri-peptides was also detected. Neither enzyme possessed leucine aminopeptidase or endopeptidase activity.

Gel filtration of cell-free extracts of *L. cremoris* SK11 on Sephadex G-200 indicated six peaks of peptidase activity.[119] One of the peptidases could hydrolyse Leu–Gly–Gly but did not possess general aminopeptidase activity.

A tripeptidase from *L. cremoris* Wg2 has been purified to homogeneity.[120] It had broad substrate specificity for tripeptides but could not hydrolyse tripeptides containing proline in the penultimate position. The molecular weight of this tripeptidase was ~104 000 and it was inhibited by EDTA, reducing agents (dithiothreitol and β-mercaptoethanol) and Cu^{2+}. The apoenzyme could be activated completely by Zn^{2+} or Mn^{2+} and partially by Co^{2+}. A partially purified tripeptidase from *L. lactis* CNRZ 267 (APII), which was also inhibited by EDTA, had a molecular weight of 75 000.[100]

A proline iminopeptidase is a peptidase which cleaves a peptide bond when proline is the amino terminal residue (see below). Such an enzyme from *L. cremoris* HP has been purified to homogeneity.[121] It was a 130 kDa metalloenzyme with a preference for tripeptides containing proline as the N-terminal residue [Pro.X.Y]. This tripeptidase was optimally active at 35°C, pH 8·0 and in 100 mM NaCl. The enzyme did not hydrolyse tetra- or pentapeptides or Pro–Pro or Pro–Pro–Pro.

3.2.13 Endopeptidases

The term 'endopeptidase' is applied to enzymes that hydrolyse interior bonds of peptides, but not of proteins. Exterkate[122] described two endopeptidase activities with different temperature optima, 37°C and 50°C, and different activities on N-glutaryl-L-phenylalanine-4-nitroanilide in stationary phase cells of *L. cremoris*. One endopeptidase, with an optimum temperature of 37°C, was detected in *L. cremoris* E8.[122] Yan *et al.*[123,124] identified two endopeptidases, LEP I and LEP II, in *L. cremoris* H61, which, since they were inhibited by EDTA and activated by Mn^{2+}, were classified as metalloenzymes. LEP I had a MW of 98 kDa and a high affinity for Gly–Asn peptide bonds. Its pH and temperature optima were 7·0 and 40°C, respectively. LEP II was an 80 kDa endopeptidase which hydrolysed peptide bonds involving amino groups of hydrophobic amino acids. It had pH and temperature optima at 6·0 and 37°C, respectively. These endopeptidases participate in the degradation of α_{s1}-casein (f1–23), the first peptide produced from α_{s1}-casein by chymosin. These enzymes were not inhibited by serine proteinase inhibitors and therefore differed from lactococcal proteinases.

Muset et al.[125] identified an intracellular endopeptidase activity in a mutant of L. lactis NCDO 763, which lacked the cell envelope-associated proteinase. This enzyme had a MW of 93 000 and, like other peptidases described here, was also inhibited by EDTA but not by PMSF or DFP. It degraded β-casein very slowly but readily hydrolysed the oxidized B-chain of insulin.

3.2.14 Proline Specific Peptidases

Since β-casein has a high content of proline, its hydrolysis by the cell wall proteinase leads to proline-rich peptides which can be hydrolysed only by proline-specific peptidases; many researchers have sought to determine the nature and occurrence of these enzymes in lactococci. Five types of exopeptidase are known which can cleave peptide bonds involving proline (Table IV).[97]

Mou et al.[119] observed three of these activities, aminopeptidase P, prolinase and proline iminopeptidase, in L. cremoris SK11, L. lactis C2 and L. lactis subsp. lactis var. diacetylactis DRC1.

A metal-independent X-prolyl-dipeptidyl-aminopeptidase (X-PDAP) has been purified from L. cremoris P8-2-47[126] and L. lactis NCDO 763.[127] Both enzymes, which had optima of 45–50°C and pH 7·0, were inhibited by DFP and were dimers of identical subunits (MW ~90 000). This was the first peptidase isolated from a Lactococcus which was a serine rather than a metalloenzyme. X-PDAP hydrolysed β-casomorphin (Tyr–Pro–Phe–Pro–Gly–Pro–Ile) to Tyr–Pro, Phe–Pro, Gly–Pro and Ile.[126] The X-PDAP purified from the cytoplasm of L. cremoris AM2 had a MW of 117 000, a broad pH optimum (6–9) and was inhibited by PMSF.[128]

The products of X-PDAP activity [X.Pro] can be hydrolysed by a prolidase. This latter enzyme was detected in L. cremoris H61;[129] it had a MW of 43 kDa and was completely inhibited by EDTA. A prolidase detected in a cytoplasmic

Table IV
Proline Specific Peptidases in the Lactococci[97]

Enzyme	Substrate
Aminopeptidase P	X Pro↓Y---
Proline iminopeptidase	Pro↓X---
Proline iminodipeptidase (prolinase)	Pro↓X
Imidodipeptidase (prolidase)	X↓Pro
Dipeptidylaminopeptidase	X Pro↓Y---

preparation of *L. cremoris* AM2 had a MW of 42 kDa and a pH optimum in the range 7·3 to 8·25. The enzyme, which was inhibited by EDTA, was active on all amino acylproline substrates tested, except Gly–Pro and Glp–Pro and also showed activity on Pro–Pro.[130]

3.2.15 Carboxypeptidase

As yet, there are no reports on carboxypeptidase activity in the lactococci.

It will be apparent from the information currently available that the lactococci possess a formidable proteolytic system. The range of peptides that might be produced from β-casein (in solution) by the concerted action of the known lactococcal proteinases and peptidases has been described by Smid *et al.*[131]

Genetics of lactococcal peptidases. Unlike the proteinase genes, peptidase genes appear to be chromosomally encoded in lactococci, which makes targeting and isolating them considerably more difficult. The gene for X-PDAP from *L. cremoris* P8/2/47[132] and *L. lactis* NCDO 763[133] has been cloned and sequenced.

Research on the genetics of lactococcal peptidases, although still in its infancy, is a rapidly developing field.

3.2.16 Contribution of Starter Proteinases to Cheese Ripening

It is generally assumed that the proteolytic system of the starter is important in cheese ripening, but little definitive information is available, although progress has been made in recent years. The literature has been comprehensively reviewed.[61,62,97,134,135] It is generally assumed that the proteolytic enzymes of the starter cells, described in the preceding section, are released into the cheese matrix when the cells lyse after death. However, there are very few reports (e.g. Ref. 136) on the microbial proteinases actually present in cheese.

Work on several varieties of cheese made with a controlled microflora indicates that starter proteinases/peptidases are primarily responsible for the formation of small peptides and free amino acids, i.e. TCA-soluble N.[9,14,18,19,70,137,138] Likewise, these studies indicate that starter proteinases contribute little to the formation of the larger, pH 4·6 or water-soluble peptides. However, the lactococcal proteinases are capable of hydrolysing intact caseins in solution, especially β-casein; relatively few strains are capable of hydrolysing α_{s1}-casein, i.e. only those with P-III-type proteinase, but this is probably not significant since this protein is rapidly hydrolysed by chymosin and other rennets. However, the cell wall-associated proteinases can readily hydrolyse α_{s1} f1–23 (produced by chymosin) at several sites; some of the resultant peptides have been demonstrated in Gouda cheese.[139,140] In Dutch and Cheddar cheeses, the concentration of β-casein decreases slowly during ripening with the formation of little, if any, β-I (suggesting the lack of chymosin activity) but with the formation of γ-caseins (indicating plasmin activity). The cell-wall proteinase of *L. lactis* NCDO 763 cleaves five bonds in β-casein, i.e. Ser–Gln (166–167), Gly–Lys (175–176), Gln–Arg (182–183), Tyr–Gln (193–194) and Ile–Ile (207–208).[141] These bonds are in a very hydrophobic region of β-casein, i.e. the region of β-casein which in

TABLE V
Water-soluble N (WSN), Phosphotungstic Acid Soluble N (PTA N), or 10 000 D-permeate N in
Cheddar Cheese made with Proteinase-positive or Proteinase-negative Starter (from Ref. 147)

Starter type	Age of cheese	WSN		PTA N		10 000 D-Permeate N	
		(% of total cheese N)					
		X	SE	X	SE	X	SE
L. lactis ssp. cremoris US 317 (Prt$^+$)	1 d	5·90	0·05	1·02	0·03	—	—
	1 m	9·67	0·03	1·27	0·10	—	—
	3 m	17·94	0·27	1·59	0·02	—	—
	6 m	25·89	1·26	2·30	0·09	5·04	0·32
L. Lactis ssp. cremoris UC 041 (Prt$^-$)	1 d	6·10	0·10	0·72	0·04	—	—
	1 m	10·01	0·66	0·88	0·02	—	—
	3 m	18·99	0·49	1·08	0·03	—	—
	6 m	25·04	0·78	1·28	0·16	2·52	0·49

solution is cleaved by chymosin. Since chymosin does not hydrolyse β-casein in cheese, possibly because of intermolecular hydrophobic interactions (as already discussed), it is probable that the starter proteinases are also unable to hydrolyse this region of β-casein in cheese.

Starter bacteria reach maximum numbers in Cheddar and Dutch cheeses at or shortly after the end of manufacture, and viable numbers decline quickly thereafter.[8,142] It is generally assumed that the cells lyse after death, releasing intracellular enzymes that diffuse into the surrounding environment. However, electron photomicrographs show that the bacterial cells remain largely intact in cheese, presumably due to the gel-like structure in which they are embedded and the high solute concentration in the aqueous environment.[143] This suggests that intracellular enzymes, e.g. peptidases, may not have ready access to substrates. An intracellular dipeptidase was extracted from cheese, suggesting that the cells had lysed and that intracellular enzymes were free.[144,145] However, it is claimed[97] that the extracting solution used in those studies would have caused lysis of the cells.

Oberg et al.[146] claimed that cheese made using a Prt$^-$ starter contained significantly higher concentrations of pH 4·6–soluble N than cheese made using normal Prt$^+$ starter, possibly due to greater carry-over of soluble N from the medium used to propagate the Prt$^-$ starter. However, the concentrations of soluble N increased at similar rates during ripening, suggesting that starter proteinases play no significant role in proteolysis.

These results were more or less confirmed by Farkye et al.[147] using Prt$^+$ and Prt$^-$ variants of the same Lactococcus strain. There were only slight differences between the electrophoretograms of the cheeses and in the levels of WSN but electrophoretograms of the WSN were markedly different and significantly more of the N in Prt$^+$ cheese was ultrafiltrable and soluble in PTA (Table V).

All investigators (see Fox[62] for Refs) agree that only very low concentrations of small peptides and free amino acids are formed in aseptic starter-free cheese. It is generally assumed that this indicates that intracellular starter peptidases are active in cheese during ripening.

3.2.17 Proteolytic System of Thermophilic Starters

In addition to the mesophilic lactic acid bacteria used as starters in dairy and other food fermentations, thermophilic starters, i.e. *S. salivarius* subsp. *thermophilus* and one of several *Lactobacillus* spp., capable of surviving at temperatures up to 60°C, are used in a number of cheese and fermented dairy products. *L. delbrueckii* subsp. *bulgaricus* and *S. thermophilus* are used in the manufacture of Parmesan and Romano, while *Lb. bulgaricus, Lb. lactis* or *Lb. helveticus* and *S. thermophilus* are used in the production of Swiss varieties in addition to *Propionibacterium shermanii* which are necessary for eye formation via CO_2 production. Thermophilic lactic acid bacteria are also used in the manufacture of Mozzarella, Provolone, Brick and Limburger.

S. thermophilus grows optimally at 44°C. Lactose is hydrolysed via a β-galactosidase.[148] A significant feature of *S. thermophilus* is its inability to utilize a wide variety of carbon substrates, e.g. most strains ferment as few as three substrates: lactose, glucose and sucrose.[149] *S. thermophilus*, like the mesophilic lactococci, is nutritionally fastidious and has complex amino acid requirements. Despite this trait, *S. thermophilus* is weakly proteolytic and its growth in milk resembles that of Prt⁻ variants of *L. cremoris*.[97] Therefore, *S. thermophilus* is usually coupled with a highly proteolytic *Lactobacillus* as combined-strain starters for many milk fermentations. The starters for yoghurt fermentation are *S. thermophilus* and *Lb. bulgaricus*, where *S. thermophilus* is stimulated by the free amino acids produced in the milk by the proteolytic activity of *Lb. bulgaricus* while the production of formic acid and CO_2 by *S. thermophilus* stimulates *Lb. bulgaricus* to produce acid.[150,151] Acetaldehyde, a distinctive flavour component of yoghurt, is derived from threonine via the threonine aldolase of *S. thermophilus*.[152] The weakly proteolytic nature of *S. thermophilus* was further highlighted by Desmazeaud & Hermier[153] who showed that enzymatic hydrolysates of whole casein stimulated its growth in milk. It has also been observed that the growth of *S. thermophilus* in mastitic milk is stimulated due to the presence in this milk of high levels of indigenous milk proteinase.[154] Therefore, evidence is now accumulating for cell wall-associated proteinase and peptidase activity in *S. thermophilus*, capable of hydrolysing proteins to the amino acids essential for their nutrition and acid production.[155,156]

Meyer *et al.*[157] reported on moderately strong proteinase activity at the cell wall and in the cytoplasmic membrane of *S. thermophilus*. However, three types of peptidase activity, aminopeptidase, dipeptidylpeptidase and dipeptidase, were found exclusively in the cytoplasm and became active in cheese ripening only after cell lysis. Desmazeaud & Juge[158] also identified intracellular amino- and dipeptidase activities while Rabier & Desmazeaud[159] purified an EDTA-sensitive dipeptidase (MW 50 000) and an aminopeptidase which hydrolysed oligopeptides, dipeptides and amino acid amides.

Perhaps the best characterized enzyme of the proteolytic system of *S. thermophilus* is an X-prolyl-dipeptidyl-aminopeptidase (X-PDAP). This enzyme was purified by Meyer & Jordi[160] and shown to have a MW of 165 kDa and to consist of two subunits. It had an isoelectric point ~4·5 and a broad pH optimum (6·5–8·2). Inhibition of X-PDAP by diisopropylfluorophosphate indicated that it was a serine enzyme which was slightly sensitive to EDTA.

In most dairy fermentations, a *Lactobacillus* is used in combination with *S. thermophilus* for the aforementioned reasons. *Lb. lactis* and *Lb. helveticus* are the main starters for Swiss-type cheeses due to their ability to grow at the high temperatures employed during the fermentation, while *Lb. bulgaricus* is used primarily in yoghurt production. Similar to *S. thermophilus*, β-galactosidase is the dominant enzyme in lactose hydrolysis by these strains.[161] Like the mesophilic lactococci and *S. thermophilus*, lactobacilli are nutritionally fastidious and possess an elaborate proteolytic system capable of degrading milk proteins to the amino acids essential for growth. In addition, these amino acids have a role as precursors for cheese flavour.

Despite one report on the purification of an extracellular proteinase from *Lb. bulgaricus* (see Ref. 162), most other reports indicate that the proteinases of *Lactobacillus* spp. are cell wall-associated[163,164] or intracellular.[165] The proteinase from *Lb. helveticus* was isolated by incubation of whole cells in a Ca-free buffer, similar to the method employed to liberate the cell wall-associated proteinase from the mesophilic lactococci.[166] This enzyme had broad specificity on both α- and β-caseins and was inhibited by PMSF and DFP, indicating that it was a serine proteinase. A proteinase from *Lb. acidophilus* was also isolated and characterized and shown to have a MW of 145 kDa, similar to the lactococcal proteinases.[167] The pH optimum of the *Lb. acidophilus* proteinase was ~5·7 but the *Lb. helveticus* proteinase had an optimum pH of 7·5–8·0. The substrate specificity of several lactobacilli proteinases has been examined; it appears that, in general, they degrade $α_{s1}$-casein and, in addition, either β- or κ-casein. The polypeptides thus produced are further degraded by peptidases.

Compared with the peptidases of the lactococci, relatively little is known about the peptidases of the lactobacilli. In general, they possess a broad range of peptidase activities.[162,168–170] Several peptidases have now been isolated and characterized, including X-PDAP from *Lb. lactis*,[160] *Lb. helveticus*,[171] *Lb. bulgaricus*,[172] and *Lb. acidophilus*; aminopeptidases from *Lb. lactis*,[173] *Lb. bulgaricus*,[174] *Lb. acidophilus*,[175] *Lb. helveticus*[176] and a dipeptidase from *Lb. bulgaricus*.[177] Many of the aminopeptidases are specific for proline, the significance of which lies in the fact that β-casein contains a high level of proline residues.

The location of the peptidases in the lactobacilli is not clearly defined and reports such as that by El-Soda *et al.*[178] indicate a cytoplasmic membrane location for several peptidases in lactobacilli while intracellular aminopeptidase, dipeptidase, endopeptidase and, interestingly, carboxypeptidase activities were reported for *Lb. casei*. *Lb. lactis* also produces a carboxypeptidase which hydrolyses N–benzoyl–Gly–Arg.[173] However, *Lb. plantarum* possessed intracellular aminopeptidase and dipeptidase activity but lacked endopeptidase and carboxypeptidase activity.[179]

Dipeptidase and tripeptidase activity was also identified in *Lb. casei* and *Lb. plantarum* in a subsequent study by Abo El-Naga & Plapp[180] who also showed that the dipeptidase of *Lb. casei* hydrolysed dipeptides when proline was at the N-terminal position. Thus, according to the classification of Thomas & Pritchard,[97] this enzyme could be referred to as a prolinase.

Lb. helveticus strains have high aminopeptidase activity compared with other *Lactobacillus* strains and crude preparations of aminopeptidase have been used to accelerate the ripening of Gouda.[181]

At the time of writing, there are no published reports on the cloning of proteinase or peptidase genes from *S. thermophilus* or lactobacilli. However, lysylaminopeptidase (AP II) and X-PDAP deficient mutants of *Lb. bulgaricus* have been isolated by classical mutagenesis techniques.[181a,181b]

3.2.18 Contribution of Starter Lactobacilli to Proteolysis

As far as we are aware, the contribution of thermophilic starters to proteolysis in cheese has not been studied as extensively or systematically as that of the mesophilic starters, e.g. no studies using controlled-flora high-cooked cheeses, similar to those on Cheddar and Gouda, have been reported. However, extensive proteolysis does occur in Swiss-type and hard Italian varieties, e.g. Parmesan and Grana; in fact the flavour of these cheeses is dominated by the products of proteolysis (see Chapters 3 and 7, Volume 2).[182]

As discussed previously, the coagulant is extensively or fully denatured by the high cooking temperatures used in the manufacture of these cheeses. However, plasmin is very active, due to activation of plasminogen, and is a major proteolytic agent in high-cooked cheeses. Although quantitative information is lacking, it can be safely assumed that the proteolytic system of the thermophilic starter makes a major contribution to proteolysis, probably at all levels, but certainly in the formation of small peptides and free amino acids.

It would be interesting to undertake studies on high-cooked cheeses with controlled microflora but the method of manufacture militates against this, e.g. Emmental and Gruyère cheeses must be large (typically 40–60 kg) for proper eye formation and Parmesan and Grana are manufactured by traditional methods. It would be very difficult or impossible to manufacture these four varieties, and probably others, under conditions that would give a controlled microflora. In some respects, high-cooked cheeses are simpler than mesophilic cheeses since the coagulant has been denatured, the high cooking temperature prevents the growth of most contaminating bacteria and the action of plasmin can be inhibited by specific inhibitors, e.g. 6-aminohexanoic acid or 3,4-dichloroisocoumarin, as discussed earlier. Studies using such model systems should provide useful information. As is obvious from the previous discussion, research activity on the proteolytic system of thermophilic bacteria has intensified in recent years. While knowledge on these bacteria lags behind that on mesophiles, it is likely that in a few years, considerable information on the action of the various proteinases and peptides of thermophilic lactic acid bacteria on the caseins and casein peptides will become available, permitting their contribution to cheese ripening to be assessed more accurately. Concomitantly, fractionation and characterization of the peptides in many high-cooked cheeses is progressing which will integrate with work on the proteolytic system of the thermophilic starters.

3.2.19 Non-Starter Lactic Acid Bacteria

Non-starter lactic acid bacteria (NSLAB) typically reach populations of 10^7–10^8 during the ripening of many cheeses. Lactobacilli usually dominate the NSLAB microflora (according to Jordon & Cogan (unpublished), the NSLAB in Irish Cheddar are almost exclusively lactobacilli) but pediococci, especially *P. pentosaceus*, may also be significant (according to Turner & Thomas[29] the NSLAB are predominantly pediococci). The contribution of NSLAB to cheese ripening and quality is a vexed question. It is generally agreed that Cheddar and other varieties produced from raw milk ripen faster and develop a more intense flavour than those made from pasteurized milk, although the quality may be more variable (see Ref. 183). This suggests that NSLAB do play a role in cheese ripening, although they may have negative effects on at least some occasions. At least some of the problems may be caused by the heterofermentative lactobacilli and pediococci. The rate at which the cheese is cooled after hooping is the principal factor that determines the growth of NSLAB[29] and hence cheese quality. This latter point has been reconfirmed in a recent study[184] in which, unfortunately, the microbiological and biochemical parameters were not investigated. Undoubtedly, the rate of cooling varies between factories and within a factory, especially if the blocks of cheese are palletized before cooling. The significance of rapid cooling of cheeses to be ripened at elevated temperatures is discussed in Chapter 14, Volume 1.

The significance of non-starter lactobacilli in Cheddar cheese has been reviewed by Peterson & Marshall.[185] These authors also reviewed work on the use of active and modified lactobacilli to accelerate cheese ripening. This subject is also reviewed in Chapter 14, Volume 1 of this book. The enzymes and significance of lactobacilli in cheese ripening were also reviewed by Khalid & Marth.[162] To a large extent these two reviews cover much the same area and the distinction between starter and non-starter lactobacilli is rather blurred. It appears superfluous to review the subject again.

Pasteurization causes a number of changes in addition to killing off the indigenous microorganisms, e.g. indigenous enzymes are inactivated, whey proteins may be denatured and may interact with the caseins, salts equilibria may be altered and vitamins and other growth factors may be destroyed. Until recently it was not possible to distinguish between and quantify the significance of these heat-induced changes in cheese ripening. However, the development of microfiltration makes it possible to remove the indigenous microorganisms (>99·9% removal) without other concomitant heat-induced changes. The results of a recent study by McSweeney et al.,[186] in which the quality and biochemical parameters of Cheddar cheeses made from raw, pasteurized or MF milk were compared, indicated that the indigenous microorganisms were the most significant factor. The pasteurized and MF cheeses were indistinguishable with respect to quality, rate, extent and pattern of proteolysis, and extent of lipolysis. In contrast, the raw milk cheeses underwent more rapid and extensive proteolysis and the products of proteolysis were markedly different, especially when analysed by RP-HPLC. The flavour of the raw milk cheese was much more intense than that of the pasteurized or MF cheeses but commercial graders regarded the raw cheeses as atypical and unacceptable.

The pasteurized and MF milks were free of NSLAB at the start of manufacture while the raw milk contained ~200 NSLAB/ml. NSLAB grew in all cheeses during ripening, reaching maximum numbers after about 11 weeks when the pasteurized and MF cheeses contained ~10^7 NSLAB/g while the raw milk cheese contained ~10^8 NSLAB/g.

Possibly the principal difference between the starter and non-starter lactobacilli in cheese is the greater variability of the latter. The starter lactobacilli are exclusively thermophilic, homofermentative strains while the non-starter lactobacilli are a heterogeneous population and include thermophilic and mesophilic, heterogermentative as well as homofermentative species. In pasteurized milk cheese the non-starter lactobacilli originate mainly as post-pasteurization contaminants from the factory environment. It is very likely that a unique, characteristic population of *Lactobacillus* species and strains will evolve in each factory. The population may vary over time and will be strongly influenced by the standards of hygiene in the factory. There is little detailed information on such variability, which is probably responsible for variations in the quality of cheese. In the case of raw milk cheeses, the non-starter lactobacilli originate from the milk supply as well as from the factory environment. Hence, the species of lactobacilli in raw and pasteurized milk cheeses might be expected to vary markedly. This can be illustrated by the data of McSweeney *et al.*:[186] 50 colonies were isolated from plates inoculated with raw or pasteurized milk Cheddar. For both cheeses, all the colonies were lactobacilli but the species differed considerably: 88% of the isolates from the pasteurized milk cheese were *L. casei* subsp. *casei*, 8% were *L. casei* subsp. *pseudoplantarum* and 4% were unidentified; the raw milk cheese contained a more heterogeneous population of lactobacilli—38% *L. casei* subsp. *casei*, 30% *L. casei* subsp. *pseudoplantarum*, 16% *L. curvatus*, 8% unidentified.

It appears that NSLAB can have a significant impact on cheese quality; however, variability in the number and type of NSLAB population, whether due to the use of raw or thermized milk, variations in the factory environment or to variations in cooling is a problem. As discussed in Chapter 14, Volume 1, there is considerable interest in inoculating milk with selected lactobacilli, either viable or attenuated. It is likely that there will be considerable activity in this area in the immediate future; it may become normal practice to inoculate pasteurized or MF cheesemilk with a culture of selected NSLAB, probably *Lactobacillus* spp., in addition to the regular starter.

3.2.20 Micrococcus

The occurrence and significance of *Micrococcus* spp. in milk and cheese have been reviewed by Bhowmik & Marth.[187] These authors cite several references indicating that *Micrococcus* spp. are major components of the microflora of raw milk and of cheese made from it and that they also occur in significant numbers in cheese from pasteurized milk. They report that *Micrococcus* grow during the early stages of cheese manufacture and ripening. However, lactobacilli and pediococci appear to be more numerous than micrococci in cheese. Bhowmik & Marth[187] also cite many references claiming that at least some *Micrococcus* spp. contribute positively to cheese quality and accelerate ripening, as shown in the recent study by Bhowmik *et al.*[187a]

Some *Micrococcus* spp. are quite proteolytic; they produce extracellular proteinases and also possess intracellular proteinase(s) and peptidases. Production of extracellular proteinase has been studied by McDonald,[188] Prasad *et al.*[189] and Garcia de Fernando & Fox.[190] Desmazeaud & Hermier[191,192] isolated and characterized an extracellular proteinase from *M. caseolyticus*. Two extracellular proteinases of a *Micrococcus* strain isolated from a farmhouse raw-milk blue cheese were isolated and characterized by Garcia de Fernando & Fox.[193] Both proteinases were optimally active at ~45°C; one was optimally active at pH 8·5 and the other at pH 9–11. Both enzymes were inactivated by EDTA, one of them irreversibly but the other could be reactivated by Ca^{2+}, Ba^{2+}, Mg^{2+}, Sr^{2+} or Zn^{2+}. The MW of proteinases I and II were 23·5 and 42·5 kDa, respectively. Proteinase I hydrolysed β-casein preferentially over α_{s1}-casein while proteinase II hydrolysed α_{s1}- and β-caseins at approximately equal rates. Nath & Ledford[194] reported that the extracellular proteinase of some *Micrococcus* spp. preferentially hydrolysed α_s-casein.

The proteinase activity of cell-free extracts of a number of *Micrococcus* spp. was studied by Bhowmik & Marth.[195] The proteinase of all species hydrolysed β-casein preferentially over α_{s1}-casein. Nath & Ledford[194] also reported that the intracellular proteinases of *Micrococcus* spp. preferentially hydrolysed β-casein. All the species/strains studied by Bhowmik and Marth[195] possessed intracellular aminopeptidase activity, Lys-p-NA being the best substrate for most strains, and most possessed iminopeptidase and dipeptidase activities. As far as we are aware, micrococcal peptidases have not been isolated and characterized.

3.2.21 Pediococci

Pediococci spp. grow in Cheddar cheese;[196,197] in some cases they are the predominant NSLAB[198] but in other studies they were a minor component of NSLAB.[199,200] Pediococci have also been reported in Grana,[201] Manchego[202] and Domiati.[203] The significance of pediococci in cheese ripening was reviewed by Bhowmik & Marth.[187]

There is little information on the significance of pediococci to cheese quality. Robertson & Perry[204] found that inoculation of cheesemilk with pediococci improved the quality of Cheddar made from it in some but not all experiments. Law *et al.*[205] reported that when pediococci were used as the only NSLAB they had no effect on cheese quality but were beneficial when used as a component of NSLAB. Bhowmik *et al.*[187a] concluded that pediococci may be useful for accelerating the ripening of low-fat Cheddar.

Nuñez[202] reported that *Pediococcus* spp. were weakly proteolytic and lipolytic. Several strains of *P. pentosaceus* were reported to have aminopeptidase activity and to have weak lipolytic/esterase activity, as well as a number of glycosidases.[206] Bhowmik & Marth[207] reported that *P. pentosaceus* and *P. acidilactici* possessed intracellular aminopeptidase, dipeptidase, dipeptidyl aminopeptidase and proteinase activity; the latter could hydrolyse both α_s- and β-caseins.

A partially purified cell-free extract from *Pediococcus* sp.LR was active on the nitroanalides of Arg, Leu, Lys and Ala but not of Pro or Gly; it also had strong X-PDAP activity on Arg. Pro-p-NA and Gly. Pro-p-NA.[208] The aminopeptidase had pH and temperature optima at 6·5 and 40°C and the X-PDAP at 7·5 and 50°C. The aminopeptidase was strongly inhibited by 1,10-phenanthroline and pCMB, while the X-PDAP was strongly inhibited by PMSF.

The results of these studies suggest that *Pediococcus* spp. probably contribute to proteolysis in Cheddar and other internal bacterially ripened cheeses. As discussed in a later section, they are also weakly lipolytic but they may be important in cheese flavour owing to their ability to oxidize lactate to acetate, as discussed earlier. They are also capable of reducing acetaldehyde and propionaldehyde to the corresponding alcohols and can produce small amounts of diacetyl.[209]

3.2.22 Proteolytic Systems of Other Microorganisms Important in Cheese Ripening

Propionibacteria.   *Propionibacteria shermanii* is an essential component of the microflora of Swiss-type cheeses. As discussed previously, its principal function in these cheeses is the metabolism of lactate to propionate, acetate and CO_2, the latter being responsible for typical eye formation. However, *Propionibacteria* also contribute to proteolysis. Emmental cheese contains a higher concentration of proline than any other variety[210] and this is considered to contribute to the sweet taste of Emmental.[211] It seems reasonable to conclude that *Propionibacteria* are primarily responsible for the formation of proline.

Langsrud *et al.*[212] showed that *P. shermanii* could liberate large amounts of proline from casein hydrolysates but little from intact casein. The organism could also biosynthesise proline. The ability of *Propionibacteria* to produce proline varies considerably between species and between strains of the same species; *P. shermanii* generally produced the highest concentrations of proline in a sodium lactate broth containing Trypticase and yeast extract.[213]

Propionibacteria appear to have very weak proteolytic activity and no proteinase activity was found in the soluble cell wall fraction of *P. shermanii* α or *P. shermanii* ATCC 9614.[214] However, the cell wall, membrane and intracellular fractions contained one or two, two or three and six or seven peptidases, respectively, determined by a zymogram technique. These peptidases were active on a range of dipeptides, including Pro-containing dipeptides, and some tripeptides; aminopeptidase, carboxypeptidase, X-prodipeptidyl aminopeptidase or other peptidase activities were not reported.

The leucine aminopeptidase and proline iminopeptidase activities of the cell-free extracts of *P. freudenreichii* subsp. *shermanii*, *P. freudenreichii* subsp. *freudenreichii*, *P. acidipropionici*, *P. jensenii* and *P. thoenii* were studied by Chaia *et al.*[215] Extracts of *P. freudenreichii* strains were more active than those of the other species on both Leu-p-NA and Pro-p-NA. For all cultures except *P. jensenii*, proline iminopeptidase activity was considerably higher than the leucine aminopeptidase, 60:1 in some strains of *P. acidipropionici*. The temperature optima of most of the leucine aminopeptidases was 37–45°C but that of *P. acidipropionici* was 30–37°C. The temperature optima of the proline iminopeptidase ranged from 21°C for the enzyme from *P. jensenii* to 55°C for that from *P. thoenii*. The pH optima of the leucine aminopeptidases and proline iminopeptidases of all the strains studied were in the range 6·4–7·2. Zymograms showed that the proline iminopeptidases of all strains had similar electrophoretic mobilities.

A partially purified cell-free extract of *P. shermanii* NZ showed weak activity on the nitroanalides of Leu, Lys, Pro, Gly, Ala, Arg, Arg.Pro, Gly.Pro and Gly.Phe.[208] The proline iminopeptidase activity was considerably higher than those of *L. casei*, *Pediococcus* sp. *Br. linens* and Leu *mesenteroides* reported in the same study. The iminopeptidase and X-PDAP were optimally active at 40°C and 7·5. The iminopeptidase was strongly inhibited by pCMB and weakly by 1,10-phenanthroline and the X-PDAP by PMSF.

Brevibacterium linens. *Brevibacterium linens* is a major component of the surface microflora of smear-ripened cheeses and to a lesser extent of traditional surface mould-ripened cheese. These varieties are discussed in Chapter 5 and 4, respectively, of Volume 2. The precise role of *Br. linens* in the ripening of these cheeses is not known but they certainly make a major contribution. *Br. linens* has an active proteolytic system, especially with respect to peptidase activity.

The proteolytic activity of *Br. linens* 450, which was mostly extracellular, was optimal at pH 7·2 and 38°C and was greater on casein than on whey proteins.[216] It hydrolysed α_s-casein more rapidly than β-casein and its activity was not significantly affected by any of several metal ions or reducing agents. An extracellular proteinase from *Br. linens* was purified and partially characterized by Tokita & Hosono.[217] This enzyme was optimally active at pH 7·0 and 25°C; it was completely inactivated in 10 min at 50°C. It hydrolysed casein and haemoglobin readily but had little activity on ovalbumin.

Using a zymogram technique, Foissy[218] showed that the proteinase profile of 15 *Br. linens* isolates differed considerably—the number of extracellular proteinases varied from two to six and two to five intracellular proteinases were also present, although the intracellular activity was considerably weaker than the extracellular.

Five extracellular proteinases of *Br. linens* were purified to homogeneity by Hayashi *et al.*[219] using chromatography on DEAE Sephadex, DEAE-Trisacryl M and Mono Q. The five enzymes were serine proteinases with temperature optima ranging from 40 to 55°C and pH optima at ~11, although their pH stability varied significantly. Molecular weights ranged from 37 to 325 kDa. The possibility of using partially purified preparations of some of these enzymes to accelerate cheese ripening was assessed by Hayashi et al.,[220] apparently with a promising outcome.

Apparently, the intracellular proteinases of *Br. linens* have not been isolated.

The peptidases of *Br. linens* have received more attention that the proteinases. Torgersen & Sørhaug[221] showed, using a zymogram technique, that *Br. linens* ATCC 9174 possessed at least six intracellular peptidases active on dipeptides. Using a similar technique, Sørhaug[222] demonstrated that six strains of *Br. linens* contained a total of 18 intracellular peptidases (assayed on dipeptides), one strain showing 14 peptidase bands.

An extracellular aminopeptidase from *Br. linens* ATCC 9174 was purified by Foissy.[223] This enzyme had a MW of ~48 kDa, a pH of 4·3, and was optimally active at pH 9·6 and 28°C. It was inactivated rapidly above 50°C and at pH <3·0 or >11·5. It was active on a range of dipeptides as well as amino acid imides, with a strong preference for leucine residues.[224] The aminopeptidase was activated by Co^{2+} and inhibited by metal chelators, heavy metals, reducing agent, hydrophobic molecules but not by NEN pCMB, DFP or TLCK.[225] Reactivation of the metal-free protein was not reported. The very high sensitivity of the enzyme to heavy metals and the low concentration of Co^{2+} in cheese suggested that the enzyme might not be fully active in cheese. The K_m for the aminopeptidase on Leu NH_2, Leu.Leu, Leu-4-NA and Leu-2-NA were $1·5 \times 10^{-2}$, $1·5 \times 10^{-2}$, $1·2 \times 10^{-3}$ and $0·75 \times 10^{-3}$, respectively.[226] Michaelis–Menten plots were sigmoidal which was interpreted to indicate an enzyme structure consisting of subunits.

Hayashi & Law[227] purified to homogeneity two aminopeptidases, A and B, from the culture filtrate of *Br. linens* F, a strain used commercially in cheesemaking. Both enzymes were optimally active at pH 9·3 and 40°C and were completely inhibited by 0·1 mM EDTA and several heavy metals. EDTA-inactivated enzymes were reactivated by Ca^{2+}, Co^{2+}, Mg^{2+}, Zn^{2+} or Mn^{2+}. The MW of aminopeptidases A and B were 150 and 110 kDa, respectively, by gel filtration and 36 and 26 kDa by SDS-PAGE, supporting the view of Foissy[226] that the native enzymes are tetramers. The K_m on Leu-p-NA were 16·1 and 15·2 mM for peptidases A and B, respectively. The aminopeptidases were capable of hydrolysing several dipeptides and tripeptides but much more slowly than nitroanalides or amino acid imides; the specificities of both enzymes were similar.

The partially purified cell-free extract of *Br. linens* HS was very active on Gly-p-NA and weakly on Ala-p-NA but had no activity on the nitroanalides of Leu, Lys, Pro, Arg, Arg.Pro Gly.Arg or Gly.Phe.[208] The aminopeptidase was most active at 40°C and pH 7·5 and was strongly inhibited by 1,10-phenanthroline and pCMB.

Br. linens actively catabolises amino acids with the production of amines, volatile acids, NH_3 and CO_2. Some of these products have a major influence on the flavour and aroma of surface-ripened cheeses (see Refs 228–230). As far as we know, the enzymes responsible for these transformations have not been isolated and characterized.

Penicillium roqueforti *and* Penicillium caseicolum. Both of these moulds possess very potent proteolytic systems that have been studied extensively; the literature has been reviewed in Chapter 4, Volume 2. The moulds secrete acid and neutral metalloproteinases and several aminopeptidases and carboxypeptidases. The mycelia also contain intracellular proteinases and peptidases that are probably released when the mould dies and lyses.

Early proteolysis in blue cheeses is due to coagulant with the rapid hydrolysis of α_{s1}-casein to α_{s1}-I; presumably, plasmin and starter proteinases also contribute. However, following mould growth and sporulation, fungal enzymes dominate proteolysis in these cheeses, in which very extensive proteolysis occurs (see Chapter 4, Volume 2).

Extensive proteolysis also occurs in surface mould-ripened cheeses. It was generally assumed that the extracellular proteinases secreted by *P. caseicolum* were responsible for this proteolysis, but it has been demonstrated that these proteinases diffuse only a very small distance from the surface. Most proteolysis within the cheese is due to coagulant, probably with a contribution from plasmin, especially when the pH rises during ripening (see Chapter 4, Volume 2). However, peptides produced by the action of the fungal proteinases have been identified in the core of Camembert cheese, presumably having been produced at the surface from which they diffuse into the cheese. Probably the most conspicuous result of fungal action is the increase in pH from the surface to the centre as discussed previously.

Amino acid catabolism. Catabolism of free amino acids probably plays some role in the ripening of most cheese varieties but is particularly important in mould and smear ripened varieties. The principal products that may arise from amino acid catabolism are (1) amines, resulting from decarboxylation, (2) ammonia, acids, ketoacids, carbonyls, alcohols, resulting from deamination, (3) other amino acids, resulting from transaminations, (4) H_2S, $(CH_3)_2S$, methanethiol, thioesters and other sulphur compounds arising from desulphurylation and demethiolation. Many of the products of these reactions are important in pH changes (NH_3), flavour (especially carbonyls, alcohols and sulphur compounds) and nutritional/toxicological aspects (e.g. biogenic amines). The significance of some are discussed elsewhere in this chapter while the toxicological aspects are discussed in Chapter 15.

Amino acid catabolism in cheese was reviewed by Hemme *et al.*[229] and Law.[61,231]

3.3 Lipolysis

In most cheese varieties, relatively little lipolysis occurs during ripening and is considered undesirable; most consumers would consider Cheddar, Dutch and

TABLE VI
Typical Concentrations of Total Free Fatty Acids (FFA) in some Cheese Varieties (Adapted from Refs 234, 235)

Variety	FFA, mg/kg	Variety	FFA, mg/kg
Sapsago	211	Gjetost	1658
Edam	356	Provolone	2118
Mozzarella	363	Brick	2150
Colby	550	Limburger	4187
Camembert	681	Goats' milk	4558
Port Salut	700	Parmesan	4993
Monterey Jack	736	Romano	6754
Cheddar	1028	Roqueforti	32 453
Gruyère	1481	Blue (US)	32 230

Swiss-type cheeses containing even a moderate level of free fatty acids to be rancid. Bills & Day[232] failed to find any significant differences, qualitatively or quantitatively, in the free fatty acids in Cheddar cheese of widely different flavour. The ratios of individual free fatty acids from $C_{6:0}$ to $C_{18:2}$ in cheese to the corresponding esterified acids in milk fat were very similar, indicating that these acids were released non-selectively. However, free butyric acid was always about double that in glycerides, suggesting that it is selectively liberated or synthesized by microorganisms. Reiter *et al.*[233] suggested that volatile fatty acids may contribute to the background flavour of Cheddar but felt that longer chain fatty acids ($>C_{4:0}$) are not important.

In extra-mature cheeses, fatty acids probably make a positive contribution to flavour when properly balanced by the products of proteolysis and other reactions. Exceptions to the above general situation are the Blue cheeses and certain hard Italian varieties, e.g. Romano and Parmesan (Table VI).

3.3.1 Lipolytic Agents in Cheese
Milk contains a very potent lipoprotein lipase which normally never reaches its potential in milk. This lipase has been isolated and well characterized at the molecular and biochemical levels; the literature has been reviewed by Olivecrona & Bengtsson–Olivecrona.[236,237] Indigenous milk lipase probably causes significant lipolysis in raw milk cheese and probably makes some contribution in pasteurized milk cheese, especially if the milk was heated at sub-pasteurization temperatures since heating at 78°C for 10 s is required to completely inactivate milk lipase.[238] Milk lipase is highly selective for fatty acids on the Sn3 position; since most of the butyric acid in milk fat is esterified at the Sn3 position, this specificity probably explains the disproportionate concentration of free butyric acid in cheese.

Good quality rennet extract contains no lipase activity. In contrast, the rennet paste used in the manufacture of some Italian cheeses contains a potent lipase, pregastric esterase (PGE), which in some countries is added to the cheesemilk in partially purified form. The literature on PGE has been reviewed by Nelson *et al.*[239] PGEs show high specificity for short-chain fatty acids esterified at the Sn3 position. Since the short-chain acids in milk fat are predominantly at the Sn3 position, the action of PGE results in the release of high concentrations of short and medium chain acids which are responsible for the characteristic piquant flavour of the hard Italian cheeses.

PGEs from calf, kid and lamb are commercially available. Although these enzymes have generally similar characteristics, they show subtle differences in specificity which facilitate the manufacture of Italian cheeses with different flavour profiles.

Most other lipases are unsatisfactory for the manufacture of Italian cheeses due to incorrect specificity. However, it has been claimed that the lipase from *Mucor miehei* and perhaps a lipase from *P. roqueforti* give satisfactory results (see Ref. 240). It is also claimed (see Ref. 240) that the addition of PGE to milk for Cheddar, Feta, Domiati, Ras, Blue and probably other varieties improves the quality of these cheeses.

Lactic acid bacteria, both *Lactococcus* and *Lactobacillus*, have low but measurable lipolytic and esterase activity.[233,241-244] There appears to be very little work on the isolation and characterization of these enzymes. Kamaly *et al.*[245] reported some characteristics of the lipase in the cell-free extracts of a number of *Lc. lactis* and *Lc. cremoris* strains. In general, the lipases were optimally active at 37°C and pH 7 to 8·5; they were stimulated by reduced glutathione and low (~2%) concentrations of NaCl but were inhibited by high concentrations of NaCl. The lipases were most active on tributyrin.

There has been substantial interest in the lipase/esterase activity of various *Lactobacillus* species in recent years.[246-249] Oterholm *et al.*[246] partially purified an acetyl ester hydrolase from the cell free extract of *Lb. plantarum* which was optimally active at pH 6·7 and 40°C. The enzyme showed a strong preference for substrates in solution rather than emulsion, and was therefore considered to be an esterase rather than a lipase. Its activity decreased in the order triacetin > tripropionin > tributyrin; activity on longer chain esters or triglycerides was not reported. The esterase was inhibited by -SH blocking agents.

The production of esterases/lipases by *Lb. helveticus*, *Lb. bulgaricus*, *Lb. lactis* and *Lb. acidophilus* was studied by El-Soda *et al.*[247,248] All the lactobacilli studied showed activities on the p-phenyl derivatives of fatty acids up to C_5. Zymograms showed three to seven esterolytic bands. Generally, the optimum temperature for esterase production was 40–45°C.

Piatkiewicz[249] reported that *L. lactis* and *L. lactis* subsp. *diacetylactis* and *Lb. casei* produced both lipases (assayed on β-naphthyl laurate) and esterases (assayed on β-naphthyl acetate); esterase activity was higher than lipase activity in all strains. The lactococci produced higher lipase and esterase activity than the *Lactobacillus*. The enzymes were associated mainly with the cell membrane and more enzyme was produced when the cells were grown in milk than in a broth.

Lb. casei subsp. *pseudoplantarum* LE2 appears to possess a potent lipase–esterase system which was purified and partially characterized by Lee & Lee.[250] Optimum activity on p-nitrophenyl derivatives of fatty acids was at pH 7·5 and

37°C. Three active fractions which were specific for C_4, C_6 or C_8 were resolved by gel filtration on Superose 12 HR; SDS-PAGE showed that those had molecular weights of 320, 110 and 40 kDa, respectively. Using a zymogram technique, Khalid *et al.*[251] showed that *Lb. helveticus* and *Lb. delbrueckii* subsp. *bulgaricus* contain two intracellular esterases, capable of hydrolysing α- and β-naphthyl esters; the esterases were not characterized.

The significance of micrococci in cheese has been discussed by Bhowmik & Marth.[187] Two lipases with MW of 25 and 250 kDa were purified from the cell-free supernatant of *M. freudenreichii* by Lawrence *et al.*[252] In the crude state, the lipases were quite heat stable, e.g. 20% loss of activity on heating at 100°C for 5 min but they became less heat stable on purification. The lipases, which were active on long- and short-chain triglycerides and esters, were strongly inhibited by organophosphates, Zn^{2+} and Hg^{2+} and less strongly by EDTA but not by pCMB.

Bhowmik and Marth[253] studied the esterase profile of a number of *Micrococcus* spp. Zymograms stained with α-naphthyl acetate showed the occurrence of one to four esterases. The esterases were optimally active at pH 8 and 40°C and were strongly inhibited by DFP and NaF and slightly by 1,10-phenanthroline and Na taurocholate; activity was strongly inhibited by NaCl, especially at pH 5.

Pediococci are also members of the NSLAB in cheese and have been reported to grow in cheese; their significance in cheese ripening has been reviewed by Bhowmik & Marth.[187] The lipase-esterase system of pediococci has received very little study. Nuñez[202] and Tzanetakis & Lipopoulou-Tzanetaki[206] reported that pediococci possessed lipolytic-esterolytic activity. Bhowmik & Marth[254] showed that all six strains of *P. pentosaceus* examined had esterase activity on α-naphthyl acetate while the two strains of *P. acidilactici* examined did not.

A potentially very important source of potent lipases in milk and cheese are psychrotrophs which dominate the microflora of refrigerated milk. These lipases, many of which are very heat stable, adsorb on the surface of fat globules and are therefore concentrated in cheese. The lipases of psychrotrophs are probably more significant in cheese, and butter, that their proteinases, which are water-soluble and are therefore lost in the whey. Various aspects of enzymes from psychrotrophs in milk and dairy products are reviewed in the book edited by McKellar.[255]

The lipase of *Propionibacterium shermanii* was studied by Oterholm *et al.*[256] The lipases and esterases of *Brevibacterium linens* were described briefly by Sørhaug and Ordal[257] and Foissy.[218]

Mould-ripened cheeses, especially blue cheeses, undergo the highest level of lipolysis of all varieties—up to 25% of the total fatty acids may be liberated in some blue cheeses (see Chapter 4, Volume 2). However, the impact of fatty acids on the flavour of these cheeses is less than that for the hard Italian varieties, possibly due to neutralization on elevation of the pH during ripening, and also due to domination of blue cheese flavour by methyl ketones.

Lipolysis in mould-ripened cheeses is due mainly to *P. roqueforti* or *P. camemberti*, both of which secrete very active extracellular lipases although considerable inter-strain variations occur (see Ref. 258). The lipases of both organisms have been isolated and well characterized (see Chapter 4, Volume 2). *P. camemberti* appears to excrete only one lipase with an optimum pH at ~9·0 and a temperature optimum at ~35°C. *P. roqueforti* excretes two lipases, one with a pH optimum at 7·5–8·0 (or perhaps 9·0–9·5), the other at pH 6·0–6·5. The acid and alkaline lipases exhibit different specificities.

3.3.2 Catabolism of Fatty Acids

The aroma and flavour of blue cheeses is dominated by methyl ketones, a homologous series of which from C_3 to C_{17} have been identified in these cheeses, the predominant ones being 2-heptanone and 2-nonanone (see Ref. 258). The mechanism of formation of methyl ketones, which was thoroughly reviewed by Kinsella & Hwang,[258] is via the β-oxidation pathway. The concentration of methyl ketones is proportional to the level of lipolysis. Spores as well as mycelia are capable of oxidizing fatty acids to methyl ketones. Blue cheeses also contain appreciable concentrations of secondary alcohols, especially 2-pentanol, 2-heptanol and 2-nonanol, produced by the reduction of the corresponding methyl ketones by *P. roqueforti*.

3.3.3 Hydroxy Acids and Fatty Acid Lactones

Lactones are cyclic esters resulting from the intramolecular esterification of a hydroxy acid through the loss of water to form a ring structure. α- and β-lactones are highly reactive and are used or occur as intermediates in organic synthesis; γ- and δ-lactones are stable and have been identified in cheese. As a group, lactones possess strong aromas. While they do not have a specifically cheese-like aroma (descriptions of aromas include 'fruity' and 'peach-like'), they may, nevertheless, be important in the overall cheese flavour impact. A range of fatty acid lactones have been identified in cheese and other dairy products.

Boldingh & Taylor[259] demonstrated the formation of δ-lactones from δ-hydroxy acids in milk fat heated in the presence of water. Eriksen[260] concluded that γ- and δ-lactones in freshly secreted milk fat originated from the corresponding γ- and δ-hydroxy acids esterified in triglycerides. γ- and δ-hydroxy fatty acids or their corresponding lactones are not found in any plant species used as feed for ruminants; thus, the hydroxy acids must result from animal metabolism.[261] The occurrence of optical activity in the hydroxy acid precursors and the fact that they occur as a homologous series with even numbers of carbons[259] suggest that lactones in milk were not caused via autooxidation.[260] Likewise, the absence of enzymes capable of reducing γ- and δ-keto acids in the mammary gland suggest that these compounds are not the precursors of γ- and δ-hydroxy acids.[261]

Tuynenburg Muys et al.[262] demonstrated the possibility of producing optimally active γ- and δ-lactones via reduction of keto acids to hydroxy acids by a number of yeasts (*Clasdosporium* spp. and *Saccharomyces* spp.), some moulds (including *Penicillium* spp.) and some bacteria (e.g. *Sarcina lutea*). However, Dimick et al.[261] reported a δ-oxidation system for fatty acid catabolism in the mammary gland of ruminants and thus oxidation within the gland appears to be the primary mechanism for the production of γ- and δ-hydroxy acids and consequently of lactone precursors. Thus, the potential for lactone production is dependent on variables such as feed, season, breed and stage of lactation.[261] The formation of lactones in milk is spontaneous following the hydrolysis of γ- or δ-hydroxy acids from triglycerides.[263] Thus, it would appear that lactone production should correlate with the hydrolysis of triglycerides.

Ellis & Wong[264] showed a correlation between the time and temperature of heating and the concentrations of lactones in butteroil and noted that the lactone profiles of freshly-made cheese and butter are almost identical. Boldingh & Taylor[259] demonstrated the production of lactones from synthetic triglycerides containing δ-hydroxy acids on heating.

In a study on Cheddar cheese, Wong *et al.*[265] found that longer chain lactones (C_{14} and C_{16}) increased disproportionately to other lactones in rancid cheese. Two alternative methods of formation were proposed. The first involved the reduction of the corresponding keto acid, but investigations tended to disprove this hypothesis. The other possible method involved the microbial metabolism of homoricinoleic acid to shorter chain hydroxy acids and lactones.

δ-Lactones have very low flavour thresholds[266] and various authors have mentioned the possibility that lactones contribute to the flavour of cheese.[267] Jolly & Kosikowski[268] found that the concentration of lactones in Blue cheese was higher than in Cheddar and concluded that the extensive lipolysis in Blue cheese influences the formation of lactones; δ-C_{14} and δ-C_{16} were the principal lactones in Blue cheese (as found also for Cheddar).[269] A stronger typical Blue cheese flavour was found in cheeses containing added lipase, perhaps because lactones blend or modify harsher flavours.

O'Keeffe *et al.*[269a] identified γ-C_{12}, γ-C_{14}, γ-C_{16}, δ-C_{10}, δ-C_{12}, δ-C_{14}, δ-C_{15}, δ-C_{16} and δ-C_{18} lactones in Cheddar cheese. The presence of most of these (γ-C_{12}, δ-C_{10}, δ-C_{12}, δ-C_{14}, γ-C_{16}) in Cheddar cheese was confirmed by Wong *et al.*[269] who showed a correlation between the number and concentration of lactones with age and flavour, suggesting that certain lactones are significant in Cheddar cheese flavour. In a further quantitative study of lactones in Cheddar cheese, Wong *et al.*[265] failed to find a close correlation between flavour and lactone concentration. In general, the above δ-lactones were produced more quickly and to higher concentrations than γ-C_{12}. Lactone levels increased most rapidly early in the ripening period and the levels found were well above the flavour threshold; it was considered likely that they influence flavour.

3.4 Role of Acid Phosphatases in Cheese Ripening

Phosphatases are enzymes which hydrolyse the C–O–P linkage of various phosphate and phosphonate esters. They are classified into 'acid' or 'alkaline' groups depending on the effect of pH on their activity.[270] Although both acid and alkaline phosphatases are present in cheese, the former are more active due to the relatively low pH (about 5·2) of cheese.

During cheese ripening the caseins are cleaved by rennet, plasmin and bacterial proteinases into phosphorus-rich peptides. The phosphate residues exert a protective effect against further proteolytic hydrolysis of the peptides.[271] Complete casein degradation during cheese ripening can be achieved only by the combined action of proteinases and phosphatases.[272] Therefore, phosphatases may play an important role in cheese maturation and flavour development.[142,273] However, the activity of acid phosphatase(s) in cheese is probably the least-well studied of the primary hydrolytic events during ripening and their significance is mainly putative.

The level of acid phosphatase activity in cheese remains constant during ripening. Andrews & Alichanidis[274] found no change in activity during storage of Feta cheese for 9–12 months at 6°C. Similar results were obtained for Telème cheese and for softer cheeses such as Kasseri when stored for up to 18 months. Examination of Cheddar cheeses of differing ages showed that over a 12 month period no significant changes in acid phosphatase activity occurred at 13°C. The levels of enzyme activity were relatively low at $8 \cdot 1 \pm 1 \cdot 11 \times 10^{-3}$ units/g throughout ripening. These comparatively low values for a hard cheese such as Cheddar suggested the involvement of factors other than the concentration effect of the milk enzyme.[274]

The origin of active acid phosphatases in ripening cheese is controversial. The enzyme is a phosphomonoesterase and it could be derived from a number of sources. Bovine milk (as well as that of other species) contains a heat-stable acid phosphatase which is not inactivated by pasteurization[275] (95% of its activity survives heating at 75°C for 3 min).[272] This enzyme is very active on phosphoproteins such as caseins. It consists of a single glycosylated polypeptide chain (42 kDa) containing two residues of galactose, two of mannose and four of N-acetyl nuraminic acid per molecule.[276] The enzyme is strongly activated by ascorbic acid and is inhibited competitively by hexametaphosphate and noncompetitively by heavy metals, particularly Ag^+ and F^-. However, it is not inhibited by sulphydryl blocking agents, e.g. N-ethyl maleimide.

Other possible sources of phosphatases in cheese are the starter bacteria. Both primary starters, such as lactococci, and secondary organisms (fungi and yeasts) contain acid phosphatases.[271] The cellular location of the enzymes in lactococci has not been reported but it would appear to be cell wall or membrane bound.[272,277] The starter enzyme has a high molecular weight.[274] Martley & Lawrence[142] indicated that a starter culture producing a good flavoured cheese should possess high acid phosphatase activity. This view is not shared by all researchers and many consider the role of the bacterial acid phosphatase in cheese to be minimal.[274]

The lactococcal enzyme binds strongly to micellar casein but it does not dephosphorylate the caseins to a significant extent.[274] Lactococcal acid phosphatase may, however, be more active on small phosphopeptides produced from casein. Optimum activity of the starter enzyme is at pH 5·2 while the milk enzyme has a pH optimum of about 5·0. The major differences between the lactococcal acid phosphatase and the indigenous milk enzyme appear to be greater sensitivity of the former to Pb^{2+} and to SH-reagents. Unlike the milk enzyme, starter acid phosphatase is not activated by ascorbic acid.[274] There is conflict as to the role of Mg^{2+} in the activation of starter acid phosphatase. While some workers[277] claim that it has an activating effect, others[275] claim that Mg^{2+} has no effect. There are no reports that milk acid phosphatase is activated by Mg^{2+}.

Fungi such as *P. roqueforti* possess a very active acid phosphatase. The specific activity of this fungal enzyme is 7·3 times higher that that of *Lc. cremoris*.[272] The levels of acid phosphatase in blue veined cheeses are higher than those in bacterially-ripened cheeses. The apparent optimum pH (4·25) of the enzyme in an extract of Roquefort cheese was similar to that of the *P. roqueforti* enzyme.[272] The thermal stability was also similar. This suggests that in the case of mould-ripened cheeses, the microbial enzyme plays a major role in casein dephosphorylation; however, the extent of dephosphorylation in mould-ripened cheeses has not been reported.

The surface flora of Tilsiter cheese contains mainly (77%) yeasts (genus *Endomycopsis*). Considerable levels of both alkaline and acid phosphatases are found in this cheese.[271] There is higher enzyme activity in the rind than in the core of the cheese. The core enzyme may be indigenous milk acid phosphatase while the enzyme at the rind may be a mixture of the milk enzyme and that derived from the yeast.

There is disagreement as to the specific contributions of the phosphatases from the above sources to dephosphorylation of peptides during cheese ripening. Andrews & Alichanidis[274] claimed that the phosphatase action in cheese was due largely to the indigenous milk enzyme with that derived from the starter being of only minor importance although inhibitor studies (which the authors advised should be treated with caution) suggested that 20–25% of acid phosphatase activity in cheese could be of starter origin. However, Dulley & Kitchen[273] credit the starter lactococci with 50–60% of the total activity in cheese. The view that the microbial acid phosphatases play a significant role in cheese ripening is also held by other researchers.[142,272]

Although the source of the acid phosphatase in cheese is not certain, most authors[142,272-274] believe that dephosphorylation of peptides by acid phosphatases is an important reaction in ripening cheese. The only report of dephosphorylation of peptides during cheese maturation is that of Dulley & Kitchen.[277] They observed no increase in inorganic phosphorus levels but the concentration of phosphopeptides increased during storage. The role of acid phosphatases in cheese ripening is, therefore, questionable and warrants further study.

4   CHEESE FLAVOUR AND AROMA

The objective of cheese manufacture is to produce a product with the flavour, aroma and texture of the intended variety, free of defects and in the shortest time possible. At the risk of antagonizing the rheologists among us, textural changes, although very complex, are probably the least complex of these three changes. The texture of cheese is determined initially by the composition of the cheesemilk, especially by the fat:casein ratio, by the manufacturing operations which control the extent of syneresis and hence the moisture content of the cheese, and the rate of acidification which controls the extent of demineralization of the curd and which in turn has a major influence on the textural parameters of the cheese. The texture of the cheese changes during ripening due to proteolysis, especially of the α_{s1}-casein by the rennet, the decrease in a_w due to the liberation of water-binding ionic groups, redistribution of salt, and, in many cases, evaporation of water and to changes in pH due to proteolysis and catabolism of lactic acid, which is most marked in surface mould-ripened cheeses. Although the factors responsible for the finer points of cheese texture are not clearly defined, at least the principal factors are clear and generally accepted. The rheological and textural aspects of cheese are reviewed in Chapter 8, Volume 1 and by Lawrence et al.[26,278] and will not be discussed further here.

Undoubtedly, cheese aroma and flavour are influenced by cheese texture, e.g. by consumer perception and release of sapid and aromatic compounds from the cheese mass during mastication.[87] The fat of cheese plays several roles in cheese quality, e.g. by directly affecting cheese texture, acting as a source of aromatic and sapid compounds and acting as a solvent for these compounds produced from cheese lipids and other sources.

Elucidation of the chemical nature of cheese flavour is one of the most intractable problems in dairy chemistry. Research on the subject dates from the beginning of this century. During this period, especially during the last 30 years, very considerable progress has been made on the identification of sapid and aromatic compounds in cheese, through the development of new analytical techniques, notably gas chromatography and mass spectroscopy, and more recently HPLC. As discussed in the preceding sections, considerable progress has also been made in identifying the biochemical reactions responsible for production of many of the principal compounds responsible for cheese flavour and aroma. However, there has been little or no progress in correlating the flavour of cheese with the multitude of chemicals present. This failure undoubtedly reflects the complexity of the subject which is made more difficult by the lack of clear definition of cheese quality among consumers and cheese researchers.

These are exceptions to this general situation, notably the blue cheese varieties in which the flavour is dominated by methyl ketones with a major contribution of free fatty acids (see Ref. 258; Chapter 4, Volume 2) and some Italian varieties, e.g. Romano, in which free fatty acids dominate the flavour.

Research on cheese flavour and aroma expanded greatly during the 1960s and 1970s, due mainly to rapid developments in GC and mass spectroscopy which showed the presence of hundreds of compounds that could contribute to cheese flavour and aroma. Most of these are present at very low concentrations, many below their flavour thresholds, but which may still affect cheese quality. Early workers believed that cheese flavour was dominated by a single compound or class of compounds, but it was soon apparent that this was not so, and this led to the view that cheese flavour is due to the correct balance of a mixture of compounds.[279] This hypothesis has become known as the Component Balance Theory[280] which has become fairly widely accepted. The problem has been, and still is, to identify the principal contributors to flavour and aroma.

A variety of techniques, including model cheese systems, with or without controlled microflora, selective removal and trapping of compounds, synthetic mixtures and more recently discriminant analysis of flavour profiles have failed to identify unequivocally the compounds responsible for the flavour and aroma of internally bacterially ripened cheeses, such as Cheddar and Dutch types. In such cheeses, there appears to be fairly good agreement that the water-insoluble fraction (consisting mainly of proteins and large peptides) is devoid of flavour or aroma, that the water soluble non-volatile fraction (small peptides, amino acids, organic acids) contains most of the compounds responsible for flavour while the aroma is principally in the volatile fraction. There appears to be strong support for the view that products of proteolysis are the principal contributors to cheese flavour.[63,87] The principal contributors to aroma are less clear but sulphur-containing compounds and carbonyl compounds are probably important. It has long been considered that sulphur compounds are major contributors to the flavour of Cheddar and other varieties. The principal early proponents of this theory were the group of Kristoffersen (see Ref. 281) but it was extensively developed by Manning and collaborators (see Refs 282, 283). These workers demonstrated the importance of methanethiol in Cheddar cheese flavour; they considered it to be the key flavour compound, although its impact was modified by others, e.g. methanol, H_2S and 2-pentanone. One of the problems with methanethiol as the key compound in Cheddar flavour is that while it is synthesized by *Br. linens* and other coryneforms, microorganisms capable of producing it are rarely found in Cheddar,[284] suggesting that it may be of chemical origin, e.g. from methional produced from methionine by Strecher degradation.[45] Later studies have failed to show a good correlation between the quality or intensity of Cheddar flavour and methanethiol concentration (see Ref. 45). The presence of several Strecher-derived aldehydes and alcohols in Cheddar cheese has been described.[285] The principal compounds found were phenylacetaldehyde and phenethanol (from phenylalanine), p-cresol, phenol (from tyrosine), 3-methyl butanal (leucine), 2-methyl butanal (isoleucine) and 2-methyl pentanal (valine). All of these, except phenol, had distinctly unclean flavours. Interestingly, Phe, Tyr, Leu, Ile and Val are participants in peptide bonds hydrolysed by chymosin early during ripening and hence are accessible for release by bacterial aminopeptidases.

The literature on cheese flavour has been reviewed extensively, e.g. Refs 45, 56, 231, 233, 271, 279–283, 286–291. There would appear to be insufficient new information to justify a further review of the literature. Furthermore, aspects of the flavour of the principal families of cheeses are discussed in the various chapters in Volume 2. Important contributors to the flavour of Swiss cheese have been described.[292–294]

The lack of progress in describing cheese flavour in precise chemical terms, despite the considerable effort and progress, is, obviously, disappointing. It is not even clear whether the rate-limiting reaction in flavour development is enzymatic or chemical (e.g. chemical modification of amino acids by Strecher degradation). It has been suggested (see Ref. 231) that the principal role of starter bacteria is to produce the substrates, e.g. amino acids, and the conditions (pH, E_h) which favour the conversion of these substrates to the compounds responsible for the final flavour. That may very well be the case for young mild cheese but it is difficult to see how the myriad of cheese types and variations within a type could develop by purely chemical means.

There has been considerably more progress on eludicating off-flavours than desirable flavours in cheese—this is probably not too surprising since each off-flavour usually has a specific cause. Examples include fruity (ethyl hexanoate and ethyl octanoate, resulting from high concentrations of ethanol), butyric acid flavour in Swiss cheese and bitterness which is common in many cheeses. It is generally agreed that bitterness is due to the accumulation of hydrophobic peptides but there is disagreement on the cause of the problem, whether it is due to a deficiency of peptidase activity or to too much proteinase activity in certain starters (see Refs 295–299).

The focus of research on cheese flavour appears to have shifted in recent years. There has certainly been a shift from the analysis of cheese volatiles by GC to analysis and characterization of the non-volatile water-soluble fraction by HPLC. Undoubtedly, several factors have contributed to this, such as the failure to correlate cheese flavour and quality with cheese volatiles, marked developments in HPLC, evidence that non-volatile water-soluble compounds, especially peptides and amino acids, are important in cheese flavour, and in view of this the role of starter proteinases and peptidases in cheese quality. Developments in molecular genetics and genetic engineering have enabled modification of the starter proteinase system and, as discussed earlier, much attention is now focused on this area.

To attempt to predict the direction of research on cheese flavour is fraught with danger but certain aspects merit mention. Further improvements in analytical techniques are likely and will extend the range of potentially important contributors to cheese flavour. Multivariate analyses of GC and HPLC profiles of a large selection of cheeses might be fruitful; attempts at this have already been made (e.g. Refs 300, 301) but the number of samples has been too limited. There is already renewed interest in the manufacture of cheeses from raw or underpasteurized milks and in the use of secondary inocula (i.e. *Lactobacillus* spp.). Some of the parameters/factors which have persistently featured as being important in cheese flavour are several reduced sulphur compounds, and the importance of low E_h in the formation/stability of these. It has been difficult or impossible to reduce the E_h of cheese by chemical or physical means without introducing chemical interference. Further work in this area might be rewarding. Kristoffersen[281] referred to 'enzymatic ripening' of milk prior to cheesemaking, i.e. holding milk at 37°C for ~3 h after milking and before cooling. Although activation of -SH groups was the objective of this treatment, it is not clear what changes might occur or what the mechanism involved might be; perhaps the proposal merits further attention.

REFERENCES

1. Perry, K.D. & McGillivray, W.A., 1964. *J. Dairy Res.*, **31**, 155.
2. O'Keeffe, R.B., Fox, P.F. & Daly, C., 1976. *Ir. J. Agric. Res.*, **15**, 151.
3. Kleter, G. & de Vries, Tj., 1974. *Neth. Milk Dairy J.*, **28**, 212.
4. Visser, F.M.W., 1976. *Neth. Milk Dairy J.*, **30**, 41.
5. Reiter, B., Sorokin, Y., Pickering, A. & Hall, A.J., 1969. *J. Dairy Res.*, **36**, 65.
6. Chapman, H.R., Mabbitt, L.A. & Sharpe, M.E., 1966. *Proc. 17th Intern. Dairy Congr., Munich*, **D1**, 55.
7. Reiter, B., Fryer, T.F., Pickering, A., Chapman, H.R., Lawrence, R.C. & Sharpe, M.E., 1967. *J. Dairy Res.*, **34**, 257.
8. Visser, F.M.V., 1977. *Neth. Milk Dairy J.*, **31**, 120.
9. Kleter, G., 1976. *Neth. Milk Dairy J.*, **30**, 254.
10. Turner, K.W., Lawrence, R.C. & Lelievre, J., 1986. *N.Z. J. Dairy Sci. Technol.*, **21**, 249.
11. Le Bars, D., Desmazeaud, M.J., Gripon, J.-C. & Bergere, J.L., 1975. *Lait.*, **55**, 377.
12. Mabbitt, L.A., Chapman, H.R. & Sharpe, M.E., 1959. *J. Dairy Res.*, **26**, 105.
13. O'Keeffe, R.B., Fox, P.F. & Daly, C., 1975. *J. Dairy Res.*, **42**, 111.
14. O'Keeffe, R.B., Fox, P.F. & Daly, C., 1976. *J. Dairy Res.*, **43**, 97.
15. Mabbitt, L.A., Chapman, H.R. & Berridge, N.J., 1955. *J. Dairy Res.*, **22**, 365.
16. Creamer, L.K., Lawrence, R.C. & Gilles, J., 1985. *N.Z. J. Dairy Sci. Technol.*, **20**, 185.
17. Visser, F.M.W., 1977. *Neth. Milk Dairy J.*, **31**, 188.
18. Visser, F.M.W., 1977. *Neth. Milk Dairy J.*, **31**, 210.
19. Visser, F.M.W. & de Groot-Mostert, A.E.A., 1977. *Neth. Milk Dairy J.*, **31**, 247.
20. O'Keeffe, A.M., Fox, P.F. & Daly, C., 1977. *J. Dairy Res.*, **44**, 335.
21. Mulvihill, D.M., Collier, T.M. & Fox, P.F., 1979. *J. Dairy Sci.*, **62**, 1567.
22. Farkye, N.Y. & Fox, P.F., 1990. *J. Dairy Res.*, **57**, 413.
23. Farkye, N.Y. & Fox, P.F., 1990. *J. Agric. Food Chem.*, **39**, 786.
24. Harper, J.W., Hemmi, K. & Powers, J.C., 1985. *Biochemistry*, **24**, 1831.
25. Creamer, L.K., Gilles, J. & Lawrence, R.C., 1988. *N.Z. J. Dairy Sci. Technol.*, **23**, 23.
26. Lawrence, R.C., Creamer, L.K. & Gilles, J., 1987. *J. Dairy Sci.*, **70**, 1748.
27. Fox, P.F., Cogan, T.M. & Lucey, J.A., 1990. *CRC Crit. Rev. Food Sci. Technol.*, **29**, 237.
28. Huffman, L.M. & Kristoffersen, T., 1984. *N.Z. J. Dairy Sci. Technol.*, **19**, 151.
29. Turner, K.W. & Thomas, T.D., 1980. *N.Z. J. Dairy Sci. Technol.*, **15**, 265.
30. Thomas, T.D. & Pearce, K.N., 1981. *N.Z. J. Dairy Sci. Technol.*, **16**, 253.
31. O'Connor, C.B., 1974. *Irish Agric. Creamery Rev.*, **27**(1), 11.
32. Raadsveld, C.W., 1957. *Neth. Milk Dairy J.*, **11**, 313.
33. Turner, K.W., Morris, H.A. & Martley, F.G., 1983. *N.Z. J. Dairy Sci. Technol.*, **18**, 117.
34. Karahadian, G. & Lindsay, R.C., 1987. *J. Dairy Sci.*, **70**, 909.
35. Tinson, W., Radcliff, M.F., Hillier, A.J. & Jago, G.R., 1982. *Aust. J. Dairy Technol.*, **37**, 17.
36. Thomas, T.D. & Crow, V.L., 1983. *N.Z. J. Dairy Sci. Technol.*, **18**, 131.
37. Pearce, K.N., Creamer, L.K. & Gilles, J., 1973. *N.Z. J. Dairy Sci. Technol.*, **8**, 3.
38. Severn, D.J., Johnson, M.E. & Olson, N.F., 1986. *J. Dairy Sci.*, **69**, 2027.
39. Dybing, S.T., Wiegand, J.A., Brudvig, S.A., Huang, E.A. & Chandan, R.C., 1988. *J. Dairy Sci.*, **71**, 1701.
40. Thomas, T.D., McKay, L.L. & Morris, H.A., 1985. *Appl. Environ. Microbiol.*, **49**, 908.
41. Thomas, T.D., 1986. *N.Z. J. Dairy Sci. Technol.*, **21**, 37.
42. Thomas, T.D., 1987. *N.Z. J. Dairy Sci. Technol.*, **22**, 25.
43. Thomas, T.D., Ellwood, D.C. & Longyear, M.C., 1979. *J. Bacteriol.*, **138**, 109.
44. Nakae, T. & Elliott, J.A., 1965. *J. Dairy Sci.*, **48**, 287.

45. Aston, J.W. & Dulley, J.R., 1982. *Aust. J. Dairy Technol.*, **37**, 59.
46. Lenoir, J., 1984. IDF Bulletin 171, 3.
47. Noomen, A., 1983. *Neth. Milk Dairy J.*, **37**, 229.
48. Deiana, P., Fatichenti, F., Farris, G.A., Mocquot, G., Lodi, R., Todesco, R. & Cecchi, L., 1984. *Lait*, **64**, 380.
49. Crow, V.L., 1986. *Appl. Environ. Microbiol.*, **52**, 352.
50. Cogan, T.M., 1985. In *Bacterial Starter Cultures for Foods*, ed. S.E. Gilliand. CRC Press, Boca Raton, FL, USA, p. 25.
51. Cogan, T.M. & Daly, C., 1987. In *Cheese: Chemistry, Physics and Microbiology*, Vol. 1, ed. P.F. Fox. Elsevier Applied Science, London, p. 179.
52. Hickey, M.W., Hillier, A.J. & Jago, G.R., 1983. *Aust. J. Biol. Sci.*, **36**, 487.
53. Fryer, T.F., 1970. *J. Dairy Res.*, **37**, 9.
54. Manning, D.J., 1979. *J. Dairy Res.*, **46**, 523.
55. Manning, D.J., 1979. *J. Dairy Res.*, **46**, 531.
56. McGugan, W.A., 1975. *J. Agric. Food Chem.*, **23**, 1047.
57. Fryer, T.F., Sharpe, M.E. & Reiter, B., 1970. *J. Dairy Res.*, **37**, 17.
58. Cogan, T.M., 1987. *J. Appl. Bacteriol.*, **63**, 551.
59. Grappin, R., Rank, T.C. & Olson, N.F., 1985. *J. Dairy Sci.*, **68**, 531.
60. Rank, T.C., Grappin, R. & Olson, N.F., 1985. *J. Dairy Sci.*, **68**, 801.
61. Law, B. A., 1987. In *Cheese: Chemistry, Physics and Microbiology*, Vol. 1, ed. P.F. Fox. Elsevier Applied Science, London, p. 365.
62. Fox, P.F., 1989. *J. Dairy Sci.*, **72**, 1379.
63. Aston, J.W., Durward, I.G. & Dulley, J.R., 1983. *Aust. J. Dairy Technol.*, **38**, 55.
64. Amantea, G.F., Shiva, B.J. & Nakai, S., 1986. *J. Food Sci.*, **51**, 912.
65. Farkye, N.Y. & Fox, P.F., 1990. *Trends Food Sci. Technol.*, **1**, 37.
66. Haasnoot, W., Stouten, P. & Venema, D.P., 1989. *J. Chromatog.*, **483**, 319.
67. IDF Monograph on Chemical Methods for Evaluating Proteolysis in Cheese Maturation, Bulletin 261, International Dairy Federation, Brussels, 1990.
68. Holmes, D.G., Duersch, J.N. & Ernstrom, C.A., 1977. *J. Dairy Sci.*, **60**, 862.
69. Green, M.L. & Foster, P.M.D., 1974. *J. Dairy Res.*, **41**, 269.
70. O'Keeffe, A.M., Fox, P.F. & Daly, C., 1978. *J. Dairy Res.*, **45**, 465.
71. Matheson, A.R., 1981. *N.Z. J. Dairy Sci. Technol.*, **16**, 33.
72. Farkye, N.Y., Kiely, L.J., Adlshouse, R.D. & Kindstedt, P.S., 1991. *J. Dairy Sci.*, **74**, 1433.
73. Green, M.L., 1977. *J. Dairy Res.*, **44**, 159.
74. Creamer, L. K., Mills, O.E. & Richards, E.L., 1971. *J. Dairy Res.*, **38**, 269.
75. Visser, S. & Slangen, K.J., 1977. *Neth. Milk Dairy J.*, **31**, 16.
76. Carles, C. & Ribadeau-Dumas, B., 1984. *Biochemistry*, **23**, 6839.
77. Fox, P.F. & Walley, B.F., 1971. *J. Dairy Res.*, **38**, 165.
78. Creamer, L.K., 1971. *N.Z. J. Dairy Sci. Technol.*, **6**, 191.
79. Creamer, L.K., Aston, J. & Knighton, D., 1988. *N.Z. J. Dairy Sci. Technol.*, **23**, 185.
80. Phelan, J.A., Guiney, J. & Fox, P.F., 1973. *J. Dairy Res.*, **40**, 105.
81. Phelan, J.A., 1985. PhD Thesis, National University of Ireland, Dublin.
82. Mulvihill, D.M. & Fox, P.F., 1977. *J. Dairy Res.*, **44**, 533.
83. Mulvihill, D.M. & Fox, P.F., 1979. *J. Dairy Res.*, **46**, 641.
84. Godinho, M. & Fox, P.F., 1982. *Milchwissenschaft*, **37**, 72.
85. de Jong, L., 1976. *Neth. Milk Dairy J.*, **30**, 242.
86. Creamer, L.K. & Olson, N.F., 1982. *J. Food Sci.*, **47**, 631.
87. McGugan, W.A., Emmons, D.B. & Larmond, E., 1979. *J. Dairy Sci.*, **62**, 398.
88. Grufferty, M.B. & Fox, P.F., 1988. *J. Dairy Res.*, **55**, 609.
89. Le Bars, D. & Gripon, J.-C., 1989. *J. Dairy Res.*, **56**, 817.
90. Visser, S., Slangen, K.J., Alting, A.C. & Vreeman, H.J., 1989. *Milchwissenschaft*, **44**, 335.
91. Grufferty, M.B. & Fox, P.F., 1988. *N.Z. J. Dairy Sci. Technol.*, **23**, 153.

92. Madkor, S.A. & Fox, P.F., 1991. *Food Chem.*, **39**, 139.
93. Casey, M., Gruskovnjak, J. & Furst, M., 1987. *Schweiz. Milchw. Forschung*, **16**, 21.
94. Ollikainen, P. & Nyberg, K., 1988. *Milchwissenschaft*, **43**, 497.
95. Ollikainen, P. & Kivela, T., 1989. *Milchwissenschaft*, **44**, 204.
96. Farkye, N.Y. & Fox, P.F., 1992. *J. Dairy Res.*, **59**, 209.
97. Thomas, T.D., Pritchard, G.G., 1987. *FEMS Microbiol. Rev.*, **46**, 245.
98. Hugenholtz, J., Exterkate, F.A. & Konings, W.N., 1984. *Appl. Environ. Microbiol.*, **48**, 105.
99. Akuzawa, R., Ito, O. & Yokoyama, I., 1983. *Jap. J. Zootech. Sci.*, **54**, 685.
100. Kok, J., 1990. *FEMS Microbiol. Rev.*, **87**, 15.
101. Power, S.D., Adams, R.M. & Wells, J.A., 1986. *Proc. Natl. Acad. Sci. (USA)*, **83**, 3096.
102. Exterkate, F.A., 1976. *Neth. Milk Dairy J.*, **30**, 95.
103. Visser, S., Exterkate, F.A., Slangen, C.J. & de Veer, G.J.C.M., 1986. *Appl. Environ. Microbiol.*, **52**, 1162.
104. Fitzgerald, G.F. & Gasson, M., 1988. *Biochemie*, **70**, 489.
105. Vos, P., van Asseldonk, M., van Jeveren, F., Siezen, R., Simons, G. & de Vos, W.M., 1989. *J. Bacteriol.*, **171**, 2795.
106. Laan, H., Smid, E.J., Tan, P.S.T. & Konings, W.N., 1989. *Neth. Milk Dairy J.*, **43**, 327.
107. Vos, P., Simons, G., Siezen, R.J. & de Vos, W.M., 1989. *J. Biol. Chem.*, **264**, 13579.
108. Laan, H. & Konings, W.N., 1989. *Appl. Environ. Microbiol.*, **55**, 3101.
109. Kiwaki, M., Ikemura, H., Shimizu-Kadota, M. & Hirashima, A., 1989. *Mol. Microbiol.*, **3**, 359.
110. Kok, J., Leenhouts, K.J., Haandrikman, A.J., Ledeboer, A.M. & Venema, G., 1988. *Appl. Environ. Microbiol.*, **54**, 231.
111. Exterkate, F.A. & de Veer, G.J.C.M., 1987. *Appl. Environ. Microbiol.*, **53**, 577.
112. Geis, A., Bockelmann, W. & Teuber, M., 1985. *Appl. Microbiol. Biotechnol.*, **23**, 79.
113. Tan, P.S.T. & Konings, W.N., 1990. *Appl. Environ. Microbiol.*, **56**, 526.
114. Neviani, E., Boquien, C.Y., Monnet, V., Phan Thanh, L. & Gripon, J.-C., 1989. *Appl. Environ. Microbiol.*, **55**, 2308.
115. Kaminogawa, S., Ninomiya, T. & Yamauchi, K., 1984. *J. Dairy Sci.*, **67**, 2483.
116. Hwang, I.K., Kaminogawa, S. & Yamauchi, K., 1981. *Agric. Biol. Chem.*, **45**, 159.
117. Van Boven, A., Tan, P.S.T. & Konings, W.N., 1988. *Appl. Environ. Microbiol.*, **54**, 43.
118. Kolstad, J. & Law, B.A., 1985. *Appl. Bacteriol.*, **58**, 449.
119. Mou, L., Sullivan, J.J. & Jago, G.R., 1975, *J. Dairy Res.*, **42**, 147.
120. Bosman, B.W., Tan, P.S.T. & Konings, W.N., 1990. *Appl. Environ. Microbiol.*, **56**, 1839.
121. Baankreis, R. & Exterkate, F.A., 1990. *Proc. 3rd Neth. Biotechnol. Congr.*, p. 336.
122. Exterkate, F.A., 1975. *Neth. Milk Dairy J.*, **29**, 303.
123. Yan, T.-R., Azuma, N., Kaminogawa, S. & Yamauchi, K., 1987. *Appl. Environ. Microbiol.*, **53**, 2296.
124. Yan, T.-R., Azuma, N., Kaminogawa, S. & Yamauchi, K., 1987. *Eur. J. Biochem.*, **163**, 259.
125. Muset, G., Monnet, V. & Gripon, J.-C., 1989. *J. Dairy Res.*, **56**, 765.
126. Kiefer-Partsch, B., Bockelmann, W., Geis, A. & Teuber, M., 1989. *Appl. Microbiol. Technol.*, **31**, 75.
127. Zevaco, C., Monnet, V. & Gripon, J.-C., 1990. *J. Appl. Bacteriol.*, **68**, 357.
128. Booth, M., Ni Fhaolain, I., Jennings, P.V. & O'Cuinn, G., 1990. *J. Dairy Res.*, **57**, 89.
129. Kaminogawa, S., Azuma, N., Hwang, I.K., Susuki, Y. & Yamauchi, K., 1984. *Agric. Biol. Chem.*, **48**, 3035.
130. Booth, M., Jennings, P.V., Ni Fhaolain, I. & O'Cuinn, G., 1990. *J. Dairy Res.*, **57**, 245.

131. Smid, E.J., Poolman, B. & Konings, W.N., 1991. *Appl. Environ. Microbiol.*, **57**, 2447.
132. Mayo, B., Kok, J., Venema, K., Bockelmann, W., Teuber, M., Reinke, H. & Venema, G., 1991. *Appl. Environ. Microbiol.*, **57**, 38.
133. Nardi, M., Chopin, M.-C., Chopin, A., Cals, M.M. & Gripon, J.-C., 1991. *Appl. Environ. Microbiol.*, **57**, 45.
134. Law, B.A. & Kolstad, J., 1983. *Antonie Leeuwenhoek, J. Microbiol.*, **49**, 225.
135. Thomas, T.D. & Mills, O.E., 1981. *Neth. Milk Dairy J.*, **35**, 255.
136. Igoshi, K., Kaminogawa, S. & Yamauchi, K., 1986. *J. Dairy Sci.*, **69**, 2018.
137. Gripon, J.-C., Desmazeaud, M.J., Le Bars, D. & Bergere, J.L., 1975. *Lait*, **55**, 502.
138. Desmazeaud, M.J., Gripon, J.-C., Le Bars, D. & Bergere, J.L., 1976. *Lait*, **56**, 379.
139. Kaminogawa, S., Yan, T.R., Azuma, N. & Yamauchi, K., 1986. *J. Food Sci.*, **51**, 1253.
140. Exterkate, F.A., Alting, A.C. & Slangen, C.J., 1991. *Biochem. J.*, **273**, 135.
141. Monnet, V., Le Bars, D. & Gripon, J.-C., 1986. *Fed. Eur. Microbiol. Soc. Microbiol. Lett.*, **36**, 127.
142. Marley, F.G. & Lawrence, R.C., 1972. *N.Z. J. Dairy Sci. Technol.*, **7**, 38.
143. Umemoto, Y., Sato, Y. & Kito, J., 1978. *Agric. Biol. Chem.*, **42**, 227.
144. Law, B.A., Sharpe, M.E. & Reiter, B., 1972. *J. Dairy Res.*, **41**, 137.
145. Law, B.A., Castanon, M.J. & Sharpe, M.E., 1976. *J. Dairy Res.*, **43**, 301.
146. Oberg, C.J., Davies, L.H., Richardson, G.H. & Ernstrom, C.A., 1986. *J. Dairy Sci.*, **69**, 2975.
147. Farkye, N.Y., Fox, P.F., Fitzgerald, G.F. & Daly, C., 1990. *J. Dairy Sci.*, **73**, 874.
148. Somkuti, G.A. & Steinberg, D.H., 1979. *J. Appl. Biochem.*, **1**, 357.
149. Mercenier, A. & Lemoine, Y., 1989. *J. Dairy Sci.*, **72**, 3444.
150. Veringa, H.A., Galesloot, Th.E. & Davelaar, H., 1968. *Neth. Milk Dairy J.*, **22**, 114.
151. Driessen, F.M., Kingma, F. & Stadhouders, J., 1982. *Neth. Milk Dairy J.*, **36**, 135.
152. Lees, G.J. & Jago, G.R., 1976. *J. Dairy Res.*, **43**, 75.
153. Desmazeaud, M.J. & Hermier, J.H., 1972. *Eur. J. Biochem.*, **28**, 190.
154. Marshall, V.M. & Bramley, A.J., 1984. *J. Dairy Res.*, **51**, 17.
155. Shankar, P.A., 1977. PhD Thesis, University of Reading.
156. Shahbal, S., Hemme, D. & Desmazeaud, M., 1991. *Lait*, **71**, 351.
157. Meyer, J., Howald, D., Jordi, R. & Fürst, M., 1989. *Milchwissenschaft*, **44**, 678.
158. Desmazeaud, M.J. & Juge, M., 1976. *Lait*, **56**, 241.
159. Rabier, D. & Desmazeaud, M.J., 1973. *Biochimie*, **55**, 389.
160. Meyer, J. & Jordi, R., 1987. *J. Dairy Sci.*, **70**, 738.
161. Premi, L., Sandine, W.E. & Elliker, P.R., 1972. *Appl. Microbiol.*, **24**, 51.
162. Khalid, N.M. & Marth, E.H., 1990. *J. Dairy Sci.*, **73**, 2669.
163. Ezzat, N., El-Soda, M., Bouillanne, C., Zevaco, C. & Blanchard, P., 1985. *Milchwissenschaft*, **40**, 140.
164. Ezzat, N., Zevaco, C., El-Soda, M. & Gripon, J.-C., 1987. *Milchwissenschaft*, **42**, 95.
165. Ohmiya, K. & Sato, Y., 1969. *Agric. Biol. Chem.*, **33**, 669.
166. Exterkate, F.A. & de Veer, G.J.C.M., 1987. *Syst. Appl. Microbiol.*, **9**, 183.
167. Bockelmann, W., 1987. PhD Thesis, University of Kiel.
168. Peterson, S.D., Marchall, R.T. & Heymann, H., 1990. *J. Dairy Sci.*, **73**, 1454.
169. Arora, G., Lee, B.H. & Lamoureux, M., 1990. *J. Dairy Sci.*, **73**, 264.
170. Arora, G. & Lee, B.H., 1990. *J. Dairy Sci.*, **73**, 274.
171. Khalid, N.M. & Marth, E.H., 1990. *Appl. Environ. Microbiol.*, **56**, 381.
172. Bockelmann, W., Fobker, M. & Teuber, M., 1991. *Intern. Dairy J.*, **1**, 51.
173. Eggimann, B. & Bachmann, M., 1980. *Appl. Environ. Microbiol.*, **40**, 876.
174. Bockelmann, W., Schulz, Y. & Teuber, M., 1992. *Intern. Dairy J.*, **2**, 95.
175. Machuga, E.J. & Ives, D.H., 1984. *Biochim. Biophys. Acta*, **789**, 26.
176. Cholette, H. & McKellar, R.C., 1990. *J. Dairy Sci.*, **73**, 2278.

177. Schulz, Y. & Bockelmann, W., 1992. *Intern. Dairy J.*, in press.
178. El-Soda, M., Bergere, J.L. & Desmazeaud, M.J., 1978. *J. Dairy Res.*, **45**, 519.
179. El-Soda, M., Said, H., Desmazeaud, M.J., Mashaly, R. & Ismail, A., 1983. *Lait*, **63**, 1.
180. Abo El-Naga, I.G. & Plapp, R., 1987. *J. Basic Microbiol.*, **27**, 123.
181. Bartels, H.J., Johnson, M.E. & Olson, N.F., 1987. *Milchwissenschaft*, **42**, 139.
181a. Atlan, D., Laloi, P. & Portalier, R., 1990. *Appl. Environ. Microbiol.*, **56**, 2174.
181b. Atlan, D., Laloi, P. & Portalier, R., 1989. *Appl. Environ. Microbiol.*, **55**, 1717.
182. Anon, 1988. Ricerca triennale sulla composizione e su alcune peculiari caratteristiche del formaggio Parmigiano-Reggiano. Consorzio del Formaggio Parmigiano-Reggiano, Reggio Emilia.
183. Lau, K.Y., Barbano, D.M. & Rasmussen, R.R., 1991. *J. Dairy Sci.*, **74**, 727.
184. Grazier, C.L., Bodyfelt, F.W., McDaniel, M.R. & Torres, J.A., 1991. *J. Dairy Sci.*, **74**, 3656.
185. Peterson, S.D. & Marshall, R.T., 1990. *J. Dairy Sci.*, **73**, 1395.
186. McSweeney, P.L.H., Lucey, J.A., Jordan, K., Cogan, T.M. & Fox, P.F., 1991. *J. Dairy Sci.*, **74** (suppl 1), 91.
187. Bhowmik, T. & Marth, E.H., 1990. *J. Dairy Sci.*, **73**, 859.
187a. Bhowmik, T., Riesterer, R., van Bockel, M.A.J.S. & Marth, E.H., 1990. *Milchwissenschaft*, **45**, 230.
188. McDonald, I.J., 1961. *Can. J. Microbiol.*, **6**, 251.
189. Prasad, R., Malik, R.K. & Mathur, D.K., 1984. *Asian J. Dairy Res.*, **3**, 25.
190. Garcia de Fernando, G.D. & Fox, P.F., 1991. *Lait*, **71**, 359.
191. Desmazeaud, M. & Hermier, J., 1968. *Ann. Biol. Anim. Biochim. Biophys.*, **8**, 565.
192. Desmazeaud, M. & Hermier, J., 1971. *Eur. J. Biochem.*, **19**, 51.
193. Garcia de Fernando, G.D. & Fox, P.F., 1991. *Lait*, **71**, 371.
194. Nath, K.R. & Ledford, R.A., 1972. *J. Dairy Sci.*, **55**, 1424.
195. Bhowmik, T. & Marth, E.H., 1988. *J. Dairy Sci.*, **71**, 2358.
196. Dacre, J.C., 1958. *J. Dairy Res.*, **25**, 409.
197. Dacre, J.C., 1958. *J. Dairy Res.*, **25**, 417.
198. Fryer, T.F. & Sharpe, M.E., 1966. *J. Dairy Res.*, **33**, 325.
199. Elliott, J.A. & Mulligan, H.T., 1966. *Can. Inst. Food Technol. J.*, **1**, 61.
200. Litopoulou-Tzanetaki, E., Graham, D.C. & Beyatli, Y., 1989. *J. Dairy Sci.*, **72**, 854.
201. Bottazzi, V., 1960. *Ann. Microbiol. Milano*, **10**, 57.
202. Nuñez, M., 1976. *Ann. Inst. Nac. Invest. Agar. Ser. Gen.*, No. 4, 75.
203. El-Gendy, S.M., Abdel-Galil, H., Shahin, Y. & Hegazi, F.Z., 1983. *J. Food Prot.*, **46**, 429.
204. Robertson, P.S. & Perry, K.D., 1961. *J. Dairy Res.*, **28**, 245.
205. Law, B.A., Castanon, M. & Sharpe, M.E., 1976. *J. Dairy Res.*, **43**, 117.
206. Tzanetakis, N. & Lipopoulou-Tzanetaki, E., 1989. *J. Dairy Sci.*, **72**, 859.
207. Bhowmik, T. & Marth, E.H., 1990. *Microbiol.*, **62**, 197.
208. El-Soda, M., Macedo, A. & Olson, N.F., 1991. *Milchwissenschaft*, **46**, 223.
209. Keenan, T.W., Parmelee, C.E. & Branen, A.L., 1968. *J. Dairy Sci.*, **51**, 1737.
210. Antila, V. & Antila, M., 1968. *Milchwissenschaft*, **23**, 597.
211. Hintz, P.C., Slatter, W.L. & Harper, W.J., 1956. *J. Dairy Sci.*, **39**, 235.
212. Langsrud, T., Reinbold, G.W. & Hammond, E.G., 1977. *J. Dairy Sci.*, **60**, 1.
213. Langsrud, T., Reinbold, G.W. & Hammond, E.G., 1978. *J. Dairy Sci.*, **61**, 303.
214. Sahlstrom, S., Espinosa, C., Langsrud, T. & Sørhaug, T., 1989. *J. Dairy Sci.*, **72**, 342.
215. Chaia, A.P., de Ruiz Holgado, A.P. & Oliver, G., 1990. *J. Food Prot.*, **53**, 237.
216. Friedman, M.E., Nelson, W.O. & Wood, W.A., 1953. *J. Dairy Sci.*, **36**, 1124.
217. Tokita, F. & Hosono, A., 1972. *Jap. J. Zootech. Sci.*, **43**, 39.
218. Foissy, H., 1974. *J. Gen. Microbiol.*, **80**, 197.
219. Hayashi, K., Cliffe, A.J. & Law, B. A., 1990. *Intern. J. Food Sci. Technol.*, **25**, 180.

220. Hayashi, K., Revell, D.F. & Law, B.A., 1990. *J. Dairy Sci.*, **73**, 579.
221. Torgersen, H. & Sørhaug, T., 1978. *FEMS Microbiol. Lett.*, **4**, 151.
222. Sørhaug, T., 1981. *Milchwissenschaft*, **36**, 137.
223. Foissy, H., 1978. *Milchwissenschaft*, **33**, 221.
224. Foissy, H., 1978. *FEMS Microbiol. Lett.*, **3**, 207.
225. Foissy, H., 1978. *Z. Lebensm. Unters-Forsch.*, **166**, 164.
226. Foissy, H., 1978. *Proc. 20th Intern. Dairy Congr. (Paris)*, **1E**, 479.
227. Hayashi, K. & Law, B.A., 1989. *J. Gen. Microbiol.*, **135**, 2027.
228. Sharpe, M.E., Law, B.A. & Pitcher, D.G., 1977. *J. Gen. Microbiol.*, **101**, 345.
229. Hemme, D., Bouillanne, C., Metro, F. & Desmazeaud, M.J., 1982. *Sci. Aliment*, **2**, 113.
230. Boyaval, P. & Desmazeaud, M.J., 1983. *Lait*, **63**, 187.
231. Law, B.A., 1984. In *Advances in the Microbiology and Biochemistry of Cheese and Fermented Milk*, ed. F.L. Davies & B.A. Law. Elsevier Science Publishers, London, p. 187.
232. Bills, D.D. & Day, E.A., 1964. *J. Dairy Sci.*, **47**, 733.
233. Reiter, B., Fryer, T.F. & Sharpe, M.E., 1966. *J. Appl. Bacteriol.*, **29**, 231.
234. Woo, A.H., Kollodge, S. & Lindsay, R.C., 1984. *J. Dairy Sci.*, **67**, 874.
235. Woo, A.H. & Lindsay, R.C., 1984. *J. Dairy Sci.*, **67**, 960.
236. Olivecrona, T. & Bengtsson-Olivecrona, G., 1991. In *Food Enzymology*, Vol. 1, ed. P.F. Fox. Elsevier Applied Science Publishers, London, p. 63.
237. Olivecrona, T., Vilaro, S. & Bengtsson-Olivecrona, G., 1992. In *Advanced Dairy Chemistry—1: Proteins*, ed. P.F. Fox. Elsevier Applied Science Publishers, London, p. 292.
238. Driessen, F.M., 1989. In *Heat-induced Changes in Milk*, Bull. 238. International Dairy Federation, Brussels.
239. Nelson, J.H., Jensen, R.G. & Pitas, R.E., 1977. *J. Dairy Sci.*, **60**, 327.
240. Fox, P.F., 1988/89. *Food Biotechnol.*, **2**, 133.
241. Lawrence, R.C., 1967. *Dairy Sci. Abstr.*, **29**, 1.
242. Lawrence, R.C., 1967. *Dairy Sci. Abstr.*, **29**, 59.
243. Morichi, T., Sharpe, M.E. & Reiter, B., 1968. *J. Gen. Microbiol.*, **53**, 405.
244. Fryer, T., Reiter, B. & Lawrence, R.C., 1967. *J. Dairy Sci.*, **50**, 388.
245. Kamaly, K.M., Takayama, K. & Marth, E.H., 1990. *J. Dairy Sci.*, **73**, 280.
246. Oterholm, A., Witter, L.D. & Ordal, Z.J., 1972. *J. Dairy Sci.*, **55**, 8.
247. El-Soda, M., Fathalla, S., Ezzat, N., Desmazeaud, M.J. & Abou Donia, S., 1986. *Sci. Aliment*, **5**, 545.
248. El-Soda, M., El-Wahab, H.A., Essat, N., Desmazeaud, M.J. & Ismail, A., 1986. *Lait*, **66**, 431.
249. Piatkiewicz, A., 1987. *Milchwissenschaft*, **42**, 561.
250. Lee, S.Y. & Lee, B.H., 1989. *Biotechnol. Appl. Biochem.*, **11**, 552.
251. Khalid, N.M., El-Soda, M. & Marth, E.H., 1990. *J. Dairy Sci.*, **73**, 2711.
252. Lawrence, R.C., Fryer, T.F. & Reiter, B., 1967. *J. Gen. Microbiol.*, **48**, 401.
253. Bhowmik, T. & Marth, E.H., 1990. *J. Dairy Sci.*, **73**, 33.
254. Bhowmik, T. & Marth, E.H., 1989. *J. Dairy Sci.*, **72**, 2869.
255. McKellar, R.C., ed. *Enzymes of Psychrotrophs in Raw Foods.* CRC Press, Boca Raton, FL, USA, 1989.
256. Oterholm, A., Ordal, Z.J. & Witter, L., 1970. *Appl. Microbiol.*, **20**, 16.
257. Sørhaug, T. & Ordal, Z.J., 1974. *Appl. Microbiol.*, **25**, 607.
258. Kinsella, J.E. & Hwang, D.M., 1976. *CRC Crit. Rev. Food Sci. Nutr.*, **8**, 191.
259. Boldingh, J. & Taylor, R.J., 1962. *Nature (London)*, **194**, 909.
260. Eriksen, S., 1975. *Milchwissenschaft*, **31**, 549.
261. Dimick, P.S., Walker, N.J. & Patton, S., 1969. *Biochem. J.*, **111**, 395.
262. Tuynenburg Muys, G., van der Ven, B. & de Jonge, A.P., 1962. *Nature (London)*, **194**, 995.

263. Christie, W.W., 1983. In *Developments in Dairy Chemistry—2: Lipids*, ed. P.F. Fox. Elsevier Applied Science Publishers, London, p. 1.
264. Ellis, R. & Wong, N.P., 1974. *J. Am. Oil Chem. Soc.*, **51**, 278A.
265. Wong, N.P., Ellis, R. & La Croix, D.E., 1975. *J. Dairy Sci.*, **58**, 1437.
266. Kinsella, J.E., Patton, S. & Dimick, P.S., 1965. *J. Am. Oil. Chem. Soc.*, **44**, 202.
267. Jolly, R.C. & Kosikowski, F.V., 1974. *J. Dairy Sci.*, **57**, 597.
268. Jolly, R.C. & Kosikowski, F.V., 1975. *J. Agric. Food Chem.*, **23**, 1175.
269. Wong, N.P., Ellis, R., La Croix, D.E. & Alford, J.A., 1973. *J. Dairy Sci.*, **56**, 636.
269a. O'Keeffe, P.W., Libbey, L.M. & Lindsay, R.C., 1969. *J. Dairy Sci.*, **52**, 888.
270. Stauffer, C.E., 1979. *Enzyme Assays for Food Scientists*, Van Nostrand, Reinhold, p. 198.
271. Schormüller, J., 1968. *Adv. Food Res.*, **16**, 231.
272. Larsen, R.F. & Parada, J.L., 1988. *Sci. Aliment*, **8**, 285.
273. Dulley, J.R. & Kitchen, B.J., 1972. *Aust. J. Dairy Technol.*, **27**, 10.
274. Andrews, A.T. & Alichanidis, E., 1975. *J. Dairy Res.*, **42**, 327.
275. Andrews, A.T., Anderson, M. & Goodenough, P.W., 1987. *J. Dairy Res.*, **54**, 237.
276. Andrews, A.T., 1991. In *Food Enzymology*, Vol. 1, ed. P.F. Fox, Elsevier Applied Science Publishers, London, p. 90.
277. Dulley, J.R. & Kitchen, B.J., 1973. *Aust. J. Dairy Technol.*, **28**, 114.
278. Lawrence, R.C., Gilles, J. & Creamer, L.K., 1983. *N.Z. J. Dairy Sci. Technol.*, **18**, 175.
279. Mulder, H., 1952. *Neth. Milk Dairy J.*, **6**, 157.
280. Kosikowski, F.V. & Mocquot, G., 1958. In *Advances in Cheese Technology*, FAO, Rome, p. 133.
281. Kristoffersen, T., 1985. *Milchwissenschaft*, **40**, 197.
282. Manning, D.J., Ridout, E.A. & Price, J.C., 1984. In *Advances in the Microbiology and Biochemistry of Cheese and Fermented Milk*, ed. F.L. Davies & B.A. Law. Elsevier Applied Science Publishers, London, p. 229.
283. Manning, D.J. & Neursten, H.E., 1985. In *Developments in Dairy Chemistry—3*, ed. P.F. Fox. Elsevier Applied Science Publishers, London, p. 217.
284. Law, B.A. & Sharpe, M.E., 1978. *J. Dairy Res.*, **45**, 267.
285. Dunn, H.C. & Lindsay, R.C., 1985. *J. Dairy Sci.*, **68**, 2859.
286. Marth, E.H., 1963. *J. Dairy Sci.*, **46**, 869.
287. Fryer, T.F., 1969. *Dairy Sci. Abstr.*, **31**, 471.
288. Forss, D.A., 1969. *J. Dairy Sci.*, **52**, 832.
289. Law, B.A., 1981. *Dairy Sci. Abstr.*, **43**, 143.
290. Adda, J., Gripon, J.-C. & Vassal, L., 1982. *Food Chem.*, **9**, 115.
291. Kristoffersen, T., 1973. *J. Agric. Food Chem.*, **21**, 573.
292. Biede, S.L. & Hammond, E.G., 1979. *J. Dairy Sci.*, **62**, 227.
293. Biede, S.L. & Hammond, E.G., 1979. *J. Dairy Sci.*, **62**, 238.
294. Kowalewska, J., Zelazowska, H., Babuchowski, A., Hammond, E.G., Glatz, B.A. & Ross, F., 1985. *J. Dairy Sci.*, **68**, 2165.
295. Lowrie, R.J., Lawrence, R.C. & Pearce, L.E., 1972. *N.Z. J. Dairy Sci. Technol.*, **7**, 44.
296. Lowrie, R.J. & Lawrence, R.C., 1972. *N.Z. J. Dairy Sci. Technol.*, **7**, 51.
297. Stadhouders, J., Hup, G., Exterkate, F.A. & Visser, S., 1983. *Neth. Milk Dairy J.*, **37**, 157.
298. Visser, S., Hup, G., Exterkate, F.A. & Stadhouders, J., 1983. *Neth. Milk Dairy J.*, **37**, 169.
299. Visser, S., Slangen, K.J., Hup, G. & Stadhouders, J., 1983. *Neth. Milk Dairy J.*, **37**, 181.
300. Pham, A.-M. & Nakai, S., 1984. *J. Dairy Sci.*, **67**, 1390.
301. Mohler Smith, A. & Nakai, S., 1990. *Can. Inst. Food Sci. Technol. J.*, **23**, 53.

11

Water Activity in Cheese in Relation to Composition, Stability and Safety

A. Marcos

Department of Food Science and Technology, University of Córdoba, E-14005 Córdoba, Spain

1 INTRODUCTION

The influence of water activity, a_w, on the physical, chemical and biological changes that occur in foods was recently established on a scientific basis.[1] At present, water activity is regarded as one of the most important developments in the field of food science in the past 50 years.[2] The interest of food scientists and technologists in water activity is clearly reflected in a series of monographs[3-12] and book chapters[13-22] published since 1975 on fundamental concepts, its measurement and calculation in foods, moisture sorption isotherms of foods and food components, the interaction of water molecules with food components and the state of water in foods, the influence of a_w on the physical, chemical, biochemical and microbiological properties of foods and the relationship between water activity and the quality, stability and safety of foods. A reference list[10] compiled in 1985 included over 2000 papers and reviews on these topics.

Next to temperature, a_w is currently regarded as probably the most important single parameter in food technology.[18]

Water activity plays a central role in cheese, as shown in the scheme below, and influences the quality, stability and safety of the final product.[14,16,17,23,24]

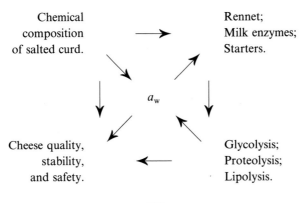

| Chemical composition of salted curd. | → | Rennet; Milk enzymes; Starters. |

a_w

| Cheese quality, stability, and safety. | ← | Glycolysis; Proteolysis; Lipolysis. |

In two book chapters, Rüegg & Blanc,[14] and Rüegg[16] reviewed available information on the relationship between a_w and cheese composition, and the influence of this parameter on starter cultures used in cheese manufacture and on proteolysis and cheese defects. The well-known relationship between chemical composition and quality of Cheddar cheese was discussed by Lawrence & Gilles[25] (see also Chapter 1, Volume 2). The significance of a_w to various aspects of cheese manufacture and the quality, stability and safety of the end product has also been discussed by other authors.[26-32]

There follows a brief discussion on the concept of water activity and underlying principles for its measurement, the experimental methods used to determine a_w in cheese and to calculate it from chemical compositional data, and a tentative scheme of the dynamics of a_w depression during cheese ripening. Due mention is also made of the interactions between a_w and other physical, chemical and microbiological parameters in relation to cheese stability and safety.

2 WATER ACTIVITY AND UNDERLYING PRINCIPLES OF ITS MEASUREMENT

According to Scott,[33] water activity (a_w) is a thermodynamic concept defined as the ratio between the vapour pressure of the water present in a system (p) and that of pure water (p_o) at the same temperature:

$$a_w = p/p_o$$

The vapour pressure of water in a given system is always lower than that of pure water, which is assumed to be unity by convention. Calculating the ratio of manometric readings taken simultaneously on the system and on pure water under identical conditions is the most direct method for obtaining experimental a_w measurements.

In so far as a_w is a colligative property, it can also be measured indirectly in dilute aqueous solutions and liquid foods (low solute concentration) by cryoscopy from the depression of the freezing point (f.p.) with respect to that of pure water, according to the Clausius–Clapeyron equation:

$$a_w = \gamma[n_2/(n_1 + n_2)]$$

where the activity coefficient is assumed to be unity ($\gamma = 1$), n_1 and n_2 are the number of moles of solute and water, respectively, and $n_1 = n_2$ (f.p./1·86), f.p. being expressed in °C.

If the system in question is at equilibrium with it gaseous atmosphere (i.e. if there is no moisture gain or loss because the vapour pressure of the water in the system is the same as that of its atmosphere), then:

$$a_w = \text{e.r.h.}/100$$

or

$$\text{e.r.h.} = a_w \times 100$$

This quantitative relation allows the water activity of the system concerned to be determined by measuring the equilibrium relative humidity (e.r.h.) in the headspace of a small closed container hygrometrically or psychrometrically, or by directly measuring the water in air by gas-liquid chromatography. The e.r.h. can also be estimated on the basis of its induction of moisture-related colour changes in paper impregnated with cobalt thiocyanate, changes that can be matched to those of paper exposed over standards of known a_w.[34]

Another alternative for a_w-range finding is provided by the hygroscopicity differences within a series of salt crystals exposed to the sample e.r.h.; those salt crystals in the series with a_w values greater than the sample e.r.h./100 will remain dry while those with lower values will be subject to surface absorption of water vapour until dissolution; thus, the sample a_w will lie between that of the dissolved (liquified) or moistened crystals of the highest a_w and the dry crystals of the lowest a_w.[35,36]

A less direct alternative to the determination of water activity involves isopiestic equilibration, whereby a dehydrated sorbent (e.g. microcrystalline cellulose) with a well-known moisture sorption isotherm is exposed in the headspaces over sorbate (test) samples until the water vapour pressure in the sorbent equals that of the sorbate samples. The moisture content of the equilibrated sorbent, measured gravimetrically, allows the a_w of the samples to be determined by reference to the moisture sorption isotherm of the sorbent through a mathematical relationship.

Under non-equilibrium conditions, the rate of moisture exchange between the system and it gaseous environment is directly proportional to the water vapour-pressure gradient (i.e. the driving force). Thus, one can determine the water activity of samples by exposing them to different atmospheres of constant known relative humidities for gravimetric measurement of moisture gains or losses and subsequent interpolation to the hypothetical e.r.h., at which the sample weight should remain constant (a_w). Conversely, if the environmental e.r.h. is unknown, it can be determined gravimetrically on the basis of vapour-pressure gradients (VPG) from moisture gains or losses ($\pm$mg) of saturated salt solutions of different water activities exposed to the same atmosphere and under the same temperature, time and surface area conditions from the linear relationship (coefficient of determination $r^2 > 0.999$):

$$a_w = (\text{r.h.}/100) - [b \times (\pm\text{mg})]$$

where r.h./100 is the intercept and b the slope, which halves as the time or surface area is doubled, and is markedly dependent on the temperature and air flow-rate—it decreases significantly with an increase in either. The above equation also allows calculation of the a_w of test samples (aqueous solutions or liquid foods containing no volatiles other than water) assayed in parallel with the reference solutions of known a_w, from the weight changes ($\pm$mg) of the test samples; the principle was used in developing a new straightforward VPG method for the simultaneous determination of the environmental r.h. and water activity of aqueous solutions or liquid foods. The temperature-dependence in static air—uniform air flow-rates are difficult to accomplish in practice—was determined empirically to derive the following general equation:

$$a_w = (\text{r.h.}/100) + [(-0.0927 + 0.0025°C^{-1}) \times (\pm \text{mg.h}^{-1}.\text{cm}^{-2})]$$

which takes account of the dependence of the slope on the temperature, time and surface area (Esteban & Marcos, unpublished results).

3 EXPERIMENTAL MEASUREMENT OF WATER ACTIVITY IN CHEESE

The water activity of cheese is usually in the range 0.70–1.00, although most varieties have water activities above 0.90 (Table I). This should always be taken into account in choosing a method for determining a_w in cheese, as should the high metabolic activity of some cheeses and the occasional release of volatile aroma compounds, which may interfere with the determination of a_w by some methods.

The vapour pressure manometer (VPM), first reported by Makower & Meyers,[37] uses a low-density, low-vapour pressure oil (Apiezon B) as manometric fluid or a capacitance manometer.[38] The measuring method involved, the accuracy of which is dependent on how accurately the temperature is controlled, is, on the other hand, rather cumbersome, so it is better suited to research (as a primary standard method) than for routine work.[39] Labuza's VPM technique[13] was applied by Leung et al.[40] to the determination of water activity in assorted cheese samples; although the method proved to be accurate in the high a_w range, it was not applicable to some types of cheese (Camembert, Blue and others) which

TABLE I
Water Activity (a_w) of Some Cheese Varieties[a]

a_w	Cheese
1.00	Cheese curd, Whey Cheese
0.99	Beaumont, Cottage, Fresh, Jackie, Quarg
0.98	Belle des Champs, Münster, Pyrénées, Processed, Taleggio
0.97	Brie, Camembert, Emmental, Fontina, Limburger, Saint Paulin, Serra da Estrela[b]
0.96	Appenzeller, Chaumes, Edam, Fontal, Havarti, Mimolette, Norvegia, Samso, Tilsit
0.95	Bleu de Bressel Cheddar, Gorgonzola, Gouda, Gruyère, Manchego
0.94	Idiazábal, Majorero, Mozzarella, Norzola, Raclette, Romano, Sbrinz, Stilton
0.93	Danablu, Edelpilzkäse, Normanna, Torta del Casar
0.92	Castellano, Parmesan, Roncal, Zamorano
0.91	Provolone, Roquefort
0.90	Cabrales, Gamalost, Gudbrandsdalost, Prim

[a] Compiled from various sources.[14,40,42–50]
[b] Predicted from its chemical composition.
N.B. The average standard deviations are about ±0.02 at the 0.90 level. They tend to decrease with increase in a_w and vice versa. Some varieties such as Mahón, over-ripe blue cheeses, some extra-hard cheeses and grated cheeses have water activities between 0.90 and 0.70, whereas activities below 0.70 are rather rare.

did not meet the equilibrium conditions. A gradual increase in manometer readings was obtained as a result of microbial respiration or the release of volatiles from the cheeses in question.[41]

Cryoscopic methods, based on the colligative properties of aqueous solutions, rely on the f.p. depression undergone by liquid foods. In 1970, Strong et al.[51] reported a technique for exclusive application to dilute liquid foods rather than solid foods; however, a cryoscopic procedure adaptable to non-liquid foods such as cheeses was also reported recently.[52,53] Cryoscopic methods can be effectively applied to liquid systems containing large amounts of volatile substances or undergoing respiration processes, which are arrested by low temperatures. Although reportedly accurate for $a_w > 0.85$,[54] cryoscopic methods yield the a_w at the freezing temperature rather than the room temperature—the difference, however, is usually fairly small (about 0.01 a_w units at 25°C). Nevertheless, the cryoscopic procedure as applied to cheese[52,53] provides the water activity at room temperature. In dilute solutions, its accuracy increases in the high activity range ($a_w > 0.98$), where the effect of temperature is negligible, and allows a_w to be expressed to the nearest fourth decimal place, if required.[55]

As f.p. measurements are routinely made in quality control programmes in the dairy industry to detect adulteration of milk with added water by using high-performance, accurate instruments (cryoscopes), the cryoscopic approach of Esteban et al.[52] to the determination of a_w in cheese may be applicable to cheese-making quality control. The procedure involved relies on the relationship between the water activity at 20°C of cheeses with $a_w \geq 0.90$ and the freezing points of their aqueous extracts. The preparative technique used is very simple: 1 part (w) of cheese is blended with 3 parts (v) of distilled water and the mixture centrifuged at 1325 g for 5 min. The centrifuge tubes are then allowed to stand in an ice bath until the supernatant fat has hardened. At that point, the incipient f.p. of the intermediate aqueous phase is measured and the a_w of the original cheese sample is calculated from the f.p. (in °C) of the cheese extract according to:

$$a_w = 1.0155 + 0.1068 \text{ f.p.}$$

This linear regression equation was arrived at by evaluating 139 pairs of a_w and f.p. measurements.[53] On application to soft cheeses, this method yielded similar results to those obtained in parallel by hygrometric, psychrometric and isopiestic techniques.[42] A similar cryoscopic procedure applicable to processed cheeses only was developed in the former USSR[56] on the basis of the more complex equation:

$$\log a_w = -0.004\,221 - 0.000\,002\,2\Delta t^2$$

where Δt is the f.p. depression of processed cheese melted at 70°C and then kept at 10–12°C for 24 h.

As stated above, hygrometric methods involve making measurements on the atmosphere in equilibrium with the sample. Hair hygrometers based on dimensional (length) changes in natural filaments—the oldest technique, reported by Leonardo da Vinci (see Ref. 13)—or synthetic (polyamide) fibres are quite affordable; some (e.g. the Lufft and Abbeon hygrometers) are used for quality control purposes by the German meat industry and have also been tested on cheese.[57,58] In fact, they are not very sensitive and are only adequate for range finding and obtaining rough estimates of higher a_w (0·70–0·95) levels.[57] On the other hand, electric hygrometers based on resistance or conductance changes experienced by hygrosensors coated with a hygroscopic salt (e.g. LiCl), electrical impedance or capacitance, are quite precise, though also expensive. They require waiting for a few hours before equilibrium is reached and are thus inadequate for numerous samples. Their major drawbacks include sensor fatigue and poisoning by volatiles such as ammonia, acetate ions and other compounds; however, the former pitfall can be avoided, while the latter can be lessened or even fully overcome with suitable filters.[59] Electronic hygrometers have been used to measure water activity in cheese by Rüegg & Blanc,[43] and by Heskedtad & Steinsholt,[44] among others.

In dew point hygrometers, an air stream in equilibrium with the sample is allowed to condense on the surface of a cooled mirror. In modern instruments, the dew point temperature of the mirror and the sample temperature—the latter is measured on the sample surface by means of an infrared sensor that avoids the need for temperature equilibration—are used to compute the a_w of the sample, which is displayed together with the sample temperature. Activity measurements by this procedure usually take less than 5 min and are rather accurate ($\pm 0\cdot003$ a_w units) throughout the a_w range (0·030–1·000). Volatile condensation may occasionally pose some problems that can be overcome. A recently reported device of this type was successfully applied to cheese.[42] Its unprecedented speed, ease of use, accuracy and precision make it ideal for the immediate determination of a_w on a moderate number of samples per working session.[45]

Thermocouple psychrometers also based on wet-bulb temperature depression and dry-bulb temperature, accurate to $1\cdot7 \times 10^{-5}$ and 0·1°C, respectively, were recently developed for small samples and used at the USDA to measure a_w of soils, foods and other agricultural products. Wet-bulb readings are accomplished by means of a thermocouple consisting of chromel and constantan wires ($\varnothing = 25$ μm), the junction of which is covered with a wetted tiny ceramic bead placed in a small headspace over the sample. A nanovolt thermometer system measures the sample temperature and wet-bulb depression (in μV) and the readings are used to calculate the psychrometric constant from a_w standards and subsequently to compute the activities of samples. In the high a_w range (1·00–0·97), readings are extremely accurate ($\pm 0\cdot0001$ a_w units), but thermal and vapour equilibria are reached slowly (in about 30 min); on the other hand, readings on samples with $a_w \leq 0\cdot97$ take only a few minutes, but are rather less accurate ($\pm 0\cdot005$ a_w units). This type of instrument, the operational range of which is 0 to 50°C, has been applied as a reference (standard) method to a variety of cheeses.[46,52,53,58,60–62] Neither microbial metabolism nor volatile compounds appear to detract from its performance.

The graphical interpolation method of Landrock & Proctor[63] for the determination of water activity is still in use owing to its simplicity and reliability. It is based on the moisture gains or losses by samples exposed to different known constant relative humidities provided by saturated aqueous solutions of salts[64,65] or dilute solutions of sulphuric acid,[66] glycerol[67] or some salts[68,69] which are placed in desiccators or, for individual samples, in small proximity equilibration cells.[70] The weight changes undergone by a sample after a few hours of thermal equilibration are plotted against the corresponding relative humidities, and the e.r.h. (no-weight-change point or a_w) is then determined by graphical interpolation. Modern analytical balances and temperature-control devices have improved on the accuracy of this procedure. Over the usual a_w range of cheeses, where the slope of the moisture sorption isotherms is very high,[17,71–73] the procedure is very sensitive since small a_w changes result in substantial moisture gains or losses. This procedure had been extensively applied to the determination of water activity in cheese.[47–50]

The isopiestic equilibration method, reported as early as 1934 by Robinson & Sinclair,[74] involves equilibrating a dehydrated sorbent material (a protein or microcrystalline cellulose) of known adsorption isotherm with a sorbate of unknown a_w (test samples) in a series of closed headspaces and subsequently measuring the moisture content of the equilibrated sorbent gravimetrically to calculate the water activity of the samples by reference to the sorbent isotherm. The Fett-Vos technique, developed by Fett[75] and by Vos & Labuza,[76] uses dry microcrystalline cellulose as absorbing material and has been applied to the determination of a_w in cheeses.[40,57]

An outstanding innovation to the isopiestic equilibration method was the introduction in 1981 by McCune et al.[77] of dehydrated filter paper as absorbent material and proximity equilibration cells (PECs) as sorbostats[70] to shorten the time required for equilibration (24 h at 20°C against a few weeks).[70,77] Over the a_w range 0·40–0·98, these authors[77] found the moisture content of the equilibrated absorbent (on a wet basis) to be linearly related to $\log(1 - a_w)$. A number of technical modifications in relation to the sorbostatic cells, sorbent filter paper and expression of the moisture of the equilibrated filter papers have been subsequently reported.[42,58,60,61,78–82] The statistical analysis of moisture absorption data —and isotherms—of both dehydrated and untreated filter paper[83–86] showed that pre-drying of the absorbent filter paper is a technical complication which does not result in improved accuracy. In any case, a loss of linearity occurs at $a_w < 0·70$ that may significantly affect the accuracy of the method in determining lower a_w values. As the mathematical linearization of the absorption moisture isotherm of the filter paper used to compute a_w is based on the Smith equation:[87]

$$\ln(1 - a_w) = -a - bm$$

where $-a$ and $-b$ are the regression constants (intercept and slope, respectively), and according to which the moisture content (m) can be expressed either on a percent wet[77] or a percent dry basis,[58] so the moisture can be expressed on an initial weight basis[42,61] if the sorbent used is in moisture equilibrium (pre-equilibrated at an a_w below the operational range[79] or untreated[42,61]) and hence, for paper of uniform quality (constant moisture content and weight per unit of surface area), the simple moisture gain of the equilibrated paper[42,82] is directly related to log

$(1 - a_w)$. By using circles of filter paper with a large enough surface area ($\emptyset = 9$ cm), the isopiestic equilibration method can be used to determine water activities between 0·70 and 1·00,[61,79] a range which encompasses both the a_w of most cheeses and the minimum a_w limits for microorganisms relevant to public health to grow and produce toxins.[88] Both the accuracy and the precision of the method increase with the a_w value, which is of microbiological and technological significance. Although the equilibration time involved is somewhat long, the isopiestic method can be simultaneously applied to a large number of samples, thus resulting in higher sample throughputs than alternatives. PEC techniques have been used to determine the a_w of hundreds of cheese samples.[42,46,60,61,89,90]

4   CALCULATION OF THE WATER ACTIVITY OF CHEESE FROM ITS CHEMICAL COMPOSITION

An alternative approach to more or less direct a_w measurements involves calculating this parameter from chemical and/or physical data. According to van den Berg,[18] some of the a_w values that are currently determined experimentally could be more readily, and at least as accurately, estimated from such data.

The depression of the water vapour pressure in a given system is directly proportional to the concentration of dissolved particles (i.e. to the number of molecules and ions) in the 'solvent' water and, for ideal aqueous solutions, though not for foods, water activity can be quantitatively calculated from Raoult's law:

$$a_w = p/p_0 = n_2/(n_2 + n_1)$$

where n_1 and n_2 denote the number of moles of solute and solvent water, respectively, as stated above for the Clausius–Clapeyron equation. This law does not apply to cheese, which is a far from ideal system in this respect.

Although there are some general alternatives (e.g. the Ross equation[91]) to the calculation of the vapour pressure depression ($a_w > 0.75$) brought about in 'formulated' foods and solutions by the occurrence of different types of 'known' solutes that result in deviations from Raoult's law, none is valid for cheese, the physico-chemical complexity and dynamic (non-steady) state of which hinder any possibility of characterization (quantitation of molecular species).

However, a search for statistical relationships between a_w and chemical composition revealed the major factors that influence the water activity of cheese and allowed the establishment of both complex and straightforward equations, which are of academic interest and practical use, respectively. Both can be used to estimate the a_w of cheese from its physico-chemical composition. Some of these equations can be applied to literature data (compositional tables) to obtain a rough estimate of the a_w of different types of cheese. Although this subject was reviewed early in 1990 by Esteban & Marcos,[92] new, more accurate, equations have been reported since then.[93–98]

TABLE II
Selected Equations for Estimation of the Water Activity (a_w) of Various Types of Cheese from their Chemical Composition

Type of cheese	Equation	Reference
(1) Unripened	$a_w = 0.9719 - 0.0044[NaCl] + 0.0041$ pH	93
(2) Bacterial ripened	$a_w = 1.0234 - 0.0070[Ash]$	48
(3) Mould ripened	$a_w = 1.0058 - 0.0045[Ash] - 0.0107[NPN]$	98
(3a) Soft	$a_w = 0.9960 - 0.0029[Ash] - 0.0106[NPN]$	94
(3b) Blue	$a_w = 0.9808 - 0.0058[Ash]$	50
(4) Processed	$a_w = 0.9951 - 0.0032[Ash]$	99
(5) All with $a_w > 0.90$	$a_w = 0.9450 - 0.0059[NaCl] - 0.0056[NPN] -$	
	$0.0019([Ash] - [NaCl]) + 0.0105$ pH	16

NPN = non-protein nitrogen (N soluble in 12% TCA).
The concentrations of all chemical components are expressed in g/100 g moisture.

Selected equations for the calculation of the water activity of different types of cheese from ordinary physico-chemical parameters are listed in Table II. Non-protein nitrogen (NPN) denotes the nitrogen soluble in 12% trichloroacetic acid (TCA) and, unless stated otherwise, all chemical components in these and other equations throughout this chapter are expressed on a moisture basis (viz. in g/100 g H_2O).

The a_w of fresh cheese[47] is determined almost solely by the aqueous concentration of NaCl, so the data reported by Robinson & Stokes,[100] which relate the molality (m) of aqueous solutions of NaCl to water activity, can be used through the relation:

$$a_w = 1 - 0.033[NaCl_m]$$

to calculate the upper maximum of a_w of unripened cheeses and related products with rather good accuracy provided the NaCl molality does not exceed 1.2. The actual a_w value is always slightly lower owing to the presence of other solutes. The above equation was used to construct a nomograph[101] for direct estimation of the a_w of fresh cheeses from the percentages of cheese moisture and salt (Fig. 1). For simplicity, the equation can be rewritten as:

$$a_w = 1 - 0.00565[NaCl]$$

which applies to salted curds and fresh and young cheese with salt-in-moisture below 7 g/100 moisture. López et al.[93] recently reported an improved equation [eqn (1) in Table II] for more accurate calculation of the actual a_w of coagulated unripe dairy products which also takes into account the effect of pH on a_w depression.

The a_w of one bacterial ripened cheese variety[48] was found to be directly related to its moisture content through:

$$a_w = 0.7662 + 0.0046 \, (\text{g } H_2O/100 \text{ g cheese})$$

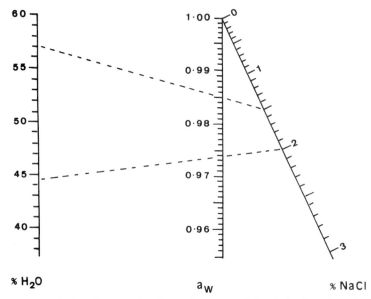

Fig. 1. Nomograph for direct estimation of water activity (a_w) of unripe cheeses from percentages of moisture (% H_2O) and salt (% NaCl). Examples: If % H_2O = 57·0, and % NaCl = 1·5, then a_w = 0·985; if % H_2O = 44·5 and % NaCl = 2·0, then a_w = 0·974. (Reproduced from *J. Dairy Sci.*[101] by courtesy of the American Dairy Science Association.)

but also to be more closely (though inversely) related to the ash content on a moisture basis [see eqn (2) in Table II]. This latter equation was also found to apply to many other bacterial ripened cheese varieties.[46] However, it has not yet been tested on some major varieties of this group (e.g. Cheddar, Dutch, Swiss type and Italian hard cheeses), where the regression constants may vary as a result of some of their features (salting, size and shape, wrapping or waxing, scalding, propionic acid fermentation and lipolysis by added lipases).

Another similar empirical equation was proposed for the estimation of water activity in mould ripened cheeses:[32]

$$a_w = 1.0076 - 0.0079[Ash]$$

According to Rüegg & Blanc,[43] the NPN content of a cheese also influences its water activity, but in searching for correlations between a_w and the ash content (on a moisture basis), for groups of cheeses undergoing proteolysis to a different extent, but similar within each group (e.g. those ripened by bacteria or moulds), the NPN content—lower in bacterial than in mould-ripened cheeses—would proportionally affect the depression of a_w in all the cheeses within a group and would therefore not substantially affect the correlations with the ash contents in each group. Equation (3) in Table II, more accurate than the previous expression, was arrived at by taking account of the NPN content and performing a multiple regression analysis of 70 data sets obtained from 40 blue cheese samples

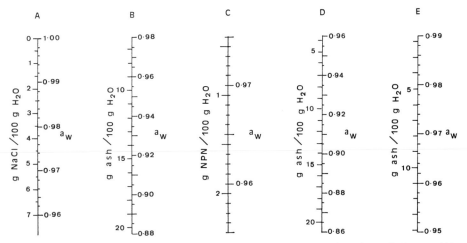

Fig. 2. Graphs for chemical estimation of water activity (a_w) of unripened cheeses (A), cheese varieties ripened by bacteria (B), white-surfaced mould (C) and blue internal moulds (D), and processed cheeses (E).

of six different varieties and 30 surface mould ripened soft cheeses of three varieties.[98]

The water activity of white mould-ripened cheese can be estimated not only from eqn (3a) in Table II, but also from other relations established by simple linear regression, namely:[94]

$$a_w = 0.9769 - 0.0019[\text{Ash}]$$
$$a_w = 0.9793 - 0.0101[\text{NPN}]$$

Likewise, the a_w of blue mould-ripened cheese can be estimated from eqn (3b) in Table II and from the alternative multiple regression relation:[50]

$$a_w = 1.0013 - 0.0051[\text{Ash}] - 0.0056[\text{SN}]$$

where SN denotes the soluble nitrogen content at pH 4.6.

Equation (5) in Table II, which was established by Rüegg & Blanc[14,43] and reported by Rüegg,[16] is applicable to most cheeses with water activities higher than 0.90. It is of greater academic than practical interest because although it provides an overview of the four major chemical parameters and the chief physical parameter that influence water activity in cheese, it demands prior knowledge of some data that cannot be found readily in the literature (compositional tables) and must therefore be determined experimentally by more complex procedures than those involved in measuring a_w values. In any case, it has been successfully applied to hundreds of cheese samples of many varieties,[55] but it is inapplicable to some blue-veined cheeses (with $a_w < 0.90$) and, in its present form, overestimates a_w values for surface mould-ripened cheeses.[94]

The above simple linear equations can be readily converted into graphs[102] (see Fig. 2) for a direct visual estimation of a_w with fewer arithmetical calculations.

Obviously, predictive equations must be based on the use of reliable a_w values and chemical parameters. This is particularly true and essential whenever nitrogen fractions such as NPN and, especially, SN are involved since they may be obtained by a variety of procedures[103] which yield conceptually different water activities because of the heterogeneity in M_r. An attempt at standardizing and characterizing such fractionation procedures was made by Kuchroo & Fox[104] and is still under way.[105]

5 TENTATIVE SCHEME OF THE DYNAMICS OF WATER ACTIVITY DEPRESSION DURING CHEESE RIPENING

The term 'water activity', as defined above, applies only to equilibrium (static) conditions, so the basic concept of a_w must be used cautiously in relation to foods such as cheese, since this is a very dynamic system. Therefore, to understand the dynamics of a_w in a real system (e.g. cheese subject to various physical and biochemical processes acting as a whole under non-equilibrium conditions), it seems useful first to analyse the theoretical mechanisms by which the water vapour pressure (a_w) of equilibrium (static) systems can be depressed.[22,106]

5.1 Foundation of Water Activity-depressing Mechanisms

The water activity of static systems can be depressed by three basic mechanisms, namely: (a) interactions between solvent water and solutes and small ions, which can be assessed quantitatively from Raoult's law for solute effects in ideal systems (*loc. cit.* equation); (b) interactions between water and polar macromolecules (e.g. proteins, polysaccharides) whereby water molecules are tightly bound to active sites such as ionized and polar groups, chiefly via water-ion or water-dipole (hydrogen bonding) associations that in theory provide a water monolayer coverage of accessible, strongly hydrophilic groups on biopolymers amenable to quantitation by the Brunauer, Emmett & Teller (BET) equation:[107]

$$a_w/[m/(1 - a_w)] = (1/m_0 C) + a_w[(C - 1)/m_0 C]$$

where the BET constants C and m_0 (the monolayer value, which allows the corresponding critical a_w value to be calculated) can be readily obtained from a linearization of the above equation; and (c) in porous systems, the a_w depression arises mainly from capillary condensation which can be estimated by using the Kelvin equation:

$$a_w = p_c/p_o = \exp(-2\sigma v \cos \Theta/r_c RT)$$

where p_c is the water pressure above the meniscus in the capillary, σ the surface tension of water, Θ the contact angle (for thorough wetting, $\Theta = 0$, so $\cos \Theta = 1$), v the molar volume of the liquid, r_c the capillary radius, R the gas constant and T the absolute temperature. According to this equation, the a_w depression increases with decreasing capillary radius, so significant predictions can only be made for capillaries with sizes of the same order of magnitude as the absorbing molecules. Most capillaries in foods are larger than 10 μm in size, so, assuming thorough wetting, the predicted a_w values would be quite high,[8] but as water is removed, the remaining water present in microcapillaries (a small fraction of the total pores[8]) represents a significant proportion of the total water,[13] and the depressing effect on a_w becomes significant as well.[8]

To fill the gap between the extreme mechanisms (a) and (b) for lowering the water vapour pressure ranging from free water filling macrocapillaries, microcapillaries and more loosely bound water in multilayers above the first single-layer, European food researchers have frequently used the three-parameter equation derived independently by Guggenheim,[108] Anderson[109] & de Boer[110] (the so-called 'GAB model'), which is expressed mathematically as:

$$m/m_0 = kCa_w/[(1 - ka_w)(1 - ka_w + kCa_w)]$$

where C and k are constants related to the energies of interaction between the first and distant sorbed molecules at the individual sorption sites. Parameter k is included on the assumption that multilayer molecules interact with the sorbent at energy levels somewhere between those of the monolayer molecules and the bulk free water.

5.2 Dynamics of Water Activity Depression in Cheese

Cheese is not a static system inasmuch as the amount of solvent water it contains is continuously evaporated and new solutes of decreasing M_r gradually arise from biochemical processes. These dynamics enhance both water-solute and water–small ion interactions and, as a further consequence of moisture loss, the ratio of water to non-solute solids decreases, thereby promoting water-casein interactions that in turn decrease the ratio of unbound to bound non-solvent water, and also perhaps influence capillary forces.

The two effects (moisture loss and solute formation) are very small in fresh cheeses; however, as they ripen, the combined effects of drying and chemical breakdown become increasingly significant and large deviations in sorption phenomena from theoretical predictions are to be expected.

The information gathered from studies on the relationship between the chemical composition and water activity of cheese has fostered the formulation of hypotheses and the development of speculative generic approaches to integrating the processes and the physico-chemical factors involved in the depression of the a_w of cheese throughout its maturation in order to account for some of the relations found.

Strictly, the dynamics of a_w depression in cheese involve both space and time dimensions or variables; however, greater emphasis is normally placed on either, depending on the particular objective. Thus, some studies focus on zonal changes in a_w within the body of a large cheese at a given stage of ripening or simply within its outer and inner parts during ripening. Others are concerned with the changes in a_w at a preset point (e.g. core or rind) or in the average a_w in the whole edible mass of a given cheese batch or, preferably, on a_w changes at different points within a cheese from manufacture to some stage of the ripening process. Rüegg & Blanc[14] determined zonal changes in a_w for various aged (cured) cheese varieties of different size; Román[111] monitored the a_w of both the core and sub-cortical regions of Manchego cheese over one year's ripening. The last approach was applied by Guinee[112] to Romano-type cheese. The discussion below, how-ever, is focused on the processes involved in the gradual decrease in the average a_w of cheeses in general after the manufacturing phase, i.e. as a result of salting the curd and of subsequent events during cheese drying, curing, storage and market-ing. In fact, most of the experimental evidence partly supporting this tentative scheme was obtained from many assorted cheese samples (whole edible portion) purchased at market within their commercial shelf-life but at different (unknown) stages of ripening.

The decrease in the initially high a_w of freshly salted curd compared to that of the cured finished cheese is unarguably due to an increase in the solute concen-tration in water. As noted earlier, this results both from the loss of solvent water through physical dehydration and from the formation of new solute species of low M_r through breakdown of the original components. In the course of cheese ripening, macromolecular species undergo hydrolytic cleavage processes such as glycolysis, proteolysis and lipolysis, effected by biochemical and microbiological agents.

5.3 The Initial Water Activity of Salted Curd

Salted cheese curd can be regarded, in an oversimplified way, as a physico-chemical system consisting of a structural matrix of macromolecular polar compounds, filled partly with hydrophobic triglycerides and partly with water, which in turn contains dissolved solutes of low M_r. Some of the water[72,113] (nearly 10%)[31] is bound to macromolecular para-casein as non-solvent water, while the rest, the bulk phase of free water physically entrapped within the porous matrix, pre-serves its solvent capacity but has a lower vapour pressure than pure water (a_w) because of the presence of the NaCl added at salting and of residual lactose and milk salts (from whey).

In theory, if the interactions between proteins and water and small ions that occur in cheese curd do not significantly influence—owing to compensation or neutralization—the vapour pressure of the free aqueous phase, the 'maximum' a_w value at any zonal locus of salted curd could be calculated very accurately, as stated above, from the data reported by Robinson & Stokes,[99] which relate the molality of aqueous NaCl solutions to their water activity. During the initial salting period, the water activity of salted curd or fresh cheese is neither static— as a result of dynamic changes arising from initial outward diffusion of water and glycolysis—nor uniform—owing to salt/moisture diffusion gradients.[31] The a_w values reported for fresh cheeses are in fact average values obtained from pre-sumably representative (homogenized) samples of the product taken at a known time after manufacture or at an unknown time during its commercial shelf-life.

These considerations are of undeniable academic interest as they enable 'prediction' of the maximal a_w at any point in the cheese after any period of brining according to the relations reported by Guinee and Fox[114] for salt/moisture diffusion in cheese during the initial salting period.

In practice, the interactions of proteins with water and ions should roughly cancel one another as the water activities of fresh cheeses with moisture contents above 40% calculated from the NaCl concentration in 'total' moisture are only slightly higher than the measured a_w values of the cheese samples.[71] The actual a_w of fresh cheese is somewhat lower than that thus calculated (NaCl in cheese moisture), most probably as a result of the additional vapour pressure depression caused by solutes of low M_r other than NaCl (e.g. lactose and salts from residual whey).

Salting is thus the most relevant process influencing the initial vapour pressure depression of the water in cheese curd.

5.4 The Major Role of Dehydration in Water Activity Depression

As stated by Fox,[115] cheese manufacture is essentially a dehydration process that may continue afterwards over a long period of ripening, lasting over a year.

If milk is concentrated four-fold, then its initial a_w ($ca.\,0.995$) drops to about 0.990, which is equivalent to the addition of 25 g of salt (NaCl) per litre of milk. Drying affects salted cheese even more markedly in relation to a_w depression; thus, as noted earlier, part of the overall moisture content is bound to casein as non-solvent water,[72,113] so that the loss of solvent water by evaporation alone will increase the proportion of bound non-solvent water in the residual total moisture and the solute concentration in the solvent water will be much greater than that expected from the solute/total moisture ratio. According to Ross,[91] the estimation of water activity involves a critical ratio of non-solute solids to moisture above which non-soluble solids must be taken into account since their a_w-depressing effect, exerted through water binding, in not negligible.

The influence of drying on the water activity of cheese, however, can be assessed from the correlation coefficients of simple linear regression analyses between a_w and the total moisture of the cheese or the aqueous concentration (in g/100 g moisture) of some inert components taken as references. Thus, data from 20 samples of two versions of a bacterial ripened cheese variety yielded a highly significant ($P < 0.001$) linear relationship between their moisture contents (ranging from 46 to 20%) and water activities from 0.98 to 0.85, with a correlation coefficient $r = 0.98$.[48] The correlation coefficient between the ash content on a moisture basis (g ash/100 g moisture) and water activity was higher ($r = -0.99$), as expected, for samples of the same population because the ash content accounted for the variable amount of salt added in the manufacturing process. Similarly, the results obtained for 82 different cheeses with a_w values in the range 0.994–0.87 showed the water activity to correlate with the total nitrogen and ash contents, which undergo no quantitative (absolute) changes during ripening and account for the differences in salting, respectively, with correlation coefficients of -0.8 and -0.9, respectively, i.e. higher than that of any other component or variable (NaCl, NPN, pH) when expressed on a moisture basis.[14] Seventeen sets of data pairs obtained during one year of ripening of a batch of Manchego cheese subjected to the same initial salting (brining) treatment resulted in coefficients of correlation of the water activity with the moisture content (g H_2O/100 g cheese) and ash content (g/100 g moisture) of equal magnitude but opposite sign (0·94 and −0·94, respectively).[111] Finally, 125 samples of 20 different varieties yielded a coefficient of correlation between a_w and moisture in fat-free cheese of 0·9 (unpublished results).

The above findings allow one to conclude that the loss of solvent water from cheese upon salting (during the drying phase and throughout ripening) is the most significant single process accounting for the increase in the solute concentration and concomitant decrease in a_w in the vast majority of cheese varieties, particularly in bacterial, long-ripening hard cheeses. Exceptions in this respect are fresh cheeses, those coated with a moist-proof barrier and mould-ripened varieties that tend to isopiestic equilibration with the high r.h. of the air currents during curing periods in cellars (preliminary unpublished results obtained for Cabrales cheese), a tendency that must be further enhanced in surface mould-ripened soft varieties (e.g. Brie and Camembert), the flat shape and small size of which result in larger surface area/volume ratios.

Hence, the concentration of all chemical constituents involved in the equations for estimation of water activity must be expressed on a moisture basis.

5.5 Contribution of Glycolysis to Water Activity Depression in Fresh Cheese

Glycolysis is the earliest biochemical event contributing to water vapour depression in fresh cheese by increasing the number of dissolved molecules and ions. The small differences found between a_w values experimentally measured on fresh cheeses and the slightly higher values calculated from the NaCl molality were subjected to linear regression analysis[89] in search of a relationship with the aqueous concentrations of reducing sugars, lactic acid and several other compounds; only the lactic acid concentration yielded a highly significant ($P < 0.001$) correlation coefficient ($r = 0.89$). This is because the lactic acid produced on the glycolysis of lactose represents a four-fold increase in the number of molecules, and thus is about four times more efficient than lactose in lowering the water vapour pressure.

However, López et al.[93] failed to find a close correlation between the lactic acid concentration and pH in fresh cheese, probably because of the buffering capacity of caseins; yet, they found a positive correlation between a_w and pH. Rennet curd contains some inorganic compounds, chiefly colloidal calcium phosphate and substantial amounts of micellar magnesium and some citrate, which may be dissolved during glycolysis, thereby causing a further decrease in the water activity. The lower pH resulting from the lactic acid produced on glycolysis affects the calcium phosphate equilibrium by increasing the amount of both soluble and ionic calcium and soluble phosphate at the expense of colloidal $Ca_3(PO_4)_2$. The role of pH is thus more relevant than that of lactic acid. The relationship between water activity, the NaCl concentration and pH in this type of cheese, viz. eqn (1) in Table II, has a square of multiple correlation coefficient $R^2 = 0.95$.

5.6 Contribution of Proteolysis to Water Activity Depression in Mould Ripened Cheese

After the initial a_w depression in cheese curd brought about by salting and the more significant effect of the gradual loss of moisture during the ripening of many varieties, proteolysis is the most important biochemical event lowering a_w as the cheese ripens (especially through mould activity) by forming new, low-M_r solutes through gradual (extensive) and progressive (in depth) protein break-down.

The major agents of proteolysis in cheese are rennet, milk enzymes and starter bacteria.[103] In all cheese varieties, α_{s1}-casein is the main target of proteolysis by residual rennet, which releases a polypeptide consisting of the first 23–24 amino acid residues from N-terminus as the primary soluble hydrolytic product. Plasmin, the main indigenous proteinase in milk, acts preferentially on β-casein releasing, also from N-terminus, soluble polypeptides of the proteose-peptone (PP) fraction, namely PP8F (28 amino acid residues), PP8S (77–79 residues) and PP5 (105–107 residues). All of these primary proteolytic products produced by rennet and plasmin from α_{s1} and β-casein, respectively, are relatively high-M_r polypeptides (SN at pH 4·6) that have little effect on a_w, as shown by the lack of a significant correlation.[14]

On the other hand, although lactic acid bacteria used as cheese starters are only weakly proteolytic, they contain a variety of proteinases and peptidases that, as a whole, are presumably capable of hydrolysing chiefly β-casein, as well as large and medium polypeptides, to oligopeptides and amino acids,[116] thus being primarily responsible for the formation of NPN (12% TCA SN), which in turn has a strong influence on water activity.[14]

The amount of SN compounds (in terms of peptides within several molecular weight ranges and amino acids) produced in aseptically made Gouda cheese by the combined and separate actions of rennet, milk proteinase and starter bacteria was determined by Visser.[117]

Strong evidence of the prominent role of proteolysis in a_w depression in some types of cheese is the lack of correlation found in high-cooked cheeses (Emmental) and soft surface ripened cheeses (Camembert and Münster) between a cryoscopic ripening index (CRI), proposed for monitoring proteolysis, and their NaCl or moisture contents—which is indicative of little variations in the two components in each type of cheese—and the positive correlation found between the CRI and NPN ($r = 0.84$), NH_3 ($r = 0.82$) and SN ($r = 0.70$).[118] Since the procedure used to prepare the cheese extract for freezing point measurement[118] was basically identical to that used for the cryoscopic determination of a_w in cheeses,[52,53] it follows that the a_w of such cheeses must be reduced primarily as a result of the formation of NPN by proteolysis.

Mould ripened cheeses undergo proteolysis to a much greater extent than bacterially ripened cheeses. In surface mould-ripened and blue-veined cheeses, SN accounts for 30–60% and 50–75%, respectively, of total nitrogen,[46] but even more significant to a_w than the extent of proteolysis is its depth, arising from the action of fungal enzymes (extracellular acid and neutral proteinases secreted by *Penicillium camemberti* and *P. roqueforti*, and, in blue cheeses, intracellular proteinases and peptidases released by lysis and/or leakage of the mould mycelium) that may convert to NPN up to 20–40% of the total nitrogen in soft cheeses and up to 40–75% of the total nitrogen in blue varieties,[46] in which free amino acids may account for about 10% of total nitrogen. Therefore, low-M_r nitrogen compounds must play a more significant role than dehydration in a_w depression in mould-ripened cheese since moisture is lost only to a small extent after removal from cellars. In fact, the NPN content must be taken into account if a_w is to be estimated accurately in mould-ripened cheeses.[94,98]

5.7 Effect of Lipolysis

The relationship between lipolysis during cheese ripening and a_w depression is still obscure. The ripening of most cheeses is accompanied by a low level of lipolysis, usually involving less than 2% of the triglycerides; however, as with proteolysis, lipolysis in mould-ripened cheese is quite extensive and may amount to 20% of the triglycerides.[119] In blue-veined cheeses, lipolysis results chiefly from the action of lipases secreted by *P. roqueforti*, although indigenous milk lipase may also contribute to some extent, particularly in raw milk cheese.

Italian hard cheese varieties (Romano, Parmesan) also undergo extensive lipolysis during ripening caused by pregastric esterase from rennet paste, which preferentially hydrolyses the soluble (short-chain) fatty acids from triglycerides esterified at the sn-3 position to yield relatively large amounts of butyric acid.[120] Although a_w influences lipolysis in Romano cheese—the free fatty acid (FFA) content increases with a_w[102]—it is uncertain whether the opposite holds true, i.e. whether high levels of short-chain FFAs significantly decrease a_w.

No doubt, the dipolar structure of the amphiphilic molecules, FFAs and partial glycerides released on hydrolysis of lipids, interact in different ways with the water molecules (ion-dipole and water-to-water hydrogen bonding) and thus affect the normal structure and properties of water; however, as a rule, the practical significance of such interactions to the water vapour pressure depression (a_w) is most probably negligible compared to that of drying, salting, proteolysis and glycolysis.

On the other hand, although glycerol is a very effective depressant of a_w and is used as humectant—antifreeze or cryoprotective—agent,[67] the amount of free glycerol resulting from total hydrolysis of the three ester linkages in the triglyceride molecules (e.g. by pancreatic lipase) must be too small to have a detectable effect—indicated by linear regression analysis—in most cheese varieties.

Although lipolysis is likely to have little effect on the a_w of many bacterial ripened varieties (e.g. Gouda, Cheddar and Gruyère, in which FFAs account for less than 3% of total fatty acids) and even mould-ripened varieties resulting from the action of surface moulds (e.g. Brie and Camembert, with 3–10% FFAs) or internal moulds (e.g. Roquefort, with 10–15% FFAs), it remains to be seen whether the effect of lipolysis on some blue cheese (Danablu and Cabrales, with 15–20% FFAs) and, especially, on lipase-supplemented cheeses, in which FFAs account for over 20% of total fatty acids, must be taken into account to refine the predictive equations developed so far for blue cheeses or to develop new expressions applicable to Italian hard cheeses (e.g. Parmesan, Provolone and Romano).

Research aimed at clarifying the potential influence of the total FFA content (as an index of lipolysis) on the a_w depression in Cabrales cheese ripened for long periods is currently under way. However, it might be better to choose Romano (with 30–45% FFAs) as a model system in which to establish the role of lipolysis in a_w depression and to develop multiple linear regression relations of the following type:

$$a_w = a - b[\text{Ash}] - c[\text{FFA}]$$

that perhaps may be applicable to hard Italian varieties in so far as (see Fox & Guinee[120]): (a) unlike blue cheeses, their level and type of proteolysis are normal and similar to those for most semi-hard and hard bacterial ripened cheeses; (b) they lose much more moisture than mould-ripened varieties, thus providing a more variable basis for expressing the FFA and ash concentrations; and (c) last, but not least, their high esterase activity results in the hydrolysis of triglycerides and consequent release of water-soluble short-chain (C_4–C_{10}) fatty acids, particularly butyric acid, to a much greater extent than other cheese varieties.

5.8 Major Factors Influencing Water Activity Depression in Cheese

Once the main physical and chemical processes involved in a_w depression during cheese ripening have been dealt with, due mention should be made of the principal individual physico-chemical factors that influence the a_w of cheese, namely: (a) the moisture content for all varieties of cheese; (b) the total moisture/non-fat solid ratio in hard to extra-hard varieties; (c) the sodium chloride content; (d) the pH in the case of fresh unripe cheeses; (e) the ash content in all varieties of cheese; (f) the non-protein nitrogen (NPN) content of mould-ripened cheeses; and (g) perhaps the total FFA or water-soluble FFA (C_2–C_{10}) content of blue cheeses and, most probably, of hard varieties containing added lipase.

6 WATER ACTIVITY INTERACTIONS WITH OTHER FACTORS IN RELATION TO CHEESE STABILITY AND SAFETY

The so-called 'biokinetic zone', which enables life through biochemical meta-bolism of useful, spoilage and pathogenic microorganisms, is determined by a favourable micro-environmental ecosystem dependent on an interactive series of physical, chemical and biological conditions. Empirically, throughout the long history of Man, and technologically in the present century, these conditions have been instinctively used or purposely manipulated and controlled to produce selective hostile habitats for undesirable organisms and to preserve foods.

6.1 Major Physical Parameters

The essential physical parameters in relation to the stability of foods are tem-perature (T), water activity (a_w) of solid and liquid substrates (and the relative humidity, e.r.h. $= a_w \times 100$, of gaseous environments). pH and redox potential (E_h), in addition to other, somewhat less influential factors, such as pressure or electromagnetic radiation.

Microbial activity is possible only over a narrow temperature range of about 100°K, set by the phase transitions of liquid water (from ca. −10°C for cryopro-tected organisms to +90°C). Although microorganisms generally adapt to the temperature of their natural ecosystems, there are narrower temperature sub-ranges for different microbial groups (e.g. psychrophiles, psychrotrophs, mesophiles and thermophiles), all of which have their own maximal, optimal and minimal values for growth which, however, vary within each genus, species and strain.

According to van den Berg,[18] water activity is the second most critical physical parameter in relation to food microbiology. The water activity spectrum ranges between 0 and 1, but microbial metabolism is restricted to the upper half, from above 0·6,[121] where there is free (useful) unbound water, to very near 1 (0·999), where enough nutrient solutes are yet available, since most microorganisms need more than plain water (i.e. some nutrients) to develop and survive. In so far as the maximal and optimal a_w values for most microorganisms are very close to unity (pathogenic bacteria and many other microorganisms grow most rapidly at a_w values in range 0·995–0·980), the minimal a_w value for growth and toxin production is the most relevant parameter for preservation technology and public health protection. The minimal a_w limits for the growth of, and toxin production by, microorganisms of public concern were compiled by Beuchat.[88] The water activity of most common bacteriological culture media lies within the range 0·999–0·990 (Chirife et al.[55]), although there are some extremophilic organisms capable of growing in some foods at low a_w values (e.g. from halophilic bacteria, osmotolerant and osmophilic yeasts to moderately or extremely fastidious xerophilic fungi), which call for special culture media with a_w values below 0·98 (Hocking & Pitt[122]). The water activity of special culture media appropriate for such organisms ranges from 0·9 to 0·7 (Esteban et al.[123]).

The hydrogen ion concentration, expressed on a pH scale, is very well known to affect the life and death of microorganisms, the growth limits of which vary over a wide range (pH 1–11). As a rule, the maximum pH values for growth are similar among bacteria, yeasts and moulds, but the latter two also grow at much lower pH values than do bacteria. Although optimal bacterial growth usually occurs at a pH close to 7, lactic acid bacteria, among others, grow optimally between pH 5·5—or lower—and 6·0.[124]

Unlike the above-mentioned parameters, the microbiological significance of the redox potential (E_h), is not yet clearly understood.[124] The standard reduction potentials of the electrochemical series range within about ±3 V,[125] but bacterial growth is restricted to a range within about ±0·5 V (Jacob[126]). While some microorganisms have only one terminal energy-producing mechanism and are metabolically active over a relatively narrow E_h range, others have alternative systems that can be switched on or off by the redox potential or the presence or absence of oxygen. Strict aerobes use oxygen as the terminal electron acceptor in respiration, while obligate anaerobes can grow only at low (negative) E_h values and some grow exclusively in the absence of molecular oxygen; on the other hand, facultative anaerobes (e.g. *Lactobacillaceae*) can use oxygen as a terminal electron acceptor but, in its absence, can also use a variety of electron acceptors. In different microbial cultures, E_h may range from about +300 mV for aerobes to less than −400 mV for anaerobes.[124]

Although each parameter may act individually on microbial activity and even be lethal at extreme levels, all usually act concomitantly and interact mutually in affecting the life and death of microorganisms.

6.2 Interactions Between Physical Parameters

Theoretically, as noted earlier, each of the above four physical parameters can be used technologically to arrest chemical and enzyme-catalysed reactions if one of them reaches an inhibitory limit even if the other three are kept at their optimal values. These parameters, however, are interactive, so the same inhibitory effect can be achieved by using suboptimal values for two or more parameters.

Exploitation of the interactive effects of a_w and pH to preserve foods are most typically expressed by cheeses and other fermented products such as semi-dry sausages. The combined inhibitory effects of these two parameters on the survival of microorganisms are clearly additive (i.e. at any given a_w value, microbial activity decreases with decreasing pH and vice versa), or, rather, synergistic (i.e. the effects of decreasing a_w and pH on the survival of spoilage or pathogenic microorganisms are mutually enhanced—Troller;[127] Lenovich[128]). Salmonellae proliferate at $a_w = 0·971$ and pH 5·8; however, if the pH is reduced to 5·0, then the minimal a_w increases to 0·986 or higher. A similar interdependent behaviour has been found for *Staphylococcus aureus*.[129,130]

Thus, in a plot of the average and standard deviation of a_w and pH of different cheese varieties (as crosses in which the intercept points and branches denote mean values and standard deviations, respectively) in a coordinate system where

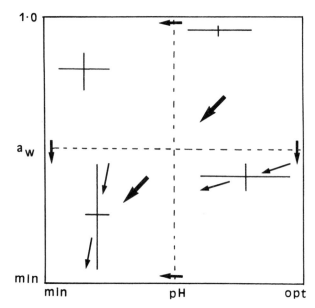

Fig. 3. Stability coordinates (a_w versus pH) of four hypothetical cheese varieties (crosses) of high a_w–high pH, high a_w–low pH, low a_w–high pH and low a_w–low pH. The intercepts of the crosses denote mean values; their branches denote standard deviations. If more cheese varieties were plotted, the stability and safety of the mapped cheeses would tend to decrease from higher to lower values as shown by the thick arrows. For individual cheese varieties, the effect of a_w or the pH may prevail, according to the greater variability of one or the other parameter, as shown by the thin arrows (for real cheese varieties, see Fig. 2 in Chapter 6, Volume 2 or Ref. 32).

the axes range equally from optimal to minimal values (see Fig. 3), the stability and safety of the cheese varieties mapped will increase from the top down and from right to left as shown by the large arrows (i.e. those varieties with high a_w and pH values are less stable than those with low values for these two parameters). For individual varieties, however, the effect of a_w or pH may prevail according to the greater variability of one or the other parameter as shown by the small arrows in Fig. 3. The effect of the pH usually prevails in fresh and mould-ripened cheeses, while that of water activity is normally prevalent in hard and grated cheeses.[32]

There are many other known possible binary combinations, of which only a few are cited here by way of examples. Thus, limit temperature and pH ranges for the survival and growth of major pathogens in cheese have been determined (see Table III); if either parameter is optimal, although the other is sub-optimal, then the rate of growth of the microorganism concerned will be higher than if both parameters are sub-optimal. Some binary combinations with a_w have also been associated with the temperature (e.g. in the study by Northolt on mycotoxin production)[131] and the removal of oxygen or E_h reduction (e.g. the minimal a_w for growth of *Staphylococcus aureus* is 0·86 aerobically and 0·90

TABLE III
Temperature and pH Ranges for the Survival and Growth of Some Pathogens in Cheese[a]

Risk ranking for cheese	Pathogen	T (°C)		pH	
		Min.	Max.	Min.	Max.
High	Salmonella paratyphy	6·5	57	4·6	10·0
	S. senftenberg	7·0	47	4·7	10·0
	Listeria monocytogenes	1·0	45	4·8	9·6
	Escherichia coli (EEC)	2·5	45	4·6	9·5
Medium	Yersinia enterocolitica	1·0	44	4·4	9·0
	Mycobacterium tuberculosis	30·0	38	6·3	7·8
Low	Staphylococcus aureus	7·0	48	4·0	9·8
	Clostridium botulinum (A and B)	10·0	50	4·7	9·0

[a] Adapted from Mitscherlich & Marth[132] and Johnson et al.[133]

anaerobically)[134] while *Clostridium botulinum* at its optimal a_w, which is very high, grows well over the range +60 to −400 mV, but at the lower potential (−400 mV), a small decrease in the a_w (0·98) on addition of NaCl significantly decreases the probability of sporal outgrowth.[135]

An example of ternary combinations of interdependent physical factors on the potential for microbial food spoilage or poisoning is reflected in Table IV, in which the minimum values for three physical parameters (T, a_w and pH) that allow the growth of major food-borne disease-causing organisms are listed. If two parameters are optimal, growth will occur at extreme values of the third, but the growth range in the last will be narrowed when the other two parameters are sub-optimal.

TABLE IV
Minimum Values of Three Major Physical Parameters Allowing Growth in Cheese of Some Pathogens Responsible for Food-borne Diseases[a]

Risk	Pathogen	T (°C)	a_w	pH
High	Salmonella spp.	5·3	0·94	4·0
Medium	Vibrio parahaemolyticus	5·0	0·94	4·8
	Yersinia enterocolitica	0·0	—	4·4
Low	Staphylococcus aureus	6·7	0·86	4·5
	(toxin production)	(10·0)	(0·90)	—
	Clostridium botulinum			
	(A and B)	10·0	0·94	4·7
	(E)	3·0	0·97	—
	Clostridium perfringens	12·0	0·93	5·0
	Bacillus cereus	7·0	0·95	4·3

[a] Adapted from the *International Commission on Microbial Specifications for Food* (ICMSF),[136] with permission of Academic Press.

At 37°C, the growth of *C. botulinum* type A occurs at $a_w = 0.94$ and pH 7·0, whereas at pH 5·3, the minimal a_w value increases to 0·99 (Baird-Parker[137]).

Quaternary combinations are also common in many food products. The possible interactions between four major physical parameters in relation to food stability and safety are illustrated in the following square diagram reported by Leistner.[138]

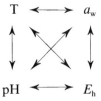

6.3 Physical, Chemical and Microbiological Interactions in Cheese

In addition to physical aspects, chemical (preservatives) and microbiological agents (competitive flora) must also be considered in relation to the stability and safety of foods in general[124] and of cheese in particular.[136] The interactions between these are schematized as follows:

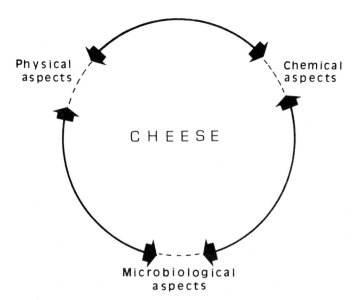

Various chemical additives commonly used in cheesemaking (e.g. H_2O_2, $CaCl_2$, NaCl, sorbic, propionic and other acids, nitrates and nitrites, etc.) may modify physical parameters (a_w, pH, E_h) through interaction, although some also exert specific effects and selective influences on certain microorganisms. Thus, H_2O_2 increases the E_h and preserves raw milk; addition of CO_2 to raw milk, which decreases its pH and reduces phsychrotroph growth, also affects starter cultures and eye development in cheese by aroma producers; $CaCl_2$ improves milk rennetability and curd syneresis, and lowers water activity; NaCl lowers a_w, but also influences the growth of starter and non-starter microorganisms, the activity of indigenous milk, coagulant and microbial enzymes, all of which in turn affect the ripening, composition, quality, stability and safety of the cheese;[31] sorbic acid and its calcium, sodium and potassium salts are very effective in preventing the growth of surface yeasts and moulds, and mycotoxin production by the latter; sorbate, propionate and acidulants (sorbic, acetic, lactic and citric acids), in addition to lowering the pH, influence the survival of *Listeria monocytogenes* in cheese; nitrates and nitrites, the antimicrobial activity of which depends on the pH, E_h and a_w—like that of NaCl—inhibit the growth of *Clostridium* spp., thereby preventing the late bloating of cheese. These are but a few examples of interdependences between chemical, physical and biological factors.

The physical and chemical microenvironmental conditions that favour the development of particular species or strains in a mixed microbial population in food may exert selective effects through competition and mutual interactions between species leading to changing dominances of microbial associations, the whole sequence of which is referred to as 'succession' by Mossel & Ingram.[139-141]

Outstanding, comprehensive and in-depth coverage of the individual physical, chemical and biological factors that affect the survival and growth of microorganisms in foods, with special emphasis on the complexity of the mutual interrelations between them is provided in *Microbial Ecology of Foods*, published by the ICMSF.[124] Numerous factors that may influence the presence and survival of pathogens in cheese were reviewed by Marth.[142,143]

Cheese is a dairy product traditionally obtained by empirical combinations of chemical, physical and biological agents while avoiding drastic treatments after milk curdling as far as possible.

Cheesemaking is a good example of the hurdle technology, a concept coined by Leistner & Rödel[144] and based on the interactive effect of several mild suboptimal physical (temperature, a_w, pH, E_h), chemical (preservatives) and microbiological parameters (competitive flora) or hurdles to control microbial spoilage and food poisoning. Thus, in cheese manufacture and ripening, an intricate and complex series of hurdles operates, in which a_w is of great relevance at the beginning and is essential at the end. A perishable raw material (i.e. milk with an indigenous flora, a high a_w value, a near-neutral pH and a relatively high E_h), usually stored at refrigeration temperature (T), is heated to lethal pasteurization temperatures (F) to destroy unwanted indigenous microorganisms (chiefly pathogens) and then inoculated with a competitive flora (lactic acid cultures) to exert a selective pressure through competition, decrease in pH and E_h in order to control the growth of surviving undesirable (spoilage) microorganisms of the initial flora, or of contaminants; then, the a_w is lowered by removing whey by syneresis and pressing the coagulum, salting the curd and drying and curing the cheese at refrigeration temperature until the water activity is low enough to endow the cheese with stability under non-refrigeration conditions. During the manufacturing process, preservatives other than salt are usually added to some varieties; some cheese

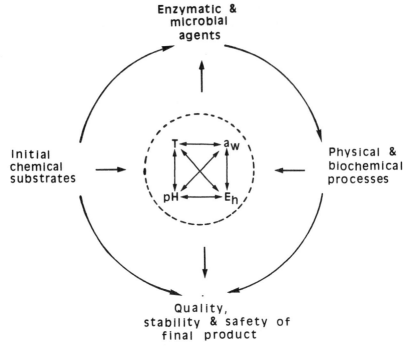

Fig. 4. Outline of major interactive pathways in the network of physical, chemical and biological parameters for an integrated approach to cheese science and technology.

varieties are smoked after manufacture, which incorporates smoke preservatives on the cheese surface while the hot air lowers its water activity.

The only hurdle that strengthens with time is a_w; therefore, this is the main agent responsible for the stability and safety of cheese ripened over a long period. Most mature hard cheeses have good keeping qualities at ambient temperature and many are self-stable products simply because of their low water activities. Only cheeses with high a_w values are markedly perishable and have been involved in some outbreaks of food poisoning; they require refrigerated storage and delivery.

Potential pathogens in cheeses of high a_w (e.g. surface-ripened soft cheeses such as Brie and Camembert), fresh or ripened for less than two months, which have been involved in several outbreaks of food-borne diseases, are those causing brucellosis (*Brucella abortus, B. melitensis*), listeriosis (*L. monocytogenes*), and shigellosis (*Shigella* spp.), mainly as a result of the use in cheesemaking of raw or improperly pasteurized milk, and those producing salmonellosis (*Salmonella* spp.), botulism (*C. botulinum*), staphylococcal food poisoning (*S. aureus*) and gastroenteritis (entheropathogenic *Escherichia coli (EEC), Yersinia enterocolitica*) that occur in cheese chiefly from recontamination or improper pasteurization of milk.[136] *Salmonella, L. monocytogenes* and EEC have been prioritized as high risks in cheese, while *S. aureus* is thought to be of low risk because it can be readily controlled in cheesemaking. In fact, mild heat-treatment (65–66°C for 16–18 s) is lethal for virtually all pathogenic microorganisms present in milk that pose major threats to cheese safety; however, neither pasteurization (due to recontamination) nor any other single factor can assure the complete safety of cheese. Thus, it is mandatory to study the relationships between physical, chemical and microbiological parameters as a whole (Fig. 4) in order to enable articulation of an integrated system of cheesemaking technology together with appropriate programmes for hazard assessment and critical control points (HACCP).[133]

Some semi-hard and hard cheeses, particularly Italian varieties (e.g. Romano, Parmesan and Provolone), with relatively low a_w values, have rarely been reported to cause food poisoning,[133] but the formation of tyramine and histamine through decarboxylation of the amino acids tyrosine and histidine, respectively, by bacterial decarboxylase may pose some problems; on the other hand, these amines can be catabolized by some starter cultures with mono- and diamine oxidase activity.[136]

Even in blue cheeses, which have rather low a_w values, mycotoxigenic moulds of concern are xerophilic and, although little is still known about the influence of water activity on mycotoxin production,[124,145] *P. roqueforti* spp. are potentially able to produce roquefortin, PR toxin and patulin.[146] However, consumption of blue cheeses is believed by many researchers not to represent a health hazard.[146]

In a very comprehensive discussion[133] on microbial safety of cheese containing nearly 400 references and encompassing almost a century (1893–1990), no mention was made of the term 'water activity'. This is a rather surprising omission since Labuza,[8] in his review of 1984, listed over 400 references dealing exclusively with water activity and microbiology; since then, this parameter has been one of the most extensively studied by food microbiologists.[17]

ACKNOWLEDGEMENTS

This chapter is dedicated to Pascual López Lorenzo, a precociously retired Professor who taught on the relevance of water in foods in the 1960s, and to whom the author is greatly indebted. The Spanish *Plan Nacional de Investigación Científica y Desarrollo Tecnológico* is acknowledged for financial support of Research Project ALI89–0037, concerned with work on water activity cited in the text or to be published in the future. The great patience and devotion of our busy Editor in supervising and polishing the rough manuscript is also gratefully acknowledged.

REFERENCES

1. Goldblith, S.A., 1989. *Food Technol.*, **43**, 88.
2. Institute of Food Technologists Staff Report, 1989. *Food Technol.*, **43**, 308.
3. Duckworth, R.B. (ed.), 1975. *Water Relations of Foods*. Academic Press, London.
4. Davies, R., Birch, G.G. & Parker, K.J. (eds), 1976. *Intermediate Moisture Foods*. Elsevier Applied Science, London.

5. Troller, J.A & Christian, J.H.B., 1978. *Water Activity and Food.* Academic Press, New York.
6. Rockland, L.B. & Stewart, G.F. (eds), 1981. *Water Activity: Influences on Food Quality,* Academic Press, New York.
7. Iglesias, H.A. & Chirife, J., 1982. *Handbook of Food Isotherms,* Academic Press, New York.
8. Labuza, T.P., 1984. *Moisture Sorption: Practical Aspects and Use.* American Association of Cereal Chemists, St. Paul.
9. Simatos, D. & Multon, J.L. (eds), 1985. *Properties of Water in Foods in Relation to Quality and Stability.* Martinus Nijhoff Publishers, Dordrecht.
10. Wolf, W., Spiess, W.E.L. & Jung, G., 1985. *Sorption Isotherms and Water Activity of Food Materials.* Science and Technology Publishers Ltd., Essex.
11. Rockland, L.B. & Beuchat, L.R. (eds), 1987. *Water Activity: Theory and Applications to Food.* Marcel Dekker, Inc., New York.
12. Hardman, T.M. (ed.), 1989. *Water and Food Quality.* Elsevier Applied Science, London.
13. Labuza, T.P., 1975. *Theory, Determination and Control of Physical Properties of Food Materials.* ed. C. Rha. D. Reidel Publishing Co., Dordrecht, p. 197.
14. Rüegg, M. & Blanc, B., 1981. In *Water Activity: Influences on Food Quality,* ed. L.B. Rockland & G.F. Stewart. Academic Press, New York, p. 791.
15. Jowitt, R., Escher, F., Hallström, B., Meffert, H.F.Th., Spiess, W.E.L. & Vos, G. (eds), 1983. *Physical Properties of Foods.* Elsevier Applied Science, London.
16. Rüegg, M., 1985. In *Properties of Water in Foods,* ed. D. Simatos & J.L. Multon. Martinus Nijhoff Publishers, Dordrecht, p. 603.
17. Luquet, M. (ed.), 1985. *Laits et Produits Laitiers.* Technique et Documentation (Lavoisier), Paris.
18. van den Berg, C., 1986. In *Concentration and Drying of Foods,* ed. D. MacCarthy. Elsevier Applied Science, London, p. 11.
19. Rao, M.A. & Rizvi, S.S.H. (eds), 1986. *Engineering Properties of Foods.* Marcel Dekker, Inc., New York.
20. MacCarthy, D. (ed.), 1986. *Concentration and Drying of Foods.* Elsevier Applied Science, London.
21. Seow, C.C. (ed.), 1988. *Food Preservation By Moisture Control.* Elsevier Applied Science, London.
22. Aguilera, J.M. & Stanley, D.W., 1990. *Microstructural Principles of Food Processing and Engineering.* Elsevier Applied Science, London.
23. Fox, P.F. (ed.), 1987. *Cheese: Chemistry, Physics and Microbiology,* Vols 1 and 2. Elsevier Applied Science, London.
24. Eck, A. (ed.), 1987. *Cheesemaking: Science and Technology.* Lavoisier Publishing Inc., New York.
25. Lawrence, R.C. & Gilles, J., 1987. In *Cheese: Chemistry, Physics and Microbiology,* Vol. 2, ed. P.F. Fox. Elsevier Applied Science, London, p. 1.
26. Hardy, J., 1984. In *Le Fromage,* ed. A. Eck. Technique et Documentation (Lavoisier), Paris, p. 37.
27. Choisy, M., Desmazeaud, M., Gripon, J.C., Lamberet, G., Lenoir, J. & Tourneur, C., 1984. In *Le Fromage,* ed. A. Eck. Technique et Documentation (Lavoisier), Paris, p. 62.
28. Weber, F. & Ramet, J.P., 1984. In *Le Fromage,* ed. A. Eck. Technique et Documentation (Lavoisier), Paris, p. 291.
29. Veillet-Poncet, L., 1984. In *Le Fromage,* ed. A. Eck. Technique et Documentation (Lavoisier), Paris, p. 476.
30. Hardy, J., 1985. In *Laits et Produits Laitiers,* Vol. 2, ed. M. Luquet. Technique et Documentation (Lavoisier), Paris, p. 603.
31. Guinee, T.P. & Fox, P.F., 1987. In *Cheese: Chemistry, Physics and Microbiology,* Vol. 1, ed. P.F. Fox. Elsevier Applied Science, London, p. 251.

32. Marcos, A., 1987. In *Cheese: Chemistry, Physics and Microbiology*, Vol. 2, ed. P.F. Fox. Elsevier Applied Science, London, p. 185.
33. Scott, W.J., 1953. *Aust. J. Biol. Sci.*, **6**, 549.
34. Solomon, M.E., 1957. *Bull. Ent. Res.*, **48**, 489.
35. Kvaale, O. & Dalhoff, E., 1963. *Food Technol.*, **17**, 659.
36. Northolt, M.D. & Heuvelman, C.J., 1982. *Voedingsmiddelentechnologie*, **15**, 11.
37. Makower, B. & Meyers, S., 1943. *Proc. Inst. Food Technol.*, Fourth Annual Meeting, p. 165.
38. Troller, J.A., 1983. *J. Food Sci.*, **48**, 739.
39. Bizot, H. & Multon, J.L., 1978. *Ann. Technol. Agric.*, **27**, 441.
40. Leung, H., Morris, H.A., Sloan, A.E. & Labuza, T.P., 1976. *Food Technol.*, **23**, 42.
41. Labuza, T.P., Acott, K., Tatini, S.R., Lee, R., Flink, J. & McCall, W., 1976. *J. Food Sci.*, **41**, 910.
42. Marcos, A., Esteban, M.A. & Alcalá, M., 1990. *Food Chem.*, **38**, 189.
43. Rüegg, M. & Blanc, B., 1977. *Milchwissenschaft*, **32**, 193.
44. Heskedtad, R. & Steinsholt, K., 1978. *Meieriposten*, **11**, 327.
45. Richard, J. & Labuza, T.P., 1990. *Sci. Aliments.*, **10**, 57.
46. Marcos, A., Fernández-Salguero, J., Esteban, M.A., León, F., Alcalá, M. & Beltrán, F.H., 1985. *Quesos españoles: Tablas de composición, valor nutritivo y estabilidad.* Servicio de Publicaciones de la Universidad, Córdoba.
47. Marcos, A., Alcalá, M., León, F., Fernández-Salguero, J. & Esteban, M.A., 1981. *J. Dairy Sci.*, **64**, 622.
48. Marcos, A., Esteban, M.A., Alcalá, M. & Millán, R., 1983. *J. Dairy Sci.*, **66**, 909.
49. Marcos, A., Millán, R., Esteban, M.A., Alcalá, M. & Fernández-Salguero, J., 1983. *J. Dairy Sci.*, **66**, 2488.
50. Fernández-Salguero, J., Alcalá, M., Marcos, A. & Esteban, M.A., 1983. *J. Dairy Res.*, **53**, 639.
51. Strong, D.H., Foster, E. & Duncan, C., 1970. *Appl. Microbiol.*, **19**, 980.
52. Esteban, M.A., Marcos, A. & Fernández-Salguero, J., 1987. *Food Chem.*, **25**, 31.
53. Cabezas, L., Marcos, A., Esteban, M.A., Fernández-Salguero, J. & Alcalá, M., 1988. *Food Chem.*, **30**, 59.
54. Lerici, C.R., Piva, M. & Dalla Rosa, M., 1983. *J. Food Sci.*, **48**, 1667.
55. Chirife, J., Favetto, G. & Scorza, O.C., 1982. *J. Appl. Bacteriol.*, **53**, 219.
56. Kairyukshtene, I., 1983. *Sbornik Nauchnykh Trudov*, **17**, 75.
57. Labuza, T.P., Kreisman, L.N., Heinz, C.A. & Lewicki, P.P., 1977. *J. Food Process. Preserv.*, **1**, 31.
58. Lenart, A. & Flink, J.M., 1983. *Lebensm. Wiss. u. Technol.*, **16**, 84.
59. Troller, J.A., 1983. *J. Food Protect.*, **46**, 129.
60. Esteban, M.A. & Marcos, A., 1989. *Alimentaria*, **203**, 27.
61. Esteban, M.A., Marcos, A., Fernández-Salguero, J. & Alcalá, M., 1989. *Intern. J. Food Sci. Technol.*, **24**, 139.
62. Cabezas, L., Marcos, A., Alcalá, M. Esteban, M.A. & Fernández-Salguero, J., 1987. *Alimentación*, **6**(3), 167.
63. Landrock, H.A. & Proctor, B.E., 1951. *Food Technol.*, **5**, 332.
64. Rockland, L.B., 1960. *Anal Chem.*, **32**, 1375.
65. Greenspan, L., 1977. *J. Res. Nat. Bureau Standards—A Physics and Chemistry*, **81A**(1), 89.
66. Rüegg, M., 1980. *Lebensm. Wiss. u. Technol*, **13**, 22.
67. Grover, D.W. & Nicol, J.M., 1940. *J. Soc. Chem. Ind. (London)*, **59**, 175.
68. Chirife, J. & Resnik, S.L., 1984. *J. Food Sci.*, **49**, 1486.
69. Kitic, D., Resnik, S.L. & Chirife, J., 1986. *Lebensm, Wiss. u. Technol*, **19**, 272.
70. Lang, K.W., McCune, T.D. & Steinberg, M.P., 1981. *J. Food Sci.*, **46**, 936.
71. Wolf, W., Spiess, W.E.L. & Jung, G., 1973. *Lebensm. Wiss. u. Technol.*, **6**, 94.
72. Geurts, T.J., Walstra, P. & Mulder, H., 1974. *Neth. Milk Dairy J.*, **28**, 46.

73. Berlin, E. & Anderson, B.A., 1975. *J. Dairy Sci.*, **58**, 25.
74. Robinson, R.A. & Sinclair, D.A., 1934. *J. Am. Chem. Soc.*, **56**, 1830.
75. Fett, H.M., 1973. *J. Food Sci.*, **38**, 1097.
76. Vos, P.T. & Labuza, T.P., 1974. *J. Agr. Food Chem.*, **22**, 326.
77. McCune, T.D., Lang, K.W. & Steinberg, M.P., 1981. *J. Food Sci.*, **46**, 1978.
78. Marcos, A., Beltrán, F.H., Alcalá, M., Fernández-Salguero, J., Esteban, M.A., León, F. & Sanz, B., 1982. *Alimentaria*, **134**, 29.
79. Marcos, A., Fernández-Salguero, J., Esteban, M.A., & Alcalá, M., 1985. *J. Food Technol.*, **20**, 523.
80. Palacha, Z. & Flink, J.M., 1987. *Intern. J. Food Sci. Technol.*, **22**, 485.
81. Esteban, M.A. & Marcos, A., 1989. *Alimentaria*, **206**, 83.
82. Marcos, A. & Esteban, M.A., 1989. *Alimentaria*, **207**, 57.
83. Esteban, M.A. & Marcos, A., 1989. *Alimentaria*, **201**, 31.
84. Marcos, A., Esteban. M.A. & López, P., 1990. *Alimentación*, **9**(4), 165.
85. López, P., Sanz, B., Burgos, J., Marcos, A. & Esteban, M.A., 1990. *Anal. Bromatol.*, **42**, 139.
86. Gómez, R., Esteban, M.A. & Marcos, A., 1991. *Proc. III World Congr. Food Technol.*, Barcelona, p. 94.
87. Smith, S.E., 1947, *J. Am. Chem. Soc.*, **69**, 646.
88. Beuchat, L., 1981. *Cereal Foods World*, **26**, 345.
89. Marcos, A., Alcalá, M., Esteban, M.A., Fernández-Salguero, J., León, F., Beltrán, F.H. & Sanz, B., 1983. *Industrias Lácteas Españolas*, **53–54**, 57.
90. Alcalá, M., Millán, M.J., Marcos, A., Fernández-Salguero, J. & Esteban, M.A., 1986. *Alimentaria*, **175**, 41.
91. Ross, K.D., 1975. *Food Technol.*, **29**, 26.
92. Esteban, M.A. & Marcos, A., 1990. *Food Chem.*, **35**, 179.
93. López, P., Marcos, A. & Esteban, M.A., 1990. *J. Dairy Res.*, **57**, 587.
94. Esteban, M.A., Marcos, A., Alcalá, M. & Gómez, R., 1991. *Food Chem.*, **40**, 147.
95. Marcos, A., Esteban, M.A., Espejo, J. & Marcos, I., 1990. *Alimentación*, **9**(6), 97.
96. Espejo, J., Marcos, I., Esteban, M.A. & Marcos, A., 1991. *Proc. III World Congr. Food Technol.*, Barcelona, p. 31.
97. Alcalá, M., Esteban M.A. & Marcos, A., 1991. *Proc. III World Congr. Food Technol.*, Barcelona, p. 21.
98. Marcos, A. & Esteban, M.A., 1991. *Intern. Dairy J.*, **1**, 137.
99. Esteban, M.A. & Marcos, A., 1989. *J. Dairy Res.*, **56**, 665.
100. Robinson, R.A. & Stokes, R.H., 1970. *Electrolyte Solutions*, 2nd edn., Butterworths, London.
101. Marcos, A. & Esteban, M.A., 1982. *J. Dairy Sci.*, **65**, 1795.
102. Esteban, M.A., Marcos, A. & López, P., 1990. *Alimentaria*, **213**, 21.
103. Fox, P.F., 1989. *J. Dairy Sci.*, **72**, 1379.
104. Kuchroo, C.N. & Fox, P.F., 1982. *Milchwissenschaft*, **37**, 331.
105. O'Sullivan, M. & Fox, P.F., 1990. *J. Dairy Res.*, **57**, 135.
106. Rizvi, S.S.H., 1986. In *Engineering Properties of Foods*, ed. M.A. Rao & S.S.H. Rizvi, Marcel Dekker Inc., New York, p. 133.
107. Brunauer, S., Emmett, P.H. & Teller, E., 1938. *J. Am. Chem. Soc.*, **60**, 309.
108. Guggenheim, E.A., 1966. *Applications of Statistical Mechanics*, Clarendon Press, Oxford.
109. Anderson, R.B., 1946. *J. Am. Chem. Soc.*, **68**, 686.
110. de Boer, J.H., 1953. *The Dynamical Character of Adsorption*, Clarendon Press, Oxford.
111. Román, M.L., 1990. PhD Thesis, University of Córdoba, Córdoba.
112. Guinee, T.P., 1985. PhD Thesis, National University of Ireland, Dublin.
113. Geurts, T.J., Walstra, P. & Mulder, P., 1974. *Neth. Milk Dairy J.*, **28**, 102.
114. Guinee, T.P. & Fox, P.F., 1983. *J. Dairy Res.*, **50**, 511.

115. Fox, P.F., 1987. In *Cheese: Chemistry, Physics and Microbiology*, Vol. 1, ed. P.F. Fox. Elsevier Applied Science, London, p. 1.
116. Mou, L., Sullivan, J.J. & Jago, G.R., 1975. *J. Dairy Res.*, **42**, 147.
117. Visser, F.M.W., 1977. *Neth. Milk Dairy J.*, **31**, 210.
118. Courroye, M., 1987. *Revue Laitière Française*, **462**, 53.
119. Gripon, J.C., 1987. In *Cheese: Chemistry, Physics and Microbiology*, Vol. 2, ed. P.F. Fox, Elsevier Applied Science, London, p. 121.
120. Fox, P.F. & Guinee, T.P., 1987. In *Cheese: Chemistry, Physics and Microbiology*, Vol. 2, ed. P.F. Fox, Elsevier Applied Science, London, p. 221.
121. Pitt, J.I. & Christian, J.H.B., 1968. *Appl. Microbiol.*, **16**, 1853.
122. Hocking, A.D. & Pitt, J.I., 1987. In *Water Activity: Theory and Application to Food*, ed. L. B. Rockland & L.R. Beuchat, Marcel Dekker Inc., New York, p. 153.
123. Esteban, M.A., Alcalá, M., Marcos, A., Fernández-Salguero, J., García de Fernando, G.D., Ordóñez, J.A. & Sanz, B., 1990. *Intern. J. Food Sci. Technol.*, **25**, 464.
124. International Commission on Microbial Specifications for Foods (ICMSF), 1980. *Microbial Ecology of Foods: Factors Affecting Life and Death of Microorganisms.* Academic Press Inc., New York.
125. Hunsgerger, J.F., 1981–82. In *Handbook of Chemistry and Physics*, 62nd edn, ed. R.C. Weast & M.J. Astle. CRC Press Inc., Boca Raton, p. D-135.
126. Jacob, H.E., 1970. In *Methods in Microbiology*, Vol. 2. Academic Press, London, p. 91.
127. Troller, J.A., 1987. In *Water Activity: Theory and Application to Food*, ed. L.B. Rockland & L.R. Beuchat. Marcel Dekker Inc., New York, p. 101.
128. Lenovich, L.M., 1987. In *Water Activity: Theory and Application to Food*, ed. L.B. Rockland & L.R. Beuchat. Marcel Dekker Inc., New York, p. 119.
129. Genigeorgis, C. & Sadler, W.W., 1966. *J. Bacteriol.*, **92**, 1383.
130. Genigeorgis, C., Martin, S., Franti, C.E. & Reinmann, H., 1971. *Appl. Microbiol.*, **21**, 934.
131. Northolt, M.D., 1979. PhD Thesis, Rijks Instituut voor de Volksgezondheid, Belthoven, The Netherlands.
132. Mitscherlich, E. & Marth, E.H., 1984. *Microbial Survival in the Environment*, Springer-Verlag, Berlin.
133. Johnson, E.A., Nelson, J.H. & Johnson, M., 1990. *J. Food Protect.*, **53**, 441, 519, 610.
134. Scott, W.J., 1957. *Avd. Food Res.*, **7**, 83.
135. Lund, B.M. & Wyatt, G.M., 1984. *Food Microbiol.*, **1**, 49.
136. International Commission on Microbial Specifications for Foods (ICMSF), 1980. *Microbial Ecology of Foods: Food Commodities*, Vol. 2. Academic Press Inc., New York, p. 507.
137. Baird-Parker, A.C., 1971. *J. Appl. Bacteriol.*, **34**, 181.
138. Leistner, L., 1987. In *Water Activity: Theory and Applications to Food*, ed. L.B. Rockland & L.R. Beuchat. Marcel Dekker, Inc., New York, p. 295.
139. Mossel, D.A.A. & Ingram, M., 1955. *J. Appl. Bacteriol.*, **18**, 232.
140. Mossel, D.A.A., 1971. *J. Appl. Bacteriol.*, **34**, 95.
141. Mossel, D.A.A., 1975. *Crit. Rev. Environ. Control*, **5**, 1.
142. Marth, E.H., 1968., *J. Dairy Sci.*, **52**, 283.
143. Marth, E.H., 1987. In *Standard Methods for the Examination of Dairy Products*, American Public Health Association, Washington.
144. Leistner, L. & Rödel, W., 1976. In *Intermediate Moisture Foods*, ed. R. Davies, G.G. Birch & K.J. Parker. Applied Science Publishers, London, p. 120.
145. Northolt, M.D., Verhulsdonk, C.A.H., Soentoro, P.S.S. & Paulsch, W.E., 1976. *J. Milk Food Technol.*, **39**, 170.
146. Renner, E., 1987. In *Cheese: Chemistry, Physics and Microbiology*, Vol. 1, ed. P.F. Fox. Elsevier Applied Science, London, p. 538.

12

Growth and Survival of Undesirable Bacteria in Cheese

EDMUND A. ZOTTOLA & LORRAINE B. SMITH

Department of Food Science and Nutrition, University of Minnesota, St Paul, Minnesota 55108, USA

1 INTRODUCTION

1.1 The Problem of Food-borne Disease

The growth and survival of bacteria in foods, particularly in dairy products such as cheese, have been studied extensively; however, changes in dietary habits, the development of new technologies and formulations to meet the demands for cheeses of lower fat, cholesterol and calorie content, may create conditions which render cheese microbiologically unsafe for human consumption. To understand fully the concern for the safety of these foods, one must first be aware of the potential hazards, of how, why and where they occur and how they can be controlled or eliminated.

Food-borne illness or disease is a direct result of the growth and survival of undesirable bacteria in food and can be classed as a food intoxication, a food infection or a food toxicoinfection. **Food intoxication** is also known as true food poisoning and is the direct result of ingesting food containing a toxin produced during bacterial growth in the food. The toxin is present in the food although the organism that produced it may be dead.[1] The occurrence of a food intoxication depends on the presence of the bacteria in the food, the suitability of the food for microbial growth, the ability of the bacteria to grow to high numbers in the food, production of toxin during the growth cycle of the bacteria, consumption of the food and the susceptibility of the consumer to the action of the toxin. *Bacillus* spp. *Clostridium botulinum* and *Staphylococcus* spp. are involved in food intoxications.[2] These microorganisms have been responsible for cases of cheese-implicated food-borne illness reported in Europe during the period 1971–1987.[3,4]

When foods containing pathogenic bacteria are consumed and the bacteria are able to colonize the intestinal tract of the host, grow and cause tissue damage, the resultant symptoms are referred to as a **food infection**. Again, the microorganism must be present in the food, conditions must be suitable for the survival of the microorganism, although it may or may not grow, the food must

471

be consumed, the organism must be able to colonize the gut and grow, and the host must be susceptible to the action of the organism. In this type of food-borne infection, the numbers of causative microorganisms are often very low, sometimes as low as one cell per gramme of food.[2] The organisms responsible for food infections are *Salmonella, Shigella, Listeria, Yersinia, Vibrio parahaemolyticus, Vibrio vulnificus* and *Campylobacter*; some of these have been involved in cheese-related food-borne outbreaks in the US and Europe.[3,4]

A third type of food-borne disease is essentially a combination of food intoxication and food infection and is classified as a **food toxicoinfection**. High numbers of the responsible bacteria are present in the food and are ingested by the host. The bacteria continue to grow in the gut, toxin is released and subsequently causes symptoms of the disease. Again, the bacteria must be present in the food, growth of the microorganism (which depends on the suitability of the food) must proceed to high populations (10^6/g), the food must be consumed, the organisms must grow in the gut, release toxin and the host must be susceptible to the toxin.[2] *Clostridium perfringens, Vibrio cholerae* and enterotoxigenic *Escherichia coli* are examples of microorganisms that are responsible for food toxicoinfections. Enteropathogenic *E. coli* were responsible for more than 737 cases of food-borne illness in Europe from 1971 to 1987.[2,3]

While a variety of foods, e.g. shellfish, meats, salads, fruits, vegetables, poultry, milk, dairy products, baked goods, beverages and ethnic specialities have been involved in outbreaks of food-borne illness, our concern in this review is focused on the bacterial contamination of cheeses. Outbreaks of food-borne illness in the US, Canada and Europe have been related mainly to the consumption of soft cheeses such as Brie, Camembert, Vacherin and 'Mexican style' soft cheese but also of Cheddar, Kuminost and Monterey.[3,4]

Cheese constitutes an important staple in the daily diet of many peoples. Today, whether consumed at commercial restaurants, in the home, at fast food chains, or in school lunch programmes, cheese, in one form or another, is often the basis for fulfilling the protein requirements of both children and adults. Because of the dietary importance of cheese, it is necessary to examine the significance of milk as a vector for bacterial contamination of cheese.

1.2 Milk as a Vector of Food-borne Disease

As early as 400 BC, Hippocrates recorded that 'milk is the most nearly perfect food'.[5] All of the fat, calcium, phosphorus and riboflavin, one-half of the protein, one-third of the thiamine, ascorbic acid and Vitamin A, one-fourth of the calories and most of the minerals are furnished to an individual by the daily consumption of one litre of cows' milk.[5] However, milk may easily become contaminated with bacteria and is a good medium for the growth of many species.

1.2.1 Microorganisms in Milk

Several factors affect the microorganisms present in milk as it is delivered from the cow. A healthy dairy herd is essential for a safe milk supply. Pathogenic microorganisms may be present in the milk of infected cows but often go undetected until the manifestation of clinical symptoms. Consumption of infected milk by humans can result in human infections.

Health of the cow. *Mycobacterium bovis* is the bacterium responsible for bovine tuberculosis and is transmissible to humans by milk from infected cows. The organism often originates in the sputum, passing through the digestive tract into manure which then may contaminate the milk from bedding or the coat of the infected animal. However, in the US, tuberculosis-free herds are assured due to the mandated tuberculosis test-and-slaughter method which began in 1917 after the development in 1890 by Robert Koch of the tuberculin test for the detection of tuberculosis. Control of tuberculosis is governed by the Animal and Plant Health Inspection Service of the US Department of Agriculture.[5]

The Recommended Uniform Methods and Rules for Brucellosis Eradication administered by the same agency serve as control measures for brucellosis which is caused by *Brucella abortus*. This organism causes abortion in cattle and undulant fever in adult humans who are infected either through the consumption of milk from an infected animal or via open scratches or wounds on the hands of dairy workers. The 'ring' test on milk and/or the agglutination test on the animal's blood serum serve as detection methods.

Udder infections. The udder of the cow is another source of possible contamination of milk. Mastitis, an infection of the udder, is caused by *Streptococcus agalactiae*, which is non-pathogenic to Man, by *Streptococcus pyogenes* or *Staphylococcus aureus*, both of which can infect humans. The toxin produced by *Staphylococcus aureus* is not inactivated by pasteurization temperatures. If the organism is present in milk and has sufficient time to grow in the milk with subsequent production of toxin, humans may be exposed to food-poisoning syndrome.[6] Several other organisms are implicated in mastitis, including *Escherichia coli, Streptococcus uberis, Streptococcus dysgalactiae* and *Pseudomonas aeruginosa*. Infection enters the mammary gland via the teat orifice and streak canal. Prevention methods include disinfection of the teats after milking by use of a bacteriocide, disinfection and sterilization of milking machines, as well as antibiotic therapy for infected animals.[5]

Husbandry practices. The prevention of disease, the isolation and treatment of infected animals, routine veterinarian assessment and vaccination of the herd are only part of the overall practices which ensure a healthy herd. The dairyman must take the initiative to provide clean bedding and food for his animals. Routine grooming of the animals and removal of waste are needed to prevent infection of the herd, and disinfection and sanitization of milking equipment, pipelines and storage tanks will reduce the potential for bacterial contamination of the milk.[6]

1.2.2 Role of Pasteurization in Controlling Microorganisms in Milk
   Basis for milk pasteurization.   The US *FDA Grade A Pasteurized Milk Ordinance* (1989 revision) indicates that pasteurization of milk shall mean the process of heating every particle of milk or milk product in properly designed and operated equipment to one of the temperatures given below and holding the milk continuously at or above that temperature for at least the corresponding specified time:

Temperature	Time
145°F (63°C)	30 min
161°F (72°C)	15 s
191°F (89°C)	1·0 s
204°F (96°C)	0·05 s
212°F (100°C)	0·01 s

If the dairy ingredient has a fat content of 10% or more, or if it contains added sweeteners, the specified temperature shall be increased by 5°F (3°C). Other time/temperature combinations which have been demonstrated to be equivalent thereto in microbial destruction may be used and are approved by the Food and Drug Administration.[7]

The objective for pasteurization of milk is to destroy pathogenic microorganisms which may be present initially and to enhance the keeping quality of the product. In the late 1800s, basic research reportedly began on the pasteurization of milk for cheesemaking using exceptionally crude equipment and without any evident control measures. The concept of pasteurization was based on the early (1860–1864) studies of Louis Pasteur in which it was demonstrated that abnormal souring and fermentations of wine could be controlled by heating at 57·2°C (135°F) for several minutes. Application of the same concept to milk many years later resulted in the development of the following requirements for the pasteurization of milk: low-temperature, long time (LTLT) at 145°F (63°C) for 30 min or the high-temperature, short-time (HTST) process at 161°F (72°C) for 15 s. Initially, the organism of concern was *Mycobacterium tuberculosis* which, because of its heat-resistant characteristics, was studied extensively. Subsequently, it was established that *Coxiella burnetti* was more heat resistant than *Mycobacterium tuberculosis* and current requirements are based on the destruction of *Coxiella burnetti*.

Enzyme inactivation.   Milk contains many indigenous enzymes, while others are produced in milk or milk products by contaminating microorganisms. These enzymes are important because of their ability to modify the constituents of milk with resultant changes in smell, taste or appearance; they also serve as indicators of quality or confirmation of product treatment. Alkaline phosphatase, which is present in raw milk, specifically in microsomal particles adsorbed on fat globules, is almost completely inactivated by heating at the minimum times and temperatures for pasteurization but even a slight decrease in either the time or temperature leaves some residual active enzyme. If, after pasteurization, alkaline phosphatase activity is detected, its presence is evidence of inadequate heating or recontamination with raw milk.

From a spoilage viewpoint, milk lipase is potentially the most significant indigenous enzyme but it is almost totally inactivated by pasteurization.[8]

1.2.3 Relative Safety of Cheese Made from Raw and Pasteurized Milk
Disease-causing microorganisms may be found in milk used for cheesemaking. Regulations were introduced in the US in 1944 requiring cheese either to be made from pasteurized milk or the product to be held for 60 days after production before being released for sale. Thirty-nine natural cheese standards were published by the FDA in 1949 which provided two options for the safe production of cheese: (1) pasteurization of milk for cheesemaking, or (2) holding the finished cheese at 35°F (2°C) for not less than 60 days.[9,10] Cheeses produced from raw milk have the potential to be unsafe. Although these cheeses are required to be held for 60 days and may then be free of undesirable microorganisms, these same microorganisms may produce toxins that remain in the product. Strict control of pH and starter activity will serve to suppress the growth of, or toxin production by, food-poisoning microorganisms. Post-pasteurization contamination of cheeses manufactured from pasteurized milk is the most frequent contributory factor to reported outbreaks of food-borne illness. Faulty pasteurization and manufacturing practices, as well as gross contamination of the production facilities, can jeopardize the safety of cheeses made from pasteurized milk.[4]

2 OUTBREAKS OF FOOD-BORNE DISEASE INVOLVING CHEESE

2.1 Historical

As early as 1884, Vaughn[11] reported a cheese-associated food poisoning. He alluded to the observations of illness throughout the world during the previous 300 years that were due to the consumption of cheese. His reported included information about the destruction of the reputation of an Ohio cheese factory due to illnesses attributed to cheese manufactured at that factory. He also reported on 300 cases of food poisonings due to cheese in Michigan at the time of writing of his article.

In 1917, Cheddar cheese was implicated as the aetiological agent responsible for 64 cases and 4 deaths in Michigan, and in 1925, the state of Minnesota reported 29 cases of typhoid due to the consumption of contaminated Cheddar cheese. In 1939, Cheddar cheese was cited as the source of typhoid illness involving 100 cases and 11 deaths in Canada. From 1932 to 1939, six epidemics of typhoid involving 760 cases and 71 deaths, which undoubtedly included those previously mentioned, were documented in Canada. In New York state, typhoid illness, involving 23 cases and one death, due to the consumption of contaminated

Cheddar curd was reported in 1941. Some other reports of typhoid illnesses due to cheese are shown below:

Year	Country	Origin	Cases	Deaths
1943	Canada	Cheddar	40	6
1944	Canada	Cheddar	83	7
1944	California/Nevada	Romano-Dolce	80	—
1944	Indiana	Green Cheddar	246	13
1945	Tennessee	Colby	484	—

From 1917 to 1944, staphylococcal and streptococcal outbreaks involving more than 265 cases, but no deaths, were also reported. A variety of cheeses were implicated, including Asiago, Colby, NY Herkimer, Cheddar, Cottage and imported Albanian.[12]

Outbreaks of food poisoning in Europe involving more than 670 cases were attributed to cheeses, including Italian, cream, Wensleydale, Canadian Cheddar, Gorgonzola, and cheeses made from milk of sheep or goats, contaminated with *Brucella*, were implicated in more than 62 cases of food-borne illness.[13]

Fabian[14] published an extensive review citing 59 epidemics of food poisonings involving 2904 cases and 117 deaths for the period 1883 to 1946 which implicated cheese as the causative agent.

2.2 Recent Outbreaks

During the period 1948 to 1988, only six outbreaks of food-borne illness attributed to contaminated cheese have occurred in the US since the Federal Standards of Identity for cheese were published in 1949. This is an exceptional record when one considers that more than 50 million tonnes of natural cheese were produced during that 40-year period. Post-pasteurization contamination was reported to be the factor most frequently responsible, together with faulty pasteurization procedures or equipment and the use of raw milk. Soft surface-ripened cheeses, such as Brie and Camembert, rather than Cheddar and Swiss, appear to be most often involved in outbreaks in the US, Canada and Europe.[3,4,15,16]

2.2.1 Salmonella

Salmonella has been involved in several cheese-implicated outbreaks of food poisoning. Its widespread presence in the environment places the organism in a high risk category for the cheese industry; however, only one of the six previously confirmed outbreaks of food-borne illnesses during the reported 40-year period referred to above involved Cheddar cheese contaminated with *Salmonella heidelberg*. In Europe during the period 1971 to 1987, *Salmonella* was responsible for 250 cases of cheese-associated food-borne illness. From 1982 to 1984, contamination of Cheddar and other types of cheese produced in Canada with *Salmonella* was responsible for more than 2700 cases of food-borne illness.[4]

In 1989, the Minnesota Department of Health investigated a multi-state outbreak of salmonellosis and established a confirmed case-list of 136 incidents of *Salmonella javiana* and 11 of *Salmonella oranienburg* covering the period from April to July of that year. Estimates of 1500 to 15 000 cases of salmonellosis may have occurred and similar incidents of this infection were also noted in the state of Wisconsin. Epidemiological evidence showed that Mozzarella cheese was the vector of contamination. The cheese, produced at a single manufacturing plant, was sent to four shredding plants and subsequently led to cross-contamination of additional cheese products. *S. javiana* was the serotype identified as being responsible for the contamination.[4,17]

2.2.2 Staphylococcus

In a paper by Johnson *et al.*,[3] three outbreaks involving *Staphylococcus aureus* as the contaminant were reported which involved Cheddar, Kuminost and Monterey cheese. Two of the outbreaks occurred in 1958 with the third in 1965. Inadequate performance of the starter culture was reportedly the cause of *S. aureus* contamination in the Cheddar cheese. Contamination of lactic cultures with *S. aureus* was the cause of Swiss cheese-associated food-borne illness in Canada in 1977, and the same organism was the source of contamination of cheese curds in an outbreak in 1980.[3]

2.2.3 Listeria

A more recently recognized contaminant of cheese products is *Listeria monocytogenes* which has been responsible for outbreaks involving cheese, both in the US and Europe. In California between January and June 1985, the mysterious deaths of a number of infants were investigated. Mexican-style soft white cheese manufactured by a Californian firm was implicated as the source of *L. monocytogenes*, the organism responsible for 29 deaths, including 13 still-births and 8 neonatal fatalities. The implicated cheese had been consumed by the mothers of the infants. Of the 86 cases of listeriosis reported in epidemiological data, 58 were Hispanic patients.[4,18,19]

Vascherin Mont d'Or cheese was responsible for 122 cases and 33 deaths from listeriosis in Vaud, Switzerland, during the period 1983 to 1987.[4,20,21] Because of a recently confirmed case of listeriosis in the UK attributed to contaminated soft cheese, the British Food Hygiene Laboratory examined 222 British and imported soft cheeses of which 10% were found to be contaminated with *L. monocytogenes* at levels of <100 to >10 0000 cfu/g. Additional results indicated that 16% of cheeses from Italy, 14% from France, 10% from Cyprus and 4% from the UK were contaminated with the organism.[4,22] Recalls of Camembert and Brie cheeses contaminated with *Listeria* have been reported in the literature.[23,24] Griffiths[25] indicated that from February to March 1986, approximately 60% of French Brie cheese was recalled from US markets because of contamination with *L. monocytogenes*. Of 864 cheeses examined for *L. monocytogenes* by Cantoni *et al.*,[26] the incidence of infection was 8·45 and 13·7% for Gorgonzola and Taleggio cheese, respectively.

It is evident from these reports that *L. monocytogenes* is present in the environment as well as in milk supplies. Although pasteurization of the milk used in cheese manufacture should eliminate this organism in the finished product, post-pasteurization and post-processing contamination is possible and could contribute to potential health hazards. The economic losses due to recall of product, potential liability for loss of life and medical costs, as well as the reputation of the manufacturer, must be considered in a discussion of outbreaks of food-borne illness due to microbial contamination of products.

3   CHEESE/CHEESEMAKING—INFLUENCE ON PATHOGEN CONTROL

3.1   Milk for Cheesemaking

3.1.1   Raw Milk
Bacteria are by far the most important microorganisms present in milk. Prevention of opportunities for contamination and subsequent growth of microorganisms is foremost in the production, handling, transportation and processing of good quality milk. Depending on their presence and the characteristics of their specific activities, some organisms can be either beneficial or detrimental, but most are considered to be part of the normal flora of milk. Several factors must be considered in the assessment of the ideal conditions for the production of good quality milk. Environmental conditions such as dust from sweepings, silage, manure and bedding materials may contribute to the microbial load found in or on the udder of the cow. The lactiferous ducts of the udder often harbour bacteria that are secreted into the milk. Reports indicate that bacterial populations are often high at the onset of milking and decrease gradually as milking progresses.[6] The normal flora of milk from a healthy cow usually consists of micrococci, streptococci and lactobacilli. Organisms responsible for mastitis are present in milk.

Other sources of contamination of raw milk are the exterior of the udder of the cow which may be soiled with bedding materials and manure, the coat of the cow which may harbour microorganisms from stagnant pools of water, manure and air, utensils used in the milking process and handling by workers, inadequate sanitization of machines and permanent pipelines. Raw milk must be cooled adequately to prevent or minimize the growth of organisms present in the milk.

Although viruses, yeasts and moulds are also present in milk, the families of bacteria of most concern are the *Lactobacteriaceae, Micrococcaceae, Pseudomonadaceae, Bacillaceae, Enterobacteriaceae* and *Achromobacteriaceae*.[6] Milk used for the manufacture of cheese must be from healthy animals as well as of good bacteriological quality. Milk quality is usually judged by the number of microorganisms present as determined by any one of several available methods. It must be free of inhibitory substances such as residual antibiotics used in mastitis control and must be supportive of active growth of acid-producing microorganisms. Lactic acid bacteria are necessary for the production of cheeses and fermented milks but some species may be undesirable since they can be responsible for the discolouration in Cheddar cheese and souring of market milk and cream.[6]

3.1.2 Pasteurized Milk

As was indicated earlier, the objective of the pasteurization of milk is to destroy non-sporeforming pathogenic bacteria that may be present in the milk. The heat treatments used to pasteurize milk for cheesemaking may have an effect on the quality of the resultant cheese. Prior to the pasteurization of milk for cheese-making, the process was markedly influenced by the native flora of the milk. If lactic acid bacteria were present, acid development was faster than anticipated and resulted in high-acid cheese. If coliform or other lactose-fermenting organisms were present, gaseous cheese or cheese with an undesirable unclean flavour resulted. Cheese made from unpasteurized milk is more likely to serve as a vector for food-borne illness.

Wilster[27] described the advantages of using pasteurized milk for cheesemaking as: (1) better control of flavour development; (2) destruction of pathogenic bacteria that might be in the milk; (3) destruction of coliform bacteria; (4) better control of acid production during manufacture; (5) more uniform cheese on a day-to-day basis; (6) cheese can be ripened at a higher temperature; (7) a slight increase in yield; and (8) better financial return as high quality cheese is produced with less second grade or undergrade cheese. The disadvantages might include: (1) a slight increase in cost; (2) cheese made from pasteurized milk ripens more slowly; (3) flavour development in pasteurized milk cheese is slower and the cheese never develops the same flavour as that made from raw milk. The advantages of making cheese from pasteurized milk strongly outweigh the disadvantages. The cheesemaking process is more easily controlled when pasteurized milk is used. In the large cheese factories common in the cheese industry today, where hundreds of thousands of litres of milk may be processed in a single day, it is imperative that milk is pasteurized to maintain control of the cheesemaking process so that high quality cheese that is safe, wholesome and of premium grade is produced.

3.1.3 Heat-Treated Milk

The term 'heat-treatment' is defined as the process of heating milk at time-temperature conditions less rigorous than pasteurization. Also referred to as 'sub-pasteurization' or 'thermization', heat-treatment has no established criteria or standards. Thermization at 63–65°C (145–150°F) for 15–20 s kills psychrotrophic bacteria, without inactivation of enzymes.[15]

According to Johnson et al.,[15] microbiologically safe cheese may, generally, be produced, from heat-treated raw milk. They reported that a heat treatment of 65·0–65·6°C (149–150°F) for 16–18 s will destroy almost all pathogenic microorganisms that pose a potential threat to a safe cheese product. The same authors summarized the results of D'Aoust et al.[28,29] and Farber et al.[30] concerning the effect of heat treatment of raw milk on the survival of pathogens. Heating raw milk inoculated at populations of 10^5/ml in a commercial HTST pasteurizer at 65°C (149°F) for a mean holding time of 17·6 s (minimum 16·2 s) destroyed all strains of *Yersinia enterocolitica, E. coli* 0157:H7, *Campylobacter* sp., and all except one *Salmonella* species. Although *Salmonella senftenberg* is, reportedly, rarely found in cheese, it was inactivated at 69°C (156·2°F). In milk naturally contaminated with *Listeria monocytogenes* at populations of 10^4 organisms/ml, inactivation occurred at 66°C (150·8°F). When milk was inoculated with this bacterium at a level of 10^5/ml, a temperature of 69·0°C (156·2°F), was required for inactivation. It was also reported by Johnson *et al.*[15] that data collected over several hundred production days from a large cheese factory, where the milk for cheesemaking is normally heated at 64·4°C (148–149°F) for 16 s, indicated a reduction of aerobic plate counts from $1·4 \times 10^6$/ml in raw milk to $2·8 \times 10^4$/ml in milk as it entered the vat and a reduction of coliform counts from $1·2 \times 10^5$/ml to <10/ml. Zottola *et al.*[31] studied the heat resistance of 236 strains of *Staphylococcus aureus* isolated from raw milk and Cheddar cheese and demonstrated that a heat treatment of 152°F (67°C) was sufficient to inactivate all isolates. These results, coupled with the others described above, imply that the sub-pasteurization heat treatments frequently used for milk for cheese manufacture will control many of the pathogenic bacteria of concern. Some advantages of the heat-treatment of cheesemilk include better control of the cheesemaking process and a more stable quality. Flavour development during curing is more rapid than in pasteurized milk cheese due to only partial inactivation of indigenous enzymes, microorganisms and other biological components of the raw milk.

3.2 Growth and Survival During Cheesemaking

3.2.1 Effect of Starter Activity and Acid Development on the Growth and/or Survival of Pathogens

The production of cheese depends on the formation of lactic acid by bacteria. Acid development aids curd formation by rennet, contributes to curd syneresis and facilitates the drainage of whey from the cheese mass, is responsible for preventing the growth of undesirable organisms during the make and ripening steps, influences the elasticity of the finished curd and affects the fusion of the curd into a cohesive mass, affects the types of enzymatic changes during ripening and aids in the development of the unique characteristics of the finished cheese. Acid production, with a corresponding drop in pH, is critical to cheesemaking and is necessary to inhibit the growth of undesirable organisms during the process. A failure of the starter culture, or the inability of the starter to produce sufficient acid to inhibit the growth of undesirable microorganisms, is a contributory factor to food-borne illnesses attributed to contaminated cheeses.[6] Tatini *et al.*[32] reported that when the lactic starter was inhibited, *Staphylococcus aureus* could grow and produce detectable amounts of enterotoxin in Cheddar and Colby cheeses. Similar findings were reported by Ibrahim *et al.*[33] and Reiter *et al.*[34] observed that starter inhibition by phage allowed staphylococci to multiply.

3.2.2 Salt

One of the most important effects of salting in cheesemaking is the control of undesirable microorganisms in the finished product. Proteolytic bacteria are very sensitive to the NaCl concentrations found in most cheeses. Ibrahim *et al.*[33] showed that salting the curd at the end of cheddaring was one factor in inhibiting the growth of *S. aureus*. However, Shahamat *et al.*[35] reported that *L. monocytogenes* survived for more than 100 days in cheese with salt-in-moisture concentrations of 10·5–30%. Salt also contributes to the flavour of cheese and aids in the removal of whey from curd, thereby controlling both moisture and acidity.[6]

3.2.3 Moisture or Water Activity (A$_w$)

The shelf life and type of cheese are determined by its water content. Soft cheeses with a high moisture content keep for just a few weeks, semi-soft varieties for a few months, hard cheeses for more than a year, and grating cheeses for indefinite periods. The moisture content of the curd depends on the degree of syneresis which also determines the lactose content of the cheese. The subsequent conversion of lactose to lactic acid determines the acidity of the fresh cheese; consequently, the moisture content of the curd is related directly to its acidity. Water activity influences the storage life of foods, including cheese, since it is a measure of water available to permit the growth of microorganisms.[6]

3.3 Growth and Survival of Bacteria During Ripening

3.3.1 Effect of Cheese Composition

Cheese defects or spoilage that occur during the ripening of cheese may be attributed to microbiological causes, as well as physical and chemical changes resulting from enzymes that are released from autolysed cells that grew during the manufacturing process and microorganisms that grow during the ripening period. Alterations in texture, flavour, body and appearance may result in poor quality cheese due to the growth of undesirable microorganisms. One type of spoilage is evidenced by late gas production that causes cracking or splitting of the cheese and the development of bitter flavours. Microorganisms and certain compounds, such as hydrogen sulphide, produced by the microorganisms during curing may interact with metals or metallic salts or added colouring materials and may cause discolouration of the ripened cheese. Loss of moisture during the ageing of hard cheeses encourages the formation of a protective rind on these cheeses, but the chemical products formed during ripening play only a minor role in the preservation of cheese. Some anaerobes are inhibited by the production of fatty acids, but most cheeses become more susceptible to bacterial spoilage as they age due to increased pH.[36] Ryser & Marth[37,38] reported rapid growth of *Listeria monocytogenes* in Brie and Camembert cheeses during ripening and growth increased as the pH increased. They also reported detecting this organism in Cheddar cheese inoculated with it after 154–434 days of ripening.

3.3.2 Effect of Storage Temperature

Storage of cheeses at refrigeration temperatures does not ensure a pathogen-free product. *Listeria monocytogenes* is not only low-temperature tolerant, but can grow at refrigeration temperatures. Apparently, *L. monocytogenes* gains access to cheese made from pasteurized milk through environmental contamination whereas cheese made from raw milk may contain the organism as a natural contaminant. Therefore, if present in the cheese, particularly soft, high-moisture cheeses, growth may occur during refrigerated storage. Marth & Ryser[39] cited several areas where contamination with *L. monocytogenes* can occur, one of which is the factory environment, including the ripening rooms, handling and storage areas. Dos Santos & Genigeorgis[40] reported that ripening time, as well as strain variation, were significant factors in the survival of *S. aureus* in commercially manufactured Minas cheese. While the numbers of staphylococci increased during manufacture, microbial counts decreased as the pH decreased during the early stages of ripening, e.g. by 90% after three days. Ibrahim *et al.*[33] reported a decrease in *S. aureus* populations during storage of salted and unsalted cheese at 4°C but essentially no effect on the concentration of enterotoxin was detected. However, at 11°C, growth and enterotoxin production by *S. aureus* were observed in salted Cheddar cheese but staphylococcoal populations decreased in unsalted cheese at the same temperature. Storage temperatures are not sufficiently low to prevent the growth of moulds on stored cheeses.[36]

3.3.3 Water Activity

The shelf life of cheese depends on the amount of water available for the growth of microorganisms. According to Frazier & Westhoff,[36] an increase in the moisture content of cured cheeses results in an increased susceptibility to spoilage. Soft cheeses, such as Brie and Limburger, are quite susceptible to spoilage by moulds and other microorganisms while hard cheeses, such as Swiss and Cheddar, are less susceptible. The rind on natural cheeses serves to deter spoilage of the anaerobic interior of these cheeses, but the moisture content is still sufficient to allow some mould growth. Recent work by Cabezas *et al.*[41] and Esteban & Marcos[42] indicates that the water activity of cheese can be determined by cryoscopic estimation. Esteban & Marcos[42] found that a_w is directly proportional to the moisture content of the cheese and inversely to the concentration of NaCl and other low molecular species.

3.3.4 Oxidation-reduction Potential (Eh)

Anaerobic conditions may prevent or minimize the growth or survival of some microorganisms. Aerated cultures of *S. aureus* may produce greater amounts of enterotoxin B than cultures that are held under the same conditions but incubated in an atmosphere of 95:5, nitrogen to carbon dioxide.[43] As cheese ages, particularly hard cheeses, the products of proteolysis and lipolysis may reduce the Eh of cheese. Consequently, anaerobic spore-forming organisms present in the cheese may germinate and grow, causing defects such as bitter, putrid flavours and undesirable gas splits. Moulds, which are aerobic, are often responsible for the spoilage of stored cheeses.

3.3.5 pH

A laboratory study on *S. aureus*-inoculated Cheddar cheese by Koenig & Marth[44] indicated that after pressing, the pH of unsalted Cheddar was lower than that of salted Cheddar. Lactic acid, produced by uninhibited lactic acid bacteria, was responsible for the lower pH values. They stated that the change in pH during manufacture was normal and storage temperature did not affect the pH of the cheese until after the 4th week of storage. At 10°C, the pH increased from 5·19 to 5·68 between weeks 4 and 8 but the pH of cheeses stored at 4°C for the same period remained essentially unchanged at 5·1. The level of starter inoculum appeared not to affect the pH. Rutzinski *et al.*[45] reported that Camembert made from pasteurized milk contaminated with *Hafnia* sp. showed a marked increase in the numbers of this organism during 7 weeks of ripening at 10°C, and that the growth of *Hafnia* sp. during storage was coincidental with an increase in pH. Manufacturers rely on lactic acid bacteria to compete with coliforms if they are present in the milk and expect that sufficient acid will be produced to inhibit or inactivate them. However, some coliforms are quite resistant to acidic conditions. Due to the production of NH_3 from proteins and the metabolism of lactic acid by the *Penicillium* spp., the pH of at least some regions of mould-ripened cheeses may become neutral or even alkaline and become suitable for the growth or survival of coliforms.[46] Soft-ripened cheeses have a moisture content above 50% and ripening of these cheeses involves protein degradation due to mould activity. Knoop & Peters[47] found that regions of these cheeses with elevated pH values (7·0 to 7·3) contained large numbers of coliforms. A report by Petran & Zottola[48] showed that *L. monocytogenes* Scott A was able to grow at pH values from 4·7 to 9·2 and at temperatures from 4 to 45°C. If cheese is contaminated with *L. monocytogenes* and stored at refrigeration temperatures, there is the potential for growth of the organism and the possibility of a public health hazard.

3.3.6 Inhibitors

Nisin is a bacteriocin produced by some subspecies of lactococci which occur naturally in raw milk supplies and inhibits most Gram-positive bacteria, especially spore-formers. The use of nisin in foods is permitted in 49 countries, including the US, the UK and the former USSR.[49] Somers & Taylor[50] reported on the effectiveness of nisin in delaying or preventing growth and toxin formation by inoculated *C. botulinum* strains in processed cheese spreads stored under conditions other than in chill cabinets. The level of nisin required depends on the number of clostridial spores present, the expected or required shelf life and the probable storage temperature; the usual levels of addition are 200 to 500 IU nisin/g.[51] Higher levels are recommended when formulations of processed cheese include added flavours, low fat and high moisture levels. Hirsch *et al.*,[52] who studied the effectiveness of nisin in Swiss cheese, reported that it prevented blowing due to the growth of clostridia. Many other antimicrobial agents occur naturally in foods. Certain enzymes and proteins present in milk have antimicrobial activity but have not as yet been studied extensively for their potential as antimicrobial food additives. Beuchat & Golden[53] discussed at length the role of these naturally occurring antimicrobial agents, including organic acids, fatty acids, plant oils and pigments. Some of these play a role in inhibiting the growth of undesirable microorganisms in cheese.

3.3.7 Other Microorganisms

Competition between different microorganisms depends on many factors. Poor growth of the starter culture can allow the growth of undesirable organisms; the temperature during the manufacturing process will preclude the survival of some microorganisms; pH and salt concentration will allow some organisms to proliferate but others will not survive. The manufacturing process, the type of cheese, the storage temperature, the expected shelf life, the metabolic end products of those organisms present in the cheese all interact and play a role in the safety of a specific cheese product.

4   EXTRINSIC CONTROL FACTORS

4.1 Plant Sanitation

The survival of pathogenic microorganisms in the plant environment is specifically linked to plant sanitation. Surak & Barefoot,[54] who evaluated 357 plants operating under the FDA dairy product safety initiative, reported that nine tested positive for *L. monocytogenes*. Herald & Zottola[55] showed that this organism was able to attach to stainless steel at temperatures of 10 to 35°C and at pH values between 5 and 8. An investigation by Strantz *et al.*[56] showed that cooling waters can be highly contaminated with *L. monocytogenes*. However, in 30% glycol, either alone or supplemented with 0·01% nonfat dry milk, *Listeria* did not survive at 4°C beyond eight weeks.[57] Overdahl & Zottola,[58] who studied the effectiveness of several sanitizers in controlling microbial contamination by cooling water, reported that low concentrations of sanitizers reduced the populations of the test organisms, *Pseudomonas fluorescens, Staphylococcus haemolyticus* and *Bacillus* spp., in a simulated sweet water system by more than 90% in 30 s. At both test temperatures, 25 and 4°C, 25 ppm of chlorine, 20 ppm quaternary ammonium or 12·5 ppm iodine were required to reduce the bacterial populations. Raw materials, unchlorinated water, refrigeration units, conveyors, air handling units, processing and packaging areas are all potential sites for harbouring *Listeria*. Coleman[59] reported that floor drains and floors are probably the primary source of *Listeria* spp. A study by Spurlock & Zottola,[60] using a simulated floor drain environment, showed that *Listeria* could survive changes in pH during expulsion of waste and routine sanitizing procedures. Food plant sanitizers commonly used, such as chlorine, acid anionics, quaternary ammonium compounds and iodophors, must be applied to clean surfaces in order to be effective.[61] It is apparent that effective sanitization procedures should be mandated and complied with to prevent contamination of the product from the plant environment.

4.2 Heat Process

In addition to pasteurization or heat-treatment to which cheesemilk is subjected, other factors must be considered. The heat resistance of some microorganisms and the manufacturing protocol for specific cheeses are important factors in determining the quality and safety of cheese. Those organisms that are not inactivated by the heat processes used in cheese manufacture may survive, resuscitate and grow under accelerated ripening and storage conditions. It is essential that all steps in the manufacturing process, especially time and temperature parameters, be monitored and documented to ensure a safe product.

4.3 Starter Activity

Active starter cultures are critical to the successful manufacture of cheese.[4] Inhibition of lactic acid bacteria by antibiotics will permit undesirable microorganisms to grow and proliferate and result in an unacceptable product and economic loss to the processor.[62] Lactic acid bacteria, especially mesophilic starter organisms, are susceptible to infection by bacteriophage. If this type of starter inhibition occurs, acid production ceases and pathogenic bacteria will grow and an outbreak of illness due to consumption of cheese may result.[63] Previously, lactic starters were propagated in milk to make 'mother' cultures but today selected cultures are commercially available in both liquid and frozen form that should ensure good starter activity and result in an acceptable product.[64]

4.4 Good Manufacturing Practices (GMP)

Control of environmental sources of microbial contamination of cheese is essential. Several schemes have been developed by the cheese industry and by regulatory agencies to monitor and prevent microbial contamination of cheese from environmental sources. In the UK, the Creamery Proprietors Association has developed 'Good Hygienic Practice in the Manufacture of Soft and Fresh Cheeses'. Similarly, the British Milk Marketing Board has produced 'Good Hygienic Practices in the Manufacture of Soft and Fresh Cheeses in Small and Farmed-Based Production Units'. These guidelines were developed to help prevent the contamination of soft cheeses with *L. monocytogenes* during manufacture.[65]

The Codex Alimentarius Commission, through the Joint FAO/WHO Food Standards Programme, has outlined 'Codes of Hygienic Practice' for the manufacture of many cheeses and these too are designed to prevent the possibility of contamination of cheese with pathogenic bacteria during manufacture.[66]

The Commission of the European Communities has developed a proposal for Council Regulation 'Laying down the health rules for the production and placing on the market of raw milk, of milk for the manufacture of milk-based products and of milk-based products'. These proposed regulations address, to some extent, hygienic requirements for the processing of milk and milk products. They do not specifically describe hygienic practices for cheese production, but in a general manner discuss requirements for all dairy-based foods and, consequently, apply to cheese manufacture.[67]

In the US, the Food and Drug Administration uses Chapter 21, Part 110, of 'Current Good Manufacturing Practice in the Manufacturing, Processing, Packing, or Holding Human Food' as a guide in assessing hygienic practices in the manufacture of cheese. This regulation gives, in detail, requirements for producing safe and wholesome foods. Failure to comply with GMPs by a cheese processing firm may result in litigation, product seizure and/or recall. Included in Part 110, Subpart A, are general requirements for personnel, cleanliness and disease control, education, training and supervision of personnel. Subpart B describes requirements for plants and grounds, water supply, sewage disposal, plumbing and toilet facilities, handwashing facilities, rubbish and offal disposal, general maintenance, animal and vermin control, sanitation of equipment and utensils, and handling of cleaned, portable equipment and utensils. Equipment requirements, i.e. design and cleanability, and general procedures are described in Subpart C, while production and process controls are discussed in Subpart E.[68]

It is generally accepted that if a cheese manufacturer complies with all of the tenets of the GMPs, the potential for microbial contamination of the products produced will be minimized. The use of an internal GMP inspection of all processing areas and other facilities of a cheese plant by quality control personnel will benefit the company and help assure the production of safe and wholesome food products.

Simply stated, the basic food law of the US, the 'Pure Food, Drug and Cosmetic Act of 1938, as Amended', prohibits the introduction into interstate commerce of foods that are adulterated or misbranded.[68] The FDA uses the GMPs as a guide when carrying out an inspection and evaluation of a food processing company to determine if the food products are adulterated or misbranded. Failure to comply with GMPs by a food processor may result in adulterated or misbranded products, and consequently, the processor may be liable for litigation.

4.5 HACCP

Hazard Analysis Critical Control Points (HACCP) is a systematic approach to the identification and assessment of microbial hazards and risks associated with a food process. Once identified, points in the process are developed that will control or prevent the hazards and risks. Development of the HACCP system by a cheese manufacturer involves several steps: (1) a commitment by management that safe and wholesome products will be produced; (2) identification of hazards and risks associated with the products and ingredients used to make the products; (3) identification of points in the process that will successfully control the identified hazards and risks; (4) development of mechanisms that will adequately monitor the control point; (5) corrective action in the event of failure of a control point; (6) verification that the HACCP system does in fact control the identified hazards and risks.[69]

In the initial development of the HACCP system, critical control points were developed to control only food safety hazards.[70] A critical control point can be succinctly defined as a point in a process where exerted control will positively control the identified hazard or risk. Such a point could be a microbiological test, recording of a temperature or some other means to assure positive control of the point. In recent years, as more and more companies, individuals and regulatory agencies become involved in the development of HACCP systems, extensive variations on the use of control points have developed. For example, in the 1988 publication on HACCP by ICMSF, discussion centres on the use of CCP1 and CCP2, where CCP1 is considered to be an effective and essential control point and CCP2 is not an absolute control point.[69]

In the HACCP system adopted in 1989 by the US Department of Agriculture, Food Safety and Inspection Service, three different control points (CCP, CP, MCP) are identified and required. A critical control point (CCP) is defined as any point or procedure in a specific food system where loss of control may result in an unacceptable health risk. A control point (CP) is defined as any point in a specific food system where loss of control does not lead to an unacceptable health risk, and a manufacturing control point (MCP) is defined as any measurable point in a process that may result in product of unacceptable quality.[71] The addition of other control points, i.e. CCP2, CP and MCP, confuses the user and detracts from the original intent of the system, i.e. to control food safety hazards.

A simple example of an HACCP system for the manufacture of cheese will be defined below. For the purposes of this discussion, only critical control points, CCPs, will be given. The first requirement in the development of an HACCP system is the identification of hazards and risks in the finished product and the ingredients used. The hazards and risks associated with cheese and the major ingredient, raw milk, are microbiological. The next step is the identification of points in the process at which to control these hazards and risks. There are essentially three critical points: (1) control of the microbiological quality of the raw milk, (2) pasteurization of the raw milk before cheesemaking and, (3) prevention of recontamination after pasteurization of the milk. The third step in establishing an HACCP system is the development of methods to monitor the control points. This can be achieved by using the following methods:

(1) Control of the microbiological quality of the raw milk is ensured by using microbiological standards and specifications related to hygienic practices during production, storage and transport of the raw milk. These are monitored by on-farm inspection of husbandry practices and microbiological

testing of the raw milk to ensure that standards are achieved. The third point in controlling the microbiological quality of raw milk is assurance that the milk has been cooled to less than 7°C (45°F) and maintained at that temperature throughout the process from production on the farm to storage at the processing plant.

(2) Pasteurization of the raw milk is essential to control pathogenic and other undesirable microorganisms that might be present in the milk. Pasteurization of milk is monitored by using proper and approved equipment, recording the temperature at the end of the holding time and using sealed timing pumps to ensure that proper holding times are attained in the holding tube.

(3) Prevention of recontamination of the pasteurized milk and the resultant cheese is critical to the manufacture of safe and wholesome cheese products. Control of recontamination is achieved by following Good Manufacturing Practices or Codes of Hygienic Practices or similar procedures. In the manufacture of many cheeses, rapid acid production by the starter culture is critical to the control of unwanted microorganisms. Prevention of recontamination of the products can be monitored by microbiological testing.

This brief description of a HACCP system for cheese manufacture is not complete. It was presented only to show the systematic manner in which HACCP should be applied. For a complete discussion of HACCP and applications of HACCP principles, the reader is referred to *Microorganisms in Foods. (4) Application of Hazard Analysis Critical Control Point Systems to ensure Microbiological Safety and Quality*, published by ICMSF.[69]

The proper use of Good Manufacturing Practices and Codes of Hygienic Practices, coupled with a systematic approach to ensuring food safety inherent in an HACCP programme, should result in cheese and cheese products that are safe, wholesome, nutritious and of value to the consumer.

5 REGULATION OF THE CHEESE INDUSTRY TO CONTROL PATHOGENIC MICROORGANISMS

5.1 The US

5.1.1 FDA

The major regulations affecting the manufacture of the many varieties of cheese produced throughout the world are directed primarily at the composition of the cheese, i.e. fat and moisture content, the methods of manufacture for the particular cheese type and permitted additives. In the US, CFR21, Part 133, presents in considerable detail the compositional requirements, permitted additives and the commonly used methods for the manufacture of 73 different cheese varieties. No mention is made of microbiological standards or specifications for the cheeses.[68]

If cheese or cheese products of any type are found by FDA inspection to be contaminated with pathogenic bacteria, the food is considered to be adulterated. The company involved is prosecuted under the general provisions of the US Food, Drug and Cosmetic Act of 1938, as Amended.[68] In instances where the violation occurs in a State that has applicable laws, the FDA may elect to have the offending company prosecuted under the State law and will request such action by the State. When this occurs, the FDA will assist the appropriate State agency in developing the litigation.

5.1.2 Grade A Products

The US Food and Drug Administration is also responsible for enforcing the Grade A Pasteurized Milk Ordinance.[7] This ordinance contains specific microbiological standards for products produced under the Regulation. If a company wishes to produce a Grade A cheese product, the product, the raw milk and the manufacturing plant must meet the standards set forth in the Ordinance.[7]

5.1.3 USDA

The US Department of Agriculture is another US agency that has some regulatory activity with the cheese industry, but in this instance, the activity is voluntary rather than compulsory. The USDA has the specific mission to promote the orderly marketing of wholesome, high-quality agricultural products. In the case of dairy products, the programme is administered by the Agricultural Marketing Service (AMS) and is carried out under the Agricultural Marketing Act of 1946. Qualification for the USDA inspection and grading service is dependent upon inspection and approval of processing plants (the USDA, AMS publish a list of approved plants annually), processing equipment and procedures approval, and the product must meet prescribed finished product standards for grades. These standards and grades include microbiological criteria. Qualification and approval allow the use of USDA Grades on products. Failure to meet these standards and grades results only in removal of the Grade designation from the products. If products are found to be in violation of the Food, Drug and Cosmetic Act of 1938, as Amended, enforcement of the violation is up to the Food and Drug Administration.[72]

The USDA Agricultural Marketing Service has developed and published microbiological standards for raw milk to be used for manufactured dairy products. These microbiological standards are used by many State agencies in the enforcement of their manufacturing milk programmes. In addition, the USDA AMS has developed microbiological standards for many manufactured dairy products, but none have application to cheese. The microbiological concerns of this agency are related more to the raw milk than to the finished products.[72]

5.2 European Community

The European Community (EC) is interested in establishing microbiological criteria for foods and wishes to do so as there will be a single common market for foods throughout the community from 1993. The Commission of European Communities has proposed council regulation for the 'Production and placing on the market of raw milk, of milk for the manufacture of milk-based products and of milk-based products'. Included in this proposed regulation are microbiological specifications for soft and fresh cheeses. These include: no *Listeria monocytogenes* or *Salmonella* in 25 g of sample and limits on the numbers of coliform organisms. Adoption of these suggested regulations by the EC should improve the microbiological quality of the cheese produced and sold in the Common Market.[67]

5.3 Codex Alimentarius

Codex Alimentarius, Division 12, General Standards for Cheese, gives fat and moisture parameters, and necessary and optional additions for 35 different cheese types. No microbiological standards or specifications for the 35 cheese types are given.[73] As was noted earlier, Codex Alimentarius has developed a Code of Practice on General Principles of Food Hygiene and several Codes of Hygienic Practice have also been developed, but none to date have been promulgated for the manufacture of cheese.[72] When such codes are developed, microbiological standards are usually included. Microbiological specifications for dried dairy products, caseins and caseinates have been developed, but none is applicable to the cheese industry.

It appears from this brief review of the regulations designed to control pathogens in cheese that specific criteria directed at pathogens are missing. Regulatory control is directed primarily at controlling the microbiological quality of the raw milk used in the manufacture of cheese, whereas concerns for the microbiological quality of the cheeses produced are dealt with in other ways, such as adulterated food. It is questionable whether the development of additional microbiological specifications to control the microbiological quality of cheese will have a marked effect on cheese quality or result in the reduced occurrence of outbreaks of disease caused by the consumption of cheese. Mechanisms are in place to deal with contaminated cheese and additional standards or specifications will only add more confusion to the already complex control system.

REFERENCES

1. Gravani, R.B., 1986. In *Professional Perspectives*, Division of Nutritional Sciences, Cornell University, Ithaca, NY.
2. Zottola, E.A. & Smith, L.B., 1990. *J. Food Safety*, **11**, 13.
3. Johnson, E.A., Nelson, J.H. & Johnson, M., 1990. *J. Food Prot.*, **53**, 519.
4. Zottola, E.A. & Smith, L.B., 1991. *Food Microbiology*, **8**, 171.
5. Campbell, J.R. & Marshall, R.T. (eds) 1975. *The Science of Providing Milk for Man.* McGraw-Hill Inc., NY.
6. Foster, E.M., Nelson, F.E., Speck, M.L., Doetsch, R.N. & Olson, J.C. (eds), 1957. *Dairy Microbiology*, Prentice-Hall Inc., Englewood Cliffs, NJ.

7. Grade A Pasteurized Milk Ordinance, Revision. US Public Health Service/Food and Drug Administration, Dept. Health and Human Services, Washington, DC, 1989.
8. Chandan, R.C. & Shahani, K.M., 1957. *J. Dairy Sci.*, **40**, 418.
9. Anon., 1949. Federal Register, April 22, pp. 1960–1992.
10. Anon., 1949. *Natl. Butter and Cheese J.*, **40**(7), 55.
11. Vaughn, A.C., 1884. Public Health Papers and Reports. *Am. Publ. Health Assn.*, **10**, 241.
12. Anon., 1947. Federal Security Agency. Exhibit 54, dated May 14, 1947.
13. Anon., 1947. Federal Security Agency. Exhibit 55, dated May 14, 1947.
14. Fabian, F.W., 1947. *Am. J. Public Health*, **37**, 987.
15. Johnson, E.A., Nelson, J.H. & Johnson, M., 1990. *J. Food Prot.*, **53**, 441.
16. Johnson, E.A., Nelson, J.H. & Johnson, M., 1990. *J. Food Prot.*, **53**, 610.
17. Anon., 1990. Acute Disease Epidemiology Section, Minnesota Department of Health, January 1990.
18. James, S.M., Fannin, S.L., Agree, B.A., Hall, B., Parker, E., Vogt, J., Run, G., Williams, J., Lieb, L., Salminen, C., Pendergast, J., Werner, S.B. & Chin, J., 1985. *Morbid. Mortal. Weekly Report*, **34**, 357.
19. Kvenberg, J.E., 1988. *Microbiol. Sci.*, **5**, 355.
20. Bille, J.M. & Glauser, P., 1988. *Bull. Off. Fed. Sante Publ.*, **3**, 28.
21. Gilbert, R.J., Hall, S.M. & Taylor, A.B., 1989. *PHLS Microbiol. Digest.*, **6**, 34.
22. Pini, P. & Gilbert, R.J., 1988. *Intern. J. Food Microbiol.*, **6**, 317.
23. Anon., 1985. FDA Enforcement Rep., September 4, 1985.
24. Anon., 1986. *Food Chem. News*, February 17, 1986.
25. Griffiths, M.W., 1989. *J. Sci. Food Agric.*, **47**, 133.
26. Cantoni, C., d'Aubert, S. & Valenti, M., 1989. *Industrie Alimentari.*, **XXVIII**, 1068.
27. Wilster, G.H., 1977. *Practical Cheesemaking*, 12th edn, DSU Bookstores Inc., Corvallis, OR.
28. D'Aoust, J.-Y., Emmons, D.W., McKellar, R., Timbers, G.E., Todd, E.C.D., Sewell, A.M. & Warburton, D.W., 1987. *J. Food Prot.*, **50**, 494.
29. D'Aoust, J.-Y., Park, C.E., Szabo, R.A., Todd, E.C.D., Emmons, D.B. & McKellar, R., 1988. *J. Dairy Sci.*, **71**, 3230.
30. Farber, J.M., Sanders, G.W., Speirs, J.I., D'Aoust, J.-Y., Emmons, D.B. & McKellar, R., 1988. *Intern. J. Food Microbiol.*, **7**, 277.
31. Zottola, E.A., Jezeski, J.J. & Al-Dulaimi, A.N., 1969. *J. Dairy Sci.*, **52**, 1707.
32. Tatini, S.R., Jezeski, J.J., Morris, H.A., Olson, J.R. Jr. & Casman, E.P., 1971. *J. Dairy Sci.*, **54**, 815.
33. Ibrahim, G.F., Radford, D.R., Baldock, A.K. & Ireland, L.B., 1981. *J. Food Prot.*, **44**, 189.
34. Reiter, B., Fewins, B.G., Fryer, T.F. & Sharpe, M.E., 1964. *J. Dairy Res.*, **31**, 261.
35. Shahamat, M., Seaman, A. & Woodbine, M., 1980. *Zentralbl. Bacteriol. Hyg. Abt. I. Orig. A.*, **246**, 506.
36. Frazier, W.C. & Westhoff, D.C., 1988. *Food Microbiology*, 4th edn. McGraw-Hill Inc., New York, NY.
37. Ryser, E.T. & Marth, E.H., 1987. *J. Food Prot.*, **50**, 372.
38. Ryser, E.T. & Marth, E.H., 1987. *J. Food Prot.*, **50**, 7.
39. Marth, E.H. & Ryser, E.T., 1990. In *Foodborne Listeriosis*, ed. A.J. Miller, J.L. Smith & G.A. Somkuti, Soc. Indus. Microbiol., p. 151.
40. Dos Santos, E.C. & Genigeorgis, C., 1981. *J. Food Prot.*, **44**, 172.
41. Cabezas, L., Marcos, A., Esteban, M.A., Fernandez-Salguero, J. & Alcala, M., 1988. *Food Chem.*, **30**, 59.
42. Esteban, M.A. & Marcos, A., 1989. *J. Dairy Res.*, **56**, 665.
43. Carpenter, D.F. & Silverman, G.J., 1974. *Appl. Microbiol.*, **28**, 628.
44. Koenig, S. & Marth, E.H., 1982. *J. Food Prot.*, **45**, 996.
45. Rutzinski, J.L., Marth, E.H. & Olson, N.F., 1979. *J. Food Prot.*, **42**, 790.

46. Keogh, B.P., 1971. *J. Dairy Res.*, **38**, 91.
47. Knoop, A.M. & Peters, K.H., 1971. *Milchwissenschaft*, **26**, 193.
48. Petran, R.L. & Zottola, E.A., 1989. *J. Food Sci.*, **54**, 458.
49. Delves-Broughton, J., 1990. *J. Soc Dairy Technol.*, **43**, 73.
50. Somers, E.B. & Taylor, S.L., 1987. *J. Food Prot.*, **50**, 842.
51. Fowler, G.G., 1979. *Food Manufacture*, **54**, 57.
52. Hirsch, A., Grinsted, A., Chapman, H.R. & Mattick, A.T.R., 1951. *J. Dairy Res.*, **18**, 205.
53. Beuchat, L.R. & Golden, D.A., 1989. *Food Technol.*, **43**, 134.
54. Surak, J.G. & Barefoot, S.F., 1987. *Vet. Hum. Toxicol.*, **29**, 247.
55. Herald, P.J. & Zottola, E.A., 1987. *J. Food Prot.*, **59**, 894.
56. Strantz, A.A., Zottola, E.A., Petran, R.L., Overdahl, B.J. & Smith, L.B., 1989. *J. Food Prot.*, **52**, 799.
57. Petran, R. & Zottola, E.A., 1988. *J. Food Prot.*, **51**, 172.
58. Overdahl, B.J. & Zottola, E.A., 1991. *J. Food Prot.*, **54**, 305.
59. Coleman, W.W., 1986. *Dairy Food Sanit.*, **6**, 555.
60. Spurlock, A.T. & Zottola, E.A., 1991. *J. Food Prot.*, **54**, 910.
61. Anon., 1988. *Dairy Food Sanit.*, **8**, 52.
62. Jezeski, J.J., Morris, H.A., Zottola, E.A., George, E. Jr. & Busta, F.F., 1961. *J. Dairy Sci.*, **44**, 1160.
63. Lawrence, R.C., Thomas, T.D. & Terzaghi, B.E., 1976. *J. Dairy Res.*, **43**, 141.
64. Price, W.V., 1971. *J. Milk Food Technol.*, **34**, 329.
65. McLauchlin, J., Greenwood, M.H. & Pini, P.N., 1990. *Intern. J. Food Microbiol.*, **10**, 255.
66. International Commission on Microbiological Specifications for Foods (ICMSF) revised, 1985. *Microorganisms in Foods. 2.* University of Toronto Press, Toronto, Canada, p. 157.
67. Commission of the European Communities, 1990. COM (89) 667 Final. Brussels, Belgium.
68. Code of Federal Regulations, 1984. 21, Parts 100–169. Office of Federal Register, US Government Printing Office, Washington, DC.
69. International Commission on Microbiological Specifications of Foods (ICMSF), 1988. *Microorganisms in Foods. 4.* Blackwell Scientific Publishers, Oxford, UK.
70. Committtee on Salmonella. 1969. *An Evaluation of the Salmonella Problem.* Natl. Academy Sci., Washington, DC, p. 108.
71. National Advisory Committee on Microbiological Criteria for Foods, 1989. HACCP, US Department of Agriculture, Food Safety and Inspection Serv., Washington, DC.
72. Subcommittee on Microbiological Criteria for Foods and Food Ingredients, 1985. *An Evaluation of the Role of Microbiological Criteria for Foods and Food Ingredients.* National Academy Press, Washington, DC.
73. Codex Alimentarius Commission, Abridged Version, 1990. Joint FAO/WHO Food Standards Programme, Rome, Italy.

13

Application of Membrane Separation Technology to Cheese Production

V.V. Mistry

Dairy Science Department, South Dakota State University,
Brookings, SD 57007, USA

&

J.-L. Maubois

Laboratoire de Recherches Laitières, Institut National de la Recherche
Agronomique, Rennes Cedex, France

1 INTRODUCTION

It has been said that 'a revolution in cheesemaking as a result of ultrafiltration is coming soon'.[1] This revolution has been in the making for the past 25 years and has encompassed not only ultrafiltration but, more recently, microfiltration as well. Indeed, more than 400 000 tonnes of cheese were made using ultrafiltration technology in 1989.[2] The history of cheesemaking using membranes commenced in the late 1960s with the invention of the MMV process.[3-5] This process, named after its inventors (Maubois, Macquot and Vassel) opened up new avenues for significant advances in cheesemaking, including improvements in plant efficiencies, increases in cheese yield, development of continuous process, and possibilities of creating new cheese varieties. As a result, numerous plants all over the world, but mainly Europe, now use this process to manufacture a wide range of cheeses.[6]

Since the introduction of the MMV process, commercial applications of membranes in the cheese industry as well as research efforts aimed at developing new applications and understanding and improving current applications have expanded all over the world. It was reported, for instance,[7] that during the period 1979–1983, a total of 213 scientific papers were published dealing with membrane separations in food processing. Of these, publications dealing with cheese formed the largest category at 25%. Publications dealing with cheese and whey combined represented 50% of the total. In a more recent literature search, it was found that more than 1000 publications on the application of membranes in food processing appeared between 1984 and 1990. These data clearly illustrate the magnitude of effort that has been invested in developing and understanding applications of membranes in the food industry.

Since 1969, cheese-related applications of membranes have expanded into numerous areas, including the manufacture of fresh, soft, semi-hard and hard cheeses from milks of cows, goats, ewes, and water buffaloes; production of milk powders with good cheesemaking properties;[8] restoration of the rennet coagulation properties of UHT treated milk;[9,10] on-farm concentration of milk;[11] removal of bacteria from cheese milk by microfiltration[12] and casein enrichment of cheese milk by microfiltration.[13] These developments were catalysed by improvements in membrane components such as the development of mineral and ceramic membranes; by studies on physico-chemical equilibria of UF retentates; by characterization of the rheological behaviour of protein-enriched milks; by studies on the growth and activity of cheese starters in liquid pre-cheeses and in the resulting cheeses; and more importantly by the generation of new ideas and the acceptance of new cheesemaking concepts in laboratories and in cheese plants around the world. In this chapter, cheesemaking using ultrafiltration, reverse osmosis and microfiltration will be discussed as well as other cheese-related applications using these processes. It would be appropriate first to define some membrane terms and discuss in brief, membrane design and configuration.

2   MEMBRANE DESIGN AND CONFIGURATION

Membrane technology is a broad term that encompasses several molecular separation processes. Each process requires its own specialized equipment and has its own characteristics that make it suitable for some applications but not for others. To date, two of the most commonly used membrane processes in the dairy industry have been reverse osmosis and ultrafiltration.[14,15] Another process, microfiltration, has emerged in the dairy industry over the past few years and shows tremendous potential for the future.[2,16] Yet another process, nanofiltration or loose reverse osmosis, has recently been introduced to the dairy industry and is now being used for whey processing in some cheese plants.[17–19]

2.1 Definitions

2.1.1 Ultrafiltration (UF)
Ultrafiltration is a process which selectively separates macromolecules having molecular weights of 1000 to 200 000 daltons (Fig. 1) from solvent and dissolved solutes. With cross-flow over a membrane surface at relatively low pressures (less than 1000 kPa), UF produces from milk a permeate (also called ultrafiltrate) containing water, lactose, soluble minerals, non-protein nitrogen and water-soluble vitamins and a retentate in which proteins, fat and colloidal salts content increase in proportion to the amount of permeate removed.[3,4,14]

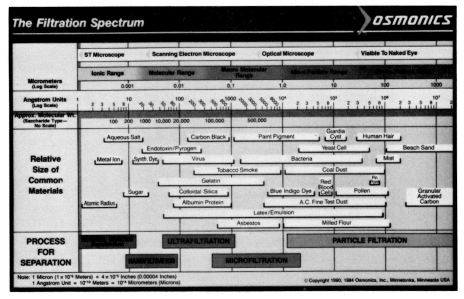

Fig. 1. Filtration spectrum showing the range of molecular separation. (Courtesy of Osmonics, Inc., Minnetonka, MN, USA.)

2.1.2 Reverse Osmosis (RO)

Reverse osmosis (hyperfiltration) is basically a dewatering process which operates at pressures at least five to ten times higher than those employed for UF.[14,20,21] Reverse osmosis membranes separate solutes with a molecular weight of approximately 150 daltons. Hence, fat, proteins, lactose and all undissociated minerals are retained and concentrated by the membrane and only water and some ionized minerals are allowed to pass through.

2.1.3 Microfiltration (MF)

Microfiltration is a process which selectively separates particles with molecular weights of greater than 200 000 daltons (Fig. 1). According to the membrane pore size, milk materials removed by MF include whey components,[12] β-casein,[22] β-lactoglobulin[23] and skim milk components.[2,11,12,24]

2.1.4 Nanofiltration (NF)

Nanofiltration, also known as loose RO,[25] falls between RO and UF. It removes particles of molecular weight less than 300–1000 daltons and retains the rest. Small molecules, such as NaCl, are removed along with water, whereas other material such as lactose, proteins and fat are retained, making it suitable for desalting cheese whey.

Fig. 2. Commercial Tubular Membrane System. (Courtesy of PCI Membrane Systems.)

2.2 Membrane Configuration

As the membrane industry expanded, a wide range of membrane configurations became available. Four basic configurations currently available for UF, RO, MF and NF applications are: (1) tubular, (2) hollow fibre, (3) plate and frame, and (4) spiral-wound.[26,27] Each of these configurations has its own advantages and disadvantages.

2.2.1 Tubular

In this configuration (Fig. 2), feed flows through a tube 85–600 cm long and 3–25 mm inside diameter. The inside wall of the tube is lined with the membrane, the outside consists of support material. Several tubes may be connected in series or in parallel as a bundle and are housed in a stainless steel casein. Recently, multi-channel geometry was developed in France[28] and in the USA[29] for MF and UF mineral membranes.

Tubular membranes are easy to clean and allow recirculation of liquids with a high level of solids and viscosity.[26,27] However, they have the lowest surface area-to-volume ratio and therefore require a high feed flow rate and consequently, a high running energy (0.6–1.0 kW/m^2).[26] Reverse osmosis operations are conducted at high pressures; hence tubular membranes for RO require additional support material to withstand the high pressures.

2.2.2 Hollow Fibre

Hollow fibre membranes (Fig. 3) can be thought of as the tubular-type except that they are self supporting. Hollow fibres also have a much smaller diameter than the tubular-type membranes. The diameter of each fibre ranges from 0.19 to 1.25 mm.[27] From 50 to 3000 such fibres may be bundled together in parallel

Fig. 3. Multi-stage commercial hollow fibre UF system. (Courtesy of Alfa-Laval Food and Dairy Group, Inc., Pleasant Prairie, WI, USA.)

in a see-through casing. Each such unit is referred to as a cartridge. Hollow fibre membranes have the highest surface area-to-volume ratio, providing for very low floor space requirements.

As in the tubular design, feed flows through the inside of the fibres and permeate is collected outside in the casing. A disadvantage with this system is that even if only one fibre fails, the entire cartridge must be replaced. Replacement costs of membranes are, therefore, high. On the other hand, since hollow fibres are self-supporting, operating pressures are low.[26] Trans-membrane pressure is limited to 170–270 kPa. While this is an advantage in terms of energy consumption (0·2 kW/m^2), this configuration may not be suitable for applications requiring high pressures. One of the greatest advantages of hollow fibre membranes as bundled tubular membranes is the ability to backflush. This aids in cleaning the membrane as well as in preventing a build-up of fouling material on the surface.

Hollow fibres such as those described above are used for UF and MF applications. For RO applications, even smaller fibres, known as hollow fine fibres, are used.[27] In these fibres, feed flows from the outside of the fibre to the inside.

2.2.3 Plate and Frame
This configuration consists of a stack of plates and flat sheet membranes, much like a filter press arrangement (Fig. 4). The flat sheet membrane and its support are sandwiched together in large numbers to form a module. Feed flows parallel to the membrane surface and permeate is channelled out of the module. Plate

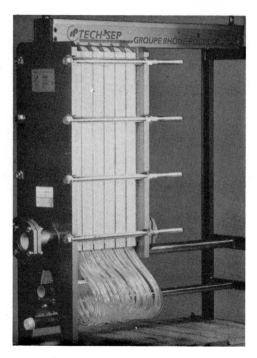

Fig. 4. Commercial plate and frame UF system. (Courtesy of Tech-Sep, S.A. Rhône-Poulenc, St. Maurice, France.)

and frame configurations are available in horizontal as well as vertical designs. The surface area-to-volume ratio is between hollow fibre and tubular designs. The required pumping energy is around 0·5 to 0·7 kW/m².[26]

2.2.4 Spiral-wound
This configuration is widely used in the dairy and food industries and is also the most inexpensive (Fig. 5). Spiral-wound membranes consist of two flat sheet membranes along with spacers wrapped around a perforated permeate-collecting tube.[27] As feed passes over the membrane surface, permeate spirals its way to the centre of the tube. Spacers are included in the assembly to promote turbulence, thereby minimizing fouling, but they can cause cleaning difficulties when highly viscous retentates are recirculated. Spiral-wound membranes are available for UF, RO, MF and NF applications. The nature of the membrane support and the general design permits operation at high trans-membrane pressures without damaging the membrane.

Good quality membrane material is critical for the proper operation of UF, RO, MF or NF plants. Cellulose acetate was the most common material for UF and RO membranes but these have now been almost completely replaced by polysulphone membranes, especially for UF applications. Numerous other materials have been tested, e.g. polyamide, polyimide, polyvinylidene fluoride, etc.

Fig. 5. Multi-stage spiral wound UF system. (Courtesy of Koch Membrane Systems, Inc., Wilmington, MA, USA.)

Mineral membranes, specifically zirconium oxide, and ceramic membranes are now being used increasingly for UF and MF. These materials have high mechanical strength, and tolerate wider pH and temperature ranges than polymeric membranes. They are more expensive but have a substantially longer life (at least five years compared to 18 months).

The above is only an overview of membrane processes used in cheese applications. Other membrane processes such as dialysis and electrodialysis and details of UF, RO, MF and NF, such as flux rates, thermodynamics of operation, fouling and concentration polarization have been discussed in depth elsewhere[27] and will not be addressed here.

3 MEMBRANE APPLICATIONS IN CHEESEMAKING

Ultrafiltration is currently the most widely used membrane process for cheesemaking and is fairly well advanced. Microfiltration for removal of bacteria or for enrichment of micellar casein is just beginning to enter industrial operations.

Before attempting to make cheese by UF or MF, specific properties of the protein-enriched products must be well understood because they strongly determine the quality of the end-products, as well as economy of the use of the membrane technology.

3.1 Properties of UF Retentates

3.1.1 Buffering Capacity of UF Retentates
If milk is ultrafiltrated at its normal pH (6·7), mineral salts (Ca, Mg, P) bound to casein micelles are concentrated in the same proportion as proteins. This results in an increase in the buffering capacity of UF retentates which will consequently modify the basic parameters of the cheesemaking process: acidification kinetics by lactic acid bacteria, ultimate pH value, rennet coagulation kinetics

and rheological characteristics of the curd, activity of ripening enzymes, growth and rate of survival of spoilage flora and water-holding capacity of the cheese mass during ripening.

According to the volumetric concentration factor (F) (ratio of milk to retentate volumes), higher lactic acid production by lactic starter bacteria is required to obtain optimal pH in cheese: usually 5·2 in hard cheeses and 4·6 in soft and fresh cheeses. For this latter category, the increase in required lactic acid production was quantified by Brulé et al.[30] and was expressed as:

$$Q_L = 4·4 \, F + 1·5$$

where Q_L is expressed in g of lactic acid per kg of pH 6·7 UF retentate.

Consequently, for most cheese varieties, use of pH 6·7 retentates results in acid-tasting products.[31] On the other hand, a large quantity of calcium salts is released into the aqueous phase of cheese curd during acidification. Ionic strength is strongly increased and casein micelle aggregation is modified. Cheese texture is crumbly or sandy[32] and spreadability and stretching properties are poor.[33] The buffering effect of pH 6·7 UF retentates leads to higher numbers of lactic starter bacteria in curd and resulting cheeses than in non-UF curd and cheeses.[34,35] While this may result in bitterness in some cheeses such as Quarg,[36] it has been used to advantage in the development of a new bulk starter.[37] This starter is manufactured by fermenting a whole or skim milk retentate containing 12% protein with a mesophilic lactic culture at 22°C for 12–15 h (Fig. 6). It has a built-in internal pH control mechanism due to the buffering capacity of the UF retentate which maintains the pH of the starter steady at 5–5·2. It, therefore, has a greater activity than traditional bulk starters made from pasteurized milk and maintains its activity for 10–12 h at room temperature. Additionally, the UF retentate starter

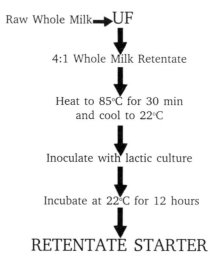

Fig. 6. Flow diagram for the production of retentate starter. (Developed from Ref. 37.)

also has a high protein content which provides for increased cheese yield,[38,39] making it very suitable for either traditional or UF cheesemaking.

The increased buffering capacity of pH 6·7 UF milk also offers a favourable environment for the growth and survival of certain bacteria such as entero-pathogenic *E. coli*.[40] This underscores the importance of adjusting the mineral content of UF retentates to avoid the unfavourable consequences in cheese-making due to the increase in buffering capacity of pH 6·7 UF milk. This aspect was emphasized as early as 1974 by Brulé *et al.*[30] who suggested several suitable ways for adjusting the mineral salts content of UF retentates, which is specific for each cheese variety. The first method involves a reduction in milk pH before or during ultrafiltration by the growth of lactic starters or by any ap-proved acidifying agent (glucono-δ-lactone or organic acids in some countries). Acidification leads to solubilization of colloidal calcium and magnesium phos-phate salts which pass into the permeate. Reduction of milk pH from 6·6 to 6·0 and 5·6 increases the Ca content of UF permeate from 0·38 to 0·50 and 0·80 g per kg. Consequently, a 5× UF retentate obtained at pH 5·6 has a Ca content 2·6 times that in milk instead of 3·8 times for the 5× UF retentate obtained at pH 6·6.[30] The second method, which eventually can be combined with the first one, is the addition of NaCl (0·5–0·9%) to UF retentate during or after ultra-filtration. The increase of ionic strength resulting from NaCl addition reduces the ionization of casein phosphoseryl groups and consequently leads to solu-bilization of colloidal calcium in the permeate or in the aqueous phase of UF retentate (up to 15–18% depending on the pH and amount of NaCl added).[30] An increase of ionic strength also lowers the isoelectric pH of casein which may offer an increased security margin for handling acidified UF retentate by the cheesemaker.

Addition of NaCl to milk or reduction of the pH by acidification reduces UF fluxes because of increased membrane fouling but it is obvious that any cheese-maker will prefer to have satisfactory cheeses even at the expense of reduced performance of UF equipment rather than defective cheeses resulting from a process involving the highest UF fluxes. It must also be remembered that milk having a pH lower than 5·0, when ultrafiltered, leads to a higher UF flux than that observed at pH 6·7 because of the weak texture of the polarization layer at the isoelectric pH of casein.[41]

3.1.2 Rheological Behaviour of UF Retentates

Milk can be considered as a Newtonian liquid, while UF retentates behave differently. The higher the protein content and/or the lower the temperature, the more pseudoplastic is their behaviour.[42] Such a rheological behaviour must be taken into consideration in the design and in the operating parameters of UF equipment (for example, in the restarting procedure after an electrical failure to avoid an hydraulic ram).

The viscosity of UF retentates increases markedly with an increase in their protein content. At 30°C, at a shear rate of $437·4 \, s^{-1}$, the observed viscosity is 45 cP for a protein content of 19·6% and 370 cP for a protein content of

20·6%.[42,43] The manufacture of semi-hard or related cheese from these highly viscous UF retentates requires removal of all dissolved gases which are entrapped in the liquid and the use of special mixing devices (static and dynamic) to enable thorough blending of rennet and lactic starters.[16] If dissolved gases are not removed by application of a vacuum, a spongy curd is obtained and the appearance and taste of the cheese are poor. If rennet is not mixed satisfactorily, the resulting curd will be flaky due to localized coagulation.

3.1.3 Rennet Coagulation

If the same amount of rennet is added to equal volumes of milk or UF retentate, the rennet clotting time is not affected by the increase in protein content (%P) but the time from clotting to cutting is reduced.[4,31,44–46] This is the net result of numerous phenomena: there is an increase in the velocity of the enzyme reaction as the protein content is increased[46] but the degree of proteolysis at gelation decreases as %P increases. At pH 6·6 and at the normal casein content of milk, coagulation occurs when 80 to 90% of the κ-casein has been hydrolysed. However, in a 4× UF retentate, hydrolysis of only 50% is necessary for curd formation.[47] Because the secondary phase of rennet action is a diffusion-controlled process, an increase in protein content leads to a sharp increase in the rate of aggregation.[46] The final firmness of rennet curd is generally assumed to be directly related to the casein content.[4–10] This is of particular importance when low-concentrated retentates (LCR) are used to make cheeses because traditional equipment is employed. This would require stronger knives and agitators to handle the firmer, stronger curd.[48]

If the primary phase of κ-casein hydrolysis by rennet is slightly affected in UHT milk,[10] coagulation does not occur owing to the increased electronegativity of the casein micelles resulting from the covalent binding of β-lactoglobulin with κ-casein.[49] Increasing the protein content by UF before or after UHT treatment restores curd forming ability.[9] According to Ferron et al.,[10] such a phenomenon would result from lowering the zeta potential of casein micelles on UF. This hypothesis, which must be confirmed by direct observations, agrees with the fact that UF retentates coagulate at a lower degree of κ-casein hydrolysis than normal milk.

3.2 Applications of UF in Cheesemaking

Cheesemaking using UF can currently be divided into three main categories.[1,2,16] These are: (1) use of protein-standardized cheese milk, (2) intermediate or medium concentrated retentates, and (3) liquid pre-cheeses,[3,4] i.e. UF retentates having the composition of the cheese variety to be made. Cheesemaking using each of the above categories will be discussed, starting with the use of protein-standardized cheese milk.

3.2.1 Use of Protein-standardized Cheese Milk

The protein content of milk collected by dairy plants varies according to season due to multiple factors: stage of lactation, weather, feeding and breed of lactating cows. Such a variation in the composition of the incoming milk requires adjustment of processing parameters by cheesemakers. Moreover, at a low protein content rennet curds are weak and lead to relatively high losses of caseins as fines in whey. A slight increase in the protein content by UF eliminates these difficulties. In many cheese plants, generally those using highly mechanized equipment, the protein content of cheese milk is increased to 3·7–4·5% throughout the year.[6,50] Protein-standardized milk is used in Europe for the manufacture of Camembert cheeses[6] using Alpma coagulator or similar equipment. It is also used for semi-hard and hard cooked cheese. In the USA, the acronym LCR (low-concentrated retentates) was proposed[1] to characterize this use of ultrafiltration in cheesemaking. Several pilot plant and industrial studies have reported on the use of the LCR concept for Cheddar cheese using either direct concentration or supplementation.[51–55] These studies concluded that the optimum degree of concentration for Cheddar cheesemaking is between 1·7:1 and 1·8:1. In the LCR or protein standardization process, cheese is made using conventional equipment and a cheese plant can easily adapt this application of UF. Manufacturers of UF equipment have now proposed specially designed ultrafiltration systems that are equipped with in-line protein and fat sensors.[56] This will make it possible to determine that fat and protein content of the incoming milk and to standardize the cheese milk for fat and protein simultaneously.

The cost of UF for this application is balanced by a slight increase in manufacturing efficiency due to increased production of cheese per vat per day, reduced rennet requirements, improved quality of marginal cheese[1] and a slight increase in yield (generally less than 1% for most varieties). This yield increase results from reduced losses of fat and casein particles in whey and better retention of whey proteins in the aqueous phase of cheese. The effect of this on cheese yield can be estimated according to the formula proposed by Vandeweghe & Maubois.[57] Another advantage is the possible added value of the resulting whey which has an increased content of protein/total solids. However, it must be said that in industrial situations these advantages are minor. Therefore, it is somewhat surprising that a large number of UF plants have been installed for protein-standardization in Europe during the last few years. An indirect but important advantage of the LCR/protein-standardization concept is the utilization of permeate to reduce the protein content of fluid UHT milk to the minimum required by law: 2·8% in most EC countries but 3·0% in France.[58] Such a practice has led to considerable profit for many UHT milk processors. The pay-back of UF investment is less than six months. While it is possible to detect dilution of milk with water, it is impossible to detect dilution with permeate.

A number of cheese varieties have been made using the LCR concept. These include cottage,[59–63] Mozzarella,[64,65] Saint Paulin[66] and others. Industrial and pilot-plant trials with cottage cheese indicate that 1·2–1·7:1 concentration ratios are optimum for yield, flavour and body characteristics. Above these levels, the texture becomes firm and the cheese has a flat flavour. Thermization of milk (74°C for 10 s) prior to UF gives the highest increase in cottage cheese yield compared with thermization after UF or no thermization.[63]

Good quality low-moisture Mozzarella cheese with excellent stretching and melting properties can be produced from low concentration retentates at 1·75:1·0 ratios.[64,65] Cheese from higher concentrations was firmer and had greater fat losses in the brine. Using LCR it is possible to produce both starter-acidified and directly-acidified Mozzarella.

The LCR concept has also been applied to Brick and Colby cheeses.[67] LCR Brick cheese had a lower pH and higher fat losses in whey than in controls. The cheese was firmer and mealy and scored lower in overall preference than control cheeses. For Colby, use of UF made it possible to eliminate the curd washing step. Sensory scores were similar to those of controls. Reduction in cooking time and rennet usage were reported. In studies with Edam cheese,[68] 2:1, 4:1 and 6:1 retentates were used. LCR (2:1 concentration) produced the best cheese with fewest defects. In this cheese, the rate of proteolysis of α_{s1}-casein was similar to that in control cheese but that of β-casein was slower. With UF Danbo cheese, made from 2:1 diafiltered UF milk, slight increases in yield, reduction in rennet requirements by 50%, and a 40% increase in the cheesemaking capacity of vats was possible.[69]

3.2.2 Use of Medium or Intermediate Concentrated Retentates

Numerous cheese varieties, ranging from soft to hard, have been made from medium-concentrated retentates. In this approach, cheese is made by using specially designed equipment able to cut and to handle firm curd resulting from the coagulation of 2:1 to 5:1 concentrated retentates, eventually diafiltrated with pure, salted or acidified water. The main applications, which are in industrial operation, are the manufacture of UF Cheddar cheese according to the APV-SiroCurd process[70,71] and the production of structured Feta cheese.[72]

APV-SiroCurd Process. This process was developed by an Australian dairy research team in collaboration with the APV firm.[71] A commercial plant has been in operation in Cobram, Australia, for over four years, and another was recently started in Perham, Minnesota, by Land O'Lakes.[73] The Minnesota facility is reportedly capable of handling approximately 800 000 kg milk per day in a plant spread over 12 000 m². The APV-SiroCurd process (Fig. 7) is a continuous process utilizing a rennet coagulator and rotating curd tumbler drums for syneresis. Both these devices replace conventional cheese vats. Milk is concentrated by UF to 40–45% solids at 50–55°C. The calcium and lactose content of milk are adjusted by diafiltration. A small portion of the retentate is fermented with lactic starters to pH 5·4. This fermented retentate is used as a bulk starter for cheesemaking and is dosed at 10–12% into the larger portion of the retentate. At pH 6·3–6·5, temperature is adjusted to 30–35°C and rennet is added. The mixture is pumped into a rennet coagulator drum consisting of six barrels, each more that 6 m long. A gel forms in 14–18 min and, as it comes out of the barrels, it is cut into cubes and transferred to large rotating drums where it is heated to 38°C. The combined retention time in the two drums is 48–52 min. After the proper pH is attained, curds are transferred to a continuous draining, matting and cheddaring machine where they mat in about 110 min. Thereafter, the curd

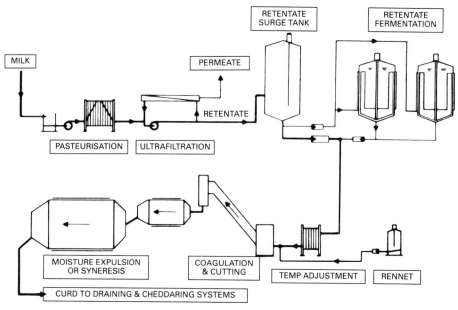

Fig. 7. The APV-SiroCurd process for Cheddar cheese (Ref. 73).

is milled and salted, pressed and packed. Increases in cheese yield of 6–8% are realized with this process and cheese of uniform composition and quality is obtained. The process is continuous and completely automated. About 100 installations of this process are anticipated over the next 10–15 years.[70]

Structured Feta cheese. This approach to making Feta cheese was developed in Denmark in response to consumer demand from many Mediterranean countries where people desired cheeses having the appearance and texture (presence of mechanical holes) similar to those of traditional products, characteristics they did not find in UF Feta cheese made from liquid pre-cheeses.[36] Pasteurized, fat-standardized milk, generally homogenized at 18 MPa and 60°C, is ultrafiltered at 50°C. The final concentrate contains 28·5% TS, which corresponds to a concentration factor of 3:1. Lipase, starter cultures or glucono-δ-lactone are added to the UF retentate, previously homogenized at 5 MPa at 65°C, heat-treated to 80°C for 60 s and cooled to 34°C. After a short storage period in a tank, the UF retentate is pumped to specially designed Alfa-Laval Alcurd or Pasilac equipment. Rennet is added in line. In both types of equipment, rennet is thoroughly mixed and UF retentates coagulate in tubes. The resulting curd is removed from the tubes, cut into cubes, moulded and drained (16 to 24 h at 10–14°C) until the pH has decreased to 4·8.[74] The product of this process is virtually indistinguishable from the traditional product. A yield increase of about 14% on a solids basis is claimed over the traditional process,[36] a far smaller value than that obtained with the process using liquid pre-cheese (30%). Such a difference explains why only a few plants produce structured Feta cheese.

Other cheeses produced from medium concentration retentates. Several experiments on the use of UF retentates of up to 5:1 concentration have been reported for making Havarti, a semi-soft cheese of Danish origin, containing approximately 26% fat and 56% solids.[75-78] Cheese milk used in these experiments was not pre-acidified and diafiltration was not performed with acidified water. Consequently, the buffering capacity of UF retentates was high and it was reported that more starter was required than with the traditional process or the use of specially selected cultures.[74] However, the taste and flavour of UF cheeses were similar to the traditional product. The texture was, nevertheless, softer and the melting properties poorer. A 10% saving in skim milk cheese manufacture was claimed resulting in a net profit of US $42 000 per year for a production of 600 tonnes of cheese.[74]

In experiments with Gouda cheese,[79,80] whole milk was first ultrafiltrated to 3·3:1 and then diafiltered to 3·6:1–5:1 concentration. Gouda from 5:1 retentate produced cheese similar in moisture, hardness and proteolysis to controls produced from non-UF milk. Savings of 33% in rennet costs were reported. Flavour development in UF Gouda cheese could be accelerated by using a combination of liposome-entrapped enzyme and freeze-shocked *Lactobacillus helveticus* cells.

Attempts have also been made to manufacture Blue cheese from UF milk.[31,81-84] According to French studies,[81] the organoleptic qualities of cheeses made from UF retentates with a protein content ranging from 3·2 to 10% and treated in traditional cheese vats were similar to reference cheeses. Above 12% protein, modifications of cheesemaking parameters and new cutting and handling equipment were required to produce satisfactory blue cheeses. Egyptian studies describe the use of recombined ultrafiltered milks for making blue cheese.[82-84]

General considerations on the use of intermediate UF retentates. Benefits accruing from the use of intermediate UF retentates for making any cheese variety must be substantial enough to justify substitution of traditional cheesemaking technology. Moreover, the organoleptic quality must also be acceptable to the consumer. Investments involve not only UF equipment, as in the LCR concept, but also additional equipment, such as curd makers. Increased cheese yield is strongly related to the volume concentration factor (F) and to the difference between the composition of the UF retentate and the final cheese.[85] The saving of skim milk increases logarithmically as the difference becomes smaller but UF operating cost also increases logarithmically with F. The economic study to be made by cheese plants to assess investment must also take into consideration the potential value of the two by-products obtained: drained whey which contains more protein and fat than normal whey, and permeate; each is of interest to a different downstream industrial network. Minor benefits also result from reduced rennet consumption and reduced requirement in volume and floor space.

3.2.3 Use of Liquid pre-cheeses (LPC concept)

In this approach, cheese milk is concentrated by UF to the composition of the drained curd being made, before addition of rennet. There is minimal whey drainage, and there is no need for cheese vats.[3,4] This principal was first applied to Camembert cheese[3,31] but many applications have been successfully developed for the manufacture of other cheese varieties, ranging from 'fromages frais' or Quarg to semi-hard cheeses such as Saint Paulin.

Fresh unripened cheeses. In early attempts to apply UF for the manufacture of cheese varieties belonging to this category, milk was preconcentrated prior to starter and rennet addition. The cheese produced had a highly acid and metallic taste, frequently associated with bitterness. These defects were attributable to the high mineral content of the curd and consequently its high buffering capacity.[32,41,86] Some reduction of acid flavour was observed when pre-acidified (pH 6·0) milk was ultrafiltered or when milk was concentrated to a higher degree than necessary and subsequently diluted with water.[32] Introduction of new membranes, such as mineral membranes and specially designed membrane supports that permit the UF of high-viscosity products,[11,87] has made it possible to solve this organoleptic defect completely by using the process initially proposed by Stenne,[88] i.e. first fermenting the milk to pH 4·6 with conventional mesophilic cultures, adding rennet and then ultrafiltering to remove lactose and mineral salts but retaining whey proteins. Unusual initial flux rates and decrease in flux rates with concentration were observed when pH 4·6 milk was ultrafiltered,[41] both phenomena being attributable to the highly porous structure of the polarization layer.[11] Because of the relatively high protein content (12%), this use of UF was successfully developed for the manufacture of Quarg,[89-95] a German cheese variety. For application to similar French cheeses which contain much less protein, it was necessary to develop specially designed UF equipment which minimized the mechanical shear stress applied to the retentate (Fig. 8). The viscosity of pH 4·6-acidified curds decreases markedly with the increase of mechanical treatment imposed during centrifugal drainage or UF concentration.[96]

Many other fresh unripened cheese varieties are now made according to the LPC approach or the MMV process. Some examples include Saint-Maure goats' milk cheese,[97] Ricotta,[98,99] Cream cheese[100,101] and Mascarpone.[102,103]

The manufacture of Ricotta presents special problems because of the complexity of precipitation, and requirements for suitable texture and flavour.[98,104] In one UF process,[98] whole milk is acidified to pH 5·9 with lactic starter, acid whey powder or food-grade acid, and ultrafiltered at 55–60°C to 12% protein. The acidified liquid pre-cheese is heated in a scraped-surface heat exchanger at 80°C and filled directly into packages. In another process,[99] milk or whey is pasteurized, acidified to pH 6·3, and ultrafiltered to 30% solids at 50°C. The retentate is heated to 90°C at a pressure of 1–1·5 bar, following which the pressure is reduced to atmospheric to aid curd formation. The product is cooled to 70°C, packaged and chilled to 10°C. No whey drainage occurs.

Another interesting application of UF for the manufacture of cheese varieties belonging to this high moisture category is the procedure developed for the production of 'Faisselles' or country cheese. Traditionally, this cheese is made from whole milk curds inoculated at 18–22°C with mesophilic starters and rennet.

Fig. 8. Commercial carbosep UF system for the production of fresh cheese from pH 4·6 milk. (Courtesy of Tech-Sep, S.A. Rhône Poulenc, St. Maurice, France.)

After overnight cooling to 12–16°C, pieces of curd are scooped by hand into moulds for slight whey drainage. The drained curd is then removed, always by hand, from the moulds and gently laid down in the retail cups. This production was disappearing because of the increasing cost of labour. Use of UF has allowed the draining step to be eliminated and consequently labour requirements are reduced substantially. The production of this cheese variety has now reached its former maximum level.[105]

Soft cheese. Camembert, a French surface-moulded cheese variety, was the first to be made according to the MMV principle.[3,4] Several recipes were proposed[5,11,31] to optimize the use of UF with industrial cheesemaking constraints (24 h production of UF retentate with 16 moulding working hours) and to obtain the very delicate equilibrium in calcium salts in the curd required to get texture and flavour similar to traditional cheeses. The procedure used with the continuous moulding and demoulding equipment, 'Camatic' (Fig. 9), developed by Alfa-Laval,[6,106,107] is the following: HTST pasteurized milk is ultrafiltered at 50°C to a pre-cheese concentration of 5:1 and a total solids content of 35%. The resulting pre-cheese is cooled to 30°C and 2% mesophilic lactic starter and 0·75% NaCl added. Then, the mixture is allowed to acidify to pH 5·5 and is automatically filled into forms with on-line inoculation with rennet. Curd wheels develop rapidly and are continuously and gently moved in the Camatic equipment for

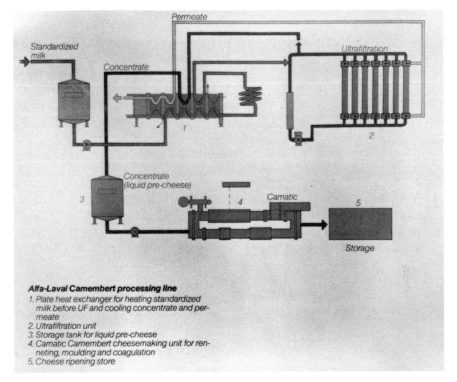

Fig. 9. The Camatic method for producing Camembert cheese by the MMV process. (Courtesy of Alfa-Laval Food and Dairy Group, Inc., Pleasant Prairie, WI, USA.)

45 min. After being turned over once, a continuous electric current is applied to each cheese between the air-exposed surface in contact with a carbon electrode and the stainless steel cup holding the cheese. Limited electrolysis of whey occurs and the use of an air injector allows perfect demoulding of the wheels onto cheese trays. Then, the fresh Camembert cheeses are brined for about 30 min, removed, sprayed with *Penicillium candidum* spores and held for 12 days at 11–12°C to permit development of the white mould covering. A yield increase between 12 and 15% is obtained. Several units of Camatic equipment have been sold, mainly in Germany. In France, UF Camembert cheeses have encountered 'psychocommercial' difficulties. The organoleptic qualities of UF Camembert were indistinguishable from those of traditional cheeses but the bulk density of the UF cheese paste is much higher than that of traditional cheese because there are no mechanical openings. Since French consumers are accustomed to buying Camembert cheese by the piece and not by weight, they are conscious of the volume of this cheese variety and get the impression that they receive less cheese for their money when buying UF Camembert.[16,69]

The commercial failure of UF Camembert has led French cheesemakers to develop new varieties. Most of them have achieved a very rapidly growing production such as 'Pavé d'Affinois' developed in 1982 and which reached a production of 2500 tonnes in 1990. This cheese is made from 4:1 UF whole milk retentate fermented with thermophilic lactic starters and rennet, then poured into rectangular plastic trays, 5 cm high. The trays are set in an incubator at 43°C for 6 h to allow acidification and coagulation to occur. After cooling to room temperature, the cheese slabs are removed from the trays and cut with an automatic dividing knife in 96 pieces, each having the size and form of a small rectangular paving stone (approximately 7×5 cm). The resulting fresh cheeses are then ripened for 10 days, as for Camembert.[6]

The greatest success world-wide of the MMV process is unquestionably the manufacture of Feta cheese.[36,86,108,109] It is believed that today Feta accounts for 35% of the cheese produced in Denmark and more than 90% of it is produced by UF. The LPC approach for making Feta has made feasible an old dream of cheesemakers: to make the cheese in its retail package. Yield increases of 30% were reported, higher than could be expected from the retention in the cheese of whey proteins (22% at most). The difference must be related to the total elimination of curd particle losses arising from the coagulation and curd cutting inside the retail tins. The same concept was followed for the manufacture of Domiati, an Egyptian cheese variety: 5:1 UF whole milk retentates were homogenized, 5% NaCl, 2% lactic starter and lipase-rennet mixture were added prior to pouring into 18 kg tins[110] or Tetra pack packages.

UF processes for Mozzarella have been reported since the mid-1970s.[111,112] In one of the first attempts to use the MMV principle for Mozzarella cheese,[111] retentates were adjusted to 33·6% solids with freeze-dried retentates, and then blended with 69% fat cream to 45–50% solids. This mixture was fermented and rennet curd was obtained. It was concluded that diafiltration was required to produce good flavour, stretch and melting characteristics. In a subsequent continuous process for low-moisture Mozzarella,[1,102] pasteurized skim milk was pre-acidified to pH 6·0 to remove calcium. It was then ultrafiltered/diafiltered at 54°C to approximately 10:1 concentration. The retentate was blended with cream to obtain 20% fat, 28% protein, and then dosed with starter and rennet. Coagulation occurred continuously, followed by conventional stretching and moulding at pH 5·2. Some problems encountered with this process included poor stretching characteristics of the cheese. Lack of proper stretching of UF Mozzarella cheese may be attributed to the incorporation of large quantities of whey proteins and their denaturation during cooking, improper calcium ratios,[86,113] inadequate removal of dissolved gases and incomplete blending of rennet.[16]

Semi-hard cheeses. Saint Paulin is a bacterial surface-ripened semi-hard cheese of French origin and contains approximately 47% moisture and 2·5% salt.[104] In the manufacture of this cheese by UF, it is necessary to obtain at least 21% protein (45% solids).[11] This is more easily attainable with mineral membranes[43] rather than with polymeric membranes. Procedures have been developed for both brine-salted and added-salt cheese. Increases in cheese yield up to 19% may be realized, with 85% savings in rennet.[43] Acid flavour and slow ripening of UF Saint Paulin cheese can be controlled by reducing the lactose and ash content of the retentate to less than 1·9%.[114] The flavour of UF Saint Paulin may be improved by adding lysozyme at 0·5–1·0 g/litre[35] which increases the proteolytic count and reduces the mesophilic count of the cheese.

A new cheese variety, with propionic bacteria fermentation, has been studied in France. The procedure includes the preparation of 7·5:1 retentate in two steps: first with continuous diafiltration at 3·0:1 concentration regulated by a refractometric sensor inserted in the permeate line, followed by heat treatment at 4·0:1 concentration, and second, with continuous ultrafiltration to 7·5:1 using specially designed equipment (short cartridges and positive displacement recirculation pumps) for handling highly viscous products. Original mixing devices for starters and rennet addition was used. The moulding equipment includes a vacuum step for removing dissolved gases and a special injection head for pouring renneted LPC into two-part spherical or cylindrical moulds.[16,115]

Other applications of the LPC concept. An original approach of UF in processed cheesemaking was pioneered by Jolly & Kosikowski[116] who proposed substitution of aged cheeses by UF skim milk retentate previously incubated with blue mould spores. Some interesting results were reported in 1979[117] for replacement of Cheddar cheese: fully acceptable processed cheese was obtained by substituting 40% of aged Cheddar cheese by enzyme-treated UF retentates containing up to 30% solids. On the other hand, cheese base made by UF can satisfactorily replace the young cheese component in the manufacture of conventional processed cheese. An Australian process[118] was supposedly commercialized in Arizona.[1] In this process, whole milk or whole milk acidified to pH 5·7 was ultrafiltered to 40% of its original weight and then diafiltered to 20% of its original weight. The product was fermented with a lactic starter for 16 h at 30°C and then vacuum evaporated to 64% solids. A similar process was developed in Denmark.[119]

By means of ultrafiltration and drying, a pre-cheese powder can be produced for subsequent reconstitution and conversion into cheese.[8,14] The primary use is for export to countries with low milk production or where the milk supply is very seasonal. In the importing country, the user needs only to add water, starter and rennet to make cheese.[120,121] Such powders could also be used in dairy countries for home cheesemaking.[8,122] Pre-cheese powders offer many advantages for both exporting and importing countries: cheesemaking abilities are better than those of even low-heat normal milk powders,[8,122–124] cheesemakers in the importing country have no whey problem, economy of production is favourable for both countries since both spray drying and transport costs are cheaper than those for normal milk powder.[120] However, this approach of UF has found very few applications, mainly because of the regulations of dairy exporting countries, such as the USA and the EC which subsidize the export of milk solids regardless of the amount of liquid milk used to make one kg of these milk solids.[120]

3.2.4 Cheese Quality

Texture. Although UF cheeses offer moderate to significant yield benefits and have been well accepted by consumers, they do possess some inherent characteristics that make them unique with respect to composition, ripening characteristics and texture qualities. It has even been suggested[86,125] that separate standards of identity for UF cheeses would be advisable, and that a new range of cheese varieties should be developed rather than duplicating traditional varieties.

One of the most notable characteristics of UF cheese is the incorporation of whey proteins in the cheese. The quantity of whey proteins retained depends on the variety and on the degree of UF concentration. If all the whey proteins of milk are retained, they will represent approximately 20% of the total protein in the cheese. Lower quantities will be retained when the LCR method is used. Part of the casein is replaced by whey proteins which act as an inert filler and may soften the cheese.[126] On the other hand, the water-binding capacity of whey proteins is much higher than those of casein and UF cheeses are less susceptible to drying during retailing than traditional cheeses.

Proteolysis and ripening characteristics. It has been commonly observed that UF cheese ripens more slowly than traditional cheese.[125-130] Generally, the larger the amount of whey proteins incorporated, the slower the flavour development. Large variations in the flavour quality of UF cheese have also been observed and these have been attributed to the varying contents of immunoglobulin and proteose-peptone in the whey proteins.[125] The effect of whey proteins on flavour development is less pronounced in LCR cheeses due to the smaller quantities of whey proteins present but is more significant in cheeses made from higher retentate concentrations and those that are ripened for long periods.

The retarded maturation could be due to several reasons. The high content of β-lactoglobulin in UF cheeses could inhibit to some extent the general proteolytic activities of rennet[127] and plasmin.[131] Undenatured whey proteins found in UF cheeses are resistant to proteolysis by these proteases as well as starter-derived enzymes. The high buffering capacity of UF cheeses prepared from pH 6·7 ultrafiltered milk retards the rate of lactic starter autolysis[35] and consequently hydrolysis of the casein network. The rate of α_{s1}-casein breakdown, as well as of β-casein, has been found to be decreased in UF cheese[127] (Fig. 10). The rate of flavour development of UF cheese may be improved by the addition of flavour-producing enzymes[80] which will be totally retained in the retentate contrary to what happens in traditional cheesemaking where 80 to 90% of the added enzyme is lost in the whey. Non-starter bacteria or slow-acidifying lactic microorganisms may be used for their proteolytic and flavour production potential because UF cheesemaking permits separate management of acidification, drainage and ripening flora.

The consumer is the final judge of cheese quality and the success of any new cheese or traditional cheese made with new technology, such as membrane separations, will depend to a large extent on acceptance by consumers.

percentage α_{s1}-casein

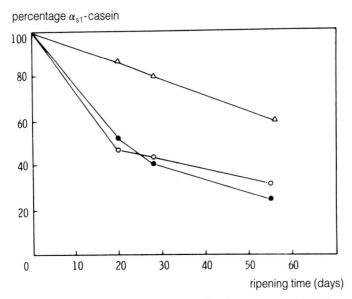

Fig. 10. Degradation of α_{s1}-casein during cheese ripening. ● = standard cheese, ○ = UF cheese with normal rennet concentration, △ = UF cheese with half normal rennet concentration (Ref. 126).

3.3 Reverse Osmosis in Cheesemaking

The use of evaporated milk in cheesemaking was proposed by Richardson[132] as early as 1929. This idea was revived some 30 years later by Stenne,[133] but within the framework of an original combination with the observations of Berridge[134] on the separation of the primary and secondary phases of rennet action at low temperatures. The same basic idea lies behind increasing the solids content of cheese milk by adding dried milk or by concentration using reverse osmosis.

For Cheddar cheesemaking, reverse osmosis was proposed for preconcentrating whole milk to 20 to 25% solids.[135–139] Cheesemaking is conducted in traditional equipment, and the gross composition of the resulting cheese is identical to that of cheese from unconcentrated milk. The amount of starter and rennet required are reduced by 50 and 60%, respectively,[135] and with a 20% milk volume reduction by RO, a 2–3% increase in cheese yield can be expected.[136] However, fat losses in the whey increase with increasing solids concentration due to a partial homogenization effect during processing. A sudden release of pressure during RO can induce lipolysis in the milk and cheese. A 15% volume reduction, representing 1·18 : 1 concentration, has been reported to be optimum.[140] For Cottage cheese, a 5% increase in yield can be realized with an 8% skim milk volume reduction by RO.[141] Such yield increases result from the entrapment of concentrated whey within the network formed by calcium paracaseinate in the cheese. Depending on the degree of concentration, the same consequences for cheese quality result from a greater mineral retention as described for UF retentates. In RO cheese, there is a high concentration of residual lactose which may lead to a resumption of lactic acid fermentation after several days in the ripening room, when sufficient lactic acid has been consumed by ripening micro-organisms. This is almost always very detrimental to organoleptic qualities.[142]

Reverse osmosis is widely used for processing whey but it is doubtful whether it will find widespread acceptance in cheesemaking. However, there are several plants in the USA that already use thermal evaporators for preconcentrating cheese milk.[143,144] Excess lactose and minerals in the curd are removed by washing with water. Benefits of the overall process must therefore be examined closely.

3.4 On-farm Concentration

Interest in on-farm concentration of milk started in 1974 in France,[11] and in 1977 the Alfa-Laval company developed an on-farm UF processing unit.[145] It was believed at that time that on-farm concentration of milk by UF would reduce milk transportation costs due to reduction in milk volume. Increases in cheese yield were also anticipated. At four French farms, milk was concentrated to 2:1 and then delivered to a cheese plant for the manufacture of Emmental and St Paulin cheeses.[146] Permeate produced at the farms was fed to cows, resulting in savings in feed. The French on-farm operation utilized thermization of the 2:1 UF milk prior to cooling and delivery. The microbiology of on-farm UF milk was favourable.[147] The economics of on-farm ultrafiltration were studied in the US by Slack et al.[148,149] using milk concentrated 2:1 or 3:1. Economic advantages were possible when the on-farm concentration concept was used for farms with 100–1000 cows. In a year-long study in California, a 900-cow herd was used to study the feasibility of on-farm UF.[150] Retentate from this farm was delivered to cheese plants for cheesemaking. The regulatory aspect of this operation was not fully resolved.[151] Use of reverse osmosis for concentration of milk on the farm has been evaluated in Australia.[152,153]

While on-farm concentration of milk showed much promise in its early years, experience and evaluation in three countries over the past 15 years indicate that before this technology can have widespread use, problems such as regulatory considerations, reduction of investment and membrane replacement costs, and safety of farm UF equipment need to be resolved. The anticipated cost savings have not been realized. It has been stated recently[151] that the cheese industry is not ready for this technology. However, UF thermization equipment has been installed recently on several Wisconsin farms and a new generation of equipment called 'Thermicon' TM was developed in 1988 by Alfa-Laval.[154]

Another approach to treating milk by UF and thermization at the producer level was studied in France.[155] During two years, milk collected from 22 farms located on an island south of Brittany, was concentrated 2:1 every second day. The retentate was HTST pasteurized and cooled to 2°C. Pooled retentates were shipped to a dairy plant twice a week. The mesophilic flora of the UF retentate was on average 7700 cfu/ml. A net benefit of 0·0482 FF/litre of milk was achieved.[155]

3.5 Applications of Microfiltration in Cheesemaking

Microfiltration, curiously often referred to as crossflow microfiltration whereas the terms 'crossflow UF' and 'crossflow RO' are never used, is a relatively new processing technique in the dairy industry. Introduction really started with the development of the mineral generation of MF membranes made from alumina[28] or from zirconium oxide supported by carbon.[156] In 1990, the total area installed in the world dairy industry was less than 750 m²,[15] but promises are very great, as indicated by numerous studies which project a potential market six to seven times higher than that for UF. In dairying, MF applications have gained increasing attention because of the wide range of available pore size, which makes it possible to separate and fractionate milk.

3.5.1 Microbial Epuration of Raw Milk by MF

Decontamination of raw milk is generally achieved through heat treatment. Various combinations of time–temperature treatments can be used, depending on the desired bactericidal effect. While heat treatment is necessary to ensure the safety of milk and milk products, it almost always induces irreversible modifications of milk components, physicochemical equilibria are disrupted, and the organoleptic quality and cheesemaking properties are also adversely affected.

As with bactofugation, MF allows the heat treatment for decontaminating milk to be minimized, but MF appears to be more efficient than bactofugation. Removal of bacteria contained in milk by MF was first suggested by Holm *et al.*[157,158] Initially, permeation and retention of bacteria were very satisfactory, but serious fouling of the MF membrane occurred rapidly. To overcome this, a new hydraulic concept, developed by Alfa-Laval[12] in 1974, could be applied because of the development of a new MF ceramic membrane with a highly permeable structure and a multichannel geometry.[28] It includes a recirculation loop of microfiltrate which permits a constant low trans-membrane pressure all along the MF tubular membrane in spite of a high retentate recirculation velocity (7 m/s). Commercialized equipment employing this concept for removal of bacteria named 'Bactocatch', is used as follows:[2,157] raw skim milk is continuously microfiltered using 1·4 μm pore size membranes at temperatures ranging between 35°C and 50°C. Retentate flow extracted from the corresponding loop generally represents 5% of the entering milk flow but it can be reduced to 3%. This retentate, which contains the milk bacteria, may be used for animal feed when raw milk cheesemaking is envisaged, or it may be continuously blended with cream for fat standardization. A moderate UHT treatment (115–120°C for 3 s) may be applied to the cream-retentate mixture, which, after cooling, is incorporated in the microfiltrate (Fig. 11).

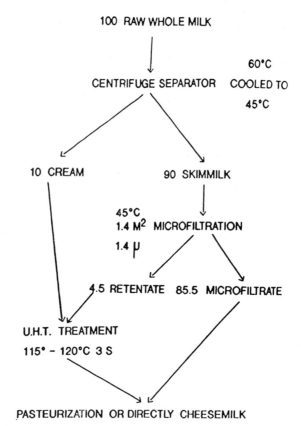

Fig. 11. Flow digram for the treatment of raw milk by the Bactocatch process (Ref. 2).

MF fluxes ranging from 500 to 700 litre/h^{-1}.m^{-2} are obtained for 10 h, with an average bacterial removal of 99·6%[24,159-162] regardless of the initial count in the raw skim milk. This means that MF is not acting as a screen filter but as a depth filter.[24] Morphology of bacterial cells and cellular volume slightly influence membrane retention. High retention rates (greater than 99·98%) observed for spore-forming bacteria, such as *Bacillus cereus*[161] or *Clostridium tyrobutyricum*,[24] are likely due to binding of bacterial spores to a part of the cell wall, consequently resulting in an apparent larger cell size. Recent observations made on the retention of *Listeria* during the Bactocatch process, showing a decimal reduction varying between 1·3 and 2·3 depending on whether milk was previously micro-filtered or not, indicate that this bacterial species may be attached to somatic cells which are totally retained by the MF membrane.[2]

While treatment of milk by the Bactocatch process will tremendously improve the hygienic quality and shelf-life of manufactured dairy products,[2,159,160,163] it will also increase the problem of how to make good quality cheeses from ultra-clean cheese milk, often described by cheesemakers as 'dead milk'. Extensive research must be conducted to understand fully the nature, role and development of ripening starters in these ultra-clean cheese milks. It appears from French studies,[2] that each cheese variety requires an independent study. For example, satisfactory distribution of eyes in Emmental cheese made from Bactocatch-treated milk requires incorporation of specific heterolactic strains along with mesophilic, thermophilic and propionic starters added to this milk at the beginning of the cheesemaking process.[2]

3.5.2 Casein Enrichment of Cheese Milk by UF

Microfiltration of skim milk with $0.2~\mu m$ pore size membranes allows a selective separation of micellar casein, i.e. native calcium phosphocaseinate.[13] Depending on the amount of extracted microfiltrate (the composition of which is similar to that of an 'ideal whey'), the casein content of the retentate increases. As soon as commercial production of casein-enriched cheese milk becomes a reality, applications such as those with UF will become possible. The first one will likely be casein-standardized cheese milk for which the acronym LCC (low casein concentrate) has already been proposed.[2] Medium and fully concentrated casein retentate concepts as in UF can also be envisioned. Such future approaches will depend on a dual strategy followed by the cheese plant: (1) to obtain the highest possible yield for cheese with the required organoleptic and functional qualities, and (2) to maximize the value of the whey protein-containing microfiltrate. For example, use of MF may be a preferred alternative to the use of medium concentrated UF retentate for Mozzarella cheesemaking. Addition of spray dried micellar casein powder to cheese milk is another approach which can be envisaged in the future for improving cheesemaking.

3.5.3 Modifications of α_s/β Casein Ratio by MF

Most of the β-casein is entrapped in the micelle structure by hydrophobic binding. On cooling milk or caseinate suspension to a temperature lower than 5°C, β-casein is solubilized. This soluble β-casein can be separated by $0.2~\mu m$ pore size MF membranes[22] or by 100 000 daltons cut-off UF membranes.[164] Both retentates have a casein content with an increased α_s/β-casein ratio and the microfiltrate is a solution of almost pure β-casein.[22] Such a separation process may allow in the future, if the economics are favourable, cheesemaking from milk with variable α_s/β-casein content.[22,165] Existing knowledge on cheese made from goats' or ewes' milk and on the role played by the degradation of each individual casein in the development of cheese flavour allows one to think that a large range of new cheese varieties might be possible from β-casein-adjusted milks.

4 CONCLUDING REMARKS

Membrane technologies have, during the last 20 years, opened new avenues for improving traditional cheesemaking procedures and consequently increased not only the overall quality of a number of cheese varieties but also net profit resulting from this transformation of milk. They have also allowed the survival of cheese varieties requiring unacceptable and tedious manual labour by traditional process. In addition, they have at least led to the creation of new cheeses in response to constant consumer demand.

On the other hand, because of difficulties encountered by all those trying to make cheeses with satisfactory organoleptic qualities through membrane technologies, an impressive amount of knowledge has been acquired in numerous fields of dairy science: protein biochemistry and physicochemistry, interrelationships between protein and minerals, dairy microbiology, rheology, etc.

The future of the use of these technologies in the world dairy industry is very promising. Many new cheese varieties might be prepared by combining the properties of mineral-adjusted UF retentates and enzymic abilities of lactic starters. Microfiltration has already opened new and very diversified avenues for research and technology. In the future, numerous ideas for cheese technologists may also originate from nanofiltration.

REFERENCES

1. Kosikowski, F.V., 1986. *Food Technol.*, **40**, 71.
2. Maubois, J.L., 1990. *Proc. XXIII Intern. Dairy Congr.*, Montreal, Canada, p. 1775.
3. Maubois, J.L., Mocquot, G. & Vassal, L., 1969. French Patent 2,052,121.
4. Maubois, J.L. & Mocquot, G., 1971. *Lait*, **51**, 495.
5. Maubois, J.L. & Mocquot, G., 1975. *J. Dairy Sci.*, **58**, 1001.
6. Korolczuk, J., Maubois, J.L. & Fauquant, J., 1986. *Proc. XXII Intern. Dairy Congr.*, The Hague, The Netherlands, p. 123.
7. Kosikowski, F.V., 1986. In *Membrane Separation in Biotechnology*, ed. W.C. McGregor, Marcel Dekker, Inc., New York, p. 201.
8. Maubois, J.L., Mocquot, G. & Vassal, L., 1973. French Patent 2,218,821.
9. Maubois, J.L., Mocquot, G. & Vassal, L., 1972. French Patent 2,166,315.
10. Ferron, C., Quiblier, J.P., Garric, G. & Maubois, J.L., 1991. *Lait*, **71**, 423.
11. Maubois, J.L., 1979. *Proc. Symp. on Ultrafiltration Membranes*, Amer. Chem. Soc. Polymer Sci. Tech., Vol. 13, Plenum Press, New York, p. 305.
12. Sandblom, R.M., 1974. Swedish Patent 7,416,257.
13. Fauquant, J., Maubois, J.L. & Pierre, A., 1988. *Tech. Lait. Market.*, **1028**, 21.
14. Glover, F.A., 1985. *Tech. Bull.*, no. 5, NIRD, Reading, UK.
15. Van der Horst, H.C. & Hanemaaijer, J.H., 1989. *Membr. Techn. Symp.*, Tylosant, Sweden.
16. Maubois, J.L., 1987. In *Proc. Symp. Cheese Biotech. Intern. Food Dev.*, Cornell Univ., Ithaca, NY, p. 24.
17. O'Shea, D.P., 1990. *Proc. ADPI/CDR Dairy Prod. Tech. Conf.*, Chicago, April 25–26, 1990.
18. Gregory, A.G., 1987. *Intern. Dairy Fed. Bull.*, **212**, 38.
19. Eriksson, P., 1988. *Environmental Progress*, **7**, 58.
20. Dziezak, J.D., 1990. *Food Technol.*, **44**, 108.
21. Glover, F.A., Skudder, P.J., Stothart, R.H. & Evans, E.W., 1978. *J. Dairy Res.*, **45**, 291.
22. Terré, E., Maubois, J.L., Brulé, G. & Pierre, A., 1987. French Patent, 2,592,769.
23. Maubois, J.L., 1988. *8th Intern. Biotech. Symp.*, Paris, France, p. 814.

24. Trouvé, E., Maubois, J.L., Piot, M., Madec, M.N., Fauquant, J., Rouault, A., Tabard, J. & Brinkman, G., 1991. *Lait*, **71**, 1.
25. Horton, B.S., 1986. *Proc. XXI Intern. Dairy Congr.*, The Hague, The Netherlands, p. 527.
26. Maubois, J.L. & Brulé, G., 1982. *Lait*, **62**, 484.
27. Cheryan, M., 1986. *Ultrafiltration Handbook*, 1st edn, Technomic Publishing Co., Lancaster, PA, USA.
28. Gillot, J., Brinkman, G. & Garcera, D., 1986. *Proc. Congr. Filtra.*, Paris, France.
29. Renner, E. & Abd El-Salam, M.H., 1991. *Application of Ultrafiltration in the Dairy Industry*, 1st edn, Elsevier Appl. Sci., England.
30. Brulé, G., Maubois, J.L. & Fauquant, J., 1974. *Lait*, **54**, 600.
31. Maubois, J.L., 1989. *Génie Rural*, **3**, 1.
32. Brulé, G., Maubois, J.L., Vandeweghe, J., Fauquant, J. & Goudédranche, H., 1975. *Rev. Lait. Fr.*, **328**, 117.
33. Green, M.L. & Grandison, A.S., 1987. In *Cheese, Chemistry, Physics, and Microbiology, Vol. 1, General Aspects*, 1st edn, ed. P.F. Fox. Elsevier Appl. Sci., London, p. 97.
34. Mistry, V.V. & Kosikowski, F.V., 1985. *J. Dairy Sci.*, **68**, 1613.
35. Goudédranche, H., Ducruet, P., Vachot, J.C., Pannetier, R. & Maubois, J.L., 1986. *Lait*, **66**, 189.
36. Mortensen, B.K., 1984. In *Milk Protein '84*, Proc. Intern. Congr. Milk Proteins, Luxembourg, ed. T.E. Galesloot & B.J. Tinbergen, p. 109.
37. Mistry, V.V. & Kosikowski, F.V., 1986. *J. Dairy Sci.*, **69**, 945.
38. Mistry, V.V. & Kosikowski, F.V., 1986. *J. Dairy Sci.*, **69**, 1484.
39. Mistry, V.V., 1990. *Milchwissenschaft*, **45**, 702.
40. Rash, K.E. & Kosikowski, F.V., 1982. *J. Food Sci.*, **47**, 47.
41. Mahaut, M., Maubois, J.L., Zink, A., Pannetier, R. & Veyre, R., 1982. *Tech. Lait.*, **961**, 9.
42. Culioli, J., Bon, J.P. & Maubois, J.L., 1974. *Lait*, **54**, 481.
43. Goudédranche, H., Maubois, J.L., Ducruet, P. & Mahaut, M., 1980. *Desalination*, **35**, 243.
44. Garnot, P., Rank, R.C. & Olson, N.F., 1982. *J. Dairy Sci.*, **65**, 2267.
45. Lucisano, M., Peri, C. & Donati, E., 1985. *Milchwissenschaft*, **40**, 600.
46. Garnot, P., 1988. *Intern. Dairy Fed. Bull.*, **225**, 11.
47. Dalgleish, D.G., 1980. *J. Dairy Res.*, **47**, 231.
48. Kosikowski, F.V., Masters, A.R. & Mistry, V.V., 1985. *J. Dairy Sci.*, **68**, 547.
49. Dalgleish, D.G., 1990. *Milchwissenschaft*, **45**, 491
50. Mietton, B., 1990. In *Proc. XXIII Intern. Dairy Congr.*, Montreal, Canada, p. 1838.
51. Rousseaux, P., Maubois, J.L. & Mahaut, M., 1978. In *Proc. XX Intern. Dairy Congr.*, Paris, France, p. 805.
52. Chapman, H.R., Bines, V.E., Glover, F.A. & Skudder, P.J., 1974. *J. Soc. Dairy Technol.*, **27**, 151.
53. Kosikowski, F.V., Masters, A.R. & Mistry, V.V., 1985. *J. Dairy Sci.*, **68**, 547.
54. Kealey, K.S. & Kosikowski, F.V., 1985. *J. Dairy Sci.*, **68**, 3148.
55. Sharma, S.K., Ferrier, L.K. & Hill, A.R., 1989. *J. Food Sci.*, **54**, 573.
56. Friis, T.L., 1985. *North Eur. Dairy J.*, **9**, 266.
57. Vandeweghe, J. & Maubois, J.L., 1987. In *Cheesemaking*, ed. A. Eck. Lavoisier Pub. Inc., New York, p. 469.
58. Maubois, J.L., 1989. *Intern. Dairy Fed. Bull.*, **244**, 26.
59. Athar, I.H., Spurgeon, K.R., Gilmore, T.M., Parsons, J.P. & Seas, S.W., 1983. *J. Dairy Sci.*, **66**(suppl. 1), 69 (Abstr.).
60. Kealey, K.S. & Kosikowski, F.V., 1986. *J. Dairy Sci.*, **69**, 1479.
61. Kosikowski, F.V., Masters, A.R. & Mistry, V.V., 1985. *J. Dairy Sci.*, **68**, 541.
62. Mattews, M.E., So, S.E., Amundson, C.H. & Hill, C.G., 1976. *J. Food Sci.*, **41**, 619.
63. Zall, R.R. & Chen, J.H., 1986. *Milchwissenschaft*, **41**, 217.

64. Fernandez, A. & Kosikowski, F.V., 1986. *J. Dairy Sci.*, **69**, 643.
65. Fernandez, A. & Kosikowski, F.V., 1986. *J. Dairy Sci.*, **69**, 2011.
66. Abrahamsen, R.K., 1988. *Milchwissenschaft*, **43**, 27.
67. Bush, C.S., Caroutte, C.A., Amundson, C.H. & Olsen, N.F., 1983. *J. Dairy Sci.*, **66**, 415.
68. Pahkala, E., Turunen, M. & Antila, V., 1984. *Meijeritiet. Aikak.*, **43**, 47.
69. Qvist, K.B., Thomsen, D. & Jensen, G.K., 1985. *Beret. Statens Mejerifors.*, **266**, 48.
70. Garrett, N.L.T., 1987. *North Eur. Dairy J.*, **5**, 135.
71. Leeuwen, J. van, Freeman, H., Sutherland, J. & Jameson, W., 1987. European Patent No. EP 0 120 879 B1.
72. Hansen, R., 1985. *North Eur. Dairy J.*, **6**, 153.
73. Honer, C., 1989. *Dairy Foods*, **90**(12), 54.
74. Skovhauge, E., 1987. *North Eur. Dairy J.*, **3**, 61.
75. Bundgaard, A.G., Olson, O.J. & Madsen, R.B., 1972. *Dairy Ind.*, **37**, 539.
76. Qvist, K.B., Thomsen, D. & Jensen, G.K., 1986. *Beret. Statens Mejerifors.*, **268**, 55.
77. Qvist, C.B., 1987. *Scand. Dairy Ind.*, **1**, 30.
78. Qvist, K.B., Thomsen, D. & Hoier, E., 1987. *J. Dairy Res.*, **54**, 437.
79. Spangler, P.L., El-Soda, M., Johnson, M.E., Olson, N.F., Amundson, C.H. & Hill, C.G. Jr., 1989. *Milchwissenschaft*, **44**, 199.
80. Spangler, P.L., Jensen, L.A., Amundson, C.H., Olson, N.F. & Hill, C.G., 1990. *J. Dairy Sci.*, **73**, 1420.
81. Mahaut, M. & Maubois, J.L., 1978. *Proc. XX Intern. Dairy Congr.*, Paris, France, p. 804.
82. Abdou, S.M., Abd-El-Salam, M.H., Abd-El-Hady, S.M., Dawood, A.H. & Montasser, E.A., 1988. *Egypt. J. Dairy Sci.*, **16**, 249.
83. Dawood, A.H., Abdou, S.M., Abd-El-Hady, S.M., Abd-El-Salam, M.H. & Montasser, E.A., 1988. *Egypt. J. Dairy Sci.*, **16**, 157.
84. Abd-El-Salam, M.H., Dawood, A.H., Abd-El-Hady, S.M., Adbou, S.M. & Montasser, E.A., 1988. *Egypt. J. Dairy Sci.*, **16**, 148.
85. Jacobsen, M.K., 1985. *North Eur. Dairy J.*, **2**, 38.
86. Lawrence, R.C., 1987. *Intern. Dairy Fed.*, Document B136.
87. Herbertz, G., 1984. *Dtsch. Milchwirtsch.*, **35**, 1004.
88. Stenne, P., 1973. French Patent 2,232,999.
89. Patel, R.S., Reuter, J., Prokopek, D., 1986. *J. Soc. Dairy Technol.*, **39**, 27.
90. Koch International GmbH, 1987. *North Eur. Dairy J.*, **53**, 75.
91. Herberts, G., 1985. *Dtsch. Milchwirtsch.*, **36**, 1042.
92. Anon., 1984. *Dtsch. Milchwirtsch.*, **35**, 1108.
93. Herbertz, G., 1984. *Dtsch. Milchwirtsch.*, **35**, 1790.
94. Anon, 1984. *Dtsch. Milchwirtsch.*, **35**, 1000.
95. Baeurle, H.W., Walenta, W. & Kessler, H.G., 1984. *Dtsch. Molk.-Ztg.*, **105**, 356.
96. Mahaut, M., 1990. PhD Thesis, Ecole Nationale Supérieure Agronomique de Rennes (France).
97. Mahaut, M., Korolczuk, J., Pannetier, R. & Maubois, J.L., 1986. *Tech. Lait. Mark.*, **1011**, 24.
98. Maubois, J.L. & Kosikowski, F.V., 1978. *J. Dairy Sci.*, **61**, 881.
99. Skovhauge, E., Aarhus, Denmark, Pasilac-Danish Turnkey Dairies Ltd., 1988. *APV Pasilac A/S*, **2**, 7.
100. Covacevich, H.R. & Kosikowski, F.V., 1978. *J. Food Sci.*, **42**, 1362.
101. Santo Neves, B. Dos & Ducruet, P., 1988. *Rev. Inst. Lacticinios Candido Tostes*, **43**, 3.
102. Resmini, P., Pagani, M.A. & Prati, F., 1984. *Sci. Tec. Latt.-casearia*, **35**, 213.
103. Sordi, F., 1984. *Latte*, **9**, 290.
104. Kosikowski, F.V., 1978. *Cheese and Fermented Milk Foods*, 2nd edn. Edwards Bros., Inc., Ann Arbor, MI.

105. Maubois, J.L., 1985. In *Laits et Produits Laitiers*, ed. F.-M. Luquet. Lavoisier, Paris, p. 121.
106. Hansen, F., 1981. *North Eur. Dairy J.*, **5**, 147.
107. Guetter, H., 1984. *Dtsch. Molk.-Ztg*, **105**, 1500.
108. Hansen, R., 1980. *North Eur. Dairy J.*, **46**, 149.
109. Hansen, M., 1984. *Cult. Dairy Prod. J.*, **19**, 16.
110. Al Khamy, A.F., 1988. PhD Thesis, Al-Azhar University (Egypt).
111. Covacevich, H.R. & Kosikowski, F.V., 1978. *J. Dairy Sci.*, **61**, 701.
112. Swientek, R.J., 1984. *Food Proc., USA*, **45**, 103.
113. Hansen, R., 1987. *North Eur. Dairy J.*, **53**, 21.
114. Delbeke, R., 1987. *Milchwissenschaft*, **42**, 222.
115. Ducruet, P., Maubois, J.L., Goudédranche, H. & Pannetier, R., 1981. *Tech Lait.*, **957**, 13.
116. Jolly, R.C. & Kosikowski, F.V., 1975. *J. Dairy Sci.*, **58**, 1272.
117. Sood, V.K. & Kosikowski, F.V., 1979. *J. Dairy Sci.*, **62**, 1713.
118. Ernstrom, C.A., Sutherland, B.J. & Jameson, G.W., 1980. *J. Dairy Sci.*, **63**, 228.
119. Madsen, R.F. & Bjerre, P., 1981. *North Eur. Dairy J.*, **5**, 135.
120. Maubois, J.L. & Fauconneau, G., 1977. C.R. ITEB.
121. Madsen, R.F. & Bjerre, P., 1981. *North Eur. Dairy J.*, **47**, 120.
122. Le Graet, Y. & Maubois, J.L. 1979. *Rev. Lait. Fr.*, **373**, 23.
123. Lablée, J., 1982. *Intern. Dairy Fed. Bull.*, **142**, 123.
124. Mahaut, M. & Maubois, J.L., 1988. In *Proc. Intern. Dairy Fed. Seminar*, special issue 9001, p. 298.
125. Lawrence, R.C., Creamer, L.K. & Gilles, J., 1989. *J. Dairy Sci.*, **70**, 1748.
126. de Koning, P. J., de Boer, R., Both, P. & Nooy, P.F.C., 1981. *Neth. Milk Dairy J.*, **35**, 35.
127. Creamer, L.K., Iyer, M. & Lelievre, J., 1987. *N.Z. J. Dairy Sci. Technol.*, **22**, 205.
128. Harper, J., Iyer, M., Knighton, D. & Lelievre, J., 1989. *J. Dairy Sci.*, **72**, 333.
129. Hickey, M.W., Van Leeuwen, H., Hillier, A.K. & Jago, G.R., 1983. *Aust. J. Dairy Technol.*, **38**, 110.
130. Furtado, M.M. & Partridge, J.A., 1983. *J. Dairy Sci.*, **71**, 2877.
131. Visser, S., 1981. *Neth. Milk Dairy J.*, **35**, 65.
132. Richardson, W.D., 1929. US Patent 1,711,032.
133. Stenne, P., 1964. *Tech. Lait.*, **463**, 13.
134. Berridge, N.J., 1970. In *The Enzymes I. Milk Coagulation*, ed. P.D. Boyer. Acad. Press, New York, p. 1079.
135. Agbevavi, T., Rouleau, D. & Mayer, R., 1983. *J. Food Sci.*, **48**, 642.
136. Barbano, D.M. & Bynum, D.G., 1984. *J. Dairy Sci.*, **67**, 2839.
137. Bynum, D.G. & Barbano, D.M., 1985. *J. Dairy Sci.*, **68**, 1.
138. Mayes, J.J., 1985. *Aust J. Dairy Technol.*, **40**, 100.
139. Schmidt, D., Fedrick, I.A. & Donovan, H.M., 1986. *N.Z. J. Dairy Sci. Technol.*, **21**, 125.
140. Barbano, D.M., Bynum, D.G. & Senyk, G.F., 1983. *J. Dairy Sci.*, **66**, 2447.
141. Barbano, D.M., 1986. In *New Dairy Products via New Technology*, Intern. Dairy Fed. Brussels, p. 31.
142. Maubois, J.L., 1987. In *Cheesemaking*, ed. A. Eck. Lavoisier Pub. Inc., New York, p. 156.
143. Honer, C., 1984. *Dairy Record*, **85**(7), 68.
144. Sandfort, P., 1983. *Dairy Record*, **84**(10), 16.
145. Kosikowski, F.V., 1985. *J. Dairy Sci.*, **68**, 2403.
146. Anon, 1984. *Tech. Lait.*, **983**, 7.
147. Benard, S., Maubois, J.L. & Tareck, A., 1981. *Lait*, **61**, 435.
148. Slack, A.W., Amundson, C.H., Hill, C.G. & Jorgensen, N.A., 1982. *Process Biochem.*, **17**, 6.
149. Slack, A.W., Amundson, C.H. & Hill, C.G., 1982. *Process Biochem.*, **17**, 23.

150. Zall, R.R., 1987. *Milchwissenschaft*, **42**, 98.
151. Honer, C., 1990. *Dairy Field Today*, **173**, 52.
152. Cox, G.C., Kyle, W.S.A., Versteeg, K., Marshall, S.C., Kitchen, B.J., Muller, L.L. & Jacob, P.H., 1985. Grant Report No. NERDDP.EG., **377**, pp. 9–15, National Energy Research Development and Demonstration Program, Department of Resources and Energy, Canberra, Australia.
153. Cox, G.C. & Langdon, I.A., 1985. *Aust. J. Dairy Technol.*, **40**, 113.
154. Alfa-Laval, 1988. Thermicon Commercial Brochure.
155. Maubois, J.L. & Maugas, J.J., 1985. *EEC Report 8010*, VI PF4, 179.
156. Cacciola, A. & Leung, P.S., 1980. *European Patent*, no. 0 040 282.
157. Holm, S., Malmberg, R. & Svensson, K., 1986. *Intern. Patent PCT* no. WO 86/01687.
158. Piot, M., Vachot, J.C., Veaux, M., Maubois, J.L. & Brinkman, G.E., 1987. *Tech. Lait. Mark.*, **1016**, 42.
159. Malmberg, R. & Holm, S., 1988. *North Eur. Food Dairy J.*, **54**, 30.
160. Meersohn, M., 1989. *North Eur. Food Dairy J.*, **55**, 108.
161. Olesen, N. & Jensen, F., 1989. *Milchwissenschaft*, **44**, 476.
162. Vincens, D. & Tabard, J., 1988. *Tech. Lait. Mark.*, **1033**, 62.
163. Kosikowski, F.V. & Mistry, V.V., 1990. *J. Dairy Sci.*, **73**, 1411.
164. Murphy, J.M. & Fox, P.F., 1991. *Food Chem.*, **39**, 27.
165. Murphy, J.M. & Fox, P.F., 1991. *Food Chem.*, **39**, 211.

14

Acceleration of Cheese Ripening

MARTIN G. WILKINSON

National Dairy Products Research Centre, Moorepark, Fermoy, Co. Cork, Republic of Ireland

1 INTRODUCTION

Cheese ripening is a complex process of concerted biochemical changes, during which a bland curd is converted into a mature cheese having the flavour, texture and aroma characteristics of the intended variety.[1-3] The gradual breakdown of carbohydrates, lipids and proteins during ripening is mediated by several agents, including: (a) residual coagulant, (b) starter bacteria and their enzymes, (c) non-starter bacteria and their enzymes, (d) indigenous milk enzymes, especially proteinases, and (e) secondary inocula with their enzymes.[1-4] Ripening is an expensive and time-consuming process, depending on the variety, e.g. Cheddar cheese is typically ripened for 6–9 months while Parmesan is usually ripened for two years. Owing to the cost of ripening cheese, there are obvious economic advantages to be gained by accelerating the process.

Greater control of ripening may also be gained by manipulating the process whereby end product quality may be predicted with greater certainty.[1] Acceleration of cheese ripening is, therefore, of benefit to the producer from both the economic and technological points of view, provided, of course, that the final product has the same flavour profile and rheological attributes as conventional cheese.[1-7]

The principal biochemical events involved in cheese ripening are: (a) glycolysis of residual sugars, (b) lipolysis, and (c) proteolysis involving the degradation of the caseins to lower molecular weight peptides and free amino acids. Acceleration of glycolysis, which occurs rapidly, is considered to be of no benefit in most or all cheese varieties. Acceleration of lipolysis may be of benefit in Blue or some Italian types where lipolysis plays a major role in the generation of characteristic flavour. The contribution of lipolysis to the flavour of Cheddar or Dutch cheeses is unclear, and acceleration of lipolysis in these types is not usually undertaken as a means of enhancing flavour development.

Proteolysis occurs in all cheese varieties and is considered to be a prerequisite for good flavour development. It is effected by a number of agents, including

TABLE I
Principal Methods for Accelerating Cheese Ripening[1,3]

Method	Advantages	Disadvantages
Elevated temperature	No legal barriers, technically simple, possible cost saving	Non-specific action; potential for microbial spoilage increased
Exogenous enzyme	Low cost, specific action, choice of flavour options	Limited choice of useful enzymes; over-ripening; difficult to incorporate uniformly; possible legal barriers
Modified starters	Easy to incorporate, natural enzyme balance retained	Technically complex
Slurries	Very rapid flavour development	High risk of microbial spoilage; final product requires processing

(1) residual coagulant, (2) indigenous milk proteinase, and (3) the proteinases and peptidases of starter and non-starter bacteria.

Ripening involves the production, via various pathways, of a pool of sapid compounds which give the flavour typical of the intended variety. The aim of accelerating the various biochemical pathways is to reduce the ripening time without adversely affecting flavour or texture. The biochemistry of cheese ripening and the important contribution by proteolysis to flavour development are reviewed in Chapter 10 of this book; in this chapter, the means by which maturation can be accelerated will be discussed.

2   METHODS FOR ACCELERATING CHEESE RIPENING

The principal methods used to accelerate the ripening of cheese are summarized in Table I, and are discussed in Sections 3–6.

3   ELEVATED RIPENING TEMPERATURE

Enzymatic, as well as chemical, reactions generally occur at faster rates as the reaction temperature is increased. Therefore, it can be reasonably assumed that the biochemical reactions which generate flavour compounds or flavour precursors in cheese will be accelerated by increasing the temperature at which the cheese is matured.

Many cheese varieties are now ripened at low temperatures, e.g. 6–8°C for Cheddar.[1,8] In a study on the influence of various factors, such as starter type, level of non-starter lactic acid bacteria (NSLAB) and ripening temperature, i.e. 6 or 13°C, on the flavour intensity of Cheddar cheese after six or nine months ripening. Law et al.[9] found that ripening temperature was the single most important factor. After a maturation period of six months, cheeses ripened at 13°C scored 4·4 on a 0 to 8 scale for flavour intensity (corresponding to medium/mature cheese) while cheeses stored at 6°C scored 3·2, corresponding to mild Cheddar. At 6°C, bitterness was more marked, either due to a lower flavour intensity or because degradation of bitter peptides by peptidases was not favoured at this temperature.

A comprehensive study on the influence of various ripening temperature regimes on proteolysis and flavour development in Cheddar cheese was undertaken by Aston et al.[10-14] and Fedrick et al.[8,15] who showed that the development of cheese flavour correlated well with the formation of soluble N, especially with 5% PTA-N (i.e. $r = 0.6-0.9$)[12] which consists primarily of free amino acids and small peptides of $M_r < 600$ D.[16] The following ripening temperatures and starter systems were investigated: (1) control cheese, with normal starter stored at 8°C (C8), (2) cheese supplemented with mutant (lac⁻ prt⁻) starter, stored at 8°C (M8), (3) cheese stored at 20°C for one month and then at 8°C for up to nine months (C20), (4) cheese supplemented with added mutant starter stored at 20°C for one month and then at 8°C for up to nine months (M20). The levels of PTA-N decreased in the order: M20 > C20 > M8 > C8. Levels of TCA-soluble tyrosine were highest in the M20 and C20 cheeses after one month but at nine months were quite close in all cheeses. The initial increases in TCA-soluble tyrosine and PTA-soluble N in the M20 and C20 cheeses were attributed to early stimulation of starter proteinases/peptidases and rennet in these cheeses compared with those ripened at 8°C (i.e. C8 and M8).

Although the levels of TCA-soluble tyrosine for all cheeses converged after nine months, the ratio of free amino acids to peptides was higher in the cheeses ripened at the elevated temperature, especially those supplemented with the mutant starter. It would appear that ripening initially at an elevated temperature, especially in combination with a mutant culture, either accelerated the production of peptide substrates for the starter peptidases or increased peptidolysis. Taste panel assessment confirmed the trends found for proteolysis, i.e. M20 > C20 > M8 > C8. In agreement with previous work by these authors,[12] a good correlation was found between levels of PTA-soluble N and flavour scores at all stages of ripening over nine months; however, flavour was strongly correlated with TCA-soluble tyrosine levels only up to three months.[10] A noteworthy finding in this and other studies is that supplementation with a lac⁻ prt⁻ mutant starter caused an atypical, undesirable flavour.

The production of the volatile sulphur compounds, hydrogen sulphide, carbonyl sulphide, methanethiol and dimethyl sulphide, in the above cheeses was also monitored[13] and correlated with flavour intensity. Concentrations of H_2S and CH_3SH generally followed the trends noted for proteolysis,[10] i.e. M20 > C20 > M8 > C8 and concentrations increased up to six months of ripening after which they decreased. Concentrations of carbonyl sulphide increased with ripening temperature and throughout ripening, but levels of dimethyl sulphide remained fairly constant up to six months but then decreased in C20, M20 and M8 cheeses. Correlations between flavour development and the concentration of H_2S were low, with only C8 and M8 cheeses showing a statistically significant correlation. The use of volatile sulphur compounds as an indicator of flavour development in Cheddar cheeses subjected to accelerated ripening was considered to be unreliable.

The influence of elevated temperature on proteolysis during the initial stages of ripening was investigated more thoroughly by Aston et al.[11] using the following protocol:

Treatment	Code
Control, 8°C for 32 weeks	C1
13°C for 4 weeks, then 8°C for 28 weeks	T13-1
13°C for 8 weeks, then 8°C for 24 weeks	T13-2
13°C for 32 weeks	T13-3
Control, 8°C for 32 weeks	C2
20°C for 4 weeks, then 8°C for 28 weeks	T20-1
20°C for 4 weeks, then 13°C for 4 weeks, then 8°C for 24 weeks	T20-2
20°C for 4 weeks, then 13°C for 28 weeks	T20-3

Proteolysis, as measured by the formation of PTA-soluble N, increased with temperature and duration of ripening, with highest levels of PTA-N in the T20-3 cheeses, values for which at 8, 12 or 24 weeks corresponded to the controls at 16, 24 or 32 weeks, respectively. Cheeses were assessed on the basis of total flavour (taste and aroma), mature flavour (denoted as a component of total flavour), estimated age and preference for each cheese. T20-3 cheeses received highest flavour scores and estimated ages, followed in order by T13-3, T20-2/T20-1 and T13-1/T13-2. T20-3 cheeses also showed the greatest advancement in age over the controls, especially after four to eight weeks, when T20-3 cheeses had a gain of six weeks over the controls but the difference decreased to four weeks after 16, 24 or 32 weeks. T13-3 cheeses had levels of advancement similar to T20-3 up to 16 weeks but were only two weeks ahead of the controls after 24 or 32 weeks. In this study, the maximum reduction in time required to achieve mature flavour development was found to be four weeks, i.e. in the T20-3 cheeses after 32 weeks. This apparently small gain may have been due to a tendency for panellists to underestimate the age of older cheeses.[12] Advancement in ripening as assessed by PTA-N was more significant, e.g. T20-3 cheeses were 16 weeks ahead of control cheeses after 32 weeks. Ripening according to regime T13-3, i.e. at 13°C for 32 weeks, was considered to be the most suitable commercially as it did not require transfer of cheeses between ripening rooms.

The effect of elevating the temperature midway through ripening on proteolysis and flavour development of Cheddar was studied by Fedrick et al.[8] using the following protocol:

Treatment	Code
8°C for 32 weeks (Control)	C1
8°C for 12 weeks, 13°C for 4 weeks, 8°C for 16 weeks	T13-1
8°C for 12 weeks, 13°C for 8 weeks, 8°C for 12 weeks	T13-2
8°C for 12 weeks, 13°C for 20 weeks	T13-3
8°C for 32 weeks (Control)	C2
8°C for 12 weeks, 20°C for 4 weeks, 8°C for 16 weeks	T20-1
8°C for 12 weeks, 20°C for 2 weeks, 13°C for 4 weeks, 8°C for 12 weeks	T20-2
8°C for 12 weeks, 20°C for 4 weeks, 13°C for 16 weeks	T20-3

Proteolysis generally increased with increasing temperature and duration of exposure and was highest in T20-3 cheeses. The levels of proteolysis (5% PTA-N and 12% TCA-N) in the T20-3 cheeses at 13 weeks were similar to the control cheeses at 16 weeks, and at 16 or 21 weeks corresponded to the control at 24 or 32 weeks, respectively. Grading scores for the T13 cheeses were similar to the controls while the T20 cheeses received higher scores and appeared to be more mature than the controls. Maximum advancement in ripening (estimated age) was three weeks at 16 weeks for T20-3 cheeses, although the level of proteolysis in these cheeses indicated advancements of 3, 8 and 11 weeks after ripening for 16, 24 or 32 weeks, respectively. The results of these studies[8,11,12] indicate that ripening at elevated temperatures may be a useful mechanism for accelerating ripening in response to market demands; however, it appears that initial storage at elevated temperatures is more effective than increasing the temperature during ripening. Significantly, no adverse effect on cheese quality was noted for any of the treatments in these studies.[12,13,15]

The effect of higher ripening temperatures for longer periods on proteolysis and flavour was investigated by Aston et al.[14] using the following protocol:

Treatment	Code
8°C for 32 weeks (Control)	C1
15°C for 8 weeks, 8°C for 24 weeks	T15-1
17·5°C for 8 weeks, 8°C for 24 weeks	T17-1
20°C for 8 weeks, 8°C for 24 weeks	T20-1
8°C for 32 weeks (Control)	C2
15°C for 32 weeks	T15-2
17·5°C for 32 weeks	T17-2
20°C for 32 weeks	T20-2

Proteolysis (PTA-N or TCA-soluble tyrosine) decreased in the order T20-2 > T17-2 > T15-2 > T20-1 > T17-1 > T15-1. At 32 weeks, T17-2 cheeses received significantly lower scores than the control of the same age while T20-2 cheeses received lower scores than the controls at 16, 24 or 32 weeks, the former being described as strongly off-flavoured, e.g. rancid, burnt or unclean.

In this study,[14] the maximum temperature at which cheeses could be stored for 32 weeks without significant deterioration in quality was 15°C, at which cheeses showed levels of proteolysis after 12 weeks similar to control cheeses (8°C) after 32 weeks. The estimated age of cheeses ripened at 15°C for 20 weeks was similar to that of the controls after 32 weeks. T20-1 cheeses (20°C for 8 weeks, followed by 24 weeks at 8°C) had levels of proteolysis and flavour scores similar to the cheeses stored at 15°C throughout ripening. The most optimistic estimate of the time saving which may be gained by elevating ripening temperature without concomitant reduction in cheese quality was approximately 12 weeks for the cheeses ripened at 15°C (control cheese ripened for 32 weeks at 8°C).

The influence of ripening temperature on cheese texture was studied by Fedrick & Dulley[15] using the following protocol:

Treatment	Code
8°C for 16 weeks (Control)	C1
15°C for 8 weeks, 8°C for 8 weeks	T15-1
17·5°C for 8 weeks, 8°C for 8 weeks	T17-1
20°C for 8 weeks, 8°C for 8 weeks	T20-1
8°C for 16 weeks (Control)	C2
15°C for 16 weeks	T15-2
17·5°C for 16 weeks	T17-2
20°C for 16 weeks	T20-2

Parameters measured after 8 and 16 weeks included hardness, springiness and fracturability. Cheeses became less springy and more fracturable with increasing storage temperature and time. Control cheeses softened with ripening time while experimental cheeses softened earlier; the rate of softening decreased markedly with age. After eight weeks ripening, the texture of the experimental cheeses appeared to be more mature than that of the control after 16 weeks. Changes in texture decreased in the order T20-2 > T17-2 > T20-1 > T15-2 > T17-1 > T15-1.

Levels of proteolysis correlated negatively with cheese hardness and springiness; hardness was negatively correlated with the levels of salt, moisture, salt-in-moisture (S/M) and moisture-in-fat-free-solids (MFFS).

The effects of ripening temperature on the microflora of cheeses stored under the time/temperature regimes used by Aston et al.[14] were reported by Cromie et al.[17] Counts of total bacteria, 'undesirable lactobacilli' (i.e. producers of off-flavours and CO_2) and total lactic acid bacteria (LAB) were generally highest in cheeses stored at 15–20°C. Counts of starter lactococci decreased in control cheeses during the first four weeks of ripening, with no further change at eight weeks; cheeses stored at 15–20°C had higher starter counts during the initial eight weeks of ripening. Non-starter lactic acid bacteria (NSLAB) decreased slightly in all cheeses during ripening and appeared not to be significantly affected by storage temperature. It was concluded that elevated ripening temperature had no clear-cut effects on cheese microflora, supporting the theory that elevated temperature accelerates flavour development by increasing the rate of flavour-forming reactions independent of the cheese microflora.[9] Off-flavours noted in T17-2 and T20-2 cheeses did not appear to be due to the dominance of any one bacterial species but may have been due to an imbalance of bacterial species.[17]

TABLE II
Specifications for Cheddar Cheese to be Ripened at Elevated Temperature[19]

Moisture	<37%
Moisture-in-the-fat-free-solids	<55%
Fat-in-dry-matter	>50%
Salt-in-moisture	3·8–5·0
pH	<5·30
Coliforms	$<10^{-1}$ g
E. coli	Absent in 0·1 g
S. aureus	Absent in 0·1 g
Yeasts and moulds	$<10^{-1}$ g
NSLAB	$<1000^{-1}$ g
Anaerobic spores	Absent in 5 ml raw milk

In an attempt to control the growth of NSLAB and subsequent spoilage in cheeses subjected to elevated ripening temperatures, rapid cooling of the cheeses to <10°C and holding for 14 days <10°C was recommended by Fryer[18] who proposed that counts of NSLAB should be $<10^3$/g at hooping and $<10^6$/g after 14 days. It was claimed that after cooling, the cheeses could be ripened at elevated temperatures without the risk of off-flavours due to secondary bacteria.

Compositional and microbiological specifications (Table II), which should curtail the growth of NSLAB and other spoilage organisms and result in balanced flavour development, were recommended by Fedrick[19] for cheeses to be ripened at elevated temperatures.

3.1 Elevated Temperature Combined with Exogenous Enzymes or Mutant Starter

Elevated ripening temperatures in combination with exogenous proteinase and/or mutant starter cultures have also been assessed.[20,21] The use of a neutral proteinase (Neutrase), a lactose and proteinase negative (lac⁻ prt⁻) supplementary starter and elevated temperatures, singly or in combinations, to accelerate flavour development and proteolysis in cheese was assessed by Fedrick et al.[20] using the following protocol:

Treatment	Code
Control starter only	
ripened at 8°C	C8
or 15°C	C15
with Neutrase, ripened at 8°C	CN8
or at 15°C	CN15
Control plus mutant starter	
ripened at 8°C	M8
or 15°C	M15
with Neutrase, ripened at 8°C	MN8
or at 15°C	MN15

Proteolysis (TCA-soluble tyrosine or PTA-soluble N), rheological changes and flavour development were assessed as previously described.[8,11,13,14] The efficacy of the treatment, i.e. storage time required for experimental cheeses to reach a level of maturity equivalent to the control cheese (C8) after six months, decreased in the order MN15 > M15 > CN15 > MN8 > M8 > CN8 > C8, corresponding to 1·4, 1·7, 2·0, 2·6, 2·8, 3·2, 4·3, and 6·0 months, respectively. Proteolysis and rheological measurements indicated a greater degree of maturity than taste panel assessment.[8,10-12,14] Storage at 15°C was the most effective single treatment and reduced maturation time by more than 50%. Increased starter populations or Neutrase were less effective and in some cases (e.g. CN8 after 24 or 32 weeks) Neutrase produced bitterness.

The use of Neutrase, FlavourAge-FR (fungal proteinase/lipase blend from Chr. Hansen's Ltd) and increased rennet levels in combination with elevated ripening temperatures was evaluated by Guinee et al.[21] for their effect on proteolysis, rheology and flavour development in Cheddar cheese according to the following protocol:

Treatment	Code
Control starters only	
ripened at 5, 10 or 15°C	CO5, CO10, CO15
with added FlavourAge-FR ripened at 5, 10 or 15°C	CFL5, CFL10, CFL15
with increased rennet levels (R1, R2, R3) ripened at	CR1, 2, 3 (5)
5, 10 or 15°C	CR1, 2, 3 (10)
	CR1, 2, 3 (15)
with added Neutrase ripened at 5, 10 or 15°C	CN5, CN10, CN15
Control plus added mutant starters	
ripened at 5, 10 or 15°C	MO5, MO10, MO15
with added FlavourAge-FR, ripened at 5, 10 or 15°C	MFL5, MFL10, MFL15
with increased rennet levels (R1 R2 R3), ripened at	MR1, 2, 3 (5)
5, 10 or 15°C	MR1, 2, 3 (10)
	MR1, 2, 3 (15)
with added Neutrase, ripened at 5, 10 or 15°C	MN5, MN10, MN15

Throughout the 240 day storage period, the level of PTA-N decreased in the following order for cheeses ripened at all temperatures: CN > CFL > CR1 > CR2 ~ CR3 > CO ~ MN > MFL > MR1 ~ MR2 ~ MR3 ~ MO. The increases (expressed as % of the control value) in PTA-N at 5, 10 or 15°C at 128 days, caused by different enzyme treatments in cheeses made with control starter only, were, ~37, 30 and 24% for Neutrase; ~13, 10 and 9% for FlavourAge-FR; 15, 3 and 1% for CR1; ~1, ~3 and ~6% for CR2 and 4, ~8 and 1% for CR3. Similar trends were noted for cheeses made with control plus mutant starter. Increasing the ripening temperature from 5 to 10 to 15°C resulted in an increase in the PTA-N for all treatments used, with the level of increase depending on the treatment interval and the sampling time.[21]

The available scientific literature indicates that increasing the ripening temperature from 5–8°C to 13–15°C offers the producer a technologically simple method, unrestricted by legal barriers, by which to achieve a significant acceleration of cheese ripening which should not be adversely affected by off-flavour development provided that the cheese is of a good compositional and microbiological quality before exposure to the higher temperature.

4 USE OF MODIFIED STARTER CULTURES

Most evidence to date indicates the important role played by the enzymes of starter and non-starter bacteria in the generation of flavour in various cheese varieties. The objective of using modified/attenuated starters is to increase the number of starter cells without detrimental effects on the acidification schedule during manufacture so that the cells contribute only to proteolysis and other changes during ripening.

A number of approaches have been adopted to augment the contribution of these cultures in an attempt to accelerate ripening, i.e. the use of (a) lysozyme-treated starters, (b) heat or freeze-shocked cells and (c) mutant cultures.

Modified starter cultures, with attenuated acid-producing abilities, are added with the normal starter culture during cheese manufacture and contribute to proteolysis during ripening.

Law[22] and Fox[1] comment that starter cultures have traditionally been selected for their ability to produce acid at a consistent rate during manufacture and that the selection protocol, e.g. the Heap–Lawerence test, does not include measurement of proteolytic activity or autolytic ability, even though certain strains, such as *L. lactis* subsp. *cremoris* AM2, SK11 or AM1, may lyse early during ripening and produce flavourful, close-textured cheeses, perhaps through early release of the intracellular enzymes.

Cheddar cheese produced using defined-strain starters, while having a clean, consistent flavour, does not satisfy the market for extra-mature cheeses. Indeed, it now appears that the focus of accelerated ripening may switch from the use of added proteinases/lipases, which is subject to strict legislation in some countries, to the selection of starter strains with enhanced autolytic properties and increased peptidase activity. This would provide a more balanced enzyme complement than that obtained through the addition of exogenous enzymes.

4.1 Stimulation of Starter Cells

The growth of starter cells may be stimulated by the addition of enzymes or hydrolysed starter cells to cheese milk. Maturation in Sovietski cheese, an Emmental-type cheese made in Russia, has been accelerated by using starters grown to high cell numbers in media supplemented with protein hydrolysates and at elevated temperatures.[3,22,23] Other Soviet varieties have been accelerated by addition to cheesemilk of starter supplements containing a high population of cells. Vassal et al.[24] used Rulactine, a metalloproteinase from *Micrococcus caseolyticus*, to accelerate the ripening of Emmental, Gouda and Carré de l'est without the development of bitter flavour. Rulactine was added to a small volume of milk prior to cheesemaking to produce small peptides and amino acids which appeared to stimulate the growth of the starter and secondary flora which, along with residual enzyme, may contribute to proteolysis. Casein or whey protein hydrolysates[25] or trace elements[26] have been used to accelerate the ripening of Ras cheese (a hard Egyptian variety).

4.2 Lysozyme Treatment

The significance of lysis of bacterial cells in proteolysis in cheese is largely unproven. In an attempt to accelerate the ripening of Cheddar, Law et al.[27] added lysozyme-sensitized cells to cheesemilk at a level equivalent to 10^{10} cells/g of cheese. Lysis of these cells, as indicated by the release of an intracellular dipeptidase in cheese (assayed in an extract prepared using a hypotonic buffer), appeared to occur on salting of the curd and resulted in a significant increase in the concentration of free amino acids at six months over the control. However, no effect on the rate of flavour development or intensity of flavour over the control was noted. It was concluded that intracellular enzymes may not play a direct role in flavour development.

Doubts have been expressed[28] about the possible overestimation of lysis of lysozyme-treated cells (owing to the use of a hypotonic extraction buffer) and replication of this study using an alternative extraction procedure may be warranted. Economically, the use of lysozyme-treated cells may not be viable for large-scale cheese manufacture owing to the cost of the enzyme. Addition of lysozyme encapsulated in a dextran matrix to cheesemilk at renneting has been suggested as a possible method to effect large-scale lysis of starter bacteria. The dextran matrix would break down at the pH of Cheddar cheese (5·2–5·4) and release the lysozyme which should degrade the bacterial cell wall, releasing cell wall enzymes and leading to lysis of the cells with release of intracellular peptidases. This method should facilitate good distribution of the lysozyme in the cheese matrix[29] but, as yet, no results are available.

The cloning and expression in E. coli of the lysin gene of the lytic phage, ø VML3,[30] probably provides the best opportunity to establish unequivocally whether increased autolysis of starter bacteria may be a serious alternative to other methods of accelerated ripening. Addition of lysin to milk-grown cultures of L. lactis subsp. cremoris in laboratory-scale experiments caused a rapid loss in cell viability, e.g. after 72 h, only 0·04% of the cells were viable compared to 44% for the control. It is envisaged that for Cheddar cheese, lysin would be added with the salt after milling. As yet, no cheesemaking trials using lysin have been reported.[31]

4.3 Heat- and Freeze-Shock Treatments

Bie & Sjostrom[32,33] subjected mixed strain starters or an L. helveticus culture to various heat-shock treatments in an attempt to reduce their acid-producing ability but to enhance their rate of autolysis. While earlier autolysis, as measured by the decrease in cell-bound DNA, of the heat-treated cells was noted, increasing the severity of the heat treatment did not result in increased proteolysis in the cheese; it was concluded that the proteinases were denatured at the temperatures used to retard the acid-producing ability of the culture, but this caused increased autolysis.

Addition of moderate amounts of heat-treated *L. helveticus* (69°C for 20 s) combined with untreated mixed-strain starters accelerated the ripening of a Swedish hard cheese.[33] Acid production by the *L. helveticus* culture appeared to have been minimized while its proteolytic activity remained unaffected by this treatment.

Pettersson & Sjostrom[34] used heat-shocked cultures to attain large numbers of a mixed-strain starter, containing *Lactococcus, Leuconostoc* or *L. helveticus* strains which were cultivated at a constant pH, followed by heating to 59 or 69°C for 15 s. Using this methodology, experimental vats were supplemented with a suspension of 10^{11}–10^{14} cells per vat which were incorporated into the cheese without adversely affecting the pH schedule. The proteolytic ability of the treated organisms was reduced by only 10–30%. Proteolysis, as measured by PTA-N, increased with increasing numbers of entrapped heat-treated *L. helveticus* cells, but only slightly with heat-shocked mixed-strain starters. Flavour scores increased with increasing numbers of heat-shocked cells and ripening was accelerated more in cheeses supplemented with *L. helveticus* than with *Lactococci*. A 50% reduction in the ripening time was possible using this treatment and bitterness was not noted in any of the cheeses.[34]

A combination of a neutral proteinase (Neutrase) with heat-shocked *L. helveticus* to a final level of 4×10^6 cfu/g curd also accelerated the ripening of the Swedish hard cheese, Svecia.[35] Levels of 12% TCA and 5% PTA-N were higher when Neutrase and heat-shocked lactobacilli were used in combination than when either was used alone, suggesting a synergistic relationship between production of peptides by Neutrase and their degradation by peptidases of the attenuated starter culture. Addition of Neutrase alone resulted in a crumbly body and a bitter taste; bitterness was much reduced when Neutrase was combined with the heat-shocked cells, probably through breakdown of the bitter peptides by their peptidases.

Addition of heat-shocked lactobacilli has also been found to increase peptidolysis and produce good flavour in low-fat (10%) Swedish semi-hard cheese which otherwise was bitter and had little typical flavour.[36] While the addition of heat-treated lactobacilli gave significant increases in proteolysis, addition of equivalent numbers of viable, untreated lactobacilli did not produce similar increases. Gel permeation studies indicated that the control cheese, which was bitter, had considerably higher levels of intermediate-sized peptides, but lower levels of small peptides and free amino acids, than the cheese containing heat-treated culture. Heat treatment may have caused early lysis of the cells, as indicated by the presence of a high level of an intercellular aminopeptidase in cheese extracts, and may have played a role in debittering the cheese.[36] More recently,

Lopez-Fandino and Ardö[37] showed that heat-shocking of *L. bulgaricus* leads to the early release of intracellular peptidases.

Ardö & Mansson[38] noted that the flavour intensity of low-fat Swedish hard cheese could be increased by increasing levels of added heat-treated lactobacilli. Debittering, which was again noted, required addition of much lower levels of heat-shocked cells than flavour enhancement and appeared to be due to the rapid breakdown of medium-sized peptides.

The effect of adding heat-shocked thermophilic lactobacilli and lactococci to Gouda cheese has also been reported.[39] Several heat shock treatments were evaluated, with best results being obtained at 70°C for 18 s. Proteolysis, as measured by PTA-N, free amino acid levels and small peptides (<1000 Da), was considerably higher in cheese supplemented with a heat-shocked *L. helveticus* strain but not with heat-shocked strains of *L. bulgaricus* or *S. thermophilus*. Heat-shocked *L. helveticus* gave considerable flavour enhancement, and again, debittering was noted.

A freeze-thaw treatment has also been used to produce cells with active proteolytic systems but diminished lactose-utilizing ability. Freeze-shocked *L. helveticus* cells autolysed twice as fast as control, untreated cells.[32,33] Addition of freeze-shocked *L. helveticus* CNRZ 32 cells to Gouda cheesemilk accelerated proteolysis and flavour development compared to a control cheese or a cheese to which untreated *L. helveticus* cells were added, the greatest difference being observed after ripening for five weeks.[40] Although untreated *L. helveticus* cells accelerated proteolysis, they produced off-flavours. The debittering effect of attenuated *L. helveticus* cells was confirmed.

Addition of a heat-shocked culture of *S. thermophilus* or *L. bulgaricus* to milk for Feta cheese produced changes in flavour, body, 12% TCA or 5% PTA-soluble N and amino acid-N comparable to an acid or neutral proteinase (Neutrase).[41] Flavour development was considerably advanced in all experimental cheeses, with mature flavour being noted after 40 days ripening compared with 80 days for the control cheeses. Cheeses with added heat-shocked lactobacilli had the best organoleptic characteristics, followed by those treated with neutral or acid proteinase. Cheeses supplemented with acid proteinase were slightly bitter initially but bitterness had disappeared by 80 days. All enzyme-treated cheeses developed flavour more slowly than cheese with added heat-shocked starters but after 80 days ripening, flavour scores for all experimental cheeses were similar.

Heat-shocked *L. helveticus* CNRZ 32 cells have also been used to accelerate the ripening of a variant of Saint Paulin cheese.[42] Optimum conditions for heat-shocking were 18 s at 64°C which destroyed 90% of the acid-producing ability but only 60% of cell wall proteinase and 10% of the aminopeptidase activity; cell viability was reduced to 0·5%, although electron microscopy showed that the treated cells had not lysed. Nitrogen soluble at pH 4·6 increased to the greatest extent in cheeses to which untreated lactobacilli were added, while PTA-N was highest for cheeses with added heat-treated lactobacilli. Although cheeses supplemented with heat-shocked lactobacilli had a somewhat atypical, sweet flavour, they received highest grading scores (mature flavour and overall quality) at all stages during the 60-day ripening period. Cheeses to which untreated cells were added showed marked textural and flavour defects.[42]

Freeze-shocked lactobacilli in combination with free or liposome-encapsulated proteinase (Corolase) have been used to accelerate the ripening of Gouda cheese made from ultrafiltered (UF) or conventional milk.[43] Cheeses containing the free Corolase had higher levels of PTA-N than those with liposome-entrapped enzyme. However, the former cheeses were markedly bitter, while the latter showed low levels of bitterness, similar to the controls. The combination of liposome-entrapped proteinase and freeze-shocked lactobacilli resulted in large increases in TCA and PTA soluble-N levels, and in the development of a more intense flavour but without the bitterness, which was found in UF control cheese and UF cheese with added entrapped enzyme. The more intense bitterness compared to the controls and the cheeses with free Corolase in the UF cheeses containing liposome-entrapped Corolase was attributed by the authors to a high retention of liposomes in the UF cheese.

The ripening of Gouda cheese was accelerated by adding heat-shocked cells (56·5°C for 17 s) of a mixed-strain starter culture.[44] An inoculum of up to 5% (v/v) could be included in the cheese milk with very little loss of the cells in the whey. The level of amino nitrogen was increased through the action of the heat-shocked cells but 4% NaCl-soluble N was not significantly increased over the controls. A reduction of 25% in the ripening time was noted and, as in other reports,[35,36,38] the level of bitterness was reduced.

4.4. Mutant Starter Cultures

Lactose-negative (lac⁻) mutants of starter strains have also been used to provide 'packets' of uniformly distributed proteinases/peptidases, enhancing the production of peptides and free amino acids. These cells do not interfere with acid production during manufacture.

Grieve et al.[45,46] added concentrates containing 10^{11} cfu/ml of L. lactis subsp. cremoris C2 (lac⁻ prt⁻) cells to milk for Cheddar cheese at levels varying from 0·15 to 0·005%, together with normal starter. The cheeses were ripened at 20°C for 1 month followed by 5 months at 8°C, or at 8°C for 6 months. At 6 months, cheeses ripened at 8°C and supplemented with mutant starter received the highest flavour scores. Cheeses containing mutant starter and ripened at 8°C throughout showed an advancement in ripening over the control of from 1–8 weeks after 3 months or 4 to 12 weeks after 6 months' storage. TCA (12%)-soluble N levels were similar in control and experimental cheeses but cheeses supplemented with mutant starter had higher levels of free amino acids (PTA-N) as noted previously by Aston et al.[10] Bitterness did not develop in the experimental cheeses, probably due to the absence of a cell surface proteinase in the mutant (prt⁻) strain.

The use of the same mutant (C2, added as a concentrate containing 2×10^{11} cfu/ml) in combination with a neutral proteinase or elevated ripening temperature was studied by Fedrick et al.[20] Formation of PTA-N was more rapid in the mutant-containing cheeses, but mean flavour scores were the same as, or slightly lower than, the control cheeses throughout ripening. A slight fruity flavour was noted in cheeses containing mutant starter which was attributed to heterolactic fermentation of lactose by the mutant cells. Debittering of bitter peptides produced by Neutrase was noted in cheeses supplemented with lac⁻, prt⁻ mutant starters.[20] This will be discussed further in Section 6.1 on exogenous proteinases.

The manufacture of Cheddar cheese using prt⁻ mutants of *L. lactis* subsp. *cremoris* was also investigated by Oberg *et al.*[47] and Farkye *et al.*[48] Oberg *et al.*[47] found higher levels of pH 4·6-soluble N in experimental cheeses compared to controls at all stages of ripening, which was attributed to a carry-over of N from the yeast extract used to grow the bulk prt⁻ culture. The quality of the experimental and control cheeses was similar, with only slightly lower values for flavour noted in the prt⁻ cheeses. Scores for body and texture were higher for the prt⁻ cheeses after 180 days ripening.

Farkye *et al.*[48] using prt⁺ or prt⁻ starters to manufacture Cheddar cheese, showed that the level and rate of formation of WSN were similar in the experimental and control cheeses, but PTA-soluble N was significantly higher in the controls (i.e. prt⁺ starters). Electrophoretograms of the WSN fractions of the cheeses were also significantly different; ultrafiltrates of the WSN of the prt⁺ cheeses contained more large peptides (retained by UF membranes). The results of this study suggest that starter proteinases contribute to the degradation of casein to small peptides. Cheese quality was not adversely affected by the use of prt⁻ starters and scores for body and flavour were only slightly lower than those for the control (prt⁺) cheese.[48] Levels of 12% TCA soluble-N were similar in both sets of cheeses.

The use of prt⁻ starters in Gouda cheese gave levels of 4% NaCl-soluble N similar to the control (i.e. prt⁺ starter).[49] However, amino nitrogen levels were lower in cheeses made with 100% prt⁻ culture. Flavour development was also lowest in the cheeses made with 100% prt⁻ starter and they lacked typical Gouda flavour. About 20% prt⁺ cells in the culture appeared to be sufficient for maximal formation of amino-N.

It is possible that the differences in proteolysis between the above studies[47–49] may be due to differences in proteinase and peptidase activities of prt⁺ parent strains and their prt⁻ derivatives, as suggested by Kamaly *et al.*[50,51]

The effect of the level of starter proteinase on the development of bitterness in Cheddar cheese was investigated by Mills & Thomas[52] using cultures containing varying proportions of prt⁺ and prt⁻ cells for cheese manufacture. Cheeses made with 45–75% prt⁻ cells were significantly less bitter than cheeses made using only prt⁺ cells; this implies that the cell wall proteinase has a role in the production of bitter peptides which may be removed by the action of intracellular peptidases.

Modified starter cultures with enhanced complements of proteinase and/or peptidases which could be released early and evenly distributed in the curd would be an ideal method of accelerating cheese ripening and recent cloning of the genes for a dipeptidyl aminopeptidase[53,54] and phage lysin[30,31] would indicate that research is being directed to this end.

Starter culture technology, as mentioned previously, offers the advantage of uniform distribution of increased amounts of proteolytic activity in cheese. Provided that the techniques used to produce these enhanced starter qualities are not prohibited, a more efficient means of accelerating ripening will be achieved.

5 NON-STARTER LACTIC ACID BACTERIA AS ADJUNCT CULTURES

Non-starter lactic acid bacteria (NSLAB) are considered to make a significant contribution to proteolysis and flavour development in cheese.[55,56] Therefore, the addition of selected NSLAB, along with the normal starter, to increase the rate of casein degradation and flavour development has begun to receive research attention. Much of the recent work has emanated from Canadian and Australian researchers where milk for Cheddar cheese has been inoculated with cultures of lactobacilli.

The influence on proteolysis and flavour development of adding different *Lactobacillus* spp. to Cheddar cheesemilk to a level of 10^5 cfu/ml was investigated by Puchades *et al.*[57] Levels of free amino acids, especially glutamic acid, leucine, phenylalanine, valine and lysine, in the water soluble fraction were increased in the experimental cheeses over the control (no lactobacilli added) throughout a 9-month ripening period. Cheeses supplemented with *L. casei* subsp. *casei* L2A graded highest for flavour after ripening at 6 or 10°C for 7 months, followed by cheeses inoculated with *L. plantarum*, *L. casei* subsp. *casei* 119 and the control. Cheeses inoculated with added *L. brevis* scored lowest after 6 or 9 months.

In a similar study,[58] the effects of using *L. casei* subsp. *casei, L. plantarum* and *L. brevis* as adjunct cultures on the bitter and astringent N fractions of cheese was investigated using size exclusion HPLC. Proteolysis of the astringent fraction was highest for *L. casei* subsp. *casei* and *L. brevis*. Proteolysis of the bitter fraction was also highest for cheeses inoculated with *L. casei* subsp. *casei*. Free amino acid levels were increased in the experimental cheeses and the added lactobacilli appeared to hydrolyse the caseins independently of rennet-producing low molecular weight products. Inoculation with *L. casei* subsp. *casei* gave best results: a high quality cheese, free of bitterness, was produced after seven months at 6°C which graded better than the control after 9 months; bitterness was detected in the control cheese. However, off-flavour developed in cheeses inoculated with *L. brevis* or *L. plantarum*.[58]

A high quality cheese was produced using *L. casei* subsp. *casei* and *L. casei* subsp. *pseudoplantarum* with more intense flavours than the controls after 6, 8 and 10 months (cheeses were ripened for two months at 15°C followed by 10 months at 7 or 15°C); flavour and body defects were noted for other strains of added lactobacilli.[59]

Inoculation of milk for Cheddar cheese with heterofermentative lactobacilli, *L. brevis* or *L. fermentum*, consistently caused flavour and body defects, e.g. fruity flavour, openness and late gassing. When combined with homofermenters, e.g.

L. casei subsp. *casei* or *L. casei* subsp. *pseudoplantarum*, heterofermenters caused no significant downgrading compared to the control cheeses, but neither did they improve the flavour over the control.[60,61] The addition of a cell concentrate or lyophilised or liquid cell homogenates of *L. casei* subsp. *casei* L2A either to the cheesemilk prior to renneting or at salting caused a 40% increase in flavour intensity over the controls (starter only) after 6 months ripening. Best results were obtained when untreated lactobacilli plus a lyophilised cell homogenate were added to milk at renneting.[62]

Broome *et al.*[56] also reported that inoculation of cheesemilk with strains of *L. casei* accelerated the ripening of Cheddar cheese. Strains were added at a level of 10^6/ml milk and reached $>10^7$/g cheese during the first 12 weeks of ripening. After 12 weeks, no utilization of citrate or formation of biogenic amines was noted in the experimental cheeses but proteolysis (PTA-soluble N and free amino acids) was significantly higher than in the control cheeses. Flavour development was more advanced after 36 and 48 weeks in the experimental cheeses and no body or textural defects were noted.

The addition of *Micrococcus* or *Pediococcus* strains to low-fat Cheddar cheese has been reported to enhance proteolysis and flavour development over control cheese (starter only) after three months ripening; after 6 months, cheeses inoculated with pediococci graded highest, while off-flavour development was noted for cheeses supplemented with micrococci.[63]

6   ADDITION OF EXOGENOUS ENZYME

Since cheese ripening is essentially an enzymatic process, it should be possible to accelerate ripening by augmenting the activity of key enzymes. Addition of enzymes has the advantage of more specific action for accelerating flavour development compared to elevated temperatures which can accelerate off-flavour development just as much as flavour-forming reactions. Enzymes may be added to generate specific flavours in cheese, e.g. lipase addition for Parmesan or blue-type cheese flavours.[1,3] On the negative side, enzyme addition is not permitted in all countries and the range of useful enzymes available is quite limited. Uniform distribution of enzymes in the curd can be difficult to achieve and may give rise to 'hot spots' if the enzyme is added with the salt. If enzymes are added to the cheesemilk, 90% of the enzyme may be lost in the whey, proteolysis can occur in the vat during manufacture, generating small peptides which are lost in the whey, leading to reduced yield, and the whey may be rendered unsuitable for further processing. Over-ripening, with flavour and body defects, may occur because of the inability to control enzyme activity during ripening.

Varieties such as Gouda, Edam or Swiss have a curd washing or partial drainage step and are surface-salted (dry or brine) so that enzyme must be added either to the cheesemilk or by injection into the cheese. The cost of enzymes such as proteinases, peptidases or lipases can be considerable and their use must be justified by a significant acceleration of flavour development.

Fox[1] was of the view that addition of a single enzyme which accelerates one particular reaction is unlikely to be successful unless that reaction is clearly rate limiting. Hence, the acceleration of proteolysis without a concomitant increase in the rate of peptidolysis, and perhaps lipolysis, may result in the development of unbalanced flavour.

To date, few, if any, attempts to accelerate the multiple secondary flavour-forming reactions, e.g., Strecker degradation, have been reported. At present, the pathways leading to the formation of flavour compounds are largely unknown[22] and hence the use of exogenous enzymes to accelerate ripening is largely an empirical process.

6.1 Proteinases/Peptidases

Proteolysis is characteristic of most cheese varieties and is a *sine qua non* for good flavour and textural development. Proteinases in cheese include (1) plasmin, (2) rennet and (3) proteinases (i.e. cell wall and/or intracellular) of the starter and non-starter bacteria. Peptidases originate from cell wall, cell membrane and intracellular locations of the starter and non-starter bacteria.[1,4,22,64]

Approximately 6% of the rennet added to cheesemilk remains in the curd after manufacture[65] and contributes significantly to proteolysis during ripening.[66] Increasing the concentration of residual coagulant in Dutch cheese increased the rate of formation of WSN without a corresponding increase in the formation of amino N.[67] It was concluded that sufficient peptides were produced by the normal level of rennet used and increasing the rennet level did not further stimulate the production of amino acids and flavour development but did lead to bitterness.

Guinee *et al.*[21] added freeze-dried rennet to Cheddar curd at milling to give 1, 1·5 and 3 times the normal level of residual rennet, in combination with a freeze-shocked culture of *L. lactis* subsp. *cremoris* added at the rate of 0·05% (w/v) to cheesemilk (which increased the number of cells by $1·2 \times 10^7$ cfu/ml). Increased rennet levels increased the level of WSN, but levels of PTA soluble-N were only slightly higher than in control cheeses. Addition of modified starter in combination with additional rennet produced slightly greater increases in both WSN and PTA soluble-N than in cheeses with increased rennet levels alone. The latter cheeses were bitter but this defect was less intense in cheese to which freeze-shocked starter was also added.

The proportion of chymosin retained in cheese curd increases as the pH of the curd at whey drainage is reduced but the retention of microbial rennet is independent of pH.[68,69] Increased retention of rennet in the curd led to faster proteolysis but reduced flavour scores for all cheeses, with bitterness being the most prevalent defect. In general, addition of increased rennet levels to cheesemilk or as a freeze-dried preparation to curds or by increased retention in the curd (through lower pH) leads to bitterness which precludes its use as a viable method for accelerating ripening.[1,22,67]

Various enzymes (rennets, acid proteases, neutral proteases, acid protease-peptidases, lipases and decarboxylases) were screened by Sood and Kosikowski,[70] using Cheddar cheese slurries, for their ability to accelerate cheese ripening at 20 or 32°C, as assessed by proteolysis, lipolysis and flavour development. Fungal rennets and bacterial decarboxylases contributed little to cheese flavour. Individual microbial acid proteinases in combination with fungal rennets produced medium cheese flavour in slurries but caused pronounced bitterness. Combinations of individual neutral proteinases and microbial peptidases intensified cheese flavour and when used in combinations with microbial rennets reduced the intensity of bitterness caused by the latter. Acid proteases caused intense bitterness. Various animal or microbial lipases gave pronounced cheese flavour, low bitterness and strong rancidity, while lipase in combination with proteinase and/or peptidases gave good cheese flavour with low levels of bitterness.[70]

The addition of combinations of various fungal proteases and lipases to Cheddar cheese has been reported[71] to reduce ripening time by 50%, with good quality, medium-sharp Cheddar being produced in 3 months at 10°C; when matured at 4·5°C, little acceleration of ripening was observed. Enzyme-treated cheeses showed higher levels of free amino acids and increased levels of β-casein breakdown compared to the controls. The inclusion of up to 60% of enzyme-treated UF retentates (incubated with fungal protease or lipase preparations at 45°C for 24 h) in a processed cheese mix improved the flavour compared to commercial processed cheeses.[72] Enzyme treatment of the retentate allowed greater replacement of Cheddar cheese by retentate in the formulation.

The effects of adding a proteinase/intracellular peptidase(s) preparation from a *Pseudomonas* sp. to cheese were described by Law.[73] Experimental cheeses showed increased levels of soluble-N, peptide-N and amino acid-N over the controls. At a low level of enzyme addition, the quality and intensity of typical cheese flavour were highest in 8 weeks-old cheeses, equivalent to 16-week controls, but at 22 weeks little difference in flavour intensity was noted between control and experimental cheeses; at higher levels of addition, typical Cheddar flavour was masked by bitterness and other off-flavours.[73]

Law & Wigmore[74] assessed the suitability of various microbial proteinases for accelerating the ripening of Cheddar cheese. Commercial preparations of a neutral proteinase from *Bacillus subtilis* (Neutrase) at a level of 0·0125 mg/kg significantly enhanced Cheddar cheese flavour without defects, but higher levels caused bitterness. The aspartyl proteinase of *Aspergillus oryzae* was very proteolytic in cheese and always led to bitterness, even when added at a level of proteolytic activity equivalent to only 4% of the optimum level for Neutrase. The alkaline protease of *Bacillus licheniformis* (Subtilisin) also produced a very bitter cheese at the same level of addition as Neutrase (i.e. 0·0125 mg/kg). At the same level of addition as Neutrase, Pronase, a broad specificity proteinase from *Streptomyces griseus*, with some aminopeptidase activity, produced strong flavours in cheese but also caused bitterness, while lower levels of Pronase gave flavour enhancement without off-flavour development.[74]

In a more detailed study on the use of Neutrase or a proteinase from *A. oryzae* to accelerate cheese ripening, Law & Wigmore[75] achieved significant (two-fold) acceleration of flavour development. The acid proteinase of *A. oryzae* caused more extensive proteolysis than in the control or Neutrase-treated cheeses but bitterness was noted at all levels of acid proteinase used, due perhaps to its greater stability and activity at the pH of Cheddar cheese. All the proteinase-treated cheeses appeared more crumbly, less elastic and less firm than the controls, but these problems were not significant at low levels of Neutrase. The effect on cheese texture was attributed to excessive breakdown of β-casein in enzyme-treated cheeses.[75]

Temperature control of ripening may be used to increase the effectiveness of Neutrase, e.g. initial storage at 12°C for 2 months or 18°C for 1 month, followed by storage at 6°C for 2 months, gave little flavour enhancement while storage at 18°C for 3 months gave marked flavour development without flavour defects; no acceleration of ripening was observed at 6°C.[74] The effect of different levels of a neutral proteinase from *A. oryzae* on the ripening of Cheddar cheese indicated[76] that while proteolysis increased with the level of enzyme added (the concentration of free amino acids in cheeses supplemented with 0·01–0·1% proteinase reached those in a 6 months-old control cheese after 2–4·5 months), low levels of proteinase (0·001 and 0·005%) had no detectable effect on cheese flavour and the lowest levels (0·25–0·5%) which gave an advancement in flavour also caused bitterness, the intensity of which increased *pro rata* with enzyme level. The maximum advancement in estimated age without a significant reduction in quality was 1·7 months for cheese with 0·05% added protease after 4 months at 10°C. Slightly higher levels of bitterness were noted with the *A. oryzae* proteinase than with comparable levels of proteolysis effected by Neutrase. Texture analysis confirmed the findings of Law & Wigmore[75] for Neutrase, i.e. enzyme-treated cheeses became softer, more fracturable, less springy and less cohesive than controls during ripening.[76]

Ripening of Feta cheese was accelerated using a neutral proteinase from *B. subtilis* or an acid proteinase from *A. oryzae* or a heat-shocked mixed culture of *S. thermophilus* and *L. bulgaricus*.[41] 12% TCA soluble-N levels were similar for all experimental cheeses but levels of 5% PTA-N were highest in cheeses with added heat-shocked bacteria, while flavour development in all treated cheeses was accelerated over the controls, with a 50% reduction in ripening time (80 to 40 days). Bitterness was noted only in the case of cheeses treated with the acid proteinase. A lipase/proteinase preparation derived from *A. oryzae* was reported to release C6–C10 fatty acids, producing a flavour typical of aged Cheddar cheese, when added at salting to Colby cheese.[77] Addition of a neutral proteinase, P-11, at salting produced a texture similar to an aged Cheddar cheese without bitterness.

FlavourAge-FR (lipase/proteinase blend from *A. oryzae*), which accelerates the release of middle chain fatty acids, has been reported[77,78] to accelerate the production of good-quality Cheddar cheese. Guinee *et al.*[21] found that addition of 48 mg FlavourAge-FR per kg curd at salting gave large increases of water-soluble, ethanol soluble and 5% PTA soluble N. At a low ripening temperature (5°C), FlavourAge-FR improved the flavour and aroma of cheese, although some bitterness was noted, but at 10 or 15°C, flavour scores were lower than for the control and bitterness was observed. FlavourAge-FR treated cheeses had a soft or brittle body.[21]

A more balanced approach to the acceleration of ripening using mixtures of proteinases and peptidases, attenuated starter cells or cell-free extracts is now favoured. Intracellular cell-free extracts (CFE) of cheese starter bacteria in combination with Neutrase resulted in significant acceleration of flavour development.[79] CFE caused no increase in the level of primary proteolysis or 12% TCA soluble N but caused the rapid release of small peptides and free amino acids. In combination with Neutrase, the levels of small peptides and free amino acids were greater than when Neutrase was used alone. Cheeses treated with CFE and Neutrase had a stronger flavour after 2 months than those treated with Neutrase alone. Increasing the level of CFE at a constant Neutrase level increased proteolysis, but did not increase the rate of flavour formation *pro rata*, suggesting that the transformation of free amino acids to flavour compounds may be rate-limiting and that these changes are not catalysed by starter enzymes.[79] Development of an enzyme system for accelerated ripening, known as 'Accelase' (Imperial Biotechnology Ltd), and containing a CFE of starter lactococci and a proteinase has emerged from this work.[79] The debittering effect and flavour development on addition of 'Accelase' to Neutrase-treated Cheddar cheese slurries was studied by Cliffe & Law[80] using the analytical methods of Cliffe *et al*.[81] 'Accelase' sequentially produced and hydrolysed bitter hydrophobic peptides giving, in the most intensely flavoured cheeses, a predominance of hydrophilic di- or tri-peptides. Starter peptidases would appear to be directly involved in the breakdown of large peptides produced by Neutrase and/or chymosin to smaller peptides and free amino acids.[80]

Addition of CFEs of *L. helveticus, L. bulgaricus* or *L. lactis* to aseptic, chemically-acidified cheese curds was reported to increase the levels of pH 4·6 soluble N (i.e. primary proteolysis).[82] Ripening was fastest for *L. lactis* CFE, which increased the rate of β-casein breakdown but after two months bitterness was noted in all CFE-treated cheeses.

A comparative study on the effects of Neutrase, calf lipase (Miles), Neutrase plus calf lipase or NaturAge (a culture-enzyme mixture from Miles) on proteolysis and textural changes was reported by Lin *et al*.[83] Rapid production of 12% TCA-soluble N, with slower production of 5% PTA-soluble N, was noted for all enzyme treatments throughout a 16-week ripening period at 7 or 13°C. A combination of Neutrase and calf lipase produced higher levels of 12% TCA-soluble N than either lipase or Neutrase alone. Cheeses treated with NaturAge had levels of TCA-soluble N similar to control or lipase-treated cheeses but had considerably more 5% PTA-soluble N and free amino acids. Enzyme treatment caused textural defects: lipase caused a softening of the cheese, neutrase and lipase/protease treated cheeses were softer than the control cheeses, while NaturAge gave significant softening over the controls only at the highest level used. Fracturability of all cheeses correlated well with 12% TCA soluble-N levels during ripening. Fedrick *et al*.[76] found a significant correlation between both 12% TCA and 5% PTA soluble-N and many of the textural parameters measured during ripening. Organoleptic assessment of cheese flavour or estimated advancement in ripening were not reported by Lin et al.[83]

Fedrick et al.[20] investigated the effects of Neutrase in combination with a lac⁻/prt⁻ mutant starter and elevated ripening temperature. The effectiveness of the treatment decreased in the order MN15 > M15 > CN15 > C15 > MN8 > M8 > CN8 > C8. As in previous studies[22,74,75] treatment with Neutrase resulted in bitterness in some cases: however, in combination with the lac⁻/prt⁻ mutant, bitterness was less pronounced than with Neutrase alone, suggesting a role of the intracellular peptidases of the mutant starter. A 25% reduction in ripening time was found for Neutrase alone. Fedrick et al.[20] noted that secondary proteolysis or peptidolysis, assessed as 5% PTA soluble-N, appeared a rate-limiting step in ripening.

Primary and secondary proteolysis in a Swedish hard cheese were significantly accelerated when Neutrase and heat-shocked lactobacilli were added to cheesemilk.[35] Addition to cheesemilk of heat-shocked lactobacilli (without Neutrase), to give 10^6 cfu/g cheese after 24 h, significantly increased secondary proteolysis. Proteolysis by Neutrase alone increased the levels of 12% TCA soluble-N but bitterness developed. Neutrase plus heat-shocked lactobacilli did not increase gross proteolysis further, but amino N was increased significantly and the bitterness generated by Neutrase was eliminated.

Aminopeptidases of *Brevibacterium linens*, a major component of the surface microflora in surface-ripened cheeses, have recently been used in combination with Neutrase to accelerate the ripening of Cheddar.[84] The level of free amino acids was considerably higher in cheeses treated with Neutrase plus the aminopeptidase of *B. linens* than in cheeses treated with Neutrase plus a cell free extract of *L. lactis* subsp. *lactis*. A synergistic effect on flavour development was noted with the aminopeptidase and Neutrase or with Neutrase and CFE of *L. lactis* subsp. *lactis*. Maturation of Cheddar cheese could be reduced from four to six months to two months by addition of Neutrase plus *B. linens* aminopeptidase. The aminopeptidase was stable in cheese but its distribution in the curd was not uniform. Strain improvement would be necessary for the production of sufficient quantities of this aminopeptidase for large-scale cheese trials, e.g. the aminopeptidase produced in five litres of culture was sufficient to treat only 1 kg of curd.[84]

Partially purified extracellular serine proteinases from *B. linens* have been used to accelerate the ripening of Cheddar.[85] Enzymes were added at salting at various levels, i.e. 2·9, 8·6 and 26 enzyme units per kg curd, and compared with Neutrase-treated cheese (added at 8·6 units per kg) and untreated controls. Increased levels of TCA-soluble N were found in all enzyme-treated cheeses *pro rata* with the level of enzyme added: *B. linens* proteinase (26 units/kg curd) gave the highest levels of proteolysis followed by Neutrase. Degradation of β-casein was considerable in all enzyme-treated cheeses but no significant differences in the level of amino acids were found between enzyme treatments. Flavour development was significantly advanced in cheese treated with *B. linens* proteinase at 26 units after 2 months at 12°C.

Ripening of Ras cheese was accelerated by addition of heat-shocked lacto-bacilli,[86] proteinases/lipases[87] or Neutrase (at 50 mg/kg milk) plus intracellular enzymes from either *L. bulgaricus, Propionibacterium freudenreichii* or *Brevibacterium linens*.[88] Intracellular enzymes plus Neutrase did not accelerate gross proteolysis compared to Neutrase alone. Cheeses treated with Neutrase plus CFEs had more intense flavour than control cheeses or cheeses treated with Neutrase alone. However, bitterness increased in the cheeses with added Neutrase and CFE after two months, in contrast to other studies where CFE reduced the bitterness caused by Neutrase.[20,75,84] The addition of intracellular enzymes reduced ripening time by up to 25% with best results obtained with the CFE from *L. bulgaricus*.[88]

Addition of cell free extracts to accelerate ripening generally leads to the accumulation of small peptides and free amino acids without an increase in gross proteolysis.[88,89,75] A greater understanding of the specificity and debittering effect of these intracellular enzymes is necessary for further development of accelerated ripening systems.

6.2 Liposome Entrapment of Proteinases

Addition of enzymes to cheesemilk leads to a partitioning of the enzyme between curds and whey, representing an economic loss of enzyme, e.g. only 5–10% of Neutrase remains in the curd after pitching.[22] Alternative methods for enzyme addition to the cheese include the addition of a dried enzyme preparation with the salt at milling but this approach precludes its use for surface or brine-salted varieties. Methods of enzyme addition to cheese are shown in Table III.

Addition of encapsulated enzymes, enclosed in liposomes (artificial lipid membrane vesicles) to cheesemilk offers an alternative method for enzyme-supplementation of cheeses. Liposomes are composed of materials which are normal food components, and degradation of the vesicle membrane in cheese releases the enzyme into the curd. Liposome encapsulation also ensures uniform distribution of the enzyme in the curd since the encapsulated enzyme is added to the cheesemilk and entrapped in the gel at renneting.

Law & King[90] reported the successful use of multilamellar liposomes for entrapment of Neutrase added to milk for Cheddar cheese in laboratory-scale trials. Caseins were not hydrolysed during manufacture by liposome-entrapped

TABLE III
Methods of Enzyme Addition to Cheese[3]

Criteria of effectiveness	Direct to milk	Encapsulated	Direct to curd
Distribution	Good	Good	Poor
Curd texture	Poor	Normal	Normal
Yield	Reduced	Normal	Normal
Whey contamination	Complete	None	Press only

Neutrase, in contrast to free Neutrase which had hydrolysed ~20% of β-casein after cheddaring. Only 1–2% of the Neutrase originally present in a solution of the enzyme was entrapped in the liposomes, while 17% of added liposomes were retained in the curd. Release of the encapsulated Neutrase was inferred since proteolysis in the cheese was similar whether Neutrase was added in liposomes or in free form. This study showed that the use of liposomes was feasible, but encapsulation efficiency and retention in the curd were too low for large-scale cheese production.

Using a dehydration-rehydration technique for liposome preparation, higher entrapment efficiency (34%) and retention in the curd (90%) was reported.[91] Electron microscopy showed that the liposomes occurred as clusters, randomly distributed throughout the cheese. Proteolysis, as detected by gel electrophoresis or TCA-soluble N, occurred more gradually in the liposome-treated cheeses than when free Neutrase was added. The texture of cheese treated with free Neutrase deteriorated after three months while cheese treated with encapsulated enzyme was normal after 8 months. Cheeses treated with liposome-entrapped Neutrase plus the CFE of *L. lactis* subsp. *lactis* (added to curds at salting) were less crumbly than cheeses treated with free Neutrase plus intracellular peptidases. Acceleration of flavour development was noted for all experimental cheeses.[91]

Liposomes prepared using three different methods which produced multilamellar vesicles (MLV), small unilamellar vesicles (SUV) or reverse phase evaporation vesicles (REV) were used to encapsulate Corolase for addition to milk for Gouda cheese.[43] The proportion of enzyme encapsulated was highest for REV and lowest for SUV. Proteolysis was lower in cheeses containing liposome-entrapped enzyme than in those with free enzyme but was slightly greater than for the control. When added to milk for the manufacture of Saint Paulin-type cheese, REV-entrapped neutrase gave a significant advancement in ripening with no textural defects or bitterness.[92]

El Soda *et al.*[93] added temperature-sensitive liposomes (containing entrapped Corolase), designed for the early release of encapsulated enzyme, to milk for Domiati cheese. Cheeses were held at high temperatures for short periods to release the encapsulated proteinase; however, due to the high temperatures employed for encapsulation (50°C for 2 h), denaturation of the proteinase may have occurred. The authors state that liposomes engineered to have a phase transition within the pH range of most cheese varieties (5·0–5·5), so as to release the encapsulated enzymes, are currently under study and should provide a simpler method for controlled enzyme release.[93]

6.3 Lipases

Lipolysis plays a major role in the generation of flavour in certain cheese types such as Romano, Blue cheeses and Feta, but its importance in varieties such as Cheddar or Gouda is unclear. The volatile fraction of Cheddar cheese, containing fatty acids, contributes significantly to cheese aroma but not to taste; the water-insoluble fraction has no taste or aroma while the non-volatile water-soluble fraction contributes most to flavour intensity.[94,95] Lipolysis makes an important contribution to Swiss cheese flavour,[96,97] due mainly to the lipolytic enzymes of the starter cultures.

Pre-gastric esterase (PGE) activity in the rennet pastes used in the manufacture of some Italian cheese varieties is mainly responsible for the release of high levels of short chain fatty acids which produce the 'piccante', peppery flavour characteristic of these cheeses.[98,99] It has been reported[100,101] that an esterase from *Mucor miehei* can produce flavour in Romano and Fontina, very similar to that obtained with PGE. The flavour of these Italian cheeses was further enhanced by addition of a commercial lipase to the rennet paste.[102]

The characteristic peppery flavour of Blue cheese is due to short chain fatty acids and methyl ketones.[103] Most of the lipolysis in blue cheese is catalysed by *P. roqueforti* lipase, with a lesser contribution from indigenous milk lipase.[104,105] Blue cheese ripening was accelerated by the addition of lipases from *Aspergillus* sp.; however, the lipase preparation contained a proteinase which led to increased levels of free amino acids.[106] Inhibition of proteolysis has been associated with the use of an added animal lipase, but no inhibition of proteolysis was associated with the use of a fungal lipase; in this case, acceleration of Blue cheese ripening was due to both lipolysis and proteolysis.[103]

Production of a 'blue cheese food' from a blend of skim milk, UF retentate and cream (previously treated with a lipase from *A. oryzae*) has been described.[107] Using coconut fat in the formulation, a product having a characteristic blue cheese flavour was manufactured. Accelerated development of Blue cheese flavour has also been achieved by mixing acid or sweet whey with butterfat or coconut oil followed by fermentation by *P. roqueforti* and addition of *A. oryzae* lipase.[108] A continuous submerged fermentation process for the production of Blue cheese flavour by *P. roqueforti* has also been developed.[109]

Acceleration of flavour development in Ras and Domiati cheeses by addition of commercial animal lipases has been reported.[110,111] Ras cheese flavour was improved by addition of capalase K (from goat gastric tissues). Enzyme-treated cheeses acquired mature flavour after 45–60 days compared to 90 days for control cheeses.[110] In the case of Domiati cheese with added lipases (i.e. kid–goat–lamb or lamb PGE), the flavour intensity of four-week-old experimental cheeses was more pronounced than that of an 8 weeks-old untreated cheese; however, at higher levels of enzyme, rancidity developed after eight weeks.[111] Improvement of the flavour of Cheddar and Provolone was noted when rennet pastes were used or when gastric lipase was added with rennet extracts;[112] however, increased proteolysis occurred, indicating significant contamination of this lipase by proteinases.

The effect of lamb PGE on the flavour of Parmesan, Romano and Cheddar cheeses has been evaluated.[113] Treatment with gastric extract alone improved flavour compared to the controls, but cheeses with added gastric and pregastric enzyme extracts were considered to have best flavour. PGE extract alone produced an unbalanced lipolytic flavour in slurries of Pizza cheese or Cheddar, while addition of PGE to Cheddar cheese caused rancidity.[113]

Addition of lipase to American Cheddar has been reported[71,72] to accelerate flavour development but the value of adding lipases to accelerate ripening of English Cheddar is questionable.[114] Law & Wigmore[114] evaluated various commercial lipases of animal (lamb or calf) or microbial origin (*Mucor* sp.), with or without the inclusion of Neutrase (proteinase) at a level which accelerated the ripening of Cheddar.[75] None of the other lipases, either alone or in combination with Neutrase, appeared to enhance flavour after two months. Free fatty acid concentrations did increase markedly over the control but not *pro rata* with enzyme dosage. Rancidity was evident after two months, even at low dosage levels, leading to 'sweaty' or soapy off-flavours, depending on the enzyme added.[114]

The lipase of FlavourAge-FR (from *A. oryzae*), has a high specificity for C_6–C_{10} fatty acids, considered very important in correct flavour generation (unlike some other lipases, it did not produce either butyric acid which causes rancidity or dodecanoic acid, which causes soapy flavour).[78] Addition of this enzyme to cheese curds at salting resulted in significantly enhanced flavour over the controls. When added to cheese milk, this enzyme forms microscopic particles, >90% of which are entrapped in the cheese. In a comparison with calf PGE, which causes off-flavours, Flavourage-FR produced an FFA profile similar to the control cheese but at much greater concentrations.[78]

Addition of *L. lactis* subsp. *lactis* or *L. casei* cultures along with kid, lamb or kid/lamb PGEs to pasteurized milk for Feta cheese produced a cheese with characteristic body, aroma and texture.[115] Best flavour developed on addition of kid/lamb or lamb PGEs; excessive rancidity in some cheeses was associated with the production of C_{12} or higher fatty acids.[115] The use of animal or microbial lipases to accelerate the ripening of Feta cheese made from ultrafiltered milk has also been described with best results obtained for the microbial enzyme; unfortunately, the origin of these enzymes was not given.[116] Addition of various animal lipases has been reported to improve the flavour characteristics of Latin American white cheese.[117]

6.4 β-Galactosidase (Lactase)

β-Galactosidase (lactase) hydrolyses lactose to glucose and galactose. Treatment of milk with β-galactosidase prior to the manufacture of yogurt, buttermilk or Cottage cheese shortened the manufacturing time by 20%.[118] Yoghurts were 'sweeter' and 'less acid' and were generally regarded as being more acceptable than controls. Cottage cheese yields were increased by pre-treatment with lactase.[118] Stimulation of lactococci occurred in lactase-treated milk. The primary effect of β-galactosidase was to shorten the lag period of growth of lactococci; enzyme treatment did not result in an increase in cell numbers.[119]

Addition of lactase to Cheddar cheese milk has been reported to reduce manufacturing time, improve flavour and accelerate ripening by about 50%.[120,121] Proteolysis (12% TCA soluble N) was significantly increased in experimental cheeses over the control.[121] The enzyme used in the latter trial was from *Kluyveromyces lactis*, available as Maxilact (Gist-Brocades). In a further study[122] on the use of

Maxilact to accelerate the ripening of Cheddar cheese it was found that the preparation contained a proteinase, which was responsible for the increased levels of peptides and free amino acids and the improved flavour of experimental cheeses. No increase in starter cell numbers could be attributed to the action of this enzyme.[122]

Treatment of milk with Maxilact did not stimulate acid production by *L. helveticus*, *L. lactis* or *L. bulgaricus*, but did stimulate *S. thermophilus*,[123] apparently due to proteolytic products of the contaminating proteinase(s) in the β-galactosidase preparation rather than to hydrolysis of lactose. Grieve *et al.*[124] found that Maxilact contained acid endopeptidase, serine endopeptidase and serine exopeptidase activities. Caseinolytic activity was demonstrated over the pH range 5·0–8·0. The rate of hydrolysis of the individual caseins was in the order κ-, >αs-, >β; no bitterness was noted in any of the hydrolysates.

Gooda *et al.*[125] claimed that the increased rate of proteolysis and ripening in cheese from lactase-treated milk are not due entirely to a contaminating protease. These workers argue that Maxilact contains very weak proteolytic activity and that the increased proteolysis effected by its addition to Cheddar cheese milk is due to the stimulation of the starter and secondary flora, with numbers of lactococci and lactobacilli reaching 10 to 6 times those in the control cheeses after 1 day. These workers argued that increased proteolysis is due primarily to the increased level of starter peptidases resulting from such high numbers of bacteria in the lactase-treated cheeses. However, Law[3] reported that high starter cell numbers do not necessarily lead to increased flavour development and suggested that available evidence supports the proposal that accelerated ripening reported for Maxilact-treated cheese is due to contaminating proteinase(s).

7   CHEESE SLURRIES

Rapid development of cheese flavour has been reported in slurries of fresh curd with various added co-factors or enzymes[126,127] and ripened at elevated temperatures for short periods. Kristoffersen *et al.*[126] reported that mild Cheddar flavour developed in unpressed Cheddar curd dispersed in 5·2% NaCl during storage for nine days at 30°C. Addition of reduced glutathione intensified the flavour and appeared to stimulate proteolysis, lipolysis and bacterial growth. High concentrations of free fatty acids (C_4 or longer) were produced. It was concluded that reduced glutathione may contribute directly to flavour or cause the exposure of the sulphydryl groups of enzymes to exert an enzymatic effect on ripening. Addition of animal lipase and glutathione produced a characteristic Romano-type flavour in 5–10 days, depending on the level of lipase and reduced glutathione added. Prolonged storage without added lipase, but including reduced glutathione, resulted in the development of Brick-type cheese flavour.[126]

Further work on cheese slurries examined the effect of lactococcal cultures and curd milling acidity on flavour development; best results were obtained with a commercial mixed-strain starter, a milling acidity of 0·55 and inclusion of reduced glutathione.[128] Optimum flavour development occurred in slurries containing 100 ppm reduced glutathione, 3% NaCl, Na citrate, Mn^{2+}, riboflavin or cobalt, incubated at 30–35°C and agitated daily. This study also provided evidence that flavour development was related to the formation of active sulphydryl groups and free fatty acids.[129]

Acceptable Cheddar flavour developed in directly-acidified curds (with added rennet) on inclusion of reduced glutathione, cobalt, riboflavin, diacetyl, lactic starter bacteria and regular adjustment of the pH to 5·3, which retarded the growth of undesirable bacteria. For slurries heated to 62 or 70°C, rennet addition was necessary to restore the capacity to develop desirable flavours, which was found to be related to the degradation of α_{s1}-casein.[127]

The production of mature Swiss cheese flavour by the slurry technique was possible in 5–6 days at 30°C;[130] as in previous studies,[126–129] addition of reduced glutathione enhanced flavour development. Curd slurries have also been used in the manufacture of processed cheese.[131] Slurries were made with various inclusions: rennets with pH adjustment to 6·1 (giving a blue type cheese flavour), reduced glutathione (giving a 'sulphury' flavour), lipases, or storage with an initial head space volume of oxygen equivalent to twice the volume of slurry. When these slurries were added singly or in combination with a processed cheese mix at levels not exceeding 20% of the total formulation, a range of different flavours could be generated in the slurries, depending on treatment. The final product also differed considerably in flavour, depending on the slurry type, combinations of slurries and their level of inclusion in the formulation. Taste panel assessments of the final products were favourable for all slurries.

Acceleration of the ripening of natural Cheddar cheese by the inclusion of cheese slurries, either to the cheesemilk, to the curd prior to cheddaring or to the salted curd prior to pressing, has been reported.[132] Inclusion of 0·3% potassium sorbate in the slurry controlled yeast growth and the development of off-flavours. The preferred method of addition was to curd before hooping since it resulted in lower losses into the whey. Experimental cheeses showed an advancement in flavour and maturity over the controls; although they had a higher moisture content than control cheeses, the effect on ripening was considered to be due mainly to the added slurry rather than to the increased moisture content. Proteolysis (12%-TCA soluble N) was not increased over the control, as in other reports on accelerated ripening. However, the inclusion of slurries led to a marked increase in the numbers of lactobacilli which may have been responsible for increased flavour development.[132]

Rapid development (7 days at 30°C) of mature Ras cheese flavour occurred in slurried curds treated with either a proteinase/lipase mixture, a solution of trace elements or sodium citrate; the flavour was similar to that of conventional Ras cheese ripened for 2 months.[133] The effectiveness of the treatments on flavour development, proteolysis or lipolysis decreased in the order: proteinase/lipase mixture > trace elements > sodium citrate. The proteinase/lipase blend enhanced both proteolysis and lipolysis and also led to increased numbers of lipolytic or proteolytic organisms. The use of Ras cheese slurries to accelerate the ripening of Ras cheese, either by addition to the cheese milk or to the curd before hooping, was reported to reduce the ripening time by 50% (two as compared to four months.[134] Addition of slurry to the curd prior to hooping produced best results, with increased proteolysis and lipolysis and higher bacterial numbers.

For a Swedish hard cheese, addition of a slurry of mature cheese to cheese-milk prior to manufacture increased the numbers of lactobacilli in the cheese and produced a stronger flavour than normal after three months' ripening.[135]

The effect of adding an aseptic curd slurry inoculated with *P. roqueforti* and ripened at 25°C for 15 days at a level of 1 or 2% to blue cheese curd was compared to the effect of proteinases or lipases from *P. roqueforti* on the development of blue cheese flavour. Addition of slurry at 2% significantly stimulated proteolysis, lipolysis and flavour development after 45 days, with levels of proteolysis or lipolysis similar to those in untreated cheese after 60 days.[136] Results similar to those for the addition of 2% slurry were found on addition of *P. roqueforti* enzymes to the blue cheese curd.

The rapid production of blue cheese flavours by fermentation has been described.[102,106–109] More recently, accelerated production of blue cheese flavours by addition of lipase to cheesemilk or by coating granular or loose curds with lipase or with an enzyme-lipolysed cream has been reported.[137] The use of treated granular curd in this manner allowed ripening to be completed within 4–6 days. Increased levels of free fatty acids and methyl ketones were noted in all experimental cheeses.

A Camembert cheese flavour concentrate has also been prepared in granular curd form by addition of veal oral lipase to homogenized milk; a highly-flavoured product was produced in 12 days.[138] These flavour concentrates can be included in other foods, e.g. processed cheese dressings or dips, as a substitute for natural cheese.[139,140]

The mechanism of flavour development in slurries is unclear and the whole process is difficult to control; yeast growth causes off-flavours but this may be controlled by adding potassium sorbate to the slurry.[3] The slurry technique has been used to assess the suitability of proteinases, lipases or peptidases for use as agents to accelerate flavour development, offering the advantage of time and cost saving. Results are obtained rapidly and large-scale cheese manufacture is not required.[70,72] Bacterial cultures can also be screened in cheese slurries for their ability to produce low molecular weight nitrogenous compounds without off-flavour development.[22]

Enzyme-modified cheeses (EMC) are variants of the slurry system and involve enzyme addition to the cheese after manufacture or after ageing.[139,140] The EMC produced may have a texture similar to or slightly modified from the intended variety or may be produced as a paste. Moskowitz[139] described the manufacture of an Italian cheese spread which included EMC Parmesan or Romano at levels of 0·6% (w/w).

The major steps in the production of EMC include preparation of a cheese slurry with addition of a particular enzyme, homogenization to form an emulsion, incubation to reach the desired flavour level, and inactivation of the enzyme, usually by heating. This product may be used alone or added to other foods to provide a strong cheesy note at a low cost. In general, the level of addition of EMC's to foods ranges from 0·1 to 0·5%, reducing the requirement for a mature cheese in these formulations. The flavour intensity of EMC is 5–25 times that of the intended variety.

The NOVO process for production of EMC involves using medium-aged cheese which is emulsified, homogenized and pasteurized, after which 'palatase' (a lipase from *M. miehei*) is added, with or without a proteinase, and the blend ripened at a high temperature for one to four days. The mixture is reheated, producing a paste which is suitable for inclusion in soups, dips, dressings or snack foods. EMC technology has been developed to produce a range of cheese types and flavour intensities including Swiss, Blue, Cheddar, Provolone or Romano, suitable for inclusion at low levels in many products.[140]

8 CONCLUSION

Much research has been directed towards the acceleration of cheese ripening because of the obvious economic and technological advantages to be gained in shortening the ripening process. While the simplest method available to the producer is still elevation of the ripening temperature, provided that the cheese is of good compositional and microbiological quality, claims in the scientific literature that exogenous enzymes are effective in accelerating ripening have not led to their widespread use, possibly due to their high cost, difficulties in distributing them uniformly in the curd and the possible danger of over-ripening the cheese.

The use of starters with enhanced flavour-producing ability through an increase in their proteinase/peptidase complement by genetic manipulation, together with increased autolytic properties to effect an early release of these enzymes, may offer the best method of accelerating cheese ripening. However, this research effort should be paralleled by studies on the mechanism of cheese flavour development so that the more important enzymes in ripening can be identified, and the exact ways in which the various end products of glycolysis, lipolysis and proteolysis interact with each other to produce characteristic cheese flavour can be understood.

ACKNOWLEDGEMENT

I wish to acknowledge the valuable assistance of Professor P.F. Fox in the preparation of this chapter, and also the advice and comments of Dr T.P. Guinee.

REFERENCES

1. Fox, P.F., 1988–89. *Food Biotechnol*, **2**, 133.
2. Fox, P.F., 1987. In Cheese: Chemistry, Physics and Microbiology, Vol. 1. Elsevier Applied Science, London, p. 1.
3. Law, B.A., 1986. *Ann. Rept.*, Food Research Institute, Reading, p. 111.
4. Fox, P.F., 1989. *J. Dairy Sci.*, **72**, 1379.
5. Wilkinson, M.G., 1990. In *Proc. 2nd Moorepark Cheese Symposium*, p. 111.
6. El-Soda, M. & Pandian, S., 1991. *J. Dairy Sci*, **74**, 2317.
7. Walstra, P., Noomen, A. & Geurts, T.J., 1987. In *Cheese: Chemistry, Physics and Microbiology*, Vol. 2. Elsevier Applied Science, London, p. 45.
8. Fedrick, I.A., Aston, J.W., Durward, I.G. & Dulley, J.R., 1983. *N.Z. J. Dairy Sci. Technol.*, **18**, 253.
9. Law, B.A., Hosking, Z.D. & Chapman, H.R., 1979. *J. Soc. Dairy Technol.*, **32**, 87.
10. Aston, J.W., Grieve, P.A., Durward, I.G. & Dulley, J.R., 1983. *Aust. J. Dairy Technol.*, **38**, 59.
11. Aston, J.W., Fedrick, I.A., Durward, I.G. & Dulley, J.R., 1983. *N.Z. J. Dairy Sci. Technol.*, **18**, 143.
12. Aston, J.W., Durward, I.G. & Dulley, J.R., 1983. *Aust. J. Dairy Technol.*, **38**, 55.
13. Aston, J.W. & Douglas, K., 1983. *Aust. J. Dairy Technol.*, **38**, 66.
14. Aston, J.W., Giles, J.E., Durward, I.G. & Dulley, J.R., 1985. *J. Dairy Res.*, **52**, 565.
15. Fedrick, I.A. & Dulley, J.R., 1984. *N.Z. J. Dairy Sci. Technol.*, **19**, 141.
16. Jarrett, W.D., Aston, J.W. & Dulley, J.R., 1982. *Aust. J. Dairy Technol.*, **37**, 55.
17. Cromie, S.J., Gilles, J.E. & Dulley, J.R., 1987. *J. Dairy Res.*, **54**, 69.
18. Fryer, T.F., 1982. *Proc. XXI Intern. Dairy Congr. (Moscow)*, **1**, 485.
19. Fedrick, I.A., 1987. Paper presented to Annual Conference of the Victorian Division of the Dairy Industry Association of Australia, Melbourne, April 8, 1987.
20. Fedrick, I.A., Cromie, S.J. & Dulley, J.R., 1986. *N.Z. J. Dairy Sci. Technol.*, **21**, 191.
21. Guinee, T.P., Wilkinson, M.G., Mulholland, E.D. & Fox, P.F., 1992. *Ir. J. Food Sci. Technol.*, **15**, 27.
22. Law, B.A., 1990. *Proc. XXIII Intern. Dairy Congr. (Montreal)*, **2**, 1616.
23. Dilanyan, Z.Kh., 1980. Principles of Cheese Making, *Moskva: Pischevaya promyshlennost.*
24. Vassal, L., Desmazeaud, M.J. & Gripon, J.-C., 1982. In *Use of Enzymes in Food Technology*, International Symposium 1982, Versailles, p. 315.
25. Hofi, A.A., Mahran, G.A., Abd El Salam, M.H. & Rifaat, I.D., 1973. *Egypt. J. Dairy Sci.*, **1**, 33.
26. Abd El Salam, M.H., Rifah, E.D., Hofi, A.A. & Mahran, G.A., 1973. *Egypt. J. Dairy Sci.*, **1**, 171.
27. Law, B.A., Castanon, M.J. & Sharpe, M.E., 1976. *J. Dairy Res.*, **43**, 301.
28. Thomas, T.D. & Pritchard, G.G., 1987. *FEMS Microbiol. Rev.*, **46**, 245.
29. Wright, P.M., 1990. *Cultured Dairy Products J.*, **25**(2), 11.
30. Shearman, C., Underwood, H., Jury, K. & Gasson, M., 1989. *Mol. Gen. Genet.*, **218**, 214.
31. Gasson, M.J., 1990. International Patent Application. Application no. PCT/4B89/OO791.
32. Bie, R. & Sjostrom, G., 1975. *Milchwissenschaft*, **30**, 653.
33. Bie. R. & Sjostrom, G., 1975. *Milchwissenschaft*, **30**, 730.
34. Pettersson, H.E. & Sjostrom, G., 1975. *J. Dairy Res.*, **42**, 313.
35. Ardö, Y. & Pettersson, H.E., 1988. *J. Dairy Res.*, **55**, 239.
36. Ardö, Y. & Larsson, P-O., 1989. *Milchwissenschaft*, **44**, 485.
37. Lopez-Fandino, R. & Ardö, Y., 1991. *J. Dairy Res.*, **58**, 469.
38. Ardö, Y. & Mansson, H.L., 1990. *Scand. Dairy Information*, **4**, 38.
39. Bartels, H.J., Johnson, M.E. & Olson, N.F., 1987. *Milchwissenschaft*, **42**, 83.

40. Bartels, H.J., Johnson, M.E. & Olson, N.F., 1987. *Milchwissenschaft*, **42**, 139.
41. Vafopoulou, A., Alichanidis, E. & Zerfidis, G., 1989. *J. Dairy Res.*, **56**, 285.
42. Castaneda, R., Vassal, L., Gripon, J.-C. & Rousseau, M., 1990. *Neth. Milk Dairy J.*, **44**, 49.
43. Spangler, P.L., El Soda, M., Johnson, M.E., Olson, N.F., Amundson, C.H. & Hill, C.G. Jr., 1989. *Milchwissenschaft*, **44**, 199.
44. Exterkate, F.A., de Veer, G.J.C.M. & Stadhouders, J., 1987. *Neth. Milk Dairy J.*, **41**, 307.
45. Grieve, P.A., Lockie, B.A. & Dulley, J.R., 1983. *Aust. J. Dairy Technol.*, **38**, 10.
46. Grieve, P.A. & Dulley, J.R., 1983. *Aust. J. Dairy Technol.*, **38**, 49.
47. Oberg, C.J., Davis, L.H., Richardson, G.H. & Ernstrom, C.A., 1986. *J. Dairy Sci.*, **69**, 2975.
48. Farkye, N.Y., Fox, P.F., Fitzgerald, G.F. & Daly, C., 1990. *J. Dairy Sci.*, **73**, 874.
49. Stadhouders, J., Teepoll, L. & Wouters, J.T.M., 1988. *Neth. Milk Dairy J.*, **42**, 183.
50. Kamaly, K.M., Johnson, M.E. & Marth, E.H., 1989. *Milchwissenschaft*, **44**, 343.
51. Kamaly, K.M. & Marth, E.H., 1988. *J. Dairy Sci.*, **71**, 2349.
52. Mills, O.E. & Thomas, T.D., 1980. *N.Z. J. Dairy Sci. Technol.*, **15**, 131.
53. Nardi, M., Chopin, M.C., Chopin, A., Cals, M.M. & Gripon, J.-C., 1991. *Appl. Environ. Microbiol.*, **57**, 45.
54. Mayo, B., Kok, J., Venema, K., Bockelmann, W., Teuber, M., Reinke, H. & Venema, G., 1991. *Appl. Environ. Microbiol.*, **57**, 38.
55. Peterson, S.D. & Marshall, R.T., 1990. *J. Dairy Sci.*, **73**, 1395.
56. Broome, M.C, Krause, D.A. & Hickey, M.W., 1990. *Aust. J. Dairy Technol.*, **45**, 67.
57. Puchades, R., Lemieux, L. & Simard, R.E., 1989. *J. Food Sci.*, **54**, 885.
58. Lemieux, L., Puchades, R. & Simard, R.E., 1989. *J. Food Sci.*, **54**, 1234.
59. Lee, B.H., Laleye, L.C., Simard, R.E., Holley, R.A., Emmons, D.B. & Giroux, R.N., 1990. *J. Food Sci.*, **55**, 386.
60. Laleye, L.C., Simard, R.E., Lee, B.H. & Holley, R.A., 1990. *J. Food Sci.*, **55**, 114.
61. Lee, B.H., Laleye, L.C., Simard, R.E., Munsch, M.H. & Holley, R.A., 1990. *J. Food Sci.*, **55**, 391.
62. Trepanier, G., Simard, R.E. & Lee, B.H., 1991. *J. Food Sci.*, **56**, 1238.
63. Bhowmik, T., Riesterer, R., van Bockel, M.A.J.S. & Marth, E.H., 1990. *Milchwissenschaft*, **45**, 230.
64. Olson, N.F., 1990. *FEMS Microbiol. Rev.*, **87**, 131.
65. Holmes, D.G., Duersch, J.W. & Ernstrom, C.A., 1977. *J. Dairy Sci.*, **60**, 862.
66. O'Keeffe, R.B., Fox, P.F. & Daly, C., 1976. *J. Dairy Res.*, **43**, 97.
67. Stadhouders, J., 1960. *Neth. Milk Dairy J.*, **14**, 83.
68. Creamer, L.K., Lawrence, R.C. & Gilles, J., 1985. *N.Z. J. Dairy Sci. Technol.*, **20**, 185.
69. Creamer, L.K., Aston, J. & Knighton, D., 1988. *N.Z. J. Dairy Sci. Technol.*, **11**, 30.
70. Sood, V.K. & Kosikowski, F.V., 1979. *J. Dairy Sci.*, **62**, 1865.
71. Kosikowski, F.V. & Iwasaki, T., 1975. *J. Dairy Sci.*, **58**, 963.
72. Sood, V.K. & Kosikowski, F.V., 1979. *J. Dairy Sci.*, **62**, 1713.
73. Law, B.A., 1980. *Dairy Ind. Intern.*, **45**, 15, 17, 19, 20, 22, 48.
74. Law, B.A. & Wigmore, A.S., 1982. *J. Soc. Dairy Technol.*, **35**, 75.
75. Law, B.A. & Wigmore, A.S., 1982. *J. Dairy Res.*, **49**, 137.
76. Fedrick, I.A., Aston, J.W., Nottingham, S.M. & Dulley, J.R., 1986. *N.Z. J. Dairy Sci. Technol.*, **21**, 9.
77. Frick, C.M., Hicks, C.L. & O'Leary, J., 1984. *J. Dairy Sci.*, **67** (suppl. 1), 89.
78. Arbige, M.V., Freund, P.R., Silver, S.C. & Zelko, J.T., 1986. *J. Food Technol.*, **40**(4), 91.
79. Law, B.A. & Wigmore, A.S., 1983. *J. Dairy Res.*, **50**, 519.
80. Cliffe, A.J. & Law, B.A., 1990. *Food Chem.*, **36**, 73.
81. Cliffe, A.J., Revell, D.F. & Law, B.A., 1989. *Food Chem.*, **34**, 147.

82. El Soda, M., Desmazeaud, M.J., Abou Donia, S. & Badran, A., 1982. *Milchwissenschaft*, **37**, 325.
83. Lin, J.C.C., Jeon, I.J., Roberts, H.A. & Milliken, G.A., 1987. *J. Food Sci.*, **52**, 620.
84. Hayashi, K., Revell, D.F. & Law, B.A., 1990. *J. Dairy Res.*, **57**, 571.
85. Hayashi, K., Revell, D.F. & Law, B.A., 1990. *J. Dairy Sci.*, **73**, 579.
86. Abd El-Baky, A., El-Neshawy, A., Rabie, A. & Ashour, M., 1986. *Food Chem.*, **21**, 201.
87. Abd El Salam, M.H., Mohamed, A., Ayad, E., Fahmy, N. & El Shibiny, S., 1979. *Egypt. J. Dairy Sci.*, **7**, 63.
88. Ezzat, N., 1990. *Le Lait*, **70**, 459.
89. Gripon, J.C., Desmazeaud, M.J., Le Bars, D. & Bergere, J.L., 1977. *J. Dairy Sci.*, **60**, 1532.
90. Law, B.A. & King, T.S., 1985. *J. Dairy Res.*, **52**, 183.
91. Kirby, C.J., Brooker, B.E. & Law, B.A., 1987. *Int. J. Food Sci. Technol.*, **22**, 355.
92. Alkhalaf, W., Paird, J.-C., El Soda, M. Gripon, J.-C., Desmazeaud, M. & Vassal, L., 1988. *J. Food Sci.*, **53**, 1674.
93. El Soda, M., Johnson, M. & Olson, N.F., 1989. *Milchwissenschaft*, **44**, 213.
94. McGugan, W.A., Emmons, D.B. & Larmond, E., 1979. *J. Dairy Sci.*, **62**, 398.
95. Aston, J.W. & Creamer, L.K., 1986. *N.Z. J. Dairy Sci.*, **21**, 229.
96. Biede, S.L. & Hammond, E.G., 1979. *J. Dairy Sci.*, **62**, 227.
97. Biede, S.L. & Hammond, E.G., 1979. *J. Dairy Sci.*, **62**, 238.
98. Fox, P.F., 1981. *J. Soc. Dairy Technol.*, **33**, 118.
99. Arnold, R.G., Shahani, K.M. & Dwivedi, B.K., 1975. *J. Dairy Sci.*, **52**, 1127.
100. Moskowitz, G.J., Shew, T., West, I.R., Cassaigne, R. & Feldman, L.I., 1977. *J. Dairy Sci.*, **60**, 1260.
101. Peppler, H.J., Dooley, J.G. & Huang, H.T., 1976. *J. Dairy Sci.*, **59**, 859.
102. Botazzi, V., 1965. *Scienza e Tecnica Lattiero-Casearia*, **16**, 229.
103. Jolly, R.C. & Kosikowski, F.V., 1975. *J. Agric. Food Chem.*, **23**, 1175.
104. Lambert, G. & Lenoir, J., 1972. *Le Lait*, **52**, 175.
105. Monassa, A. & Lambert G., 1982. *Proc. XXI Intern. Dairy Congr. (Moscow)*, Vol. 1, Book 1, p. 509.
106. Jolly, R.C. & Kosikowski, F.V., 1978. *J. Dairy Sci.*, **61**, 536.
107. Jolly, R.C. & Kosikowski, F.C., 1974. *J. Dairy Sci.*, **58**, 1272.
108. Jolly, R.C. & Kosikowski, F.V., 1975. *J. Food Sci.*, **40**, 285.
109. Dwivedi, B.K. & Kinsella, J.E., 1974. *J. Food Sci.*, **39**, 620.
110. Abd El Salam, M.H., El-Shibiny, S., El-Bogoury, E., Ayad, E. & Fahmy, N., 1978. *J. Dairy Res.*, **45**, 491.
111. El-Neshawy, A.A., Abd El Baky, A.A. & Farahat, S.M., 1982. *Dairy Ind. Intern.*, **47**(2), 29.
112. Richardson, G.H., Nelson, J.H. & Farnham, M.G., 1971. *J. Dairy Sci.*, **54**, 643.
113. Chaudhari, R.V. & Richardson, G.H., 1971. *J. Dairy Sci.*, **54**, 467.
114. Law, B.A. & Wigmore, A.S., 1985. *J. Soc. Dairy Technol.*, **38**, 86.
115. Efthymiou, C.C. & Mattick, J.F., 1964. *J. Dairy Sci.*, **47**, 593.
116. Mahmoud, M.M. & Kosikowski, F.V., 1980. *J. Dairy Sci.*, **63**(suppl 1), 47.
117. Torres, N. & Chandan, R.C., 1981. *J. Dairy Sci.*, **64**, 2161.
118. Thompson, M.P. & Gyuricsek, D.M., 1974. *J. Dairy Sci.*, **57**, 584.
119. Gilliland, S.E., Speck, M.L. & Woodard, J.R., 1972. *Appl. Microbiol.*, **23**, 21.
120. Thompson, M.P. & Brower, D.P., 1976. *Cultured Dairy Prod. J.*, **11**, 22.
121. Marschke, R.J. & Dulley, J.R., 1978. *Aust. J. Dairy Technol.*, **33**, 139.
122. Marschke, R.J., Nickerson, D.E.J., Jarrett, W.D. & Dulley, J.R., 1980. *Aust. J. Dairy Technol.*, **35**, 84.
123. Hemme, D., Vassal, L., Foyen, H. & Auclair, J., 1979. *Le Lait*, **59**, 597.
124. Grieve, P.A, Kitchen, B.J., Dulley, J.R. & Bartley, J., 1983. *J. Dairy Res.*, **50**, 469.
125. Gooda, E., Bednarski, N. & Poznanski, S., 1983. *Milchwissenschaft*, **38**, 83.

126. Kristoffersen, T., Mikolajcik, C.J. & Gould, I.A., 1967. *J. Dairy Sci.*, **50**, 292.
127. Singh, S. & Kristoffersen, T., 1972. *J. Dairy Sci.*, **55**, 744.
128. Singh, S. & Kristoffersen, T., 1971. *J. Dairy Sci.*, **54**, 1589.
129. Singh, S. & Kristoffersen, T., 1970. *J. Dairy Sci.*, **53**, 533.
130. Singh, S. & Kristoffersen, T., 1971. *J. Dairy Sci.*, **54**, 349.
131. Sutherland, B.J., 1975. *Aust. J. Dairy Technol.*, **30**, 138.
132. Dulley, J.R., 1976. *Aust. J. Dairy Technol.*, **31**, 143.
133. Abd El Baky, A.A., El Neshewy, A., Rabie, A.H.M. & Farahat, S., 1982. *J. Dairy Res.*, **49**, 337.
134. Abd El Baky, A.A., El Fak, A.M., Rabie, A.M. & El Neshawy, A.A., 1982. *J. Food Protect.*, **45**, 894.
135. van Bockelmann, I. & Lodin, I.O., 1974. *Proc. XIX Intern. Dairy Congr. (New Delhi)*, **IE**:441.
136. Revah, S. & Lebeault, J.M., 1989. *Le Lait*, **69**, 281.
137. Rabie, A.M., 1989. *Le Lait*, **69**, 305.
138. Furtado, M.M., Chandan, R.C. & Wishnetsky, T., 1984. *J. Dairy Sci.*, **67**, 2850.
139. Moskowitz, G. & Noerck, S.S., 1987. *J. Dairy Sci.*, **70**, 1761.
140. Kilara, A., 1985. *Process Biochem.*, **20**(2), 35.

15

Nutritional Aspects of Cheese

E. RENNER

Dairy Science Section, Justus Liebig University, Giessen, Germany

1 CHEESE CONSUMPTION

Although the consumption of liquid milk and most dairy products is decreasing, there has been a steady increase in the consumption of cheese in most countries; worldwide, the annual growth rate in cheese consumption is about 3%. In Germany, cheese consumption has risen since 1962 from 8 to 18 kg per caput per annum. Per caput cheese consumption is highest in France while consumption is very low in most South American, African and Asian countries (Fig. 1).

2 NUTRIENTS OF CHEESE

2.1 Milk Fat

By adjusting the fat content of cheese milk to different values, cheeses of widely different fat contents (usually expressed as % fat-in-dry matter) are produced. Fresh cheeses have an absolute fat content of up to 12%, while ripened cheeses, in general, contain between 20 and 30% fat (Table I). Consumers generally prefer high-fat cheeses because a high fat content contributes significantly to flavour quality. The typical aroma of some types of cheese, for instance Cheddar, develops only when the fat-in-dry matter content is at least 40–50% because the aroma is due mainly to the breakdown products of fat formed during cheese ripening.

The production of low-fat cheeses offers great opportunities to market new products which are perceived as 'healthy' due to their low fat content.[4] It is, however, essential that low-fat cheeses are organoleptically acceptable.[5]

Lipolysis during cheese ripening is caused primarily by microbial lipases because the indigenous lipase in milk is largely inactivated by pasteurization (except in cheeses in which rennet paste or pregastric esterase is used). As a result of lipolysis, the concentration of free fatty acids in cheese is usually 1–5 g/kg. There is a close link between the content of free volatile fatty acids in a number of cheese varieties and their flavour.[6]

557

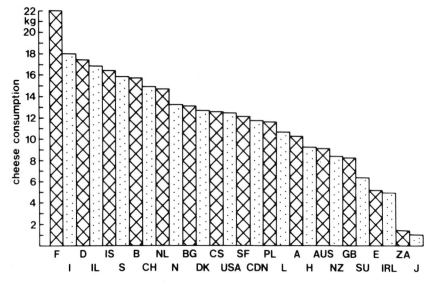

Fig. 1. Consumption (kg per caput per annum) of cheese in some countries (1988).[1]

TABLE I
Average Content of Fat and Protein of a Number of Cheese Varieties[2,3]

Cheese variety	Moisture content (%)	Fat content		Protein content (%)
		Fat-in-dry matter (%)	Absolute (%)	
Parmesan	31	35	26·0	37·5
Emmentaler	36	45	30·0	28·9
Tilsiter	46	45	25·4	24·1
Cheddar	37	50	32·4	25·4
Edam/Gouda	46	45	25·4	24·1
Butter cheese	49	50	26·5	19·4
Blue cheese	43	50	29·0	22·4
Brie	52	50	25·5	21·1
Camembert	54	45	21·8	21·0
Limburger	53	40	19·7	23·2
Romadur	57	30	14·1	24·8
Feta	63	40	16·0	18·4
Cottage cheese	80	20	5·0	10·0
Fresh cheese	75	40	10·3	9·0
	82	skimmed	0·2	12·3

In Germany, cheese contributes only 4·7% of total fat intake (128 g per caput per day). As to the cholesterol content of cheese, it must be emphasized that:

—The cholesterol content of cheese is rather low (0–100 mg/100 g, depending on the fat content).[2]

—Therefore, cheese contributes only 3–4% of total cholesterol intake.

—Cholesterol in the diet has only a limited effect on the level of blood cholesterol. The body has a control mechanism which ensures that the synthesis of cholesterol by the body is reduced when the amount of cholesterol consumed increases.[7,8]

The coefficient of digestibility of the fat of different varieties of cheese is reported to be 88–94%.[2]

2.2 Protein

The nutritional importance of cheese arises from its high content of biologically valuable proteins. Table I shows that the protein content of different varieties of cheese varies between 20 and 35%. Within any one type of cheese, the protein content varies inversely with the fat content. The nutrient density (related to the energy content) for the protein content of different types of cheese is 2·9–6·2. A 100 g portion of soft cheese will provide 30–40% of the daily protein requirements of an adult and 100 g of a hard cheese will supply 40–50%.

In cheese manufacture, the casein of milk is incorporated into the cheese while most of the biologically valuable whey proteins pass into the whey. Thus, 75–80% of the total protein and about 95% of the casein are transferred from the milk to the fresh cheese.[9,10] Therefore, in cheeses produced from pasteurized milk, the content of whey protein in total cheese protein is 4–6%.[11] Since the whey proteins are nutritionally superior to casein, which is somewhat deficient in sulphur amino acids, the biological value of the proteins in cheese is somewhat lower than that of the total milk protein, but is still higher than that of casein alone. In Cheddar cheese, the PER values were found to be significantly higher than for casein (3·7 vs 2·5).[12] The NPU values of Camembert, Cheddar and Gouda were found to be between 75·0 and 76·3, compared with 78·0 for whole milk.[13] If the essential amino acid index of total milk protein is given a value of 100, then the corresponding values of the proteins in a number of cheese varieties range from 91 to 97. The biological value of the proteins is not impaired by the action of rennet or of other enzymes active during cheese ripening, nor is it affected by acid formation. The Maillard reaction does not occur during cheese manufacture, so that the availability of lysine in cheese is almost the same as in milk. Ripening periods of 16–20 weeks produce no significant changes in the NPU and PER values of the proteins of Tilsiter and Gouda cheeses; in fact, in some cases the NPR and PER values of cheese proteins are higher even than those of milk proteins.[14-16]

When ultrafiltration is used to concentrate cheese milk up to the dry matter content of cheese so that no whey is produced, the whey proteins are also incor-

Table II

Concentration of Essential Amino Acids in Milk and Cheese Protein, Compared to the Reference Protein[18]

Essential amino acid	Content (g/100 g protein)		
	Reference protein FAO/WHO	Milk protein	Cheese protein
Tryptophan	1·0	1·4	1·4
Phenylalanine + tyrosine	6·0	10·5	10·9
Leucine	7·0	10·4	10·4
Isoleucine	4·0	6·4	5·8
Threonine	4·0	5·1	4·8
Methionine + cystine	3·5	3·6	3·2
Lysine	5·5	8·3	8·3
Valine	5·0	6·8	6·8
Total	36·0	52·5	51·6

porated into the cheese, thereby improving the nutritive value of the protein. The whey proteins in such a cheese represent about 15% of the total protein.

Cheese can contribute significantly to the supply of essential amino acids. In Table II, where the amino acid composition of milk and cheese proteins are compared to the reference protein, which indicates the ideal concentration of essential amino acids in a dietary protein, it can be seen that cheese protein meets the requirements to the same extent as milk protein, except those for methionine plus cystine.

During cheese ripening, part of the water-insoluble casein is converted into water-soluble nitrogenous compounds which include the intermediate products of protein hydrolysis as well as free amino acids. Cheese ripening can be looked upon as a sort of predigestion whereby the digestibility of the proteins is increased. The true digestibility of a number of cheese varieties is almost 100%;[17] the protein digestibility of cheese is considered to be higher than that of whole milk (96·2–97·5% vs 91·9%).[13] Small peptides can pass through the walls of the intestine and it is possible that they penetrate even cell membranes so that they become directly available to the cell. An experiment with rats demonstrated that the rate of utilization of cheese protein was higher than the rate for casein. The mean degree of utilization of the essential amino acids in cheese protein is 89·1%, i.e. greater than the corresponding value for milk protein (which is 85·7%) and almost equal to the value for egg protein, which is 89·6%. The free amino acids of cheese, particularly aspartic and glutamic acids, are said to promote the secretion of gastric juices. It should be noted that a food allergy to cheese protein has never been described.[17]

5-Aminovaleric acid, a degradation product of casein, could not be detected in cheese; a physiological significance was attributed to this compound by inhibiting the binding of γ-aminobutyric acid to brain cortex receptors.[19]

Free D-amino acids are common in ripened cheeses (mg/100 g and % of total free amino acids, respectively): Gouda, 18·5 and 2·6; Trappist cheese, 44·0 and 4·7;

TABLE III
Average Concentration of Tyramine and Histamine in Some Cheese Varieties[23,28,30-38]

Cheese variety	Content of	
	Histamine, mg/100 g	Tyramine, mg/100 g
Cheddar	16	97
Emmentaler, Gruyère	17	18
Blue cheese	9	25
Edam, Gouda	21	17
Camembert, Brie	13	16

Emmentaler, 83·1 and 3·7; Parmesan, 403 and 5·5. D-Amino acids are considered to be of no nutritional value for humans. Therefore, their concentration has, particularly in the case of essential amino acids and those present in high amounts, to be subtracted from the total amount of amino acids used for nutritional calculations. Since D-amino acid oxidase is present in the kidney, liver and brain of almost all vertebrates, low amounts of D-amino acids may not be harmful to man.[20]

The decarboxylation of free amino acids during cheese ripening produces amines. The principal amines found in cheese are histamine, tyramine, tryptamine, putrescine, cadaverine and phenylethylamine. The concentrations of individual amines in cheese show great variations and depend on the ripening period, the quality of the base milk, the storage conditions, the salt content and the microbial flora.[21,22] Average values for the contents of tyramine and histamine in different types of cheese are shown in Table III. It is evident that Cheddar cheese contains an astonishingly high concentration of tyramine.[23] In contrast to earlier findings, blue cheeses do not have substantially elevated concentrations of amines, although tyramine values up to 420 mg/100 g have been reported by Evans et al.[24] There is little difference between hard, semi-hard and soft cheeses.[25-28] As a rule, ripened cheeses contain 10–30 mg histamine per 100 g and 15–40 mg tyramine.[29] The concentration of amines is very low in cottage cheese, quarg, cream cheese, curd cheese and processed cheese.[24,30]

The phenylethylamine content of cheese is clearly below 10 mg/100 g;[29] the concentration of the amines tryptamine, putrescine and cadaverine is usually in the range between 0 and 5 mg/100 g.

Physiologically active amines can affect the blood pressure, with tyramine and phenylethylamine having a hypertensive and histamine a hypotensive effect. However, mono- and diamine oxidases convert the biogenic amines that are consumed in food relatively quickly into aldehydes and finally into carboxylic acids by oxidative deamination. Although opinions on the toxicity threshold values of amines vary widely (for tyramine they are 10–80 mg, for histamine, 70–100 mg), it is concluded that healthy persons are able to metabolize the biogenic amines ingested, even when large amounts of cheese are consumed, without adverse physiological reactions.[31,39,40]

Some sensitive people may be subject to attacks of migraine as a result of eating cheese. It is possible that such persons suffer from a genetically determined lack of monoamine oxidase. The consumption of 100 mg of tyramine produces severe headaches in a large number of these patients. Also, when patients suffering from high blood pressure or similar disorders are treated with drugs containing a monoamine oxidase inhibitor, the normal breakdown of amines in the body is prevented. Cases have been reported in which the consumption of cheese by such patients has led, within 30 min to 2 h, to hypertensive reactions which are therefore called the 'cheese syndrome'. However, such drugs are rarely used nowadays. In those cases where drugs containing monoamine oxidase inhibitors are prescribed, the patient should be warned not to eat cheese or other tyramine-containing foods during the period of treatment, although the great variation in the tyramine content of different types of cheese does not make an adverse reaction inevitable.[41-43] Biogenic amines may also be potential producers of pseudoallergies, i.e. intolerance reactions which are not coupled to an immune mechanism.[44,45]

2.3 Lactose and Lactic Acid

There is no lactose in many cheeses or only a very low concentration (1–3 g/ 100 g) because most of the lactose of the milk passes into the whey and that retained in the cheese curd is partly or fully converted to lactic acid during cheese ripening. Therefore, cheese is suitable for the diets of persons suffering from lactose malabsorption and of diabetics.[46]

The average lactic acid content of a number of cheeses is as follows: Parmesan, 0·7%; Cheddar, 1·3%; Tilsiter, 1·0%, Quarg (curd cheese), 0·7% Blue cheese, 0·6%; Emmentaler, 0·4%; Cottage cheese, 0·3%; Camembert, 0·2%.[47] Cheese usually contains both lactic acid isomers, L(+) and D(−), the relative proportion of the D-isomer depending on the type of starter culture used and on some other ripening factors. During storage, a conversion of L(+) lactate to D(−) lactate occurs by non-starter bacteria; therefore, after storage, D(−) lactate may be present at substantial concentrations.[48] The content of D(−) lactic acid in different types of cheese can be very different (fresh cheese 4–14%; ripened cheeses, 10–50%).[49-51] D(−) lactic acid can be metabolized by humans only to a certain extent by the non-specific mitochondrial enzyme D-2-hydroxy acid dehydrogenase,[52] but from the data available in the literature, a toxic effect of D(−) lactic acid cannot be derived for the adolescent or the adult. As a logical conclusion, the WHO, in a revised statement, has not limited the admissible intake for adults, while for infants (up to one year of age), a D(−) lactic acid-free diet is recommended.[48,53]

2.4 Minerals

The calcium and phosphorus contents of cheese are as important as those of milk, since 100 g of soft cheese will supply 30–40% of the daily Ca requirement and 12–20% of the daily P requirement and 100 g of a hard cheese will meet the

TABLE IV
Average Content of Minerals in Various Cheese Varieties[3,66-76]

Cheese variety	Content (mg/100 g) of				
	Calcium	Phosphorus	Sodium	Potassium	Magnesium
Parmesan	1300	850	1200	100	44
Cheddar	760	500	640	90	30
Emmental	1080	730	250	90	43
Edam/Gouda	800	600	800	100	40
Tilsit	800	500	750	100	40
Blue cheese	420	350	1200	110	50
Mozzarella	400	340	450	100	16
Camembert	350	300	930	150	20
Processed cheese	600	600	135	100	24
Cottage cheese	80	140	380	75	8
Fresh cheese skimmed	90	190	30	120	9

daily Ca requirement completely and contribute 40–50% of the P requirement. In the USA, a serving of cheese is said to contribute on average about 25% and up to 42% of the recommended dietary allowance of calcium.[54] In Germany and France, adults ingest about 150 mg Ca per caput per day from cheese, i.e. about 20% of the daily Ca intake. The average concentrations of Ca and P, as well as of some other minerals, in a number of cheese varieties are shown in Table IV. It should be noted that where one variety of cheese is made with different fat contents, the higher fat cheese contains less Ca and P. The nutrient density for Ca in different types of cheese is 1·3–7·0 and for P, 2·6–5·7. Cheeses produced by rennet coagulation usually have higher calcium contents than those made from acid-coagulated milk.[55] About 60–65% of the calcium and 50–55% of the phosphorus of milk are retained in Tilsiter and Trappist cheeses. The calcium, phosphorus and magnesium in cheese are utilized as well by the body as those in milk.[56] The physicochemical changes occurring during cheese manufacture and ripening do not affect calcium bioavailability; Ca absorption by rats from Cheddar cheese averaged 76·8%.[57] While the phosphorus absorption by human subjects was 64% from milk and 62% from cheese, it was only 29% from wholemeal rye bread.[58] The ratio of calcium to phosphorus in cheese is also thought to be desirable nutritionally. Cheese is one of the foods that is not cariogenic.[59]

The wide range of the Na contents is due to the different amounts of NaCl added to cheeses; the following are average values for the NaCl content (%) of different cheeses:[60-65] fresh cheese, 0·1; cottage cheese, 0·8; processed cheese, 2·5; Feta, 3·7; Camembert, 2·4; Limburg, 3·0; Edam/Gouda, 1·9; Tilsit, 1·6; Mozzarella, 1·6; blue cheeses, 4·1; Gruyère, 1·5; Emmental, 0·5; Cheddar, 1·7.

A minimum intake of less than 500 mg and a maximum of 4 g sodium per caput per day is suggested by the German Nutrition Association. There is still a controversial discussion whether high dietary intakes of sodium are harmful to the general population or whether high Na intake is the major dietary factor causing problems with hypertensive individuals.[77] Nevertheless, a restricted sodium intake is often recommended to accommodate the diets of consumers under medical management for hypertension. Cheese contributes to the total sodium intake only to a small extent, even in countries with a high cheese consumption, e.g. 0·12–0·23 g Na per caput per day in Australia, Switzerland and the United Kingdom.[78] Although cheese contributes only about 5% to the total sodium intake, the manufacture of low-sodium cheese is recommended by using a brine containing potassium or magnesium chloride.[79,80] When, for example, a solution of $MgCl_2$ was used instead of NaCl, the Na content of Gruyère cheese was lowered by 80% (50 vs 250 mg/100 g) and the Mg content increased two-fold.[81] According to Lindsay et al.,[77] consumer-acceptable low-sodium Cheddar cheese and processed cheeses (less than 500 mg/100 g) can be manufactured with substitutions of NaCl/KCl (1:1) and combinations of potassium-based emulsifiers with delta-gluconolactone. While in some cases taste panel results showed that cheeses prepared to contain up to 75% less sodium than traditional cheese are acceptable to consumers,[82,83] it is reported that replacement by KCl produced a bitter taste and that consumers preferred the fuller salty taste of Cheddar cheese with higher NaCl concentrations.[84] According to the German regulations, sodium-reduced cheese may contain a maximum of 450 mg sodium per 100 g; this can be obtained by keeping Edam cheese in brine (18% NaCl) for only 6 h; the resulting product was sensorially acceptable.[85–87]

It should be considered also that hypertension may be due to a deficiency of dietary calcium rather than to an excessive intake of sodium, since it has been observed that patients suffering from hypertension consume about 25% less Ca than normotensive persons, because of a low consumption of milk and dairy products.[88]

2.5 Trace Elements

The concentrations of some trace elements in various cheese varieties are given in Table V. The following aspects are also noteworthy:

—Processed cheeses are considered as major sources of aluminium as they contain higher Al levels than other cheese varieties. The estimated average Al intake is about 10 mg/caput/day, less than 10% being derived from cheese.[89,90] On peptic and pancreatic digestion, a high proportion of Zn (91·0%) is released into the soluble digesta of cheese, which may be an indication of good availability of Zn from cheese.[91]

—As a rule, the Se content of cheese is in the range between 5 and 12 μg/ 100g.[92,93] In the USA, the recommended dietary intake range is 50–200 μg/ day; milk and dairy products contribute about 5% of the total Se intake, cheese about 1%.[92,94]

—In the Netherlands, the average zinc intake from cheese is 1·1 mg per caput per day.[95]

TABLE V

Literature Data on the Concentration of Some Trace Elements in Several Cheese Varieties[3,90,94,102-108]

Cheese variety	Concentration (per 100 g) of						
	Zn (mg)	Fe (mg)	I (μg)	Mn (μg)	Se (μg)	Cu (μg)	Al (mg)
Cheddar	3·8	0·6	52	40	11	50	0·02
Parmesan	3·6	0·6	40	40	12	200	—
Emmental	5·0	0·5	40	40	11	200	0·2
Edam/Gouda	4·0	0·5	35	40	—	100	0·3
Mozzarella	3·5	0·4	45	30	3	60	—
Processed cheese	3·4	0·35	48	22	10	50	1·4
Cottage cheese	0·5	0·2	20	6	5	17	0·1

—The nickel content of cheese is in the range between 2 and 34 μg/100 g.[96]

—The values found for the mercury content in cheese (0·04–0·16 μg/100 g) are far below the suggested allowable amounts.[97]

Milk protein does not affect iron absorption. Cottage cheese made from iron-enriched milk retains 58% of the iron of the milk so that 100 g of such a cheese will supply one-third of the recommended daily iron intake for a female adult.[98] Cheese fortified with ferric ammonium citrate is effective in restoring low haemoglobin; in rat experiments, 66–75% of the iron consumed from fortified cheeses was incorporated into haemoglobin, which was approximately equivalent to beef and egg.[99,100] The quality of fortified Cheddar cheese (up to 4 mg/100 g) was found to be as good as unfortified cheese.[101]

2.6 Vitamins

The concentration of fat-soluble vitamins in cheese depends on its fat content. Most (80–85%) of the vitamin A contained in milk passes into the cheese. The figure is naturally lower for the water-soluble vitamins. The values for thiamine, nicotinic acid, folic acid and ascorbic acid are 10–20%, for riboflavin and biotin, 20–30%, for pyridoxine and pantothenic acid, 25–45% and for cobalamin, 30–60%; the rest remains in the whey.[109,110] However, milk contains such high concentrations of some B vitamins that cheese still contributes significantly to the supply of these vitamins. This is especially true of the vitamins B_{12} and B_2 with a supply of up to 100 and 27%, respectively, by 100 g of cheese (folic acid up to 15%).[111,112] On the other hand, the contribution of cheese to the vitamin B_6 intake is only 1–2% in the Netherlands.[113]

Table VI lists the average concentrations of vitamins in a number of cheese types. Some mould-ripened cheeses contain more of the B vitamins than other types of cheese. Examples are the high contents of vitamins B_2, B_6 and niacin in Camembert.

TABLE VI
Average Vitamin Content of Some Cheese Varieties[3,112,117–122]

Cheese variety	Content (per 100 g) of vitamin						
	A (mg)	B_1 (μg)	B_2 (mg)	B_6 (μg)	B_{12} (μg)	folic acid (μg)	toco-pherol (mg)
Emmental	0·33	35	0·30	105	2·7	12	0·9
Cheddar	0·36	35	0·40	75	1·0	16	1·0
Edam/Gouda	0·21	35	0·35	70	1·9	25	0·7
Tilsit	0·25	45	0·35	65	2·3	16	0·8
Blue cheese	0·40	36	0·50	100	1·2	45	0·9
Camembert/Brie	0·30	40	0·52	150	1·8	60	0·7
Cream cheese	0·44	27	0·26	45	1·0	15	0·9
Cottage cheese	0·08	28	0·24	55	1·0	15	0·2
Quarg skimmed	—	35	0·28	50	1·0	30	—
Processed cheese	0·30	23	0·30	55	1·0	15	0·4

The concentration of total vitamin B_2 in Emmental, Greyerzer and Sbrinz cheeses is reported to be 0·42–0·55 mg/100 g with a proportion of about 25–35% in the form of flavin adenine dinucleotide.[114]

There are reported the following other vitamin contents in cheese: vitamin D, 0·4–0·5 μg/100 g;[115,116] niacin, about 70 μg/100 g; biotin, 0·6–3·2 μg/100 g; pantothenic acid, 0·11–0·71 mg/100 g.

All-trans retinol and 13-cis retinol are the major retinoids in dairy products, only small amounts of 9-cis, 11-cis and 9,11-cis retinols are found: 0·18–0·30 μg all-trans retinol per 100 g, and 23–56 μg 13-cis retinol per 100 g; the β-carotene content is given as 0·09–0·13 μg/100 g.[123,124]

The concentrations of B vitamins change during ripening since these are both used and synthesized by the cheese microflora. The concentration of several of the B vitamins depends on the type of starter culture used and increases with time of storage. After a long ripening period, the concentrations of these vitamins in cheese are therefore increased.[17,125] By isolating individual microorganisms from cheese it could be shown that they are able to synthesize niacin, folic acid, biotin and pantothenic acid. The synthesis of vitamin B_{12} by propionic acid bacteria in hard cheese, especially in Emmentaler, has aroused great interest. Propionic acid bacteria have, therefore, been added experimentally to cheesemilk for the manufacture of Edam, Tilsiter and a number of other types of cheese with the result that in some cases, especially with *Propionibacterium freudenreichii*, the cobalamin content was doubled.[126] Most of the ascorbic acid, on the other hand, is degraded during cheese ripening.

3 FRESH CHEESE (QUARG)

Tables I, IV, V and VI include values for the concentrations of protein, minerals, trace elements and vitamins in Quarg (fresh cheese). Milk destined for Quarg production is nowadays often strongly heat-treated (at 95°C for 10 min). This leads to complex formation between casein and whey proteins so that a large part of the whey proteins is precipitated with the casein on acidification and is included in the Quarg. The percentage of the total nitrogen coagulated increases from 77–79% to 88–89%. This product has, therefore, a higher content of essential amino acids and a higher protein biological value.[127,128] The complete incorporation of the whey proteins in the fresh cheese can be achieved by ultrafiltering the cheesemilk.[129]

From the point of view of nutrition, Quarg, which is usually produced by means of a lactic acid culture, is similar to other cultured milk products. Because low-fat Quarg is rich in biologically valuable proteins, calcium and phosphorus, and because its calorie content is relatively low, it is recommended for all sections of the population, but particularly for older people and as part of slimming diets. Quarg is also easily digestible and this makes it valuable in therapeutic diets, especially in cases of liver disease.[130]

4 PROCESSED CHEESE

In the production of processed cheese, the casein is hydrated and peptidized by the action of the emulsifying salts and the proportion of water-soluble protein therefore increases considerably. Polyphosphates have the widest range of application as emulsifying salts, but citrates and lactates are used also. During the storage of the processed cheese, the polyphosphates are converted, either partly or wholly, into di- and monophosphates. Processed cheeses contain roughly the same proportions of nutrients as the cheeses from which they were made. The fat content varies between 9 and 31% and the protein content between 8 and 24%. Except for the Na and K contents, which are higher, the mineral concentrations are also similar to those in the original cheeses. The addition of polyphosphate does not increase the phosphate content significantly; the natural variation in the phosphate content of cheese is 0·4–2·7% and of processed cheese, 0·8–2·7%. Some losses of vitamins B_1, B_2, niacin, pantothenic acid and vitamin B_{12} occur during the manufacture of processed cheese. The free amino acid content of the cheese and the in-vitro digestibility of the proteins are increased by processing and the utilization of the proteins of processed cheese is thought to be better than that of the proteins of natural cheese. No change in the availability of lysine could be detected.[131,132] According to Dupuis et al.,[133] Ca utilization is better with a processed cheese diet than with milk and yoghurt diets.

Polyphosphates ingested with food do not exert a physiological effect because they are quickly converted by enzymes to monophosphates which are then absorbed. They are, therefore, no danger to health. Experiments with rats have shown that polyphosphates are well tolerated, even when administered over long periods of time. The phosphate ingested as part of processed cheese must be considered in the context of the total phosphorus intake which consists of the natural phosphorus content of the diet plus any additional mono- and polyphosphates. The phosphate contained in processed cheese might even contribute to meeting the P requirement. Because the recommended long-term daily intake of polyphosphate is 40 mg/kg of body weight, on average, there is no danger that the consumption of processed cheese could lead to an overdose of polyphosphates. It has been calculated that the amount of additional phosphorus ingested through the consumption of processed cheese and other phosphate-enriched foods is about 1·2 g/day and this is well within the range of variation of the normal phosphate intake. Processed cheese may be regarded as a valuable food because phosphates are said to inhibit the formation of dental caries. There are no objections to the use of citrates as emulsifying salts because citric acid and its salts occur in many foods and are normal metabolic products of the human body.[134,135]

5 SPECIFIC NUTRITIONAL EFFECTS

5.1 Anticariogenic Effect of Cheese

It was observed in experiments with animals and humans that cheese may protect against dental caries. In a chewing test, cheese produced considerable quantities of lactic acid which, however, was largely degraded after 1–10 min; the pH after 90 min was about 6·70 and was never lower than the critical pH value of 5·5.[136] Cheese eaten immediately after six sucrose rinses a day reduced the demineralization caused by sucrose by an average of 71% in five subjects,[137] it resulted in a smaller pH drop and a more rapid return to the resting pH (Fig. 2). In-vivo intraoral tests in humans confirmed the cariostatic nature of cheese.[138] In rats fed standard diets containing 5 or 20% sucrose, there was significantly greater caries activity than in rats fed cheeses containing 20% sucrose.[138a] Rats developed fewer and less severe caries when fed cheese snacks in addition to a cariogenic diet; the effects were particularly dramatic on root-surface caries.[139]

A number of possible mechanisms have been offered to explain this beneficial effect of cheese:

—The release of calcium and its diffusion into the plaque might be regarded as the most important effect. When the most active anti-cariogenic components of aqueous cheese extract were isolated and identified, the results indicated that most of this effect is due to the Ca and P contents which probably influence the demineralization–remineralization process.[137,140,141]

—Eating cheese causes an increased flow of saliva, which is slightly alkaline and thus acts as a buffer.[140]

—A substantial proportion of the protection may be related also to textural influences.[142]

—Casein or particular peptides may also play a role in counteracting the development of caries.[143]

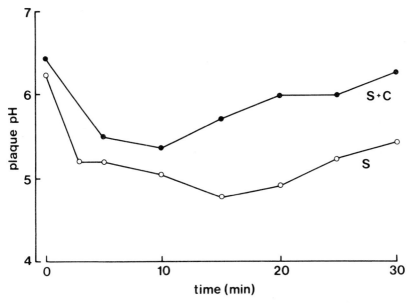

Fig. 2. Mean plaque pH response with sucrose only (S) and sucrose followed by cheese (S + C).[137]

It should be ensured that any cheese eaten by individuals especially prone to caries is taken at the time that it can exert its maximum effect, this is immediately after meals when acid production is at its greatest intensity and neutralization by saliva and an increase in plaque calcium would have their greatest impact.[140]

5.2 Tumour Incidence

In a case-control study, the risk of breast cancer was found to be positively associated with the frequency of cheese consumption: 68% of the cases and 63% of the controls consumed cheese daily.[144] As the difference found between both groups was very small, more research work should be conducted before conclusions are drawn.

An epidemiological study revealed a positive correlation between the consumption of fava beans and the incidence of gastric cancer; it could, however, be shown that cheese has an inhibitory effect on the direct-acting mutagens that are generated in fava beans after treatment with nitrite.[145]

6 ADDITION OF NITRATE

6.1 Nitrite

In most cheese varieties which undergo a long ripening period there is the danger that anaerobic spore-forming clostridia, particularly *Clostridium tyrobutyricum*, which are not destroyed by pasteurization, may produce considerable butyric acid fermentation resulting in bloating of the cheese, which would make it unfit for consumption. The addition of a maximum of 15 g of $NaNO_3$ or KNO_3 per 100 litres of cheese milk is therefore permitted in the manufacture of some types of cheese (semi-hard cheese), because during the ripening period the nitrates are reduced to nitrites which inhibit the growth of clostridia and thus prevent the so-called late bloating of cheese. Nitrites have no effect on the growth of lactic acid bacteria. No nitrate, or only very little of it is used during the manufacture of Emmentaler cheese, otherwise the propionic acid fermentation may be disturbed. The application of microfiltration is suggested for making cheese without the addition of nitrate, or the use of lysozyme or bactofugation to reduce the addition of nitrate.

Part of the nitrate which is added during cheesemaking passes into the whey or diffuses into the brine, another part is reduced to gases so that only a small part remains as nitrite.[146] It has to be considered that in cheeses made without the addition of nitrate, the naturally derived nitrate occurs in the range 1–8 mg/kg, while values between 1 and 41 mg/kg are found in cheeses to which nitrate has been added.[147–150] The Dutch regulations include a value for the residual nitrate in cheese which must not exceed 50 mg/kg of cheese. This value is only very rarely exceeded.[151,152] In an investigation in Germany, nitrite was not detectable in 95% of the hard cheese samples and 88% of the semi-hard cheese samples.[153] The nitrate content of cheese falls progressively during ripening.

Nitrite is a toxic compound and cheese should therefore not contain any harmful amounts of it at the end of ripening. This is actually the case, because the nitrite formed during cheese ripening is destroyed rapidly so that the finished product contains only traces of nitrite. When 20 g of nitrate are added per 100 litres of cheese milk, generally no nitrite is detectable.[151,152,154,155] The majority of cheeses do not need the addition of nitrate during their manufacture and they are free of nitrite. In Holland, the maximum permitted amount of nitrite in cheese is 2 ppm but the actual values found in an extensive investigation were always lower.[147] Figure 3 shows the changes in the nitrite content of Gouda cheese made from milk to which 20 g of nitrate per 100 litres of milk had been added. The nitrite content of the cheese increases to a maximum of 0·7 mg/kg and then falls considerably during ripening. The interaction of lipids with nitrite is considered to cause a considerable reduction in the nitrite content of cheese.[156]

On the basis of the results obtained from animal experiments and including a safety factor of 1:1000, the maximum safe daily dose of nitrite for humans has been worked out to be 46 µg/kg body weight. Preliminary ADI (acceptable daily intake) values suggested by WHO are 5 mg of nitrate or 0·2 mg of nitrite/kg body weight/day. The low concentrations of nitrate and nitrite in cheese do not in any way present a health hazard for the consumer. The average daily intake of nitrate in the diet in different countries varies between 50 and 100 mg, to which vegetables contribute 70–80%, while the nitrate intake from milk and milk products, including cheese, is only 0·2–0·3 mg/day which is 0·2–0·6% of the total nitrite intake.[157–160] The amount of nitrite ingested with milk and dairy products

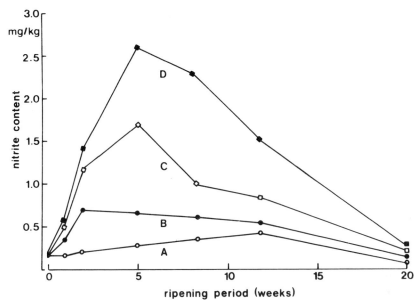

Fig. 3. Changes in the nitrite content of Gouda cheese made from milk to which various amounts of nitrate had been added (A = 10 g, B = 20 g, C = 40 g, D = 60 g per 100 litres).[164a]

is even less (0·01–0·02 mg/day or 0·1% of the total nitrite intake).[160,161] The nitrate intake is only 0·10% of the suggested ADI values, the nitrate intake 0·07%.[162–164]

6.2 Nitrosamines

Nitrosamines can be produced by a reaction between secondary amines and nitrite. Sixty different nitrosamines are known and the majority of them have been found to be relatively strongly carcinogenic in rats. The formation of nitrosamines depends on the amount of nitrite present, but is unrelated to the amine concentration.[165] Histamine and tyramine, which are the chief amines occurring in cheese, are not among those that can be converted to nitrosamines. The reaction is pH dependent, occurring preferentially in the pH range of 2–4·5. Cheese has a higher pH value and this prevents the reaction leading to the formation of nitrosamines. Some moulds, such as *Penicillium camemberti*, are able to synthesize nitrosamines in the pH range of cheese. However, those cheese varieties usually made from nitrated milk do not contain this mould culture. Nitrosamines can be formed in the stomach of animals and humans from nitrites and secondary amines; the low pH value of the gastric juices favours the reaction. However, the kinetics of the reaction make it improbable that it should occur either in cheese or in the stomach. This is why only trace amounts of nitrosamines are found in the stomach. Moreover, the reaction can be completely inhibited by ascorbic acid.[166] It is also thought possible that nitroso compounds in cheese might be degraded by enzymes.[167]

These are the reasons why nitrosamines are found only rarely and then at very low concentrations in cheeses that have been made with the addition of the permitted amounts of nitrates.[147,168–170] In Germany, values of up to 1·2 μg/kg have been reported; in 96% of the cheese samples, the concentration of the nitrosamines was below 0·5 μg/kg.[160,171] The nitrosamine content of most cheese samples examined was below the detectable limit, which is very low, namely 0·01 μg/kg. There is no correlation between the nitrate content of a cheese and its nitrosamine content. It has been pointed out that cheeses made without the addition of nitrate often contain nitrosamines. For instance, when nitrite derived either from nitrite-containing cheese or from meat products is present during the manufacture of processed cheese or of products based on processed cheese, nitrosamines may be formed. This can, however, be prevented by the addition of ascorbic acid. No nitrosamines could be detected in dishes made in the home from nitrite-containing cheeses or meats (e.g. ham and cheese on toast).

The levels of apparent total N-nitroso compounds were found to be below the detection limit in 28 out of 31 different cheese varieties analysed; these substances were detected in most of those cheese samples manufactured with added nitrate, including Edam, Gouda and Havarti.[172]

The compound most often found in cheese is dimethyl nitrosamine: in the Netherlands, concentrations of up to 0·15 μg/kg are reported for this compound, and up to 0·03 μg/kg for diethyl nitrosamine.[151]

Nitrosamines belong to a class of compounds which are highly carcinogenic for animals. It is not known for certain, but strongly suspected, that they are also carcinogenic for man. The 'maximum acceptable dose' of these compounds for humans is at present under discussion. On the basis of toxicity studies on animals, a value of 5–10 μg/kg of foodstuff has been proposed. This includes a certain safety factor. It is concluded from relevant studies that individuals ingest less than 50 μg of nitrosamines/year with their food. The average daily intake in the United Kingdom is said to be about 1 μg, of which cheese contributes 4%; the corresponding figures for Japan are 0·5 μg and 0·1%.[170,173] The small amounts of nitrosamines sometimes found in cheese can therefore be regarded as quite unimportant. In view of the fact that the body itself produces nitrites and that there is therefore an endogenous nitrosamine synthesis, these small amounts derived from cheese can be neglected.[174] As a rule, nitrosamine formation occurred at pH values that are common in laboratory situations, but that are rare during the preparation of food in the household.[175]

7 MICROBIOLOGICAL TOXINS

Because moulds, particularly strains of *Penicillium*, are used in the manufacture of blue cheese as well as of cheeses with a surface mould, the question arises whether mycotoxins could be formed. The following substances are degradation products formed by the action of *Penicillium roqueforti*:

1. The alkaloid, roquefortin, a degradation product typical of *P. roqueforti*, has been detected in blue cheese at a concentration of 0·05–6·8 ppm. According to the currently available toxicity data, these concentrations are too low to be toxic. The consumption of blue cheese is therefore not a health hazard.[176]

2. The so-called PR-toxin is formed only by a few *P. roqueforti* strains and then only on artificial nutrient media; cheese is an unsuitable medium for the formation of PR-toxin. Moreover, PR-toxin is unstable, and even if present in cheese it would react with amino groups and be converted very quickly to non-toxic substances. This toxin has, therefore, never been found in cheese, not even when the cheese was made with PR-toxin-producing organisms.[177,178] *P. roqueforti* isolates from the Spanish blue cheese, Cabrales, could produce 1·6–1·9 mg/100 ml of PR-toxin and 0·09–0·18 mg/100 ml of roquefortin.[179]

3. The toxic mould product, patulin, which is carcinogenic for mice, and which causes effects on the kidney and gastrointestinal tract of rats, is not produced by those strains of *P. roqueforti* used in cheesemaking. Moreover, the formation of patulin in cheese is strongly inhibited, probably by a reaction with sulphydryl groups, so that this substance would disappear rapidly from the cheese even if it had been present at the beginning. No patulin was detected in Tilsiter cheese which had been deliberately infected with the patulin-producing organisms.[180–182]

4. Furthermore, *P. roqueforti* is able to form mycophenolic acid (0·01–15 mg/kg) and penicillic acid.[183]

In subchronic toxicity tests with mould-ripened cheese, no signs of adverse effects by cheese mycotoxins could be detected after 28 days of feeding high amounts of mycelium to mice; no still unknown toxic metabolites could be demonstrated.[184]

None of the mycotoxins under investigation could be detected in cultures of *P. caseicolum* or *P. camemberti*.[185,186] Cyclopiazonic acid, which is formed by some *P. camemberti* strains and which may be found sporadically in cheese samples with a white surface mould (mainly in the rind), does not have a mutagenic potential.[187,188]

When mould cultures used in cheesemaking were fed to animals or when crude extracts of these moulds were injected, no harmful effects could be observed.[189,190] Taking into consideration all the results on the occurrence and biological activity of mycotoxins occurring naturally in mould-ripened cheeses, a health hazard for Man can be excluded, even with a considerable intake of cheese.[191]

Of 22 *Aspergillus* species under investigation, one was capable of producing aflatoxins after direct growth on the Greek cheese variety, Teleme.[192] In special circumstances, *Aspergillus versicolor*, a mould which usually belongs to the contaminating flora of Gouda cheese, can produce the mycotoxin, sterigmatocystin; therefore, growth of this mould on cheese should be avoided by technological means.[193,194]

Benzoic acid may also be formed by microbial activity during cheese manufacture. The following concentrations are reported (mg/kg): Cheddar and Quarg, 20–25; Gouda, cottage cheese and processed cheese, about 10; Camembert and Edam, about 7; Brie and blue cheeses, traces.[195,196] The carry-over of hippuric and benzoic acids, which are naturally present in milk, to cheese only occurs at a level of less than 10%.[197]

8 PRESERVATION OF CHEESE

Sorbic acid and its calcium, sodium and potassium salts are very effective in preventing the growth of yeasts and moulds. Sorbic acid, therefore, might be used for the surface treatment of hard and semi-hard cheeses to prevent the growth of moulds during ripening and storage and thus to preserve the quality of the cheese. This method of preservation derives a special importance from the fact that it also prevents the growth of aflatoxin-producing moulds: for instance, a sorbate concentration of 200–400 ppm inhibits the growth of mycotoxin-producing moulds.[198,199] Sorbate also greatly reduces or prevents the production of patulin by *P. patulum*.[200]

Sorbic acid is quite harmless because the body uses it as it uses any other dietary fatty acid. It is metabolized like other fatty acids with the same number of C-atoms such as, for example, caproic acid. Experimental animals suffered no ill effects when their diet contained 5% sorbic acid. In a number of countries, the use of sorbic acid and its salts is, therefore, permitted for the treatment of a range of foods, especially of cheese. Sorbic acid is one of the most commonly used preservatives because it is harmless and very effective.[201,202]

Natamycin (Pimaricin) is an antibiotic produced by *Streptomyces natalensis*. Like sorbic acid, it suppresses the growth of yeasts and moulds, but has very little effect on bacteria; *Aspergillus flavus*, in particular, is very sensitive to natamycin. It has been suggested that natamycin should be used in the same way as sorbic acid for the surface treatment of cheeses. Natamycin remains on the cheese surface for a relatively long time. While sorbate in the cheese coating dispersion migrates into the whole cheese, in the case of natamycin no migration can be observed further than 2 mm into the rind.[203–205] Because the depth of penetration is so small, cheeses treated with natamycin are protected against mould infection for about eight weeks. Aflatoxin formation is also prevented.[206,207] At the time of sale to the consumer, the permitted level of natamycin should not exceed 2 mg/dm^2 of cheese surface and the penetration depth into the cheese should not exceed 5 mm. Although natamycin has now been used for several years, yeasts and moulds have not developed resistance against this antibiotic. Natamycin has no physiological effects and is non-toxic. Doubts have been expressed whether it is advisable to permit the use of an antibiotic that has been used successfully in medicine for the preservation of foods, although investigations have shown that the development of resistant organisms can practically be ruled out. Neither are allergic reactions expected to occur. The acceptable daily intake of natamycin is said to be 0·25 mg per kg body weight.[208]

REFERENCES

1. International Dairy Federation, 1988. *IDF Bulletin*, **246**.
2. Renner, E., 1983. *Milk and Dairy Products in Human Nutrition*, Volkswirtsch. Verlag, München.
3. Renner, E. & Renz-Schauen, A., 1986. *Nährwerttabellen für Milch und Milchprodukte*, Verlag B. Renner, Giessen.
4. Stanton, R., 1984. *Aust. J. Dairy Technol.*, **39**, 99.
5. Jameson, G.W., 1990. *Aust. J. Dairy Technol.*, **45**, 93.
6. Biede, S.L., Paulsen, P.V., Hammond, E.G. & Glatz, B.A., 1979. *Dev. Industr. Microb.*, **20**, 203.
7. Finegan, A., Hickey, N., Maurer, B. & Mulcahy, R., 1968. *Am. J. Clin. Nutr.*, **21**, 143.
8. Flaim, E., Ferreri, L.F., Thye, F.W., Hill, J.E. & Ritchey, S.J., 1981. *Am. J. Clin. Nutr.*, **34**, 1103.
9. Antila, V., Hakkarainen, H. & Lappalainen, R., 1982. *Milchwissenschaft*, **37**, 321.
10. Rommel, G., 1983. PhD Thesis, University of Giessen.
11. Resmini, P., Mazzolini, C. & Pelegrino, L., 1982. *Latte*, **7**, 548.
12. Kotula, K.T., Nikazy, J.N., McGinnis, N., Lowe, C.M. & Briggs, G.M., 1987. *J. Food Sci.*, **52**, 1245.
13. Dreyer, J.J., 1984. *S. Afr. J. Dairy Technol.*, **16**, 139.
14. Korolczuk, D., Cieslak, D., Luczynska, A. & Bijok, F., 1978. *Proc. 20th Intern. Dairy Congr., Paris*, Vol. **E**, p. 1073.
15. Staub, H.W., 1978. *Food Technol.*, **32**(12), 57.
16. Blanc, B. & Sieber, R., 1978. *Alimenta*, **17**, 59.
17. Dillon, J.C., 1984. In *Le Fromage*, ed. A. Eck, Lavoisier, Paris, p. 497.
18. Renner, E. (ed.), 1991. *Dictionary of Milk and Dairying*, Volkswirtsch. Verlag, München.
19. Karabelnik, D., Amado, R., Arrigoni, E. & Solms, J., 1982. *Lebensm. Wiss. Technol.*, **15**, 245.
20. Brückner, H. & Hausch, M., 1990. *Milchwissenschaft*, **45**, 421.
21. Joosten, H.M.L.J., 1988. *Neth. Milk Dairy J.*, **41**, 329.
22. Antila, P., Antila, V., Mattila, J. & Hakkarainen, H., 1984. *Milchwissenschaft*, **39**, 400.
23. Klein, D.F. & Sandler, M., 1983. *Psychopharm. Bull.*, **19**, 496.
24. Evans, C.S., Gray, S. & Kazim, N.O., 1988. *Analyst*, **113**, 1605.
25. Wortberg, B. & Zieprath, G., 1981. *Lebensm. Chem. Ger. Chem.*, **35**, 89.
26. Feldman, J.M., 1983. *Arch. Intern. Med.*, **143**, 2099.
27. Pechanek, U., Pfannhauser, W. & Woidich, H., 1983. *Z. Lebensm. Unters. Forsch.*, **176**, 335.
28. Antila, P., Antila, V., Mattila, J. & Hakkarainen, H., 1984. *Milchwissenschaft*, **39**, 81.
29. Sieber, R. & Lavanchy, P., 1990. *Mitt. Gebiete Lebensm. Hyg.*, **81**, 82.
30. Reuvers, T.B.A., de Pozuelo, M.M., Ramos, M. & Jimeney, R., 1986. *J. Food Sci.*, **51**, 84.
31. Binder, E. & Brandl, E., 1984. *Österr. Milchwirtsch., Wiss. Beilage 1*, **39**, 1.
32. Ingles, D.L., Back, J.F., Gallimore, D., Tindale, R. & Shaw, K.J., 1985. *J. Sci. Food Agr.*, **36**, 402.
33. Vidaud, Z.E., Chaviano, J., Gonzales, E. & Garcia Roché, M.O., 1987. *Nahrung*, **31**, 221.
34. Laleye, L.C., Simard, R.E., Gosselin, C., Lee, B.H. & Giroux, R.N., 1987. *J. Food Sci.*, **52**, 303.
35. Valletrisco, M., Azzi, A. & de Clemente, I.M., 1989. *Ind. Alim.*, **28**, 1084.
36. Bütikofer, U., Fuchs, D., Hurni, D. & Bosset, J.O., 1990. *Mitt. Gebiete Lebensm. Hyg.*, **81**, 120.
37. Mehanna, N.M., Antila, P. & Pahkala, E., 1989. *Egypt. J. Dairy Sci.*, **17**, 19.

38. Sieber, R., Collomb, M., Lavanchy, P. & Steiger, G., 1988. *Schweiz. Milchw. Forsch.*, **17**, 9.
39. Taylor, S.L., Keefe, T.J., Windham, E.S. & Howell, J.F., 1982. *J. Food Protect.*, **45**, 455.
40. Chang, S.-F., Ayres, J.W. & Sandine, W.E., 1985. *J. Dairy Sci*, **68**, 2840.
41. Terplan, G., Wenzel, S. & Grove, H.-H., 1973. *Wien. Tierärztl. Mschr.*, **60**, 46.
42. Kaplan, E.R., Sapeika, N. & Moodie, I.M., 1974. *Analyst*, **99**, 565.
43. Sattler, J. & Lorenz, W., 1987. *Münch. Med. Wochschr.*, **129**, 551.
44. Häberle, M., 1987. *Ernähr. Umschau*, **34**, 287.
45. Häberle, M., 1990. *Lebensm. Technik*, **22**, 632.
46. Blanc, B., 1982. *Alimenta*, **21**, 125.
47. Florence, E., Milner, D.F. & Harris, W.M., 1984. *J. Soc. Dairy Technol.*, **37**, 13.
48. Thomas, T.D. & Crow, V.L., 1983. *N.Z. J. Dairy Sci. Technol.* **18**, 131.
49. Pahkala, E. & Antila, M., 1976. *Finn. Chem. Lett.*, **1**, 21.
50. Puhan, Z., 1976. *Schweiz. Milchw. Forsch.*, **5**, 55.
51. Krusch, U., 1978. *Kieler Milchw. Forsch. Ber.*, **30**, 341.
52. Giesecke, D., 1988. *Ärztezeitschr. Naturheilverf.*, **29**, 67.
53. Barth, C.A. & De Vrese, M., 1984. *Kieler Milchw. Forsch. Ber.*, **36** 155.
54. Tunick, M.H., 1987. *J. Dairy Sci.*, **70**, 2429.
55. Lagrange, V., 1982. *Méd. Nutr.*, **18**, 200.
56. Kansal, V.K. & Chaudhary, S., 1982. *Milchwissenschaft*, **37**, 261.
57. Buchowski, M.S. & Miller, D.D., 1990. *J. Food Sci.*, **55**, 1293.
58. Strobel, C. & Kluthe, R., 1986. *Akt. Ernähr.*, **11**, 96.
59. Andlaw, R.J., 1977. *J. Human Nutr.*, **31**, 45.
60. Donovan, S., 1983. *Proc. Nutr. Soc.*, **42**, 375.
61. Kindstedt, P.S. & Kosikowski, F.V., 1984. *J. Dairy Sci.*, **67**, 879.
62. Tröger, J., 1985. *AID-Verbraucherdienst*, **30**, 135.
63. Sieber, R., Collomb, M. & Steiger, G., 1987. *Mitt. Gebiete Lebensm. Hyg.*, **78**, 106.
64. Klostermeyer, H., 1988. *Deut. Molk. Ztg.*, **109**, 570.
65. Zangerl, P. & Tschager, E., 1990. *Milchw. Ber.*, **105**, 214.
66. Juarez, M. & Martin-Hernandez, M.C., 1983. *Rev. Agroquim. Tecnol. Aliment.*, **23**, 417.
67. Farrer, K.T.H., 1984. *Aust. J. Dairy Technol.*, **39**, 108.
68. Tschager, E., 1984. *Milchw. Ber.*, **81**, 335.
69. Bruhn, J.C. & Franke, A.A., 1988. *J. Dairy Sci.*, **71**, 2885.
70. Pennington, J.A.T., Young, B.E., Wilson, D.B., Johnson, R.D. & Vanderveen, J.E., 1986. *J. Am. Diet. Assoc.*, **86**, 876.
71. Miller, D.D. & Bisogni, C.A., 1987. *N.Y. Food Life Sci. Quart.*, **17**(3), 5.
72. Petik, S., 1987. *Cult. Dairy Prod. J.*, **22**(1), 12.
73. Van Binsbergen, J.J., Hulshof, K.F.A.M., Egger, R.J., Wedel, M. & Hermus, R.J.J., 1987. *Voeding*, **48**, 351.
74. Kindstedt, P.S. & Kosikowski, F.V., 1988. *J. Dairy Sci.*, **71**, 285.
75. Hill, A.R. & Ferrier, L.K., 1989. *Can. Inst. Food Sci. Technol. J.*, **22**, 75.
76. Kitts, D.D., 1990. *Nutr. Quart.*, **14**(2), 28.
77. Lindsay, R.C., Karahadian, C. & Amundson, C.H., 1985. *Proc. IDF. Sem.*, Atlanta/USA, p. 55.
78. Edwards, D.G., Kaye, A.E. & Druce, E., 1989. *Eur. J. Clin. Nutr.*, **43**, 855.
79. Greenfield, H., Smith, A.M., Maples, J. & Wills, R.B.H., 1984. *Human Nutr. Appl. Nutr.*, **38A**, 203.
80. Karahadian, C., Linday, R.C., Dillman, L.L. & Deibel, R.H, 1985. *J. Food Protect.*, **48**, 63.
81. Lefier, D., Grappin, R., Grosclaude, G. & Curtat, G., 1987. *Lait*, **67**, 451.
82. Demott, B.J., Hitchcock, J.P. & Sanders, O.G., 1984. *J. Dairy Sci.*, **67**, 1539.
83. Karahadian. C. & Lindsay, R.C., 1984. *J. Dairy Sci.*, **67**, 1892.

84. Lindsay, R.C., Hargett, S.M. & Bush, C.S., 1982. *J. Dairy Sci.*, **65**, 360.
85. Barth, C., Krusch, U., Meisel, H., Prokopek, D., Schlimme, E. & de Vrese, M., 1989. *Kieler Milchw. Forsch. Ber.*, **41**, 105.
86. Prokopek, D., Barth, C., Klobes, H., Krusch, U., Meisel, H., Schlimme, E., de Vrese, M. & Wotha, H.J., 1990. *Kieler Milchw. Forsch. Ber.*, **42**, 565.
87. Barth, C., Krusch, U., Meisel, H., Prokopek, D., Schlimme, E. & de Vrese, M., 1990. *Lebensm. Ind. Milchw.*, **111**, 1276 and 1312.
88. McCarron, A., Morris, C.D. & Cole, C., 1982. *Science*, **217**, 267.
89. Treptow, H. & Askar, A., 1987. *Ernähr. Umschau*, **34**, 364.
90. Pennington, J.A.T., 1987. *Food Add. Contam.*, **5**, 161.
91. Ikeda, S., 1990. *J. Sci. Food Agr.*, **53**, 229.
92. Robberecht, H.J. & Deelstra, H.A., 1985. *Voeding*, **46**, 262.
93. Elmadfa, I., 1986. *AID-Verbraucherdienst*, **31**, 183.
94. Schubert, A., Holden, J.M. & Wolf, W.R., 1987. *J. Am. Diet. Assoc.*, **87**, 285.
95. Keijbets, M.J.H., van Boekel, M.A.J.S., van der Meer, M.A., Reijenga, M.C. & Warnaar, F.M., 1985. *Voeding*, **46**, 306.
96. Flyvholm, M.-A., Nielsen, G.D. & Andersen, A., 1984. *Z. Lebensm. Unters. Forsch.*, **179**, 427.
97. Koops, J. & Westerbeek, D., 1984. *Neth. Milk Dairy J.*, **38**, 241.
98. Sadler, A.M., Lacroix, D.E. & Alford, J.A., 1973. *J. Dairy Sci.*, **56**, 1267.
99. Wong, N.P., McDonough, F.E., LaCroix, D.E. & Vestal, J.H., 1984. *Nutr. Rep. Int.*, **29**, 135.
100. Zhang, D. & Mahoney, A.W., 1989. *J. Dairy Sci.*, **72**, 2845.
101. Zhang, D. & Mahoney, A.W., 1990. *J. Dairy Sci.*, **73**, 2252.
102. Greger, J.L., 1985. *Food Technol.*, **39**(5), 73.
103. Häberle, M., 1987. *Ernähr. Umschau*, **34**, 48.
104. Fairweather-Tait, S.J., Faulks, R.M., Fatemi, S.J.A. & Moore, G.R., 1987. *Human Nutr. Food Sci. Nutr.*, **41F**, 183.
105. Bester, B.H., 1988. *S. Afr. J. Dairy Sci.*, **20**, 1.
106. Intrieri, F., Cavaliere, A. & Ferolla, B., 1988. *Act. Med. Vet.*, **34**, 219.
107. Gabrielli Favretto, L., 1990. *Food Add. Contam.*, **7**, 425.
108. Koops, J., Klomp, H. & Westerbeek, D., 1986. *Neth. Milk Dairy J.*, **40**, 337.
109. Rolls, B.A. & Porter, J.W.G., 1973. *Proc. Nutr. Soc.*, **32**, 9.
110. Reif, G.D., Shahani, K.M., Vakil, J.R. & Crowe, L.K., 1976. *J. Dairy Sci.*, **59**, 410.
111. Tan, S.P., Wenlock, R.W. & Buss, D.H., 1984. *Human Nutr. Appl. Nutr.*, **38A**, 17.
112. Laukkanen, M., Antila, P., Antila, V. & Salminen, K., 1989. *Finn. J. Dairy Sci.*, **47**(1), 10.
113. Brug, J., Löwik, M.R.H., Kistemaker, C. & Wedel, M., 1991. *Voeding*, **52**, 4.
114. Bilic, N. & Sieber, R., 1990. *Schweiz. Milchw. Forsch.*, **19**(4), 71.
115. Jackson, P.A., Shelton, C.J. & Frier, P.J., 1982. *Analyst*, **107**, 1363.
116. Hulshof, P.J.M. & Katan, M.B., 1990. *Voeding*, **51**, 234.
117. Esha Research, The Food Processor II, Nutrition Analysis System, 1987. Salom, Oregon/USA.
118. Scott, K.J. & Bishop, D.R., 1988. *J. Sci. Food Agr.*, **43**, 187.
119. Ashoor, S.H., Seperich, G.J., Monte, W.C. & Welty, J., 1983. *J. Food Sci.*, **48**, 92.
120. Ashoor, S.H., Knox, M.J., Olsen, J.R. & Deger, D.A., 1985. *J. Assoc. Off. Analyt. Chem.*, **68**, 693.
121. Wills, R.B.H., Wimalasiri, P. & Greenfield, H., 1985. *J. Micronutr. Anal.*, **1**, 23.
122. Rao, D.R. & Shahani, K.M., 1987. *Cult. Dairy Prod. J.*, **27**(2), 6.
123. Sivell, L.M., Bull, N.L., Buss, D.H., Wiggins, R.A., Scuffam, D. & Jackson, P.A., 1984. *J. Sci. Food Agr.*, **35**, 931.
124. Ollilainen, V., Heinonen, M., Linkola, E., Varo, P. & Koivistoinen, P., 1989. *J. Dairy Sci.*, **72**, 2257.
125. Zehren, V., 1982. *Proc. 21st Intern. Dairy Congr., Moscow*, Vol. **2**, p. 177.

126. Janicki, J., Pedziwilk, F. & Kisza, J., 1963. *Nahrung*, **7**, 406.
127. Puhan, Z. & Flüeler, O., 1974. *Milchwissenschaft*, **29**, 148.
128. Renner, E., Karasch, U., Renz-Schauen, A. & Hauber, A., 1983. *Deut. Milchwirtsch.*, **34**, 1410.
129. Thomasow, J. & Hardung, C., 1978. *Proc. 20th Intern. Dairy Congr., Paris*, Vol. E, p. 750.
130. Halden, W., 1978. *Milch und Milchprodukte in der Ernährung*, Facultas-Verlag, Vienna.
131. Bijok, F., 1974. *Proc. 19th Intern. Dairy Congr., New Delhi*, Vol. 1E, p. 574.
132. Lee, B.O. & Alais, C., 1981. *Lait*, **61**, 140.
133. Dupuis, Y., Gambier, J. & Fournier, P., 1985. *Sci. Alim.*, **5**, 559.
134. Cremer, H.-D. & Büttner, W., 1962. *Ernähr. Umschau*, **9**, 68.
135. Fingerhut, M., Ruf, F. & Lang, K., 1966. *Z. Ernährungswiss.*, **6**, 228.
136. Lembke, A., 1987. *Milchwissenschaft*, **42**, 573.
137. Silva, M.F. de A., Jenkins, G.N., Burgess, R.C. & Sandham, H.J., 1986. *Caries Res.*, **20**, 263.
138. Thomson, M.E., 1988. *Caries Res.*, **22**, 246.
138a. Rosen, S., Min, D.B., Harper, D.S., Harper, W.J., Beck, E.X. & Beck, F.M., 1984. *J. Dent. Res.*, **63**, 894.
139. Krobicka, A., Bowen, W.H., Pearson, S. & Young, D.A., 1987. *J. Dent. Res.*, **66**, 1116.
140. Jenkins, G.N., 1989. *Nutr. Quart.*, **13**, 33.
141. Silva, M.F. de A., Burgess, R.C. & Sandham, H.J., 1987. *J. Dent. Res.*, **66**, 1527.
142. Harper, D.S., Osborn, J.C., Hefferren, J.J. & Clayton, R., 1986. *Caries Res.*, **20**, 123.
143. Sieber, R. & Graf, H., 1990. *Ernähr. Nutr.*, **14**, 63.
144. Lê, M.G., Moulton, L.H., Hill, C. & Kramar, A., 1986. *J. Nat. Cancer Inst.*, **77**, 633.
145. Jongen, W.M.F., van Boekel, M.A.J.S. & van Broekhoven, L.W., 1987. *Food Chem. Toxic.*, **25**, 141.
146. Munksgaard, L. & Werner, H., 1987. *Milchwissenschaft*, **42**, 216.
147. Stephany, R.W., Elgersma, R.H.C. & Schuller, P.L., 1978. *Neth. Milk Dairy J.*, **32**, 143.
148. Heeschen, W. & Nijhuis, H., 1985. *Molk. Ztg. Welt der Milch*, **39**, 961.
149. Walker, R., 1990. *Food Add. Contam.*, **7**, 717.
150. Glaeser, H., 1989. *Dairy Ind. Int.*, **54**(11), 19.
151. Van Faassen, A. & Pieters, J.J.L., 1983. *Voeding*, **44**, 412.
152. Collet, P., 1983. *Deut. Lebensm. Rundschau*, **79**, 370.
153. Beyer, F., 1984. *Deut. Milchwirtsch.*, **35**, 2061.
154. Sen, N.P. & Donaldson, B., 1978. *J. Assoc. Off. Analyt. Chem.*, **61**, 1389.
155. Zerfiridis, G.K. & Manolkidis, K.S., 1981. *J. Food Protect.*, **44**, 576.
156. Kurechi, T. & Kikugawa, K., 1979. *J. Food Sci.*, **44**, 1263.
157. Biedermann, R., Leu, D. & Vogelsanger, W., 1980. *Deut. Lebensm. Rundschau*, **76**, 149.
158. Tremp, E., 1980. *Mitt. Gebiete Lebensm. Hyg.*, **71**, 182.
159. Garcia Roche, M.O., Del Pozo, E., Izquierdo, L. & Fontaine, M., 1983. *Nahrung*, **27**, 125.
160. Hofmann, K., 1986. *AID-Verbraucherdienst*, **31**, 98.
161. Gray, J.I., Irvine, D.M. & Kakuda, Y., 1979. *J. Food Protect.*, **42**, 261.
162. Luf, W. & Brandl, E., 1986. *Ernähr. Nutr.*, **10**, 683.
163. Luf, W. & Brandl, E., 1986. *Österr. Milchw.*, **41** (wiss. Beil. 7), 57.
164. Luf, W. & Brandl, E., 1987. *Deut. Milchwirtsch.*, **38**, 116.
164a. Sieber, R. & Blanc, B., 1978. *Deut. Molk. Ztg.*, **99**, 240.
165. Askar, A., 1982. *Ernähr. Umschau*, **29**, 143.
166. Fritz, W. & Uhde, W.-J., 1980. *Ernährungsforschung*, **25**, 17.
167. Huynh, C.-H., Huynh, S. & Boivinet, P., 1980. *Ann. Nutr. Alim.*, **34**, 1069.
168. Gough, T.A., McPhail, M.F., Webb, K.S., Wood, B.J. & Coleman, R.F., 1977. *J. Sci. Food Agr.*, **28**, 345.

169. Pedersen, E., Thomsen, J. & Werner, H., 1980. In *N-Nitroso Compounds: Analysis, Formation and Occurrence*, ed. E.A. Walker, M. Castegnaro, L. Griciute & M. Boerzsoenyi, IACR Scient. Publ. No. 31, Lyon, p. 493.
170. Gray, J.I. & Mortin, I.D., 1981. *J. Human Nutr.*; **35**, 5.
171. Elgersma, R.H.C., Sen, N.P., Stephany, R.W., Schuller, P.L., Webb, K.S. & Gough, T.A., 1978. *Neth. Milk Dairy J.*, **32**, 125.
172. Massey, R. & Key, P.E., 1989. *Food Add. Contam.*, **6**, 453.
173. Yamamoto, M., Iwata, R., Ishiwata, H., Yamada, T. & Tanimura, A., 1984. *Food Chem. Toxic.*, **22**, 61.
174. Terplan, G., Bucsis, L. & Heerdegen, C., 1980. *Arch. Lebensm. Hyg.*, **31**, 1.
175. Groenen, P.J., Busink, E. & van Wandelen, M., 1987. *Z. Lebensm. Unters. Forsch.*, **185**, 24.
176. Ware, G.M., Thorpe, C.W. & Pohland, A.E., 1980. *J. Assoc. Off. Analyt. Chem.*, **63**, 637.
177. Engel, G. & Prokopek, D., 1979. *Milchwissenschaft*, **34**, 272.
178. Moreau, C., 1980. *Lait*, **60**, 254.
179. Medina, M., Gaya, P. & Nuñez, M., 1985. *J. Food Protect.*, **48**, 118.
180. Bullerman, L.B., 1981. *J. Dairy Sci.*, **64**, 2439.
181. Harwig, J., Blanchfield, B.J. & Scott, P.M., 1978. *Can. Inst. Food Sci. Technol. J.*, **11**, 149.
182. Speijers, G.J.A., Franken, M.A.M. & van Leeuwen, F.X.R., 1988. *Food Chem. Toxic.*, **26**, 23.
183. Pfleger, R., 1985. *Milchw. Ber.*, **85**, 297.
184. Schoch, U., Lüthy, J. & Schlatter, C., 1984. *Z. Lebensm. Unters. Forsch.*, **179**, 99.
185. Engel, G. & von Milczewski, K.E., 1977. *Milchwissenschaft*, **32**, 517.
186. Krusch, U., Lompe, A., Engel, G. & von Milczewski, K.E., 1977. *Milchwissenschaft*, **32**, 713.
187. Schoch, U., Luethy, J. & Schlatter, C., 1983. *Mitt. Gebiete Lebensm. Hyg.*, **74**, 50.
188. Sieber, R., 1988. *Schweiz. Landw. Forsch.*, **27**, 251.
189. Frank, H.K., Orth, R., Ivankovic, S., Kuhlmann, M. & Schmaehl, D., 1976. *Experientia*, **33**, 515.
190. Schoch, U., Luethy, J. & Schlatter, C., 1984. *Z. Lebensm. Unters. Forsch.*, **178**, 351.
191. Schoch, U., Luethy, J. & Schlatter, C., 1984. *Milchwissenschaft*, **39**, 583.
192. Zerfiridis, G.K., 1985. *J. Dairy Sci.*, **68**, 2184.
193. Veringa, H.A., van den Berg, G. & Daamen, C.B.G., 1989. *Neth. Milk Dairy J.*, **43**, 311.
194. Van Egmond, H.P., Speijers, G.J.A. & Wouters, R.B.M., 1990. *Voeding*, **51**, 82.
195. Sieber, R., Bütikofer, U., Bosset, J.O. & Rüegg, M., 1989. *Mitt. Gebiete Lebensm. Hyg.*, **80**, 345.
196. Sieber, R., Bütikofer, U., Baumann, E. & Bosset, J.O., 1990. *Mitt. Gebiete Lebensm. Hyg.*, **81**, 484.
197. Sieber, R., Bütikofer, U., Baumann, E. & Bosset, J.O., 1990. *Mitt. Gebiete Lebensm. Hyg.*, **81**, 722.
198. Wallhäusser, K.H. & Lück, E., 1978. *Z. Lebensm. Unters. Forsch.*, **167**, 156.
199. Lück, E., 1990. *Food Add. Contam.*, **7**, 711.
200. Bullerman, L.B., 1984. *J. Food Protect.*, **47**, 312.
201. Corradini, C. & Battistotti, B., 1981. *Scienza Tec. Latt.-Casear.*, **32**, 173.
202. Sofos, J.N. & Busta, F.F., 1981. *J. Food Protect.*, **44**, 614.
203. Engel, G., Rohmann, G. & Teuber, M., 1983. *Milchwissenschaft*, **38**, 592.
204. Riedl, R., 1984. *Milchw. Ber.*, **81**, 293.
205. De Ruig, W.G. & van den Berg. G., 1985. *Neth. Milk Dairy J.*, **39**, 165.
206. Kiermeier, F. & Zierer, E., 1975. *Z. Lebensm. Unters. Forsch.*, **157**, 253.
207. De Boer, E. & Stolk-Horsthuis, M., 1977. *J. Food Protect.*, **40**, 533.
208. Cerutti, G. & Battisti, P., 1972. *Latte*, **46**, 1.

Index

Note: Figures and Tables are indicated by *italic page numbers.*

581